Quantitative Finance with Case Studies in Python

Quantitative Finance with Case Studies in Python: A Practical Guide to Investment Management, Trading and Financial Engineering bridges the gap between the theory of mathematical finance and the practical applications of these concepts for derivative pricing and portfolio management. The book provides students with a very hands-on, rigorous introduction to foundational topics in quant finance, such as options pricing, portfolio optimization and machine learning. Simultaneously, the reader benefits from a strong emphasis on the practical applications of these concepts for institutional investors.

This new edition includes brand new material on data science and AI concepts, including large language models, as well as updated content to reflect the transition from Libor to SOFR to bring the text right up to date. It also includes expanded material on inflation and mortgage backed securities, more trade ideas embedded in each chapter and also via a dedicated chapter analyzing a set of derivatives trades. There are additional examples throughout based on recent market dynamics, including the post-Covid inflation shock and its impact on risk parity strategies.

Overall, the new edition is designed to be even more of a practical tool than the first edition, and more firmly rooted in real-world data, applications, and examples.

Features

- Useful as both a teaching resource and as a practical tool for professional investors
- Ideal textbook for first year graduate students in quantitative finance programs, such as those in master's programs in Mathematical Finance, Quant Finance or Financial Engineering
- Includes a perspective on the future of quant finance techniques, and in particular covers concepts of Machine Learning and Artificial Intelligence
- Free-to-access repository with Python codes available at www.routledge.com/ 9781032014432 and on https://github.com/lingyixu/Quant-Finance-With-Python-Code.

Chapman & Hall/CRC Financial Mathematics Series

Series Editors

Recently Published Titles

Foundations of Quantitative Finance
Book II: Probability Spaces and Random Variables
Robert R. Reitano

Foundations of Quantitative Finance
Book III: The Integrais of Riemann, Lebesgue and (Riemann-)Stieltjes
Robert R. Reitano

Foundations of Quantitative Finance
Book IV: Distribution Functions and Expectations
Robert R. Reitano

Foundations of Quantitative Finance
Book V: General Measure and Integration Theory
Robert R. Reitano

Computational Methods in Finance, Second Edition
Ali Hirsa

Interest Rate Modeling
Theory and Practice, Third Edition
Lixin Wu

Data Science and Risk Analytics in Finance and Insurance
Tze Leung Lai and Haipeng Xing

Foundations of Quantitative Finance
Book VI: Densities, Transformed Distributions and Limit
Robert R. Reitano

Malliavin Calculus in Finance
Theory and Practice, Second Edition
Elisa Alòs and David Garcia Lorite

Arbitrage and Rational Decisions
Robert Nau

Equity Release Products
Radu S. Tunaru and Enoch B. Quaye

Quantitative Finance with Case Studies in Python
A Practical Guide to Investment Management, Trading and Financial Engineering
Chris Kelliher

For more information about this series please visit: Chapman and Hall/CRC Financial Mathematics Series - Book Series - Routledge & CRC Press

Quantitative Finance with Case Studies in Python

A Practical Guide to Investment Management, Trading and Financial Engineering

Second Edition

Chris Kelliher

CRC Press
Taylor & Francis Group
Boca Raton London New York

CRC Press is an imprint of the
Taylor & Francis Group, an **informa** business
A CHAPMAN & HALL BOOK

Designed cover image: Shutterstock images

Second edition published 2025
by CRC Press
2385 NW Executive Center Drive, Suite 320, Boca Raton FL 33431

and by CRC Press
4 Park Square, Milton Park, Abingdon, Oxon, OX14 4RN

CRC Press is an imprint of Taylor & Francis Group, LLC

ISBN: 978-1-032-86800-4 (hbk)
ISBN: 978-1-032-87041-0 (pbk)
ISBN: 978-1-003-53062-6 (ebk)

DOI: 10.1201/9781003530626

Typeset in Latin Modern font
by KnowledgeWorks Global Ltd.

Publisher's note: This book has been prepared from camera-ready copy provided by the authors.

Access the support material at: www.routledge.com/9781032014432

To my amazing daughter, Sloane, my light and purpose.
To my wonderful, loving wife, Andrea, without whom none of my achievements would be possible.
To my incredible, supportive parents and sister and brother, Jen and Lucas.

Contents

About the Author

Chris Kelliher is a multi-asset portfolio manager and senior quantitative researcher with over 20 years of investment experience at asset management firms and hedge funds. In addition, Mr. Kelliher is an adjunct professor in the Master's in Mathematical Finance and Financial Technology program at Boston University's Questrom School of Business where he has also held the role of Executive Director. In these roles, he has taught graduate-level courses on computational methods in finance, fixed income, credit risk and programming for quant finance. He is also the author of "Quantitative Finance with Python: A Practical Guide to Investment Management, Trading and Financial Engineering" and was named among the top 20 US Finance Professors in 2024 by Rebellion Research. Mr. Kelliher earned a BA in economics from Gordon College, where he graduated Cum Laude with departmental honours and an MS in mathematical finance from New York University's Courant Institute.

Foreword

In March 2018, the Federal Reserve ("Fed") was in the midst of its first hiking cycle in over a decade, and the European Central Bank ("ECB"), still reeling from the Eurozone debt crisis, continued to charge investors for the privilege of borrowing money. US sovereign bonds ("Treasuries") were yielding 3% over their German counterparts ("Bunds"), an all-time high, and unconventional monetary policy from the two central banks had pushed the cost of protection to an all-time low.

Meanwhile, across the pond, a sophisticated Canadian pension flipped a rather esoteric coin: A so-called digital put-on Euro/Dollar, a currency pair that trades over a trillion dollars a day. On this crisp winter morning, the EURUSD exchange rate ("spot") was 1.2500. If the flip resulted in heads and spot ended below 1.2500 in 2 years, the pension would receive \$10 million. If the flip were tails and spot ended above 1.2500, the pension would have to pay \$2.5 million. Naturally, the 4 to 1 asymmetry in the payout suggests that the odds of heads were only 25%. Interestingly, the flip yielded heads, and in 2 years, spot was below 1.2500.

After the trade, I called Chris, reiterated the pitch, and explained that since January 1999, when EURUSD first started trading, the market implied odds of heads had never been lower. As macroeconomic analysis and empirical realizations suggest that the coin is fair, and there is about 50% chance of getting heads, should the client perhaps consider trading the digital put in 10x the size? In his quintessentially measured manner, Chris noted, "We must have a repeatable experiment to isolate a statistical edge." Ten separate flips, for instance, could reduce the risk by over 2/3. Moreover, negative bund yields, which guarantee that investors will lose money, incentivize capital flows to Treasuries, and the anomalous rates "carry is a well-rewarded risk premium." Furthermore, as "investors value \$1 in risk-off more than \$1 in risk-on," does the limited upside in the payout of the digital put also harness a well-rewarded tail risk premium?

I wish I were surprised by Chris's nuance, or objectivity, or spontaneity. Having known him for 8 years, though, I have come to realize that he is the most gifted quant I have had the privilege to work with, and this book is a testament to his ability to break complex ideas down to first principles, even in the treatment of the most complex financial theory. The balance between rigor and intuition is masterful, and the textbook is essential reading for graduate students who aspire to work in investment management. Further, the depth of the material in each chapter makes this book indispensable for derivative traders and financial engineers at investment banks, and for quantitative portfolio managers at pensions, insurers, hedge funds and mutual funds. Lastly, the "investment perspectives" and case studies make this an invaluable guide for practitioners structuring overlays, hedges and absolute return strategies in fixed income, credit, equities, currencies and commodities.

In writing this book, Chris has also made a concerted effort to acknowledge that markets are not about what is true, but rather what can be true, and when: With negative yields, will Bunds decay to near zero in many years? If so, will $1 invested in Treasuries, compound and buy all Bunds in the distant future? Or will the inflation differential between the Eurozone and the US lead to a secular decline in the purchasing power of $1? One may conjecture that no intelligent investor will buy perpetual Bunds with a negative yield. However, even if the Bund yield in the distant future is positive, but less than the Treasury yield, market implied odds of heads, for a perpetual flip, must be zero. As the price of the perpetual digital put is zero, *must* the intelligent investor add this option to her portfolio?

Since the global financial crisis, the search for yield has increasingly pushed investors down the risk spectrum, and negative interest rates, and unconventional monetary policy, are likely just the tip of the iceberg. This book recognizes that unlike physics, finance has no universal laws, and an asset manager must develop an investment philosophy to navigate the known knowns, known unknowns and unknown unknowns. To allow a portfolio manager to see the world as it was, and as it can be, this book balances the traditional investment finance topics with the more innovative quant techniques, such as machine learning. Our hope is that the principles in this book transcend the outcome of the perpetual flip.

– *Tushar Arora*

Contributors

Tushar Arora
INSEAD
Paris, France

George Kepertis
Boston University
Boston, Massachusetts

You Xie
Boston University
Boston, Massachusetts

Lingyi Xu
Boston University
Boston, Massachusetts

Maximillian Zhang
Boston University
Boston, Massachusetts

Acknowledgments

Thank you to the many talented mathematical finance graduate students at Boston University who have taken my courses and motivated me to write this book. Special thanks to my exceptional working team of Lingyi Xu, George Kepertis, You Xie and Max Zhang, who contributed to the coding and in-practice examples in the book, as well as provided invaluable feedback throughout the process. My sincerest appreciation to Eugene Sorets, who was a tremendous colleague and mentor for many years, and who co-taught multiple courses with me that helped shaped the courses that served as the baseline for this text. I am also eternally grateful to the many mentors that I have had throughout the years, most notably Mitch Livstone, Bill Irving, Neville McCaghren, Ali Hirsa and Tom Knox. My deepest gratitude to Tushar Arora for his significant contributions to the book and his continued feedback along the way. I am thankful to Tinghe Lou, Alan Grezda, Yincong He, Boyu Dong, Tim Lee, Ziyuan Zhang, Jipeng Li and Maksym Naumenko for their feedback on the book and for their work in finding errors in the text. Thanks to my family for their support throughout the process of writing this text, and their patience with my perpetual focus on the book during the last year. Any typos and errors are entirely my responsibility.

I

Foundations of Quant Modeling

CHAPTER 1

Setting the Stage: Quant Landscape

1.1 INTRODUCTION

QUANTITATIVE finance and investment management are broad fields that are still evolving rapidly. Technological and modeling innovations over the last decades have led to a series of fundamental shifts in the industry that are likely just beginning. This creates challenges and opportunities for aspiring quants who are, for example, able to harness the new data sources, apply new machine learning techniques or leverage cutting-edge derivative models. The landscape of these fields is also fairly broad, ranging from solving investment problems related to optimal retirement outcomes for investors to providing value to firms by helping them hedge certain undesirable exposures.

In spite of the innovations, however, quants must remember that models are, by definition, approximations of the world. As George Box famously said, "All Models are Wrong, Some are Useful," and this is something all quants should take to heart. When applying a model, the knowledge of its strengths, weaknesses and limitations must be at the forefront of our mind. Exacerbating this phenomenon is the fact that finance, unlike hard sciences like Physics or Chemistry, does not allow for repeatable experiments. In fact, in finance we don't even know for sure that consecutive experiments come from the same underlying distribution because of the possibility of a regime change. Making this even more challenging is the potential feedback loop created by the presence of human behavior and psychology in the market. This is a key differentiating factor relative to other, harder, sciences. As a result, quants should aim to make models that are parsimonious and only as complex as the situation requires and to be aware and transparent about the limitations and assumptions.[1]

The goal of this book is to bridge the gap between theory and practice in this world. This book is designed to help the reader understand the underlying financial theory but also to become more fluent in applying these concepts in Python. To achieve this, there is a corresponding coding repository with a bevy of practical

[1] Along these lines, Wilmott and Derman have created a so-called modeling manifesto designed to emphasize the differences between models and actual, traded markets [200].

DOI: 10.1201/9781003530626-2

applications and Python code designed to build the reader's intuition and provide a coding baseline. These examples are designed to provide the reader with coding examples that they can take and tweak to fit their own applications.

It should be emphasized, however, that this book assumes some working knowledge of Python, and a solid background in single and multi-variate calculus, probability, statistics and linear algebra. A review of many of these quantitative underpinnings is provided in the book's supplementary materials.

In this chapter we aim to orient the reader to the landscape of quant finance. In doing so, we provide an overview of what types of firms make up the financial world, what some common quant careers tend to look like, and what the instruments are that quants are asked to model. In later chapters, we discuss the specifics of modeling each instrument mentioned here, and many of the modeling techniques that are central to the quant careers discussed. Before diving into the modeling techniques, however, our goal is to provide more context around why the techniques and instruments matter and how they fit into the larger picture of quant finance and investment management.

We then discuss what a typical quant project looks like, with an emphasis on what is common across all roles and seats at different organizations. Of critical importance is not only being able to understand the models and the mathematical theory but also being fluent with the corresponding data and having the required tools to validate the models being created. Lastly, we try to provide the reader with some perspective on trends in the industry. Unlike other less dynamic and more mature fields, finance is still rapidly changing as new data and tools become available. Against this backdrop, the author provides some perspective on what skills might be most valuable going forward, and what the industry might look like in the future.

1.2 QUANT FINANCE INSTITUTIONS

The fields of finance and investment management contain many interconnected players that serve investors in different ways. For some, such as buy-side hedge funds and asset managers, their main function is to generate investments with positive returns and create products that are attractive to investors. For others, such as dealers, market makers and other sell-side institutions, their main function is to match buyers and sellers in different markets and create customized derivative structures that can help their clients hedge their underlying risks. Lastly, the field includes a lot of additional service providers. These providers may provide access to data, including innovative data sources, or an analytics platform for helping other institutions with common quant calculations. This space includes financial technology companies, whose primary role is to leverage technology and new data sources to create applications or signals that can be leveraged by buy or sell-side institutions to make more efficient decisions. In the remainder of this section, we provide the reader some context on what functions these organizations provide. We then proceed to discuss where quants might fit in at these various entities in the following section.

1.2.1 Sell-Side: Dealers & Market Makers

Sell-side institutions facilitate markets by providing liquidity through making markets and by structuring deals that are customized to meet client demands. On the market-making side, sell-side institutions provide liquidity by stepping in as buyers or sellers when needed and subsequently looking to offload the risk that they take on quickly as the market returns to an equilibrium. Market makers are compensated for their liquidity provision by collecting the bid-offer spread on these transactions. In many markets, market makers use automated execution algorithms to make markets. In other cases, traders may fill this role by manually surveying the available order stack and stepping in when the order book becomes skewed toward buys or sells, respectively.

Dealers also facilitate markets by creating structures for clients that help them to hedge their risks, or instrument a macroeconomic view, efficiently. This often involves creating derivative structures and exotic options that fit a client's needs. When doing this, a dealer may seek another client willing to take on the offsetting trade that eliminates the risk of the structure completely. In other cases, the dealer/market maker may warehouse the risk internally, or may choose to hedge certain sensitivities of the underlying structure, but allow certain other sensitivities to remain on their books. We will discuss this hedging process for various derivative structures in more detail in chapter 16.

1.2.2 Buy-Side: Asset Managers & Hedge Funds

Buy-side institutions, such as hedge funds and asset managers, are responsible for managing assets on behalf of their clients. The main goal of a buy-side firm is to deliver strong investment returns. To this end, hedge funds and asset managers may employ many different types of strategies and investment philosophies, ranging from purely discretionary to purely systematic. For example, hedge funds and asset managers may pursue the following types of strategies:

Global Macro: Global macro strategies may be discretionary in nature, where views are generated based on a portfolio manager's assessment of economic conditions, or systematic, where positions are based on quantitative signals that are linked to macroeconomic variables.

Relative Value: Relative value strategies try to identify inconsistencies in the pricing of related instruments and profit from their expected convergence. These relative value strategies may be within a certain asset class, or may try to capture relative value between two asset classes, and may be pursued via a discretionary approach, or systematically.

Event Driven: Event-driven strategies try to profit from upcoming corporate events, such as mergers and acquisitions. These strategies tend to bet on whether these transactions will be completed, benefiting from subsequently adjusted valuations.

Risk Premia: Risk premia strategies try to identify risks that are well-rewarded and harvest them through consistent, isolated exposure to the premia. Common risk premia strategies include carry, value, volatility, quality and momentum. Some of these risk premia strategies are discussed in chapter 27.

Statistical Arbitrage: Statistical arbitrage strategies are a quantitative form of relative value strategies where a quantitative model is used to identify anomalies between assets, for example, through a factor model. In other cases, a pairs trading approach may be used, where we bet on convergence of highly correlated stocks that have recently diverged. These types of statistical arbitrage models are discussed further in detail in chapter 27.

A key differentiating factor between hedge funds and asset managers is the level and structure of their fees. Hedge funds generally charge significantly higher fees and have sizable fees linked to their funds performance. Additionally, most hedge funds are so-called absolute return funds, meaning that their performance is judged in absolute terms. Asset managers, by contrast, often do not collect performance fees and measure their performance relative to benchmark indices with comparable market exposure. Hedge funds also tend to have considerably more freedom in the instruments and structures that they can trade and require less transparency[2].

1.2.3 Financial Technology Firms

Generally speaking, financial technology firms leverage data and technology to create products that they can market to buy-side and sell-side institutions. The proliferation of available data over the last decade has led to a large increase in FinTech companies. Many FinTech companies at their core solve big data problems, where they take non-structured data and transform it into a usable format for their clients. As an example, a FinTech company might track traffic on different companies websites and create a summary signal for investment managers. In another context, FinTech companies might leverage technological innovations and cloud computing to provide faster and more accurate methods for pricing complex derivatives. Buy-side and sell-side institutions would then purchase these firms services and then incorporate them into their processes, either directly or indirectly.

1.3 MOST COMMON QUANT CAREER PATHS

Aspiring quants may find themselves situated in any of these organizations and following many disparate career paths along the way. In many ways, the organization that a quant chooses will determine what type of modeling skills will be most emphasized. In **buy-side** institutions, such as hedge funds and asset managers, a heavy focus will be placed on econometric and portfolio construction techniques. Conversely, while working on the **sell-side**, at dealers, or investment banks, understanding of stochastic processes may play a larger role. Even within these institutions, a quant's role may vary greatly depending on their group/department. To provide additional context, in the following sections, we briefly describe what the most common quant functions are at buy-side, sell-side and FinTech companies.

[2]Because of this, there are tighter restrictions on who can invest in hedge funds.

1.3.1 Buy Side

At buy-side institutions, building investment products and delivering investment outcomes are at the core of the business. As such, many quants join these shops, such as asset management firms, pension funds, and hedge funds, and focus on building models that lead to optimal portfolios, alpha signals or proper risk management.

Examples of roles within buy-side institutions include:

Desk Quants: A desk quant sits on a trading floor at a hedge fund or other buy-side firm and supports portfolio managers through quantitative analysis. The function of this support can vary greatly from institution to institution and may in some cases involve a great deal of forecasting models and have a heavy emphasis on regression methods, machine learning techniques and time series/econometric modeling, such as those described in chapter 5. In other cases, this support may involve more analysis of derivative valuations, and identifying hedging strategies, as discussed in section 12 of this text. Desk quants may also help provide portfolio managers with quantitative analysis that supports discretionary trading process.

Asset Management Quants: A quant at an asset manager will have a large emphasis placed on portfolio construction and portfolio optimization techniques. As such, understanding of the theory of optimization, and the various ways to apply it to investment portfolios is a critical skill set. Asset management quants may also be responsible for building alpha models and other signals, and in doing so will leverage econometric modeling tools. Asset management quants will rely heavily on the material covered in section 23 of this text.

Research Quants: Research quants tend to focus more on longer term research projects and try to build new innovative models. These may be relative value models, proprietary alpha signals or innovative portfolio construction techniques. Research quants utilize many of the same skills as other quants but are more focused on designing proprietary, groundbreaking models rather than providing ad-hoc analysis for portfolio managers.

Quant Developers: Quant developers are responsible for building production models and applications for buy-side shops. In this role, quant developers must be experts in programming but also have a mastery of the underlying financial theory. Quant developers will rely on the programming skills described in chapters 7 and 8 and also must be able to leverage the financial theory and models described in the rest of the book.

Quant Portfolio Managers: Generally these quants are given a set risk-budget that they use to deploy quantitative strategies. These roles have a heavy market-facing component but also require an ability to leverage quant tools such as regression and machine learning to build systematic models. As such, these roles also require a strong background in finance in order to understand the dynamics of the market and uncover attractive strategies to run quantitatively. Simplified versions of some of the strategies that might be employed by quantitative portfolio managers are described in more detail in chapter 27.

1.3.2 Sell Side

At sell-side institutions, making markets and structuring products for clients are crucial drivers of success, and quants at these institutions can play a large role in both of these pursuits. Many quants join sell-side shops and are responsible for creating automated execution algorithms that help the firm make markets. Other quants may be responsible for helping build customized derivative products catered to clients hedging needs. This process is commonly referred to on the sell-side as structuring. Examples of quant roles on the sell-side include the following.

Desk Quants: Like a buy-side desk quant, a desk quant on the sell-side sits on a trading desk and supports traders and market makers. Sell-side desk quants will often help create structured products that are customized for clients. This may involve creating exotic option payoffs that provide a precise set of desired exposures. It may also involve creating pricing models for exotic options, modeling sensitivities (Greeks) for complex derivatives and building different types of hedging portfolios. Sell-side desk quants will leverage the concepts discussed in section 12 with a particular emphasis on the exotic option pricing topics discussed in chapter 15 and the hedging topics discussed in chapter 16.

Risk Quants: Risk quants help sell-side institutions measure various forms of risk, such as market risk, counterparty risk, model risk and operational risk. These are often significant roles at banks as they determine their capital ratios and subsequently the cash that banks must hold. Market risk quants are responsible for determining risk limits and designing stress tests. Many of the topics relevant to risk quants are discussed in chapter 26. Risk quants also often need to work with the modeling concepts presented in the rest of the book, such as the time series analysis and derivatives modeling concepts discussed in chapter 5 and section 12, respectively.

Model Validation Quants: Many sell-side institutions have separate teams designed to validate newly created production models and production model changes. The quants on these teams, model validation quants, are responsible for understanding the financial theory behind the models, analyzing the assumptions, and independently verifying the results. The principles of model validation are discussed in detail in chapter 11. Additionally, model validation quants gain exposure to the underlying models that they are validating through implementing them independently and the verification process.

Quant Trader/Automated Market Maker: Quant traders and quant market makers are responsible for building automated market-making algorithms. These algorithms are used to match buyers and sellers in a highly efficient manner, while collecting the bid-offer spread. Unlike a quantitative portfolio manager, a market-making quant tends to build higher frequency trading models and will hold risk for very short time periods. This leads to an increased emphasis on coding efficiency and algorithms. The foundations for a quant trader's coding background are discussed in more detail in chapter 8, and some examples of the types of trading algorithms that they rely on are presented in chapter 27.

Quant Developer: Much like quant developers on the buy-side, sell-side quant developers are responsible for building scalable, production code leveraged by sell-side institutions. This requires mastery of coding languages like Python, and also strong knowledge of financial theory. In contrast to buy-side quant developers, sell-side quant developers will have a larger emphasis on the options modeling techniques discussed in section 12.

1.3.3 Financial Technology

Data Scientist: Financial technology is a burgeoning area of growth within the finance industry and is a natural place for quants. Technological advances have led to a proliferation of data over the last decade, leading to new opportunities for quants to analyze and try to extract signals from the data. At these firms, quants generally serve in data scientist type roles, apply machine learning techniques and solve big data problems. For example, a quant may be responsible for building or applying Natural Language Processing algorithms to try to company press releases and trying to extract a meaningful signal to market to buy-side institutions.

1.3.4 What's Common between Roles?

Regardless of the type of institution a quant ends up in, there are certain common themes in their modeling work and required expertise. In particular, at the heart of the majority of most quant problem is trying to understand the underlying distribution of assets. In chapters 4 and 5 we discuss the core mathematical tools used by buy- and sell-side institutions for understanding these underlying distributions. Additionally, solving quant problems in practice generally involves implementing a numerical approximation to a chosen model, and working with market data. As a result, a strong mastery of coding languages is central to success as a quant.

1.4 STAGES OF A QUANT PROJECT

Quant projects can vary greatly by organization and role, as we saw earlier in this chapter, however, there is a great deal of commonality in how these varying quant tasks are structured. In particular, quant projects, generally speaking consist of the following four main steps: data collection, data cleaning, model implementation and model validation. In the following sections, we briefly describe what each of these steps entails. Throughout the book, we will highlight approaches to tackling these steps efficiently, both from a technical and quantitative perspective.

1.4.1 Data Collection

Data collection is the process of identifying the proper source for our model and gathering it in a desired format for use in the model. This part of a quant project requires being able to interact with different types of data sources, such as databases, flat/CSV files, ftp sites and websites. As such, we need to be familiar with the libraries in Python that support these tasks. We discuss this stage, and some of the more commonly used financial data sources, in more detail in chapter 9.

1.4.2 Data Cleaning

Once we have obtained the relevant data for our model the next step is to make sure that it is in proper order for our model to use. This process is traditionally referred to as data cleaning, where we analyze the set of data that we are given to make sure there are no invalid values, missing data or data that is not in line with the assumptions of the model. Depending on the type of data we are working with, this process can range from very simple to extremely complex. For example, with equity data we may need to verify that corporate actions are handled in an appropriate manner. In contrast, when working with derivatives, we need to check the data for arbitrage that would violate our model assumptions. Completing this stage of a project will require a level of mastery of Python data structures and an ability to use them to transform and manipulate data. These data structures are discussed in more detail in chapter 8, and more details on the typical cleaning procedure for different types of financial data is discussed in chapter 9.

1.4.3 Model Implementation

Model implementation is the core task we face as quants. It is the process of writing a piece of code that efficiently implements our model of choice. In practical quant finance applications, closed form solutions are rarely available to solve realistic problems. As a result, quant projects require implementation of an approximation of a given model in code. The underlying models and techniques vary by application, ranging from using simulation of an Stochastic Differential Equation to estimate an option price to using econometric techniques to forecast stock returns. As models become more complex, this step becomes increasingly challenging.

The vast majority of this book is dedicated to model implementation and the most common models and techniques used to solve these problems. We also focus heavily on methods for implementing models in a robust and scalable way by providing the required background in object-oriented programming concepts. Further, in section 18 we work through different model implementations across asset classes and discuss the key considerations for modeling across different markets.

1.4.4 Model Validation

Once we have implemented our model, a separate process must begin that convinces us that the model has implemented correctly and robustly. This process is referred to as model validation, and it is designed to catch unintended software bugs, identify model assumptions and limitations of the model. For simple models, this process may be relatively straight-forward, as we can verify our model against another independent third party implementation. As models get increasingly realistic, this process becomes much less trivial as the true model values themselves become elusive. In this context, we need to rely on a set of procedures to ensure that we have coded our models correctly. This model validation step is discussed in detail in chapter 11.

1.5 TRENDS: WHAT HAS CHANGED IN THE LAST FEW YEARS, AND WHERE IS QUANT FINANCE GOING?

In this chapter we have tried to provide somewhat of a roadmap to the quant finance and investment management industry. As a relatively new, younger field, quant finance and investment management is still in a very dynamic phase. In some cases, these changes are driven by technological innovation, or groundbreaking models. In other cases, however, these changes are driven by changes in the underlying dynamics of the market itself. In this section, we attempt to highlight a few of the most important changes that we have seen in the past few years as well as areas of potential innovation and evolution.

1.5.1 Change in Inflation Regime

Fixed income markets have been the home of some of the most striking changes to the market ecosystem in recent years, beginning with the introduction of negative rates several years ago to the recent inflation shock which led to a significant re-pricing of interest rates higher. This change in the inflationary environment led to generational volatility in the rates and credit markets, and has led to increased popularity of inflation hedges, such as commodities and inflation linked bonds, and inflation derivatives.

Whether this trend of higher inflation, and higher inflation volatility persists, is and will continue to be an open topic among economists, market practitioners and the Fed. As a result, the inflation landscape will be one of the key themes for quants and other market participants alike and knowledge of the underlying drivers and principles will continue to be of the utmost importance for those entering the field.

1.5.2 Automation

Automation is a key trend in the finance industry that is likely to continue for the foreseeable future. On the one hand, technological advances have led to the ability to streamline and automate processes that used to require manual intervention or calculations. Automation of these processes generally requires strong programming knowledge and often also requires a solid understanding of the underlying financial concepts. In some cases, automation may involve writing a script to take the place of a manual reconciliation process a portfolio manager used to do. More substantively, automation may also involve replacing human based trading algorithms with automated execution algorithms. This is a trend that we have seen in most exchange traded markets, that is, execution has gotten significantly lower-touch. In the future, there is the potential for this to extend to other segments of the market.

1.5.3 Rapid Increase of Available Data

Over the past few decades, a plethora of new data sources have become available, many of which are relevant for buy-side and sell-side institutions. This has created a dramatic rise in the number of big data problems and data scientists in the quant finance industry. Many of these data sources have substantially different structure than standard financial datasets, such as text data and satellite photographs. The ability to parse text data, such as newspaper articles, could be directly relevant to buy-side institutions who want to process news data using a systematic approach. Similarly, image data of store parking lots may provide insight into the demand for different stores and products that leads balance sheet data.

This trend in the availability of data is likely to continue. It has been said that something like 90% of the financial data available to market participants is from the last decade and further that proliferation of data is likely to make this statement true in the next few decades as well. This is a welcome trend for quants, FinTech firms and data scientists looking to apply their skills in finance as it provides a richer opportunity set. This data is not without challenges, however, as the fact that these datasets are new makes it challenging to thoroughly analyze them historically in different regimes. It stands to reason, however, that the reward for being able to process these new data sources robustly should also be quite high before other market participants catch on.

1.5.4 Commoditization of Factor Premia

Another key trend in the quant finance and investment management community has been the evolution of the concepts of **alpha** and **beta**. Traditionally, investment returns have been viewed against a benchmark with comparable *market exposure*. This means that for most hedge funds, if they arguably take minimal market exposure, or beta, over long periods of time, then their performance[3] would be judged in absolute terms. For asset management firms who take large amounts of beta, their performance would be judged against a balanced benchmark with the appropriate beta, such as a 60/40[4]. This ensures that investment managers are compensated for the excess returns, or alpha that they generate but are not compensated for their market exposure, which could easily be replicated elsewhere more cheaply.

More recently, there has been a movement toward identification of additional factors, or risk premia, such as carry, value, momentum, (low) volatility and quality. This has created a headwind for many investment firms as the returns in these premia have become increasingly commoditized, leading to lower fees and cheaper replication. Further, returns from these premia, which used to be classified as **alpha**, have become another type of **beta**. Although in some ways this has been a challenge for the buy-side, it also creates an opportunity for quants who are able to identify and find robust ways to harvest these premia.

1.5.5 Increased Popularity of Multi-Strategy Hedge Funds

While the increased presence of factor premia has generally served as a headwind for hedge funds, with it providing investors an alternative lower fee way to gain consistent exposures to these factor premia, it has not necessarily lead to a decrease

[3] And consequently their ability to charge performance fees.

[4] A 60/40 portfolio has a 60% allocation to equities and 40% allocation to fixed income.

in interest in the hedge fund industry. On the contrary, there is one type of hedge fund that has continued to attract significant interest among clients, and justify high fee structures: a multi-strategy fund. These funds typically allocate risk to a large number of uncorrelated portfolio managers across investment strategy and asset class. Investing in many underlying managers allows these funds to benefit from strategy and manager diversification while harvesting small pockets of the market that large AUM strategies may not be incentivized to search. Funds like Millennium and Citadel are the most famous examples of this underlying hedge fund model.

1.5.6 Proliferation of Artificial Intelligence, Machine Learning and Large Language Models

In recent years, there has been a push toward using machine learning techniques and artificial intelligence to help solve quant finance and portfolio management problems. This trend then further accelerated with the promise of the Large Language Models. This trend is likely just reaching an inflection point, with many seminal works in machine learning discussing the potential applications of these techniques in a financial setting[5]. For example, Machine Learning may help us with many quantitative tasks, such as extracting meaning from unstructured data, building complex optimal hedging schemes and creating higher frequency trading strategies. Most recently, work has been done connecting the use of LLMs to solving quant problems, such as the estimation of expected returns [267].

While this trend toward AI and Machine Learning is likely to continue its acceleration in the future it is important to know the strengths and weaknesses of these different techniques and keep in mind Wilmott's quant manifesto [200]. No model or technique will be a perfect representation of the world or perfect in all circumstances. In the context of AI & machine learning, this may mean that there are certain instances where application of these techniques is natural and leads to significant improvement. Conversely, it is important to keep in mind that in other cases where the data is less plentiful, or we face other challenges, these techniques are likely to struggle to add value and may not be worth the added complexity. For example, in some cases, such as lower frequency strategies, sufficient data is generally not available to warrant use of sophisticated machine learning techniques with large feature sets. In chapter 28 and 29 we discuss the potential challenges and opportunities of leveraged these techniques in finance.

1.5.7 Increasing Prevalence of Required Quant/Technical Skills

Taken together, these trends lead to a larger overarching trend in favor of the importance of quantitative techniques that can help us uncover new data sources, explain the cross-section of market returns via a set of harvestable premia, and help us automate trading and other processes. Over the past few decades, many roles in the financial industry have begun to require more technical skills and a more quantitative inclination. Knowledge of a coding language such as Python and fluency with

[5]Such as Halperin [141] and Lopez de Prado [74] [76].

basic quant techniques has become more widespread, leading to an industry where substantial quantitative analysis is required even for the some of the most fundamentally oriented institutions.

1.5.8 Increased Regulatory Complexity and the Potential for Democratization

As the financial industry continues to evolve and incorporate alternative data sources and sophisticated techniques, it adds layers of both complexity and promise. On the one hand, as the data and models become more complex, so does the need for a regulatory environment that ensures that these techniques are implemented responsibility. When leveraging alternative data, for example, validating the data becomes more challenging and there are legitimate question regarding accountability and accessibility. At the same time, as a technique gets more precise, the impact of undetected bad data can actually be larger. With many of these datasets and techniques being newly introduced, and the space continuing to evolve, it is likely that this will be a continued point of emphasis in markets for the foreseeable future.

On the other hand, if these new techniques and datasets are leveraged responsibly, it has the potential to create an opportunity for better investment outcomes for all investors by improving the overall efficiency of the market. Importantly, quants stand to play a key role in both parts of this process, first by being at the forefront of harvesting these new technologies and data sources, and secondly by playing a role in helping strike a balance between fostering innovation and ensuring models are implemented sensibly.

CHAPTER 2

Setting the Stage: Landscape of Financial Instruments

THERE are two main types of instruments in financial markets, cash instruments, such as stocks and bonds, and derivatives, whose value is contingent on an underlying asset, such as a stock or a bond.

Analysis of different types of financial instruments requires a potentially different set of quantitative tools. As we will see, in some cases, such as forward contracts, simple, static, replication and no arbitrage arguments will help us model the instrument. Other times we will need more complex replication arguments, based on dynamically replicated portfolios, such as when valuing options. Lastly, in some circumstances these replication arguments may fail and, we may need to instead invoke the principles of market efficiency and behavioral finance to calculate an expectation.

The following sections briefly describe the main financial instruments that are of interest to quants and financial firms.

2.1 EQUITY INSTRUMENTS

2.1.1 Overview & Mechanics

Equity instruments enable investors to purchase a stake in the future profits of a company. Some equities may pay periodic payments in the form of dividends, whereas others might forego dividends and rely on price appreciation in order to generate returns for investors. Equity instruments may include both public and private companies and arise when companies issue securities (i.e. stocks). Companies may issue these securities in order to help finance new projects, and in doing so are sharing the future profits of these projects directly with investors. Equity investments embed a significant amount of risk as they are at the bottom of the capital structure, meaning that they are last to be re-paid in the event of a bankruptcy or default event for the firm. Because of this, it is natural to think that equity investors would expect to be paid a premium for taking on this risk. Market participants commonly refer to this

DOI: 10.1201/9781003530626-3

as the equity risk premium. The equity market also has other equity like products such as exchange traded products and exchange traded notes. Common models for equities are discussed in section 23 and the underlying techniques that tend to belie these models are discussed in chapter 5.

2.1.2 Public Equity

Publicly traded companies issue shares of stock that are traded on exchanges on a secondary market where these shares can change hands many times in a short period of time. These shares of public equities are highly liquid and enable investors to participate in the future profits of public companies. These profits may come in the form of dividends, or price appreciation. Therefore, when modeling an equity, we must model the future value of this dividend stream as well as estimate the future value of the stock price. In the next section, we introduce one such model for valuing an equity in this context as a collection of future dividends.

2.1.3 Modeling Equity via a Discounted Dividend Growth Model

As discussed in the last section, when modeling an equity security we need to factor in the potential upside, or price appreciation of the stock as well as any potential dividends. A standard approach to valuing equities is via a Discounted Dividend Growth model, or Gordon Growth Model [116]. In this model, we assume that any price appreciation is eventually returned to investors in the form of a future dividend. Thus, in this model we assume that there is no possibility of default and treat an equity investment as an infinite series of future dividends, with dividends being the only mechanism for returning money to shareholders. While this is no doubt an oversimplification, it provides an intuitive, tractable model and in general companies achieving high levels of profit and corresponding stock price increases are likely to increase their dividends, whereas struggling companies are likely to see their stock price and dividends both decrease.

As a result, in this type of model valuing an equity can be done by finding the present value of all future dividends, as shown in the following equation:

$$P_0 = \sum_{t=1}^{\infty} D_0 \frac{(1+g)^t}{(1+r)^t} \tag{2.1}$$

where r is the firm's cost of capital, discount rate, or required rate of return, and g is the firm's dividend growth rate. If we factor out a $\frac{1+g}{1+r}$, then this equation can be rewritten as:

$$P_0 = D_0 \frac{1+g}{1+r} \left(1 + \frac{1+g}{1+r} + \frac{1+g}{1+r}^2 + \frac{1+g}{1+r}^3 + \cdots \right) \tag{2.2}$$

The series inside the parenthesis is a standard geometric series which, if $\frac{1+g}{1+r} < 1$ sums to $\frac{1}{1-\frac{1+g}{1+r}}$

Plugging this in and simplifying,[1] we are left with the following:

$$P_0 = \frac{D_0\,(1+g)}{r-g} \tag{2.3}$$

This is the well-known equation for valuing an equity in a Discounted Dividend Growth Model.

An astute reader may notice several key assumptions in the model. Notably, we are assuming a constant dividend growth rate, g and discount rate, r. In practice, this is unlikely to be realistic as they are likely to vary over time due to economic conditions and company profitability. Additionally, as we are modeling an equity as an infinite stream of dividends we are neglecting the possibility of a jump into default. Readers may also notice that for the model to generate sensible valuations, the discount rate must be above dividend growth rate.[2] In spite of these simplifying assumptions, these models are commonly relied on in practice.

2.1.4 Private Equity

If a company is not publicly listed on a stock exchange, then it may be harder for investors to access shares in the firm. To access these shares would require a private investment in the company. Private equity firms, however, specialize in these types of investments, and provide exposure to pools of these investments, thus enabling investors to access this potentially attractive return stream and benefit from diversification.

Unlike public equities, where there is a liquid and active secondary market with deep liquidity, private equity securities may require locking up an investors capital for longer and are characterized by significantly less liquidity. This illiquid profile means that investors are not able to withdraw their capital when they want to, and they may demand a premium, often referred to as a liquidity premium, to bear this risk [147]. Conversely, private equity and other illiquid assets are not required to mark their assets to market daily, creating a return smoothing profile that some have speculated as desirable [24]. In the next section, we compare the performance of investments in public and private equities, analyzing the presence of a liquidity premium in private assets.

2.1.5 Private Equity vs. Public Equity Returns in Practice

Private equity investments have gained popularity in recent years as they have experienced strong returns and proved to be relatively resilient to market volatility. This has led many investors to question whether an additional risk premium exists in private assets. A full examination of this phenomenon is beyond the scope of this text, however, in the following chart we analyze the returns over last ten years of the S&P 500 and an index of private equity firms:

[1] To simplify we multiply the numerator and denominator by $1 + r$.

[2] In fact, when they are close the denominator approaches zero leading to unrealistic valuations.

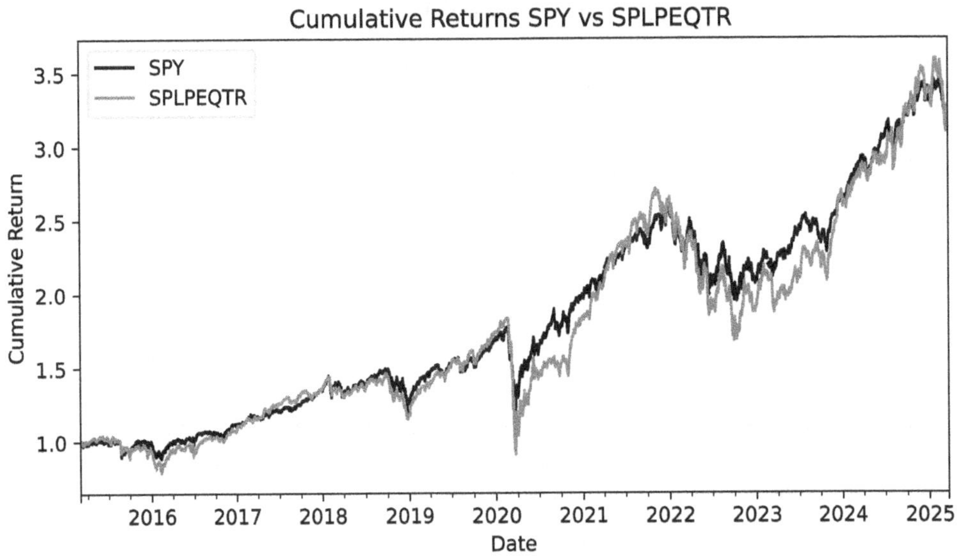

As we can see, the return profiles of these two indices over this time period are quite similar. Some have speculated that any potentially higher returns in private assets are a reflection of the fact that these instruments do not have to mark to market in the presence of market stress, unlike typical equity investments. Excess returns in private assets could similarly be due to size and other premia embedded in these assets, or compensation for taking on larger risks associated with the lower transparency in these investments. Interested readers are encouraged to consult [16] for more information on these phenomenon.

2.1.6 Preferred Stock

Preferred Stock is a special kind of equity instrument that, as the name implies, is further up the capital structure. Because of this, Preferred Stock has more protection than common equity in the event of a default event, and has first right to the firm's cashflows to pay dividends. This gives Preferred Stock an added layer of protection relative to common equity.

Preferred Stock generally pays a higher dividend than traditional public equity securities, however, investors in preferred stock do not get to participate in the upside, or price appreciation of a firm's equity. As a result, Preferred Stock is generally viewed as a hybrid of equity and bond instruments, with some features, like its fixed coupon aligning with bonds, and others, like the ability for a firm to increase or decrease its dividend and the absence of a maturity date, matching more with equity investments.

2.2 DEBT INSTRUMENTS

2.2.1 Overview & Mechanics

Governments and private companies often gain access to capital by borrowing money to fund certain projects or expenses. When doing so, they often create a debt security, such as a government or corporate bond. These bonds are the most common type of fixed income or debt instrument, and are structured such that the money is repaid at a certain maturity date. The value that is re-paid at the maturity date is referred to as a bond's principal. In addition, a periodic stream of coupons may be paid between the initiation and maturity dates. In other cases, a bond might not pay coupons but instead have a higher principal relative to its current price.[3]

A bond that does not pay any coupons is referred to as a zero-coupon bond. Pricing debt instruments, such as zero-coupon bonds, relies on present value and time value of money concepts, which are further explored in chapter 5.[4] As an example, the pricing equation for a zero-coupon bond can be written as:

$$V_0 = \exp\left(-yT\right)P \tag{2.4}$$

where P is the bond's principal and V_0 is the current value or price of the bond. Further, y is the yield that is required in order for investors to bear the risks associated with the bond. In the case of a government bond, the primary risk would be that this yield would change, which would change the market value of the bond. This is referred to as interest rate risk. In the case of corporate bonds, another critical risk would be that the underlying corporation might go bankrupt and fail to repay their debt. Investors are likely to require a higher yield, or a lower initial price, in order to withstand this risk of default. It should be noted, however, that debt holders are above equities in the capital structure, meaning they will be re-paid, or partially re-paid, prior to any payment to equity holders. In practice this means that in the event of a default bond holders often received some payment below the bond's principal, which is known as the recovery value.

Astute readers may notice that this implies that both equity and corporate bond holders share the same default risk for a particular firm. That is, they are both linked to the same firm, and as a result the same underlying earnings and future cashflows, however, are characterized by different payoffs and are at different places in the firms capitalization structure. This creates a natural relationship between equities and corporate bonds, which we explore further in chapter 20.

There are many different types and variations of bonds within fixed income markets. Some bonds are linked to nominal rates whereas others are adjusted for changes in inflation. Additionally, bonds may contain many other features, such as gradual repayment of principal[5] or have coupons that vary with a certain reference rate. More information is provided on modelling debt instruments in chapter 19.

2.2.2 Sovereign Bonds

Sovereign bonds, or government bonds, are debt instruments issued by Federal government entities. Sovereign bonds can have many underlying features, including different coupon structures and callability properties. Sovereign bonds issued by some government entities, such as the United States are considered risk-free, and when

[3] Assuming positive interest rates.

[4] More detail on these concepts can also be found in [251].

[5] Which is referred to as an Amortizing Bond.

they are valued the prospect of a default is not taken into account. Other sovereign bonds, such as Emerging Market Sovereign bonds, are not considered risk-free. In fact, in these markets, there is a significant historical precedent for default and as a result default risk[6] must be modeled. In practice, this means that investors will require a higher yield to be compensated for bearing the default risk embedded in the sovereign bond.

2.2.3 Corporate Bonds

Corporate bonds are issued by individual companies instead of government agencies. These bonds may be issued when firms are looking to finance a new project, and choose to do so by taking on debt instead of issuing new equity shares. Like Emerging Market bonds, when we model corporate debt we must account for the possibility of default. Similarly, corporate bonds are likely to offer higher yields to compensate investor for a potential loss in the event of default. The excess spread embedded in a corporate bond over the risk-free rate, is generally referred to as a credit spread, and in practice will vary based on the implied credit quality of the firm issuing the bond.

2.2.4 Municipal Bonds

Municipal bonds, or Munis, are bonds issued by local governments instead of Federal governments. Because municipalities typically fund their bonds via tax revenues, the perceived risk of default is low. Nonetheless, municipal bonds may have embedded credit risk. Another defining feature of municipal bonds is their tax treatment. Unlike standard corporate or sovereign bonds, interest on Munis is often tax-exempt. This means that, all else equal, an investor should prefer a municipal bond to a corporate bond with the same yield if they are holding it in a taxable account.[7] In fact, in general investors might be willing to invest in Munis with lower yields to obviate the tax bill of standard corporate bonds. In the next section, we explore the hypothetical after tax returns for investors between Munis and treasuries.

2.2.5 Municipal Bond Valuation in Practice

While a full examination of municipal bonds relative to nominal bonds is beyond the scope of this text, in this section we attempt to motivate the topic by examining the difference in yield between a municipal bond and a treasury bond. As highlighted in the last section, municipal bonds, unlike treasury are exempt from certain taxes. Therefore, all else equal, investors should prefer Munis in taxable accounts. In fact, this preferential tax treatment leads Munis to have lower yields, reflecting this advantage. To see this, we examine two ETFs with otherwise similar exposure,[8] MUB a municipal bond ETF and GOVT a treasury bond ETF. In the following table, we provide a comparison of the yield of the two instruments:

---

[6]More information on default risk modeling can be found in chapter 20.

[7]Some retirement and other accounts may be tax deferred or tax exempt reducing an investors incentize to invest in Munis.

[8]For example, they have similar duration risk.

ETF	Yield
MUB	3.92%
GOVT (before tax)	4.42%

As expected, we can see that the yield on the Muni ETF is lower, reflecting the advantageous tax position of these bonds. In fact, once we account for a reasonable tax the investor would pay on the treasury bond ETF, it becomes the lower yielding security. This implies that an investor in municipal bonds may be able to achieve a higher yield while taking on the same risks as the treasury investor. This higher tax adjusted yield of a municipal bond relative to a treasury, however, may be due to liquidity reasons, and the flight to quality properties inherent in US treasuries. Interested readers are encouraged to explore this phenomenon more with actual data from interest rate and Muni markets.

2.2.6 Inflation Linked Bonds

Any of the bonds we have introduced so far will lose value when inflation rises. Mechanically, this happens as increases in inflation cause the yield on nominal bonds to increase, and the corresponding bond price will decrease. This means that these bonds are subject to inflation risk, among other things.

Inflation linked bonds, such as Treasury Inflation Protected Securities (TIPS) remove this inflation risk by adjusting the principal in the bond as inflation increases. This means that, the traditional decrease in price that we observe from an increase in the bond yield due to inflation increasing will be offset by the principal increasing. The mechanics of TIPS are discussed further in section 15.6.

While inflation linked bonds are not sensitive to changes in inflation, because of their underlying structure, they still have embedded interest rate risk. This is because they will lose value in the event that inflation stays the same but the remaining risk free rate, often referred to as the real interest rate, changes. This real interest rate, which is calculated as the difference between the nominal yield of a risk-free bond and the inflation rate, is a quantity of great interest to market participants, and is also the main driver of risk in an inflation linked bond.

Inflation linked instruments and the relationship between real rates and inflation are described in more detail in 15.6.

2.2.7 Convertible Bonds

Convertible Bonds can be viewed as a hybrid of equity and bond securities. Convertible Bonds, in particular, are debt instruments that can be converted to equity at a predetermined price per share. This leads to some interesting dynamics for convertible bonds. On the one hand, if the value of the equity is significantly above the conversion price, then the convertible will tend to trade like an equity instrument, as investors will be incentivized to convert their bonds to shares of stock. Conversely, if the stock is trading significantly below the conversion price, then the convertible bond will trade more like a bond as there will be little value in a potential conversion.

This conversion decision within a Convertible Bonds is an example of embedded optionality because the holder of the bond gets to decide when and if to convert their bonds to shares, and will only do so if it is economically advantageous. When the conversion price is significantly above the current stock price, then this embedded option has little value because it is out-of-the-money. On the other hand, when the conversion price is above the current stock price, there is a lot of value in the conversion and we say the option is in-the-money. The mechanics of options and the definition of in and out-of-the-money options are presented in 2.4. More information on modeling convertible bonds can also be found [247].

2.2.8 Green Bonds

Green bonds are a class of bonds where the proceeds of the bonds are used to finance projects related to sustainability, climate and other green initiatives. These bonds are conceptually similar to the other bonds discussed in this section, however, like Munis, they may also have preferential tax treatment, depending on the project being invested in. Additionally, because green bonds support "green" initiatives, they may have an embedded discount or premium in them depending on investors collective desire to invest in such projects. As an example, if these types of projects are perceived as contributing to the greater societal good,[9] then an investor may be willing to forego some of the yield they would be able to obtain elsewhere because of the utility that they gain from investing in bonds with a net positive social impact. More information on green bonds and their effectiveness can be found in [170].

2.3 FORWARDS & FUTURES

Forwards and futures are, generally speaking, the simplest derivatives instrument. A forward or futures contract is an agreement to buy or sell a specific asset or security at a predetermined date. Importantly, in a forward or futures contract it is *required* that the security be bought or sold at the predetermined price, regardless of whether it is economically advantageous to the investor. Forwards and futures contracts themselves are quite similar, with the main differentiating factor being that forwards are over-the-counter contracts and futures are exchange traded.

Forward and futures contracts can be used to hedge against price changes in a given asset by locking in the price today. As an example, an investor with foreign currency exposure may choose to hedge that currency risk via a forward contract rather than bear the risk that the currency will move in an adverse way.

The payoff for a long position in a forward contract can be written as:

$$V_0 = S_T - F \tag{2.5}$$

where F is the agreed upon delivery price and S_T is the asset price on the delivery date. Similarly, a short position in a forward contract can be written as:

$$V_0 = F - S_T \tag{2.6}$$

[9]Which is a conclusion that is left to the reader.

F is typically chosen such that the value of the forward contract at initiation, V_0, is equal to zero [235].

Replication arguments can be used to find the relationship between spot (current asset) prices and forward prices. To see this, consider the following two portfolios:

- Long Position in Forward Contract
- Borrow S_0 dollars at Risk Free Rate & Buy Asset for S_0 dollars.

The following table summarizes the payoffs of these respective portfolios both at trade initiation ($t = 0$) and expiry ($t = T$):

Time	Portfolio 1: Long Forward	Portfolio 2: Borrow and Long Stock
0	0	0
T	$S_T - F$	$S_T - S_0 \exp(rT)$

Note that we are assuming a constant interest rate, no transactions costs and that the underlying asset does not pay dividends. As you can see, the value of both of these portfolios at $t = 0$ is zero. The value at expiry in both cases depends on S_T, and the other terms, such as F, are known at trade initiation.

These portfolios have the same economics, in that they both provide exposure the underlying asset on the expiry/delivery date. To see this, consider a portfolio that is short portfolio 1 and long portfolio 2. In that case, the payoff for this investor becomes:

$$V_0 = 0 \tag{2.7}$$
$$V_T = F - S_0 \exp(rT) \tag{2.8}$$

This portfolio has zero cost, therefore, in the absence of arbitrage, it's payoff must also be equal to zero. This leads to a forward pricing equation of:

$$F = S_0 \exp(rT) \tag{2.9}$$

If the forward prices diverges from this, it can easily be shown that this results in an arbitrage opportunity. For example, if $F - S_0 \exp(rT) < 0$, then we can buy a forward contract, sell the stock at the current price and lend the proceeds. Conversely, if $F - S_0 \exp(rT) > 0$, then we can sell the forward contract and borrow money to buy the stock at the current price.

It should be emphasized that this replication is **static**. This means that we built a single replicating portfolio and were able to wait until expiry without adjusting our hedge. Later in the book we will use replication arguments to value option payoffs, and in this case **dynamic replication** will be required.

These forward pricing replication arguments can easily extended to include dividends [139]. Incorporation of dividends[10] leads to the following formula for forward pricing:

$$F = S_0 \exp((r - q)T) \tag{2.10}$$

[10]In the form of a continuous dividend yield.

where q is an asset's dividend yield. Under certain assumptions, such as constant or deterministic interest rates, it can also be shown that futures and forward prices will be the same.

Forward contracts are particularly common in foreign exchange markets. Equity and commodity markets, in contrast, have liquid markets for many index futures. In interest rate markets, both forwards and futures contracts are traded, and the role of a stochastic discount factor create another level of complexity in valuing and differentiating between the two.[11] Futures and forwards contracts in these different markets are discussed in more detail in section 18. Additionally, more detailed treatment of forwards and futures contracts can be found in [139].

2.4 OPTIONS

2.4.1 Mechanics

An option provides an investors with the right, rather than the obligation to buy or sell an asset at a predetermined price on a specified date. This contrasts with a forward contract, where that exchange is required. Thus, as the name implies, an option provides the investor with a choice of whether to engage in the transaction depending on whether it is economically favorable.[12] It turns out that this right to choose whether to exercise leads to some interesting and subtle mathematical properties.

The most common options are **call** and **put** options which are often referred to as vanilla options. A call option provides an investor the right to buy a security or asset for a given price at the option's expiry date. Clearly, situations when the asset price at expiry are highest are best for holders of call options.

A put option, conversely, gives an investor the right to sell a security for a pre-specified price at expiry. Put options will be most valuable when the asset price at expiry is lowest. The agreed upon transaction price specified in an option is referred to as the **strike price**.

The payoff for a call option, C and put option, P, respectively, can be written as:

$$C = \max(S_T - K, 0) \tag{2.11}$$

$$P = \max(K - S_T, 0) \tag{2.12}$$

where the max in the payoff functions reflects that investors will only exercise their option if it makes sense economically.[13] We can see that a call option payoff looks like a long position in a forward contract when the option is exercised, and will only be exercised if $S_T > K$. Similarly, a put option's payoffs are similar to a short position in a forward contract conditional on the option being exercised, which will only happen if $S_T < K$.

In the following chart, we can what a payoff diagram for a call and put option look like:

---

[11] More details on futures modelling in an interest rate setting, and the convexity correction that arises, can be found in chapter 19.

[12] This is referred to as an exercise decision.

[13] Because the payoff is greater than zero.

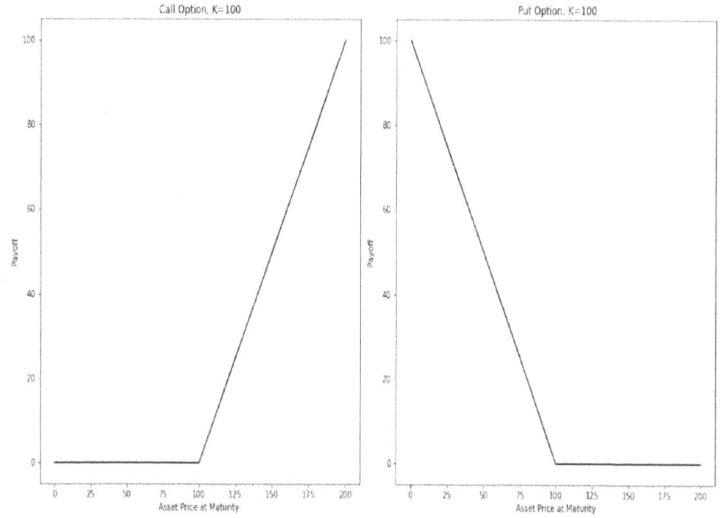

As we can see, the payoff of a long position in a call or a put option is always greater than or equal to zero. As a result, in contrast to forwards, where the upfront cost is zero, options will require an upfront payment to receive this potential positive payment in the future. Additionally, we can see that the inclusion of the max function, created by an option holders right to choose, leads to non-linearity in the payoff function.

A significant portion of this book is dedicated to understanding, modelling and trading these types of options structures. In particular, chapter 4 and section 12 provide the foundational tools for modelling options.

2.4.2 Option Straddles in Practice

A commonly traded option strategy is a so-called straddle, where we combine a long position in a put option with a long position in a call option at the same strike. In the following chart, we can see the payoff diagram for this options portfolio:

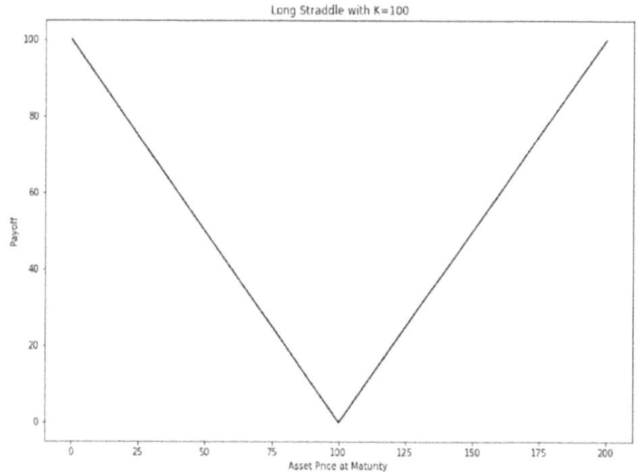

From the payoff diagram, we can see an interesting feature of this strategy is that the payoff is high when the underlying asset moves significantly in either direction. It is also defined by a strictly non-negative payoff, meaning that we should expect to pay a premium for it. At first glance, this appears to be an appealing proposition for investors. We are indifferent in how the asset moves, we just need to wait for it to move in order for us to profit. The truth, however, turns out to be far more nuanced. Nonetheless, straddles are fundamental options structures that we will continue to explore throughout the book, most notably in chapter 16.

2.4.3 Put-Call Parity

A relationship between call and put options can also be established via a static replication argument akin to the one we saw earlier for forward contracts. This relationship is known as put-call parity, and is based on the fact that a long position in a call, combined with a short position in a put, replicates a forward contract in the underlying asset.

To see this, lets consider an investor with the following options portfolio:

Portfolio 1

- Buy a call with strike K and expiry T
- Short a put with strike K and expiry T

The payoff at expiry for an investor in this portfolio can be expressed via the following formula:

$$V_T = \max\left(S_T - K, 0\right) - \max\left(K - S_T, 0\right) \tag{2.13}$$

If S_T is above K at expiry, then the put we sold will expiry worthless, and the payoff will simply be $S_T - K$. Similarly, if S_T expires below K, then the call expires worthless and the payoff is $-\left(K - S_T\right) = S_T - K$. Therefore, the payoff of this portfolio is $V_T = S_T - K$ regardless of the value at expiry.

Now lets consider a second portfolio:

Portfolio 2

- Buy a single unit of stock at the current price, S_0
- Borrows K dollars in the risk-free bond

At time T, this second portfolio will also have a value of $S_T - K$ regardless of the value of S_T. Therefore, these two portfolios have the same terminal value at all times, and therefore must have the same upfront cost in the absence of arbitrage. This means we must have the following relationship:

$$C - P = \left(S_0 - Ke^{-rT}\right) \tag{2.14}$$

where C is the price of a call option with strike K and time to expiry T, P is the price of a put option with the same strike and expiry, S_0 is the current stock price, and r

is the risk-free rate. It should be emphasized that the left-hand side of this equation is the cost of portfolio 1 above, a long position in a call and a short position in a put. Similarly, the right hand side is the cost of the second portfolio, a long position in the underlying asset and an appropriately sized short position in the risk-free rate. This formula can be quite useful as it enables us to establish the price of a put option given the price of a call, or vice versa.

2.5 SWAPS

2.5.1 Mechanics

A swap is an agreement to a periodic exchange of cashflows between two parties. The swap contract specifies the dates of the exchanges of cash flows, and defines the reference variable or variables that determine the cashflow. In contrast to forwards, swaps are characterized by multiple cash flows on different dates.

Swaps define at least one so called **floating leg**, whose cashflow is dependent on the level, return or change in some underlying reference variable. It may also include a **fixed leg** where the coupon is set at contract initiation. The most common types of swaps are fixed for floating swaps, where one leg is linked to an interest rate, equity return or other market variable, and the fixed leg coupon is set when the contract is entered. Other swaps may be floating for floating swaps, with each leg referencing a different market variable.[14]

The cashflows for the fixed leg of a swap can be calculated using the present value concepts detailed in chapter 5.[15] In particular, the present value for each leg of a swap is calculated by discounting each cashflow to today. The present value for the cashflows of the fixed leg of a swap can then be written as:

$$\text{PV(fixed)} = \sum_{i=1}^{N} \delta_{t_i} C D(0, t_i) \tag{2.15}$$

where C is the set fixed coupon and $D(0, t_i)$ is the discount factor from time t_i to today, and δ_{t_i} is the time between payments. The reader may also notice that, because C is a constant set when the trade is entered, it can be moved outside the summation. We can see that the present value of the fixed leg of a swap is the sum of the discounted cashflows weighted by the time interval. While this part may seem trivial, it turns out there is actually some ambiguity in how we calculate the time between payments, δ_i. For example, do we include all days or only business days? Do we assume each month has 30 days or count the actual number of days in each month? In practice a **daycount convention** is specified to help us measure time intervals precisely. The most common daycount conventions are discussed in more detail in chapter 19.

To calculate the present value of the cashflows of a floating leg in a swap we can

[14]With a potential fixed basis, or spread.

[15]For more information see [251].

use the following equation:

$$\text{PV(float)} = \sum_{i=1}^{N} \delta_{t_i} F_{t_i} D(0, t_i) \tag{2.16}$$

where F_{t_i} is the value of the reference rate, index or return at time t_i. Unlike the fixed leg, this value is not known at trade initiation and requires knowledge of the expected future value of the reference variable.[16] However, in many cases, a forward or futures contract may directly tell us the expected value of the reference variable, $\mathbb{E}\left[F_{t_i}\right]$.

The present value of the swap then becomes:[17]

$$\text{PV} = \text{PV(float)} - \text{PV(fixed)} \tag{2.17}$$

The most common types of swaps are interest rate swaps and credit default swaps.[18] In interest rate swaps, the most common products are fixed for floating swaps where SOFR[19] is the reference rate for the floating leg. In credit default swaps, a fixed coupon is paid regularly in exchange for a payment conditional on default of an underlying entity.[20] In the next section, we look at an example of finding the fair swap rate for a swap contract and then look at a market driven example: equity index total return swaps.

2.5.2 Equity Index Total Return Swaps in Practice

In this section we leverage the coding example found in the coding repository to value a total return swap on the S&P 500 where the return of the index is defined as the floating leg and the fixed leg is a constant, financing leg. The rate for this financing leg is set by market participants via hedging and replication arguments similar to those introduced in this chapter for forwards. In particular, in the following chart we show the cumulative P&L to an investor who receives the floating S&P return and pays the fixed financing cost:

[16] It should be noted that only the expected value of the reference variable is needed to value a swap contract, in contrast to options structures where the entire distribution will be required.

[17] Note that this is from the perspective of the buyer of the floating leg.

[18] Interest Rate swaps are discussed in more detail in chapter 19. Credit Default Swaps are discussed in more detail in chapter 20. More information on swaps can also be found in [139] and [134].

[19] Or previously Libor.

[20] Given their structure, credit default swaps can be viewed as analogous to life insurance contracts on corporations or governments.

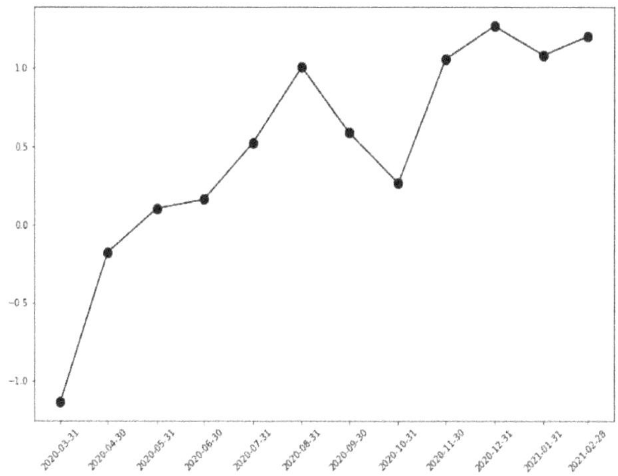

We can see that, in this example the investor who receives the S&P return makes a substantial profit, leading to a loss on the other side of the trade. This should be unsurprising, however, as the period of observation was defined by particularly strong equity returns as the drawdown from the Covid crisis eased, resulting in initially negative returns but a steep uptrend in the above chart.

2.5.3 Over-The-Counter vs. Exchange Traded Products

In many markets, instruments trade on an exchange. These instruments tend to be liquid and are characterized by standardized terms, such as contract size and expiration date. Exchange traded markets naturally lend themselves to automated trading and low-touch execution.[21] The majority of the products in equity and equity derivatives markets are exchange traded.

In other cases, over-the-counter (OTC) contracts are the market standard. In these cases, an intermediary such as a dealer will customize a structure to cater to the needs of a client. These markets are far less standardized with customizable terms/features, and are characterized by high-touch execution. Execution in OTC markets often happens via phone or chat requiring interaction and potential negotiation with a sell-side counterpart. In these markets, clients, such as buy-side institutions often reach out to dealers with a desired structure, and the dealer responds with pricing. This process may then include several iterations working with the dealer to finalize the trades terms and negotiate the price. Many derivatives, including the vast majority of exotic options, are traded OTC.

The primary benefit of an OTC contract is that an investor may customize the terms and exposure to meet their exact needs. This is not possible in exchange traded products which consist of only standard instruments with preset features. As an example, a client looking to hedge currency risk that is contingent on future sales

[21] Low touch execution refers to the ability to execute trades in an automated manner with minimal dealer contact.

of a product might want to enter a customized contract that hedges this currency risk conditional on positive equity returns, when sales are likely to be strongest. This type of OTC product is commonly referred to as a **hybrid** option, as it is contingent on returns in both the foreign exchange and equity markets. One of the drawbacks of OTC contracts are their less automated, higher touch execution. and the corresponding lower levels of liquidity that can lead to higher bid-offer spreads. This leads to a situation where investors must solicit pricing from a dealer to start the execution process, rather than observing a set of market data and choosing the pockets where pricing looks most competitive.

In section 18, we will highlight which markets trade which instruments on exchanges vs. OTC, and discuss the implications for investors looking to leverage those products, and for quants looking to model them. We will find that the distinction is most important in interest rate markets, where we use non-deterministic interest rates.[22]

[22]See chapter 19 for more details.

CHAPTER 3

Theoretical Underpinnings of Quant Modeling: Modeling the Risk Neutral Measure

3.1 INTRODUCTION

PERHAPS the canonical mathematical technique associated with quant finance is that of stochastic calculus, where we learn to model the evolution of stochastic processes. As this subject requires a mastery of calculus, probability, differential equations and measure theory, many if not all graduate students find these concepts to be the most daunting. Nonetheless, it is a critical tool in a quants skillset that will permeate a quant's career. In this chapter we provide a broad, high level overview of the techniques required in order to tackle the financial modeling techniques detailed in later chapters. This treatment is meant to augment, and not replace by any means the standard and more rigorous treatment that dedicated stochastic calculus texts and courses provide. Instead, we look to provide an intuitive, accessible treatment that equips the reader with the tools required to handle practical derivatives valuation problems.

As we will see, stochastic processes are the evolution of sequences of random variables indexed by time. Modeling the evolution of the entire sequence then requires a different set of tools than modeling a single probability distribution. In a stochastic process, we need to model a series of interconnected steps, each of which is random, and drawn from a probabilistic distribution. This is in contrast to other applications of probability theory where we look at successive, often independent, single draws from a distribution. Of course, this adds complexity, and means that we need a new set of techniques. For example, it requires a different type of calculus, stochastic calculus, that incorporates the behavior of the entire sequence in a stochastic process instead of working on deterministic functions.

In this chapter, we provide the framework, and mathematical machinery, for dealing with these stochastic processes in the context of options pricing. As we will soon see, this is fundamentally distinct from the pursuit of forecasting or prediction models. In this context, instead of building models based on our estimates of future

DOI: 10.1201/9781003530626-4

probabilities, we will instead develop arguments that are based on no-arbitrage, hedging and replication. If we can perfectly replicate an option, or contingent claim, for example, with a position in the underlying asset, then we know we have found an asset that mimics its exposure and payoff regardless of the true future probabilities. No arbitrage will then further guide us by telling us that these two claims that are economically equivalent must also have the same price. Thus, to the extent that we are able to do this, which we will soon judge for ourselves, we will then be able to build models that are agnostic to these future probabilities and instead built on our ability to hedge the options risk, or replicate the claim. Instead, we will build a different set of probabilities, under the assumption that investors are risk neutral. Importantly building this set of probabilities will not require forecasting the drift of the asset in the future. Later in the text, in chapter 17, we will generalize this to a market implied distribution, based on the set of risk-neutral probabilities extracted from options prices, using these replication arguments.

In this text a working knowledge of calculus and probability concepts is assumed. Readers looking to review these topics should consult [237] or [64]. In the following sections, we then extend these concepts to the case of a stochastic process, and highlight many of the challenges that arise. It should again be emphasized that the treatment of stochastic calculus here is meant to be light and intuitive. Those looking for a more rigorous treatment should consult [232], [35] or [193].

3.2 RISK NEUTRAL PRICING & NO ARBITRAGE

3.2.1 Risk Neutral vs. Actual Probabilities

Before delving into the theory of risk-neutral probabilities and no-arbitrage theory, it is worth first understanding it intuitively. Along those lines, Baxter and Rennie [35] present an exceptional example highlighting the difference between risk-neutral and actual probabilities that the reader is strongly encouraged to review. At the core of this example is the idea of replication. If we are able to synthetically replicate a given bet, or contingent claim, then the actual probabilities of the underlying sequence of events are no longer important in how we value the claim. Instead we can rely on the relative pricing of the contingent claim and our replication strategy.

When working with actual probabilities, we look for trades that are significantly skewed in our favor. That is, we want a high probability of favorable outcomes and a relative low probability of unfavorable outcomes. The idea is then, if we can engage in this strategy many times, we should profit over time. On each individual bet, we have no idea what is going to happen, it is random, but over time we should be able to accumulate a profit if the odds are skewed in our favor. In fact this is the basis for most, if not all, systematic investing strategies, and is a major focus on 23. But, importantly, as we will see in the remainder of this chapter, there are times when we are able to be agnostic to the actual probabilities. That is, of course, because we have built a replicating portfolio. If we are able to do this, and the replicating portfolio is priced differently than the contingent claim, then we want to buy (or sell) the contingent claim and sell (or buy) the replicating portfolio. This will be true regardless of the actual probabilities. Even if the contingent claim is in isolation a great investment strategy, that is, even if the probabilities are skewed in our favor, we are better off trading the replicating portfolio against the contingent claim. This is because, in that case, we are agnostic to the outcome, we can simply wait for the claim to expire and collect our profit. The pricing models that we develop in this text will be based on this phenomenon, and the fact that this relationship should keep the pricing of derivatives, or contingent claims, and their replicating portfolios, in line.

Throughout the text, we will emphasize the fundamental difference between risk-neutral probabilities, and the risk-neutral pricing measure and the actual set of probabilities, which we will refer to as the physical measure. In the risk-neutral measure, we work in the context of a replicating portfolio. This makes us indifferent to what happens in the future, as long as our replication strategy holds. In the physical measure, we will not use these replication arguments but instead rely on forecasts of the actual probabilities. This will become necessary in the case where replication arguments do not exist.

3.2.2 Theory of No Arbitrage

The concept of no-arbitrage is a fundamental tenet of quant finance [41]. Arbitrage refers to the ability to engage in a trading strategy that obtains excess profits without taking risk. For example, if we were able to buy and sell the same security at different prices at the same time, then we could simultaneously buy the lower priced and sell the higher priced, leaving us a risk-free profit. Economic theory posits that these types of opportunities are not present in markets, and, to the extent that they appear, market participants are happy to make the appropriate trades that force prices to re-align toward their equilibrium, no-arbitrage state. This means that, if we are able to replicate a claim, whether it is a position in a stock or a complex derivative, then the price of the replicating portfolio should be the same as the price of the original asset. If not, an arbitrage would occur, where we could buy the cheaper of the replicating portfolio and the original asset, and collect a profit without bearing any risk. This type of no-arbitrage and replication argument is at the heart of so-called risk-neutral pricing theory. Instead of attempting to forecast the asset price and use that to value our derivative, we can instead look to replicate it using simpler instruments. The price of the replicating portfolio, will then help us value the original, more complex derivative.

In the last chapter, we saw an example of this when we derived the pricing formula for a forward or futures contract under certain conditions. In this case we were able to build a replicating portfolio for a forward or a future that involved a position in the underlying asset, and a position in a risk-free bond. Notably, this position was static: we put on our position at trade origination and simply waited for the expiration date. Importantly, we didn't need to re-balance. As the derivatives that we work with get more complex, so too will the replication schemes. In this chapter we will consider options, or contingent claims, which will require a dynamic hedge in order to replicate.

3.2.3 Complete Markets

Another pivotal concept in the pursuit of valuing complex derivative securities is that of a complete vs. incomplete market [207]. In a complete market, any contingent claim, regardless of complexity can be replicated. This ability to replicate these contingent claims will be central to our modeling efforts, and therefore this is an important distinction. An incomplete market, by contrast is one in which not every claim can be replicated. In this case, valuation of claims that we cannot replicate will be more complex, and will not be able to be done using the standard risk-neutral valuation framework that we develop.

3.2.4 Risk Neutral Valuation Equation

The first fundamental theorem of asset pricing [207] states that there is no arbitrage if and only if there exists a risk-neutral measure. In this risk-neutral measure, valuations of all replicable securities can be computed by discounting prices at the risk-free rate. This means that, in this case, asset prices can be written as:

$$S_0 = \tilde{\mathbb{E}}\left[e^{-\int_0^T r_u\,du} S_T\right] \tag{3.1}$$

for a given risk-neutral measure $\tilde{\mathbb{P}}$ and an interest rate process r. Under this measure, $\tilde{\mathbb{P}}$, the current stock price reflects the discounted risk-neutral expectation of all future cashflows.

Similarly, the price today, p_0, of a contingent claim with payoff $V(S_T)$ can be written as:

$$p_0 = \mathbb{E}\left[e^{\int_0^T -r_\tau d\tau} V(S_T)\right] \tag{3.2}$$

In the case of a constant interest rate, this equation simplifies to:

$$p_0 = e^{-rT}\mathbb{E}\left[V(S_T)\right] \tag{3.3}$$

This risk-neutral valuation formula, built on replication and hedging arguments, will be one of the key pillars in our derivative modeling efforts, and as such, will be used frequently throughout the text.

3.2.5 Risk Neutral Discounting, Risk Premia & Stochastic Discount Factors

The use of risk-neutral discounting, and the corresponding risk-neutral measure is a noteworthy part of our framework that is worth emphasizing. As we can see from equation (3.2), discounting in a risk-neutral context is done via a deterministic function, which can easily be obtained from the current interest rate curve. Importantly, these discount factors are not state dependent, or stochastic. They are also based on the assumption that investors are risk-neutral. Said differently, they are built independently of investors varying risk preferences. This brings a great deal of convenience and simplicity to the risk-neutral valuation framework. If we were forced to incorporate investor risk preferences into our model, it would result in a different more complex discount or drift term, that would vary not only from state to state, but also potentially from investor to investor. We would also then need to model the correlation structure between our stochastic discount factors, and the underlying asset that our derivative is based on.

In the next chapter, we transition to the physical measure where the assumption of risk-neutrality no longer holds. Instead, we will then have to incorporate investor preferences. Further, these investor preferences may be different from investor to investor, which will manifest itself in different utility functions. A share of Apple stock, for example, may provide one investor with more utility than another, if the payoffs for Apple are high in a particular state that the first investor values greatly. The amount of utility for a given security will be dependent on how risk averse they are which gives rise to the concept of a risk premia. Risk premia is a foundational concept in quantitative finance, and refers to the fact that investors may earn a premia for bearing risks that others are averse from. In this context, the discount factors that we compute will themselves be stochastic, and be functions of state and risk preferences. The fact that we are able to obviate these challenges associated with risk premia and risk preferences in the context of derivatives valuation is a fundamental point, and one that the reader is encouraged to think through.

3.3 BINOMIAL TREES

3.3.1 Discrete vs. Continuous Time Models

The study of stochastic processes is broken into discrete and continuous models. Continuous models rely on the same tools that we rely on in ordinary calculus problem, albeit updated to handle the stochastic nature of the processes we are working with. They assume that asset prices themselves are continuous processes, that can be traded and move instantaneously, rather than at set, discrete increments. In a discrete model, conversely, we might assume that prices move every d days, or every m minutes, but importantly the movements would be finite.

Of course, in practice markets actually trade more in alignment with discrete time models. Yet, in spite of this, quants have a tendency to gravitate toward continuous time models, and more work has been done building continuous models that closely resemble market dynamics. A significant reason for this is that, as quants, we have a more robust toolset for solving continuous problems (calculus) relative to discrete time problems (tree methods). Thus, working in a continuous time model is more mathematically convenient. This is analogous to calculus itself, where the assumption of the infinitesimal limit may likewise be unrealistic, however, it still provides a useful framework and accurate approximation for solving the problem of interest even as we relax the relevant assumptions. Nonetheless, in this section we detail the canonical discrete time tool for modeling a stochastic process, a binomial tree. We then move on to continuous stochastic processes, which are the focus of the remainder of the text.

3.3.2 Scaled Random Walk

Let's begin with our first building block for stochastic calculus, a discrete random walk. A random walk is a stochastic process that is a martingale, meaning its expected future value is equal to the current value of the process. A random walk can be motivated by considering the example of flipping a fair coin with equal probability of heads and tails. Let's assume there is a process that moves up one if a head occurs and back one if a tail occurs. Mathematically, we can write this as:

$$X_t = \begin{cases} 1 & \text{if heads} \\ -1 & \text{if tails} \end{cases} \tag{3.4}$$

where since we are tossing a fair coin, p, the probability of a head is 0.5, as is the probability of tails, $1-p$. We further assume that each increment, X_i is independent. Readers may notice that each element in this sequence of random variables, X_t, follows a Bernoulli distribution.

To construct a random walk, we can sum these independent Bernoulli increments, that is:

$$W_n = \sum_{i=1}^{n} X_i \tag{3.5}$$

Using the know expected value and variance formulas for a Bernoulli process, we can compute the mean and variance of the random walk process, W_t that we have created:

$$\mathbb{E}\left[W_n\right] = 0 \tag{3.6}$$

$$\text{var}(W_n) = \sum_{i=1}^{n} \text{var}(X_i) \tag{3.7}$$

$$= n\left(\mathbb{E}\left[X_i^2\right]\right) \tag{3.8}$$

$$= n \tag{3.9}$$

where in (3.7) we are using the independence property to eliminate the covariance terms. In the next line, we are using the definition of variance of the random variable X_i and, as we have shown the mean is equal to zero, this term can be eliminated as well. Finally, in the last line, we use the fact that $\text{var}(X_i) = (p)(1)^2+(1-p)(-1)^2 = 1$.

An unappealing feature of the random walk that we just created was that the variance is a function of the number of steps in our grid. In reality, this isn't a desirable quality, instead we will want it to be proportional to the length of time in our interval. For example, a path that spans one-year should intuitively have more variance than one that spans one-week, however, a path with 100 steps in a year should have the same variance as one with 500 steps that also spans a year.

To overcome this, let's create a grid that incorporates the timeframe that we are looking to model. Consider the grid defined by k, the number of total coin flips, or steps, and n the number of trials per period. Further, the time interval, t, can then be defined in terms of k and n via the following equation:

$$t = \frac{k}{n} \tag{3.10}$$

Using this grid, lets construct a scaled version of the previously constructed random walk which is defined as follows:

$$\hat{W_t} = \frac{1}{\sqrt{n}} W_k \tag{3.11}$$

Notice that our new grid is still based on a random walk that takes k steps, however, these steps are now rescaled to reflect the time period, by normalizing by $\frac{1}{\sqrt{n}}$. This scaled random walk has the following mean and variance:

$$\mathbb{E}\left[\hat{W}_t\right] = 0 \tag{3.12}$$

$$\text{var}(\hat{W}_t) = \text{var}(\frac{1}{\sqrt{n}}W_k) \tag{3.13}$$

$$= \frac{k}{n} = t \tag{3.14}$$

Importantly, notice that the variance is independent of k and n individually and is instead only a function of the time interval t. This rescaled random walk will become a building block for our discrete time models, and, as we transition to continuous time models, we will leverage this process and take the limit as the step size gets infinitesimally small.

3.3.3 Discrete Binomial Tree Model

The first model that we will consider is a discrete time model that relies on similar principles to a random walk in that we assume at each increment the asset moves up or down by the same amount. This model is referred to as a binomial tree [231], and is the most common discrete model applied in practice.

Let's begin with a one-period binomial tree model, with an asset that starts at S_0. As stated, the asset may then either go up and end at uS_0 or go down and finish at $dS_0 = \frac{1}{u}S_0$. Lastly, we need to define the probability of up and down moves, which will be p and $1-p$, respectively. Visually, our one-period binomial tree model can be expressed in the following chart:

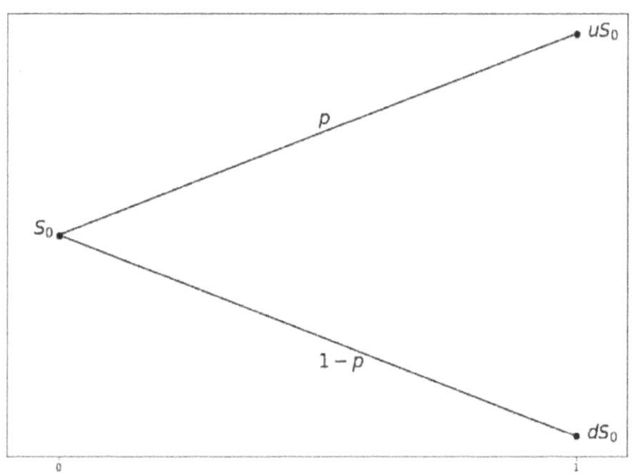

Notice that the up and down movements in the model we have created are symmetric. This ends up being an important feature as it causes the tree to recombine as we add more periods, leaving us with less nodes overall.

In order to value an option on our binomial tree we can iterate backward, beginning at the end of the tree where the payoff of the option should be known. For example, suppose the derivative of interest is a call option that expiries at the end of the first period and whose strike is equal to K. The payoffs of this structure at the end of the first period are known. If the movement of the asset is down, then the payoff is zero, otherwise it is $uS_0 - K$. Further, we know that the up movement occurs with probability p, and the down movement occurs with probability $1 - p$. Therefore, according to risk-neutral pricing theory, the expected value at time zero is this payoff, discounted by the risk-free rate, and can be written as:

$$c_0 = e^{-rt} p \left(uS_0 - K\right) \tag{3.15}$$

where we have omitted the second term for a down movement because the payoff is zero by definition. This can be seen via the following tree of the option valuation:

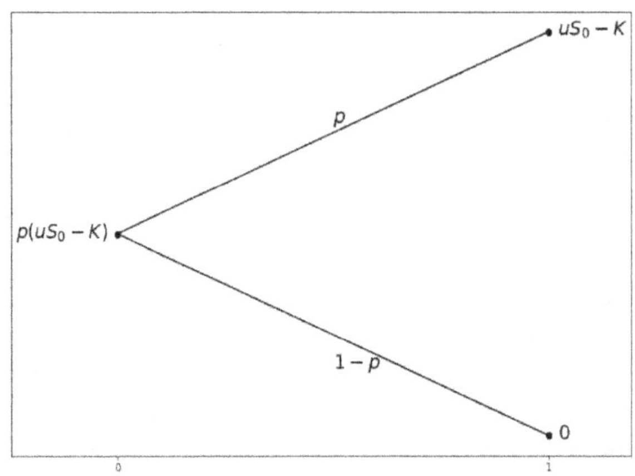

This simple one-period example highlights how we use a binomial tree in order to value a derivative. In particular, we first choose the parameters in our tree, notably u, d and p. u and d are chosen such that the movements are symmetric, resulting in a recombining tree. Additionally, we will want to choose u and d such that the variance in the underlying asset matches our desired level. p, in contrast is set such that the drift aligns with the risk-neutral valuation framework introduced above. The following equations show how u, d and p should be set if we are targeting an annualized volatility σ and have a constant interest rate, r:

$$u = e^{\sigma\sqrt{\Delta t}} \tag{3.16}$$

$$d = e^{-\sigma\sqrt{\Delta t}} = \frac{1}{u} \tag{3.17}$$

$$p = \frac{e^{r\Delta t} - d}{u - d} \tag{3.18}$$

Once we have selected the parameters, we can evolve asset prices forward along the tree until we reach the terminal nodes. We can then work backward along the tree of asset prices computing the value for the derivative of interest at each node.

At the end, the value of the derivative is known as it is just the payoff function of that derivative:

$$c_{n,i} = V(S_i) \tag{3.19}$$

We can then find the value one period sooner by leveraging the terminal values, and the probabilities of each terminal node:

$$c_{n-1,i} = e^{-r\tau}\left(pc_{n,i} + (1-p)c_{n,i+1}\right) \tag{3.20}$$

This process can be repeated until we reach the beginning of the tree, where we obtain a price for the derivative.

In the following visualization, we show how a recombining binomial tree works over multiple periods:

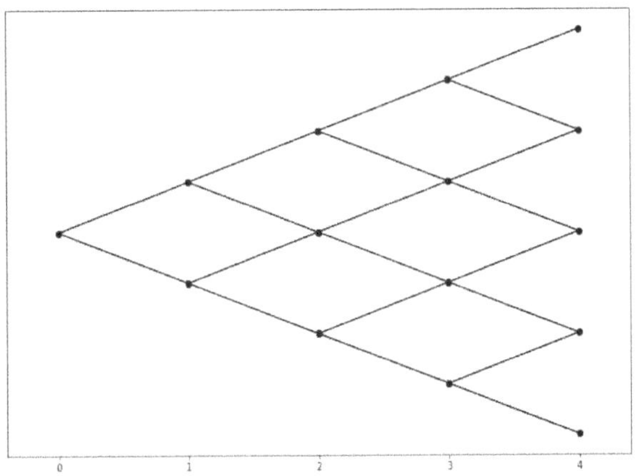

Readers may recall that the increments in a binomial tree come from a Bernoulli distribution. Therefore, when we aggregate the steps we obtain a Binomial distribution. We can see this by examining the probabilities of the asset prices at the end of the tree. At the extremes, the probability of achieving the highest and the lowest values of S_{T_i} will be p^n and $(1-p)^n$, respectively. That is, in order for that sequence to happen, we need to have observed only up or down movements. Similarly, for another point on the grid of S_T, it is also a combination of k up movements and $m = n - k$ down movements, the probability of which is defined by the following PDF of the binomial distribution:

$$f(S_{T_i}; p) = \binom{n}{k} p^k (1-p)^m \tag{3.21}$$

3.3.4 Limiting Distribution of Binomial Tree Model

The Central Limit theorem says that, in a Binomial distribution defined with n independent events, that the distribution of $f(x;p)$ converges to normal as n increases.

Therefore, as we increase the nodes in our binomial tree, the distribution becomes closer to a continuous time model that is defined by the normal distribution. In the next section, we build a continuous time model that is based on Brownian Motion, or normally distributed increments. According to the central limit theorem this continuous model should closely approximate the model presented here, and, in practice, we can see an equivalence between the binomial tree method here and the implementation of the Black-Scholes dynamics via a continuous process in the next section.

3.4 BUILDING BLOCKS OF STOCHASTIC CALCULUS

3.4.1 Deterministic vs. Stochastic Calculus

Before going through the building blocks of stochastic calculus it is worth highlighting how and why it differs from the ordinary calculus techniques readers learned about in their undergraduate courses. Generally speaking, in calculus we compute derivatives of functions that are smooth and well behaved. Taking a derivative in that context works by doing so locally, that is, treating the function as linear over infinitesimal horizons. As the process becomes stochastic, the random nature of the process will force us to deviate from this approach. To see this, consider the following chart of a deterministic function and stochastic process:

As the chart clearly highlights, there is a fundamental difference in the behavior of the two series. While both are continuous, the stochastic process is defined by bursts, or sudden movements in either nature that are driven by its random nature. These bursts can be thought of as up or down movements in a random walk. In practice, this difference in behavior creates challenges when we are trying to calculate a derivative of the process with respect to time. As we will soon see, these processes are not differentiable, and the presence of these jagged movements means that we cannot integrate them the way we could an ordinary function. Instead, we will need a new set of techniques, and a new set of building blocks, which we detail in the remainder of this section.

3.4.2 Stochastic Processes

A stochastic process is a time sequenced collection of random variables. The random nature of the increments makes the process stochastic.

$$S_t = \{S_1, \ldots, S_t\} \tag{3.22}$$

The study of stochastic processes such as S_t are the focus of stochastic calculus and derivatives modeling. The random nature of each element in the stochastic process, coupled with the dependency between them because they are in fact a time ordered sequence, leads to challenges when modeling stochastic processes. In the rest of this section, we develop the building blocks that might be contained with S_t. We then formulate a generic stochastic differential equation consisting of drift and random components, and show how we can mathematically analyze these types of processes.

3.4.3 Martingales

Examples of martingales abound in financial applications. The discrete random walk that we just introduced, is an example of a martingale. Similarly, we introduce Brownian Motion in the next section, which is also a martingale. Intuitively, we can think of a martingale as a stochastic process where the future conditional expectation is equal to the present value of the process [232]. Financial theory is built on concepts of efficient markets, which we discuss in the next chapter, which states that investors should not be able to predict asset prices. Additionally, replication and hedging arguments enable us to replicate complex derivatives structures. Both of these concepts naturally motivate the use of martingales in asset pricing.

Informally, for our purposes we will write a martingale as:

$$\mathbb{E}\left[S_T | S_t\right] = S_t \tag{3.23}$$

We can interpret this as a statement that, conditional on the information in the asset price until time t, which is measurable at time t, our best guess, or expectation of the future value at time T is the current price, S_t. When we say S_t is measurable at time t, we mean that the values of the process up to and including time t are observed and therefore known. In probability measure theory, there is additional notation and terminology that we neglect here for simplicity and brevity. Readers interested in a more rigorous treatment of martingales and measure theory should consult [86].

3.4.4 Brownian Motion

Brownian motion, or a Weiner process, is perhaps the most important fundamental building block of stochastic calculus [232] [144]. It is a continuous time stochastic process that will be at the core for defining uncertainty in many of the stochastic models that we build. It can be motivated by taking the limit of the scaled random walk that we introduced in the previous section. It is a stochastic process that starts at zero, and is then defined by independent, Gaussian increments:

$$B_0 = 0 \tag{3.24}$$

$$B_t - B_s \sim N(0, t-s) \tag{3.25}$$

Equivalently, we often write an increment of Brownian Motion in the following form:

$$dB_t \sim N(0, \delta t) \tag{3.26}$$

where dB_t is an increment from a Brownian Motion and δt is the time interval. As a Brownian Motion is a continuous process we are generally concerned with the limiting distribution as $\delta t \to 0$. In the next section, we delve deeper into the properties of Brownian Motion.

3.4.5 Properties of Brownian Motion

In the previous section we defined the most fundamental building block of stochastic processes, a Brownian Motion. In particular, we saw that it is itself a stochastic process that is defined by certain useful properties. In the following list, we highlight the main properties and features of Brownian Motion:

- **Independence of Increments**: The increments, dB_t in a Brownian motion process are independent by construction.
- **Normally Distributed Increments**: These increments are also assumed to be normally distributed. This normal distribution holds for arbitrary time intervals, t and s. That is, the distribution of an increment is normal regardless of its period length.
- **Continuous But Not Differentiable Paths**: It can be shown that the paths of a Brownian Motion, $B(t)$ are a continuous function of time, with probability one, however, that they are not differentiable. This is a noteworthy property that is worth understanding as it brings challenges in our modeling efforts of stochastic processes. We do not derive this result in this text as it is a fairly technical derivation, however, interested readers should refer to [144], [86] or [232].

  The reader may recall from calculus that a derivative may not exist at a given point because the function has a discontinuity, or because the function has a kink or bends such that the limit that we compute from the left hand side does not match what we compute from the right hand side. For example consider the absolute value function:

$$f(x) = |x| = \begin{cases} -x & x < 0 \\ x & x \geq 0 \end{cases} \tag{3.27}$$

  If we consider the point $x = 0$, clearly the derivative of this function coming from the left, -1, and the derivative of this function coming from the right, 1, are different. Further, we can see from the following plot of the function that the function indeed does have a kink at the point $x = 0$:

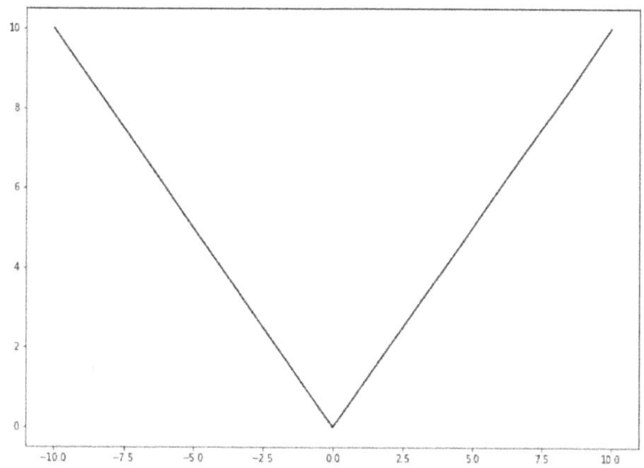

Notably, it is still continuous at this point, but it is not differentiable. To see this, imagine drawing the tangent line to this function at 0 from the right and the left. At other points, however, the absolute value function is differentiable. It is only where the kink occurs that we cannot compute the derivative. What is interesting and striking about Brownian motion then it is nowhere differentiable, yet still continuous. This means that the type of bend that we observe at $x = 0$ for the absolute value function is what we encounter at every point in a path of Brownian motion. Intuitively, we might think of a derivative operator as something that linearizes our function in order to estimate a rate of change. Brownian Motion, however, due to its nature as the infinite summation of a random walk, turns every instant and therefore cannot be linearized. This has significant implications for working with stochastic processes that are built on Brownian Motion, as we will soon see.

- **Quadratic Variation**: A defining characteristic of Brownian Motion is its quadratic variation. The presence of this quadratic variation is what makes stochastic calculus different from the standard calculus techniques that we reviewed above. As we will soon see, the presence of quadratic variation is what leads to a correction term in Ito's Lemma. A full derivation of the quadratic variation in Brownian Motion is beyond the scope of this text. Interested readers should consult dedicated stochastic calculus books such as [232], [144] or [86] for a rigorous derivation.

  Quadratic variation for a generic function, f can be informally written as:

$$(df\,df) = \lim_{||\Pi|| \to 0} \sum_{j=0}^{n-1} \left(f(t_{j+1}) - f(t_j)\right)^2 \tag{3.28}$$

where $(df df)$ is the quadratic variation of the function f and Π is the sum of the squared distance between the intervals on the time grid, t_j. Notice that as the width in between the points gets smaller, Π tends toward zero. Therefore, this limit is telling us what happens as the width of the time grid approaches zero. Ordinary calculus, most notably the mean-value theorem tell us that this quantity converges to zero for continuous, differentiable functions as our grid space shrinks to zero. Thus, for these functions, quadratic variation can be proven to be equal to zero.

Fundamentally, however, this is not true for Brownian Motion because of the properties of its paths, in particular that they are not smooth functions with differentiable paths. Instead the quadratic variation can be proven to be equal to t, which we can informally express as:

$$dB_t dB_t = dt \tag{3.29}$$

In practice, we will use this result to show that terms of $\mathcal{O}(dB_t^2)$ are equal to dt and cannot be ignored. The standard calculus equivalent, dt^2, conversely will converge to zero. This difference between the behavior of terms involving dB_t^2 and dt^2 is fundamental to stochastic calculus and Ito's Lemma.

- **Covariance Properties of Brownian Motion**: We can easily derive the covariance in a Brownian Motion process using the independent increment property stated above. To see the covariance between a Brownian Motion at times t and s let's begin with the definition of covariance reviewed earlier in the chapter:

$$\text{cov}(B_t, B_s) = \mathbb{E}\left[B_t B_s\right] \tag{3.30}$$

For simplicity, let's assume $s < t$, although this argument clearly works either way. Next, let's break down B_t into the part of the path that is shared with B_s, notably the part until time s, and the remainder. The reader should note that both of these terms can be thought of as increments of a Brownian Motion, with one spanning time 0 to s and the other spanning time s to t. That leads to the following expression for the covariance:

$$\begin{aligned} \text{cov}(B_t, B_s) &= \mathbb{E}\left[\left((B_t - B_s) + B_s\right) B_s\right] & (3.31) \\ &= \mathbb{E}\left[(B_t - B_s) B_s\right] + \mathbb{E}\left[B_s B_s\right] & (3.32) \end{aligned}$$

The independent increments property tells us that the first term in (3.32) is equal to zero as they are non-overlapping increments. Further, the second term is the variance of B_s. As we know B_s is normally distributed with mean 0 and variance s, we have:

$$\text{cov}(B_t, B_s) = \min(s, t) \tag{3.33}$$

3.5 STOCHASTIC DIFFERENTIAL EQUATIONS

3.5.1 Generic SDE Formulation

Using the building blocks defined above, our next goal will then be writing an equation that defines the movements of the stochastic process over time. Clearly, this will need to include a random component, that incorporates our uncertainty about future outcomes. Brownian motion, which we introduced as a key building block in the last section, is a natural choice to represent the uncertainty, or variance in the process. Our equation may also need to incorporate some sort of underlying drift that we tend to observe in the underlying stochastic process. Putting this together, we can write what we call a stochastic differential equation that defines the increments in a stochastic process, via the following form:

$$dS_t = \mu(S_t, t)dt + \sigma(S_t, t)dW_t \tag{3.34}$$

where $\mu(S_t, t)$ and $\sigma(S_t, t)$ are arbitrary deterministic functions that define the drift and volatility as a function of the current time, t, and asset price, S_t, and W_t is a standard Brownian Motion as introduced in the last section.

Equivalently, we can express this in integral form as the following expression:

$$S_T = S_t + \int_t^T \mu(S_t, t)dt + \int_t^T \sigma(S_t, t)dW_t \tag{3.35}$$

In these equations we can see that the underlying asset price, or stochastic process, S_t, evolves according to two terms. The first term is a drift term, that defines the deterministic movement of the asset over time. As we can see in (3.34), this drift can be dependent on time, and the current asset price, but a defining feature of this term is that it is deterministic, or certain conditional on the time and state. The second term then represents the volatility in the stochastic process, and leverages the Brownian Motion building block.

This formulation provides flexibility in how we specify the underlying SDE, however, it also makes certain assumptions. For example, (3.34) assumes that all movements in the stochastic process are continuous, as they emanate from a Brownian Motion. That is, there are no jumps. Later in the text, we will consider models with a third component that allows for jumps in the stochastic process. It should be again emphasized that while we have flexibility in choosing the functions, $\mu(S_t, t)$ and $\sigma(S_t, t)$, this formulation assumes they are deterministic and not stochastic themselves. In chapter 13 we will see some examples of models where the volatility function, $\sigma(S_t, t)$ is itself stochastic, leading to a two-dimensional system of SDEs.

3.5.2 Bachelier SDE

Perhaps the simplest SDE specification is the Bachelier model where we assume that a Brownian Motion defines the volatility in the stochastic process and also allow for drift in the underlying asset [29]. This can be seen as a specification of (3.34) with $\mu(S_t, t) = rS_t$ and $\sigma(S_t, t) = \sigma$. This leads to the following SDE:

$$dS_t = rS_t dt + \sigma dW_t \tag{3.36}$$

where σ and r are now constant model parameters. This process is known as an arithmetic Brownian Motion because the volatility term is simply summing the increments in a Brownian Motion. It can be easily shown that this SDE in (3.36) leads to normally distribution asset prices.

3.5.3 Black-Scholes SDE

Another common formulation of the generic SDE in (3.34) is the Black-Scholes SDE, which, as the name implies, leads to the well-known Black-Scholes formula [43]. Assuming that we are pricing in the risk neutral world, in the Black-Scholes SDE we again set $\mu(S_t, t) = rS_t$. The volatility function, $\sigma(S_t, t)$, however, deviates from in the Bachelier SDE in (3.36). Instead, we use $\sigma(S_t, t) = \sigma S_t$ leading to the following popular Black-Scholes SDE:

$$dS_t = rS_t dt + \sigma S_t dW_t \tag{3.37}$$

where again r and σ are constants. This process is also known as a geometric random walk.

This difference in the formulation of the $\sigma(S_t, t)$ function in (3.34) leads to a critical difference in the behavior of the model. As we will see later, specification of the Black-Scholes dynamics, as we have done in (3.37), yields asset prices, S_t that are log-normally distributed, in contrast to the normally distributed asset prices in the Bachelier model.

3.5.4 Stochastic Models in Practice

The Bachelier and Black-Scholes model are two of the most commonly employed SDEs in practice. In most markets, one of these two models provides a market standard, or baseline that all options structures are quoted in terms of. One of the main advantages of these two choices as an SDE is they naturally lead to a closed form solution for the asset price. This makes it easier to solve derivative valuation problems in the form of (3.2) for different payoff structures. For example, as we will see later, European call and put options will have closed form solutions under both of these model specifications.

Conversely, while these models are naturally appealing due to their simplicity and tractability, they don't fully explain the dynamics embedded in options markets. This is a phenomenon that we will continue to explore throughout the book, and is evidenced by the fact that a single Bachelier or Black-Scholes model, generally speaking, will not be able to explain the prices of all options for a given asset. In order to accomplish this, we will instead need to leverage more complex SDEs, that incorporate either a jump component, or allow for a stochastic rather than deterministic volatility function.

In chapter 13, we provide a more detailed survey of the most common stochastic models leveraged in practice, beginning with a more detailed treatment of the Bachelier and Black-Scholes model, and then introduce additional commonly used approaches, such as stochastic volatility models and those with a jump component.

3.6 ITO'S LEMMA

3.6.1 General Formulation & Theory

Ito's Formula, or Lemma is perhaps the most important result in stochastic calculus. It is how we calculate differentials of complex functions in stochastic calculus, and, along those lines, is the stochastic calculus equivalent of the chain rule in ordinary calculus. For example, Ito's formula can help us to write a differential equation of a function, $f(S_t, t)$, that is a function of time and a stochastic process. Of course, derivatives valuation problems will fit into this mold of functions, and, as such Ito's Lemma will be an important tool for modeling options.

A proper proof of Ito's Lemma is beyond the scope of this text, however, readers interested in one should refer to [232] or [86]. Instead, we motivate Ito's Lemma based on the Taylor series expansion of a multi-variate function. To begin, let's write this expansion for a function f which is a function of two variables, t and S_t:

$$df = \frac{\partial f}{\partial t}dt + \frac{\partial f}{\partial S_t}dS_t + \frac{\partial f}{\partial t S_t}dS_t dt + \frac{1}{2}\frac{\partial^2 f}{\partial S_t^2}dS_t^2 + \frac{1}{2}\frac{\partial^2 f}{\partial t^2}dt^2 + \cdots \tag{3.38}$$

where we are assuming that the S_t is of the form of (3.34) discussed above:

$$dS_t = \mu(S_t, t)dt + \sigma(S_t, t)dW_t \tag{3.39}$$

Notably, it is a drift-diffusion model, with a drift term characterized by dt and a diffusion term characterized by a Brownian Motion.

In ordinary calculus, differentials such as (3.38) would be computed by ignoring terms smaller than $\mathcal{O}(dt)$ and $\mathcal{O}(dS_t)$, which would leave us with only the first order partial derivatives. As we have already seen, however, calculus is different when working with stochastic processes, because of the quadratic variation in Brownian Motion. Most crucially, quadratic variation tells us that a term of order dW_t^2 is of order $\mathcal{O}(dt)$ and cannot be ignored. Therefore, we must include part of the dS_t^2 term in the above Taylor series expansion in our calculation. We can, however, ignore terms that are $\mathcal{O}(dt^2)$ or $\mathcal{O}(dtdW_t)$, meaning the second-order derivative with respect to time and the cross partial derivative are negligible. This leaves us with the following differential equation, which is a generic form of Ito's Lemma:

$$df = \frac{\partial f}{\partial t}dt + \frac{\partial f}{\partial S_t}dS_t + \frac{1}{2}\frac{\partial^2 f}{\partial S_t^2}dS_t^2 \tag{3.40}$$

Further, within the dS_t^2 term only the Brownian component will be large enough to include. All other terms can be ignored. Plugging in the generic drift-diffusion model above for dS_t, we then have the following Ito formula:indexIto's Lemma

$$df = \frac{\partial f}{\partial t}dt + \frac{\partial f}{\partial S_t}\mu(S_t, t)dt + \frac{\partial f}{\partial S_t}\sigma(S_t, t)dW_t + \frac{1}{2}\frac{\partial^2 f}{\partial S_t^2}\sigma(S_t, t)^2 dt \tag{3.41}$$

This result is one of the most important results in stochastic calculus, and, makes working with stochastic processes far more tractable. Ito's formula can be used to solve an SDE for the distribution of an asset price, S_t, as we explore in the remainder of the section. It can also be used to derive the PDE that corresponds with a stochastic process, something we detail later in the chapter.

3.6.2 Ito in Practice: Risk-Free Bond

To see how Ito's Lemma can be applied in practice, let's first consider a simple example of a risk-free bond with certain cashflows. Intuitively, as there is no variance or Brownian component to this, we would expect that it would follow the rules of standard calculus, and, as we will see, it does. The dynamics of this risk-free bond can be expressed via the following SDE:

$$dB_t = rB_t dt \tag{3.42}$$

$$\frac{dB_t}{B_t} = rdt \tag{3.43}$$

Importantly, in this process we have only have a drift term and no diffusion, or volatility term. To solve this SDE, let's apply Ito's formula to (3.42). Readers may notice that the left hand side of (3.43) looks like the derivative of $\log(x)$. As a result, let's apply Ito to the function, $f(B_t) = \log(B_t)$. This leads to the following set of partial derivatives that we will need in order to apply Ito:

$$f(B_t, t) = \log(B_t) \tag{3.44}$$

$$\frac{\partial f}{\partial t} = 0 \tag{3.45}$$

$$\frac{\partial f}{\partial B_t} = \frac{1}{B_t} \tag{3.46}$$

$$\frac{\partial^2 f}{\partial B_t^2} = -\frac{1}{B_t^2} \tag{3.47}$$

Next, recall that Ito's formula tells us how to calculate the change in a function, $f(B_t, t)$ for a given SDE, in this case dB_t as defined in (3.42). It can be written as:

$$df = \frac{\partial f}{\partial t}dt + \frac{\partial f}{\partial B_t}dB_t + \frac{1}{2}\frac{\partial^2 f}{\partial B_t^2}dB_t^2 \tag{3.48}$$

Plugging in the previously obtained derivatives into this equation, as well as the definition of dB_t, we can see that the second-order term dB_t^2 is $\mathcal{O}(dt^2)$ and is therefore negligible and can be ignored. The first term is also equal to zero because of the partial derivative of our function with respect to t. That leaves us with the following:

$$df = \frac{1}{B_t}(rB_t dt) \tag{3.49}$$

$$df = rdt \tag{3.50}$$

This expression can then be used to find the value of B_t at a given time, t. In particular, the notation df is short-hand for the value of $f_t - f_0$ over a given interval. This means we can write the value for B_t as:

$$f_t - f_0 = rdt \tag{3.51}$$

$$\log(B_t) - \log(B_0) = rdt \tag{3.52}$$

$$B_t = B_0 \exp(rt) \tag{3.53}$$

As expected, in the case of a risk-free bond we recover the standard calculus answer, without a correction term. This is because we were able to ignore the dB_t^2 term, as we would in ordinary calculus. This was because of the lack of a dW_t term, however, and generally speaking, as we incorporate these Brownian terms this will not be the case. In particular, as we will see in the next section, if we introduce a similarly defined dW_t term in the SDE, as we saw in the Black-Scholes model, this second-order term no longer goes away and a correction term emerges.

3.6.3 Ito in Practice: Black-Scholes Dynamics

Next, let's consider an example of Ito's lemma that uses a drift-diffusion process, and incorporates a random component. In particular, let's consider the widely popular Black-Scholes dynamics:

$$dS_t = rS_t dt + \sigma S_t dW_t \tag{3.54}$$

$$\frac{dS_t}{S_t} = rdt + \sigma dW_t \tag{3.55}$$

In addition to the Black-Scholes model, this model for S_t is commonly known as the log-normal model, and in this section we will see why. As in the last example, our goal will be to apply Ito in order to solve the SDE in (3.54) for S_t. This SDE appears to have similar structure as the risk-free bond in that S_t appears in both the drift and diffusion terms. This means we can divide by both sides by S_t to obtain the SDE in (3.55). As before, the left-hand side of this equation looks like the derivative of $\log(x)$, so we will again proceed by defining $f(S_t, t) = \log(S_t)$ and apply Ito's Lemma. This leads to the following set of partial derivatives:

$$f(S_t) = \log(S_t) \tag{3.56}$$

$$\frac{\partial f}{\partial t} = 0 \tag{3.57}$$

$$\frac{\partial f}{\partial S_t} = \frac{1}{S_t} \tag{3.58}$$

$$\frac{\partial^2 f}{\partial S_t^2} = -\frac{1}{S_t^2} \tag{3.59}$$

We can then apply these partial derivatives into Ito's formula for S_t, which is:

$$df = \frac{\partial f}{\partial t}dt + \frac{\partial f}{\partial S_t}dS_t + \frac{1}{2}\frac{\partial^2 f}{\partial S_t^2}dS_t^2 \tag{3.60}$$

It should be emphasized that the dS_t^2 term now contains components that are $\mathcal{O}(dt^2)$, $\mathcal{O}(dt dW_t)$ and $\mathcal{O}(dW_t^2)$. The first two types of terms can be ignored as they are negligible, however, due to the quadratic variation of Brownian Motion that we saw earlier, the dW_t^2 term is of $\mathcal{O}(dt)$ and cannot be ignored. This means, if we plug in the appropriate partial derivatives and omit only terms smaller than $\mathcal{O}(dt)$ we are left with:

$$df = \frac{1}{S_t}(rS_t dt + \sigma S_t dW_t) - \frac{1}{2}\frac{1}{S_t^2}(rS_t dt + \sigma S_t dW_t)^2 \quad (3.61)$$

$$df = rdt + \sigma dW_t - \frac{1}{2}\sigma^2 dt \quad (3.62)$$

As we did before, we can now use the above expression in order to find the distribution of S_t. Importantly, the dW_t term above can now be integrated as it is simply a Brownian Motion, or Weiner process, W_t. Following the same logic as before, we then have:

$$f_t - f_0 = rt + \sigma W_t - \frac{1}{2}\sigma^2 t \quad (3.63)$$

$$\log(S_t) - \log(S_0) = rt + \sigma W_t - \frac{1}{2}\sigma^2 t \quad (3.64)$$

$$S_t = S_0 \exp\left(rt + \sigma W_t - \frac{1}{2}\sigma^2 t\right) \quad (3.65)$$

Therefore, we have been able to express the distribution of S_t in terms of a single Brownian Motion, whose distribution is known to be normal, and a deterministic drift term. This representation in (3.65) is the closed-form solution for the asset price under the Black-Scholes model. From (3.65) and (3.64) we can see that the distribution of the log of the asset price, $\log(S_t)$ is normally distribution. Hence, S_t is log-normally distributed.

It should also be emphasized that, unlike in the previous example, a correction term in (3.65) has emerged relative to the analogous derivative in ordinary calculus. This term, $\frac{1}{2}\sigma^2 t$ emanated from the quadratic variation of Brownian Motion which prevented us from ignoring the second-order term, dS_t^2. This correction term is a defining feature of stochastic calculus relative to ordinary calculus. In this particular case, readers versed in statistics may notice that this correction term looks familiar, as it is the same correction term that emerges when converting from a normal to log-normal distribution.

One point to highlight with these examples of Ito's lemma is that, in both cases we began by specifying a function $f(x) = \log(x)$. It turned out that these functions were well chosen, as they enabled us to solve the respective SDEs. In practice, if we know how to specify this function, then, as we saw, applying Ito's Lemma becomes mechanical. Conversely, if we don't know the appropriate function, the process can become much more challenging. In these examples, we started with an observation that the left-hand sides of (3.43) and (3.55) looked like the derivative of the logarithm function. Just as important, we knew how to integrate the right-hand side of these equations in this form. In many cases with more complex SDEs, the specification of this function may become much harder, and in some cases, of course, no function will exist that solves the SDE. This makes the process difficult and often built on trial and error.

3.7 CONNECTION BETWEEN SDEs AND PDEs

3.7.1 PDEs & Stochastic Processes

Readers who have studied mathematics might be familiar with the concept of partial differential equations (PDEs). We saw earlier in this chapter the complexity that gets introduced as we make different quantities stochastic. PDEs, then, describe how the different partial derivatives evolve and defines the relationship between them. Once we have established this connection, then we are able to use the wide array of PDE techniques, both numerical and analytical, to solve our problem of interest.

In this section, we describe how we can connect a stochastic process to an ensuing PDE, beginning with a specific case of the Black-Scholes model and then presenting a general formula that works on a generic SDE of the form of (3.34).

3.7.2 Deriving the Black-Scholes PDE

To see how we can go from a stochastic differential equation to a partial differential equation let's consider the case of the Black-Scholes model. As we saw, the Black-Scholes model is defined by the following dynamics for an underlying asset, S_t:

$$dS_t = \mu S_t \, dt + \sigma S_t \, dW \tag{3.66}$$

Further let's assume there is some option, contingent claim or derivative structure $c(S_t, t)$ that we are trying to model. We also suppose that the payoff at maturity for the derivative, $c(S_T, T)$ is known. For example, in the case of a European call option, the payoff at maturity would be defined as:

$$c(S_T, T) = \max\left(S_T - K, 0\right) \tag{3.67}$$

Returning to the general form of $c(S_t, t)$, let's apply Ito's Lemma to $c(S_t, t)$ with dS_t defined by (3.66). This yields:

$$dc(S_t, t) = \frac{\partial c}{\partial s} dS_t + \frac{1}{2}\frac{\partial^2 c}{\partial s^2} (dS_t)^2 + \frac{\partial c}{\partial t} dt \tag{3.68}$$

As before only the dW_t^2 term from the expansion of dS_t^2 needs to be included as the other terms are small and can be ignored. Plugging in the dynamics in (3.66) and ignoring higher order terms, we have:

$$\begin{aligned} dc(S_t, t) &= \frac{\partial c}{\partial s} \mu S_t \, dt + \frac{\partial c}{\partial s} \sigma S_t \, dW_t + \frac{1}{2}\frac{\partial^2 c}{\partial s^2} \sigma^2 S_t^2 \, (dW_t)^2 + \frac{\partial c}{\partial t} dt \\ &= \left(\frac{\partial c}{\partial t} + \mu S_t \frac{\partial c}{\partial s} + \frac{1}{2}\sigma^2 S_t^2 \frac{\partial^2 c}{\partial s^2}\right) dt + \frac{\partial c}{\partial s} \sigma S_t \, dW_t \end{aligned} \tag{3.69}$$

This equation (3.69) defines how the option, $c(S_t, t)$ evolves over time as a function of its different components. To push forward and try to understand the behavior of this contingent claim more, let's now consider a portfolio that invests in the derivative as well as the underlying asset, S_t. More specifically, let's form a position in the underlying that tries to hedge the risk in $c(S_t, t)$. This means eliminating the dW_t term in (3.68).

There is a single term that is exposed to this randomness, $\frac{\partial c}{\partial s}\sigma S_t\, dW_t$. If we own one unit of the derivative, $c(S_t, t)$, then $-\frac{\partial c}{\partial s}$ units of S_t makes the portfolio immune from the uncertainty of dW_t. More generally, as the Black-Scholes model assumes that this is the only source of randomness, the portfolio becomes risk-free. This risk-free portfolio can be written as:

$$\Pi(S_t, t) = c(S_t, t) - \frac{\partial c}{\partial s} S_t. \tag{3.70}$$

Practitioners refer to (3.70) as a delta hedged portfolio, and $\frac{\partial c}{\partial s}$ as the delta of an option or other derivative. Next, let's look at the evolution of $\Pi(S_t, t)$ over time, $d\Pi$. The first component of $\Pi(S_t, t)$, $c(S_t, t)$ evolves according to (3.68), as we just derived via Ito's Lemma. The second part deals with a position in the underlying asset. As it is a linear component, it is defined solely by its $\frac{\partial c}{\partial s}$ term. This means that $d\Pi$ evolves according to the following equation:

$$\begin{aligned} d\Pi(S_t, t) &= \left(\frac{\partial c}{\partial t} + \mu S_t \frac{\partial c}{\partial s} + \frac{1}{2}\sigma^2 S_t^2 \frac{\partial^2 c}{\partial s^2}\right) dt + \frac{\partial c}{\partial s}\sigma S_t\, dW_t - \frac{\partial c}{\partial s} dS_t & (3.71) \\ &= \left(\frac{\partial c}{\partial t} + \mu S_t \frac{\partial c}{\partial s} + \frac{1}{2}\sigma^2 S_t^2 \frac{\partial^2 c}{\partial s^2}\right) dt + \frac{\partial c}{\partial s}\sigma S_t\, dW_t - \frac{\partial c}{\partial s}\mu S_t dt - \frac{\partial c}{\partial s}\sigma S_t dW_t & \\ &= \left(\frac{\partial c}{\partial t} + \frac{1}{2}\sigma^2 S_t^2 \frac{\partial^2 c}{\partial s^2}\right) dt & (3.72) \end{aligned}$$

where the last two terms involving dW_t cancel because of the delta hedge that we have put in place. The most noteworthy feature of the evolution of our delta-hedged portfolio is that it has no exposure to uncertainty, it is risk-less. This is because we immunized ourself from the one source of variation within the context of a Black-Scholes model, by building a hedge with the underlying asset.

Because we have now built a risk-free portfolio, that is instantaneously delta-hedged, we can then apply the type of arbitrage argument introduced earlier in the chapter to state that this portfolio must earn the same return as a position in the risk-free rate. If this portfolio were to earn a higher return than the risk-free rate, than a arbitrageur could simply buy this portfolio and sell the risk-free rate, collecting a profit without taking any risk. As a result, we can add the following condition to the evolution of our delta hedged portfolio:

$$\begin{aligned} d\Pi &= r\Pi\, dt & (3.73) \\ d\Pi &= r\left(c(S_t, t) - \frac{\partial c}{\partial s} S_t\right) dt & (3.74) \end{aligned}$$

where in the second equation we substituted the definition of $\Pi(S_t, t)$. This means that we have been able to derive two different expressions for the evolution of $\Pi(S_t, t)$, one based on Ito's Lemma and the dynamics of the Black-Scholes SDE and another based on our no-arbitrage argument which stated that, because the portfolio was risk-free, it should earn the risk-free rate. Equating these two difference expressions for $d\Pi(S_t, t)$, we get:

$$\left(\frac{\partial c}{\partial t} + \frac{1}{2}\sigma^2 S_t^2 \frac{\partial^2 c}{\partial s^2}\right) = rc(S_t, t) - r\frac{\partial c}{\partial s} S_t \tag{3.75}$$

Rearranging terms, we are left with the famous Black-Scholes PDE:

$$\frac{\partial c}{\partial t} + \frac{1}{2}\sigma^2 S_t^2 \frac{\partial^2 c}{\partial s^2} + r\frac{\partial c}{\partial s} S_t - rc(S_t, t) = 0 \tag{3.76}$$

At first glance, the fact that we were able to build a risk-free portfolio using the option and a position in underlying asset is a striking and possibly a somewhat surprising result. Given the additional complexity of options and derivatives, it is not obvious that we would be able to replicate them, or fully remove the embedded risk via a position in the underlying asset. Two points should be emphasized in this context. First, we can see from our SDE that there is a single driver of volatility in the Black-Scholes model. This means that a single hedging vehicle can hedge this single source of risk. If we were to consider models with additional sources of randomness, we would need additional hedging instruments in order to fully immunize our portfolio from risk. Secondly, this immunization from risk only holds instantaneously. Over time, the non-linearity in the behavior of an option will mean that we are no longer fully hedged.

Said differently, the term $\frac{\partial c}{\partial s}$ is itself dependent on the asset price and time meaning that the hedging portfolio that we created in (3.70) is dynamic. This is a fundamental distinction from the first chapter when we built hedging portfolios that are static and don't need to be modified over time. Here, rather than this static hedging argument we have a dynamic hedging portfolio that can be used to replicate the option by trading the underlying continuously, and in doing so removing the instantaneous risk at all time steps. The reader might also note that we are assuming we can hedge continuously, meaning we would need to able to trade continuously to adjust our hedge, and also that there are no transactions costs as we implement or modify our hedge.

As we explore later in the text, the Black-Scholes PDE tells us some important things about the underlying dynamics of options valuations. In particular, for delta-hedged portfolios, it implies a fundamental relationship between $\frac{\partial^2 c}{\partial s^2}$, called gamma in the industry, and $\frac{\partial c}{\partial t}$, which is referred to as theta. This relationship, the interpretation of the Black-Scholes PDE, and numerical methods for solving PDEs are discussed in more detail in chapters 15 and 16, respectively.

Most of the PDEs that occur in finance cannot be solved analytically, and instead we need to resort to numerical methods. The Black-Scholes model, however, is an exception, as the repeated use of the change of variables technique enables us to obtain the celebrated Black-Scholes formula from the PDE in (3.76).

3.7.3 General Formulation: Feynman-Kac Formula

In the previous section, we detailed how to use a hedging and replication argument in order to take a specific SDE and convert it into a PDE. This PDE, as we saw, then defines the evolution of different contingent claims in the context of the stochastic model. It turns out, that the Feynma-Kac formula [155] can be used to establish a more general link between the expectation of a contingent claim and a corresponding partial differential equation.

To see this, let's again consider the generic SDE formulation in (3.34), that is:

$$dS_t = \mu(S_t, t)dt + \sigma(S_t, t)dW_t \tag{3.77}$$

Further, let's assume that there is a contingent claim or derivative structure that we are interested in modeling, whose discounted payoff can be written as:

$$u(s,t) = \mathbb{E}_{S_t=s}\left[e^{-\int_t^T b(\tau,s)d\tau} V(S_T)\right] \tag{3.78}$$

where $b(s,\tau)$ is the discounting process and $V(S_T)$ is the payoff function for the derivative. Importantly, we assume that the payoff is only a function of the terminal price of the asset, as is the case for European style options. Additionally, while we have flexibility in how the payoff $V(S_T)$ is defined, we assume that it is known. For example, in the case of a European call, it would be $(S-K)^+$.

Feynam and Kac showed that the expectation on the right hand side of (3.78) can be written as:

$$\frac{\partial u}{\partial t} + \frac{1}{2}\sigma(s,t)\frac{\partial^2 u}{\partial s^2} - b(t,s)u(s,t) = 0 \tag{3.79}$$

Further, as $V(S_T)$ is a known, deterministic payoff function, the terminal value can be written as:

$$u(S_T, T) = V(S_T) \tag{3.80}$$

This will serve as a boundary condition in our PDE solution.

The reader should notice that the PDE solution in (3.79) closely resembles the PDE that we derived for the Black-Scholes model. Intuitively, we can think of the Feynman-Kac formula in a similar manner to the previously constructed replication argument above. In particular, it can be proven that the quantity inside the expectation in (3.78) is a martingale. This martingale property stems from our ability to hedge the risk in the claim instantaneously using a dynamic trading strategy, and means that the drift in the discounted expectation should equal zero. A full derivation of the result in (3.79) is beyond the scope of the text, however, readers looking for more detail should see [86] or [144].

3.7.4 Working with PDEs in Practice

In practice, PDEs such as the Black-Scholes model can be solved analytically, and solutions to PDEs is itself a widely researched field within mathematics. While we do not cover such solutions to PDEs in this text, interested readers are encouraged to consult [95]. It is far more common, however, in practical applications, to encounter PDEs with no analytical solution. Instead, in these cases, we will need to employ numerical techniques in order to solve the relevant PDE. This will be the case for the majority of models that are able to explain the behavior of the volatility surface, for example, stochastic volatility models or jump diffusion models. In chapter 15 we explore the most common approaches to solving PDEs numerically.

3.8 GIRSANOV'S THEOREM

3.8.1 Change of Measure via Girsanov's Theorem

Girsanov's theorem [112] is another foundational tool for modeling derivatives and contingent claims. Girsanov's theorem enables us to change the pricing measure that we are working in. This means working with a different set of underlying probabilities when calculating our expectation, and adjusting our expectation to account for the fact that we have changed the probabilities. A canonical example of this technique is to change from a normal distribution with mean zero to a normal distribution with mean μ and the same variance.

To see how a change of measure works, let's begin with the standard risk-neutral valuation equation in (3.3)[1]:

$$p_0 = e^{-rT}\mathbb{E}_f\left[V(S_T)\right] \tag{3.81}$$

where the $\mathbb{E}_f$ signifies that the expectation is being done under the probability measure f with a density function defined as $f(S_T)$. Using the definition of expectation, we can equivalently write this as the following:

$$p_0 = e^{-rT}\int_{-\infty}^{\infty} V(S_T)f(S_T)dS_T \tag{3.82}$$

This equation shows us that pricing a European style option, which is only dependent on the terminal value of the asset, can be achieved by calculating a single integral against a PDF, if the PDF is known. This is an important realization, and something that we return to in chapter 14. Now let's suppose that the probability measure f is inconvenient for us. Later in the text we will cover several examples of why this might be the case. Instead, we want to work in another probability measure, g.

Next, let's consider the following equivalent expression of (3.82):

$$p_0 = e^{-rT}\int_{-\infty}^{\infty} V(S_T)\frac{f(S_T)}{g(S_T)}g(S_T)dS_T \tag{3.83}$$

This equation is the same as before. We have simply multiplied and divided by the probability density function $g(S_T)$, which therefore cancels. However, it is now possible to express this equation in terms of the probability density for $g(S_T)$, albeit with an augmented payoff function, $V(S_T)\frac{f(S_T)}{g(S_T)}$ that incorporates the transformation

[1]Here we focus on the valuation equation with a constant interest rate, however, this is for simplicity and would not impact the result.

of probabilities. In expectation form, we can equivalent re-write our new integral directly as an expectation in the new measure, g:

$$p_0 = e^{-rT}\mathbb{E}_g\left[V(S_T)\frac{f(S_T)}{g(S_T)}\right] \tag{3.84}$$

$\frac{f(S_T)}{g(S_T)}$ is known as the Radon-Nikodym derivative, and intuitively, is the ratio of the probabilities between the measures f and g. So at a high level, what we have done is transformed the probabilities and then offset this transformation by this Radon-Nikodym term. Once this Radon-Nikodym term is combined with the new probability measure, we are left with the probabilities in our original probability measure f.

One way to think about the measure f and g respectively is that they are the probabilities that arise when deflating prices by different assets in our valuation equation. For example, perhaps f is the standard risk-neutral valuation framework, where the risk-free bond is used to deflate prices from the future to today. The measure g could then be the probabilities when a unit of Nike stock, or bitcoin are respectively used to deflate prices. This would result in a different set of probabilities and a different, state dependent drift component, that we saw here manifest itself in the Radon Nikodym derivative, $\frac{f(S_T)}{g(S_T)}$. The larger point however is that the choice of the asset that we use to deflate prices is a matter of convenience. In the standard pricing setup, a risk-neutral measure is naturally convenient because the drift terms are easy to model. Using a stock or bitcoin to deflate prices, alternatively, would make these terms much more difficult to estimate. However, as we will see in the next section, there are some cases where we want to leave the risk neutral world and choose another, more convenient, measure.

3.8.2 Applications of Girsanov's Theorem

Using Girsanov's theorem to change measure is something that we will return to at multiple point throughout the text. For example, in the realm of fixed income, and in particular interest rate modeling, we will find that the risk-neutral measure is not always the most convenient pricing measure. This is inherently due to interest rates being a stochastic quantity, making the discounting terms themselves stochastic. This will make it easier to work in another measure where we do not have to calculate these stochastic discount terms, which will be possible after a change of measure. More details on this application of Girsanov's theorem can be found in chapter 19. Similarly, we will find this phenomenon exists for certain credit securities, most notably swaptions, and show how changing measure lessens the complexity of their valuation in chapter 20. In FX markets, changes of measure naturally arise as we reorient a particular currency pair toward its home investor. For example, we might think about the dynamics of the Euro vs. US dollar exchange rate through the lens of a European or US investor. Changing measure provides us the tool for switching back and forth between these perspectives. Finally, in chapter 15 we will leverage Girsanov's theorem in the context of simulation, where we will use it to simulate rare events more efficiently.

CHAPTER 4

Theoretical Underpinnings of Quant Modeling: Modeling the Physical Measure

4.1 INTRODUCTION: FORECASTING VS. REPLICATION

In the last chapter we developed the core mathematical tools that are needed to model the risk-neutral measure. At the heart of this was a set of replication arguments. In the presence of these replication arguments, we were able to immunize our risk over short periods. We were then able to build a dynamic hedging strategy that kept our risk immunized, and we concluded that, since this portfolio was risk-free, it must earn the risk-free rate in the absence of arbitrage. This replication argument told us that we were able to replicate an option using this dynamic hedging strategy while trading only the underlying asset and the risk-free asset. This then enabled us to pick a probability measure where investors were risk-neutral, which was chosen as it is the measure where it is easiest to estimate discount rates and drift terms. They are simply a function of the risk-free rate.

The presence of these hedging and replication arguments, coupled with the use of the risk-neutral discounting, are defining features of working in the risk-neutral measure. When working in this measure, we rely on the techniques of stochastic calculus, build replicating portfolios to mimic complex derivatives and generate models that produce derivative valuations and hedging schemes. This risk-neutral measure then defines the distribution, or dynamics, of an asset that is implied by a set of option prices. This risk-neutral world is of great use within the context of valuing derivatives, as it enables us to build models whose prices are consistent with the most liquid options. Conversely, the risk-neutral measure does not represent a prediction of actual future outcomes. These actual future outcomes come from a different distribution, where investors are not risk-neutral but are instead risk-averse. This difference in risk preferences impacts the distribution as investors may place more value on payoffs in certain states, and less value in others.

In this chapter we transition from working in the risk-neutral measure where we can replicate a contingent claim using a combination of simpler assets, to the physical

DOI: 10.1201/9781003530626-5

measure, where investors are risk-averse, replication arguments are no longer possible and instead we must rely on our ability to forecast. In the physical measure we are no longer interested in the market implied prices, as determined by a set of options, we are interested in actual future outcomes of the assets themselves. This measure is the focus of countless buy-side institutions. With this change in perspective, comes a change in the necessary set of tools, which we introduce in the remainder of this chapter. Instead of relying on stochastic models we will rely on forecasting techniques. And instead of hedging and replication arguments, we will rely on techniques for building portfolios. In the rest of this chapter we provide an overview of the techniques required to operate within this context and also an overview of the prevailing theory, including the seminal Efficient Market Hypothesis, [1] and a set of documented market anomalies.

4.2 MARKET EFFICIENCY AND RISK PREMIA

4.2.1 Efficient Market Hypothesis

The Efficient Market Hypothesis (EMH) is a key tenet of financial market theory and states that asset prices reflect all available information. This means that, according to the EMH, investors should not be able to consistently outperform on a risk-adjusted basis. This idea is closely related to the idea prevalent in economics that there is no such thing as a free lunch. Essentially, the efficient market hypothesis is based on the idea that if all investors are rational and have access to the same information, then they should reach the same conclusions, adjusting the prices of any over or underpriced securities accordingly. Even if irrational investors exist, they should be phased out of the market over time as they suffer comparably worse returns. Any new information then, such as changes to earnings estimates, should similarly be incorporated into the price of a security immediately, as the consensus opinion of investors shifts.

The efficient market hypothesis is broken into three forms, weak, semi-strong and strong market efficiency, each of which is described below:

- **Weak**: Current prices reflect all information available in historical prices. If this form of efficiency is true, then technical analysis and quant strategies based on momentum and mean-reversion should fail to generate excess profits.
- **Semi-Strong**: Current prices reflect all publicly available information about a firm or asset. If this form of market efficiency is correct, then investors cannot generate profits by relying on fundamental data for a company, trading based on news releases, earnings statements, or any other information that is accessible to a broad set of investors.
- **Strong**: Current prices reflect all public and private information related to a firm. If this form of market efficiency is to be believed, then even insider trading cannot generate excess returns.

[1] Which states that beating the market should be impossible.

The strong form of the EMH is particularly draconian, as it is well documented that traders relying on inside information tend to earn excess profits[2]. Empirical analysis of the EMH is somewhat mixed. On the one hand, a considerable amount of evidence exists that shows the difficulty in forecasting returns and in obtaining a competitive advantage over other market participants[169]. On the other hand, many documented anomalies have been identified and seemingly persist, some of which we explore in the next section. Additionally, one could conversely argue that the EMH is against the spirit of free markets in that it doesn't provide a reward to researchers or analysts for conducting the necessary research, or developing the necessary strategy. If those doing this are not rewarded, then it begs the question why they are willing to do this, and if participants are not incentivized to bring the information to the market in the first place, then what mechanism exists to ensure that the information continues to be harvested and reflected in asset prices.

Regardless of our philosophical beliefs regarding market efficiency, it serves as a natural baseline or prior, and in practice markets are highly, if not perfectly efficient. Even if an edge is found, other participants exploring the same datasets are likely to find similar phenomenon, and an inevitable decay ensues.

Market efficiency also does not take into account the human element of market participants. Some, such as Shleifer [230] have suggested that this creates behavioral biases in the market which may cause more persistent dislocations. These behavioral phenomenon give rise to the notion of underlying risk premia. Behavioral finance, and risk premia are introduced and explored further in the next section.

4.2.2 Market Anomalies, Behavioral Finance & Risk Premia

While there is strong evidence that shows how challenging it is for professional money managers to outperform on a consistent basis, there have also been several documented, persistent anomalies. Additionally, some have argued that strict Market Efficiency may not allow for a sufficient reward for research and innovation within the field. Shleifer [230] also noted the challenges with implementing many arbitrage strategies in practice, and proposed that leads to limits on arbitrageurs, leading to potential deviations from market efficiency.

One of the more notable and persistent anomalies is the so-called momentum effect, whereby we observe assets who have previously increased continue to increase, and vice versa [150]. The presence of this phenomenon means that investors can, theoretically, achieve excess returns by forming portfolios that have higher weights in assets with previously positive returns, and lower[3] weights in assets with poor returns in the previous period. Another documented anomaly is that firms with low book-to-market values tend to outperform over longer periods than firms with high book-to-market values [98]. This phenomenon is often referred to as the value premia, or value effect. Similarly, a size effect has been identified, wherein firms with smaller market capitalizations tend to outperform larger market cap stocks. Other research has been done that has showed potential lead-lag relationships between small and

[2] And one could argue this is why it is illegal.

[3] Or negative.

large cap firms [167]. Others have studied seasonality in market returns and found a so-called January Effect where excess returns tend to occur consistently in January relative to other months [159].

Taken collectively, the documented set of market anomalies might seem to indicate that even though markets are highly efficient, they do not strictly speaking meet the standards of the cited forms of the EMH. An alternative theory is that of behavioral finance [31], which looks to incorporate cognitive biases of market participants and allows for the existence of structural premia that result from these biases. Kahneman [156], for example, postulated that investors inherently are characterized by loss aversion, meaning they are more concerned with avoiding losses than they are with achieving gains.

Through the lens of this theory, the set of documented market anomalies might in reality then be a set of harvestable risk premia that result from dislocations caused by investor biases. The so-called size effect, for example, might be a risk premium investors are paid to invest in smaller firms, who are inherently less liquid and potentially contain larger tail risks. This lens of risk premia is a prevalent theme throughout the text, and can help the reader frame many of the phenomenon that permeate financial markets.

4.2.3 Risk Premia Example: Selling Insurance

A canonical example of a risk premia is a premium earned from selling insurance. The rationale for this risk premia emanates from loss aversion as postulated by Kahneman [156]. Investors buy insurance to hedge tail risk, and are generally not price sensitive when doing so. Instead, they do so because of the value that this insurance provides in a state of the world where they are likely to suffer large losses. When buying homeowner's insurance, for example, homeowner's are happily willing to pay the prevailing market rate generally without asking questions. Fundamentally, this is to avoid large lump sum payments. This simple concept is potentially extendable to other market settings, and can be used to justify the existence of an equity risk premium[4], as well as an option's selling premium. These premia are characterized by large losses in states of the world that the remainder of investors portfolios, and career prospects are likely to suffer significantly, and thus it is reasonable to expect investors to earn a premia for bearing this risk. One could argue that the rationale for the premia is two-fold. Partly it may because of when these losses occur, and correlation of these losses to when investors most need capital. It is also partly because of a preference for convex vs. concave payoffs, characterized by larger gains and smaller losses instead of large losses and smaller gains. These are both premia that we explore later in the text.

[4]The premium on equities relative to bond's is commonly known as the equity risk premium.

4.3 LINEAR REGRESSION MODELS

4.3.1 Introduction & Terminology

When working in the physical measure, we are often tasked with finding the relationship between two or more underlying variables. Often one of these variables is a future return, meaning we are making a forecast, and there are several variables that we postulate have some bearing on that future return. That is, we are trying to model something of the following form:

$$r_{t+1} = f(r_t, x_t, \dots) + \epsilon_t \tag{4.1}$$

where r_{t+1} is the next period return, r_t is the current return in the last period, x_t is the starting value of a conditioning variable, and ϵ_t is the residual that represents the part of the forecasted return that we are unable to explain. r_{t+1} is often referred to as the dependent variable, whereas r_t and x_t are generally referred to as explanatory variables. For example, x_t could represent the price-to-earnings ratio of a stock, the seasonality of a commodity or some other relevant quantity. It could rely on traditional data sets, or it could leverage alternative data. There are many approaches to trying to handle models like (4.1), however, perhaps the simplest of which is to assume a linear relationship between the variables and use linear regression techniques. This is the technique that we explore further in this section.

As the name implies, in a linear regression model we would assume a linear relationship between the next period's return, r_{t+1} and the conditioning variables r_t, x_t, etc. This makes (4.1) become:

$$r_{t+1} = \alpha + \beta_r r_t + \beta_x x_t + \cdots + \epsilon_t \tag{4.2}$$

where β_r and β_x are the coefficients that define the linear relationship between r_t and r_{t+1}, and x_t and r_{t+1}, respectively.

In this context, these β's tell us how much higher or lower we should expect the next period's returns to be as the value of the conditioning variables, or explanatory variables, changes. If the coefficients are low, then we expect similar returns in the next period regardless of the starting values of r_t and x_t. If the coefficients are large, however, then the starting value of these variables may lead to significantly different return expectations in the next period.

Clearly, as most relationships in finance are quite complex, especially those in which we are making a prediction, there is very little reason to assume that the relationship between the dependent variable and explanatory variables should be linear. The assumption of linearity implicitly assumes that we have transformed our explanatory variables, or signals in such a way that they have a linear relationship with the dependent variable. This means that how we represent the explanatory variables can make a big difference in our success. Under the right transformation a linear relationship may appear, however, this may not be the case if we use the raw value or a different transformation.

In this context, visualization exercises that plot an explanatory variable against a dependent variable can be highly beneficial in gauging whether a linear relationship is appropriate. We should not, in contrast, just blindly assume a linear relationship in all cases. Having said that, the tractable nature or linear regression has undeniable appeal, and as a result linear regression is one of the most commonly used quant techniques on the buy-side. Recently, there has been a push to incorporate machine learning techniques into models of the form of (4.1). Later in the text, in chapter 28, we introduce how we might replace this assumption of a linear relationship with some more flexible machine learning techniques that allow for non-linearity.

To reinforce the most common terminology associated with regression models, consider the following generic regression equation:

$$Y = \alpha + \sum_{i=1}^{N} \beta_i X_i + \epsilon \tag{4.3}$$

In the following list, we highlight the associated terminology with each part of this equation. In the remainder of the chapter, we will leverage this terminology when discussing different regression models.

- Y: dependent variable
- X_i: explanatory or independent variable i
- α: regression intercept
- β_i: regression coefficient that defines the (linear) relationship between the dependent variable, Y, and the ith explanatory variable.
- ϵ: residual, or part of the dependent variable, Y that we cannot explain with the explanatory variables

In the remainder of this section, we provide the analytical framework for solving linear regression problems, and discuss how they can be applied in practice.

4.3.2 Univariate Linear Regression

The simplest type of linear regression model is one that relies on a single explanatory variable in order to predict the dependent variable. This model can be expressed as:

$$Y = \alpha + \beta X + \epsilon \tag{4.4}$$

In finance problems, the Y variable that we are looking to explain is often a future return, but it could also be a contemporaneous return or other quantity. The X variable could be a piece of fundamental data, a prior available return or a signal motivated via an alternative dataset.

In order to identify the relationship between Y and X, we need to choose values for the coefficients α, the intercept, and β. We do this by fitting the values to a set of available data. This is an example of model calibration or estimation, a concept that will repeat frequently throughout the text. We are often given a set of data, whether it be option prices, or a time series of stock returns, and are required to calibrate some sort of parametric model to the available data.

In the case of this particular estimation problem, we do so such that we can explain as much of the movement in Y as possible, which means making the residuals, ϵ as small as possible. More formally, we can formulate an optimization problem that minimizes the squared residuals, ϵ^2 as shown below[5]:

$$\min_{\alpha,\beta} \quad \sum_{t=1}^{M} (\epsilon)^2 \tag{4.5}$$

$$\min_{\alpha,\beta} \quad \sum_{t=1}^{M} (Y - (\alpha + \beta X))^2 \tag{4.6}$$

This is a standard minimization problem that can be solved by finding the first and second-order conditions by taking the appropriate derivatives of the objective function in (4.5) with respect to α and β. This is left as an exercise for the reader, however, in the next section we show how this minimization works in a more general context with multiple explanatory variables. Additionally, more context on the theory of solving optimization problems is provided in chapter 14.

In the following chart, we highlight a simple example of a univariate regression:

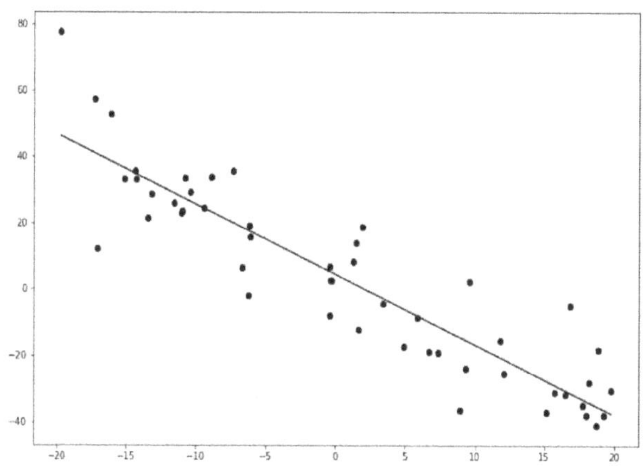

This visualization helps to provide some intuition on the two parameters that we estimated. α, the regression intercept, defines the expected value of the dependent variable assuming that the explanatory value equals zero. As we increase or decrease α, we find that the regression lines moves up or down in parallel. The second parameter, β, defines the slope of the regression line with respect to the explanatory variable X. Said differently it tells us how much we should expect Y to increase for a given shift in X. A steeper regression line indicates a stronger relationship between

[5]This is known as using an Ordinary Least Squares approach [128].

the dependent and explanatory variables, whereas a flat regression line implies that the value of X provides little information for Y.

In this chart we can see the data is scattered and doesn't fit perfectly to the regression line. We can also see, however, that as the regression line moved up or down on the y-axis, the line moves further away from the center of the data. Similarly, we can see the same phenomenon as we increase or decrease or decrease the slope of the regression line.

4.3.3 Multivariate Linear Regression

Of course, in practice we usually want to be able to rely on multiple explanatory variables to aid in our forecast of the dependent variable, Y. Along those lines, (4.4) can be generalized to include N explanatory variables using the following formula:

$$Y = \alpha + \sum_{i=1}^{N} \beta_i X^i + \epsilon \tag{4.7}$$

where as before α is the intercept of the regression and β_i is now the coefficient with respect to explanatory variable i. It should be noted that we now have N β's to estimate in our regression model in addition to the intercept, α. It can often be more convenient to express (4.7) in matrix form, which can be done via the following equivalent equation:

$$Y = X\beta + \epsilon \tag{4.8}$$

where β is now an $(N+1) \times 1$ vector of the following form:

$$\beta = \begin{pmatrix} \alpha \\ \beta_1 \\ \dots \\ \beta_N \end{pmatrix} \tag{4.9}$$

It should be emphasized that the vector β now incorporates the intercept as well as the standard regression β's. Similarly, as β has now been updated to incorporate the intercept, X will need to be updated such that it is the following $M \times (N+1)$ matrix:

$$X = \begin{pmatrix} 1 & x_{1,1} & \dots & x_{1,N} \\ 1 & x_{2,1} & \dots & x_{2,N} \\ 1 & x_{i,1} & \dots & x_{i,N} \\ \dots & \dots & \dots & \dots \\ 1 & x_{N,1} & \dots & x_{N,N} \end{pmatrix} \tag{4.10}$$

where M is the number of observations of the dependent and explanatory variables. This means that $X\beta$ is a $M \times 1$ dimensioned vector matching the dimensions of Y, as desired. Further, the values in $X\beta$ are the predicted values in our regression model for each historical date.

Just as in the univariate case, the coefficients can be estimated by minimizing the squared residuals $\epsilon^\top \epsilon$. This is known as an ordinary least squares, or OLS approach, and can be expressed via the following objective function:

$$\min_{\beta} \quad \left(\epsilon^\top \epsilon\right) \tag{4.11}$$

$$\min_{\beta} \quad (Y - X\beta)^\top (Y - X\beta) \tag{4.12}$$

$$\min_{\beta} \quad Y^\top Y - \beta^\top X^\top Y - Y^\top \beta X + \beta^\top X^\top X\beta \tag{4.13}$$

where in the second step we have substituted the definition of ϵ and in the last step we have expanded the terms in the above product. This minimum can be found by solving for the first order condition, which can be done by differentiating (4.13) with respect to β and setting the derivative equal to zero. Doing so yields:

$$\frac{\partial}{\partial \beta} = -2X^\top Y + 2X^\top X\beta = 0 \tag{4.14}$$

Finally, solving for β, the parameter that we are selecting, we get:

$$2X^\top X\beta = 2X^\top Y \tag{4.15}$$

$$\beta = \left(X^\top X\right)^{-1} X^\top Y \tag{4.16}$$

This is known as the OLS estimator for β. As we can see, calculation of the coefficients, β, requires inversion of a matrix, $X^\top X$, which is then multiplied by $X^\top Y$. This is noteworthy as it can lead to problems when the matrix is singular and therefore not invertible. Later in the chapter we discuss one technique for handling singular matrices: principal component analysis.

It should be noted that to prove this point is a minimum we also need to verify the sign of the second-order condition. In this case however it can be shown that this second derivative is positive at all points, as the objective function is a convex function of β. This means that the function forms a unique minimum at the specified first-order condition. For more details on convex functions and solving optimization problems the reader is encouraged to see the appropriate section in chapter 14. Additionally, those interested in more context on matrix derivatives should consult [162].

4.3.4 Standard Errors & Significance Tests

In the previous section, we developed OLS-based estimates for our regression coefficients for multi-variate and uni-variate regression models. Of course, in addition to the coefficient estimates themselves, of direct interest to us will be whether the coefficients are statistically and economically significant. The latter is a more subjective question that cannot be answered with a statistical test. Statistical significance, however, can be gauged using significance tests. It should be emphasized that strong statistical significance, however, does not imply a causal relationship or economic significance. When thinking about statistical significance, we ideally want to know whether our entire model, which may consist of many underlying variables, is significant, and relatedly how much of the movement in our dependent variable it explains. In addition, we want to know how statistically significant each explanatory variable, or feature is. Gauging significance for the whole model versus an individual coefficient will require a different set of tests & metrics, however, fortunately, there is a well-established toolset for analyzing this within the context of regression.

To understand the statistical significance of the underlying explanatory variables, we need to know how much certainty, or uncertainty we have on our coefficient estimates. Remember that when we are building a regression model we are trying to estimate a relationship within a population based on a sample with limited data. This sample provides some information on the regression coefficients for the population, however, there is uncertainty in the estimate of the coefficients that is, among other things, a function of the size of the sample. If we were to obtain another sample from the same population, we would be unlikely to get the exact same regression coefficients. However, it is likely that the regression coefficients would both be within a reasonable range, or confidence interval. In judging the statistical significance of each explanatory variable, we will want to build these confidence intervals.

To do this, we rely on the volatility of the coefficient estimates, which we refer to as standard errors, as well as the distributional properties of the sampling estimator, which can be proven to come from a t-distribution. To determine whether the coefficient is statistically significant we then test the null hypothesis that the coefficient $\beta = 0$ against the alternate hypothesis that it does not equal zero:

$$H_{\text{null}} : \beta = 0 \tag{4.17}$$

$$H_{\text{alternate}} : \beta \neq 0 \tag{4.18}$$

It should be noted that in some cases we might consider a one-sided alternate hypothesis, (i.e. $\beta < 0$) but the standard hypothesis test in a regression model is as defined in (4.17). To test this null hypothesis, we obtain a confidence interval, with confidence level α, for our coefficient estimate, from β_{min} to β_{max} using the standard error of our regression:

$$\beta_{\text{min}} = \beta + t_{\text{min}}\sigma_\beta \tag{4.19}$$

$$\beta_{\text{max}} = \beta + t_{\text{max}}\sigma_\beta \tag{4.20}$$

where σ_β is the standard error for the coefficient β, and β is the estimated value of the coefficient. This standard error, σ_β is a function of the size of our sample, and, as the sample size increases, we will begin to have smaller standard errors, leading us to more confidence in our coefficient estimates. Of course, in finance, this is inherently a challenge where datasets tend to be small due to short time series histories.

Additionally, t_{min} and t_{max} are the inverse of the CDF of a t-distribution evaluated at α and $1-\alpha$, respectively. Therefore, t_{min} is a negative number leading β_{min} to be below the coefficient estimate, β, and t_{max} is positive leading to $\beta_{\text{max}} > \beta$. If this confidence interval for our estimated β includes 0, then we conclude that we cannot reject our null hypothesis, meaning the explanatory variable is not statistically significant at the specified confidence level. Conversely, if the interval is strictly speaking above or below zero, then this implies that the null hypothesis can be rejected and the explanatory variable is statistically significant.

4.3.5 T-Statistics, P-Values & Statistical Significance

Another approach, which is perhaps the most common among quants, statisticians and econometricians, is to look at the so-called t-stat and p-values in a regression for each coefficient estimate. The t-stat, which is a function of the standard error, measures the normalized distance between the coefficient estimate and the null hypothesis value of zero. This can be written as:

$$\hat{t} = \frac{\beta}{\sigma_\beta} \tag{4.21}$$

A larger t-statistic indicates a larger deviation from the null hypothesis, indicating stronger statistical significance. To judge the t-stats we obtain, we might compare $\hat{t}$ to values of the t-distribution. A p-value, then, relatedly, is the inverse of the CDF[6] of this t-distribution evaluated at our t-statistic. This p-value is defined as:

$$p = P\left(\tau \geq t | H_{\text{null}}\right) \tag{4.22}$$

Importantly, this p-value then can be interpreted as the probability of obtaining this coefficient conditional on the null hypothesis being true. Said differently, it is the probability of obtaining the resulting coefficient estimate conditional on the true value of $\beta = 0$ as the null hypothesis suggests. Therefore, a small p-value implies high confidence that the results are statistically significant and not a false positive that may have occurred by chance. Each of these metrics provide a related, similar lens to judging statistical significance at a given threshold, α. T-Statistics require some knowledge of the t-distribution, in order to put in perspective. Conversely, p-values are naturally intuitive, yet closely interconnected.

High level programming languages such as Python include standard output in their regression models that includes standard errors, t-statistics and p-values for each coefficient. These quantities can then be used to judge significance of the explanatory variables, and can aid model selection by helping isolate the most pertinent variables.

At the overarching regression level, the most common metric to measure the fit of the model is R^2, which measures the percent of the variance explained by the model. Mathematically, R^2 can be expressed as:

$$R^2 = 1 - \frac{\text{SSE}}{\text{TSS}} \tag{4.23}$$

$$\text{SSE} = \epsilon_t^\top \epsilon \tag{4.24}$$

$$\text{TSS} = Y^\top Y \tag{4.25}$$

R^2 provides a meaningful and convenient measure of the fit of a regression model. In particular, as it measures the variance explained by the model relative to the

[6] Or one minus the inverse of the CDF, depending on the sign of the coefficient.

variance explained by the residuals, it is easy to interpret and use to judge whether our overarching model is successful. By contrast, it does not help us to determine which of our explanatory variables are most important, so it is less useful for feature selection. When forecasting, one should expect to obtain a relatively low R^2 as the signal to noise ratio is inevitably low. In contrast, when performing contemporaneous regressions, such as building factor models, then we are able to explain far more of the variability, leading to significantly higher R^2.

4.3.6 Assumptions of Linear Regression

So far, we have detailed some of the embedded assumptions in a linear regression model and commented on how realistic they might be. Most notably, we highlighted the assumption of a linear relationship between the dependent and explanatory variables, as well as the potential challenges that might arise when inverting a singular matrix of explanatory variables. In the following list, we summarize some of these main assumptions embedded in a linear regression model:

- **Linear relationship between dependent and explanatory variables**: Perhaps the main overarching assumption is that there is a linear relationship between the dependent and explanatory variables. This can be a challenging assumption to assess as in reality as the relationships we are trying to uncover are quite complex, if they exist at all. In practice, however, the inherit complexity makes the likelihood of the relationships being linear seemingly low. This presents a trade-off between the appealing tractable nature of linear regression vs. the potential oversimplified model of the relationship between the variables.

- **Homoscedasticity & Independence of Residuals**: In the linear regression formulation, we assume that the residuals are i.i.d. Importantly, the independence assumption means that the residuals should be independent of each and not autocorrelated. Additionally, the identically distributed assumption means that each residual comes from the same distribution. This means they should have the same mean, which we know will be zero otherwise it would be incorporated into the intercept. More fundamentally, each residual should also have the same variance. This property of equal variance is known as homoscedasticity. Series with different variances, in contrast, are known as heteroscedastic. In chapter 25 we explore the GARCH model that incorporates heteroscedasticity via a mean-reverting volatility process.

- **Lack of High Correlation Between Explanatory Variables**: One of the most challenging issues associated with multi-variate linear regression is that of multi-collinearity. This occurs when the input explanatory variables are correlated and can lead to bias in the calculated standard errors and overconfidence in the statistical significance in our results. Certain amounts of correlation between the input variables is not problematic, however, as correlation increases the explanatory variable covariance matrix may become singular, which is when problems arise. In particular, it can also lead to situations where it is hard for the model to choose between increasing the slope of different explanatory variables. As a result, the coefficient for one might end up being increased at the expense of the other. This can lead to unstable regression coefficients that can vary as we, for example, change the order of the explanatory variables. To guard against this, analysis of the rank of the covariance matrix, $X^\top X$ should be done to ensure the underlying matrix is full rank.

4.3.7 How Are Regression Models Used in Practice?

Regression techniques are possibly the most commonly used econometric technique when modeling the physical measure. Fundamentally, they allow us to establish a relationship between two or more variables. In many cases, this relationship is a forecast. A common example of this is when we build an expected return model. In this case, we will build expected returns that are conditional on a set of explanatory variables. In the equity markets, these explanatory variables may be motivated by a company's balance sheet, earnings estimates, press releases or other fundamental data. For example, a commonly used value signal uses current price-to-earnings or price-to-book as a forecast for subsequent returns. In many cases the explanatory variable may be a return from a previous period. In this case, we are implicitly working with a time-series model, which we introduce in the following section.

A defining feature of this type of regression is that the signal to noise ratio is unavoidably low. That means we are likely to find low R^2's and low explanatory power in our regressions. This obviously creates a challenge as we look to uncover any underlying relationship, as even in a best case scenario we are likely to find something with weak statistical significance. Further, as we add variables to our regression models, understanding of the underlying correlation structure between them becomes similarly challenging. This leads to questions about whether we should include interaction terms between certain variables, for example. Against this backdrop, the use of parsimonious models that are based on and underlying economic rationale can be advantageous and lead to more stable, robust models.

Additionally, we often rely on regression techniques to build factor models that are based on contemporaneous regressions. A common use of this is to decompose a set of returns on a set of underlying factors. In doing this, we can view the exposure of a given set of returns to this set of underlying factors. We consider an example of this in the following section, where we decompose returns on simple stock and bond benchmarks. Notably, a positive intercept in this type of return decomposition model implies a positive expected return assuming the other variables stay the same. This is a desirable feature, as it means that even after the factor exposures have been accounted for, we expect on average to generate a profit. Thus, in this context a positive intercept can be viewed as an ex-post alpha, or excess return.

It should also be noted that, while regression provides a convenient framework for analyzing these types of problems, we could similarly try to apply non-linear regression techniques, more advanced time series models, or machine learning techniques to these types of problems. The basic framework for this type of extension is presented partly in the remainder of the chapter, and partly in chapter 28.

4.3.8 Regression Models in Practice: Calculating High-Yield Betas to Stocks and Bonds

In this example we leverage the example in the coding repository to calculate the sensitivity of high-yield credit on broad stock and treasury indices[7]. High Yield is an interesting asset class in that it inherits certain properties from bonds and certain characteristics from the equity market. This relationship, and the broader link between equities and credit instruments is a theme throughout the text and is, for example, covered in more detail in chapter 20. In this example, however we examine this interconnection by decomposing the returns of high yield on stocks and treasuries via a contemporaneous regression. In this example we do this using Yahoo Finance data from 2012 to 2021 and obtain the following coefficient estimates:

Factor	Beta
Intercept	−0.000041
Stocks (SPY)	0.39
Treasuries (GOVT)	0.18

As we can see, there appears to be a significant relationship between high yield and both stocks and bonds, as evidenced by the relatively high coefficients. For example, the 0.39 coefficient with respect to SPY tells us that, for a one percent increase (or decrease) in the S&P we would expect an increase (or decrease) of 39 bps in High Yield. Additionally, we see that the intercept is close to zero but slightly negative. This implied that, if stocks and bonds remain unchanged, we would expect to lose money in an investment in High Yield. This intercept, then is often annualized and interpreted as an ex-post alpha. In this case, annualizing the quantity shows that, on average, we would expect to lose 1% per year in the absence of any movement in stocks and bonds. The reader is welcome to try different time periods and see how persistent these exposures and intercept are in different market conditions, and think through what this implies for investments in high yield.

4.4 TIME SERIES MODELS

4.4.1 Time Series Data

In financial models we are almost always working with time series data. That is, we are modeling the evolution of a series over time, that is indexed by time in a specific order. This is in contrast to some statistics exercises where our goal is to model the value of random variables in isolation, rather than as a time ordered sequence. In this case, subsequent events or trials may be i.i.d., however, in the case of time series, the next step in the process most often depends on its previous path. For example, perhaps the fact that a security has risen rapidly in the past impacts our expected future path for the asset. Time series models, such as the ones introduced in this section, can help us try to uncover this dependency. One key defining feature of time series data is whether it is stationary or non-stationary. In the next section,

[7]Proxied by the SPY and GOVT ETFs, respectively.

we develop this concept and describe how we can test whether our data is stationary. In practice, we will find it convenient to work with stationary processes, which may require differencing the underlying process.

In finance, we also often have time series data for many assets, creating a cross-sectional as well as a time series component. In the next section we explore techniques for dealing with such panel data that have cross-sectional and time-series elements. In this section, however, we provide the foundation for working with and analyzing time series.

4.4.2 Stationary vs. Non-Stationary Series & Differencing

Perhaps the most fundamental defining feature of time series data is whether it is stationary or non-stationary. In finance, many of the datasets that we encounter are non-stationary. This means that the distribution evolves over time, which creates challenges when modeling the process. Stationary processes, on the other hand, have the same distributional properties regardless of time. This makes stationary processes comparatively easier to model because we know all observations emanate from the same distribution, making the distribution we are looking to forecast more tractable. As a result, quants commonly find themselves analyzing time series data and performing stationarity tests to determine whether the data is likely to be stationary.

The following properties define a stationary process [250]:

- $\mathbb{E}\left[X_t\right] = \mu$
- $\text{var}\left(X_t\right) = \sigma^2 < \infty$
- $\text{cov}\left(X_t, X_s\right) = \text{cov}\left(X_{t+h}, X_{s+h}\right)$

From these properties we can see that a process is stationary if the first two moments are independent of time. There may be other components to the process, such as a seasonality component or other conditioning variable that might impact the distribution of X_t. In many cases, it is convenient to first de-trend and remove these components, such as seasonality, prior to analyzing the time series.

An important feature of stationary asset price processes is that they imply a mean-reverting component. This is of direct interest to market participants, as it means that divergences from a trend or baseline are likely to revert. Many quantitative signals are based on these types of dislocations correcting as we will explore throughout the text. For example, in chapter 27, we explore common approaches to quant trading, including those based on mean-reversion signals.

Conversely, if data is non-stationary, as it is in a random walk, as we will soon see, then it is commonplace to difference the data. The difference, then, is often stationary. This is the case in a random walk, which is non-stationary, but when differenced is stationary, as it is a simple white noise process. These building blocks are explored further in the next section. In practice, we often observe that an assets price process is non-stationary. This implies a lack of mean-reversion in the asset prices, and is in line with the EMH. In these cases, we then difference the data, and work with the return process, which is usually found to be stationary.

4.4.3 White Noise & Random Walks

The simplest building block in time series models is a white noise process, ϵ_t, which is a sequence of independent, identically distributed random variables of the form:

$$\epsilon_t \sim \phi(0, \sigma^2) \tag{4.26}$$

This white noise term is analogous to the increments in a Brownian Motion that we introduced in the last chapter, and, in fact, Gaussian White Noise is equivalent to Brownian Motion. Additionally, in a time series context, we often work with discrete rather continuous processes, and, as a result, we work with discrete white noise terms instead of their continuous Brownian increment counterparts. More generally, a white noise term does not have to be Gaussian, but can instead emanate from another distribution.

It can easily be shown that a white noise process is stationarity, as it has a constant mean equal to zero, constant, finite, variance and zero autocorrelation between successive increments by nature of the i.i.d. assumption. That is:

$$\mathbb{E}\left[\epsilon_t\right] = 0 \tag{4.27}$$

$$\operatorname{var}\left(\epsilon_t\right) = \sigma^2 \tag{4.28}$$

$$\operatorname{cov}\left(\epsilon_t, \epsilon_s\right) = 0 \tag{4.29}$$

If we then aggregate these successive white noise terms, we obtain the well-known random walk which can be written as:

$$Z_t = \sum_{i=1}^{N} \epsilon_t \tag{4.30}$$

It can easily show by analyzing the properties of stationarity in 4.4.2 that a random-walk is a non-stationary process:

$$\mathbb{E}\left[Z_t\right] = 0 \tag{4.31}$$

$$\operatorname{var}\left(Z_t\right) = N\sigma^2 \tag{4.32}$$

Notably, we can see that variance in a random walk increases with time, or the number of periods, N. This means that a random walk is a non-stationary process. Conceptually, random walks arise natural in time series analysis from the efficient market hypothesis, as it is line with the idea that current prices reflect all available information. In particular, we often assume that the price process for an asset follows a random walk, in which case the return process would be defined by white noise, and as a result will be unpredictable. As such, random walks are a staple for quant models. They are also, however, characterized by an inability to explain future price movements making them useful models in many contexts, but less useful for generating quantitatively oriented strategies. Instead, many quant strategies rely on autocorrelation in the paths of asset prices and returns, if it can be found. In the next section, we introduce these types of mean-reverting models.

4.4.4 Autoregressive Processes & Unit Root Tests

Autoregressive (AR) processes are another important tool for time series analysis, and differ from a random walk in that they allow for momentum or mean-reversion in the return process for an asset, a notable deviation from a random walk. AR processes, Y_t are a linear combination of a white noise term, ϵ_t and past values of itself (i.e. Y_{t-1}). An AR process can incorporate an arbitrary number of lags, but in its simplest for can be written as:

$$Y_t = \phi Y_{t-1} + \epsilon_t \tag{4.33}$$

where ϵ_t is a white noise process and Y_t is the AR process. This process in (4.33) is known as an AR(1) process.

More generally, an AR(n) process can be constructed by including lagged terms up to n, as in the following equation:

$$Y_t = \sum_{i=1}^{N} \phi_i Y_{t-i} + \epsilon_t \tag{4.34}$$

Returning to an AR(1) process, it should be highlighted that if $\phi = 1$, then the process simplifies to a random walk. Thus, we know that the process is non-stationary when $\phi = 1$. It turns out, however, that if $\phi < 1$, then the process is stationary, making it a useful process in our modeling efforts.

To see this, let's start by computing the expectation for our AR(1) process as defined in (4.33). This expectation can be written as:

$$\mathbb{E}\left[Y_t\right] = \phi \mathbb{E}\left[Y_{t-1}\right] + \mathbb{E}\left[\epsilon_t\right] \tag{4.35}$$

As ϵ_t is a white noise process, we know that it has zero expectation and this term can therefore be eliminated. This leaves us with $\mathbb{E}\left[Y_t\right] = \phi \mathbb{E}\left[Y_{t-1}\right]$ which we can use an induction argument to solve. Because Y_t is made up of all white noise terms, each of which has expectation zero, we can similarly argue that Y_t must have the same expectation. Therefore, we must have:

$$\mathbb{E}\left[Y_t\right] = 0 \tag{4.36}$$

It should be noted however, that, while the equation above defines the unconditional expectation, the conditional expectation, conditional on a value of Y_{t-1} is no longer zero. This is an interesting property of the process.

Next, let's consider the variance of our AR process, Y_t. This can be written as:

$$\operatorname{var}\left(Y_t\right) = \phi^2 \operatorname{var}\left(Y_{t-1}\right) + \operatorname{var}\left(\epsilon_t\right) \tag{4.37}$$

The second term on the right-hand side, $\operatorname{var}\left(\epsilon_t\right)$ is well known as it comes from the distribution of white noise and is therefore equal to σ^2. If the process is to be stationary, then that means $\operatorname{var}\left(Y_t\right) = \operatorname{var}\left(Y_{t-1}\right)$. Therefore, lets define a constant v and suppose they are equal. This leaves us with the following equation:

$$v = \phi v + \sigma^2 \tag{4.38}$$

$$v = \frac{\sigma^2}{1-\phi^2} \tag{4.39}$$

This has a solution provided $\phi < 1$, and therefore, in this case meets the property of constant variance which can be written as:

$$\operatorname{var}(Y_t) = \frac{\sigma^2}{(1-\phi^2)} \tag{4.40}$$

If $\phi \geq 1$, then this condition no longer holds and the process becomes non-stationary. Using a similar methodology, we can show that the covariance of an AR(1) process also adheres to the requirements of stationarity, and is defined as:

$$\operatorname{cov}(Y_t, Y_s) = \frac{\phi^{t-s}\sigma^2}{1-\phi^2} \tag{4.41}$$

Therefore, we can see that, if $\phi < 1$, then an AR(1) process is stationary with the following properties:

$$\mathbb{E}[Y_t] = 0 \tag{4.42}$$

$$\operatorname{var}(Y_t) = \frac{\sigma^2}{1-\phi^2} \tag{4.43}$$

$$\operatorname{cov}(Y_t, Y_s) = \frac{\phi^{t-s}\sigma^2}{1-\phi^2} \tag{4.44}$$

where $t > s$. This distinction between a non-stationary process when $\phi = 1$ and a stationary process when $\phi < 1$ is an important one and means that we can test for stationarity by estimating the coefficient ϕ and testing the null hypothesis that $\phi = 1$. This is called a unit root, or stationarity test. These tests violate the assumptions of standard OLS regressions, and therefore requires special treatment. Fortunately, several of these unit root tests exist, such as the Dickey-Fuller and Augmented Dickey-Fuller tests [80] and have implementations in Python. Later in the text we will see examples of unit root tests, and how they are applied in the context of quant finance applications.

4.4.5 Moving Average Models

Another commonly used time series process is a moving average (MA) model, which can be written as:

$$Y_t = \epsilon_t + \theta\epsilon_{t-1} \tag{4.45}$$

where as before ϵ_t is a white noise process and θ is a constant model parameter. In a moving average model, our process is a linear combination of white noise terms. In the case of (4.45), the next step in our time series is a function of two white noise terms, the current and previous realization. The weight on the previous realization of the white noise term, θ is a model parameter that we can choose to fit the available data.

It can easily be shown that the process in (4.45) is defined by the following properties and is therefore stationary:

$$\mathbb{E}[Y_t] = 0 \tag{4.46}$$

$$\operatorname{var}(Y_t) = (1+\beta)\,\sigma^2 \tag{4.47}$$

$$\operatorname{cov}(Y_t, Y_s) = \begin{cases} \beta, t-s=1 \\ 0, \text{otherwise} \end{cases} \tag{4.48}$$

where $t > s$.

More generally, an MA process can include an arbitrary number of lags, as we show in the following equation:

$$Y_t = \epsilon_t + \sum_{i=1}^{N} \theta_i \epsilon_{t-i} \tag{4.49}$$

where ϵ_t is again a white noise process, n is the number of lags and θ_i is the coefficient on the white noise term with lag i. Notice that we do not have n model parameters in $\Theta = (\theta_1, \theta_2, \ldots, \theta_n)$. Like AR models, MA models are a common building blocks in time series models. In the next section, we discuss ARMA models, which combine terms from both AR and MA models.

4.4.6 ARMA Models

In an ARMA, or Autoregressive Moving Average model, we combine parts of AR and MA models. Like AR and MA models, ARMA models are defined by the number of lags that are included in the process. In the case of an ARMA models, we need to specify the number of both AR and MA lags. An ARMA(1,1) model, for example is one that incorporates a single AR and MA lag. More generally, an ARMA(n, q) model is one that incorporates n AR lags and q MA lags.

Mathematically, an ARMA(1,1) model can be written as:

$$Y_t = \phi Y_{t-1} + \epsilon_t + \theta \epsilon_{t-1} \tag{4.50}$$

Similarly, an ARMA(n, q) model can be expressed as:

$$Y_t = \sum_{i=1}^{n} \phi_i Y_{t-i} + \epsilon_t + \sum_{i=1}^{q} \theta_i \epsilon_{t-i} \tag{4.51}$$

where n is the number of AR lags, q is the number of MA lags, and ϕ_i and θ_i are the model parameters for the ith AR or MA term, respectively.

In practice, most advanced programming languages, including python have tools for estimating ARMA models. We explore these Python tools in more detail in chapter 8. When estimating an ARMA model, or an AR or MA model, part of estimating the model is choosing the appropriate number of lags. One approach to this is to use an Akaike information criterion (AIC) [53], which is a technique for model selection that helps balance model complexity vs. model fit. Additionally, plots of the autocorrelation and partial autocorrelation function can be useful for identifying the number of significant lags. In the following list, we briefly describe these two concepts:

- **Autocorrelation Function (ACF)**: measures the total autocorrelation between the process at time t and $t+s$ without controlling for other lags.
- **Partial Autocorrelation Function (PACF)**: measures the partial autocorrelation between the process at time t and $t+s$ removing the effects of the lags at time $t+1$ through $t+s-1$.

The partial autocorrelation function is useful for model selection of AR lags, because, as we see in (4.41), the impact of autocorrelation at a given lag decays slowly at prior lags, making it challenging to distinguish the source of the autocorrelation if we look simply at the ACF. As a result, it is helpful to control for the effects of intermediate lags by using the PACF instead of the ACF. Conversely, the autocorrelation function is useful for selecting the number of MA lags because the dependence on previous lags dissipates quickly, as evidenced from (4.48).

4.4.7 State Space Models

A defining feature of the models we have introduced in this section is that the underlying process is assumed to be directly observable. In practice, however, the process may not be directly observed. Instead it might be observed with measurement error, through some filter. More advanced time series models rely on state space representations to incorporate this potential observation error. These techniques are not covered in detail in this text, however, to see the concepts consider the following generic state space representation of a time series process:

$$x_{t+1} = A_t x_t + B_t u_t \tag{4.52}$$

$$y_t = C_t x_t + D_t u_t \tag{4.53}$$

where y_t is the process that we observe, and x_t is the true, latent process, that is often referred to as the state process. Importantly, we allow some transformation to be done to this state process, represented in this equation by C_t, prior to our ability to observe it. We also allow for transformations with the latent process, x_t with the matrix A_t.

Kalman filters are a common example of a state space model that allows for measurement error in the observed series. For more information on estimation of Kalman filters within the context of quant finance problems, as well as other filtering techniques, readers are encouraged to consult [133].

4.4.8 How Are Time Series Models Used in practice?

Time series models occur naturally in quant finance problems as the data we are inherently working with are time series in nature. Time series models, for example, are commonly applied in the development of quantitative strategies. Most commonly, this is done by applying time series techniques to uncover a dependence between past and future returns. Perhaps we are able to uncover some momentum or mean-reversion in an underlying time series. Alternatively, perhaps we can extend these concepts to multiple assets and analyze how one assets prior returns impact another assets future returns. This type of cross-asset autocorrelation model is introduced in chapter 27. In another context, economists use time series analysis to try to forecast economic quantities, such as inflation, employment and GDP trends, and may use this to make predictions about changes in the macroeconomic climate or business cycle.

Later in the text, in chapter 25 we will also explore analogous time series models applied to realized volatility estimates. In this context, the use of a time series model can help us incorporate the commonly observed mean-reverting tendency of volatility and help us forecast volatility. These volatility forecasts can then be leveraged by volatility traders in creating trades. Alternatively, risk managers may use these volatility estimates to help quantify risk for their portfolio. Additionally, portfolio managers may use this estimated volatility in order to size their positions, both in absolute and relative terms.

4.5 PANEL REGRESSION MODELS

In many financial datasets we observe a cross-section at each point in time and have a subsequent time series for each asset in the cross section. That is, we are often working with panel data. When modeling this underlying data, sometimes we do so cross-sectionally in each period, and then look to draw overarching conclusions. In other cases, we may look to analyze each asset in isolation without knowledge of the other members of the cross section. However, in many cases it will be most attractive to examine the entire cross-section, and time series collectively. Using this approach we can benefit from similar relationships across different assets, while still allowing for some heterogeneity.

To see this, let's consider using the following simple lag-one time series model to a number of assets:

$$r_{i,t+1} = \alpha_i + \beta_i r_{i,t} + \epsilon \tag{4.54}$$

where $r_{i,t}$ is the return of asset i at time t and follows an AR(1) process, α_i is the intercept in the regression coefficient for asset i and β_i is the autocorrelation coefficient for asset i.

If we estimate each time series in isolation, then this means each α_i and β_i will be estimated without knowledge of the analogous coefficient for other assets. In reality, however, we may prefer the coefficients across assets to be consistent. If we are estimating a mean-reversion or momentum model, then it is reasonable to expect that we should want similar coefficients for different assets. If half of the coefficients have a positive sign, and half of the coefficients have a negative sign, it becomes hard to justify the underlying model is economically significant and will perform out-of-sample. Phrased differently, can we really explain why two similar assets might end up with vastly different coefficient estimates in our model? The specification above does not lead to this type of solution by necessity, however, as it doesn't enforce consistency across the set of assets. Therefore, this situation of inconsistent coefficients with different signs can easily arise. This approach also often leads to too many free parameters, as we will need to estimate two parameters independently per asset.

At the other end of the spectrum, we might replace the equation above to rely on a single α and β that applies to all assets:

$$r_{i,t+1} = \alpha + \beta r_{i,t} + \epsilon \tag{4.55}$$

This approach solves our previous issue of having inconsistent coefficient estimates across the set of assets, as we are now enforcing that the estimates must be the same across the cross section. However, this model will now be unable to incorporate any heterogeneity in the underlying data. Additionally, in this model specification the observations from different panel groups at given times may be correlated and our regression models must account for this otherwise our standard errors may be biased. In terms of heterogeneity in the underlying panel groups, for example, perhaps the intercept terms for the underlying assets are legitimately different. To see this, consider the following illustration of hypothetical future returns as a function of a single hypothetical explanatory variable.

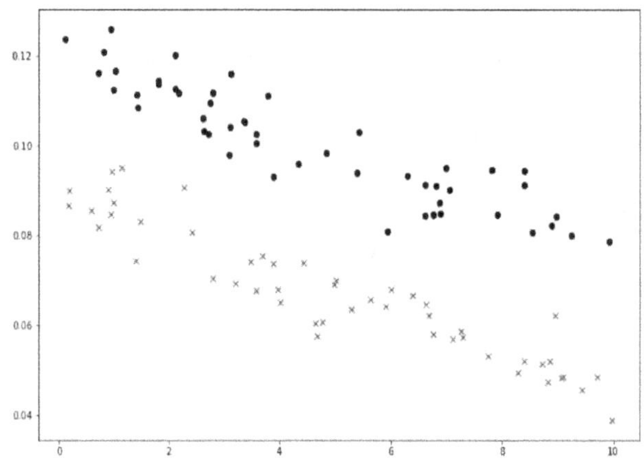

Here we can see that a (hypothetical) strong linear relationship exists, and that the slopes of the lines appear similar, characterized by a consistent sign. We can also see a difference in the underlying linear relationship between the two assets, which we may naturally want to incorporate in our model. In particular, we can see in this case that it appears the two series have different intercepts. As a result, when working with panel data such as this we often want to construct a solution that is somewhere in between the two introduced above. That is, we want to incorporate some heterogeneity while still incorporating some commonality in the underlying dynamics. One common way to do this is to allow the intercepts to vary across the panel, in this case across the asset universe, but to use a single slope parameter to define the model. This method is referred to as incorporating fixed effects. Alternatively, we could look to incorporate random effects, which allow for the impact of heterogeneity in a panel to be random. In the following list, we briefly describe these most common techniques for working with panel data:

- **Fixed Effects**: Assume a constant slope across the panel but allow the intercept to vary. This can be equivalently formulated as including a dummy or indicator variable for each asset. Mathematically, we can write this as:

$$r_{i,t+1} = \delta_i + \beta r_{i,t} + \epsilon \tag{4.56}$$

where δ_i represents the fixed effects and is interpretable as an asset or group specific intercept. Importantly, this value δ_i is a constant value that does not vary over time and is not a random variable.

- **Random Effects**: The distinguishing feature of a random effects model is that it allows for the effects in different assets, or groups, to be random, rather than a constant as in the fixed effects model. This random effect is assumed to be independent of the explanatory variables.

- **Mixed Effects**: Incorporates a combination of fixed and random effects into the panel regression model for different variables.

In chapter 25 we return to the concept of working with panel data and introduce a commonly used technique in practice for working with cross-sectional and time-series data: Fama-Macbeth. Readers looking for a more rigorous introduction to panel regression should consult [263].

4.6 CORE PORTFOLIO AND INVESTMENT CONCEPTS

One of the core concepts in investing is that of a portfolio. Few if any investors choose to invest their money in a single asset. Instead, they rely on a portfolio, or set of assets. This enables them to benefit from the seminal concept of diversification. Some have even gone so far as to describe diversification as the only available free lunch in financial markets. Other foundational concepts are the time value of money and compounding. Time value of money is built upon the idea of opportunity cost and that money in the present should be worth more than money in the future. Compounding, likewise refers to the phenomenon that investment returns tend to compound over time, and over sufficiently long periods small investments can increase in value significantly. In the remainder of this section, we introduce these concepts in more detail, and discuss how return and volatilities work in the context of a portfolio, highlighting the potential benefit of diversification.

4.6.1 Time Value of Money

The concept of time value of money is centered on the idea that a dollar today is worth more than a dollar tomorrow. This concept is closely related to the idea of opportunity costs, and is fundamentally because, if we have the dollar today we can invest it and earn interest on it, or spend it and earn utility from our purchase. Recently, negative rates have blurred this fundamental relationship, as they imply a dollar today is worth less than a dollar in the future. Nonetheless, it is a seminal relationship that is used in constructing investment portfolios, in addition to pricing derivative instruments.

We can use the concept of time value of money to establish an equivalence between present and future cashflows. Said differently, we can establish the amount that would be required in the future to make us indifferent between a specified amount today and this specified future amount. Mathematically, this relationship can be expressed via the following formula [251]:

$$P = \frac{F}{1+r} \tag{4.57}$$

where P is the present value of a cash-flow, F is the future value of the cash-flow and r is the required interest rate. We can interpret F as the amount that would be required to be received in the future, in order to forego P today. The larger the interest rate, the more future cashflows are discounted, and the larger payment we would need to receive in the future to defer consumption from the present.

Often we use the continuously compounded equivalent of (4.57), which is:

$$P = \exp(-rt) F \tag{4.58}$$

These formulas are used throughout the field of finance, notably for modeling fixed income cashflows, valuing portfolios in the future and pricing derivative securities. In some cases, a risk-neutral discount rate, r is used in the equation(4.58). In other cases, we will instead rely on a discount rate from the physical measure, μ, which incorporates investors risk preferences.

4.6.2 Compounding Returns

Another central concept in the realm of portfolio management is compounding, which refers to the idea that over time, investment gains are magnified as we not only earn interest on initial capital, but we also earn interest on previously earned interest. Over time, we continue to accumulate gains on a bigger principal amount. This is often cited as a fundamental tenet of retirement investing. Saving even small amounts of money can over time transform into a greater sum.

To see this, consider an investor who invests the value P in a project with known return r over the next year. At the end of the first year, the investor has $P(1+r)$ which they can invest in the next period. Assuming they can again invest in an investment that earns the same interest rate, r, the future value of the investments after two periods can be written as:

$$F = P(1+r)(1+r) \tag{4.59}$$

$$F = P(1+r)^2 \tag{4.60}$$

As we increase the number of periods, the impact of this phenomenon grows. In the following table, we consider a specific example of investments each of which earn 5% in a given period:

Period	Capital at Beginning of Period	Return
1	\$100	$100\,(1+0.05)$
2	\$105	$105\,(1+0.05)$
3	\$110.25	$110.25\,(1+0.05)$
4	\$115.76	$115.76\,(1+0.05)$
5	\$121.55	$121.55\,(1+0.05) = 127.63$

This leads to a terminal value of \$127.63 which exceeds the naive value of \$125 that we would have earned if we had simply earned 5% a year for 5 periods on the original principal. This difference is due to the fact that our principal balance grows in each period, enabling us to collect interest on past interest received. We can see that, over time the balance that the interest is earned on grows, leading to an increasingly large deviation from the equivalent simple interest calculation where interest is earned only on the initial principal. If we generalize this concept to t periods, with projects that are table to be reinvested every n periods, this leads to the following more general version of (4.60) can be written as:

$$F = P\left(1 + \frac{r}{n}\right)^{nt} \tag{4.61}$$

where r is an annualized interest rate. We can similarly extend this to a continuous compounding by taking the limit as $n \to 0$, which yields the following formula:

$$F = P \exp\left(rt\right) \tag{4.62}$$

More information on compounding concepts can be found in [251] and [235].

4.6.3 Portfolio Calculations

Another fundamental concept in investment management is the concept of a portfolio. Investors almost always invest in an array of securities, and in doing so, are able to benefit from diversification. Diversification emanates from the fact that the underlying securities are not perfectly correlated. Instead, when one security declines, perhaps another security is likely to appreciate, offsetting the losses. This phenomenon leads to better investment outcomes, and is therefore of utmost importance to understand and be able to model. As a result, portfolio calculations are a core part of quant modeling. This means that, as quants, we must be adept at understanding the properties of a given portfolio, regardless of the complexity or number of assets in the underlying portfolio. Two of the most common portfolio statistics that we will need to compute are the return and volatility, or standard deviation of a portfolio. In conjunction, these two metrics help us identify the risk-adjusted return of a portfolio which is perhaps the most common mechanism of gauging a portfolio's attractiveness.

To begin, let's consider a portfolio of two assets, which we can then generalize to a broader set. Further, let's suppose that we have M dollars in initial capital, with N_1 dollars invested in security one and $N_2 = M - N_1$ invested in security two. This means that we can define the weights in assets one and two as:

$$w_1 = \frac{N_1}{M} \tag{4.63}$$

$$w_2 = \frac{N_2}{M} = 1 - w_1 \tag{4.64}$$

- **Computing Portfolio Returns**: The return for this investor can then be expressed as a linear combination of the returns on assets one and two, with the weights defined by w_1 and w_2:

  $$R_p = w_1 r_1 + w_2 r_2 \tag{4.65}$$

  where r_1 and r_2 are the returns on asset one and two, respectively. r_1 and r_2 are clearly random variables, as we don't know for certain what the future returns will be. Instead we might be concerned with the expected value of (4.65). These expected returns, can then be used to help investors form a portfolio that they believe will have the most compelling future returns. As one might expect, estimating these expected returns can be a challenging process. In fact, according to the theory of market efficiency introduced earlier in the chapter, we should have very little ability, if any, to forecast returns for an asset. In chapters 24 and 25 we explore the procedure for constructing expected return estimates and also emphasize the difficulty in producing reliable estimates, and how to make our estimates robust in the presence of estimation error.

  The formula in (4.65) can be generalized to a portfolio of n assets with weights w_i in each asset:

  $$R_p = \sum_{i=1}^{N} w_i r_i \tag{4.66}$$

  Here we can see that again the returns on a portfolio can be expressed as a linear combination of the returns of the individual assets. This is because calculating portfolio return is a linear function. It is often more convenient to express this portfolio return in matrix form, where we can equivalently express the portfolio return R_p as the dot product of the vector of weights and vector of expected returns. The matrix equivalent of equation 4.66 can be written as:

  $$R_p = w^T r \tag{4.67}$$

  That is, the portfolio return is the dot product of the weights vector w and the vector of individual asset returns, r.

- **Computing Portfolio Volatility**:

  Of course, investors are not only concerned with expected returns, they should instead be concerned with the entire distribution of returns, most notably including its variance or standard deviation. Knowledge of this standard deviation, or volatility of returns, enables us to measure returns on a risk-adjusted, instead of absolute, basis. Returning to the case of two assets, we can leverage the basic properties of the variance of the sum of random variables, in addition to the scaling properties of variance, to obtain the following portfolio variance, σ_p^2, for a portfolio invested in two assets:

$$\sigma_p^2 = w_1^2\sigma_1^2 + w_2^2\sigma_2^2 + 2\rho\sigma_1\sigma_2 \tag{4.68}$$

where σ_1^2 and σ_2^2 are the variances of assets one and two, respectively, and ρ is the correlation between asset one and asset two. Note that because variance is a non-linear function of the underlying returns, we now observe that the portfolio variance is not a linear combination of the underlying component variances. Instead, it is also a function of the embedded correlation between the assets. Importantly, we can see from this equation that the variance of a two asset portfolio is highest when the underlying assets are perfectly correlated, and the portfolio variance decreases as the correlation between the asset decreases. This ends up being a critical theme in the investment management world. Fundamentally, it shows that we can reduce the variance in our portfolio if we are able to find assets with low, or negative correlations. In practice, this can make assets with less attractive return characteristics desirable because of the reduction in volatility, or variance, that they lead to.

Equation (4.68) can be generalized to n assets using the following formula:

$$\sigma_p^2 = \sum_{i=1}^{N} w_i^2\sigma_i^2 + 2\sum_{i<j} w_i w_j \rho_{i,j}\sigma_i\sigma_j \tag{4.69}$$

where σ_i^2 is the variance of asset i and ρ_{ij} is the correlation between asset i and asset j. As with returns, it is often more convenient to express this relationship in matrix form, which can be done via the following equation:

$$\sigma_p^2 = w^T \Sigma w \tag{4.70}$$

where Σ is now a covariance matrix of the following form:

$$\Sigma = \begin{bmatrix} \sigma_1^2 & \rho_{1,2}\sigma_1\sigma_2 & \dots & \rho_{1,n}\sigma_1\sigma_n \\ \rho_{2,1}\sigma_2\sigma_1 & \sigma_1^2 & \dots & \rho_{1,}\sigma_2\sigma_n \\ \dots & \dots & \sigma_i^2 & \dots \\ \rho_{n,1}\sigma_n\sigma_1 & \rho_{n,2}\sigma_n\sigma_2 & \dots & \sigma_n^2 \end{bmatrix} \tag{4.71}$$

Expansion of (4.70) shows that the off diagonal elements in Σ appear twice and represent the correlation or covariance terms in (4.69), and the diagonal elements represent the individual volatility terms expressed in the first sum of the same equation.

- **Annualizing Returns and Volatilities**:

  Another common task for quants in investment management is to convert returns from a given frequency to another. Most commonly, we work with annualized returns. If we assume that we are working with log-returns, and that the returns are i.i.d., then we can annualize a return using the following simple equation:

$$r_A = \frac{1}{\Delta_t} r \tag{4.72}$$

where Δt is the length of time in the original frequency. For example, if the returns are already annualized, then we would have $\Delta_t = 1$ and the returns would remain unchanged, as expected. Conversely, if we were starting with weekly returns, then we would have $\Delta_t = \frac{1}{52}$ and we would multiply the raw return by 52 to obtain an annualized log return.

Similarly, under the assumption of i.i.d. log returns with variance that grows at a rate of t, the annualized volatility can be expressed as:

$$\sigma_A = \sqrt{\frac{1}{\Delta_t}} \sigma \tag{4.73}$$

Therefore, if we are starting with weekly returns, then again $\Delta_t = \frac{1}{52}$ and the annualized volatility would be obtained by multiplying the original volatility by the square root of 52.

An important caveat to emphasize here is that this annualization formula for variance and volatility assumes that the returns are i.i.d. In many cases this is a reasonable and valid assumptions, however, if there is auto-correlation in the returns, for example, then the returns will no longer be i.i.d. and this formula will become less accurate. In these cases it is recommended that we first remove the autocorrelation from the data prior to annualizing the volatility.

4.6.4 Portfolio Concepts in Practice: Benefit of Diversification

Using the coding example provided in the repository allows us to illustrate one of the more fundamental concepts in finance, diversification. Diversification is often said to be the only free lunch in financial markets, and is the idea that we are able to reduce the risk in our portfolio, without sacrificing return, by introducing uncorrelated assets. To see this, we consider a simplified example of two assets, equities[8] and treasuries[9]. Historically, at least in recent decades, stocks and bonds tend to exhibit negative correlation. As we can see from (4.68), this causes the volatility of portfolios involving both treasuries and equities to decrease. In the following chart we see how the expected return and volatility of portfolios of these two assets evolve as we vary the weight in treasuries:

[8]Proxied by the S&P 500 ETF SPY.
[9]Proxied by the GOVT ETF.

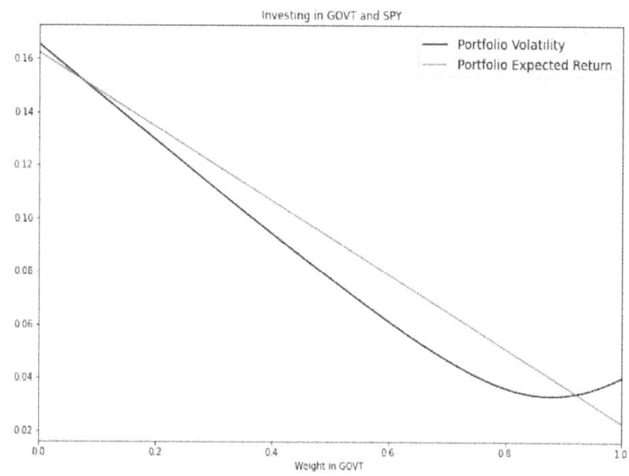

As we can see, there is a linear decrease in expected return as we increase the weight in treasuries and decrease the weight in equities. This is of course expected, as stocks are characterized by higher returns that treasuries. The portfolio volatility, however, decreases faster than the expected return does, leading to higher risk-adjusted returns when portfolios of equities and treasuries are combined. This is a powerful concept that we will keep returning to throughout the text.

4.7 BOOTSTRAPPING

4.7.1 Overview

Bootstrapping is a non-parametric technique for creating sample paths from an empirical distribution. The idea behind bootstrapping is that each realization we observe is in actuality just a draw from some unobserved density. Bootstrapping provides a method for sampling from this density with minimal assumptions. It is also a very simple and intuitive technique, but is quite powerful nonetheless. In practice it involves simulating returns, and paths of asset prices by scrambling the historical data, or sampling various historical returns and creating series of them in random order.

Bootstrapping has applications across the field of quant finance, including in risk management, portfolio optimization and trading. Essentially, bootstrapping is a technique for generating new, synthetic data which is statistically equivalent to an observed dataset. It is a topic that we will refer to throughout the book, including when covering the following topics:

- Generating Synthetic Data and Filling Missing Data in chapter 9
- Generating a Prior Distribution in Weighted Monte Carlo in chapter 17
- Re-sampling an Efficient Frontier in chapter 24.
- Using Historical Simulation to calculate VaR in chapter 26

As bootstrapping works by re-sampling paths from an observed dataset, it is inherently a method for generating synthetic data. Use of this synthetic data can be of great use in a finance world where we are generally lacking data. Additionally, bootstrapping can be useful in that it gives a broad sense of the range of potential outcomes based on limited parametric assumptions about the observed data.

It can be proven that the bootstrapping algorithm will preserve the characteristics of the empirical density. For example, if we generate enough samples, then the mean via bootstrapping will match the mean in the historical data set. In some applications this may be the desired result, however, in other cases this may mean that we want to de-mean our input return data prior to bootstrapping. This is true for other characteristics of the empirical density as well, such as variances and cross-asset correlations.

In a standard bootstrapping algorithm, at each point the probability of each historical observation is equal regardless of the previous path the asset has taken. This is an important assumption in bootstrapping. Phrased differently, we treat the historical data samples as independent and identically distributed (i.i.d.). In finance this is often a realistic assumption and is in the spirit of market efficiency.

Auto-correlation, however, is one case that would cause the i.i.d. assumption to be violated, or unrealistic. In order to overcome this, we would need to modify our bootstrapping algorithm, which could be done in a few different ways. First and perhaps simplest, we could instead rely on a block bootstrapping algorithm, where we would select blocks rather than single dates, where the autocorrelation is assumed to be embedded within the selected blocks. For example, if we knew the data was AR(2), then we generate samples consisting of three consecutive days. Note that in order to pursue this approach we need to understand the structure of the autocorrelation to know how many lags to incorporate, and the subsequent size of the blocks we should select. Alternatively, we could remove the auto-correlation from the data using standard techniques, such as a Blundell-Ward filter [45], generate the simulate paths, and then add the auto-correlation back.

Another example where this assumption may be inaccurate is if the distribution is dependent on an external conditioning variable. In this case we might want to make the probability of drawing each date a function of this conditioning variable, rather than equal. These types of extensions are possible, in a bootstrapping algorithm, however, we will want to make sure that in adding these features we don't distort the underlying statistical properties.

Putting aside these potential complications, a standard bootstrapping algorithm can be summarized as follows:

- Create a matrix of historical returns for the universe of assets.
- Generate a sequence of random integers between 1 and N, where N is the number of historical days of data.
- Look up the return data for all assets that corresponds to that randomly chosen date.

- Apply these returns to the entire asset universe.
- Continue this for many time steps until the end of a path is reached. Repeat until a large number of paths are created.
- Calculate whatever payoffs, risk measures or other analytics desired from the full set of simulated paths.

There are two key points to emphasize here. First, notice that bootstrapping works by resampling increments repeatedly, such as daily returns. This increments are then stitched together to form paths for the underlying asset. This process means that bootstrapping fills out the distribution of possible outcomes, and is not constrained by the observed realizations. Secondly, it should be emphasized that, once a given day is chosen, the returns for all assets on that day are used in the simulation algorithm. This embeds information about the correlation of different assets in the bootstrapping algorithm. If we were to select a different historical date randomly for each asset, in contrast, then we would lose this information about the correlation structure.

A final note is that bootstrapping works by applying a sequence of changes, or returns to a series. Along these lines, in a bootstrapping algorithm we will need to choose whether we believe log returns are appropriate, as they are likely to be in the case of stocks, or if we should use absolute differences, as we might in the case of yields or credit spreads. Later in the text, in chapter 13 we will see that this is equivalent to choosing between the Black-Scholes (log-normal) model, which is built on log returns, and the Bachelier (normal) model which is built on absolute deviations.[10]

4.8 PRINCIPAL COMPONENT ANALYSIS

One of the more common tasks for a quantitative researcher analyzing a set of data is to understand the co-movement of a set of assets or other variables. As we saw earlier in this chapter,, the correlation between different assets plays a large role in determining the variance of any portfolio. As such, the underlying correlation structure between the assets is of direct relevance. We also saw that portfolio construction problems inherently are based on calculations using covariance matrices. We will also see later, that certain transformations of these covariance matrices appear repeatedly, most notably inversion and taking the square root. This means that, we need tools to help us accomplish these tasks in a robust manner, and Principal Component Analysis (PCA) is one such tool. Additionally, PCA has particular appeal as it provides insight into the structure of the underlying matrix. It can help us identify, for example, what are the most important drivers of variance within the set of assets.

PCA does this by transforming the underlying matrix such that the transformed matrix is easier to work with. PCA is at heart a matrix decomposition that creates orthogonal factors, and is in reality just eigen or singular value decomposition by another name [262]. The underlying spectral decomposition for PCA, for a given matrix Σ can be written as:

$$\Sigma = X^\top X = CDC^\top \tag{4.74}$$

[10] We will also see that both of these models are a specification of the more general CEV model.

where X is a matrix of demeaned historical return data that is used to estimate the covariance matrix, D is a diagonal matrix with λ_i's on the diagonal. These diagonal elements, λ_i are known as the eigenvalues of the matrix V. Each row in the matrix C, by contrast, is an orthogonal eigenvector. The fact that these eigenvectors are orthogonal is an important and useful property of PCA. It means that when we combine them, we do not need to account for correlation terms.

Additionally, the magnitude of the eigenvalues, λ_i, in D play an important role in understanding the underlying matrix. The larger eigenvalues, in particular, are the eigenvalues that explain most of the movements in the covariance matrix. The smaller eigenvalues, by contrast, explain very little movement in the underlying matrix. The percent of variance explained by each principal component, P_i, can then be written as:

$$P_i = \frac{\lambda_i}{\sum_{i=1}^{N} \lambda_i} \tag{4.75}$$

This means that the larger eigenvalues are likely to be most important and explain the structural elements of the data. The smaller eigenvalues, conversely, are far more likely to be noise.

Once we have performed the above decomposition on Σ, matrix operations such as the square root and inverse will be easier to compute. For example, if we consider the inverse, it can be expressed as:

$$\Sigma^{-1} = CD^{-1}C^{\top} \tag{4.76}$$

Further, since D is a diagonal matrix its inverse will also be diagonal with values $\frac{1}{\lambda_i}$. Importantly, we can see that when we invert the matrix D, the smaller eigenvalues have the highest weight when inverting Σ.

This illustrates the next useful feature of PCA, which is that it can help us reduce the dimensionality of a matrix and identify a set of orthogonal main driving factors. To do this, we can use the relationship in (4.75) and choose the largest principal components, ignoring the rest which are more likely to be noise. In practice, this can either be done by choosing a fixed number of components that we think explain the majority of the movement in the underlying matrix, or setting a fixed percentage of the variance that we would like to explain, most commonly 90 or 95%. If we choose the first approach, then we have the benefit of a consistent number of principal components as we repeat the analysis over time, however, over time we may explain a substantially different proportion of the variance. In the second approach, the reverse will be true. Over time the number of principal components will vary, however, the proportion of variance explained will be consistent.

Throughout the text, we will see several examples of principal component analysis. For example, in chapter 10 we discuss how PCA can be used in exploratory data analysis. In chapter 19, we then see how PCA can be used to help better understand the behavior or a yield curve. In chapter 24, we will see how we can use PCA to increase the stability of matrix inverse operations that are used to help solve portfolio optimization problems. Additionally, in chapter 27, we will see how PCA might be used to help construct trading strategies.

4.9 CONCLUSIONS: COMPARISON TO RISK NEUTRAL MEASURE MODELING

In the last two chapters, we have detailed the foundational tools for modeling the risk-neutral and physical measure, respectively. In many ways, the skillset in the two worlds appears quite different. In the risk-neutral world, we rely on hedging arguments, replication and must be well versed in stochastic calculus. In the physical measure, by contrast, these replication arguments no longer help us. Instead, we need to rely on forecasting, which we can do leveraging some of the econometric tools discussed in this chapter, or we might seek to use machine learning techniques, which are introduced in chapter 28. This gives rise to the popular delineation of quants into so-called P-measure quants, who populate the sell-side and model exotic options and other complex derivatives, and Q-measure quants, who populate the buy-side and try to use econometrics forecast asset price movements.

Underneath these differences, however, there is also a great deal of commonality in quant modeling techniques between the two spaces. For example, in order to tackle any quant problem, a quant will by necessity rely on a simplified model of how the world works. This is true in both the risk-neutral and physical measures and is evident in the assumptions that the underlying models make. The quant will then need to identify how to calculate this expectation under the simplified model and will need to write a suitable piece of code to calculate the expectation. Against this backdrop, we will find that almost all quant problems will not have a closed form solution. As a result, experience with numerical techniques and coding becomes of critical importance for quants regardless of whether they are working on the P or Q measure.

The same stochastic processes that permeate the risk-neutral world, and are used to solve option pricing problems, may be applied in the context of a forecasting, or estimation problem as well, in the physical measure. The fundamental difference between the two approaches, then, is how we handle the drift of the asset. In one case, we assume investors are risk-neutral, and therefore we can discount all cashflows at the risk-free rate. In the Q measure, this is justified by our replication arguments. In the P measure, we must incorporate different investors risk preference, which fundamentally means we must incorporate the discount, or premium that investors place on cashflows in different states of the world. As we incorporate these risk preferences, the drift in the underlying assets changes. Generally speaking this drift in risky assets will increase in the P-measure, a reflection that investors require a premium to invest in risky assets. The drift in this measure, in the presence of risk preferences, may also be highly state dependent.

Knowledge of this subtle distinction between the two worlds, and understanding of the commonality as well as differences, can be a distinguishing feature of success for practitioners. In some cases market participants may look to take a step further and juxtapose the two measures, isolating the risk premium embedded in different markets. This approach is not without complications, but is explored in further detail in chapter 17.

II

Fundamentals of Coding and Data Analysis

CHAPTER 5

Python Programming Environment

5.1 THE PYTHON PROGRAMMING LANGUAGE

BROADLY speaking, programming languages can be broken down into low and high level languages. High Level Languages translate the source code that the programmer writes into machine code before the program can be run. Some common examples of high level languages are C++, Python, C#, R and MATLAB. Low level languages, in contrast are more raw and written more directly in terms that the machine can process without being transformed. Low level languages include C and Assembly. High level languages have more embedded functionality and can be easier to code in. Conversely, they may require more overhead and can be less efficient.

Within the scope of high-level languages, some are referred to as interpreted languages whereas others are compiled. In an interpreted language, each line is translated as it is run. In a compiled language, the entire program must be translated prior to execution, a process that is called compilation. Python is an interpreted language. This is a desirable feature because it gives coders more flexibility in debugging and modifying their code without requiring the code to be re-compiled. This is because the code is read line by line and can therefore be modified as it is debugged, or executed. Conversely, in a compiled language, for new code to take effect would require a full recompilation of the code.

As a whole, Python is an intuitive language with easy to understand syntax. Even relative to other high level, interpreted languages, most coders find Python both powerful and easy to use. When pursuing quant finance applications, the enhanced convenience of Python make it a very attractive choice. In the next section we explore the advantages and disadvantages of coding in Python relative to other high level languages.

5.2 ADVANTAGES AND DISADVANTAGES OF PYTHON

Python has become a very trendy coding language lately in part due to its ease of use and intuitive syntax, and in part for its support for more robust features. In

DOI: 10.1201/9781003530626-7

particular, most users find Python code readable and find that reading/writing a piece of code in Python is not that different from reading text. This compares favorably with C++ and other languages where the use of pointers, semi-colons and colons can make the code more difficult to follow. Python also has a set of powerful interactive development environments (IDEs) that make debugging and running Python code quite user-friendly.

Python also contains a broad library of open-source packages that are directly relevant for quantitative applications. In particular, as we will see in chapter 8, the pandas and numpy modules provide significant functionality for transforming data and performing complex numerical calculations. Python also contains many open-source packages related to Machine Learning and time series analysis that quants will rely on frequently.

Another advantage of Python is it's robust support for object-oriented programming. This is a significant advantage relative to other high level languages as it enables us to use classes to build scalable, robust code. Importantly, compared to other languages with similar mathematical and statistical functionality, Python's object-oriented support is far more integrated and far superior. As we will see in chapter 8, this is a powerful concept that enables us to write far more efficient code.

Having said that, while Python does have support for object-oriented programming that compares favorably to many other languages, such as R and MATLAB, it's support for object-oriented programming is not as strong as in C++. Additionally, C++ provides more control over things like memory management, making it more efficient in certain situations. Because of this, C++ tends to be commonly used in high frequency trading applications where the primary concern is execution speed.

5.3 PYTHON DEVELOPMENT ENVIRONMENTS

Python comes with a set of robust integrated development environments (IDEs) that make it easy for users to run and debug their code. Within each of these environments we should be able to run specific python files that we have created, run ad-hoc code via the python console, and set breakpoints that enable us to stop our code at specific places in our files. The ability to debug and set breakpoints in code is of particular importance to coders building complex models, as it enables coders to isolate and identify bugs faster. Python even enables us to modify our code while we are debugging, meaning once a bug is found we can use this feature to try different methods to fix it in-line, while debugging, until it has been fixed[1].

A few noteworthy Python IDEs are:

- Spyder
- PyCharm
- jupyter Notebook

[1]This is in contrast to compiled languages like C++ where we could debug but would need to re-build should we modify the code.

Spyder and PyCharm are powerful Windows, Mac OS and Linux based IDEs that have features for seamlessly debugging, navigating code and running it. These tools also have advanced auto-completion features, making it easier to write new pieces of code. jupyter Notebooks provide a web interface that enables us to execute code line by line and visualize the results as we step through. As such, jupyter Notebooks can be a powerful tool for analyzing datasets and building prototypes.

5.4 BASIC PROGRAMMING CONCEPTS IN PYTHON

In this section we attempt to orient the reader to the basic concepts used in Python. This book is targeted at readers who have some experience working with Python and have completed one or more projects using Python. As a result, the coverage of these basic topics is meant to serve as a review. Readers who need more detailed treatment of the basics should consult [204]. In the following sections, we briefly describe the syntax used by the Python language, how it handles data types, and then proceed to discussing built-in functions, conditional statements, loops and other more advanced concepts.

5.4.1 Language Syntax

One of the often cited advantages of Python as a coding language is its readable syntax. When we look at completed python coding samples they read like pseudo-code or text explanations of a program. However, behind the scenes is a formal syntax process that requires the code to be written in a certain way. If it is not written according to this standard, then we will receive syntax errors when we try to run our Python program. These syntax errors could arise for many reasons, for example, if a variable is defined as a reserved keyword.

One important component of this syntax is that Python uses an indentation-based syntax methodology for things like classes, user-defined functions and loops. This means that if we do not use the proper indentation in our code, the code will not run. As an example, consider the following example of a for loop:

```
x = range(1,100)

for z in x:
    print(z)

print("done with for loop")
```

If we were to change the whitespace in the line within the for loop, then the code would no longer run. Notice also the colon prior to the indentation. This is also required in Python to avoid syntax errors. Lastly, notice that once the for loop is exited, in the last line, the indentation is reset. This indentation reset is how Python knows that the for loop has existed.

Unlike some other languages, such as Java, C++ and C#, Python does not require a semi-colon or other character at the end of each line. Instead, Python knows that the line of code has completed by the new-line character.

5.4.2 Data Types in Python

In Python each value or variable that we create is stored internally as a different data type. For example, when we create a variable using text data, it is stored internally as a string object. Conversely, if we store numeric data it may be stored internally as an integer or floating point number.

Python's basic data types include:

- String (str)
- Integer (int)
- Floating Point Number (float)
- Complex Number (complex)
- Boolean (bool)

In chapter 8 we will also review the most common data structure types in Python, such as lists and tuples.

In many programming languages, such as C++, variables need to be declared to have a specific type. Python, by contrast, is a bit more flexible with its typing. This means that when you create a variable Python will dynamically figure out what type it should be. For example, if you create a variable with numeric data, it may be stored as an integer or as a floating point number depending on whether a decimal is included. We can check the type that Python has assigned to a variable using the built-in type function, as shown in the example below:

```
x = 3
type(x)

y = 3.5
type(y)

z = x / y
type(z)
```

In this example we create three variables, x, y and z. Notably, the first variable, x is created as an integer and y is created as a float. Lastly, notice that z is also created as a floating point number, even though x had been defined as an integer. This is because Python is smart enough to know that the result of the division operation will be a floating point number. This introduces one of the challenges associated with Python's flexible typing. It is important to understand how Python handles these implicit typing and type conversions in various situations to ensure there are no unintended side effects.

5.4.3 Working with Built-in Functions

The Python language consists of a wide array of built-in functions that coders can use to facilitate writing their programs. Throughout the book, we detail examples of different functions for generating random numbers, performing regression analysis, perform matrix decompositions and much more. Functions may return a value that is the result of an internal calculation, or they may not return a value but instead perform some sort of internal operation or transformation of an object. Calling functions in Python is seamless. For functions that return a value, we can assign the return value of the function to a variable that can be accessed later.

The simplest example of a function that performs an operation rather than returns a value is the print function. The print function enables us to write a given message to the console, as the following coding example shows:

```
print("calling my first function that does not return a value")
```

As quants, many of the functions that we rely on will inherently be functions that help us with calculations or data transformations, and as a result many of them will return the results of that calculation or transformation. Much of the book is dedicated to such examples. Additionally, the following example shows how to use a built-in log function to convert from a zero-coupon bond price to a discount rate, using the formula presented in chapter 2:

```
import math
T = 1.0
V_0 = 97.0
P = 100.0

yld = -math.log(V_0 / P)/T
print(yld)
```

Here the built in log function is used to compute the yield of a zero coupon bond, with the result stored in a variable named yld. It should also be noted that in some cases we need to load a module or library to access a function that we want to use, as is the case with the math module in the above example. As we can see, this can be done via an import statement, which are described in more detail in section 5.4.10.

5.4.4 Conditional Statements

One of the first features that coders need to become familiar with in Python are conditional statements. Conditional statements allow us to execute different code depending on one or more conditions. Conditional statements in Python are implemented via if elif (else if) else statements. The conditions within these statements must evaluate to a boolean, true or false, and the code will proceed inside the block if the condition is true, otherwise it will move onto the next condition.

The following Python example shows how we can use such a conditional statement to execute conditional code in Python:

```
if (condition1):
  print("call function1()")
elif (condition2):
  print("call function2()")
else:
  print("call function3()")
```

The conditions in the code above may be based on equality comparisons, using the == operator, or based on other operators such as < or >=.

Readers who are new to programming should note the important distinction between the equality operator, ==, and the assignment operator, =. The assignment operator is used to assign a value to a variable, as the name implies. The equality operator checks whether two values are the same.

Conditional statements in an if block can be based on one or more conditions. For example, in condition1 above, we could require that multiple criteria are true in order for the condition to be met. The logical operators and, or and not in Python can be used to create a condition based on multiple criteria, as shown in the following example:

```
condition1_1 = True
condition1_2 = True
condition2_1 = True
condition2_2 = True

if (condition1_1) and (condition1_2):
    print("call function1()")
elif (condition2_1) or not (condition2_2):
    print("call function2()")
else:
    print("call function3()")
```

5.4.5 Operator Precedence

When executing conditional statements it is important to know the order in which they will be executed. This is analogous to the well-known order of operations in math where we know that parenthesis are evaluated first, followed by exponents, multiplication and division and finally addition and subtraction.

In Python, the parenthesis operator will similarly be evaluated first, and has the highest precedent, followed by function calls. Additionally, as in mathematics, exponents are executed prior to multiplication and addition, which is executed prior to addition and subtraction. A full list table of Python's operator precedence levels can be found in [210].

5.4.6 Loops

Iterating or looping through data structures will be another common task when coding in Python, and will come up in many contexts. For example, one can imagine iterating through a list of positions, dates, or characters in a string.

In Python loops can be traversed using definite, for loops, or indefinite, while loops. The following code shows an example of each:

Definite Loops:

```
nums = range(0,100)
for x in nums:
  print(x)
```

Indefinite Loops:

```
x = 0
while x < 100:
  print(x)
  x += 1
```

In the first example, we are executing a for loop through a list object, which will print each element in a list. We can loop through a set of numbers or the elements in a data structure. Importantly, use of a definite loop requires us to know before entering the loop the set of values that need to be traversed. In the second example, we used an indefinite while loop in order to iterate. Here, we can specify an exit condition in the loop without having a set list of values to loop through. This gives us a bit more flexibility, however, as we can update the variables used in the condition within the while code block. The cost of this flexibility is that while loops are not as safe, as there is no guarantee that the exit condition is ever met, which would lead to an infinite loop. Because of this, whenever possible it is recommended to use for loops, and to use while loops only when we can't specify the elements to iterate over beforehand.

5.4.7 Working with Strings

Strings are a critical data type in Python that all coders will become intimately familiar with. Being able to transform, parse and create strings are all routine tasks that will come up again and again. In Python, strings are compound data types that consist of smaller strings each containing a single character. The string class in Python has a great deal of useful functionality built into it. For example, it enables us to easily compare strings, concatenate strings and search within a string. An important feature of strings is that they are immutable. This means that they cannot be modified after they have been initialized and we cannot use the [] operator to change a part of a string.

The list below summarizes some of the most common function for manipulating strings:

[] operator : enables us to select part of a string

== operator : enables us to compare strings

+ operator : concatenate two strings

in & not in : check if another string is in a string

find : find the start index of a string within another string

split : split a string by a specified character

len : determine length of a string

5.4.8 User-Defined Functions

In addition to leveraging Python's set of built-in functions in different libraries/modules, Python also has support for creating user-defined, stand-alone functions[2]. A user-defined function in Python is specified by the keyword **def** followed by a function name, and then a list of function parameters in parenthesis. A function can have an arbitrary number of parameters. Some functions may not pass in any parameters, wheras others might pass in many. Additionally, a function may return a value using the return keyword.

The following example shows a user-defined function for the Black-Scholes call price formula, which is a result of the Black-Scholes dynamics introduced in chapter 4:

```
import math
from scipy.stats import norm

def callPx(s_0, k, r, sigma, tau):
  sigmaRtT = (sigma * math.sqrt(tau))
  rSigTerm = (r + sigma * sigma/2.0) * tau
  d1 = (math.log(s_0/k) + rSigTerm) / sigmaRtT
  d2 = d1 - sigmaRtT
  term1 = s_0 * norm.cdf(d1)
  term2 = k * math.exp(-r * tau) * norm.cdf(d2)
  price = term1 - term2
  return price

px = callPx(100, 100, 0.0, 0.1, 0.25)
print(px)
```

As we can see, the first line shows the signature of the function, which is named callPx and takes five parameters. The function then implements the Black-Scholes formula over the next several lines. At the end of the function, we can see that the Black-Scholes price is then returned using the return keyword. More details on the Black-Scholes model, and a derivation of the formula implemented in the function above can be found in chapter 13. The code ends by calling the function assigning the value to a variable, px, prior to printing it. The reader should notice the change in indentation when the function definition has ended after the return line above.

5.4.9 Variable Scope

Scoping of variables is an important concept in Python as well as any other programming language. Variables can be defined to have different scopes. They can be global variables that exist throughout the entirety of our programs. Alternatively, they may

[2]Later we will also see that we can also write user-defined classes.

be specific to a function, or a class. Importantly, once a variable goes out of scope, the memory for the variable is freed up and the variable can no longer be accessed [3].

In the previous example, we showed how to implement a user-defined function that implemented the Black-Scholes model which returned an option price. If we try to reference variables created inside the function after the function has exited, such as term1 and term2, the variables will be out of scope and we will receive an error. Relatedly, if we did not return the price variable from the function than we would have no way to extract the price outside the function. The px variable which is set to the return value of the function however will have global scope, making it available throughout the program. These differences in scope are important to be cognizant of when coding.

5.4.10 Importing Modules

As mentioned earlier, one of the main advantages of using Python is the rich set of libraries, or modules that help us to perform different functionality. Python has countless useful modules that will make our lives easier as coders. Perhaps the two most common examples are numpy and pandas, however, there are others that provide functions and classes for machine learning, statistical analysis and much more. For example, if we are looking to download data from a given source such as the yahoo finance website, we will even find that there are modules in Python that have already wrapped this functionality for us. The process for accessing one of these modules in Python is seamless, as we simply need to specify an import statement, as shown below:

```
import numpy as np
import pandas as pd
```

In this example we are importing the numpy and pandas modules, and when doing so we are giving each of them an alias using the as keyword. Once this code has been executed, we can access any numpy or pandas functions or classes using the np and pd prefix, respectively, like shown below for a hypothetical function blah:

```
np.blah()
pd.blah()
```

In addition to pandas and numpy, we will also find the following modules useful in tackling many quant tasks, such as math, statsmodels, scipy, cvxpy, Matplotlib, tensorflow and scikit-learn. The reader is encouraged to consult the relevant documentation to obtain more information on each of these specific modules. The supplementary coding repository also contains several examples leveraging each of these modules. The import command can also be used to source other user-created Python files in addition to open-source modules.

[3] A destructor for the class going out of scope is also called, as we will learn in the next chapter, 8.

5.4.11 Exception Handling

Errors occur in Python programs frequently. These errors could happen for many reasons, and are bound to occur no matter how well a piece of code is written. When these errors do happen, they trigger what are called **exceptions** in the code. The process of handling these exceptions is then referred to as exception handling.

Exceptions are a type of class that exists within an inheritance hierarchy. As a result, depending on the type of error that occurs, a different type of exception object may be thrown. A few ways that an exception might occur in our code would be if we try to divide a number by zero, open a file that doesn't exist, change a single character in a string [4] or access a value in a list or array beyond its length.

Because exceptions are unavoidable and will inevitably happen regardless of the quality of the code, it is best practice to handle these exceptions explicitly in our code. Python provides a framework for doing this via **try except finally** statements. The except keyword is used to define the code that is run after an exception occurs, as shown:

```
try:
  somecode()
except:
  handle_exception()
```

In this case, the handle_exception() function is assumed to be a user-defined function that is meant to handle exceptions. Exceptions that occur anywhere in the try block of the code will be caught in the except block. In this case, that means that exceptions triggered in the somecode() function will be handled using the handle_exception() function. When we are handling exceptions, we may want to write a detailed explanatory message to the console, write a warning message to a log file, exit an application safely or modify some variables prior to re-trying the code.

As coders, the more user-friendly we can make our exceptions, the easier our code will be to use. Along those lines, it is our job to make sure that any exception that occurs in our code is easy to identify and fix not just by us but by any user. A large part of this will be creating informative error messages.

In addition to handling exceptions that are thrown in various parts of our code, we may also want to force an exception to be thrown in certain situations. We may do this to avoid the code continuing and producing a misleading or dangerous result. As an example, if we are performing an options back-test we might want to throw an exception if we encounter an option's price that is negative. Left uncaught, this could lead to significantly inflated back-test results. Therefore, it is better to throw an exception so that the user may fix the data rather than to continue processing the results.

In these cases, exceptions can be thrown using the Python **raise** command. The following coding example shows how an exception could be raised if a negative option price is encountered:

[4] Remember that strings are immutable, meaning they cannot be modified using the [] operator.

```
if (px < 0):
  err = ValueError("Negative price provided")
  raise err
```

We can see that the raise keyword is used to throw, or raise the exception. In this case, the exception type being raised is a ValueError, however, there are many other types of exceptions we could also raise.

5.4.12 Recursive Functions

Python supports recursive functions, which are functions that call themselves. When writing a recursive function, the idea is that each successive call within the recursive function should simplify the calculation until it ultimately reaches a point where the calculation is tractable and the function can return a value. A simple example of a recursive function is one that computes a number's factorial, as shown here:

```
def recur_factorial(n):
  if n == 1:
    return n
  else:
    return n*recur_factorial(n-1)
```

As we can see, the recur_factorial function is called within its own body, but that it is called with increasingly small values of n until it reaches $n = 1$ where a value is returned.

When we write recursive functions, we need to make sure that all paths of the calculation reach an end-point. In the example above, we can see this will happen when the value passed to the function reaches one. If that condition was removed from the function, then the function would never reach an exit condition and the recursion would be infinite. Most programming languages, including Python, have a limit to the depth of recursion that is allowed in order to prevent this type of infinite recursion. This means that the code would generate an exception related to maximum recursion depth rather than running indefinitely.

5.4.13 Plotting/Visualizations

As previously mentioned, one of the many advantages of Python is the wide universe of open-source built-in packages. Among these, there are libraries/modules that facilitate visualizing and plotting data in various forms. As a quant this will be a common task. Once a model has been created and calibrated to real data, the results will need to be presented. Generally, visualization and plotting tools can help us convey the message from our model and summarize the results. Additionally, these plotting tools are extremely useful when working with and validating input data. We will explore how some of the visualizations can be used in this context in chapter 9 and 10.

The matplotlib module, in particular, contains many useful plotting functions. More information on this module can be found in [178]. This module enables us to build many different types of plots, including:

- Box Plots via the **matplotlib.pyplot.boxplot** function
- Time Series / Line Charts via the **matplotlib.pyplot.plot** function
- Histograms via the **matplotlib.pyplot.hist** function
- Scatter plots via the **matplotlib.pyplot.scatter** function

5.5 WORKING IN A MULTI-PROGRAMMER ENVIRONMENT

When coding scalable models in Python quants are often working in teams. When doing so, quants need a means of coordinating their changes to the code and ensuring that the changes that they make don't negatively impact other's work. Many programs exist to facilitate this task, most notably git/github [113].

The motivation for using git, or another similar tool, is partly to help coders coordinate and merge their changes to a shared codebase. Additionally, these tools provide a robust means of version control. That is, git enables us to keep track of all historical changes made to our codebase in an organized way. It even gives us a way to revert back to a previous version if necessary. Quant teams at both buy-side and sell-side institutions generally have strict release processes for their models. Using git enables firm's to keep historical code for all previous production versions of the code and enables a quant to roll back if necessary.

Because of this, git has benefits for both large teams and quants working individually. Most notably, quants working individually will benefit from the version control aspect of git, as will larger organizations with formal release processes. Teams using git will also benefit from easier code coordination and merging of new changes to the codebase.

Git consists of the following three layers of repositories :

- Local Working Copy
- Local Repository
- Remote Repository

The first layer is a copy of the code files on each user's machine, which the user will continually update as they program. This will not be viewable by any other users. The next layer is also stored on the user's machine, and is called the local repository. The local repository is also inaccessible to other users, and is where a user might store code that has been tested and seems safe but is not necessarily production ready [5]. Moving files from a local working copy to the local repository is called a commit, and is done using the git commit command. The final layer in a git infrastructure is the remote repository, which is accessible to all users. The remote repository is meant to only contain changes to the code base that have been finalized and tested. Moving files from a local repository to a remote repository is called a push, and is done via the git push command.

[5] Or ready for others to view.

Some of the most common tasks in git are to add new files to a codebase, remove files from a codebase, to update existing files in the local or remote repository, or to get the latest changes from the remote repository.

The most common commands used in git are:

- **git status**: lists all changes you have made in your local files and local repository.
- **git add**: adds a new local file to your local repository.
- **git commit**: takes changes to all local files and commits them to your local repository (where they are still only visible to you).
- **git push**: pushes all committed changes from the local repository to the remote repository.
- **git pull**: retrieves changes from the remote repository.
- **git merge**: merges your changes with changes from a remote repository.

The merge command in particular is how a user would synchronize changes that they've made to their code if that file has been changed by another user as well. In some cases, git may be able to complete this merge automatically. In other more complex cases, git may not know how to conduct the merge, and some manual intervention may be required [6].

[6] In these situations, the git stash command is of great use.

CHAPTER 6

Programming Concepts in Python

6.1 INTRODUCTION

In the previous chapter we explored the Python programming environment and its basic components. In this chapter we go beyond these basics and explore how quants can leverage Python to build models related to the theory of mathematical finance. Python provides a great deal of open-source resources for accomplishing these tasks, and in this chapter we highlight what we believe will be some of the most commonly used functions, libraries and techniques. Additionally, we cover some more advanced programming techniques designed to provide the reader with the ability to write flexible, generic and scalable code. Of utmost importance in this pursuit is the creation of user-defined classes, and the use of design patterns, which are detailed later in the chapter.

We begin by introducing the two most important Python libraries for quants, NumPy and Pandas. These modules contain data structures that will be the foundation for building practical quant finance applications and working with real world data. The NumPy module also contains a large amount of built-in mathematical machinery/techniques. After introducing these modules, we then introduce the most relevant Python data structures and built-in mathematical tools that quants will need to be familiar with and will want to leverage often. We then turn our attention to object-oriented programming, design patterns, and algorithms, with the goal of showing how we can leverage these concepts to write more robust, scalable code.

6.2 NUMPY LIBRARY

Numpy is a critical module for quants relying on Python to build and work with models. It is an open-source library that contains a great deal of mathematical functionality, such as common statistical and linear algebra techniques. In particular, the NumPy module has implementation for many of the quant techniques that underpin many of the quant concepts discussed throughout the book, such as:

DOI: 10.1201/9781003530626-8

- Random Number Generation
- Linear Algebra Functions (e.g., Matrix Decompositions)
- Fourier Transform Calculations

Many of the quant techniques embedded in the NumPy library are detailed in section 6.5.

The NumPy library also contains one of the most useful data structures in Python, a **NumPy array**. NumPy arrays are flexible data structures that can be single or multi-dimensional. They can be used for storing numeric data, such as time series, matrices and panel data. NumPy arrays are discussed in more detail in section 6.4. More information on the NumPy library can also be found in [191].

6.3 PANDAS LIBRARY

Pandas is another Python library that will quickly become a staple for readers trying to implement quant models. It will also be of critical importance to anyone working with real financial datasets that need to be transformed, merged, or cleaned. It is of particular use for coders who are working with relational data, such as interacting with databases, reading or writing text files, or parsing a website to extract data, and provides a set of functions that facilitate the process of retrieving or writing data. These features of the Pandas library are described in more detail in chapter 9.

The Pandas library is a broad module that contains a set of data structures with some very useful accompanying functionality. These data structures, such as **Data Frames**, are ideal for time series or panel data because they give us the ability to:

- Index and slice our data
- Append to our data
- Merge our data with another dataset
- Repeat our data
- Iterate through our data with a for loop

All Pandas objects are **derived** from NumPy array objects, which allows them to **inherit** the functionality of numpy arrays while also providing new functionality. As such, they are great examples of the power of the object-oriented programming concepts that we discuss later in the chapter. In the next section, 6.4 we discuss how to leverage the pandas library to slice, merge and iterate through our data, and how we can leverage pandas data structures when working with time series and panel data. More information on the details on the pandas library can also be found in [195].

6.4 DATA STRUCTURES IN PYTHON

In this section we describe the most important and commonly used data structures in Python, such as Lists, Data Frames and Arrays. When doing so, we attempt to highlight the main differences, strengths and weaknesses in each structure.

6.4.1 Tuples

Tuples are a flexible and easy to use data structure that can handle heterogeneous data with multiple underlying types. As such, tuples provide us with a convenient data structure to store unstructured data. Tuples can be single or multi-dimensional. In the example below, we create a one-dimensional tuple whose elements each have a different underlying datatype:

```
tuple1 = ('SPY', '2018-09-06', 290.31)
```

Notice that each tuple can have an arbitrary number elements. In this case we created a tuple with three elements, one which was meant to represent a ticker string, one which represents a date, and another which represents a numeric value such as price. Once we have created a Tuple we can access a particular element in a tuple using the [] operator. However, like strings, tuples are immutable, meaning their values cannot be modified after they have been created. For example, the following code would generate an exception:

```
tuple1 = ('SPY', '2018-09-06', 290.31)
tuple1[0] = 'QQQ'
```

Tuples can be multi-dimensional and can store different data types in each element, as shown:

```
tuple1 = ('SPY', 290.31)
tuple2 = ('QQQ', 181.24)
tuples = (tuple1, tuple2)
```

Additionally, in a multi-dimensional tuple each embedded tuple can be of different lengths and contain different types, as shown in the following example:

```
tuple1 = ('SPY', 290.31, 1)
tuple2 = (181.24, 'QQQ')
tuples = (tuple1, tuple2)
```

This high level of flexibility makes tuples useful structures for dealing with unstructured data. Because of their immutable nature, Tuples are particularly useful as a return value from a function as they provide a means for returning somewhat unstructured data that the caller of the function can then unpack.

6.4.2 Lists

Lists are similar to tuples in that they also can handle generic, heterogeneous data in a convenient way. Like tuples, lists can also be nested and therefore multi-dimensional. Lists can also support data with different data types in different elements, as we saw with tuples. Lists are created using brackets rather than parenthesis:

```
list1 = ["SPY", 1, 2]
list2 = [0, "QQQ"]
lists = [list1, list2]
print(lists[0][1])
```

Like tuples, the [] operator can be used to access specific elements in a single or multi-dimensional list. In the above code, we show an example of this in a multi-dimensional list by printing the second element in the first nested list. Aside from the different syntax used by lists and tuples (the use of brackets instead of parenthesis), the biggest difference between lists and tuples is that lists are mutable whereas tuples are immutable. This means that we can modify the elements in a list once they have been set, as shown below:

```
list1 = ["SPY", 1, 2]
list1[0] = "IWM"
list2 = [0, "QQQ"]
lists = [list1, list2]
```

This ability to modify the elements in a list as done to list1 above is generally an attractive feature, making lists a more commonly used data type than tuples. A defining characteristic of lists is that they do not have mathematical operators defined on them, even if a list is created with only numeric data. The + operator, for example, is defined as *concatenation* rather than *addition*. Because of this, we will find that lists are extremely useful for unstructured data sets but we will instead want to rely on other objects for numeric, structured data. In the next section we explore a NumPy Array, which is a preferable data structure for storing numeric data and performing mathematical calculations.

6.4.3 Array

The NumPy library includes an Array data structure that is extremely useful for working with numeric datasets. Readers will find Arrays to be superior to Lists and tuples when working with numeric data because of their convenient structure for performing calculations on numeric data and built-in mathematical functionality. NumPy arrays, like the other data structures we have seen, can be multi-dimensional. NumPy arrays are mutable, meaning we can modify the elements after we have assigned them. Importantly, unlike lists and tuples, mathematical operators, such as addition, multiplication and division are defined on arrays. For example, the + operator is defined as addition, and the * operator is defined as multiplication. An example of these operations is provided in the following coding example:

```
import numpy as np

a = np.array([1,2,3])
a[1] = 3
b = np.array([4,5,6])
c = a + b
```

In addition to the standard set of mathematical operators, such as + and *, a set of convenient functions including sum() and mean() are defined for numpy arrays. It should be noted that mathematical operations are defined on arrays as element-wise operations by default. For example, the * operator is defined as element-wise multiplication rather than matrix multiplication. This is highlighted in the following coding sample:

```
a = np.array([1,2,3])
b = np.array([4,5,6])
c = a * b
```

Arrays can perform matrix multiplication and other matrix operations, however, for example using the np.dot function:

```
a = np.array([1,2,3])
b = np.array([4,5,6])
c = a.dot(b)
d = np.dot(a, b)
```

Notice that Python provides two ways to call the dot function, a.dot and np.dot. In the case above, where both inputs are arrays, the two function calls are equivalent and will return the same value, however, there are some subtle differences in how type conversions are handled between the two functions. The **matmul** function in numpy can also be used to perform matrix multiplication.

6.4.4 Differences between Lists and NumPy Arrays

The main difference between a list and a numpy array is the presence of mathematical operations such as the + operator and the * operator. As a result, lists tend to be better for handling unstructured data that doesn't require typing or have any required consistency and NumPy arrays tend to be the optimal data structure to use for numeric data sets. Using this practice for choosing between lists and arrays enables us to benefit from the built-in functionality and mathematical operations associated with Arrays, while also benefiting from the added flexibility when using lists. It should also be noted that there is nothing preventing us from creating an array with non-numeric data, however, it will lead to errors should we try to invoke any mathematical operations.

6.4.5 Covariance Matrices in Practice

In quant finance applications, working with a covariance matrix will be central to many of the tasks that we are trying to perform. These applications include portfolio optimization, where the correlation structure determines how much weight we want to invest in different assets. All else equal, the more correlated an asset is to the remaining assets in a portfolio, the less weight we are going to want in that asset because the diversification benefit will be smaller. This is a concept that we explored in chapter 5. Further, when we are building simulation algorithms that are based on multiple dimensions, we need to create a covariance matrix and simulate random numbers that are in line with that covariance matrix. These multi-dimensional problems are quite common and may arise either because we have multiple assets or a stochastic volatility process, as we will see in later chapters. One particular type of application that would use this type of multi-variate simulation is in risk management, where we would want to simulate all assets in a given portfolio using a specified covariance matrix.

As we saw in the previous section, NumPy arrays will be useful data structures for working with covariance matrices. The NumPy module also has built-in functions for calculating covariance and correlation matrices, **numpy.cov** and **numpy.corrcoef.**

Once we have a covariance or correlation matrix, we will often need to do some sort of transformation to that matrix. Most commonly, when working with portfolio optimization problems we will need to work with the inverse of a covariance matrix, and when pursuing simulation algorithms we will need to work with a square root of a covariance matrix. When doing these transformations we need to make sure and use methods that can handle matrices that are not full rank. These applications of matrix transformations are detailed in chapters 15 and 24, respectively.

6.4.6 Covariance Matrices in Practice: Are Correlations Stationary?

In practice, it is a well-known phenomenon that correlations tend to be non-stationary, meaning they tend to vary over time. In particular, we observe that correlations increase in periods of market stress. In this section we conduct a simple test of this phenomenon. To do this, we calculate rolling correlations of Apple and the S&P 500. We focus on a monthly rolling correlation[1], although the reader is encouraged to test this phenomenon with different lookback windows and asset pairings to see how prevalent it is. In the following plot, we see how these rolling correlations varied over the period beginning in March of 2020 and ending in the beginning of 2021:

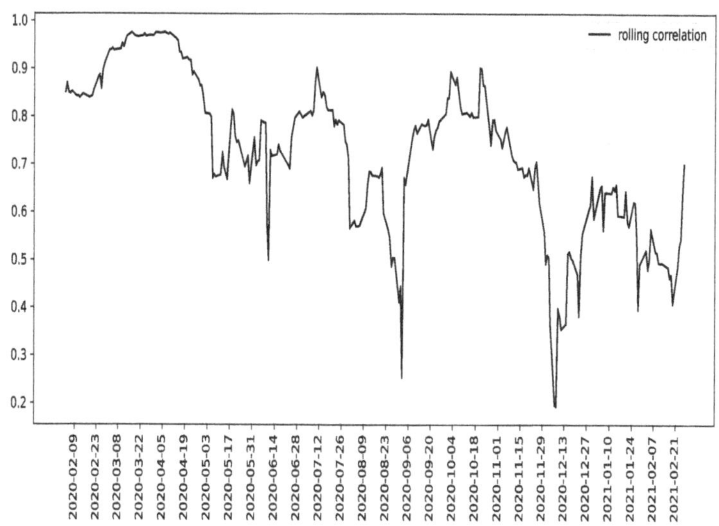

[1]For simplicity a rolling 21-day correlation is used in lieu of a true monthly correlation.

As we can see from the chart, there is clear evidence that the correlations do vary greatly over time, reaching their high during the Covid crisis in March of 2020 and then subsequently declining substantially. This potential non-stationary of asset correlations, and in particular their tendency to rise in the presence of market stress, is a common theme explored throughout the text. It is also a phenomenon that practitioners have consistently observed persists across assets.

Note that we could also use a more formal test of the stationarity of the correlation between the two assets using an augmented Dickey Fuller test [2]. This is left as an exercise for the interested reader.

6.4.7 Series

A Series is a data structure within the pandas library that extends the functionality of a NumPy Array[3]. A pandas series is a one dimensional array with an index. This index can be thought of as the name for each element, or row in the Series. Series objects are convenient for working with time-series data, where the index would naturally correspond to a date. The index can be another type, however, for example it may be an integer or a string. The index values for the elements in a series must be unique, meaning that a Series object is more appealing for time series data than panel data where the corresponding key would be less obvious.

The following coding example shows how a pandas Series object can be created from a set of synthetic data:

```
import pandas as pd
s = pd.Series(data=np.random.random(size=5),
              index=['2014','2015','2016','2017','2018'])
```

Pandas Series structures include a lot of helpful functionality for aggregating data, selecting a subset of data and performing other transformations. Much of the functionality of Series objects is shared with the functionality of the next data structure that we explore, a Data Frame. As a result, the functionality for each is presented in the next section on Data Frames.

6.4.8 DataFrame

Data Frames are possibly the most widely used and convenient data structure in Python. They are another data structure within the pandas module, and are similar to pandas Series except they allow multiple columns, or series. Data Frames are tabular data structures where both the rows and columns can be named. Data Frames are used throughout the pandas library, with many pandas data writing and reading functions returning Data Frames. As such, it is critical for an aspiring quant to be intimately acquainted with Data Frames. One of the main strengths of Data Frame objects is their built-in ability for data transformations, such as joining data frames to another data frame, or grouping or aggregating our data. This ability to use a

[2] As detailed in 6.5.4

[3] Note that pandas Series, and other data structures inherit from NumPy arrays.

Data Frame to transform, aggregate and merge our data, as we often do in SQL, is a significant benefit of the pandas library and coding in Python.

In the following coding sample, we show how to create a Data Frame that is populated with random data, with named rows and columns:

```
import pandas as pd
df = pd.DataFrame(data=np.random.random(size=(5,3)),
                  index=['2014','2015','2016','2017','2018'],
                  columns=['SPY', 'QQQQ', 'IWM'])
```

The primary features and functionality of a Data Frame object, including the ability to merge, aggregate and transform data frames, is summarized in the following list:

- Accessing Rows or Columns in DataFrames
  - We can use the **loc** method to access a group of rows and columns by label(s) or a boolean array in the given DataFrame. This is helpful to extract a useful subset from the original table. Here is a simple example:

```
# use df_ret in 6.4.6
df = df_ret.copy()
df.loc['2021-02-22']   # row access
df.loc[['2021-02-22']]   # subtable
df.loc[:,['AAPL']]   # subtable
```

  - Instead of accessing columns and rows by name/label, we can use their index numbers with the **iloc** method. Here is a simple example:

```
df.iloc[0]   # row access
df.iloc[[0]]   # subtable
df.iloc[:,[0]]   # subtable
```

- Merging DataFrames
  - An important feature of Dataframes is the ability to merge columns on an index (often a date). Here is a simple example:

```
dfSPY = pd.DataFrame(data=np.random.random(size=(5,1)),
                     index=['2014','2015','2016','2017','
    2018'], columns=['SPY'])
dfQQQ = pd.DataFrame(data=np.random.random(size=(5,1)),
                     index=['2014','2015','2016','2017','
    2018'], columns=['QQQ'])
dfSPY.join(dfQQQ)
```

  - As in SQL, we can define different inner and outer JOIN types when doing this.

- Concatenating and Appending DataFrames
  - Another important feature of Dataframes is the ability to concatenate or append rows. Here is a simple example:

```
dfSPYOld = pd.DataFrame(data=np.random.random(size=(5,1)),
                        index=['2009','2010','2011','2012','2013'], columns=['SPY'])
dfSPYNew = pd.DataFrame(data=np.random.random(size=(5,1)),
                        index=['2014','2015','2016','2017','2018'], columns=['SPY'])

dfSPY = pd.concat([dfSPYOld, dfSPYNew])
```

- Note that we passed a list of Dataframes to the **concat** function.

- Returning the first and last rows in a Data Frame
  - The **head** and **tail** functions return the first n rows and last n rows, respectively:

```
dfSPY = pd.DataFrame(data=np.random.random(size=(5,1)),
                     index=['2014','2015','2016','2017','2018'], columns=['SPY'])

dfSPY.head(3)
dfSPY.tail(2)
```

  - These functions have a natural application to computing rolling, overlapping returns with period length equal to n.
  - To do this, we can create an array of current values using the tail function (with parameter -n) and an array of previous values using the head function.

- Transforming DataFrames
  - Like making pivot tables in Excel, DataFrames enable us to generate reshaped tables using unique values from specified indices and columns. Here is a simple example of pivoting our data:

```
df = pd.DataFrame(data={'date': ['2021-2-20']*3 + ['2021-2-21']*3,
  'stock': ['SPY','AAPL','GOOG']*2,
  'forecast1': ['T','F','F','T','T','T'],
  'forecast2': ['F','F','T','T','F','T'],
  'forecast3': ['T','T','F','F','T','F']})

# make a pivot table
df.pivot(index='date', columns='stock', values='forecast1')

# make a pivot table with hierarchical columns (multiple values)
df.pivot(index='date', columns='stock', values=['forecast2', 'forecast3'])
```

  - We can also "unpivot" a DataFrame from wide to long format, where one or more columns are identifier variables. Here is a simple example:

```
df = pd.DataFrame(data={'date': ['2021-2-20','2020-2-21', '2020-2-22'],
  'SPY': ['T','F','F'],
  'AAPL': ['F','F','T'],
  'GOOG': ['T','T','F']})
df.melt(id_vars=['date'], value_vars=['SPY','AAPL','GOOG'], value_name='forecast')
```

- Performing Calculations within a Data Frame
  - We can easily access the basic statistics, or conduct simple calculations of a DataFrame. Commonly used methods include **mean**, **std**, **min**, **max**, etc. Here is a simple example:

```
df = pd.DataFrame(data=np.random.random(size=(5,3)),
                  index=['2009','2010','2011','2012','2013'],
                  columns=['SPY','AAPL','GOOG'])
print(df.mean())
print(df.std())
print(df.min())
print(df.shift(periods=2, fill_value=0))
print(df.rolling(3).mean())
```

- Applying User-defined Functions to DataFrames
  - In addition to using pre-defined functions provided by the pandas module, we are allowed to build our own function and apply it to DataFrames. Here is a simple example:

```
# use the dataframe in the basic statistics example
# this self-defined function converts simple returns to log returns
df.apply(lambda x: np.log(x+1))
```

6.4.9 Dictionary

Dictionaries are another important data structure in Python that consists of a sequence of key-value pairs. Dictionaries are extremely useful objects for keeping lookup/mapping tables. For example, we might keep a dictionary that stores instrument id's and instrument tickers, making it easy to retrieve the id for any instrument. The key in a dictionary must be unique, and the keys are also immutable, meaning they cannot be changed once they are created[4]. Values in a dictionary can be modified for a specific key using the [] operator. This operator is structured such that the key-value pair will be updated if the key exists, otherwise a new key-value pair will be added to the dictionary.

[4] Note that we saw this earlier with strings as well.

A blank dictionary can be created using the following syntax:

```
dict = {}
```

We could then use the [] operator to add items to the dictionary. Alternatively, we can initialize a dictionary with an arbitrary number of items as shown below:

```
dict = {0: "SPY", 1: "QQQ", 2: "IWM"}
```

Like other structures we have seen, dictionaries come with a set of convenient pre-built functions and methods that enable us to easily search within them (via the [] operator or in function), print their contents, and remove one or more elements (via the del function).

6.5 IMPLEMENTATION OF QUANT TECHNIQUES IN PYTHON

As quants, our jobs are generally centered on performing some sort of statistical analysis. This could mean performing a set of econometric tests on an underlying dataset, or it could also mean understanding the underlying distribution for valuation of a complex derivative instrument[5]. Regardless of the task, as quants we are likely to rely on a central set of techniques to solve these problem. Python is a convenient home for trying to solve these types of problems because it provides a set of pre-built modules or libraries that implement many quant tools that we will need. In the following sections, we briefly discuss the libraries/modules where these common quant tools reside.

6.5.1 Random Number Generation

One of the most common tasks a quant will face is generating random numbers from various distributions. This is a core part of building a simulation algorithm, and is also central to statistical bootstrapping and other applications. The NumPy library has a series of functions that can be used to generate random numbers from the most common distributions.

The simplest distribution of interest is a uniform distribution, where all values have equal probability. This can be done in Python by using the **numpy.random.rand** function. The following example shows how to create a matrix of uniformly distributed random numbers between 0 and 1:

```
rnds = np.random.rand(100, 100)
```

Later in the book, in chapter 15, we will show how to transform a set of uniform random numbers to various other distributions.

Additionally, whenever we are working with a Brownian Motion, which as we see throughout the text is quite common in quant finance applications, we will need to simulate from a normal distribution in order to simulate the increments. Simulating from a normal distribution can also be done via the NumPy library using the **numpy.random.normal** function as shown below:

[5] Recall from chapter 5 the subtle difference between the two types of models.

```
mat_shape = (100, 100)
norm_rnds = np.random.normal(0.0, 1.0, mat_shape)
```

It should be noted that the return value for both the uniform and normal random number generation functions is a multi-dimensional NumPy array. Further, while the normal distribution example above generates samples from a standard normal distribution, additional parameters are available to generate samples from normal distributions with an arbitrary mean and variance. Lastly, the numbers that we are generating will be *uncorrelated.* In some applications we will need to generate a set of multi-variate normally distributed numbers with a specified correlation structure. This can be accomplished using the **numpy.random.multivariate_normal** function, which enables us to generate multi-variate random normals with a specified set of means and covariance matrix. Later in the book we will also learn how to transform uncorrelated random normal variables to a set of correlated random normals with a specified correlation matrix.

When modeling default we may also want to be able to simulate from an exponential distribution. An example of this type of simulation is considered in chapter 15. Conveniently, NumPy provides a function for simulating from this distribution as well, **numpy.random.exponential**.

6.5.2 Linear Regression

Earlier in the book, in chapter 5, we introduced the concepts and assumptions behind linear regression. In particular, we saw how to derive an unbiased OLS estimator. In practice, employing regression techniques is a very common quant task, and unsurprisingly, Python has a set of built-in tools that we can leverage to run different kinds of regressions. In particular, Python has a machine learning module, sklearn, which has, among other things, the capability to estimate linear regression models.

In the following coding example, we show how to leverage the sklearn module to run a multi-variate linear regression model on a set of synthetic data created by generating random numbers:

```
import numpy as np
from sklearn import datasets, linear_model

Xs = np.random.rand(100, 2)
Ys = np.random.rand(100, 1)

# Create linear regression object
regr = linear_model.LinearRegression()

# Train the model using the training sets
regr.fit(Xs, Ys)
print(regr.coef_)
```

One thing that might immediately catch the readers eye is the setup of the linear regression model. In particular, to perform the regression we needed to first create an instance of a class (named regr above) and then utilize the fit function. This is in contrast to other quant techniques, such as random number generation, where a single, global function is used to perform the desired calculations. Linear regression based tasks are prevalent throughout much of the quant finance landscape, and especially at buy-side institutions. In particular, regression techniques are relied upon to solve factor modeling problems, as we will see in the next example.

6.5.3 Linear Regression in Practice: Equity Return Decomposition by Fama-French Factors

In this section we leverage the regression coding example from the previous section to conduct an experiment on the characteristics of stocks returns for Apple (ticker AAPL). In particular, we use a contemporaneous regression to determine the sensitivities of Apple returns to the three Fama-French factors, including an intercept. Doing so yields the following regression coefficients:

Factor	Exposure
Intercept	0.0199
Mkt-RF	1.27
SMB	0.2433
HML	-0.9707

From this simple analysis, we can see that Apple is characterized by high exposure to the market. This is evidenced through its market beta of ≈ 1.27 which implies that Apple is more volatile than the market, and that a 1% increase or decrease in the market would lead to a 1.27% increase or decrease in Apple, in expectation. Further, we notice that Apple has significant exposure to the HML factor. As this is a value factor this coefficient is in line with our expectations, as we know Apple has been characterized by high valuations due to its substantial popularity and growth. Additionally, the intercept in this regression is notably positive, which we can interpret as an excess return in the absence of movements in these factors. In this case, the regression is telling us that we can expect an $\approx 2\%$ increase in Apple a month, or $\approx 25\%$ a year, in the absence of an increase in the market, or other factors. This is clearly an impressive proposition that holders of Apple stock over the last two decades are likely happily familiar with.

6.5.4 Autocorrelation Tests

Earlier in the book we discussed some of the most common quant techniques that come up in modeling the physical measure[6]. Python also has built-in, open-source libraries/modules for handling the majority of these techniques. In particular, we distinguished between stationary and non-stationary time series, and showed how we can use an augmented Dickey-Fuller test to check statistically whether a time series is stationary or non-stationary. The statsmodel package can be used to perform this stationarity test using the **statsmodels.tsa.stattools.adfuller** function.

[6]See chapter 5 for more details.

Recall that in this function we are testing the null hypothesis that the data is non-stationary, meaning that higher t-stat sand lower p-values indicate higher confidence that the data is stationary. Also recall that if a data series is non-stationary, we will generally want to difference the series before proceeding to other time series tests.

In chapter 5 we also learned about autoregressive and moving average processes, and how we can combine them into a single ARMA model. ARMA models are one of the most common techniques for checking stationary time series for mean-reversion or momentum. In Python these models can be calculated using the **statsmodels.tsa.arima.model.ARIMA** function.

The resulting coefficients that we get from this function define the relationship between current values in a process and its previous values. Perhaps most commonly in quant finance this type of model is used to measure the relationship between current returns from an asset and its previous returns. We saw earlier that market efficiency principles dictate that future returns should not depend on past returns. Along these lines, these models can be used as tests of market efficiency, with statistically significant mean-reversion or momentum coefficients violating the so-called weak-form of market efficiency.

The ARMA functions support arbitrary AR and MA lengths. This means that we must choose how many of each type of term to include in our model. Information Criteria, such as AIC measures, can aid this decision by helping us measure the trade-off between better model fit and adding parameters.

Stationarity tests and ARMA models are commonly used in forecasting and alpha research. If we are able to find tradeable market quantities that exhibit mean-reversion or momentum, then the potential reward is high. As a result, these techniques, as well as variations of them, are at the heart of the work for many buy-side quants. More details on these techniques are discuss later in the book, in chapters 25 and 27. Interested readers can also see [250] for more information on time series models.

6.5.5 ARMA Models in Practice: Testing for Mean-Reversion in Equity Index Returns

In this section we apply the code in the coding repository to the returns of the S&P 500 in order to test for mean-reversion. As before, we begin with the augmented Dickey Fuller test of the price, which results in a high p-value, leading us to not be able to reject the null hypothesis that the price series is non-stationary. Next, we calibrate a mean-reversion coefficient by first differencing the data and then calibrating an AR(1) model to the returns. Doing so yields the following AR(1) coefficient and corresponding z-value:

AR Lag	Coefficient	Z-Value
1	-0.0967	-13.94

Therefore, from our stationarity test we might conclude that there is little evidence that the price series for S&P is stationary, however, once we have taken the first order difference, we do find some evidence of mean reversion in the returns by using an AR(1) model. This would indicate that an increase of 1% in the S&P today would lead to a roughly 9–10 basis point decrease the following day in expectation. The astute reader might then look to leverage this into a trading strategy where they enter long positions when the market has decreased substantially and vice versa. Whether this is a reasonable strategy to try to leverage in practice, or an untradable artifact of market microstructure is left as a thought exercise to the reader. The reader is also encouraged to see whether this phenomenon persists for other stocks, or asset classes.

6.5.6 Matrix Decompositions

As we saw earlier, working with covariance matrices will be a pivotal task in quant finance, whether we are solving portfolio optimization problems, running a simulation or building a model for risk management. More generally, there are a set of matrix decompositions that quants will find themselves performing repeatedly. Thankfully and unsurprisingly, Python has built-in libraries that calculate these decompositions for an arbitrary matrix.

The first such function is Cholesky decomposition, which can be done in Python using the following function call:

```
# cholesky
mat = np.random.rand(5, 5) # full-rank square matrix
A = mat @ mat.T  # matrix product to get positive definite matrix
L = cholesky(A)  # lower triangular matrix
```

An important caveat for Cholesky decomposition is that it requires an input matrix that is positive definite. That is, it requires an input matrix that is full rank with all positive eigenvalues. In practice we will often find this not to be the case, especially in problems of higher dimensions. In these cases, we tend to find that the matrix is not full rank and has several eigenvalues that are at or close to zero. In these cases, Cholesky decomposition should not be utilized.

A second and more robust decomposition, that works as long as the input matrix is positive semi-definite, meaning it can handle eigenvalues that are equal to zero as long as there are no negative eigenvalues, is eigenvalue decomposition[7].

```
# eigenvalue decomposition
mat = np.vstack([np.random.rand(4, 5), np.zeros(5)]) # rank-deficient
    square matrix
A = mat @ mat.T  # matrix product to get positive-semidefinite matrix
w, v = eig(A)  # eigenvalues w and normalized eigenvectors v
```

We can see that both decompositions are easily performed within the NumPy library. In practice, it is recommended the reader uses Eigenvalue decomposition instead of Cholesky decomposition because of its ability to handle positive semi-definite matrices. We will see applications of these techniques throughout the text, most notably in chapters 15 and 24. Readers looking for background on the mechanics

[7]Or the closely related, singular value decomposition.

of the underlying decompositions, and a review of linear algebra are also welcome to consult [162] or [262].

6.6 OBJECT-ORIENTED PROGRAMMING IN PYTHON

As mentioned earlier, one of the main advantages of Python versus other advanced languages, such as R and MATLAB is more robust support for object-oriented programming. Quants often have a tendency to write code either as scripts or using functional programming[8]. In many cases, this approach is completely appropriate, such as when performing exploratory analysis or building a prototype. However, a side-effect of this approach is that it doesn't scale well. As we try to add more pieces to it, it will become increasingly challenging to maintain, and we will likely have to duplicate our code in order to expand our analysis.

To avoid this, another approach that can be applied is object-oriented programming, which can lead to better, more scalable and generic code. As the name implies, when we use object-oriented programming, we create a series of objects, or classes instead of a series of functions. These classes can then interact with each other, be related to each other and define actions on themselves or others. We have already seen examples of object-oriented programming, and how it can make our lives easier as coders, in the set of pandas data structures that inherit from NumPy arrays, such as Data Frames. Fundamentally, these structures are at heart classes that have been defined in a generic, reusable manner.

The objects that we define may have different parts. They can store certain data within them, which are referred to as attributes, or member data. They can also execute certain code when they are created, or when they are destroyed. These are referred to as constructors and destructors, respectively. We can also have classes that inherit certain functionality from other classes. This is referred to as inheritance, and is one of foundational concepts in object-oriented programming that enables us to write more robust code. In the following sections, we discuss the core tenets of object-oriented programming and how they can help us write better code. We also describe how to implement classes in Pythons, including how to code the different parts of a class.

6.6.1 Principles of Object-Oriented Programming

Before presenting the coding infrastructure for building classes in Python, we first outline the principles of object-oriented design. Object-oriented programming is based on a set of underlying principles that are meant to help us understand why code that is based on classes or objects is more scalable and robust than code based on functions. There are four main tenets of object-oriented programming: inheritance, encapsulation, abstraction and polymorphism, each of which is briefly summarized in the following list:

[8] Where the coder writes a series of functions to accomplish the task they are trying to solve.

- **Inheritance**: allows us to create class hierarchies and have classes inherit functionality from other classes.
- **Encapsulation**: an object's internal representation is hidden from outside the object.
- **Abstraction**: exposing necessary functionality from an object while abstracting other details from the user of a class.
- **Polymorphism**: ability of a sub-class to take on different forms and either have their own behavior or inherit their behavior from a parent

Encapsulation and abstraction are closely related concepts, and are related to the fact that when coding objects we are able to keep the details of the class from other objects or other coders. This is a desirable quality as it means that the logic for each class is contained to within the class. That way, when the logic for a class needs to be modified, the coder knows that the single point to update will be within the class, rather than having to trace all dependencies. Abstraction of the logic from users of a class or function also enables other coders to leverage and benefit from a class without fully having to understand the details. Instead, they must only know how to interact with the class.

Inheritance and polymorphism are also critical benefits of object-oriented programming because they enable us to make class hierarchies that can be used flexibly and interchangeably. In part inheritance enables us to create code, and by extension functionality, shared by many similar classes but also to differentiate and override certain functionality. Polymorphism then enables us to dynamically choose among the class hierarchy at run-time, enabling our code to be completely generic. Inheritance and polymorphism are both seminal concepts, and are discussed further in sections 6.6.9 and 6.6.10, respectively.

6.6.2 Classes in Python

Classes in Python provide us a way to define complex user-defined data types. These data types can contain within them a variety of embedded functionality and components. For example, they often contain certain **properties or attributes** which help us define the structure of our data type. These attributes can be thought of as variables that are internal to a particular class. As coders we may decide we want these attributes to be accessible to the outside world, or we may decide that the attributes should only be visible within the class. Classes can also define **functions, or methods,** that act on themselves or other classes. Just like with attributes, we may prefer these functions be restricted from being called from outside the class, or we may want these functions to be available to users of the class.

In Python, classes are defined using the **class** keyword. Classes must have at least one constructor, which is called when the class is initialized, and may also contain any arbitrary number of attributes and class methods.

```
class Position:
  def __init__(self):
    self.shares = 0
    self.price = 0
```

We can see that in the first line we are declaring a class called Position which subsequently has a single constructor defined. Notice that the constructor takes self as a parameter. This signifies the current instance of the current object, and is passed to all constructors, class methods, operators and destructors. Constructors, as well as other class components, are described in more detail in the next few sections.

Programming via the lens of classes instead of via the lens of functions is a paradigm shift that often requires considerable upfront investment to design properly. While the actual classes that we are creating during a project vary greatly depending on the application, there are certain classes that tend to be at the heart of many quant finance applications. In particular, a few examples that come up frequently are:

- A stochastic process class that enables us to model the behavior of asset prices while switching between different SDEs.
- An option payoff class that enables us to switch between various vanilla and exotic option payoff structures.
- An instrument class that defines the various attributes of an instrument, such as a stock, bond or option
- A back-test class that enables us to take a series of signals and historical return data and generate a set of standardized summary metrics.

6.6.3 Constructors

A constructor is a piece of code that runs when an object is **initialized**. Constructors are often used to assign values to the member data, or attributes, in a given class, and may also contain additional logic. As an example, if we are working with a class that defines a Position, we may have a constructor that defines different types of positions, such as stocks, bonds and options, depending on the parameters passed to the constructor. In this case, some of the logic for setting the underlying member data might be dependent on some conditional logic based on the type of position that it is[9].

Defining a constructor in Python is accomplished by creating a function with a certain name, __init__ and a conforming signature. A constructor must take at least one parameter, self, which corresponds to the current instance of the object being created. An arbitrary number of additional parameters can be passed to the constructor as well. Following our position example, a string for instrument type, and numeric values for shares/contracts and price would be natural constructor parameters.

[9]Later we will see that inheritance is another way to approach this problem.

The following piece of code is an example of a constructor for a Position class that takes no parameters:

```
class Position:
  def __init__(self):
    print("initializing position")
```

We can see that this constructor was defined by creating a function with a single parameter, self, along with the correct name. This type of constructor, which requires no additional arguments, is referred to as a **default constructor**. Constructors are called when a new instance of an object is created. For example, the Position class constructor above would be called when the following code is executed:

```
pos = Position()
```

This would then create an instance of our user-defined Position class named pos. For each class, we may choose to define multiple constructors[10]. We can also choose to provide default values for some or all of the parameters in the constructor. For example, consider the following updated position class constructor which takes two additional arguments, shrs and px, both of which have default values:

```
class Position:
  def __init__(self, shrs=0, px=0):
    self.shares = shrs
    self.price = px
```

This constructor could then be called without passing any arguments, or could be called with the additional shares and price arguments specified, as in the following example:

```
pos = Position()
pos2 = Position(100, 100)
```

The result of this code would be two position objects, one with shares and price set to zero, and another with shares and prices set to 100.

6.6.4 Destructors

Just as we are able to define code that is executed each time an instance of an object is initialized, we are also able to define code that executes every time an object gets destroyed. These functions, known as destructors, can be used to make sure that file or database connections are released when an object is destroyed, for example. Destructors are a critical component of some programming languages, such as C++, as they are be required to make sure memory is released when objects go out of scope. In Python, however this process is handled via automatic garbage collection making destructors less commonly overridden and less important.

As was the case with constructors, in Python destructors are implemented by defining a function with a certain signature. In particular, destructors in Python are

[10]Provided they have different signatures.

specified using the name __del__ and must also take a single parameter self. Remember that this self-parameter refers to the current instance of the class being destroyed.

The following code shows what a destructor for a Position class might look like:

```
class Position:
  def __init__(self, shrs=0, px=0):
    self.shares = shrs
    self.price = px

  def __del__(self):
    print("destroying object")
```

We can see that in this simple example we are only using the destructor to notify us that the object has been destroyed, by printing a message, but in practice would generally want to include some additionally functionality in the body of the destructor. Destructors are called when objects are deleted. This might happen automatically, when an object goes out-of-scope, or might happen if an object is deleted explicitly using the del command.

6.6.5 Class Attributes

A key feature of object-oriented design is the ability to build classes that store their own data. This data for a class is referred to as a set of attributes, or member data, and is essentially a set of variables that are stored within, and are specific to, an instance of a class. These attributes may have simple built-in types or be other user-defined complex types. They may be simple string or numeric data types, or may be other classes themselves, such as a Data Frame, or the Position object defined in the previous section. Attributes may also be classes that rely on an inheritance framework, and as a result we may choose to store base or derived classes as attributes in a given class.

Attributes are referenced within a class by using the self keyword. For example, an attribute named *attribute1* would be referenced using the syntax *self.attribute1*. In Python attributes must be defined in the body of the constructor for a class. Continuing with our Position example defined in the previous section, we create a Position class with two attributes, shares and price:

```
class Position:
  def __init__(self, shrs=0, px=0):
    self.shares = shrs
    self.price = px

pos = Position(100, 100)
s = pos.shares
pos.shares = 200
```

As we can see, these attributes are accessible to for us to read and write after the class has been created. In particular, in the constructor for this Position class we created two attributes. By default these attributes will be public, meaning they are available from outside a given class. As a result, once we have initialized an instance of the Python class, we can then read and write the attribute values using the syntax *instanceName.attributeName*.

6.6.6 Class Methods

In addition to defining attributes in our classes we may want to define a set of actions, or functions for the class. We would accomplish this by creating class methods, which are functions that operate on a particular class instance. The actions within these functions could be defined to modify the attributes safely. They could also be defined to perform calculations within a class.

Class methods are defined in Python similarly to defining an isolated function[11]. The key differences are that class methods are defined within a given class and that the first argument is the self-parameter. As was the case for constructors and destructors, this self-parameter represents the current instance of a given class. Inclusion of the self-parameter is critical, and in fact, if it is omitted the function that we create will not be a class method, but instead a global function[12]. Aside from the self-parameter, the coder can pass an arbitrary number of parameters, with arbitrary type and structure, to a class methods. Because class methods operate on an instance of a class, they can access and modify class attributes and call other class methods.

The following coding example shows how a simple, market value class method could be added to our Position class:

```
class Position:
  def __init__(self, shrs=0, px=0):
    self.shares = shrs
    self.price = px

  def mktValue(self):
    return self.shares*self.price

pos = Position(100, 50)
mktval = pos.mktValue()
```

Notice that self is defined as a parameter to the mktValue function. This signifies that the function is a class method, and that it will have access to all of the attributes of the underlying Position class.

6.6.7 Class Methods vs. Global Functions

As we noted in the last section, depending on how we define our function it may be a class method that is tethered to a specific instance of a class or a global function that is comparable to the type of stand-alone functions one would create in functional programming. It turns out that we can also create a global function within a class, or at the global scope.

In the piece of code below, we explore the difference between these three types of function implementations:

[11] As we saw in chapter 7.
[12] As we explore in the next section.

- via a Class Method
- via a Global Function defined outside the Position class
- via a Global Function defined inside the Position class

```
class Position:
  def __init__(self, shrs=0, px=0):
    self.shares = shrs
    self.price = px

  def mktValue(self):
    return self.shares * self.price

  def mktValue3(pos):
    return pos.shares * pos.price

def mktValue2(pos):
  return pos.shares * pos.price

pos = Position(100, 50)
mktval = pos.mktValue()
mktval2 = mktValue2(pos)
mktval3 = Position.mktValue3(pos)
```

These three approaches enable us to accomplish the same result but there is an important subtle difference. In the first approach, the mktValue function, the class was being accessed from inside itself, whereas in the second and third approaches, the mktValue2 and mktValue3 functions, the class was being accessed externally as a passed in function parameter. A benefit of the first approach is that it is consistent with the object-oriented principles of **abstraction** and **encapsulation** in that they ensure calculations involving the Position class are contained within the class itself and that other objects & functions aren't required to know the details of the class. Additionally, if we had attributes that we wanted to be private, then we would not want to use the second approach to access them.

6.6.8 Operator Overloading

Python defines certain special class methods or functions, often referred to as operators, which a programmer is able to override in the classes that they define. The process of changing the definition of one of these built-in functions, it is called operator overloading. Examples of such operators are the assignment operator, the equality operator, and the addition and multiplication operators. A full list of operators in the Python language can be found in [258]. For some of these operators, such as the == operator, default implementations exist that are used should the function not be overridden in a class definition. For others, such as the + operator, a default implementation may not exist causing an exception if they are called without being overridden in our class design.

To override an operator we need to create a function in the body of our Python class definition that has a certain name and signature, as we did for constructors and destructors. For example, when overriding the addition operator, we simply need to define a function called __add__ that takes two parameters, *self* and *other*. Here *self* refers to the current instance of the class and other refers to another instance of the same class that we are adding to the current instance. The following code snippet shows how we might override the + operator for our Position class:

```
class Position:
  def __init__(self, shrs=0, px=0):
    self.shares = shrs
    self.price = px

  def __add__(self, other):
    return Position(self.shares + other.shares, self.price)

pos = Position(100, 50)
pos2 = Position(100,50)
pos3 = pos + pos2
```

Defining the addition operator on mathematical classes can be useful for writing clear, concise code. It should be emphasized that the addition operator returns a new instance of the Position class, rather than modifying the current instance.

The equality operator, or == operator, is also of particular interest and direct relevance as it describes how we check whether two instances of the same class are equal. The following code snippet shows how an equality comparison between two Position classes would look:

```
pos = Position(100, 50)
pos2 = Position(100,50)
print(pos == pos2)
```

Somewhat surprisingly, this comparison returns and prints false. This is because the default implementation of the == operator for complex types checks whether the two objects refer to the same instance and memory location. By default, the == does not validate whether the two instances of an object have the same attribute values. In some cases, this may be desired but in other cases it will clearly not be. In these cases, we can overcome this by overriding the == operator. This can be done by adding a definition of the __eq__ function to our class definition, as in the following example:

```
class Position:
  def __init__(self, shrs=0, px=0):
    self.shares = shrs
    self.price = px

  def __eq__(self, other):
    return (self.shares == other.shares and self.price == other.price)

pos = Position(100, 50)
pos2 = Position(100,50)
print(pos == pos2)
```

Operator overloading can be a helpful tool in writing concise, readable code. Some operator definitions might also be required in order to utilize a user-defined class in certain data structures or calculations. For example, a sorted list class might require that the less than and greater than operators for a class have been defined, in order to know how to sort.

6.6.9 Inheritance in Python

Inheritance is a fundamental part of object-oriented programming. As we saw earlier, it is one of the four pillars of object-oriented programming. In particular, it enables us to create objects that exist within a **class hierarchy** by allowing us to create classes that are modified or specialized versions of each other.

Inheritance is based on the concept of **Base** and **Derived** classes, which are defined as:

- Base Class: A higher level, more generic class with some desired basic functionality that is useful in derived classes.
- Derived Class: A lower level, child class that defines a class with more specialization and inherits the functionality in the base class.

The following coding example shows how one can implement inheritance in Python:

```
class BasePosition:
  def __init__(self):
    print("initializing from BasePosition")

  def price(self):
    print("calling base class price function")

class DerivedPosition(BasePosition):
  def __init__(self):
    super().__init__()
    print("initializing from DerivedPosition")

  def price(self):
    print("calling derived class price function")

b = BasePosition()
d = DerivedPosition()
```

We can see that, in order to have a class inherit from a base class, we specify the base class name in parenthesis after the class name. It should also be noted that the derived class explicitly calls the BasePosition class constructor by invoking the **super()** function[13]. This ability to inherit functionality from derived classes, while overriding or adding new functionality where necessary, is a critical part benefit of using object-oriented programming, and of writing scalable, re-usable, robust code. In the next section, we show an example of how inheritance can be used to accomplish this, in the context of a generic stochastic process class.

[13]This could also be achieved by calling the BasePosition class constructor by name.

6.6.10 Polymorphism in Python

In the last section, we saw how we could use inheritance within a class hierarchy to build a generic piece of code. One of the key drivers behind our ability to write generic code via inheritance is the concept of polymorphism. In the following hypothetical coding example, we highlight the importance of polymorphism by creating instances of both a Black-Scholes and base Stochastic process class:

```
bsProcess = BlackScholesProcess(0.1)
baseProcess = StochasticProcess(0.1)

my_procs = [bsProcess, baseProcess]
for stoch_proc in my_procs:
  print(stoch_proc.price())
```

Importantly, in the above code two different types of stochastic processes are stored within the same list and the price function corresponding to the appropriate place in the class hierarchy is called inside the loop. This is because Python is able to dynamically identify which type each object has been created as within a hierarchy. This powerful concept is known as polymorphism and means we can iterate through a list of all different types of stochastic process objects, knowing that they will invoke the proper pricing function. Polymorphism happens automatically in Python. In many other languages, such as C++, we need to explicitly declare functions as virtual in order to take advantage of dynamic binding. In the next section, we will see another example of polymorphism at work in the Factory Design Pattern.

6.7 DESIGN PATTERNS

Design Patterns are a specified set of coding structures or guidelines that lead to re-usable, generic code. This set of design patterns is intended to make creating objects, structuring them and communication between them more seamless. A set of analogous design patterns exist in many languages, including Python but also C++, Java and C#. In this section we discuss how we can leverage a handful of design patterns to make our quant models more efficient. We will only discuss a few of the many existing design patterns with the goal of giving the reader a sense of what these patterns can accomplish. Those interested in exploring further should see [209], [28] or [261].

6.7.1 Types of Design Patterns

The set of available design patterns are broken into three logical groups: creational, structural and behavioral. A brief definition of each type of design pattern can be found here:

- **Creational**: deals with optimal creation/initialization of objects.
- **Structural**: deals with organizing classes to form larger objects that provide new functionality.
- **Behavioral**: deals with communications between two or more objects.

6.7.2 Abstract Base Classes

Abstract Base classes are classes within an inheritance hierarchy that aren't meant to be created themselves, but instead are meant to provide a shell with pre-determined functionality for derived classes to implement. The benefit of an abstract class is that it provides a contract that all derived classes must adhere to if they want to inherit their functionality. This ensures some degree of commonality in all classes that inherit from the abstract base class.

A coding example is presented in the coding repository that shows how the previously developed Position class could be extended to an abstract base class with derived EquityPosition and BondPosition classes.

6.7.3 Factory Pattern

The **Factory Pattern** is useful for helping to create a class within a class hierarchy with many derived classes. It is a creational design pattern where we enable the user to specify the type of class within a hierarchy that they would like to create, along with some required additional arguments to create the object. The relevant object is then created based on this information. The Factory Pattern consists of a **single global function** that initializes and returns an instance of the appropriate base or derived class type based on a given input parameter.

A coding example is presented in the corresponding coding repository that shows how the factory pattern can be implemented. Importantly, as Python supports polymorphism automatically, the return type of the factory pattern will depend on the input parameter passed to the function. As a result, when common functions are called after the object is created via the factory, Python will know which level in the hierarchy should be called. This makes the Factory pattern very powerful and generic, as we are able to specify our object types as parameters via command line arguments or configuration files.

6.7.4 Singleton Pattern

The **Singleton Pattern** is also a *creational pattern* that ensures that **only one instance of a class** can be created each time the application is run. This paradigm is useful for things that we want to make sure are unique, such as connections to databases and log files. Generally speaking, this is accomplished by checking to see if the object has been created, and if it has using that instance of the object instead of creating a new instance. In order to control the number of instances of an object, the constructor for the class must be private. As a result, the Singleton Pattern is implemented by making a class's constructor **private** and defining a **static getInstance method** that checks whether the class is initialized and creates it only if it has not been initialized. An example of the singleton pattern in Python is presented in the coding repository.

6.7.5 Template Method

The template method works when we **define a sketch of an algorithm in an abstract base class** without defining certain details of the functionality. This design pattern is a specific example of using an abstract base class in that we do not define certain functions in the base class and then force the derived classes to implement them. The benefit of this method is that we only need to code the sketch of the larger algorithm once in the base class.

One example of the template method would be to define a base simulation class as an abstract base class, with a function called nextStep that is not implemented, and then define a set of derived classes that implement this method based on the specific SDE that they rely on (e.g. a Black-Scholes simulation class and a Bachelier simulation class). A corresponding coding example for the template method is presented in the text's coding repository.

6.8 SEARCH ALGORITHMS

While Python has built-in functionality to perform searching operations for different types, the mechanics of one particular underlying search algorithm, a binary search algorithm, is of direct relevance. There are many potential ways to search for a value in an array. The simplest such way would just to iterate through every element in the array and compare the value of the element to the value we are searching for. This is a naive search algorithm as it requires us to perform the comparison on all array elements. An alternative method, on an already sorted array, is **binary search**, which we discuss in more detail in the next section.

6.8.1 Binary Search Algorithm

The binary search algorithm is a way of searching for a value in an *already sorted* array. We explore methods for sorting arrays in the next section. Binary search works by splitting the sorted array in half at each step, and checking whether the item in the middle is above or below the search value. The algorithm then continues by setting the left or right end point equal to the middle, depending on the result of the comparison, and then repeating the process. This algorithm is more efficient than a naive approach because it reduces the number of comparisons by using the sort order to choose the points of comparison more judiciously.

Binary search algorithms can be written generically such that they can apply to any class or data type that has certain operators defined.[14] They are also of interest to quant finance because they are based on the same underlying approach as the root finding technique bisection.[15] A coding example for a binary search algorithm is provided in the coding repository.

[14] Such as the less than or greater than operator, for example.

[15] Bisection and other root finding and optimization techniques are explored in chapter 15.

6.9 SORT ALGORITHMS

Sorting an array in Python, or any other programming language, can be done in numerous ways, using many different underlying sorting algorithms. Some of the more naive algorithms end up being less efficient, because they require traversing through the array multiple times. While Python contains built-in functionality that encapsulates this sorting functionality, it is important for quants to know how the underlying algorithms work so that they might be able to apply to concepts to solve other more complex problems. Additionally, questions regarding sorting algorithms frequently come up in technical interviews.

6.9.1 Selection Sort

A selection sort algorithm is a naive sort algorithm that sorts by repeatedly finding the smallest value in the array and placing it at the beginning of the array. To do this, if there are n elements in the array, then we need to loop through the array n times in order to perform this sort algorithm. Please see the corresponding coding repository for an example of this sorting algorithm.

6.9.2 Insertion Sort

An insertion sort works by iterating sequentially through the array elements and placing the chosen element in the appropriate place in the already sorted portion of the array. This algorithm requires that we perform one loop through the entire array and for each iteration of the loop we must also find it's appropriate place in the smaller sorted array. The most naive implementation of this would be to simply iterate through the smaller array as well until we find the right location of the chosen element. This approach would require a nested for loop, or a while loop within a for loop. An example of this sorting algorithm is provided in the book's coding repository.

An insertion sort algorithm can be improved by finding the right location of each element in the smaller sorted array more efficiently. One way to do this is to use **Binary Search** to find the appropriate location for each element. This could be implemented by replacing the while loop in the code provided in the repository with a call to the binary search algorithm defined in section 6.8.1.

6.9.3 Bubble Sort

A Bubble Sort is a simple sorting algorithm that works by repeatedly swapping elements that are sorted incorrectly until all elements are sorted in the correct order. Because this algorithm only moves elements one place at a time, it is not an efficient algorithm and many comparisons need to be done in order for the algorithm to know it is finished. In particular, the algorithm only knows that the array is sorted if a loop finishes without swapping any elements, which can be the exit criteria for our bubble sort algorithm. Conversely, this algorithm is more efficient than the other algorithms discussed previously in the event that an already sorted array is passed to the algorithm. In the case of a bubble sort, an already sorted array will only require the loop to be traversed a single time. An example of implementing a bubble sort is provided in the coding repository.

6.9.4 Merge Sort

Lastly, we consider a merge sort, which is a more involved sorting algorithm based on a divide and conquer approach. The algorithm works by dividing the array in half, and merging the two halves after they have been sorted. This process is called recursively, that is, we break the array into smaller and smaller pieces until we are able to sort them. Once we have broken them into pieces that we are able to sort easily, we then aggregate the already sorted pieces by merging them sequentially. A merge sort algorithm consists of two functions, a merge function that defines how to merge two sorted halves (and preserve the sort order) and an outer function that is called recursively. Please see the coding repository for an example of a merge sort algorithm.

CHAPTER 7

Working with Financial Datasets

7.1 INTRODUCTION

At this point in the book, we have built the foundation for tackling quant problems, both in terms of the required programming skills as well as the quant finance theory that will underlie our models. Earlier in the book, in section 1.4, we went over the different stages in a quant project. In the subsequent chapters we laid the groundwork for the model implementation phase, which is of course at the crux of being a quant. In this chapter, we take on different stages of quant projects: data collection and data cleaning. Working with, transforming and cleaning data are imperative skills for quants, and in this chapter we aim to build the readers skillset and introduce them to these concepts. Additionally, we provide an overview of many of the more common data challenges that quants face when solving different types of problems, and also discuss where quants generally turn to find data.

Quants often find that a disproportionate fraction of their time is spent not on implementing cutting-edge models but instead working with data. Once the data has been collected, cleaned and validated, then implementing the model is often, but not always, found to be the easy part. One reason for this is that in finance we tend to work with messy datasets that might have gaps, contain arbitrage or outliers, or need to be adjusted for corporate actions or other events. We also often find ourselves working with large datasets, making it harder to validate our data without a systematic process for doing so.

The way that we engage with data depends on the type of data that we are working with. For example, conducting a quant project that relies on stock price data is inherently very different than working with options data. In one case, we will be concerned about accounting for stock splits, dividends and ensuring that our datasets are free of survivorship bias. In the case of options, we may want to instead focus on the data being arbitrage free and choosing a set of strikes & expiries that are sufficiently liquid. More generally, depending on the type of data that we are working with, we can expect a fairly different **frequency**, **size** and overall **data structure**.

DOI: 10.1201/9781003530626-9

Additionally, some types of data, such as options, are inherently messier than others and will subsequently require more rigorous cleaning & validation processes.

Some of the most common challenges that we face when working with financial data include:

- Cleaning data in a robust, consistent way
- Handling gaps in different parts of our data
- Handling extremely large data sets
- Structuring the data in an optimal way
- Building robust data integrity checks to ensure high quality data

In the remainder of this chapter, we discuss how these challenges tend to present themselves in different markets, and the potential approaches for handling each of these challenges.

7.2 DATA COLLECTION

7.2.1 Overview

The first step in the data collection process is to identify a proper source for the data that you need. In some cases this process will be trivial, and in other cases it will be quite onerous. Depending on the source and type of data that we are working with, the format of the underlying data might be quite different. For example, in some cases the data might be exposed to us through access to a database. In other cases, we might be able to download a CSV file from a website or other shared location. We will find this to be true for many open-source financial datasets, such as the Fama-French factors that are publicly downloadable on Ken French's website. In some cases we may be able to retrieve our data via a call to an API that is available either through Python, or through an external provider. One example of this is yahoo finance data, which has an API in Python that we can leverage, within the pandas module, that enables us to access the data directly. Use of an API is generally the preferred approach to accessing data for our models to process, as it is the most automated method and requires less dependence on the structure of the underlying data source. Further, it makes changes behind the scenes to the data source more likely to integrate seamlessly into our process.

Once we have identified the proper source, and analyzed the type of connection that will be required to obtain data from it, the next step is to build an automated process to retrieve the data from this source. Generally this will be done by building a piece of code in Python that retrieves the data from the source, parses it, and stores it in a friendly format for the ensuing models to process.

7.2.2 Reading & Writing Files in Python

In many situations, the data collection process will involve reading from a file in a shared location, such as a CSV or text file. Python provides tools for easily reading from these types of files, either through the use of single-line functions which can read in files in standardized formats, or via the ability to open a connection to a file, and process the contents line by line with the appropriate parsing logic. The pandas module, which we introduced in chapter 8, is extremely helpful for working with these types of files. For example, the **pandas.read_csv** function can be used to read a CSV file. This function is able to parse most CSV files and conveniently stores the result in a data frame so that it can then easily be manipulated. The pandas module also has similar functions for working with other types of files, such as the **pandas.read_table** and **pandas.read_html** functions. More information on these functions can be found in [195].

The following coding example shows how we can apply the read_csv function to parse a CSV file containing returns data:

```
df = pd.read_csv('AAPL.csv', index_col='Date')
```

In most cases, these built-in reading functions will be sufficient to handle the data that we need to process. However, in some cases the formatting of our underlying files may be too complex, and these functions may struggle to interpret and read the files. In these cases, we can instead use a file object in Python to read the file in line by line. We can then build in our own logic to handle the more complex formatting in the file.

The following coding example shows how to use an indefinite loop to iterate through and parse all lines of an underlying file using this approach:

```
f = open(fname, 'r'):
line = f.readline()
while line:
  parse_line()
  line = f.readline()
f.close()
```

In this example, we begin by opening a connection to the file in Python using the open function. It should be noted that the 'r' parameter specifies that we are opening the file read-only. If we specified 'w' instead then we would be able to write to the file. We then use the readline function of the file_object class to read a line of the file and store it as a string variable line. We then proceed to iterate through the file using a while loop, where we use a parse_line() function to process the text in each line. This approach gives us more flexibility, as we can define our own business logic, clean the data as we process it line by line, and parse the data into the appropriate types, ourselves. It does, however, require some additional overhead relative to the single-line function calls, making this approach generally only used when the standard functions fail to handle the complexity of the underlying file.

Generally speaking, for every reader function that pandas has, there is a corresponding writer function, which would enable you to write an object, such as a DataFrame to a CSV or other file type. We can also write to files using a single-line function call, or we can do so using the file_object class discussed above. For example, if we are working with a DataFrame, then we can use it to_csv function to write its contents. Alternatively, we could open a connection to the desired file and loop through the DataFrame, writing each line to the file with our desired logic.

7.2.3 Parsing Data from a Website

In many cases, it will be useful to automatically download data from a website. In some cases, we may need to parse the content of an actual website in order to obtain the data that we require. If this is the case, depending on the complexity of the site we are parsing, this task may be relatively simple or may be quite complex. In other cases the file may be available to download as a file via a URL. In this case, our job is significantly easier and we can generally use the methods described in the above section.

If we are working with a website that has relatively simple formatting, then we can use the read_html function in the pandas module to read the html of a website into a data frame, as the following example shows:

```
dfs = pd.read_html('https://en.wikipedia.org/wiki/Minnesota')
```

As was the case when working with CSV and text files, there will be some cases where we will need more control in order to parse the content effectively. This will be the case if the web formatting is more complex or should we need to embed additional logic into our parsing algorithm. In Python, this can be accomplished with the requests module which enables us to make a request to a site and download the resulting html code. This html would then be stored as a long text string that we could then iterate through the code line by line.

The following example shows how we could use this methodology to download an html page:

```
import requests
url = 'https://en.wikipedia.org/wiki/Minnesota'
response = requests.get(url)
```

Note that in this case we would then need to break the html code down by its components in order to parse the returned html response. This means that when we use this approach, we need to be familiar with common html tags in order to parse our file correctly. For example, if the result we are parsing is an html table, then we might iterate over the rows in the table by finding instances of <trow> tags in the response string. As the complexity of the website grows, and more components are added, building this logic becomes significantly more challenging.

7.2.4 Interacting with Databases in Python

Another common source for data that we will use in our models will be databases. As such, in this phase of a quant project we will need to be able to connect to a database and manipulate and retrieve data from the database. In Python, modules exist that facilitate connections with the most commonly used databases including SQL Server, Oracle and PostgreSQL. One common technique for connecting to a database is via an ODBC connection. If we use this approach, we can connect to different types of databases, and the **pypyodbc** module in Python can be used to make this type of connection seamlessly. The following code snippet can be used to connect to a SQL Server database using an ODBC connection:

```
import pypyodbc
import pandas as pd

connection_string = 'Driver={SQL Server};Server={server_name};
    Database={db_name};UID={user_id};PWD={password};'
db = pyodbc.connect(connection_string)
qry = 'select * from table'
df = pd.read_sql_query(qry, db)
```

In this coding example we import the pypyodbc module and use its connect function to create a database connection object. To call the connect function we need to pass certain information related to the database that we are connecting to, such as the database and server name, and login credentials. Next, we rely on the built-in pandas function read_sql_query to execute a SQL query, created via a string, using that connection object. Also notice that the read_sql_query function within the pypyodbc module returns a pandas data frame. This means that the results retrieved from a database will be easy to transform, merge or aggregate with other data sources. While connecting to a database is important, it is clearly only useful if we know how to build the underlying queries themselves, which we provide an introduction to in the next chapter.

7.3 COMMON FINANCIAL DATASETS

In finance we encounter many different types of datasets that we are trying to model. Perhaps most commonly, we find ourselves working with price history for a particular asset, such as a stock, commodity or currency. This will be true in many applications, such as risk management, where we are looking to model the expected risk in each particular asset. It will also be true in many other buy-side applications, such as building forecasting models in order to try to predict movements of the underlying asset. Other times, we may need to work with options data, in which case we will need to be able to assemble and make sense of option prices from different strikes and expiries. When doing this, we need to ensure that they are consistent and arbitrage free, prior to analyzing them. This will be true in options trading applications, and also in applications that require pricing exotic options. In other cases, we may need to work with fundamental or economic data. For example, we might hypothesize that some balance sheet or fundamental corporate data has predictive ability for future stock returns. Lastly, in the world of fixed income, and in particular mortgage backed securities, we may need to work with extremely large datasets that contain loan level information for all underlying borrowers. In this case we might be interested in various information that helps us understand the underlying customer's financial health, such as credit score, debt balances, etc. This will naturally be a big data problem as these types of securities often have very large universes of loans that underlie them. In the following sections, we comment on the structure of different financial datasets that we encounter in quant modeling, and describe the necessary validation steps we need to perform on these different types of data.

7.3.1 Stock Data

When working on models related to stocks or equity markets, price/returns data will be the most commonly utilized dataset. This data is widely available from multiple sources often at little to no cost. Additionally, we may want to look for clues about what future returns for a given equity might be. To do so we might look in a number of different places. One option, of course, would be to look at past returns, and try to build a momentum or mean-reversion signal. Alternatively, we could look at fundamental data, such as earnings, in order to inform our prediction. We could also look to less traditional data sources, such as algorithms that track web-clicks or parse news related to different companies. We could also look to incorporate information from the options market into our model. Generally, when building a model related to stocks we are working with time series data, and in particular are often working with panel data. The panel data aspect arises because we often have a cross-section of underlying assets within our model, and a time series for each asset in the cross-section. When we are cleaning this type of data, we need to do things like check for outliers and ensure that splits, mergers, and dividends are properly handled.

7.3.2 Currency Data

Like equity data, currency data starts with historical data for a set of exchange rates. Because we have a potential cross-section of exchange rates, working with this data also often entails analyzing panel data. Just as with equities, we may also want to bring in additional information to help explain or predict currency movements. This could include equity data itself, macroeconomic data, or potentially data related to currency options. Cleaning data for currency exchange rates tends to be less onerous than working with other types of datasets. This is because currency data tends to be fairly clean, and we don't need to worry about things like acquisitions or splits. However, we should still check our data for outliers as we would any dataset.

7.3.3 Futures Data

Analyzing futures data instead of data for the underlying asset itself brings with it a set of complications. First, there are a set of futures that exist at any given time, each with a different expiry, rather than a single asset price. As a result, instead of working with a single point, we now need to think in terms of a futures curve. Further, different points along this futures curve are clearly related to each by certain arbitrage conditions. If one point on the curve become sufficiently high or low relative to the others, then arbitrageurs will step in and move the curve back in line. As a result, these arbitrage conditions should be monitored when working with futures datasets.

It should be noted that there are two main types of futures. For many futures, the underlying itself is tradable. This leads to a very tight relationship between the set of futures and the underlying index that is bound by arbitrage. This is the relationship that we discussed in chapter 2. For some other types of futures, such as VIX futures, the underlying asset itself is not tradable. In these cases, there is more freedom in the movement of the futures curve because the arbitrage relationship breaks because the underlying asset cannot be held. There are also some futures that fall somewhere in the middle. While the underlying asset is technically attainable, it may be extremely hard and costly to store, creating some limits on arbitrageurs willing to step in. Knowing what type of futures contract you are working with can often be a pivotal part of performing the necessary analysis. In chapter 22, we explore the implications on futures curves of this difficulty to store, from an investment perspective.

A second key point regarding working with futures data is that, by the nature of their fixed expiry, futures will eventually expire and need to be *rolled* in order to maintain similar levels of exposure. The process of selling a futures position that is close to expiry and entering a new long futures positions in a future further out on the futures curve is referred to as *rolling*. One common strategy is to roll to the futures with the next closest expiry a predetermined number of days prior to expiration. When working with historical futures data, we generally want a single time series that shows the returns of buying or selling a particular asset. In this case, we need to combine positions in many underlying futures to achieve this via the same rolling strategy.

When we are working with futures data, validating the data should include checking for arbitrage between the spot and futures, keeping in mind any storage costs. It should also include a rolling strategy in order to patch together a single time series for a particular future from the individual fixed expiry futures.

7.3.4 Options Data

The structure of options data is inherently quite different than the structure of futures or stock data. This is because the presence of data for different strikes adds an additional dimension to the data. When working with futures data we have a set of futures that are for different expiries. In the case of options we have an entire surface for each historical date that we are analyzing. The first step in analyzing the set of options data is to identify which options are liquid enough to be included in our set. We also want to make sure that the data doesn't contain arbitrage[1]. The options models that we will apply in later chapter will assume that there is no arbitrage in the data, and if that assumption is violated then the model will have to stretch in order to accommodate the violation, making the results worthless. Options data for vanilla options is generally used to help model, price and build hedging schemes for exotic options. Additionally, this type of options data would be required to build and back-test an options trading strategy, as we will see later in the book. Because we are working with a multi-dimensional object at each point in history, we will find that working with options becomes more complex than working with stocks or currencies.

[1] Bearing in mind that, if the data does contain arbitrage, it is highly unlikely that the market will as well.

7.3.5 Fixed Income Data

In fixed income markets, like in futures, we have to deal with data for multiple expiries, and therefore, a curve arises for each historical point in time. This creates the same set of challenges discussed above with regard to fixed expiries and added dimensionality. The world of fixed income can be broadly broken down into interest rate markets[2] and credit markets[3]. In interest rate markets, the curve that we are referring to is called a yield curve, and it is a representation for yields of different instruments, such as bonds or swaps. In credit markets, we are instead concerned with a default rate curve, which helps us infer the probability that a firm will default at various times. Fixed income data could be used for many purposes.

Introducing derivatives in fixed income markets leads to even more complexity in our data. For example, for instrument rate derivatives a so-called volatility cube arises, as opposed to the volatility surface we generally see in options markets. This is because we are able to trade options for different strikes, option expiries as well as contract maturity.

In the cases of yield curves, and the volatility surfaces that arise in fixed income markets, certain arbitrage relationships naturally occur as well, and we should be cognizant of these when analyzing our data.

7.4 COMMON FINANCIAL DATA SOURCES

In the last section we talked about the potential different types of data that we might be interested in, and in this section we describe some of the most frequently relied upon data sources for those getting started in quant finance. Many of these sources are free and have Python API's that facilitate the data retrieval process. For certain types of data, such as stock price data, there are a plethora of free sources with reasonable quality historical data. At the other end of the spectrum, data for some segments of the fixed income market, such as mortgage backed securities, is almost impossible to find, and usually very costly to obtain. With respect to options data, it is quite hard to find free historical data sources for the data, however, there are several low-cost solutions, such as iVolatility. The quality of these low-cost solutions isn't quite as strong as other providers, such as OptionMetrics. As a result, if we rely on these lower cost solutions we need to implement the cleaning procedures ourselves. This would include, but not be limited to performing arbitrage checks on the data and isolating sufficiently liquid strikes.

The following tables lists some of the more useful data sources for quants:

- **Yahoo Finance**: Perhaps the most popular free source of data for aspiring quants. Mainly consists of stock price data with some futures data as well.
- **Ken French's Website**: Useful historical datasets of the returns for Fama-French factors.

[2]Which are discussed in chapter 19.
[3]Which are discussed in chapter 20.

- **FRED**: Federal Reserve Bank of St. Louis website. Contains a significant amount of historical data on economic variables, such as GDP, employment and credit spreads.
- **Treasury.gov**: Historical yield curve data for the US.
- **Quandl**: Contains equity market data as well as data on futures in different asset classes.
- **HistData**: Contains free data from FX markets, including intra-day data.
- **OptionMetrics**: Contains relatively clean options and futures data for equity and other markets. A great, but costly source of options data.
- **CRSP**: A broad, robust, historical database that doesn't suffer from survivorship bias. Equity prices, and other datasets available via CRSP.

Both Yahoo Finance and Quandl have API's in the Python language making interacting with these data sources completely automated and easy to setup. In other cases, like Ken French's website, the standard procedure may be to download the file manually and then to use the methods discussed in this chapter to parse the file into a data frame for us by the ensuing model.

Bloomberg is another data source that many financial professionals will rely on for retrieving data for ad-hoc analysis or for building model prototypes. Python, and other similar programming languages have an API for accessing this data although a Bloomberg terminal[4] is required to utilize this data source.

7.5 CLEANING DIFFERENT TYPES OF FINANCIAL DATA

Earlier in the chapter, we discussed the most common types of financial data that we encounter in our roles as quants, and provided some information on the general structure and potential issues associated with the various types of data. In this section, we explore a few of the most common cleaning considerations for different types of data: handling corporate actions, avoiding survivorship bias and ensuring that our data is arbitrage free.

7.5.1 Proper Handling of Corporate Actions

Corporate actions include dividends, stock splits, mergers and acquisitions. Each of these impacts the returns stream for a given stock, and as a result we will need to make sure that they are properly incorporated into any analysis that we perform. For example, it has been shown that dividends account for a significant return of equity portfolios. Therefore, if we were to ignore that component of an equity's return we would not be getting a full picture of how attractive they might be. Similarly, if we observe the price of a stock double we might naturally think we obtained a huge

[4]Which is not free.

return, but would then be very disappointed to learn that was just the result of a reverse stock split.

The easiest and most robust way to handle corporate actions is to rely on a source that tracks and adjusts for these events. For example, in Yahoo Finance there is an Adjusted Close column which is meant to account for all dividends and stock splits. Using this type of data prevents us from having to track the events ourselves, which can be quite onerous. In the absence of these datasets, then we would need to maintain tables of these corporate actions, and create adjusted price and return series that reflect these events.

Even if we are relying on a source that takes into account corporate actions, it is still recommended that we use some standard data checks to make sure that nothing has been missed. As an example, we can use the outlier detection methods discussed later in the chapter to identify stock splits that were missed.

7.5.2 Avoiding Survivorship Bias

When working in certain markets, such as equities or corporate bonds, the universe of companies will evolve naturally over time. This will be caused by new companies being created and companies ceasing to exist for a host of reasons. An important consideration for modeling these asset classes is how we handle this evolution of the universe.

In particular, there is a tendency to consider only the current universe of companies in our sample, but there is an inherent bias in this approach. For a company to continue to exist requires that they are somewhat profitable. Companies may cease to exist for many reasons, for example they could go bankrupt, or be part of a merger with another firm. However, if we eliminate all companies that no longer exist, including those that have entered bankruptcy in our analysis, it leads to a potential inflation of results. This bias is referred to as survivorship bias, and is based on this phenomenon that companies that have survived are likely to have been successful in the past. Excluding companies that have not survived excludes all bankrupt firms and may inflate the implied returns.

This effect is magnified if we get the universe of companies in our analysis via an index that selects the firms with the biggest market capitalizations, as the biggest firms today are likely to have been the most successful firms in our backward-looking period of analysis.

The impact of survivorship bias depends on the application that we are pursuing. If we are back-testing a long-only equity strategy, then survivorship bias is a serious concern. This concern would be somewhat mitigated if we were analyzing a long-short strategy, because it might inflate the results of the long parts of the trade but would similarly understate the results of the short trades.

To avoid survivorship bias and related issues we want to look at the universe at every point in time rather than only the current universe which we know may have bias. This will require us getting access to an additional dataset that is often harder to find. Secondly, we will want to make sure that positions in all firms in the sample are tracked, regardless of whether they ceased to exist at some point.

Fortunately, some of the better data sources in these markets, such as CRSP, will account for this phenomenon by providing data for all underlying assets even after they cease to exist. Using these sources combined with constituent data for all historical data points can mitigate the possibility of survivorship bias. In the absence of these sources, we need to be cognizant of potential bias in our results that might be cause by survivorship bias.

7.5.3 Detecting Arbitrage in the Data

When working with derivatives data, such as options and futures, identifying arbitrage in the data is a key part of the validation process for the data. Remember that any derivative model that we apply will require our data be arbitrage free. As a result, it is our job to ensure that the data that we feed to the model complies with its assumptions, which means verifying there is no arbitrage.

With respect to options data, there are set of arbitrage conditions that must hold for options of different strikes:[5]

- Call prices must be monotonically decreasing in strike.
- Put prices must be monotonically increasing in strike.
- The rate of change with respect to strike for call prices must be greater than -1 and less than 0.
- The rate of change with respect to strike for put prices must be greater than 0 and less than 1.
- Call and put prices must be convex with respect to changes in strike, that is:[6]

$$C(K-h) - 2C(K) + C(K+h) > 0 \tag{7.1}$$

If these relationships do not hold, it is not hard to show that there is an arbitrage opportunity. For example, with respect to the first condition, suppose $C(K) < C(K+h)$. We could then buy one unit of the call with strike K and sell one unit of the call with strike $K+h$. This would lead to us receiving an upfront payment of $C(K+h) - C(K) > 0$ and would also lead to a payout that is strictly greater than or equal to zero.

We could create a similar set of conditions that show how option prices must behave as a function of expiry. For example, we know that for both calls and puts, prices must be monotonically increasing as we increase option expiry. It is generally best practice to discard one-sided quotes (that is, with 0 bid or offer), as they usually don't represent an actual trading level.

A similar set of rules can be created for futures data using the arbitrage argument shown in chapter 2, keeping in mind that this argument requires us to be able to trade both the underlying asset, and the futures contracts.

[5] For a more detailed set of arbitrage conditions that incorporate a bid-offer spread, see [133].
[6] See chapter 17 for more details on why this convex relationship is required by arbitrage.

7.6 HANDLING MISSING DATA

Missing data is one of the most frequent data challenges that quants face. In some cases, we may just want to eliminate the data that has any component missing. Missing data may occur, for example, because a particular asset didn't exist in the given historical period, or because it ceased trading for a time. For example, if we are tracking data for 10 ETFs, we might choose to only look at the set of dates that have observations for all 10. Provided that we have a sufficient amount data, this could be a perfectly legitimate way to handle this situation. However if data is scarce, then we may want to try to fill in the gaps in a robust way such that the bias we have introduced is minimal. How we choose to fill in the gaps will vary by the nature of the missing data. For example, if we are using datasets that correspond to different regions, they might naturally have different calendars and different holidays. In this case the gaps are clearly likely to be small, lasting only one or two days, and it would be entirely appropriate to fill the data forward given that the market was closed on each holiday. Alternatively, if there are longer gaps in our data, then different methods will be required that help us preserve the distribution and are consistent with the remaining history.

The method that we use to fill missing data will depend on the specific application that we are creating. For example, for risk management calculations, less precision is often required, and we have more flexibility in how we fill the data. Conversely, if we are using the data in a back-testing framework, then improper data could have a larger impact and more care would be required.

Regardless of the method that we choose, or the application that we have in mind, a cardinal rule for filling missing data is that we should try to avoid looking ahead. If we did this in the example of back-testing, looking ahead could lead to inflated and misleading results. In this section, we discuss a handful of potential methods for filling missing data, and discuss the strengths and weakness of each method. As we do this, we attempt to highlight the potential bias each method introduces.

7.6.1 Interpolation & Filling Forward

The simplest approach to filling in missing data is either to use a constant, generally the last good value[7], or a simple interpolation scheme such as linear interpolation. These techniques are both well suited to handle small gaps in the data, however, are sub-optimal for filling longer gaps. The inappropriateness of these methods for longer periods is fairly intuitive. If we have a year long gap of stock prices that we are trying to fill, then it would be unrealistic to assume a constant return for all days in the year, as interpolation would. It would be equally unrealistic to assume the stock did not change throughout the entire year and then jumped to the new value entirely on the preceding day.

The primary benefit of using interpolation or filling the data forward is simplicity. The primary drawback, however, is the potential bias that is introduced in the distribution, in particular the dampening or heightening of volatility. This bias emanates

[7]We use the last good value to avoid looking ahead.

from the fact that the returns that we are inputting do not have the same statistical properties as the entire empirical dataset, as we illustrated in the example above. Note that forward filling is more often used than backward filling in the financial setting, essentially because we do not want to include future information in historical records.

Python has built-in capability to perform this type of data filling within a Data Frame object. This can be accomplished using the filling and interpolate functions. In the coding repository, we show how we can apply these functions to handle gaps in a data frame.

7.6.2 Filling via Regression

One feature of the previous approaches is that they don't rely on, or utilize, the performance of other potentially similar assets when filling in the data. This may be a virtue in some cases, but in many cases utilizing this information may lead to a better estimate for our missing data. For example, if we are filling in missing data for a stock that we know has high correlation with equity market indices, than if we knew the S&P had declined 5% during the missing data point that would give us strong conviction that the missing return is also likely to be highly negative. Perhaps in this case we should center our expectations for the missing data based on what happened to other, correlated assets. This is precisely what filling via regression entails.

When using a regression based filling technique we use linear regression to calibrate an expected change for each missing data point, and we use this as our best guess. The first step in doing this is identifying related assets that have a large history that includes data on the missing data points we are looking to fill. We could use a single market index, or we could rely on multiple related securities.

For simplicity, let's assume that we are operating in a world where only the market return impacts the missing return for the asset. In this case, at any point in time our best guess for the return of our stock is:

$$r_{i,t} = \beta r_{\text{mkt},t} + \epsilon_{i,t} \tag{7.2}$$

Where β is the regression coefficient between our stock and the market index, and $\epsilon_{i,t}$ is the idiosyncratic return of the stock at a given time. Importantly, $\mathbb{E}[\epsilon_{i,t}] = 0$, meaning that, in expectation, the return for our stock is a function of only the market return and the β that we calculate.

Therefore, to apply this method we simply need to estimate the β parameter, and we do so using all available data with returns for our stock and the market. We can then calculate the stock's return on each missing day using the market return according to (7.2).[8]

In this example we have focused on a single-factor approach, however, we could easily use multiple correlated factors or underlying assets instead. An attractive feature of this approach is that it retains the correlation information with the other factors that we use to build our model. In doing so, it reduces the amount of bias

[8]Eliminating the residual, or noise term because it is zero in expectation.

that we introduce into our distribution. Having said that, the success of this method is largely dependent on the fit of the regression. In cases where the explanatory power is low, this approach may lead to a dampening effect in volatility. Therefore, this method is most effective when there are longer gaps in our data and we are able to find a set of highly correlated assets.

7.6.3 Filling via Bootstrapping

As we saw in chapter 5, bootstrapping is a non-parametric technique for creating sample paths from an empirical distribution. The idea behind bootstrapping is that each realization is in actuality just a draw from some unobserved density. Bootstrapping provides a method for sampling from that density with minimal assumptions. It is an intuitive technique that is quite powerful and has many applications within the realm of quant finance. One of these applications is filling missing data, as it provides a means for generating synthetic data that is consistent with the empirical distribution, embedding information related to not only the individual asset's statistical properties, such as mean and variance, but also the correlation structure. In chapter 26 we revisit bootstrapping in the context of solving risk management problems.

In practice bootstrapping involves simulating returns, and paths of asset prices by scrambling historical returns data, or sampling various historical returns and creating series of them in random order. In the context of missing data, bootstrapping enables us to create a synthetic dataset that is statistically equivalent to the dataset with missing data. We can then use the synthetic dataset instead.

As we saw in chapter 5, a bootstrapping algorithm can be written as follows:

- Create a matrix of historical returns for the universe of assets.
- Generate a sequence of random integers between 1 and N, where N is the number of historical days of data.
- Look up the return data for all assets that corresponds to that randomly chosen date.
- Apply these returns to the entire asset universe.
- Continue this for many time steps until the end of a path is reached.

Each time this process is followed, a new bootstrapped path is created. In some applications of bootstrapping, we may want to create a sequence of bootstrapped paths and examine the properties of the paths. When filling missing data however, we simply want to create a single path that is statistically equivalent, and as a result do not need to repeat the process. Instead we can take the path that results from the process above and substitute it for the period with missing data.

The main benefit of bootstrapping is its non-parametric nature and its ability to preserve the market correlation structure and other properties. Bootstrapping is very commonly used in cases where data is sparse as a way of creating additional data. It can also be applied to periods where many assets have data missing from the dataset.

7.6.4 Filling via K-Nearest Neighbor

Using K-Nearest Neighbor (KNN) to fill missing data is similar conceptually to the regression approach we discussed earlier. In particular, it enables us to use information related to other, related assets, in order to help us inform our estimates of the missing data. The details of the K-Nearest Neighbor algorithm are discussed in more detail in chapter 28, however, in this example we provide the intuition of what the model does and show how it can be applied to filling miss data.

At a high level, the K-Nearest Neighbor works by locating the points that are the most similar to the point it is trying to estimate. It then uses an average of those closest points, or nearest neighbors, in order to make a prediction. In the case of $k = 1$ simply finds the closest point and predicts that the value will be equal to what it was at this point. In the case of $k > 1$, the closest k points will be found, and an average of their values will be used as the prediction.

For example, in the context of filling in missing data, if we are trying to estimate the most likely the return of a given commodity, say Oil, and we are trying to do so using only the broader commodity index BCOM, then the KNN algorithm would proceed by finding the k closest returns to the current BCOM index return, and averaging the return of Oil over those k periods. The average of these k returns would then be the estimate we would use to fill in our missing data for that point. This is analogous to the regression approach that we described above, however instead of assuming a linear relationship we consider a close set of neighboring and use them to estimate what the return was most likely to be.

CHAPTER 8

Data Science Techniques in Finance

8.1 INTRODUCTION: WHAT IS DATA SCIENCE?

Much of what quants do centers on their ability to understand, analyze and manipulative complex datasets and identify patterns, characteristics and defining features. Against this backdrop, it is not surprising that the fields of data science and data analysis are of the utmost importance to aspiring quants. In fact, many in quant finance have advocated for a so-called 80/20 rule, wherein quants should spend 80% of their time cleaning, validating and getting their data ready to input in a model, making sure they understand the properties, limitations and exceptions, and only 20% of their time working on the actual model. To many aspiring quants excited about the possibility of interacting with models, this is a sobering breakdown. It does, however, underscore that quants are inherently engineers and statisticians, and emphasizes the importance of data science skills for those entering the field.

While working with data has always been important, the proliferation of new available datasets [1] has led to increase in data scientists within the field, and an increased emphasis on data science skills. In this chapter we aim to outline the main components of data analysis and data science within the field of finance, highlighting the primary techniques and programming tools.

At heart, data science is about understanding our datasets, identifying any underlying patterns and any deviations from these patterns. The tools that we use to do this, are similar in spirit to the tools that we present in the rest of the book. For example, in chapter 8 we show how to transform data in Python to be consumed by quantitative models. This ability to code in Python, and to work with datasets, will likewise be a core capability for data science work. Similarly, in chapter 28 we show how machine learning is used in the context of portfolio optimization and trading. As we will see in this chapter, many of the same concepts leveraged their will also be of great use when analyzing and trying to understand structured or unstructured datasets. Additionally, the statistics and econometric concepts that we covered in

[1] As discussed in chapter 2.

DOI: 10.1201/9781003530626-10

chapter 5 are also vital to data scientists who are trying to summarize data and extract the key features.

In the remainder of this chapter we aim to highlight the most central tools for data science tasks. In the next section, we begin with an overview of Exploratory Data Analysis, an important tool for visualizing and summarizing large, often highly dimensional data. We then discuss how SQL can be used in conjunction with Python to manipulate datasets, followed by applications of dimensionality reduction and machine learning. Finally, we discuss techniques for working with large datasets and handling outliers.

8.2 EXPLORATORY DATA ANALYSIS

As data scientists in the financial world, it is pivotal that we understand the datasets that we are working with. This ensures that we make the appropriate assumptions in our models and handle special cases. Exploratory data analysis (EDA) is a set of tools first developed by Tukey that aid this process [252]. Importantly, EDA is done prior to making any assumptions about the data. Instead, it is a set of tools, many of which are visualization based, that help us understand how an underlying dataset is structured.

Postponing the traditional assumptions on our data helps us ensure the models that we build are parsimonious and only based on relationships that are optimally formulated and reveal themselves in the data. For example, linear regression is the most commonly employed technique in quant finance [2]. Prior to leveraging linear regression, however, we want to ensure that the relationships we are extracting are linear. This can be achieved by plotting, and EDA. In particular, a scatter plot with a linear trend can help us intuitively understand whether a relationship can be approximated as linear.

As discussed in [91], Exploratory data analysis defines a set of techniques and general framework for data analysis. This framework consists of:

- Maximize insight into dataset
- Identify structure of dataset
- Extract important features
- Detect anomalies or outliers
- Test underlying assumptions
- Develop parsimonious models

In the remainder of this section, we describe the main techniques used for EDA in greater detail.

[2] And even the most commonly applied machine learning technique.

8.2.1 Single vs. Multivariate Exploratory Data Analysis

One of the defining features of the datasets that we work with is the dimensionality. When we work with data that has one or two dimensions, it is inherently easier for us to visualize, and simple plotting tools usually suffice in helping us understand the data. As the dimensionality increases, so too does the difficulty in visualizing it. Within that context, dimensionality reduction and summarizing the data in a presentable way become of the utmost importance. In the following sections, we share a few visualization techniques, such as clustering, dendrograms and t-SNE that are designed to work in this type of highly dimensioned environment.

8.2.2 Data Visualization

A core tenet of EDA is that visualizing a dataset in various forms can help to bring out patterns in the data that may be otherwise unobservable. As such, at the heart of EDA is a set of plotting and other visualization tools that help us identify relationships between variables and plot highly dimensioned data using a lower dimension representation of the data [3]. In the following list, we detail some of the most import visualization tools that data scientists and other finance practitioners will find them using as the explore datasets:

- **Time Series Plots**: A time series plot is a line chart with a date or timestamp on the x-axis and the value of some variable of interest on the y-axis. These two-dimensional plots help us to identify patterns in a single variable over time.
- **Scatter Plots**: A scatter plot is a two-dimensional chart that plots the relationship between an x and y variable by plotting a dot for each x-y pair. The y-variable can be thought of as the dependent variable, as in a linear regression model, whereas the x-variable can be thought of as a single explanatory variable. If there is a relationship between the two variables, then we'd expect to see a dependence, either linear or non-linear, between values on x-axis and y-axis. Scatter plots are often overlaid with a linear [4] trend line, and can help us understand the validity of a linear relationship between the two variables.
- **Box Plots**: A boxplot is a way to visualize the range of a particular value and analyze how it differs across some sort of subset. A box plot displays the median, certain percentiles and range of the dataset and is designed to be compared across a cross-section. For example, we could define a cross-section of assets and use a box-pot to effectively examine the return distribution across the assets.
- **Histogram**: A histogram or frequency distribution shows the frequency that a given quantity occurs within certain predetermined buckets. Histograms can help us to understand the underlying distribution of our data, which is helpful prior to beginning to make model assumptions. For example, it can help us

[3] One such example is t-SNE, which is discussed later in the section.

[4] Or polynomial.

identify whether the data appears normal, has fat tails, or is positively or negatively skewed.

- **QQ Plots**: A QQ plot tests whether a given dataset matches a specified distribution. It does this by plotting the quantiles of the data, and the quantiles of the reference distribution. If the data matches the reference distribution, then we expect to see the quantiles match closely. This technique is often applied in finance to test the normality of underlying datasets, but can also be used in the context of EDA to understand additional features of a distribution, such as excess kurtosis. Later in the chapter, we explore an example of how a QQ plot can be used in the context of outlier detection.
- **Heat Maps**: A two dimensional visualization where magnitude or intensity is depicted by a color scheme. For example, a heat map might be used to visualize a large correlation matrix so that we can identify points of high and low correlation.
- **Dendrograms**: Based on an underlying clustering algorithm, dendrograms depict the distance between different quantities via a tree diagram.[5] Items that appear closest to each other in a dendrogram are the most similar, whereas those that are far apart are the most heterogeneous. Dendrograms partition the quantities into different clusters, each of which is generally represented in a different color, based on hierarchical clustering, with the most similar entities being matched in the same clusters. A dendrogram can be helpful for understanding the structure and similarity of different entities in a dataset. For example, we could view a dendrogram on a set of stocks or other assets to identify which are the most similar and can naturally be bunched together.

In section 8.2.6 we provide examples of many of these types of charts in the context of a set of country ETFs.

8.2.3 Statistical Analysis of Data

When performing EDA, in addition to visualizing the raw data in different forms we often leveraging statistical summary metrics to help us understand the data better. For example, to understand the distribution of a set of realized returns, we might start by looking at the mean, standard deviation, skewness and kurtosis of the returns. We might also be interested in seeing how these summary statistics evolve over time, and how consistent they are across different underlying assets.

The following list details a few main types of statistical analysis that are often done as part of an EDA exercise:

- **Summary Statistics**: Calculate entire period statistics such as mean, standard deviation and skewness. We could choose to calculate these summary statistics unconditionally, or conditional on some criteria. For example, to uncover

[5]Clustering algorithms and common distance metrics are discussed in more detail in 28.

seasonality we might calculate the conditional summary statistics by month or season.

- **Rolling Summary Statistics**: Calculate means, standard deviations and other summary metrics over time, observing how they evolve and whether there are any trends.
- **Cross-Sectional Summary Statistics**: Calculate summary statistics for each asset, entity or column in the cross-section of the dataset with the goal of identifying similarities and differences in their structure.
- **Density Analysis**: Perform analysis of the underlying density, or densities using plots such as QQ plots and histograms.

When performing EDA, we often attempt to visualize these statistical summary metrics to help us understand the data. For example, we might construct a box plot of the cross-sectional means of each asset in a panel. This can help us determine whether the panel we are working with is homo or heterogeneous. As we learned in section 4.5 a key feature of building regression models is determining how much differentiation, or heterogeneity across the panel we want to incorporate, and as such these types of plotting exercises can be crucial to understand our data. Similarly, plotting a rolling mean and standard deviation across time can help us to understand the stability and stationarity of the underlying data, and identify patterns like seasonality.

In exploratory data analysis we also aim to construct these visualization of the summary statistics in an optimal way so that it helps us identify patterns. For example, we could plot time series mean charts side by side for multiple assets so that we observe any clear trends that are common across the asset universe.

8.2.4 Factor & Principal Component Analysis

As we detailed above, one of the objectives of EDA is to extract the most relevant features from a dataset with minimal assumptions. This process is commonly known as factor analysis, and can be done in one of the following two ways:

- **Named Factor Analysis**: In this approach we postulate what the significant factors may be with the aid of visualization techniques. This enables us to focus on factors that are in line with our intuition but creates a risk that relationships go undetected. Once the relationships are understood via plotting exercises, then techniques like linear regression can be used to model the dependency structure of the chosen factors.
- **Statistical Factor Analysis**: In this approach instead of naming the factors we allow an underlying algorithm to identify the most important factors statistically. Perhaps the most common example of this type of factor analysis is Principal Component Analysis. PCA, which was detailed in chapter 5, has many applications across the field of finance. Recall from chapter 5 that PCA is based on the following matrix decomposition:

$$\Sigma = X^\top X = CDC^\top \tag{8.1}$$

Further recall that D is a diagonal matrix of eigenvalues with diagonal elements λ_i. The largest eigenvalues, λ_i represent the factors that explain the most variance and are thus what we are interested in when searching for underlying factors. The corresponding row in the matrix C, the eigenvector, then provides us the linear combination of the underlying data that enables us to construct the factor.

8.2.5 t-distributed Stochastic Neighbor Embedding

One of the main challenges in performing exploratory analysis on a dataset with many dimensions is that it is by construction difficult to visualize the entire data simultaneously. As such, techniques that enable us to do so are incredibly valuable. One such tool is t-distributed stochastic neighbor embedding (t-SNE) [131] which relies on creating a lower dimension representation of the data and then visualizing this more tractable representation of the data. As such, t-SNE can be useful for many purposes within EDA, including visualization, dimensionality reduction and outlier detection.

t-SNE works by creating a two or three dimensional representation of the data that is easy to visualize and also closely replicates the original dataset. Importantly, the dimensionality reduction is done is such a way that similar data points will be grouped, or clustered near each other on the resulting plot. This representation then can be used by quants in analyzing the data as it helps identify similarity among the data and natural clusters that emerge.

The algorithm for t-SNE is based on building a probability distribution of similarity between data points [6]. This probability distribution is first built using the entire dataset, and then again in a lower dimensional space. When the second, lower dimensional distribution is created, it is done such that we minimize the distance[7] to the original probability distribution, while also reducing the dimensionality as needed. This means that the visualization that results from this algorithm is both reasonably close to the original data, and also easy and intuitive to visualize and draw conclusions from.

The details of the t-SNE algorithm are beyond the scope of this text, however, interested readers can consult [132]. Other clustering algorithms, as well as some of the most common distance metrics used are also detailed in chapter 28. In the next section, we show how t-SNE can be used in the context of EDA by helping us visualize highly dimensional data.

8.2.6 EDA in Practice

In this section we leverage many of the EDA techniques introduced throughout the chapter in the context of analyzing a set of country ETFs.[8] The reader is encouraged to consult the coding repository that was leveraged in this EDA exercise and tweak the

---

[6]Where similarity is defined as Euclidean distance.

[7]Where distance is defined using entropy or Kullback-Leibler divergence.

[8]For brevity, we only show results for the first 5 country ETFs in the text.

code to fit other uses of EDA. As the name implies, EDA is a somewhat open ended task in which we try to understand our data better. This can be done via plotting exercises, computing summary statistics, outlier analysis, or density analysis.

In this case, we began by calculating summary statistics for each country ETF. We focus on the first four moments of the distribution as summary statistics, with the results shown in the following table:

ETF	Mean	Volatility	Skewness	Kurtosis
EPI	11.5%	21.6%	-1.1	14.9
EWC	8.8%	20.0%	-1.2	25.0
EWG	4.7%	21.9%	-1.0	13.7
EWH	1.0%	19.2%	-0.4	5.5

Here we can see that the returns and volatilities across the panel [9] are relatively similar. This might be something we would check when building a panel regression model to verify if the behavior of the underlying components could be treated universally or if cross sectional effects needed to be taken into account.

Next, to analyze the time series properties of the data, we might plot the rolling statistics, which we do in the following charts for the mean and standard deviation:

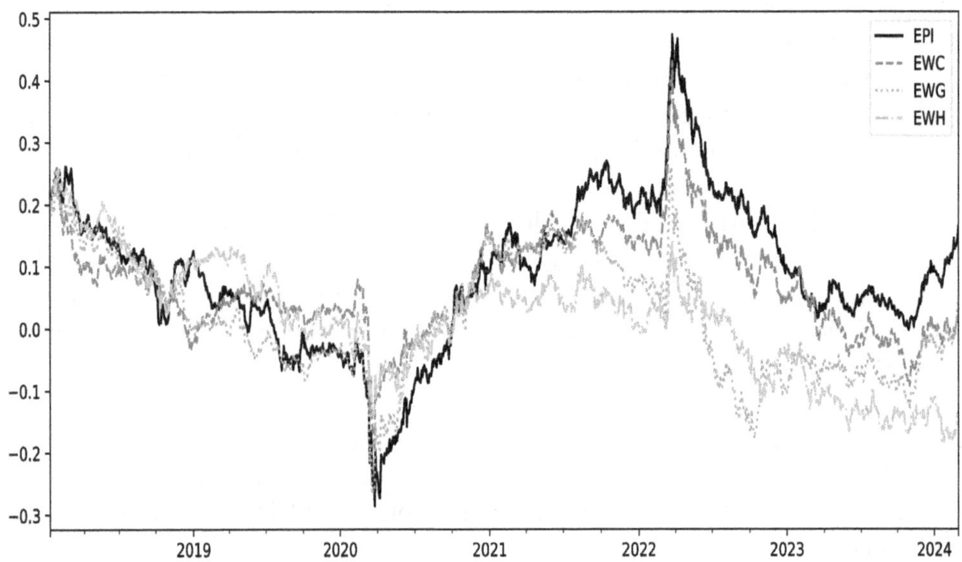

[9]That is, across the set of ETFs.

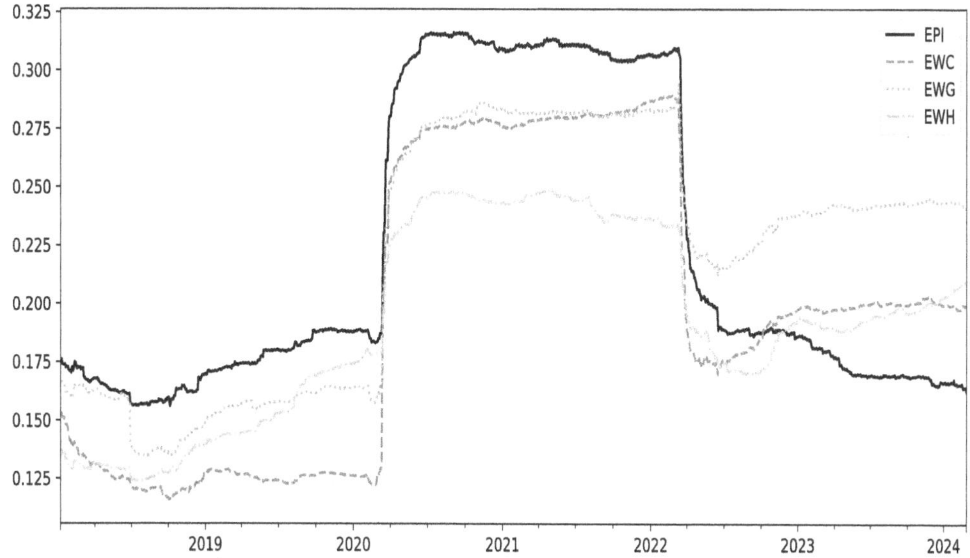

Here, we can see the non-stationarity in the data as both the mean and the volatility tend to vary significantly over time. We can also see evidence of positive correlation between the underlying assets, as both the returns and standard deviations tend to increase and decrease together, likely in response to broader market themes. To understand this correlation structure better one might naturally want to examine the correlation matrix, which is depicted in the following heat map:

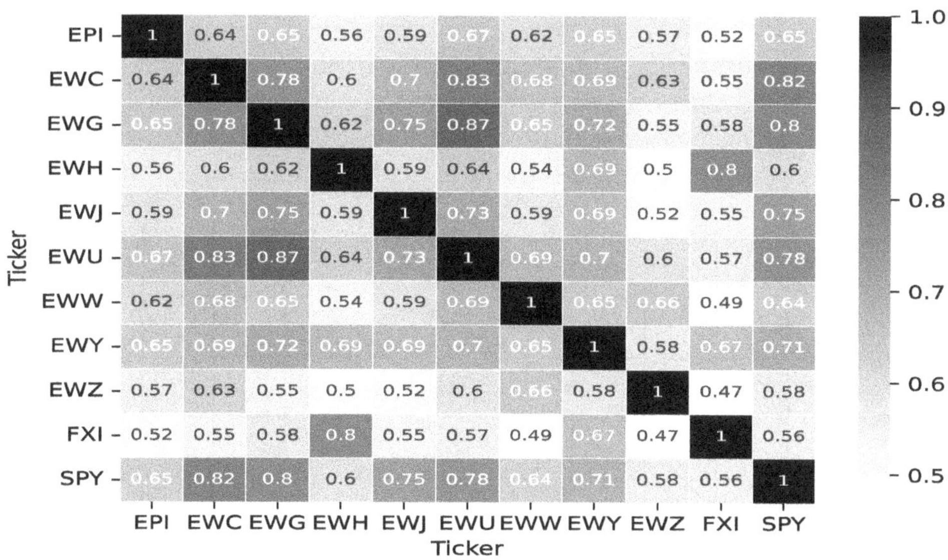

Here we can this positive correlation between the entire set of ETFs, although we do not observe that the indices are even close to perfectly correlated, evidence that each ETF is impacted by idiosyncratic risks specific to its country in addition to broader market themes.

Next, we might examine the density of returns for one or more underlying ETFs, as we do in the following density chart:

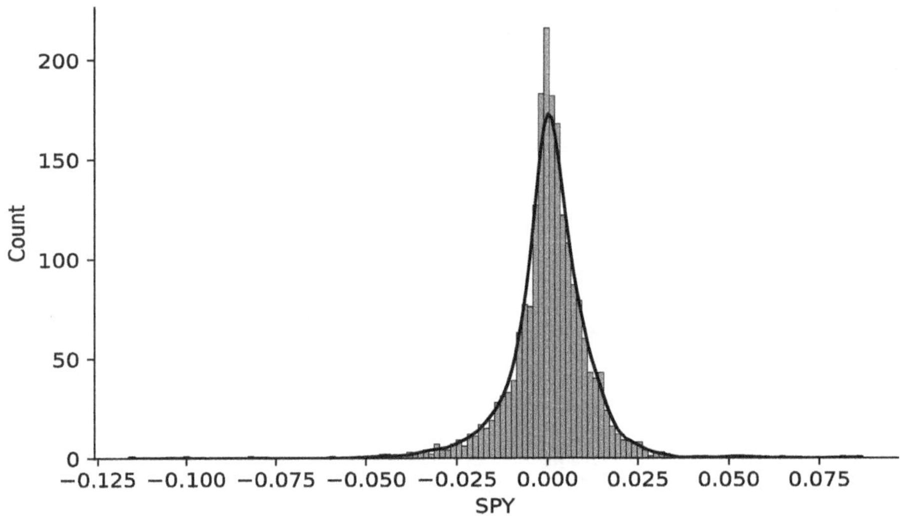

This analysis can help us uncover outliers and can also help us gauge whether the data is normally distributed [10]. Similarly, we could use a technique like a QQ plot to help understand the normality, or fat tailed nature of the return distribution. Analyzing whether these density plots display similar profiles across the set of country ETFs can also be informative and have implications for how we model the data.

Factor analysis, or identifying the main factors driving our underlying data is another critical part of EDA. This can be done via PCA, where as introduced earlier in the chapter we identify a set of principal components that explain the majority of the variance. Importantly, the eigenvectors uncovered by the PCA algorithm provide information on the definition of these statistically extracted factors.

In the following charts we first show the percentage of variance explained as we increase the number of principal components used, and then, subsequently, the corresponding eigenvectors for the top two principal components:

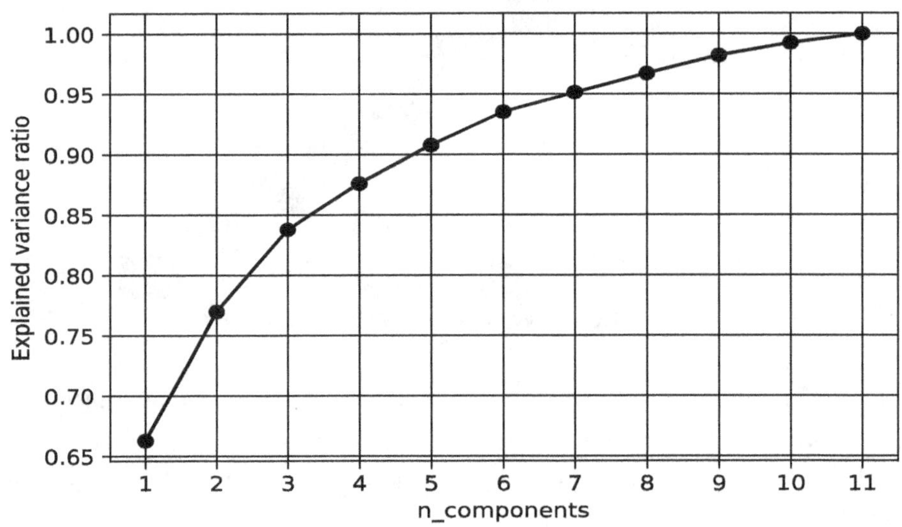

[10] Or distributed according to some other benchmark distribution.

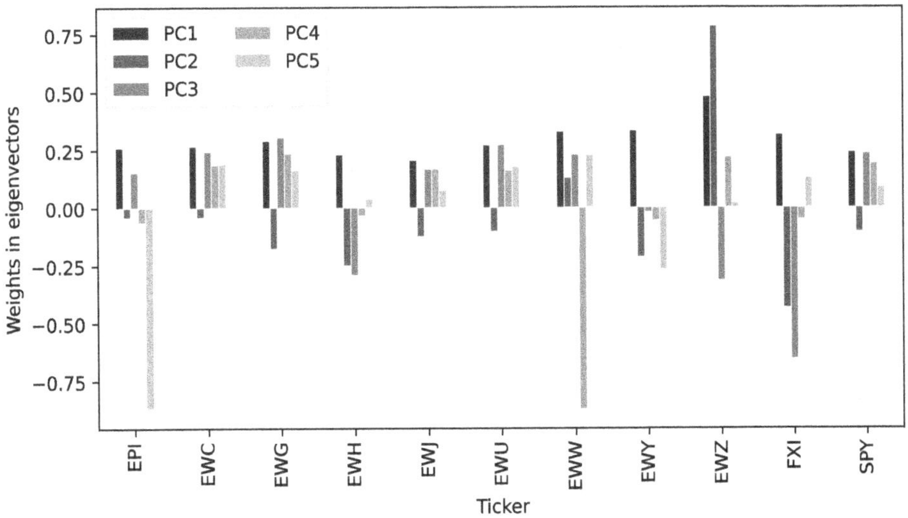

Here, we can see that explaining more than 90% of the variance would require five principal components. Another way to view this is that there are distinct five, orthogonal factors that are driving the vast majority of country ETF returns. We can also see that the weights of all country ETFs in the first principal component have the same sign, and similar magnitude. This implies that the first PC, which explains the majority of the variance, is a shift where all country ETFs move together. We might think of this as a market shock, positive or negative, which universally impacts the set of ETFs.

Finally, we attempt to use t-SNE to help us create a lower dimension representation of the data that we can visualize and closely approximates the full dataset. In this example, we do this with two t-SNE features which attempt to replicate the full set of country ETF returns. In the following plot, we show a scatter of the two t-SNE features and shade them according to whether the actual return of ACWI was positive or negative on the date they occurred:

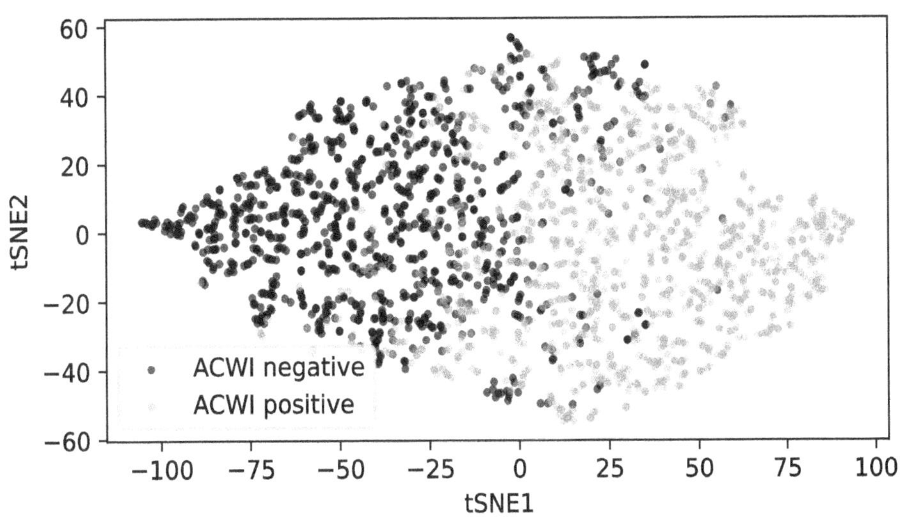

Here we can see that the algorithm fairly reliably clusters positive and negative global index returns together, while also significantly reducing the dimensionality.

8.3 MACHINE LEARNING TECHNIQUES FOR DATA SCIENCE

8.3.1 Overview of Machine Learning for Data Science

One of the main tools for data scientists is machine learning because of its ability to help us understand the underlying relationships in datasets. Machine Learning is a rich discipline that combines mathematics, statistics and programming skills, each of which are core skills for data scientists. Machine Learning algorithms and their applications in finance are covered in more detail in chapter 28 and 29. In this chapter, however, we aim to provide the reader with an understanding of how Machine Learning can be used in the context of solving data science problems.

As we saw in the last section, machine learning techniques such as PCA, Clustering and t-SNE can be used for Exploratory Data Analysis. In this context, we often employ what is called unsupervised learning to allow an algorithm to find patterns in the data without providing any structure to the model. Machine Learning techniques can similarly be used in dimensionality reduction, which is discussed later in the chapter. In this case, the algorithm is used to isolate the most important features and eliminate those that are not significant. This can also be done in an unsupervised fashion, or we can choose to provide some structure to how these factors are extracted. These algorithms are referred to as supervised learning algorithms. Machine Learning can also be used in outlier detection, as we explore later in the chapter.

8.3.2 Supervised vs. Unsupervised Learning

The ecosystem of Machine Learning Models can generally be broken down into two major classes of models: supervised and unsupervised learning models. [11] When we employ unsupervised learning we do not provide any structure to the algorithm. Instead, we simply allow the model to detect the underlying features, or patterns in the data. This has the benefit of enabling us to identify patterns that we otherwise would not have. Conversely, especially in the context of finance, it creates the risk that we find patterns that are less likely to persist. PCA, autoencoder and clustering algorithms are all examples of unsupervised learning.

Supervised learning, in contrast, does provide additional structure to the model in the form of output labels and a potential feature set of interest. Linear regression, logistic regression and support vector machine are just a few examples of supervised learning, each of which are discussed in more detail in chapter 28.

The distinction between supervised and unsupervised learning can perhaps best be seen by comparing PCA to linear regression. In the case of PCA, we simply decompose a set of input data in a way that we are able to find the most important factors. We do so without providing any insight on what these features might be. That is, it is an unsupervised learning process that is used to broadly understand the

[11] A third class of machine learning algorithm also exists called reinforcement learning.

data. When we employ linear regression, on the other hand, we do so by selecting a set of named features and providing a set of output labels (referred to the dependent variables). The structure that we provide, in particular the choice of a set of features, thus influences the supervised learning algorithm. This has the benefit of enabling us to focus on relationships that are in line with an economically motivated prior but also creates the risk that certain complex relationships would go undetected because the proper structure has not been provided.

In the context of data science, we often explore unsupervised learning techniques as they provide the most freedom for us to understand our dataset and identify patterns. In the context of the broader financial landscape and its applications of machine learning, supervised learning techniques are often preferred as they prevent us from overfitting and provide better interpretability. More information on supervised and unsupervised learning algorithms can be found in chapter 28.

8.3.3 Structured vs. Unstructured Data

Data scientists are asked to work with many disparate types of data. In finance, traditional datasets include price, return and other time series datasets. These datasets are often examples of panel data analysis, as we usually have data for a cross-section of assets in addition to a time series dimension. Nonetheless, these types of data, which are inherently numeric in nature, are highly structured datasets. This leads to certain computationally advantages as it is relatively easy to perform Exploratory Data Analysis, search for outliers and reduce dimensionality as needed. In chapter 9 we provided an overview of the different types of structured datasets that quants encounter across asset classes, including asset class price/return data, fundamental data, macroeconomic variables, and options/futures data. When working with these types of datasets, we tend to rely on a certain set of tools, such as traditional statistics and econometrics. Further, when applying machine learning in this context, we tend to focus on classical machine learning models instead of things like Large Language Models and Natural Language Processing. More information on these types of machine learning algorithms can be found in 28.

Other financial datasets are unstructured, such as when working with alternative data and Natural Language Processing algorithms. In these cases, the types of techniques that we rely on tends to be different and we often need to apply additional preprocessing to the data to be able to perform exploratory analysis or to use it in a model. Another feature of these less structured datasets is that they are inherently harder to validate and to check for outliers. More information on working with these types of alternative datasets and natural language processing algorithms can be found in chapter 29.

8.4 TRANSFORMING DATA USING SQL

In chapter 8 we provided an overview of how to connect to a SQL database within Python. Working with databases is a core competency for data scientists as they frequently need to extract data from databases, transform it into a usable format, and load related data into a database. This process is known as ETL, and more information on its principles can be found in [160].

Once we have connected to a given database, the natural next step is to be begin to execute a series of SQL statements against it. SQL is a powerful language that enables us to retrieve and aggregate data from multiple database tables. SQL can also be used to remove data, add data, or update existing data in a database. In this book we cover only the preliminary topics related to the SQL language. For more information, interested readers should consult [211].

The SQL language has four main types of commands that help us read and modify data, which are briefly summarized here:

- **SELECT**: retrieve rows from a set of tables.
- **UPDATE**: modify rows in a table.
- **INSERT**: add rows to a table.
- **DELETE**: remove rows from a table.

In general, select queries will be the most common types of queries that we write, as they enable us to access data from the database and feed it into our models. An important part of writing SQL statements is being able to retrieve data from the database in a manner that is accessible to our model or application. Generally, once the select statement is run, the results are stored in a Python data frame, as in the example in the previous chapter, and then used in broader calculations. The other three types of statements, INSERT, UPDATE and DELETE statements, enable us to modify our database rather than retrieve data. These statements may be useful in our data collection process if we choose to aggregate data from external sources and store it in a single centralized database, as is best practice in most production environments.

As a note to the reader, the SQL programming language varies slightly between database engines, such as SQL Server, Oracle and PostreSQL. In this text we focus on the syntax employed by SQL Server. Those who use other databases may need to tweak the syntax of their queries from those presented in the text, however, these differences are almost always minor. Readers are encouraged to consult the documentation for the specific database engine that they are working with for more information.

8.4.1 Modifying Data via Insert, Update and Delete Statements

In the following example, we show examples of insert, update and delete statements:

```
INSERT INTO prices (ticker, close_price)
VALUES ('MSFT', 100.0);

UPDATE prices SET close_price = 105.0
WHERE ticker = 'MSFT';

DELETE FROM prices
WHERE ticker = 'MSFT';
```

In the code above, the first SQL statement inserts a row into a table named prices, the second modifies all prices for a given ticker, again in the prices table, and the final statement deletes all rows for a given ticker. It should be noted that while insert statements are generally seen as safe operations, because they can easily be undone if anything unexpected occurs, this is not true for UPDATE and DELETE statements. Once we have deleted all rows from a table, or set all prices to the same value in a given table, the process of restoring the original data will be much more involved. Because of this, UPDATE and DELETE commands are considered risky and should be executed carefully. It is best practice to always run the UPDATE or DELETE query initially as a SELECT statement to ensure that the number of rows and content of the rows is what we expect. This should help catch situations where we would have accidentally updated or deleted an entire table, instead of a specific set of rows.

8.4.2 Retrieving Data via Select Statements

As previously mentioned, we can also use SQL queries to retrieve the data that we need from a set of underlying tables. This will be a critical tool in our toolkit when it comes to data collection. There are a handful of core components in a select statement which help us identify which tables we want to include in our query, which columns we want to return, and what conditions we want on the query. The *SELECT* clause defines the columns from our underlying tables that we would like returned in our query. Next, the *FROM* clause defines the table that we are referencing. In future examples we will show how to include multiple tables in a single query using a *JOIN*. Finally, a *WHERE* clause restricts the rows that are selected from the table or tables based on one or more conditions.

To see a select statement at work, let's start with the following query that contains only a SELECT and FROM clause:

```
SELECT ticker, price
FROM Equity_Prices
```

This query would return the ticker and price columns for **ALL** rows in the Equity_Prices table. We may instead want to restrict to only rows from the current day, for example, which we could do using the following WHERE clause:

```
SELECT ticker, price
FROM Equity_Prices
WHERE quote_date = 'today'
```

WHERE clauses such as the one above can include simple equality comparisons, but can also be arbitrarily complex. For example, we could use the results from another query in a WHERE clause, in which case that query would be referred to as a subquery.

The next relevant clause in a select statement is an *ORDER BY* clause, which can be used to sort the results of a query by one or more columns.

```
SELECT ticker, shares, mkt_value
FROM positions
WHERE quote_date = 'today'
ORDER BY mkt_value DESC
```

Order by clauses can reference the column name, or can reference the order it appears in the SELECT statement. [12] Specifically, this query will show all positions for a given quote_date, and will display them in order from highest market value to lowest. The default sort order is ascending. The **DESC** keyword will sort the results in descending order.

8.4.3 Retrieving Data from Multiple Tables via JOINs

In most cases, we will need to retrieve data from more than one table in our queries. In order to accomplish this, we can utilize the *JOIN* keyword with the *FROM* clause. JOIN's are a powerful component of SQL programming and enable us to define how we merge the tables in a select statement. As an example, if we have two tables, one for positions and one for prices, then it will be a very common task to want to select data from both table simultaneously. When doing so, we need to know which positions rows match with which rows in the prices table. In this example, we might define the matching criteria to include the date and the ticker that the position and price correspond to. The JOIN clause is where we are able to specify this relationship, as shown in the example query below:

```
SELECT px.ticker, price, shares
FROM prices px
JOIN positions pos ON px.ticker = pos.ticker
AND px.date = pos.date
```

In this query we can see that the JOIN keyword is used to specify how the prices and positions tables are merged in the query and in particular the content after the ON clause defines the JOIN columns. In this example we are specifying that rows with matching tickers and dates in the positions and prices tables will be matched.

Also note that in this query an **alias** is used for the table names in the FROM clause. The aliases are defined as px and pos respectively, and can generally be specified immediately following the table name in the FROM clause. This alias can be created as it is done in the above query, or can be specified with an optional AS keyword. When an alias is specified, that alias can then be used in other parts of the query, such as in the query above where the alias is used in the SELECT clause and to define the JOIN columns. If we had not specified an alias for the first column in the select statement, ticker, then the query would no longer work as there are multiple columns named ticker in the query and without that prefix the interpreter would not know which column to select.

The JOIN used in the above example is referred to as an inner join, meaning that this query will only return rows where the ticker and date exists in both the prices

[12] Along these lines, we could replace mkt_value in the ORDER BY clause with a 3 and the result would be identical.

and positions tables. This means that if a row is only present in the prices table, but has no corresponding positions row, then it will not be included in the query. This is often desired but in some cases we may want to return all rows in the a given table, such as the prices table, even if there is no row in the second table, in this case positions. In SQL we can accomplish this by specifying a different JOIN type, as we explore next.

SQL supports the following main types of JOIN's [13]:

- INNER: Return only rows that exist in both tables.
- LEFT: Return all rows in the left table regardless of whether they exist in the right table.
- RIGHT: Return all rows in the right table regardless of whether they exist in the left table.
- FULL: Return all rows in if they exist in either the left or right tables.

The default JOIN type in SQL is an INNER JOIN, meaning in cases where there is no keyword preceding the JOIN keyword, such as in the example above, an INNER JOIN is performed. If, in the previous example we wanted to return all prices rows whether or not a corresponding positions row existed, we could specify the JOIN as a LEFT JOIN rather than an INNER JOIN. If we were to use a RIGHT JOIN, then all rows in the positions table would be returned, regardless of whether a corresponding prices row exists.

8.4.4 Aggregating Data via a Group by Clause

Lastly, we explore how we can use SQL queries to aggregate data. When doing this, we can utilize the *GROUP BY* clause in a select statement. SQL has a set of aggregate functions that we can perform in our GROUP BY clauses, such as SUM, COUNT, MAX and MIN. The list of columns in the **GROUP BY** clause define how the results will be aggregated. We can choose to group by one or more columns, depending on the specific query we are creating.

To see how aggregate functions in SQL work, consider the following simple example of an aggregate function without a corresponding GROUP BY clause:

```
SELECT SUM(shares)
FROM positions
```

In this case we are able to aggregate the shares column, however, it aggregates the shares for all rows in the table. In many cases, we won't want a single aggregate value but instead will want an aggregate value for a subset of groups, for example by ticker, by date, or by another column. The group by clause enables us to embed this logic in our queries. In particular, if we were to add a GROUP BY clause to the above query, it would then return the sum of the shares column for each group. An example of this can be found in the next query:

[13]Note that the four types of JOIN's listed above are the most commonly used ones. There are also other types of JOIN's, such as CROSS JOIN and SELF JOIN.

```
SELECT ticker, SUM(shares)
FROM positions
GROUP BY ticker
```

In this case we are grouping by ticker, meaning that the query will return a row for each ticker in the positions table. The query will return two columns, the first of which will contain the ticker, and the second of which will contain our aggregate, that is, the sum of the shares for that specific ticker.

If we want to restrict that groups that are returned in a query that uses a GROUP BY clause, we can use the *HAVING* clause. A HAVING clause restricts the groups returned in a query. This contrasts with a WHERE clause which restricts the rows that are used in the underlying calculation, rather than the groups that are displayed after the aggregation is performed.

The following example sample SQL statement shows how we can utilize a HAVING clause to modify the groups that are returned:

```
SELECT ticker, SUM(shares) as shrs
FROM positions
GROUP BY ticker
HAVING shrs > 0
```

In particular, in this example we used the HAVING clause to only show long-only positions, that is, cases where the total positions is greater than zero.

8.5 DIMENSIONALITY REDUCTION

One of the most common tasks for data scientists is to isolate the most important factors, or features in a given dataset. Often, quants or data scientists will also look to isolate and discard features that do not appear significant, but instead are more likely to be noise. This process is known as dimensionality reduction and is something that is done in a variety of applications of data science. For example, in image compression, dimensionality reduction techniques enable us to replicate an image to a high degree of accuracy more efficiently, by only preserving the important features. In finance, we pursue this type of dimensionality reduction not only for efficiency reasons but also for improved stability. In section 8.5.3 we consider a canonical example of this in finance, that is, using dimensionality reduction on a covariance matrix.

8.5.1 Principal Component Analysis

Perhaps the most common dimensionality reduction technique in finance is PCA. The theory behind PCA was covered in chapter 5. Additionally, earlier in this chapter, we also saw that PCA could be used as a form of factor analysis, where we could use it to find a set of the most important statistically chosen factors. In the realm of dimensionality reduction, PCA can be used to remove the least important, or statistically insignificant features, enabling us to obtain a lower dimension representation of the data that is statistically equivalent and preserves its key features.

To see how this can be accomplished, recall from chapter 5 that PCA relies on the following underlying decomposition: [14]

$$\Sigma = X^\top X = CDC^\top \tag{8.2}$$

Further recall that D is a diagonal matrix of eigenvalues, λ_i. In the previous section, when focused on finding the predominant features, we were correspondingly interested in the largest eigenvalues. In this case, we will instead be focused on the smallest eigenvalues that explain the least variance and are likely to be statistically equivalent to zero. We can then replace these small eigenvalues, λ_i with 0 and recover Σ using the updated matrix D to obtain a lower rank with lower dimensionality. In chapter 25 we consider an extension of this approach, Random Matrix Theory, where we determine the threshold for setting an eigenvalue to zero based on its likelihood of being statistically significant and not to be the result of a random matrix.

At heart, PCA is a linear transformation where we decompose the original data on a set of orthogonal, linear factors. Other linear dimensionality reduction techniques, such as Linear Discriminant Analysis [181] also exist and provide an alternative to PCA. Other dimensionality reduction techniques aim to incorporate non-linear features in the data. In the next section we explore one such technique, autoencoders.

8.5.2 Autoencoder

The dynamics and relationships in finance are often quite complex and difficult to observe. Because of this, there is little reason to believe, or assume, that these relationships should be linear. [15] This makes dimensionality techniques that are able to incorporate non-linear relationships inherently appealing. One such technique for doing this in the context of dimensionality reduction is an autoencoder.

In machine learning, an autoencoder is an algorithm based on a neural network where we take a set of input data and then develop an *encoding* scheme that transforms the input data. We then build a corresponding *decoding* scheme that recreates the data from its encoded state, reproducing the original values. Comparing to PCA, we can think of the encoding scheme as the transformation from Σ to $CDC^\top$. Decoding the data, could then analogously be done by going from C and D back to Σ using the appropriate linear transformations. Of course, a unique feature of an autoencoder is the ability to incorporate non-linearity into this transformation.

In the context of dimensionality reduction, an autoencoder works by limiting the encoding layer to a small number of networks, or dimensions. This means that the autoencoder will be forced to encode the input data into a small number of dimensions but will then decode it, or project it back to its original space. In the case of PCA, this process was done by zeroing out one or more eigenvalues. Here, conversely, we limit the dimensional of the network used to transform or encode the data.

In this section, we do not cover the detailed mechanics of neural networks or autoencoders. More background, however, on neural networks can be found in chapter 28, and autoencoders are covered in more detail in 29. This includes an example of using autoencoder as an alternative to PCA to reduce dimensionality of a covariance matrix.

[14] Which is commonly known as eigenvalue decomposition or singular value decomposition.

[15] Aside from computational convenience.

8.5.3 Dimensionality Reduction in Practice: Reducing the Dimensionality of a Covariance Matrix Using PCA

In this section we leverage an example in the coding repository to show how dimensionality reduction, done via PCA can be used in the presence of potentially correlated factors and multi-colinearity. In doing this, we examine the predictive power of country ETFs return as predictors for the global market [16]. To do this, we first use a regression technique where lagged returns of the country ETFs are regressed against subsequent ACWI returns. This leads to the following table of parameters:

Variable	Coefficient	t-Statistic	p-Value
Intercept	0.0005	1.7	0.084
EPI	-0.11	-3.9	0.000
EWC	-0.08	-1.9	0.061
EWG	0.04	1.0	0.314
EWH	-0.06	-1.5	0.140
EWJ	-0.04	-0.8	0.401
EWU	0.07	1.4	0.150
EWW	0.06	2.4	0.018
EWY	0.07	2.2	0.026
EWZ	-0.05	-3.1	0.002
FXI	0.03	1.2	0.232
SPY	-0.11	-2.4	0.017

In aggregate we also find that the adjusted R^2 is 0.038 showing that even though certain features may be statistically significant, the overarching model fit, as expected, is fairly poor. We do see some potentially statistically significant relationships, with some coefficients being positive and other being negative. It is possible that these mixed signs in the results are an artificant of the high correlation in the underlying indices. In the following chart, we compare our model fit to realized returns [17]:

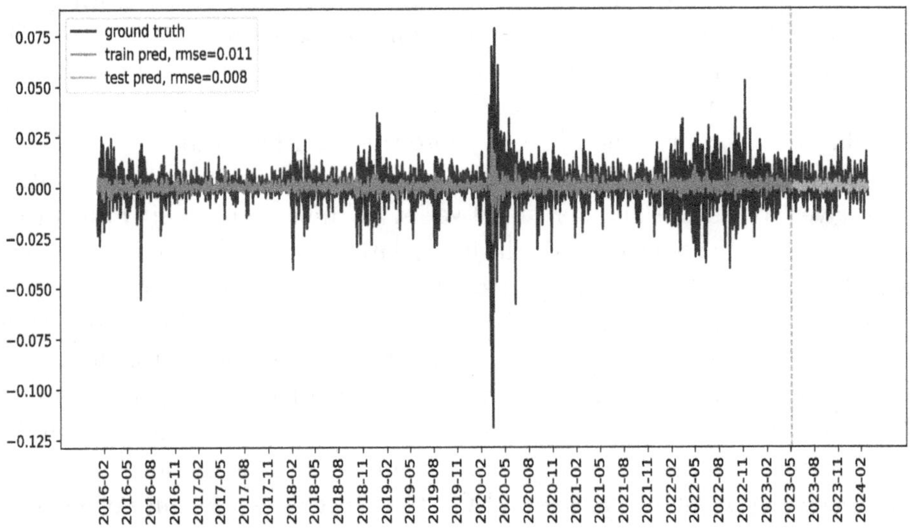

[16] Proxied via the All Country World Index (ACWI).

[17] Both using a training and test dataset.

Here we can see the poor fit clearly, as the model explains a small fraction of the overall variance.

Next, we attempt to repeat this analysis with reduced dimensionality. This has the benefit of avoid potential highly correlated explanatory variables. To do this dimensionality reduction, we leverage PCA. We could also leverage a non-linear dimensionality reduction technique, such as autoencoder, and we explore that approach in chapter 29. To do this, we perform PCA of country ETF returns, which helps us to identify the most significant factors. In this example, we found that a single principal component explained the majority of the variance in country ETF returns. This means that we can justifiably replace our prior model with each country ETF as a regressor with a simpler model with the first PC [18] as a regressor. This leads to the following regression result:

Variable	Coefficient	t-Statistic	p-Value
Intercept	0.0003	1.3	0.193
PC1	0.04	6.0	0.000

While in this case the adjusted R^2 is 0.019 indicative of a poor fit. However, while the fit is still poor, the parameter set is smaller and thus more interpretable. Further, because the principal components are by definition orthogonal, as we learned in section 4.8, this form of the regression equation will be more stable and less prone to multicolinearity issues.

In the following chart, we show the realized returns vs. our updated PCA based prediction model results:

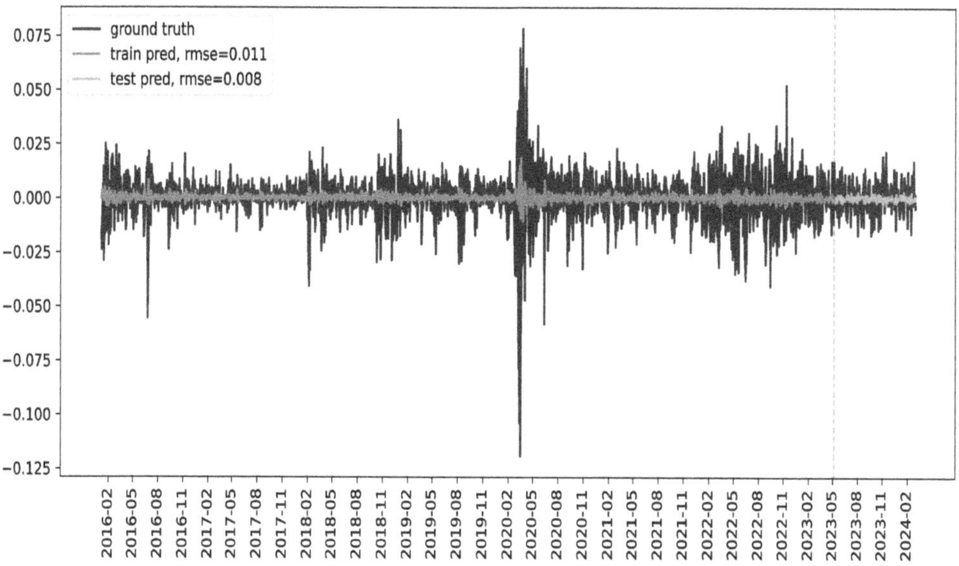

While the fit is still poor, as one might expect given our simple prediction model, one way to frame this is that we were able to achieve similar results with lower dimensionality and a more stable formulation of the problem.

[18]Or first few PCs if one desires to capture more variance.

8.6 PYTHON TOOLS FOR BIG DATA

8.6.1 Working with Large Datasets

In finance, we are often forced to work with datasets with highly degrees of complexity. Sometimes this is because of their inherent size, such as when we are working with high frequency data such as order book data or the underlying loan level data for mortgage backed securities. In other instances, the structure of the data is what makes it complex, and larger than traditional datasets. Typically, working with this type of large dataset requires different tools so that it can be done in an efficient way. As detailed in chapter 8, DataFrames are immensely valuable tools for working with, validating, and transforming data. As the size and complexity of the dataset increases, however, these standard tools will become slower and often unable to store the entire dataset. As a result, when confronted with these types of problems we need to rely on Python packages, such as Spark [17], that are specially designed to handle large datasets.

8.6.2 Memory & Large Datasets

One of the main issues that we encounter as datasets increase in size is that the memory that we need to allocate correspondingly increases, and at some point, we will reach a point where we run out of memory to allocate. To overcome this, we can use tools that use memory more efficiently, or we can partition the memory over multiple processors. The latter is referred to as parallel computing [117], and is detailed in the next section.

8.6.3 Parallel Computing

One way to increase the efficiency of working with large datasets is to split up the computational load across different processors. The idea behind parallel computing is that we distribute the load so that multiple tasks can be computed at the same time. When doing this, the load must be distributed in a sensible way such that we can then aggregate across processes to perform the final calculation. For example, consider the case of a Monte Carlo simulation. The standard approach is to create a single for loop over a desired number of draws. However, to increase efficiency we could alternatively process the draws across multiple processors, with a set of draws happening in parallel. This would result in increased efficiency as less draws would be required to be computed per processor and we would benefit from multiple paths being simulated at the same time. Similar efficiency gains can be obtained by breaking up computations on a large datasets and handling them in parallel.

Python has a set of built in tools for parallel computing, including the ***multiprocessing*** package. In the context of working with large datasets, parallel computing can help ensure that transforms and calculations on the dataset are efficient. In the coding repository, we provide an example for working with these multiprocessor tools in Python, showing how to create a new process and manage a set of tasks across processes. In the next section, we explore a distributed computation framework that naturally embeds parallel computing and in doing so enables us to work with large datasets in a highly efficient manner.

8.6.4 Spark & Hadoop

Apache Hadoop is a framework for distributed processing that is specifically designed for large datasets that are unable to be processed via standard tools. It provides efficiency by scaling across machines and processors, and does so with high availability.

Within the Hadoop ecosystem is Apache Spark [17], which provides a set of tools for working with big data. PySpark is, importantly, a Python library that makes this framework accessible in Python. PySpark includes within it a DataFrame object that mirrors a standard DataFrame except that it is designed to work with large datasets and process these datasets across multiple processors. This PySpark DataFrame object will be of utmost importance to quants when they are working with large structured datasets that cannot be handled by standard tools. PySpark also enables us to inherit the highly desirable functionality of the pandas library when working with big data through the inclusion of a pandas API within PySpark. More information on Apache Spark, and the PySpark API can be found in [18].

8.7 OUTLIER DETECTION

The last topic that we cover related to data cleaning, data manipulation and data science is outlier detection. In many classes of models, such as linear regression models, we will find that outliers will have a disproportionate impact on our solution. As a result, as part of a data validation process, it is best practice to analyze data for outliers, make sure any outliers are legitimate, and determine the optimal method of handling them. Of course, not all outliers are bad data points or can be omitted, and in fact in many cases they arise naturally due to market conditions and may represent important market tail events. But before we feed our data into a model we want to have confidence that the outliers present in our data arrived by legitimate means and are not an artifact of a bad data source. This is especially important given the large impact outliers can have on our results.

There are many different ways that we can search for outliers, from just visualizing our data to see if any anomalies present themselves, to using machine learning techniques to identify observations that look unusual.

8.7.1 Single vs. Multi-Variate Outlier Detection

One key differentiating feature when looking for outliers is whether we are scanning for outliers along a single or multiple dimensions. In the first case, we might be looking at the return series for a given foreign exchange rate, say EURUSD, and looking for returns that appears suspicious. In the second case, we may want to incorporate returns of other FX rates to help inform our understanding of whether the EURUSD return is an outlier. Alternatively, we may use this second approach to determine if all currency returns for a given day are anomalous. Depending on which of the above is our objective, we may end up pursuing different techniques/methodologies.

8.7.2 Plotting

The first tool in our toolkit when looking for outliers is to plot or visualize the data. This will be far more effective if we are trying to do a univariate outlier search than if we have many dimensions, as the data quickly becomes difficult to visualize. In the one-dimensional case, plotting a series of returns or a price series can often lead us to catch anomalies. Does the price series suddenly jump? If so, maybe that reflects an unaccounted for stock split. Does the return series contain any returns of a different order of magnitude than the rest? These are the types of questions plotting can answer for us, and as a result plotting the data is often a first step in identifying outliers.

8.7.3 Standard Deviation

The next tool we might look to in our search for outliers is to examine the data beyond a certain standard deviation threshold. For example, we might choose to isolate any data that is beyond a three standard deviation event and confirm that these data don't appear to be errors. One way to do this would be validate these isolated points against some broader market data to analyze whether such an outlier would be expected given that market context. For example, if an ETF designed to replicate an equity market displayed a -5 standard deviation return on a given day, we could see if a corresponding large standard deviation move was present in other equity instruments. If other equities indices had positive returns, this would increase our confidence that this was a bad data point.

It should be emphasized that it is always good practice to check stationarity before performing this type of standard deviation analysis. For non-stationary time series like a price process, the growing trend makes the price incomparable over time, making the standard deviation uninformative. On the other hand, a stationary time series, like a return process, would be well-prevented from such issues and that in these circumstances this analysis can be insightful.

8.7.4 Density Analysis

More generally, we can isolate outliers and obtain clues as to whether they are legitimate or may indicate data issues by analyzing the empirical density. The rationale for this is similar to that of looking at the standard deviation, however, it is reasonable to expect that their might also be clues in the higher moments, such as skewness and kurtosis as well.

The first step in this type of density analysis is to simply plot the empirical PDF, or a histogram of the data. We can then look for observations that appear as outliers in the histogram, if any exist.

Further, if we have a prior about what the distribution of our data should closely resemble, then we can use a QQ plot to see how well the actual data lines up with that distribution. A QQ plot compares the CDF of our data against a benchmark distribution, which is often a Normal distribution [19].

[19] There are several other similar tools, such as a Kolmogorov-Smirnov Test.

Additionally, the skewness and kurtosis of our dataset can help identify the presence of outliers. Both of these measures are likely to be heavily swayed by outliers, so if we find our empirical distribution has high amounts of kurtosis, that might be a clue that we should look and see if that is being driven by one or more outliers.

8.7.5 Distance from K Nearest Neighbor

Another approach for detecting outliers leverages a Machine Learning technique that we introduced in the last section, K Nearest Neighbor. In particular, one characterization of outliers is that they are far removed from the data. Said differently, they will be very far away from their neighboring data points. As a result, we can use the K Nearest Neighbor to find a set of one or more close data points, and measure the distance between the point and its neighbors. Points with large distances from their neighbors are then natural outlier candidates, and should be examined more thoroughly.

To accomplish this, we start by iterating through each data point and running the KNN algorithm to find its neighboring points. Next, we can use a metric, such as Euclidean distance [20] to measure the distance between the original data points and its neighbor.

The following equation can be used to find the distance between each point x and a single neighbor n:

$$D(x,n) = \frac{1}{d}\sum_{i=1}^{d}(x_i - n_i)^2 \tag{8.3}$$

Where d is the number of dimensions. We can then average this over the desired N closest nearest neighbors to obtain an aggregate distance matrix:

$$D(x,n) = \frac{1}{M}\sum_{j=1}^{N}\left(\frac{1}{d}\sum_{i=1}^{d}(x_i - n_{i,j})^2\right) \tag{8.4}$$

Values that have larger distance metric values in this equation would then be identified as outliers and would be investigated in further detail.

8.7.6 Outlier Detection in Practice: Identifying Anomalies in ETF Returns

In this section, we leverage the density analysis portion of the coding example presented in the repository to show how we might look for outliers in the returns of the technology ETF XLK. We can do this by looking at the histogram-density plot or the QQ plot, each of which provides a comparison between the empirical distribution and a benchmark. In this case, we assume our benchmark to be a standard normal distribution. In practice, however, this may not be a realistic benchmark, and the reader is encouraged to think through how they would choose an optimal benchmark distribution and why stock returns may or may not follow a normal distribution. This analysis leads to the following empirical density and QQ plot, respectively:

[20] Which is simply the sum of the squared distance.

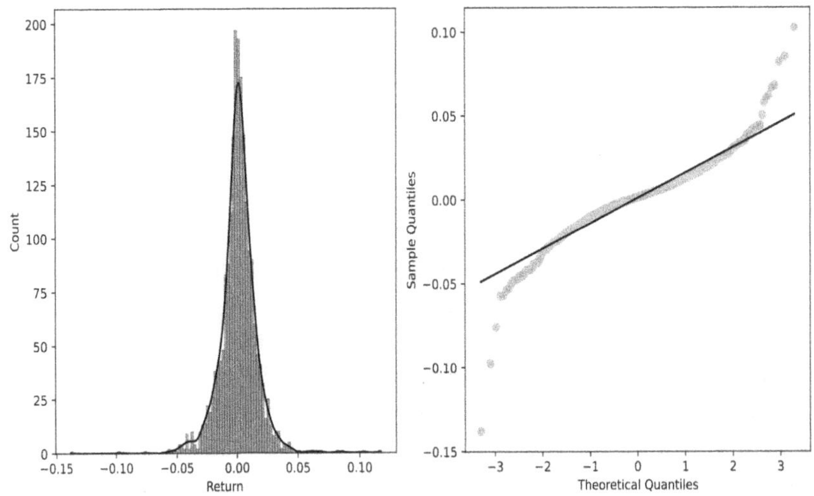

In the density plot, we see the histogram is biased from the benchmark distribution. In the QQ plot, the dots don't fully align with the reference straight line. Instead, both plots show potential evidence of outliers. We can double check this by looking at the skewness and the kurtosis of the empirical distribution, which are −0.21 and 9.21 respectively. Typically, the skewness of a "healthy" distribution should be around zero, and the kurtosis should be around 3. Obviously, the empirical distribution deviates from the benchmark, indicating potential issues that we may want to explore further. An alternate explanation, of course, is that the benchmark distribution was poorly chosen and is not representative of the distribution we are modeling, which would also lead to the behavior we see here. Fatter tails are a commonly observed market phenomenon, which we return to again in chapter 13, so this may indeed the case. In either case, however, this analysis may lead us to examine the returns at the tails of the distribution in order to validate that they don't emanate from a data error.

As detailed in the last section, another way to search for outliers in the data is to leverage a K-Nearest Neighbor algorithm. To do this, as discussed in the last section, we can measure the distance between points and their nearest neighbors, with large distances being potential outliers. In the chart below, we leverage KNN to predict the value of each piece of data using its nearest neighbors, providing a similar analysis. As we can see, the large deviations between the true data and the predicted returns, are candidates to be outliers:

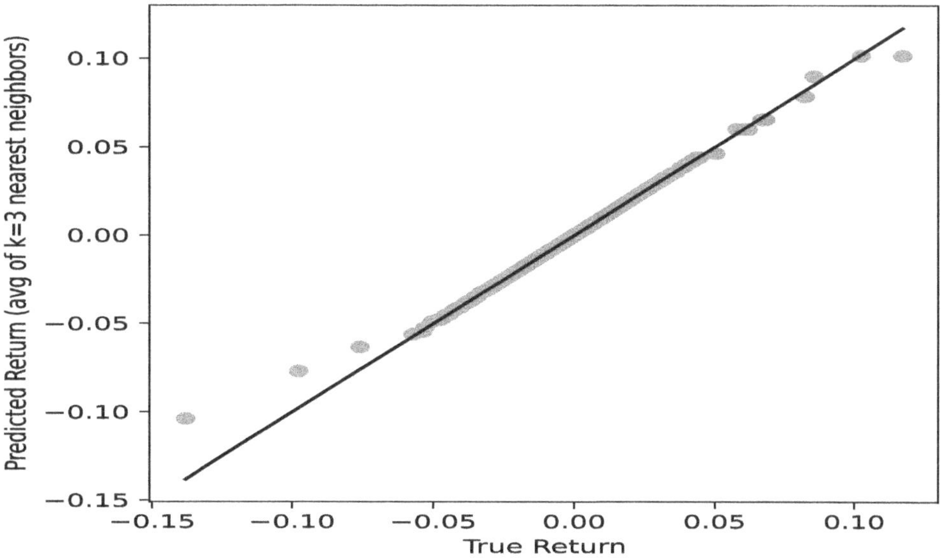
Predicted Return (avg of k=3 nearest neighbors)
0.10
0.05
0.00
−0.05
−0.10
−0.15
−0.15
−0.10
−0.05
0.00
0.05
0.10
True Return

CHAPTER 9

Model Validation

A critical component of being a successful quant is being able to implement models correctly and robustly. This involves being able to convince ourselves, and others, that the models we create have been implemented as intended. In some cases this may not be a heavy lift. As an example, when the model produces known values, such as in the case of the Black-Scholes model, then we can simply compare our model results to the known result. As models get more complex, however, this process becomes increasingly challenging. In particular, as we start to incorporate more complex stochastic processes or work with exotic options, we will not know what the result should be, making it harder to verify that our implementation is correct.

Model validation is aimed at building a process to verify our model and ensure it works as intended in different circumstances. It is also designed to provide us a baseline of what we expect from the model. We can then monitor this and catch any unintended deviations from this baseline. Further, it is designed to identify model assumptions and boundary cases that may require special consideration. We should also use model validation as an opportunity to help us understand a model, and its strengths and weaknesses, so that we can better understand when we should and should not apply the model in practice.

Model validation processes may vary greatly depending on the size and type of organization. Larger banks, for example, tend to have entire independent teams dedicated to reviewing new models, whereas smaller institutions may have less formal validation processes where quants must be responsible for reviewing their own code. Regardless, model validation is one of the most important aspects of building models and being a successful quant.

9.1 WHY IS MODEL VALIDATION SO IMPORTANT?

Depending on the financial institution a quant works in, and their level of experience, quant models may be used for different purposes. For example, some buy-side institutions are entirely quantitatively driven, meaning the models are directly responsible for investment decisions. In other cases, models may be used to determine how much risk is allocated to different portfolio managers, or how much risk a bank can have on a balance sheet.

DOI: 10.1201/9781003530626-11

In cases such as these, models will be directly connected to decisions that lead to alpha or revenue for a firm. If those decisions are based on incorrect models, improper data, or unrealistic assumptions, then it can lead to sub-optimal decision making. Additionally, in practice models that add value tend to have a high level of complexity. This puts a lot of pressure on the models and the quants that create them. In order for models to perform in these circumstances, they need to be properly stress tested and properly validated. This isn't just about finding the bugs in the code, although that is obviously critical. A model that runs with bad data or is used in inappropriate circumstances is as dangerous as a model that has bugs.

Ensuring that a model is ready for production use requires a robust set of guidelines and practices to uncover assumptions, bugs and data requirements. A set of unit tests will be a critical part of this process, as will an independent coding review designed to find any hidden risks in the code. It should also serve as an opportunity to document the model, how it was implemented, what assumptions it is based on, and any limitations.

9.2 HOW DO WE ENSURE OUR MODELS ARE CORRECT?

Model validation is the process of checking that our code is correct and that we have implemented the model in the desired way. As our models become more complex, so too does validating its results. In some models the true price is not known, making it much harder to prove their implementation is correct. Against this backdrop, it is important to have a set of tools to accomplish this.

Of utmost importance in this process is identifying bugs in our model code. However, a proper validation process goes beyond this. This involves testing the underlying building blocks with a set of unit tests to ensure each piece is implemented properly.

It also involves being able to simplify complex models so that they are more tractable and have known solutions, and testing them in these circumstances. As an example, if we are working with an SDE with no known solution, such as a stochastic volatility model, we can attempt to identify a set of parameters such that the SDE simplifies to a Black-Scholes or Bachelier model where the solution is known.

Another potential tool for validating a model is to compare the model results using a different and independently implemented technique. For example, if we are working with the Black-Scholes model and European options, we could compare the results obtained using the Black-Scholes formula to a simulation based approach using the Black-Scholes SDE. If the two methods yield very similar results, it does not necessarily mean that the model has been implemented correctly, as both could suffer from the same bug, however, it should give us confidence that the model has at least been implemented consistently.

Lastly, if a third party provider of the model, or a similar model is available, then we could use this as an additional check on our model results. As models become more customized, or more complex, third party tools will be less prevalent and we won't be able to rely as much on this as a source of validation.

9.3 COMPONENTS OF A MODEL VALIDATION PROCESS

A model validation process consists of documentation, an independent code review, a set of unit tests and a production model change process. Each of these components are critical to ensuring the robustness of initial model implementations and also that future changes are incorporated in a manner that poses minimal risk. In this section we describe what each of these pieces should entail.

9.3.1 Model Documentation

Whenever we complete a model, we should create a document explaining the underlying financial theory and corresponding code. The document should summarize the prevailing theory and relevant literature, as well as what the model was designed to do. It should also describe the algorithm that is used to fit the model to market data.

It should also discuss the set of model parameters that need to be passed to the model and any expectations regarding their values. As an example, a correlation parameter should be expected to be within -1 and 1, and a volatility parameter would be expected to be positive.

Further, our documentation should include an analysis of any data that is required to calibrate or run the model, and any assumptions related to that set of data. For example if we are calibrated an options surface, are we assuming that the data passed in is arbitrage free? Alternatively, if we are performing analysis on stock prices, are we assuming that the data has been adjusted for splits and corporate actions before being passed to the model?

Lastly, our model documentation should include a summary of both the model's theoretical and practical assumptions. Some models may assume a lack of transactions costs, or an ability to trade continuously, for example. In some cases, these assumptions may not be problematic, however, in other cases, such as back-testing a high frequency trading application, this assumption would significantly inflate the results. Uncovering and analyzing the effect of these types of assumptions is an important part of model validation. This analysis should then be used to determine the circumstances in which it is appropriate to apply this model, and the circumstances where another model would be preferred.

Model documentation should be a continuous process that is updated whenever a new version of the model is released (i.e. a model enhancement). Catalogues should be maintained that keep version histories of this documentation as well, and when production switches occurred.

9.3.2 Code Review

The next component of a model validation process should be an independent coding review. The core part of a coding review is simply looking through the code and making sure the intent is clear and as desired. Part of this process should be checking any formulas within the model against a relevant reference. Additionally, a coding review should include running different parts of the code and verifying results of the building blocks, and analyzing the input data to try to identify any hidden assumptions about the required inputs. In addition to searching for bugs, the coding review can serve as a way to add tests of the model, and to ensure the model parameters and assumptions are clear and explicit.

9.3.3 Unit Tests

The process of creating a series of tests for our model that help ensure that it has been implemented correctly is referred to unit testing. This process is designed to validate our model in a vast array of situations, including simplified and special cases, and is also meant to test the parts of the code as well as the whole. As we will see in later chapters, when working with complex financial models, it can be difficult to prove that our implementation is correct, because we don't know how the model should perform. Unit tests are an important tool to help us gain confidence that our model matches the theoretical assumptions of the model, and has been coded correctly.

The unit tests that we build should be designed to span a great deal of possible outcomes and market conditions, including both the most common and unexpected situations. For example, if we are working on a model that calibrates a volatility surface, we should make sure it can handle cases of extreme call and put skew, as well as cases where the term structure of volatility is upward sloping or downward sloping. If the model doesn't perform well in one of these situations, then unit testing can help us uncover this, and avoid using the model in practice in these conditions.

Python provides a convenient framework for unit testing via the **unittest** module. This module contains functions that make the creation of unit tests seamless and enable quants to easily catch deviations from their baseline. In particular, the module contains an **assertEqual** function which enables us to validate the results of parts of our model calculation. If the model deviates from this expected result then an exception will be triggered so that the quant knows there is an issue. In the coding repository, we provide an example of how the unittest module can be leveraged to build a series of unit tests for our code.

Unit tests are at the core of the model validation process, and quants should build a set of unit tests on their model that stress it from different angles. For example, unit tests should be created on the underlying building blocks, in addition to the overarching model result. Unit tests can also be used to ensure the model is consistent in certain known situations. For example, we might create a unittest that assert's that the value of an out-of-the-money option at expiry is zero. Should this situation ever be violated, it would indicate a problem in our model that we would need to resolve.

Additionally, when building unit tests we should try to identify special sets of model parameters that simplify the model to a known, more tractable case. As an example, if we are working with a stochastic volatility model, can we find a set of parameters that sets the volatility process to be a constant so that it simplifies to a Black-Scholes model? If that's the case, then this is a natural unit test to include as we can compare to the closed form solution for Black-Scholes.

We should also consider any special/boundary cases that arise in the model to ensure that the model calculation works in these situations. The purpose of this is two-fold. First, it can help us gain additional intuition of the model by understanding its behavior in these limiting cases while also making sure it is consistent with the theory. As an example, we might compute the price of an option with zero and infinite levels of implied volatility. It can also help us uncover any numerical issues in the coding implementation when the parameters are at or near a boundary. A common example of this would be when a parameter that shows up in the denominator of a formula converges to zero. In this case, we might catch that a model simplifies to zero over zero in this case but that this can be avoided if we apply L'Hopital's Rule [237].

Once we have built a comprehensive set of unit tests, and the results of these unit tests have been validated, the results of the these calculations can then be stored and re-run regularly in an attempt to catch any deviations from these baseline results. In particular, they can become a natural part of a model change process. That is, if we modify the model and see any deviations from our baseline tests, it will force us to determine if they are intentional as part of the model change, or due to a coding error. In the coding repository, we provide an example of model validation in the case of simulating from a Black-Scholes model.

9.3.4 Production Model Change Process

A model change process is a set of protocols that are required in order to modify production code after an original version has been agreed upon. Many technology and financial firms have strict model change processes based on independent coding reviews, replication of model results by multiple users and an explanation of all changes to unit tests. This process may vary by organization, however, is critically important in eliminating potential bugs and other issues. Git and other version control products can be used to help with model change processes by allowing us to explain our code change and assign users to review the changes before they may be incorporated.

9.4 GOALS OF MODEL VALIDATION

The main goals of a model validation process are to validate model implementation, understand the model's strengths and weaknesses, and uncover the model's assumptions (both theoretically and in practice). If done properly, this can help us avoid improper model use, such as invalid model parameters or bad data. It can also help us to uncover any software bugs that may exist. Finally, it can help raise any limitations of the model to help us with model selection.

9.4.1 Validating Model Implementation

The primary goal of a model valid process is to ensure that the model is implemented correctly. This includes validating how a model was implemented by checking the code and attempting to identify divergences between the theory in the model documentation and the implementation. As mentioned, this process will rely on understanding the code, via a coding review, and also on the creation of a set of unit tests. It will also include validating the model under simplified parameter sets and assumptions, and checking relevant boundary cases. It may also include verification against a third party or independently created model, however, as our model's become more complex these independent implementation become harder to find. We should also verify how the model behaves as a function of its different parameters. For example, the Black-Scholes models behavior should be consistent, and smooth, as we change the input volatility. If not, we likely have an error in our code. Lastly we run the model through a wide array of parameterizations and check its behavior, trying to identify anomalies and cases where the results look suspect.

9.4.2 Understanding Model Strengths and Weaknesses

In addition to proving that our model has been implemented correctly, and in line with the corresponding theory, another key goal of model validation is to analyze the model's strengths and weaknesses. When doing this we should keep in mind that no model is perfect. All models, regardless of how complex they are, rely on simplifying assumptions about markets and consequently, each model has limitations.

As an example, the Black-Scholes Model has many documented uses but also a long list of limitations that a quant or trader must be aware of when using it. We need to make judgments about the appropriateness of these assumptions in different use cases. This judgment should also incorporate information on how practical a model is. The Black-Scholes model can be replaced with a model that makes more realistic assumptions, however, this introduces other challenges, such as increased complexity and slower calibration to market data. Some more complex models, like Heston or SABR may do a better job of matching the volatility skews that we observe in markets, but may still struggle with the skew for shorted dated options. One of the goals of model validation is to uncover these types of features in each model by stressing the inputs and considering many market situations. Knowledge of this can help us know when to use each model, and ensure optimal model selection.

9.4.3 Identifying Model Assumptions

Another integral component of model validation is that it can help to understand the assumptions that our model makes. Further, it can ensure that the assumptions of our implemented model match its theoretical assumptions. As an example, all models that rely on Brownian Motions as the driver of uncertainty make certain assumptions about asset prices. Similar assumptions are made in models that rely on linear regression to explain the relationship between different variables. In many cases, further assumptions are made about transaction costs, hedging, arbitrage and continuity of trading. Depending on the model application, these assumptions may be entirely reasonable or not, and model validation can help us make that delineation.

In addition to knowing the theoretical assumptions of the model, we also want to make sure that our implementation does not make additional assumptions beyond what is required. In some cases we may only be able to approximate the assumptions in the theoretical model. In these cases, it is of critical importance to ensure that the assumptions in the theoretical model approximate, or closely resemble, the assumptions being implemented. One example of this is when using simulation, where we have to discretize the problem and in doing so are relying on an approximation.

9.5 TRADEOFF BETWEEN REALISTIC ASSUMPTIONS AND PARSIMONY IN MODELS

One prevailing theme throughout the model validation process is that as models become more complex they become harder to validate. In many cases, knowing that they are right will be no small task as the true answer is unknown. On the flip side, models that are simple and easily verified often add minimal value as they can be easily replicated. This presents an overarching tradeoff that quants face when choosing between models. If a simple model is chosen, it is less likely to make realistic assumptions about the market. It is also less likely to be valuable to have an internal implementation of the model. Conversely, bringing in a more complex model which makes more realistic assumptions about the world requires more overhead. Part of this overhead is the additional effort to validate and prove its effectiveness. These models are also more likely to have more parameters and be more prone to over-fitting and will generally be harder to calibrate to a set of market data. Given this, it is advisable to lean toward simple parsimonious models and add only the level of complexity warranted by the particular application.

III

Options Modeling

CHAPTER 10

Stochastic Models

IN this chapter we explore the most common stochastic models that are used in quant finance applications. We begin by reviewing the simplest class of models including the Black-Scholes model/Geometric Brownian Motion and the Bachelier/Normal model. These models are ubiquitous in quant finance in spite of their relative simplicity and serve as quoting conventions in many markets. We then turn our attention to different approaches to modeling volatility skew, starting with stochastic volatility, then proceeding to jump-diffusion models before finally turning to local volatility models. The goal of this chapter is to orient the reader to the features and limitations of these models and to highlight how to work with the models in practice.

As a note to the reader, in this chapter, as we introduce new models we often introduce the characteristic function associated with the model. Characteristic functions will turn out to be a key building block for efficient pricing via Fast Fourier Transform, as we explore in chapter 14. More background on characteristic functions and their importance in solving pricing problems is provided in that chapter as well.

10.1 SIMPLE MODELS

10.1.1 Black-Scholes Model

Perhaps the most famous and commonly used stochastic model in finance is the so-called Black-Scholes model. In this model, we assume that an asset price follows the dynamics below:

$$dS_t = rS_t\,dt + \sigma S_t\,dW \tag{10.1}$$

These dynamics are often referred to as Geometric Brownian motion, and it can be shown that specifying this model is equivalent to assuming that prices are lognormally distributed, and that returns are subsequently normally distributed. The model is characterized by a single parameter, σ which defines the volatility of the asset.

Importantly, volatility in the model is a function of the current asset price, and the σ parameter is interpreted as a relative or percentage volatility. As the price process increases, so will the magnitude of the dW term. That is, the volatility of the price of the underlying asset depends on the level of the underlying asset. The

DOI: 10.1201/9781003530626-13

volatility of the return of the underlying asset, $\frac{dS_t}{S_t}$, however, is constant. This is a defining feature of the Black-Scholes model.

This model is perhaps still the most fundamental model in all of quant finance, and quants will find themselves constantly working with these dynamics. Even in the case where more complex models are used, these models are generally translated back to a Black-Scholes model in order to gain intuition and present results to market participants. Consequently, both technical and conceptual mastery of this model is of paramount importance for readers aspiring to be quants.

Under Black-Scholes dynamics, European option prices can be shown to have closed form solutions that are commonly known as the Black-Scholes formula.

$$C_0 = \Phi(d_1)S_0 - \Phi(d_2)Ke^{-rT} \tag{10.2}$$

$$P_0 = \Phi(-d_2)Ke^{-rT} - \Phi(-d_1)S_0 \tag{10.3}$$

Here, S_0 is the current asset price, K is the strike, r is risk-free rate and T is the option expiry time. Further, Φ is the CDF of the standard normal distribution d_1 and d_2 are defined as:

$$\begin{aligned} d_1 &= \frac{1}{\sigma\sqrt{T}}\left(\ln\frac{S_0}{K} + \left(r + \frac{\sigma^2}{2}\right)T\right), \\ d_2 &= d_1 - \sigma\sqrt{T} \end{aligned} \tag{10.4}$$

There are multiple ways to derive the Black-Scholes formula from the underlying dynamics. Perhaps the simplest approach is to leverage our knowledge of the distribution of asset prices and log returns in the model to calculate the appropriate integrals from first principles. To see this, recall that the price of a call option can be written as:

$$C_0 = \exp(-rT)\int_{-\infty}^{\infty}(S-K)^{+}\phi(S_T)dS_T \tag{10.5}$$

As the payoff is equal to 0 when $S < K$ we can replace the integral bounds and remove the max function in the payoff and then separate into two terms:

$$C_0 = \exp(-rT)\left[\int_K^{\infty} S\phi(S_T)dS_T - K\int_K^{\infty}\phi(S_T)dS_T\right] \tag{10.6}$$

The second term is closely related to the CDF of S_T. Further, given the dynamics we specified, we know that S_T is log-normally distributed. Therefore $log(S_T)$ is normally distributed. We can use this fact to perform a change of variables to re-write the second term as an evaluation of the CDF of a standard normal distribution.

For the first term, we can use a similar change of variables and we will see that the problem becomes that of: $\mathbb{E}\left(e^{\log(S_T)}\right)$, where again $\log(S_T)$ is distributed normally. Therefore, this integral is equivalent to calculating the moment-generating function of a normal distribution, which has a known solution[1]. Performing the appropriate change of variables and re-arranging leads to the so-called Black-Scholes formula in equation (10.2).

[1]Alternatively, this integral can be by completing without knowledge of the moment-generating function by completing the square.

10.1.2 Black-Scholes Model in Practice: Are Equity Returns Log-Normally Distributed?

While the Black-Scholes model has some theoretically appealing properties, such as its analytical tractability and intuitive motivating replication argument, it is natural to question whether asset prices in practice are actually log-normally distributed. In this example we see how this assumptions holds up in the context of a popular equity market, the S&P 500. In the following chart, we show a QQ plot of log returns of the S&P 500 to that of the normal distribution, as is implied by the Black-Scholes model:

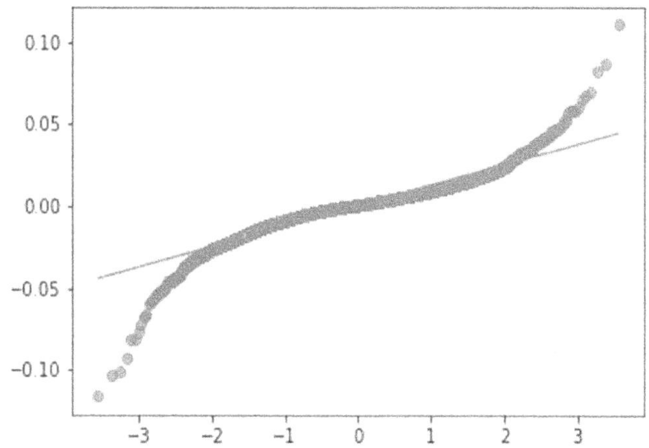

In this chart we can clearly see that equity returns do not appear to be normally distributed. Instead we observe much fatter tails in the empirical distribution of equity returns than in the normal distribution. Thus, this is an argument against leveraging the Black-Scholes model, and instead working to build a model that can incorporate this feature of fatter tails. In fact, we will see in many stochastic volatility and jump process models that we will be better able to calibrate the kurtosis in the distribution, leading to much more realistic dynamics.

10.1.3 Implied Volatility Surfaces in Practice: Equity Options

In the previous section we performed a test of the Black-Scholes model against actual movements in the underlying asset, for the S&P. Said differently, we compared a potential distribution for modeling options prices in the risk-neutral measure, to movements in the physical measure, where risk preferences are introduced. As such, deviations in distributions could theoretically be explained by differing risk preferences in the measure. Therefore, in this section we look to make a more apples to apples comparison. That is, we compare the theoretical Black-Scholes model to a set of observed option prices for the S&P. If the Black-Scholes model were to properly reflect the dynamics implied in the options prices, then we would expect a single implied volatility parameter would enable us to fit all options at different strikes[2]. In the following chart, however, we show the implied volatility required to match the option prices at different strikes:

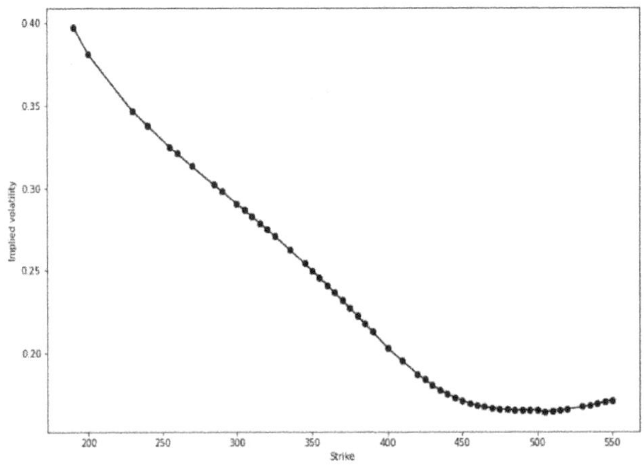

Fundamentally, we can see that as the strike varies so too does the implied volatility. In particular, it appears that lower strike options tend to have higher levels of implied volatility. This fact that the implied volatility varies by strike is clearly unsatisfactory, as it means we need to rely on a different distribution for each strike. More generally, we will find that this phenomenon often exists at different expiries as well. Using different distributions for movements in the same underlying asset is dubious at best, and is a sign that the Black-Scholes model doesn't accurately reflect the dynamics we observe in traded option prices.

Putting aside the theoretical issues with different implied volatilities at different strikes and expiries, this phenomenon gives rise to a so-called implied volatility smile and implied volatility surface, where we plot the implied volatility by strike, as is done in the above chart, or as a three-dimensional chart against both strike and expiry. This implied volatility surface is then a convenient representation for us to compare options of differing characteristics.

As such, we can look to understand the theory behind the behavior in these implied volatility surfaces. For example, in the above chart we can see that a low strike put option has a higher implied volatility than a high strike call option. This has a natural risk premia explanation, due to the demand for low strike puts as hedging vehicles. More generally, the reader is encouraged to try to understand what is driving the behavior in these implied volatility surfaces. Implied volatility surfaces, and how they vary in different markets, will also be a focal point throughout the text.

At a high level, we might conclude that the Black-Scholes model does a poor job of fitting the implied volatility surfaces that we observe in practice. As such, in the rest of the section we will look at the most common approaches to trying to explain this

[2]And different expiries.

behavior, most notably stochastic volatility models, jump process and local volatility models.

10.1.4 Bachelier Model

In the Bachelier model, or normal model, we assume asset prices are normally, rather than log-normally, distributed [29]. It is a natural alternative to Black-Scholes model that is simple and analytically tractable. The dynamics for the Bachelier model can be described as:

$$dS_t = rS_t\,dt + \sigma\,dW \tag{10.7}$$

We can see that the SDE is similar to the dynamics of Geometric Brownian Motion, although the dW term has been altered. In particular, the dW term no longer depends on S_t, which leads to different model behavior. The Bachelier model is also based on a single underlying parameter, σ. In contrast to the Black-Scholes model, volatility in the Bachelier model is not a function of the current asset price. This means that magnitude of the dW term is independent of the level of the underlying asset. This leads to a different interpretation of the σ parameter. Instead of being a relative to an asset price it is now an absolute level of volatility that is often referred to as basis point volatility.

Another differentiating feature of the Bachelier model is that it allows for the possibility of negative asset values. This is because the distribution of asset prices is normal, rather than log-normal. We can also see this intuitively by examining the SDE. In the case of the Black-Scholes model, when the asset price decreases, the volatility in the asset is damped preventing it from being negative. There is no such phenomenon here. When asset prices are close to zero, the magnitude of the changes does not adapt, enabling it to pass zero with positive probability. In some markets, such as interest rates, this potential for negative rates might be a feature[3], however in other markets such as equities is would be seen as a severe drawback. As a result, the main applications of Bachelier's model tend to be in rates markets. Recently, negative prices were experienced in oil markets [199], and, there has been an increased adoption of the Bachelier model in commodity markets as a consequence.

Under these dynamics, closed form solutions are available for the SDE as well as European option prices. This makes the model efficient and easy to use in practice. As a result, in some markets such as rates, it often serves as a standard quoting convention. The following formulas define call and put option prices under the Bachelier model:

$$C_0 = \exp(-rT)\sigma\sqrt{T}\,(d_+\Phi(d_+) + \phi(d_+)) \tag{10.8}$$

$$P_0 = \exp(-rT)\sigma\sqrt{T}\,(d_-\Phi(d_-) + \phi(d_-)) \tag{10.9}$$

[3]Although in rates markets this also used to be regarded as a drawback, prior to the emergence of negative rates.

where d_+ and d_- are defined as:

$$d_+ = \frac{S_0 e^{rT} - K}{\sigma\sqrt{T}} \tag{10.10}$$

$$d_- = -d_+ \tag{10.11}$$

Here, S_0 is the current asset price, K is the strike, r is risk-free rate and T is the option expiry time. Further, Φ is the CDF of the standard normal distribution, and ϕ is the PDF of the standard normal distribution.

The pricing formulas under the Bachelier model can be calculated directly via standard integration techniques. For example, for a put option we have the following risk neutral pricing equation:

$$P_0 = \exp(-rT) \int_{-\infty}^{\infty} (K - S)^+ \phi(S_T) dS_T \tag{10.12}$$

where given the dynamics it can easily be shown that $S_T = S_0 e^{rT} + \sigma W$ is the solution to the dynamics in (10.7). As a result, S_T is normally distributed as its lone source of randomness is an arithmetic Brownian motion that has mean zero and variance T. Substituting the solution of S_T into the pricing equation we have:

$$P_0 = \exp(-rT) \int_{-\infty}^{\infty} (K - S_0 e^{rT} + \sigma W)^+ \phi(W) dW \tag{10.13}$$

$$= \exp(-rT) \int_{-\infty}^{\frac{K - S_0 e^{rT}}{\sigma}} (K - S_0 e^{rT} + \sigma W) e^{\frac{W}{2T}} \tag{10.14}$$

where in the second step we found the point at which the max function was equal to zero and used the fact that $W \sim N(0, T)$[4]. With the non-linear max function removed, we can then split the integral into two terms, one containing the $(K - S_0 e^r T)$ piece and another with the σW_t piece and compute these integrals separately. It is straightforward to see that the first term will be a function of the CDF of the normal distribution while the second term will be a function of the PDF of the normal distribution. Simplifying and re-arranging leads to the Bachelier formulas in (10.8).

It should be noted that the Bachelier model will lead to some amount of skew in an implied volatility surface [5], which is a desirable feature. Similarly, if we defined volatility surfaces in terms of Bachelier volatilities, then the Black-Scholes model would lead to some degree of skew. We will often find, however, that the Bachelier model doesn't provide enough control of this Black-Scholes implied volatility skew to match many market settings.

10.1.5 CEV Model

The CEV, or Constant Elasticity of Variance model, is a generalization of the log-normal and normal models [36]. The model contains an additional parameter

[4] See chapter 2 for more information on the properties of Brownian Motion.

[5] Which by definition plots Black-Scholes volatilities.

controlling the exponent of the volatility term of the SDE, as shown below:

$$dS_t = rS_t\,dt + \sigma S_t^{\beta}\,dW \tag{10.15}$$

As we can see, in addition the standard σ parameter, the model contains a new parameter, β, which is the exponent of the asset price used in the volatility term. We can see this model is a generalization of the first two stochastic models that we encountered, the Black-Scholes and Bachelier models, and that these models are special cases of the CEV model. In particular, $\beta = 1$ and $\beta = 0$ reduce back to the log-normal and normal models, respectively.

The inclusion of β as a second parameter enables us to try to fit more of the volatility surface. In practice, while this model is able to account for some degree of skew due to the added β parameter, we will find that it does not provide the flexibility to match market observed skews in most asset classes.

Derivation of European option prices under the CEV model is quite involved, and beyond the scope of the book, however, closed form solutions for call and put prices do exist. The reader should consult [77], for example, for more details.

The pricing formulas for calls and puts can be written as: [13]

$$\begin{aligned}
C_0 &= e^{-rT}\left\{S_0e^{rT}\left(1-\chi^2\left(\frac{4\nu^2K^{\frac{1}{\nu}}}{\sigma^2T};2\nu+2,\frac{4\nu^2(S_0e^{rT})^{\frac{1}{\nu}}}{\sigma^2T}\right)\right)\right.\\
&\quad\left.-K\left(\chi^2\left(\frac{4\nu^2(S_0e^{rT})^{\frac{1}{\nu}}}{\sigma^2T};2\nu,\frac{4\nu^2K^{\frac{1}{\nu}}}{\sigma^2T}\right)\right)\right\}\\
P_0 &= e^{-rT}\left\{S_0e^{rT}\left(\chi^2\left(\frac{4\nu^2K^{\frac{1}{\nu}}}{\sigma^2T};2\nu+2,\frac{4\nu^2(S_0e^{rT})^{\frac{1}{\nu}}}{\sigma^2T}\right)\right)\right.\\
&\quad\left.-K\left(1-\chi^2\left(\frac{4\nu^2(S_0e^{rT})^{\frac{1}{\nu}}}{\sigma^2T};2\nu,\frac{4\nu^2K^{\frac{1}{\nu}}}{\sigma^2T}\right)\right)\right\}\\
\nu &= \frac{1}{2(1-\beta)}
\end{aligned}$$

where $\chi(x;\kappa,\lambda)$ is the CDF of a non-central chi square distribution, κ is the degrees of freedom and λ is the non-centrality parameter and it is assumed that $\beta < 1$.

10.1.6 CEV Model in Practice: Impact of Beta

In this section we explore the impact of the additional model parameter β on our ability to match market implied volatility skews. As we noted in the last section, the normal model, which corresponds to $\beta = 0$ is characterized by a (log-normal) volatility skew, however, has no parameter to control its slope. In the CEV model, the β parameter serves as this slope parameter. To see this, we show a chart of the implied volatility skews for a given expiry using different values of the parameter β, which is shown below:

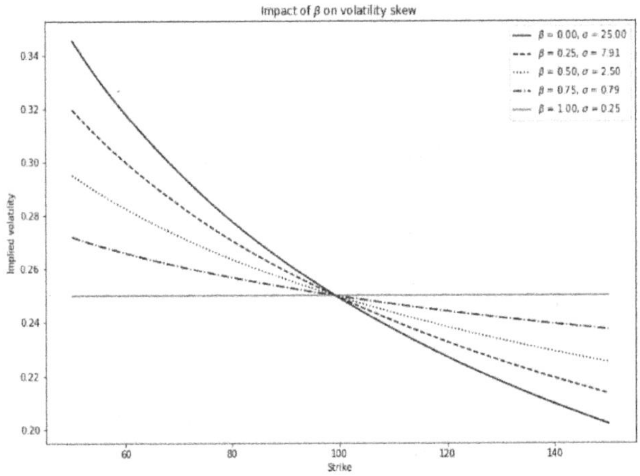

In the chart we can clearly see that as β decreases, the amount of skew increases, reaching its maximum slope at $\beta = 0$ in the Bachelier Model. At the other extreme, when $\beta = 1$ we recover the Black-Scholes model and then have a corresponding flat volatility surface. Therefore, we can see that, relative to the other models that we have covered, the CEV model has the most flexibility, enabling it to match skews with different slopes.

What it will struggle with, however, is matching options at different expiries, as we are able to control the slope of the skew but are not able to control the slope of the term structure of volatility. Additionally, the CEV model will struggle with other implied volatility shapes, such as smirks and smiles that have more complex features than the ones shown here.

It should also be noted that different levels of volatility, σ, were used to correspond to each level of β. This is because for each β the σ has a different expectation, and therefore a different required magnitude. At the extreme of this, we saw that in the Bachelier model, σ was interpreted as an absolute volatility. As such, it is by necessity higher than in the Black-Scholes model, where σ has a percentage interpretation. Therefore, in this chart we used the equivalent level of volatility for each specified β. It is left as an exercise for the reader how to calculate this conversion.

10.1.7 Ornstein-Uhlenbeck Process

An Ornstein-Uhlenbeck (OU) Process is a first order mean-reverting process where we assume deviations from some long run equilibrium revert at a specified speed, κ. The mean-reverting nature of the process makes it attractive in markets where there is empirical evidence of mean-reversion in the underlying asset, such as volatility and interest rates. This feature is less desirable in markets like equities where mean-reversion in the underlying asset would violate no-arbitrage conditions or weak-form market efficiency.

The SDE for an OU process can be written as:

$$dS_t = \kappa(\theta - S_t)dt + \sigma dW_t \tag{10.16}$$

An OU process is the continuous time equivalent of an AR(1) process that we covered in chapter 5. It contains three model parameters, κ, θ and σ, which are briefly summarized in the list below:

- κ: speed of mean-reversion
- θ: long run equilibrium asset level
- σ: volatility parameter

It can be shown that asset prices under an OU process are normally distributed with the following conditional distribution:

$$S_t = S_0 \exp(-\kappa t) + \theta(1 - \exp(-\kappa t)) + \sigma \int_0^t \exp(-\kappa(t-u))\, dW_u$$

Here, we can see that there are three terms in the above equation, a term related to the current asset price, discounted by the speed of mean-reversion, a term related to the drift of the underlying asset and a third term related to the volatility. Importantly, only the final term has a stochastic component, and is based on the evolution of Brownian motion. As a result, the distribution of S_t will be Gaussian.

Taking expectation of S_t, we can see that at any point its conditional expectation will be:

$$S_t = S_0 \exp(-\kappa t) + \theta(1 - \exp(-\kappa t)) \tag{10.17}$$

In practice, this means that, like the Bachelier model, negative values of the asset are permitted. This is another reason that this model is commonly employed in interest rate markets, as the combination of the presence of negative rates and the intuitive mean-reverting nature of the rate process are an appealing combination.

The distribution in equation 10.1.7 can be derived by applying an integrating factor to the SDE in (10.16). To see this, let's start by rearranging our SDE such that it has the following form:

$$dS_t + \kappa S_t dt = \kappa\theta dt + \sigma dW_t$$

Examination of the right-hand side of the equation shows that we know how to calculate the integral of each term. In particular, we know from chapter 4 that integrals of Brownian increment dW_t are a Weiner process W_t which are normally distributed with mean 0 and variance t. The dt term is a standard integral that we could compute from first principles.

As a result, let's work on the left-hand side of the equation multiplied by an integrating factor of $\exp(\kappa t)$. To do this, lets examine the derivative of $d(e^{\kappa t} S_t)$, or S_t times our integrating factor:

$$\begin{aligned} d(e^{\kappa t} S_t) &= dS_t e^{\kappa t} + d(e^{\kappa t}) S_t &(10.18)\\ d(e^{\kappa t} S_t) &= dS_t e^{\kappa t} + \kappa e^{\kappa t} S_t dt &(10.19) \end{aligned}$$

As we can see, the right hand side of equation 10.19 looks just like the left-hand side of our SDE times our integrating factor. As a result, we know how to integrate the left-hand side of the equation if we apply our integrating factor. Applying this integrating factor and solving for S_t leads to 10.1.7 as shown above.

10.1.8 Cox-Ingersol-Ross Model

Like an OU process, the Cox-Ingersol-Ross (CIR) model [153] is a mean reverting process with a square root in the volatility term:

$$dS_t = \kappa(\theta - S_t)dt + \sigma\sqrt{S_t}dW_t \tag{10.20}$$

Inclusion of this square root in the volatility term prevents asset prices from becoming negative, which is a desirable feature in many markets[6]. More generally, we can see that this volatility term follows a CEV model with $\beta = 0.5$. Aside from this modification, the CIR model mimics the OU process we saw above in its parameters and mean-reverting nature. In particular, the model parameters are κ, θ and σ, with κ and θ defining how the asset price mean-reverts and σ controlling the volatility.

The CIR model is most commonly applied in interest rate markets, where the mean-reverting nature of the asset price is particularly appealing. Additionally, the CIR model is often used to model volatility, as it is in the Heston model. The square root in the volatility term, however, prevents the underlying distribution from being Gaussian and makes solving the SDE significantly more challenging than in the case of an OU process.

10.1.9 Conclusions

The stochastic models that we have discussed so far are characterized by simplicity and a parsimonious set of parameters. As a result, these models will be unable to match the behavior of many different volatility surfaces that we observe in practice, such as the persistent put skew we observe in equity markets, or the smile phenomenon that we sometimes observe in FX markets.

That doesn't mean, however, that these models don't add value. In fact, the Black-Scholes model is the most common stochastic model applied in quant finance and is a standard quoting convention in many markets. In spite of their inability to match true market implied dynamics, these models do provide us with a natural means of comparing options across strikes and expiries. Additionally, these models are defined by an intuitive set of sensitivities, often referred to as Greeks. As we will see in chapter 16, this feature makes the Black-Scholes an attractive choice for measuring option Greeks.

What these models will fail to do is provide a single, unified, consistent set of dynamics for the entire volatility surface. To accomplish this, we will need to rely on a more realistic and advanced model. There are three general approaches that we can use to try to make our SDEs more accurately reflect market dynamics:

- Stochastic Volatility: allow volatility, σ to be governed by a stochastic process
- Jump Process: allow jumps in the underlying asset price.
- Local Volatility: allow volatility, σ, to be a deterministic function parameterized by state.

In the following sections, we explore each of these three approaches in more detail.

[6] But, as we will see in chapter 19 is not necessarily a desirable feature in all cases.

10.2 STOCHASTIC VOLATILITY MODELS

10.2.1 Introduction

In the models we have seen so far in this chapter, the volatility, or σ parameter has been a constant. In the real-world, this is not what we observe and thus may be unrealistic assumption. In fact, if we consider equity markets we observe heightened levels of volatility when asset prices decline. This negative correlation between volatility and asset prices is a persistent feature we observe in markets and is something we may want to incorporate into our models.

In particular, the presence of this negative correlation makes it appealing to consider the level of volatility to be a function of the current asset price rather than constant. There are two ways to achieve this. The first approach would be to create a deterministic volatility function. We explore this approach in section 10.4.

Stochastic Volatility models, instead attempt to incorporate skew into the volatility surface via the addition of another stochastic component which is followed by the asset's volatility or variance. Importantly, these two stochastic processes may be correlated. This is an important feature that we will find will be critical for helping us match the observed skew behavior in markets.

Clearly, attaching another stochastic process to our problem adds a great deal of complexity, and as a result in these classes of models we should not expect a closed form solution to be available. Fortunately, however, Fourier techniques, which are discussed in chapter 14, enable us to solve option pricing problems efficiently for many such models. In certain other cases, as we will see with the SABR model, efficient pricing algorithms exist in the form of asymptotic approximations. Additionally, in stochastic volatility models the traditional delta-hedging based replication arguments that motivate risk-neutral pricing need to be extended to account for incorporation, and hedging, of a second source of randomness. In this book this theory is taken for granted, however, interested readers should consult [108] for a rigorous treatment of this derivation.

In practice, we will find that in many cases the added complexity of a stochastic volatility model is warranted by the increased ability to match the behavior of volatility smiles and skews in practice. Most volatility models enable us to in particular control the higher moments of the underlying distributions, such as skewness and kurtosis, which we will see is critical to modeling the volatility surface accurately. In this section we discuss two of the most commonly used stochastic volatility models, Heston and SABR.

10.2.2 Heston Model

The Heston model [130] is a stochastic volatility model that is commonly used in equities, FX and commodities. Because it is a stochastic volatility model, it is defined by a two-dimensional system of SDEs. The price process is log-normal, albeit with stochastic volatility. The variance process is a mean reverting process in the form of the CIR model shown above.

The system of SDEs for Heston is defined as follows:

$$\begin{aligned} dS_t &= rS_t\,dt + \sigma_t S_t\,dW_t^1 \\ d\sigma_t^2 &= \kappa(\theta - \sigma_t^2)\,dt + \xi\sigma_t\,dW_t^2 \\ Cov(dW_t^1, dW_t^2) &= \rho\,dt \end{aligned} \tag{10.21}$$

Here we can see that the dynamics show how both the asset and its variance evolve. The model parameters are κ, θ, ρ and ξ. Additionally, σ_0, the starting level for the volatility or variance process is also a model parameter as this is a latent variable and hence we do not observe its initial value. A brief summary of the interpretations of each model parameters is in the following list:

- κ: Speed of Mean-Reversion of Volatility
- θ: Long Run Equilibrium Level of Volatility
- ρ: Correlation between the asset and the asset's volatility.
- ξ: Volatility of volatility (vol of vol)
- σ_0: initial level of volatility.

The richer parameter set of the Heston model in comparison to the previous models we investigated, combined with the power of incorporating stochastic volatility, make Heston far more effective at modeling volatility surfaces in different markets. As the price dynamics are inherently log-normal, this model is applied more in contexts where that assumption is desired, such as equities, rather than in markets such as rates, where that assumption may be less appropriate.

There is no solution to the system of SDEs underlying the Heston model, and there is not tractable representation of the PDF of the model. We can however use Fourier transform techniques to obtain efficient pricing estimates, which are detailed in chapter 14.

Of particular interest in applying the Fourier transform techniques will be the characteristic function of the underlying model. For Heston this characteristic function is known and can be expressed as:

$$\omega(u) = \frac{\exp\left(iu\ln S_0 + iu(r-q)t + \frac{\kappa\theta t(\kappa - i\rho\sigma u)}{\sigma^2}\right)}{\left(\cosh\frac{\lambda t}{2} + \frac{\kappa - i\rho\sigma u}{\lambda}\sinh\frac{\lambda t}{2}\right)^{\frac{2\kappa\theta}{\sigma^2}}} \tag{10.22}$$

$$\Phi(u) = \omega(u)\exp\left(\frac{-(u^2+iu)\nu_0}{\lambda\coth\frac{\lambda t}{2} + \kappa - i\rho\sigma u}\right) \tag{10.23}$$

$$\lambda = \sqrt{\sigma^2(u^2+iu) + (\kappa - i\rho\sigma u)^2} \tag{10.24}$$

where $\Phi(u)$ is the characteristic function evaluated at u.

10.2.3 SABR Model

Another influential stochastic volatility model is the Stochastic Alpha Beta Rho (SABR) Model [198]. Like the Heston model, it is defined by a two-dimensional system of SDEs, one for the asset price and one for its volatility. As in the Heston model, we enable these processes to be correlated. Unlike the Heston model, however, the SABR model does not necessarily assume a log-normal price process for the asset, instead relying on the more general CEV model. Another key differentiating feature is that volatility does not follow a mean-reverting process in the SABR model, instead it follows a log-normal process. The more general CEV price process makes the SABR model more attractive in rates in particular, but SABR is also often applied in FX markets.

The system of SDEs for SABR can be written as:

$$\begin{aligned} dF_t &= \sigma_t F_t^{\beta}\, dW_t^1 \\ d\sigma_t &= \alpha\sigma_t\, dW_t^2 \\ Cov(dW_t^1, dW_t^2) &= \rho\, dt \end{aligned} \tag{10.25}$$

Notice that the dynamics are expressed in terms of the forward, F_t but could equivalently be expressed in terms of the underlying asset using the identity $F_t = S_0 e^{rt}$ to add a corresponding drift term to (10.25).

Again we can see that the system of SDEs defines the evolution of the asset and its volatility. The model parameters are r, α, β, ρ and σ_0. σ_0 is a model parameter because volatility is again a hidden variable, meaning we don't observe its initial level. A brief summary of the interpretations of each model parameters is in the following list:

- α: Volatility of volatility (vol of vol)
- β: CEV Exponent
- ρ: Correlation between asset and asset volatility.
- σ_0: initial level of volatility.

As with the Heston model, the more robust set of parameters in the SABR model relative to the earlier models we saw will lead to a better fit of various volatility skews, smirks and smiles. This better fit has made SABR a popular model, with the most noteworthy applications being in rates markets. These examples of using SABR are detailed in chapter 19.

A noteworthy byproduct of the simpler non–mean-reverting dynamics of the SABR volatility process is that, while it is able to fit volatility skew very well, it is unable to fit the volatility term structure. As a result, when using SABR it is best practice to calibrate a separate set of parameters per expiry, unlike in the Heston model where we can match all expiries simultaneously.

There is no solution to the system of SDEs underlying the SABR model, however, a set of asymptotic formulas have been created that enables us to map from a set of

SABR and option parameters to a Black-Scholes implied volatility, or to a Bachelier, Normal Volatility. This asymptotic formula leads to computational efficient calculation of option prices under the SABR model. The following asymptotic formula can be used to compute the Bachelier or Normal volatility of an option under the SABR model:

$$\begin{aligned}
&\sigma_n(T, K, F_0, \sigma_0, \alpha, \beta, \rho) \\
&= \alpha \frac{F_0 - K}{\Delta(K, F_0, \sigma_0, \alpha, \beta)} \\
&\times \ldots \left\{ 1 + \left[\frac{2\gamma_2 - \gamma_1^2}{24} (\frac{\sigma_0 C(F_{mid})}{\alpha})^2 + \frac{\rho\gamma_1}{4} \frac{\sigma_0 C(F_{mid})}{\alpha} + \frac{2 - 3\rho^2}{24} \right] \epsilon \right\}
\end{aligned}$$

Where:

$$\begin{aligned}
C(F) &= F^{\beta} \\
\gamma_1 &= \frac{C'(F_{mid})}{C(F_{mid})} \\
\gamma_2 &= \frac{C''(F_{mid})}{C(F_{mid})} \\
F_{mid} &= \frac{F_0 + K}{2} \\
\Delta(K, F_0, \sigma_0, \alpha, \beta) &= \log \left(\frac{\sqrt{1 - 2\rho\zeta + \zeta^2} + \zeta - \rho}{1 - \rho} \right) \\
\zeta &= \frac{\alpha}{\sigma_0(1 - \beta)} \left(F_0^{1-\beta} - K^{1-\beta} \right) \\
\epsilon &= T\alpha^2
\end{aligned}$$

A similar expansion exists for Black's model and can be found in [198]. Once the equivalent normal or log-normal model implied volatility has been calculated, the formulas earlier in the chapter can be applied to obtain option prices. It should again be emphasized that while this expression is written in terms of the dynamics of the forward, we could easily translate it to be in terms of dynamics of the underlying asset.

10.2.4 SABR Model in Practice: Relationship between Model Parameters and Volatility Surface

In this section we explore how the underlying SABR model parameters impact the subsequent volatility surface implied by the model. In this section we focus on two model parameters of interest, α and ρ, as β was explored in a previous section on the CEV model and σ_0 has some similarity to the previously examined σ parameters in simpler models. That is, it mostly informs the level of implied volatility in the volatility surface.

First, we consider the parameter ρ. In the following chart, we display the volatility smile for a given expiry with four different values of ρ specified:

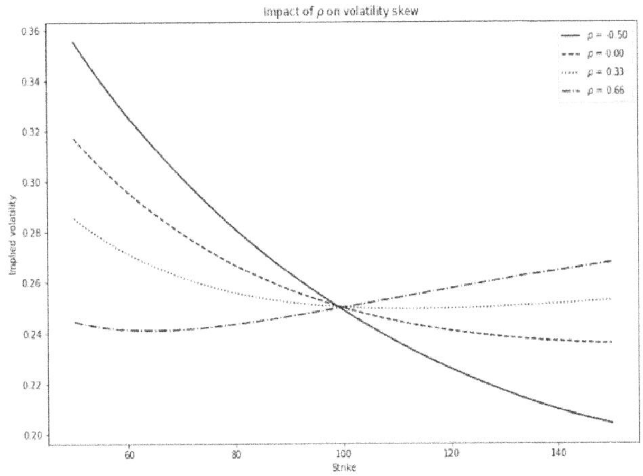

From this chart we can clearly see that, much like β in the CEV model, the ρ parameter impacts the slope of the volatility skew greatly, with more negative values leading to higher implied volatilities on lower strike options and vice versa. In fact, as β and ρ are inherently both skew related parameters, they are actually highly correlated in practice. This presents challenges in SABR model calibrations, and means, in practice, it is standard to fix one of these parameters and calibrate the remaining three. Nonetheless, we can clearly see here the relationship between ρ and the slope of the volatility skew.

Next, we consider the model parameter α, which we had previously described as the volatility of volatility. Again, to understand its impact on a volatility surface we again plot the volatility skew while varying α and leaving all other model parameters constant, as is displayed in the following chart:

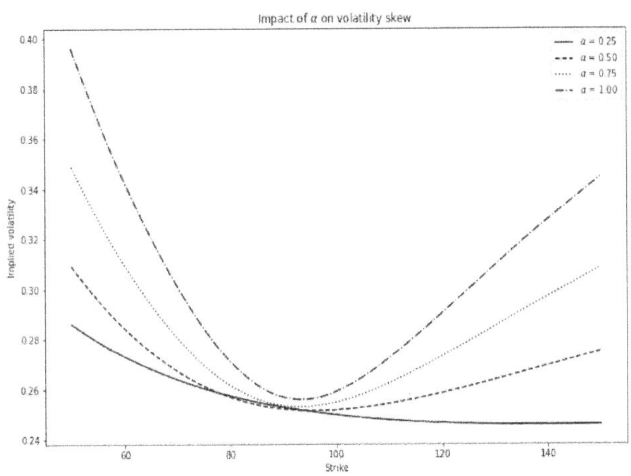

From this chart, we can see that higher levels of alpha seem to fundamentally impact the smile shape in the volatility surface. In particular, we observe that higher levels of alpha tend to lead to more pronounced smile effects. Therefore, we might conclude that the parameter α can be thought of as controlling the kurtosis in the model distributions.

10.2.5 Stochastic Volatility Models: Comments

In this section we discussed the first approach for rigorously incorporating the volatility skew in a consistent, unified manner: via stochastic volatility. The two models that we discussed, Heston and SABR are perhaps the most commonly used stochastic vol models in quant finance, and are thus worth exploring. Both of these models enable us to fit many different types of skews, whether it is put or call side skew, or a volatility smile. A key component of this is that these models each contain parameters that help us control the skewness and kurtosis of the underlying distribution. Skewness, in particular, was tightly connected to the correlation between the underlying stochastic processes. The interested reader is encouraged to think about why that might be, and try to connect that to expectations about how actual volatility should behave when an asset rises or falls in value. Another important feature of these models is that they are reducible back to the Black-Scholes model with certain sets of parameters. This proves to be a convenient feature for model validation purposes.

One particularly noteworthy difference between the Heston and SABR models was in the structure of the volatility processes. As noted previously, the Heston model contains a mean reverting volatility process, whereas SABR's volatility process is lognormal. The mean-reverting nature of volatility in Heston matches what we experience empirically. As a result, the mean-reversion related parameters in the Heston model enable us to fit the slope of the volatility surface across expiries in addition to across strikes. This means that when working with the Heston model, we are able to use a single set of parameters to define the parameters for all expiries. The SABR model, in contrast, will do an admirable job of calibrating a volatility smile across strikes, but will be unable to do so across expiries, instead requiring a separate calibration[7] for each expiry.

This is an important distinction as it means that, while Heston will create a single model and consistent set of distributions for the entire volatility surface, the SABR will require a different set of dynamics/distributions for different expiries. Having this overarching consistency is clearly desirable, and we will see the challenges created by having separate parameters at each expiry when we attempt to price exotic options in chapter 15.

One weakness of stochastic volatility models is their inability to capture extreme skew of short dated options. This is a common and persistent problem for these models, and is inherently because it is difficult to make continuous paths that reach far enough out of the money. A natural solution to this issue is to incorporate jumps into the model, which we explore in the next section.

[7]Or set of model parameters.

10.3 JUMP DIFFUSION MODELS

10.3.1 Introduction

A second way to incorporate the volatility surface behavior that we observe in practice is via jumps in the underlying asset process. Adding jumps in the asset is quite intuitive as in markets we do observe gap moves in many asset following the release of news or economic data. More generally, we tend to observe periods of different trading conditions, or market regimes, and the transitions between these regimes often display dis-continuous, jump characteristics. One particular type of jump is a so-called jump-to-default. This type of jump is at the heart of credit risk modeling but is also relevant to equity markets as clearly a company default will result in a negative jump to zero in the corresponding equity price. This relationship, and the connection between equity and credit markets, is explored in more detail in chapter 20. A major strength of models that leverage jump processes is an improved ability to match the behavior in extreme shorter dated volatility smiles. Jump models are characterized by an ability to control the higher moments in the risk neutral distribution, most notably skewness and kurtosis. Conversely, as with stochastic volatility models, we will find inclusion of a jump process increases model complexity while still not guaranteeing a fit the entire volatility surface.

Models that rely solely on Brownian motion as the driver, or drivers of randomness are inherently continuous and as a result won't allow these types of jumps that we observe in markets. To integrate jumps into our model, we will need to include an additional source of randomness. Of particular interest in this endeavor will be Poisson processes, a statistical distribution that models the number of (jump) occurrences for a given time interval and Exponential distributions, which model the time between jumps. An introduction to the theory behind these distributions was presented in the text's supplementary materials. A far more detailed treatment of the underlying theory of jump processes and stochastic modeling with jumps can be found in [240].

10.3.2 Merton's Jump Diffusion Model

Merton's Jump Diffusion model [108] is one of the more popular models that incorporates potential jumps into the underlying asset process. It combines a Brownian motion based diffusion component with a jump process.

The SDE for Merton's Jump Diffusion model can be written as:

$$dS_t = rS_t dt + \sigma S_t dW_t + (e^{\alpha+\gamma\epsilon} - 1)S_t dq \tag{10.26}$$

As we can see, this model is characterized by a single volatility parameter, σ, and a log-normal volatility process. In fact, the first two terms are identical to the Black-Scholes model, however, a third term has been introduced that will allow jumps in the model. The jump component, dq, follows a Poisson process that takes the value of zero with probability $1-\lambda$ and one with probability λ. The jump size itself, $(e^{\alpha+\gamma\epsilon}-1)S_t$, is also random in this model, following a log-normal distribution[8]. Merton's Jump

[8]Since ϵ follows a standard normal distribution.

Diffusion model is defined by the parameters σ, λ, α and γ, each of which is described briefly in the following list:

- σ: log-normal volatility of the diffusion process.
- λ: probability of a jump
- α: mean log-jump size
- γ: standard deviation of jump sizes

Merton's Jump model is attractive in that it enables us to not only incorporate jumps via a Poisson process but also control their distribution via α and γ.

As will be the case with all of the pure jump and jump diffusion processes we explore, there is no closed form solution to the SDE for Merton's Jump model. The characteristic function for the model, however, is available in closed-form [108], and can be written as:

$$\Phi(u) = \exp\left[iu\ln S_0 + (r-q) + iu\omega t - \frac{1}{2}u^2\sigma^2 t + \lambda t\left(e^{iu\alpha - \frac{u^2\gamma^2}{2}} - 1\right)\right] \quad (10.27)$$

$$\omega = r - q - \frac{1}{2}\sigma^2 - \lambda\left(e^{\alpha+\frac{\gamma}{2}} - 1\right) \quad (10.28)$$

where ω was chosen to ensure the process is a martingale after introducing the jump term. Pricing in Merton's Jump Diffusion model can be done using the Fourier Transform techniques are discussed in chapter 14.

10.3.3 SVJ Model

The Stochastic Volatility Plus Jumps (SVJ) model [33] is an extension of Merton's Jump Diffusion model that also incorporates a stochastic volatility process. The volatility process follows the same structure as the Heston model, inheriting its desirable mean-reverting nature. As a result, the SVJ model can be viewed as a Heston model with jumps in the underlying asset. Alternatively, it can be viewed as a Merton Jump Diffusion model with stochastic volatility.

The dynamics for SVJ can be written as:

$$dS_t = rS_t dt + \sigma_t S_t dW_t + (e^{\alpha+\gamma\epsilon} - 1)dq \quad (10.29)$$

$$d\sigma_t^2 = \kappa(\theta - \sigma_t^2)\,dt + \xi\sigma_t\,dW_t^2$$

$$Cov(dW_t^1, dW_t^2) = \rho\,dt \quad (10.30)$$

As we can see, the price process mimics Merton's model in the previous section and the variance process mimics that of Heston. The SVJ model is characterized by a relatively rich parameter set. Like the Heston model, the mean-reverting nature of volatility will make it an effective model for incorporating skew and the term structure of volatility. Additionally, the incorporation of jumps with control over the distribution of the jump sizes, leads to better fit of extreme short dated option skews. Conversely, in some cases addition of this jump component is unnecessary as the Heston model alone leads to a reasonable fit in many situations. The parameters in the SVJ model are κ, θ, ρ, ξ, σ_0, λ, α and γ. A brief summary of the interpretation of each parameter is detailed below:

- κ: Speed of Mean-Reversion of Volatility
- θ: Long Run Equilibrium Level of Volatility
- ρ: Correlation between the asset and its volatility.
- ξ: Volatility of volatility (vol of vol)
- σ_0: initial level of volatility.
- λ: probability of a jump
- α: mean log-jump size
- γ: standard deviation of jump sizes

As was the case for both Heston and Merton's Jump Diffusion model, the characteristic function for the SVJ model is known to be the product of the characteristic functions for the Heston Model and Merton's Jump Diffusion model. The characteristic function for the log of the asset price can therefore be written as:

$$\Phi(u) = \omega(u)\exp\left(\frac{-(u^2+iu)\nu_0}{\lambda\coth\frac{\lambda t}{2}+\kappa-i\rho\sigma u}\right) \tag{10.31}$$

$$\exp\left(iu\omega t-\frac{1}{2}u^2\sigma^2 t+\lambda t\left(e^{iu\alpha-\frac{u^2\gamma^2}{2}}-1\right)\right) \tag{10.32}$$

$$\omega(u) = \frac{\exp\left(iu\ln S_0+iu(r-q)t+\frac{\kappa\theta t(\kappa-i\rho\sigma u)}{\sigma^2}\right)}{\left(\cosh\frac{\lambda t}{2}+\frac{\kappa-i\rho\sigma u}{\lambda}\sinh\frac{\lambda t}{2}\right)^{\frac{2\kappa\theta}{\sigma^2}}} \tag{10.33}$$

$$\lambda = \sqrt{\sigma^2(u^2+iu)+(\kappa-i\rho\sigma u)^2} \tag{10.34}$$

$$\omega = \frac{1}{2}\sigma^2-\lambda\left(e^{\alpha+\frac{\gamma}{2}}-1\right) \tag{10.35}$$

Because the characteristic function is known and analytically tractable, pricing European options under the SVJ model can be done efficiently using transform techniques[9]. The SVJ model can be further extended to include jumps in both the asset and volatility processes. This model is referred to as SVJJ [14]. Interested readers should consult [108] for treatment of the SVJJ model and a comparison of its effectiveness vs. the SVJ and Heston models.

---

[9]See chapter 14 for more details.

10.3.4 Variance Gamma Model

The variance gamma (VG) model [202] is a jump process where we use a stochastic clock that determines the time steps in our process. The movements in this stochastic clock adhere to a gamma distribution and can be viewed as a pure jump approach to modeling the stochastic nature of volatility. One way to think about the random clock is as a way to incorporate different trading behavior in different market conditions. For example, in practice, we observe much of the trading volume in markets to be clustered near certain periods, such as the being or end of a trading day, or news driven events. The presence of a stochastic clock accounts for these periods of substantially different levels of trading volume by winding up or down the clock as a result.

The Variance Gamma model leverages a Geometric Brownian motion structure with the addition of the random time changes being the key differentiating feature. The dynamics for the Variance Gamma model can be written as:

$$
\begin{aligned}
X(t;\sigma,\nu,\theta) &= \theta\gamma(t;1,\nu)+\sigma W(\gamma(t;1,v)) && (10.36)\\
\ln(S_t) &= \ln(S_0)+(r-q+\omega)t+X(t;\sigma,\nu,\theta) && (10.37)\\
\omega &= \frac{1}{\nu}\log\left(1-\theta\nu-\frac{\sigma^2\nu}{2}\right) && (10.38)
\end{aligned}
$$

Here, $\gamma(t;1,\nu)$ defines the distribution of the random time changes, or stochastic clock. These time changes follow a Gamma distribution with mean 1 and variance ν. Further, as we saw in previous models, ω is chosen such that the resulting process will be a martingale. The Variance Gamma model has three underlying parameters: θ, σ and ν. A brief explanation of each parameter is provided below:

- σ: volatility of the underlying stochastic process.
- θ: drift of the underlying stochastic process.
- ν: variance of the stochastic time change

An attractive feature of the VG model is that the presence of these three parameters enables the model to fit the skewness and kurtosis that is present in options markets. Importantly, in practice we find ν to be the model parameter that is most closely related to kurtosis and θ to be the parameter that maps most closely to skewness. As a result, θ is often found to be negative in calibrations of equity markets. One feature of the variance gamma model is the independence of successive movements in the random clock. Each movement of the random clock is defined by a gamma distribution, however, there is no interconnection, or clustering between them. This is noteworthy as in practice clustering of these types of periods of high or low trading volume is observed in practice.

The characteristic function for the log of the asset price in the Variance Gamma model is known and can be expressed as:

$$
\Phi(u) = e^{iu(\log(S0)+(r-q+\omega)t)}\left(\frac{1}{1-iu\theta\nu+\frac{\sigma^2u^2\nu}{2}}\right)^{\frac{t}{\nu}} \qquad (10.39)
$$

where $\Phi(u)$ is once again the characteristic function evaluated at u, θ, ν and σ are the model parameters, and t is the time to expiry of the option. As the characteristic function is known and tractable, Fourier Transform techniques provide an efficient approach to valuing European options under the VG model.

10.3.5 VGSA Model

One of the perceived limitations of the Variance Gamma model is that successive steps in its random clock are independent. This means that the model cannot account for volatility clustering. The Variance Gamma with Stochastic Arrival model [203] was created to address this weakness.

In the VGSA model, a mean-reverting structure is given to the stochastic time changes, in the form of the CIR model presented earlier in the chapter. That is, in the VGSA model a CIR process dictates the process of instantaneous time changes. Integrals of this process are then used to determine movements in the underlying random clock.

Recall from 10.20 that the SDE for a CIR process can be written as:

$$dy_t = \kappa(\eta - S_t)dt + \lambda\sqrt{S_t}dW_t \tag{10.40}$$

$$Y(t) = \int_0^t y_u du \tag{10.41}$$

In the VGSA model, the time clock $Y(t)$ is then substituted for t in the Variance Gamma process. This leads to the following dynamics:

$$X(t;\sigma,\nu,\theta) = \theta\gamma(t;1,\nu) + \sigma W(\gamma(t;1,v)) \tag{10.42}$$

$$Z(t;\sigma,\nu,\theta,\kappa,\eta) = X(Y(t);\sigma,\nu,\theta) \tag{10.43}$$

$$\ln(S_t) = \ln(S_0) + (r - q + \omega)t + Z(t;\sigma,\nu,\theta,\kappa,\eta) \tag{10.44}$$

where as before ω is chosen to ensure that the resulting process is a martingale. We can see that this model is equivalent to the Variance Gamma with a modified time process $Y(t)$. This mean reverting time process enables us to incorporate clustering, and as a result is an important model feature. The VGSA model also has a richer parameterization than the VG model, with the standard VG parameters σ, θ and ν as well as a set of new parameters that define how the time process mean reverts (κ, η and λ). A brief description of each parameter is listed below:

- σ: volatility of the underlying stochastic process.
- θ: drift of the underlying stochastic process.
- ν: variance of the stochastic clock
- κ: speed of mean reversion of the instantaneous time change process.
- η: long run equilibrium level of the instantaneous time change process.
- λ: volatility of the instantaneous time change process.

The mean-reverting nature of the time change process in the VGSA provides an additional set of parameters for better fit to the volatility surface. In particular, the incorporation of this mean-reversion will enable us to match both the volatility skew and term structure, requiring only one set of model parameters for a surface. This contrasts with the standard Variance Gamma model, where a calibration or set of model parameters would be required per expiry.

The VGSA model has a known characteristic function enabling efficient European option pricing via FFT techniques, as we will see in chapter 14. The characteristic function for the log of the asset price can be written as:

$$\Phi(u) = e^{iu(\log(S0)+(r-q+\omega)t)} \frac{\phi(-i\Psi_{VG}(u), t, \frac{1}{\nu}, \kappa, \eta, \lambda)}{\phi(-i\Psi_{VG}(-i), t, \frac{1}{\nu}, \kappa, \eta, \lambda)^{iu}} \quad (10.45)$$

$$\phi(u) = A(t,u)e^{B(t,u)y(0)} \quad (10.46)$$

$$A(t,u) = \frac{\exp \frac{\kappa^2 \eta t}{\lambda^2}}{\left(\cosh(\frac{\gamma t}{2}) + \frac{\kappa}{\gamma}\sinh(\frac{\gamma t}{2})\right)^{\frac{2\kappa\eta}{\lambda^2}}} \quad (10.47)$$

$$B(t,u) = \frac{2iu}{\kappa + \gamma \coth(\frac{\gamma t}{2})} \quad (10.48)$$

$$\gamma = \sqrt{\kappa^2 - 2\lambda^2 iu} \quad (10.49)$$

$$\Psi_{VG}(u) = -\frac{1}{\nu}\log\left(1 - iu\theta\nu + \frac{\sigma^2 \nu u^2}{2}\right) \quad (10.50)$$

where $\Phi(u)$ is the characteristic function for the VGSA model evaluated at u. Additionally, $\Psi_{VG}(u)$ is the characteristic function of the Variance Gamma model and ϕ is the characteristic function of the integral of the CIR process $y(t)$. We will soon see how to leverage these characteristic functions to solve European option pricing problems.

10.3.6 Comments on Jump Processes

In this section we detailed a handful of relevant jump diffusion models that are commonly used in markets to explain volatility surface behavior. As we saw, jump processes are of particular use when trying to model extreme volatility skews of very short dated options. In these cases, we observe unreasonable levels of implied volatility that are required to match market prices. At heart, this is because forming continuous paths that travel sufficiently far to make the out-of-the-money options payoff in a short time requires massive levels of volatility. An alternative and perhaps better explanation is that the level of volatility is actually significantly lower and more reasonable, but that there is some positive probability of a jump occurring in the price process due to earnings, news, default, or other factors. Allowing jumps is a logical solution to this problem, and, in fact, a better fit of shorter dated skews are a strength of most jump-diffusion models.

The main drawbacks to working with jump models are the additional complexity required, and also the lack of intuitive, and potentially mean-reverting volatility processes. That is one reason that some models, such as SVJ and SVJJ have attempted to combine the desirable features of jump processes, with the mean-reverting stochastic volatility nature of Heston.

When working with jump diffusion models, we will find that the Fourier Transform techniques detailed in chapter 14 are particularly relevant. Fundamentally this is because closed-form solutions to these types of models are not available, and the PDF's of the model tend to be intractable. Fortunately, however, the characteristic function for many such jump process is known in closed-form and thus can easily be input into the FFT techniques we explore.

A key differentiating feature between jump-diffusion models is whether or not they are able to match the volatility term structure[10] as well as volatility skew. This is analogous to the comparison between SABR and Heston that we presented in the previous section. As we saw for SABR, some jump-diffusion models will be able to explain the skew behavior but not the term structure, and will thus require a separate set of model parameters of calibration per expiry. The Variance Gamma model described in this section falls into this category. On the other hand, other models are similar to Heston in that they have a rich enough parameterization to fit both the skew and term structure. Heston in particular was able to accomplish this via a mean-reverting stochastic volatility process. VGSA is analogous to the Heston model in this regard, in particular, the ability to define clustering in the jump process enables us to fit the term structure in addition to the skew that we can fit using the simpler Variance Gamma model.

10.4 LOCAL VOLATILITY MODELS

10.4.1 Dupire's Formula

Another approach to modeling the volatility smile is via a Local Volatility model, which was first created by Dupire [85] and separately by Derman and Kani [78]. In the Local Volatility model, we make volatility a deterministic function of the current state. That is, volatility may depend on time and the underlying asset price, but it is not stochastic. This dependence of the volatility on the asset price enables us to match the volatility smile, and its dependence on time helps us to match the volatility term structure.

The Local Volatility model can be written as:

$$dS_t = r(t)S_t\,dt + \sigma(S_t, t)S_t\,dW \tag{10.51}$$

where $r(t)$ is the discount process and is often assumed to be a constant and $\sigma(S_t, t)$ is a state dependent volatility function.

As we can see, the model is characterized by asset prices that are log-normally distributed at each increment, but we now have a volatility function rather than a single, constant value. An appealing feature of this model is that it imposes relatively minimal restrictions on the process of the underlying asset, because of the flexibility we have in choosing the function $\sigma(S_t, t)$. This means that, in the absence of arbitrage or data issues the model will be able to fit the volatility surface exactly.

[10]Defined as the implied volatility of options with the same strike across a set of expiries.

In fact, Dupire [85] found that there is a unique solution to this function $\sigma(S_t, t)$ that enables us to recover the volatility surface exactly:

$$\sigma(K,T)^2 = \frac{\frac{\partial C}{\partial T}}{\frac{1}{2}K^2\frac{\partial^2 C}{\partial K^2}} \tag{10.52}$$

This function is often referred to as Dupire's formula. A nice, straightforward derivation for Dupire's formula can be found in Gatheral [108]. The derivation hinges on two fundamental concepts. The first is the relationship between an asset's risk neutral distribution and the second derivative of a call option with respect to strike, which we explore further in chapter 17. The second key concept is the structure of the transition densities, that is the evolution of the call price as a function of time.

Importantly, the formula can be calculate using only current option prices as each derivative in the above equation can be approximated by a finite difference approximation. In particular:

$$\frac{\partial C}{\partial T} \approx \frac{C(K,T+h) - C(K,T-h)}{2h} \tag{10.53}$$

$$\frac{\partial^2 C}{\partial K^2} \approx \frac{c(K-h) - 2c(K) + c(K+h)}{h^2} \tag{10.54}$$

The theory behind these finite difference approximations is discussed in more detail in chapter 15. Fundamentally however, notice that each derivative can be estimated as a function of observable call option prices. As a result, if we assume that prices are available for all strikes and expiries, then this becomes an elegant solution to the problem of matching a volatility surface.

The main challenge in using a Local Volatility model then becomes incorporating it in a world with imperfect data. In practice, we do not observe data for all strikes and expiries. Further, we often observe a substantial bid-offer spread in options prices which creates uncertainty in any single traded price. Any potential error is then magnified in the formula as we are estimating first and second-order derivatives, which are far more prone to these types of errors. This creates an implementation challenge when using Local Volatility models and means that in practice they tend to only be applied in the most liquid derivatives markets.

10.4.2 Local Volatility Model in Practice: S&P Option Local Volatility Surface

In this section we apply the code presented in the repository to create a local volatility surface, or set of local volatilities as a function of time and strike, for the S&P 500:

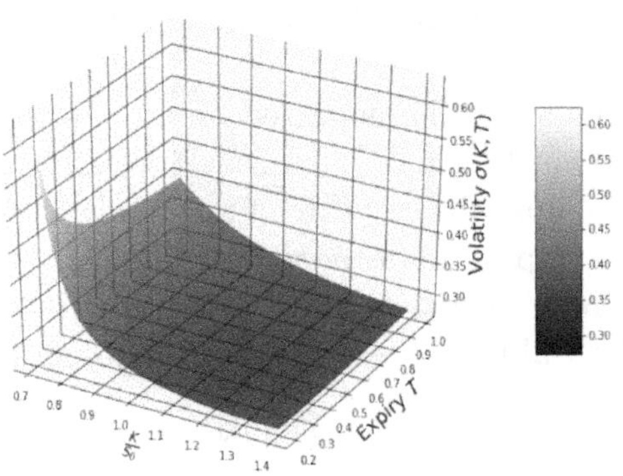

Looking at this surface, we can see the heightened levels of local volatility for low strike options. This feature is consistent with the well-known put skew observed in equity markets, where low strike options trade at higher levels of implied volatility.

10.5 STOCHASTIC LOCAL VOLATILITY MODELS

Stochastic Local Volatility (SLV) models combine a stochastic volatility component, as we saw in the Heston and SABR models, with a Local Volatility component, as we saw in the previous section. These models inherit a two-dimensional SDE from their stochastic volatility counterparts and also include a call to a local volatility function. The SDE governing the asset's volatility can be chosen in many ways, just as we saw in the Heston and SABR models, respectively. Employing a Heston approach, with a mean-reverting stochastic volatility process, would lead to the following system of SDE's:

$$\begin{aligned} dS_t &= rS_t\,dt + \sigma_t L(t, S_t) S_t\,dW_t^1 \\ d\sigma_t^2 &= \kappa(\theta - \sigma_t^2)\,dt + \xi\sigma_t\,dW_t^2 \\ Cov(dW_t^1, dW_t^2) &= \rho\,dt \end{aligned} \tag{10.55}$$

Here $L(t, S_t)$ is a local volatility function. Aside from this additional term, the reader will notice that this system of SDE's is analogous to Heston. Importantly though, and in contrast to a standard Heston model, the presence of a local volatility function gives us enough freedom to fit the entire volatility surface exactly. Further, relative to a local volatility model, inclusion of a stochastic volatility component enables us to provide additional structure to the model where the quoted market data may be insufficient. In that sense, it can be viewed as a happy medium between the stochastic volatility and local volatility approaches, and in practice often leads to more realistic exotic options pricing. Conversely, SLV models add a great deal of complexity in terms of pricing and calibration. The details of calibration of SLV models are beyond the scope of this book. Interested readers should consult [37] or [11].

10.6 PRACTICALITIES OF USING THESE MODELS

10.6.1 Comparison of Stochastic Models

In this chapter we have reviewed many of the most common models and stochastic processes used by practitioners. Some of the models that we discussed will do a reasonable job of approximating the volatility smiles that we observe in practice, such as the Heston or SABR model. Phrased in another way, these models will provide us with a single, consistent model of the dynamics of the underlying asset, and implicitly of the risk-neutral distribution in all states.

Other models are oversimplifications of markets, such as the Black-Scholes or Bachelier models. These models will fail to capture much if any of the observed patterns of skew. As a result, different dynamics will need to be used for different strikes and expiries, resulting in inconsistent distributions across a set of options with varying strikes and expiries. This is clearly not ideal, and creates challenges for using these models in many situations. However, it does not mean that these models are without use. Both the Black-Scholes and Bachelier models in particular are treated as quoting conventions in many markets, and provide a convenient and simplified way to compare options with different characteristics.

In practice to implement any of the models discussed in this chapter, we need to find an efficient pricing technique. In simple cases, such as the Black-Scholes and Bachelier models, a closed-form solution to vanilla option prices exists, making pricing tractable and computationally efficient. In other cases, we may need to resort to asymptotic approximations, as is the case for the SABR model. These approximations will still be reasonably fast, making these models easy to work with. In other cases, we may need to employ more advanced techniques such as quadrature or Fast Fourier Transforms (FFTs) to model vanilla option prices. We will explore these techniques in chapter 14, and will find FFT to be of particular interest for many stochastic volatility models and jump processes. Other techniques such as simulation and numerical solutions to PDE's will be key to our core quant skillset but are generally reserved for more advanced pricing problems such as exotic, path dependent options. These techniques are discussed in chapter 15.

10.6.2 Leveraging Stochastic Models in Practice

In markets, we are generally given a set of option prices for different strikes and expiries and need to find the optimal model for these traded options. The first part of accomplishing this is choice of a suitable model. Essentially, we need to identify the model which best represents the dynamics of the market. This decision generally involves specifying an SDE, which contains a great deal of subjectivity. Once we have specified the model, our goal will be to find the optimal model parameters that give us the best fit to the data. This process is referred to as calibration and requires solving an optimization problem that searches the space of parameters and finds the set that best matches the given market prices. Calibration techniques are discussed in more detail in chapter 14. Importantly, however, because calibration requires searching a potentially multi-dimensional space to choose the optimal parameter set, many evaluations will be required, and we will need each evaluations to be fast. This makes techniques like FFT and quadrature appealing, and will make simulation and the use of PDE's less attractive in these applications.

In most of the models specified, it should be emphasized that we have a sparse set of parameters relative to the potential data of a volatility surface. As a result, these models are overdetermined. That is, we have fewer model parameters than market prices. This means that we will, generally speaking, not be able to fit the volatility surface exactly. For these models, the degree to which we can fit the data will depend on the dynamics of a given market, and as noted will require a subjective decision by the practitioner when choosing the appropriate model. The local volatility model, notably is an exception to this as we are able to find a single, unique solution that matches the volatility surface. This model, however, is also not without challenges, in particular those related to data availability and the underlying assumption of unique option prices for every strike.

As we saw in this chapter, there are many potential approaches that we can employ to solve derivatives modeling problems, each with their own set of strengths and weaknesses. For example, the Heston model and the Variance Gamma model will contain different tail behavior and will subsequently lead to different properties for out-of-the-money options. Intimate knowledge of this behavior, and the strengths and weaknesses of different approaches, is of utmost importance when tackling these problems, and makes these problems a combination of art and science.

CHAPTER 11

Options Pricing Techniques for European Options

11.1 MODELS WITH CLOSED FORM SOLUTIONS OR ASYMPTOTIC APPROXIMATIONS

In the last chapter, we detailed many stochastic models that can be used to solve options pricing problems, ranging from simple models that were defined by computational simplicity but unrealistic assumptions to more complex models that incorporate stochastic volatility and jump processes. In this chapter, we review the pricing techniques associated with these models. When doing this, we focus on vanilla, European call and puts options whose payoff is a function only of the price of the corresponding asset at expiry. In the next chapter, we delve into non-standard, path dependent payoffs. All else equal, we would prefer models that have closed-form, analytical pricing formulas for European options. For some models, this turns out to be possible, such as:

- Black-Scholes (10.2)
- Bachelier (10.8)
- CEV (10.16)

Unfortunately, and perhaps unsurprisingly, this analytical tractability comes at a cost, and that is that these models don't match realistic market behavior. In practice we observe significant volatility smiles in many markets, which none of these models will be able to incorporate fully.

We also saw that the SABR model was defined by a convenient asymptotic formula, (10.26), that can be used to extract a corresponding normal or log-normal volatility.

In the absence of a closed-form solution or this type of asymptotic formula, other techniques will be required. In the remainder of the chapter we discuss two such techniques: quadrature and FFT. The FFT is of particular importance as it enables us to work efficiently with many stochastic vol and jump models in an efficient manner.

DOI: 10.1201/9781003530626-14

11.2 OPTION PRICING VIA QUADRATURE

11.2.1 Overview

In the absence of a closed form solution to European options pricing problems, a natural next method is to rely on quadrature to approximate the pricing integral. Said differently, in order to calculate the pricing integral, we approximate the integral numerically as the sum over a set of discretized points. For this method to work, knowledge of the risk-neutral probability density function for the stochastic process will be required. This is a somewhat limiting factor for the model, as for many more complex models this risk-neutral PDF will not be known. Assuming that this is known, however, then this is a valid and reasonably efficient method.

In the next section, we discuss some of the most common quadrature rules. These rules define how we will approximate our integral over a set of discretized nodes. For example, we might assume that the function is rectangular over that area, and is a function of the mid, left or right points. As we decrease the width of these intervals, we would naturally then expect that the approximation to get increasingly accurate, eventually converging to a solution. In the following section, we describe how to apply these quadrature rules in the setting of an options pricing problem.

11.2.2 Quadrature Approximations

In order to approximate an integral numerically, we can use one of many quadrature rules [7]. These rules define how we approximate the area of each sub-interval. Mathematically, we apply quadrature techniques when we need to approximate integrals of the form:

$$I(f) = \int_a^b f(x)dx \tag{11.1}$$

where $f(x)$ is the function that we are trying to integrate and the upper and lower bounds of the integrals are a and b, respectively. Clearly, we are assuming the anti-derivative of $f(x)$ is not known and cannot be computed analytically and therefore we must instead resort to a numerical approximation. Otherwise, we would perform this calculation analytically rather than resorting to numerical techniques.

To approximate the integral numerically, we can break the interval into N equally spaced sub-intervals[1] and then approximate the integral over each sub-interval:

$$I(f) = \sum_{i=1}^{M} \int_{x_{k-1}}^{x_k} f(x)dx \tag{11.2}$$

where $x_k = a + k\frac{b-a}{M}$ is a uniform grid of nodes and $\int_{x_{k-1}}^{x_k}$ is the integral over each sub-interval. To make the problem more tractable, we would then specify simplifying assumptions about the integral over each sub-interval. The idea is to then shrink the length of the sub-intervals such that the numerical approximation of the integral converges to the integral in (11.1).

[1] Although equal spacing is not required by any means.

For example, in a rectangular quadrature we would simplify the function by assuming that the area of the function is a rectangle in each sub-interval. This is equivalent to assuming the value of the function, $f(x^*)$, is constant over the sub-interval, as the following chart highlights:

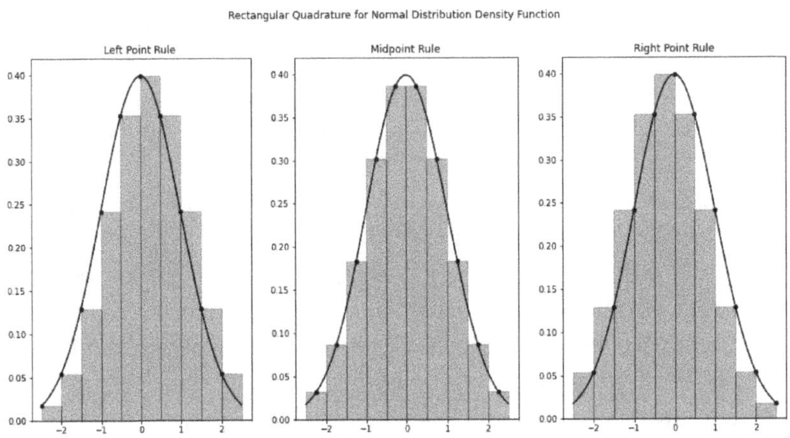

Examples of rectangular quadratures are the left-point, mid-point and right-point rules, each of which are explained below:

- **Left-Point Rule**: assume the function, $f(x)$ takes on a constant value corresponding to the value at the left hand side of the sub-interval, that is:

$$I(f) = \sum_{i=1}^{M} w_i f(x_{k-1}) \tag{11.3}$$

- **Right-Point Rule**: assumed to take on the value at the right hand side of the sub-interval, that is:

$$I(f) = \sum_{i=1}^{M} w_i f(x_k) \tag{11.4}$$

- **Mid-Point Rule**: assumed to take on the value at the mid-point of the sub-interval, that is:

$$I(f) = \sum_{i=1}^{M} w_i f\left(\frac{x_k + x_{k-1}}{2}\right) \tag{11.5}$$

Where in each equation $w_i = \frac{b-a}{M}$ is set to reflect the equally spaced node grid specified above. Clearly, as we increase M we are increasing the number of sub-intervals in our approximation, and a result we expect our approximation to become more accurate. Different quadrature rules, however, have different convergence properties.

For example, it can be proven that we can achieve a significantly higher level of accuracy with the mid-point rule relative to the left or right point rules, as a result, it is preferred in applications. More information on the approximation error of each quadrature rule can be found in [7].

An alternative to the rectangular quadrature is to assume that the function is a trapezoid, rather than a rectangle over a given sub-interval:

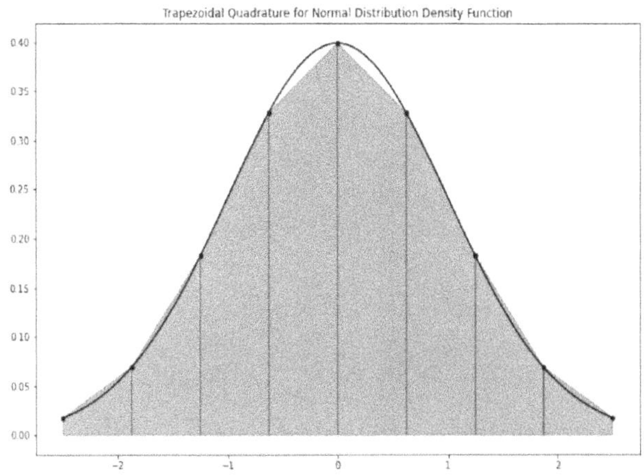

As we can see from the chart, when using the trapezoidal rule we approximate the function as the average of the left-point and right-point rules, forming a trapezoid in each sub-interval.

- **Trapezoidal Rule**

$$I(f) = \sum_{i=1}^{M} \frac{w_i}{2} f\left(f(x_k) + f(x_{k-1})\right) \tag{11.6}$$

where again $w_i = \frac{b-a}{M}$.

If we expand the sum in (11.6) we will see that values at the beginning and end of the grid for x_k will appear once and all others will appear twice.

In this book we have discussed only the simplest quadrature rules. Many other rules exist, such as Simpson's rule and those based on Lagrange Polynomials and Gauss Nodes. Interested readers should consult [115] for more information on these techniques.

11.2.3 Approximating a Pricing Integral via Quadrature

In order to apply one of the quadrature rules discussed in the previous section we begin with the risk neutral pricing formula for a call or put. That is:

$$c \;=\; \tilde{\mathbb{E}}\left[e^{-\int_0^T r_u\, du}(S_T - K)^+\right] \tag{11.7}$$

This payoff can be simplified by breaking the integral into two regions, from $-\infty$ to K and from K to ∞. Over the first interval, the payoff is equal to zero, that is, we have: $\int_{-\infty}^{K}(x-K)^{+}\phi(x)\,dx = 0$. Therefore, we can replace our lower integral limit with K, which enables us to remove the max function as well, leaving us with:

$$c = e^{-\int_0^T r_u\,du}\int_K^{+\infty}(x-K)\phi(x)\,dx \tag{11.8}$$

In order to approximate this integral numerically, we would then need to replace the upper integral limit with some upper bound B:

$$c \approx e^{-\int_0^T r_u\,du}\int_K^{B}(x-K)\phi(x)\,dx \tag{11.9}$$

Assuming the density, $\phi(x)$ is known, then, if we wanted, we could then apply our favorite quadrature rule to (11.9) in order to compute the integral numerically. This would be a legitimate approach, however, instead let's apply a change of variable in (11.9) to formulate the call price as a function of the log of the asset price. That is, let $s=\ln(x)$, $k=\ln(K)$ and $b=\ln(B)$. The pricing equation then becomes:

$$c = e^{-\int_0^T r_u\,du}\int_k^{b}(e^s-e^k)q(s)ds \tag{11.10}$$

where $q(s)$ is the density of the log of the asset price.

We could then equivalently apply our favorite quadrature method to the integral in (11.10). To proceed, we would then need to define a quadrature grid. For example, we could define n nodes that are equally spaced between the integral limits, b and k:

$$\Delta s = (b-k)/n \tag{11.11}$$
$$s_i = k + i\Delta s \tag{11.12}$$

If we were to specify the mid-point rule, our approximation would become:

$$\hat{c} \approx e^{-\int_0^T r_u\,du}\sum_0^{n-1}\left(e^{\frac{s_i+s_{i+1}}{2}}-e^k\right)q\left(\frac{s_i+s_{i+1}}{2}\right)\Delta s \tag{11.13}$$

Similarly, if we chose the trapezoidal rule, the ensuing approximation would become:

$$\hat{c} \approx e^{-\int_0^T r_u\,du}\sum_0^{n-1}\frac{1}{2}\left\{(e^{s_{i+1}}-e^k)q(s_{i+1})+(e^{s_i}-e^k)q(s_i)\right\}\Delta s \tag{11.14}$$

Assuming that the density of the log of the asset price, $q(s)$ is known, then we can apply quadrature techniques to numerically estimate the pricing integral for European options.

While this method is useful, the number of practical applications are somewhat limited by the number of stochastic processes that have a known and analytically tractable density function. In cases where this density is unknown, or intractable to evaluate, we can try an alternative technique, using Fourier Transform techniques to help us price European options in the absence of a known probability density function. This method will open up a whole new set of stochastic processes that are ability to fit the observed behavior of volatility smiles much more effectively.

11.2.4 Quadrature Methods in Practice: Digital Options Prices in Black-Scholes vs. Bachelier Model

In this section we leverage the coding example provided in the repository, using it to price digital options in the Black-Scholes and Bachelier models in order to highlight a key difference between the models. In the following illustration, we compare the price of digital calls, on the left-hand side, and digital puts, on the right hand side:

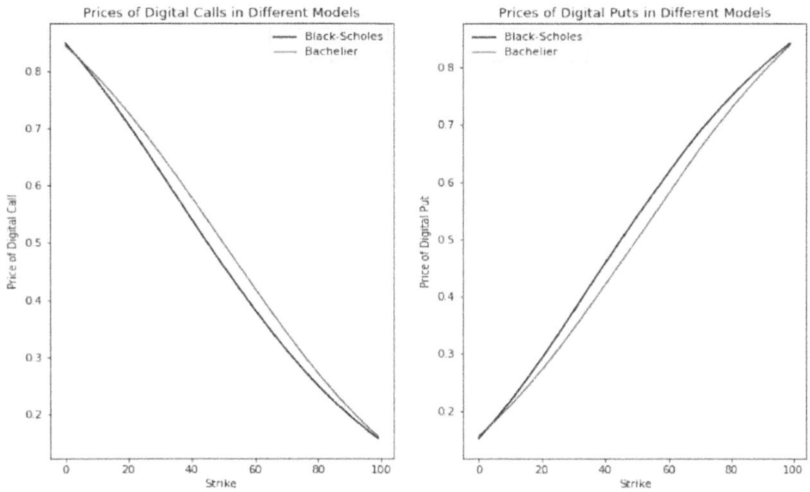

Intuitively, we can see that at the extremes the prices under both models converge to 0 and 1, respectively, when they are deeply in or out-of-the-money. In the center, however, we observe that digital call options are more expensive under the Bachelier model, whereas digit put options are more expensive under the Black-Scholes model. The reader is encouraged to think through this phenomenon and attempt to connect it with the skewness in the right tail that we observe in a log-normal distribution. Essentially, because the forward, and consequently the mean is set by arbitrage, this skewness leads to smaller probabilities closer to the mean, and lower digital call options prices.

11.3 OPTION PRICING VIA FFT

11.3.1 Fourier Transforms & Characteristic Functions

In this chapter we will utilize Fourier transforms and characteristic functions in order to solve options pricing problems. As we will see, this approach will be an attractive alternative to the quadrature based techniques discussed in the last section. Along those lines, in this section we define these foundational concepts before proceeding to show how they can be used in an options pricing setting [133] [47]:

- Fourier Transform

$$\psi(u) = \int e^{iux} h(x)\, dx \tag{11.15}$$

- Inverse Fourier Transform

$$\psi(u) = \frac{1}{2\pi} \int e^{-iux} h(x)\, dx \tag{11.16}$$

- Characteristic Function

$$\psi(u) = \mathbb{E}\left[e^{iuX}\right] = \int e^{iux} \phi(x)\, dx \tag{11.17}$$

Importantly, the **Characteristic Function** is the **Fourier transform** of a function if the function, $h(x)$, is a probability density function. Because of this, we can theoretically use **Fourier Inversion** to obtain a probability density function from a given **Characteristic Function.**

Moment Generating Functions, as defined below, are also closely related to characteristic functions:

$$\psi(u) = \mathbb{E}\left[e^{tX}\right] = \int e^{tx} \phi(x)\, dx$$

In fact, moment-generating functions are the real counterpart of the complex characteristic function. In this setting, however, we will find it preferable to work with characteristic functions as they exist for all distributions. Both moment generating and characteristic functions can be used to obtain moments of a distribution by taking its nth derivative.

Euler's formula [237] states that we can decompose the function $f(u) = e^{iu}$ into sin and cos components, that is:

$$f(u) = \cos(u) + i\sin(u)$$

Therefore, f is a periodic function with period length 2π, meaning:

$$f(u + 2\pi) = f(u)$$

We can also see the periodic nature of the function $f(u) = e^{iu}$ in the chart below:

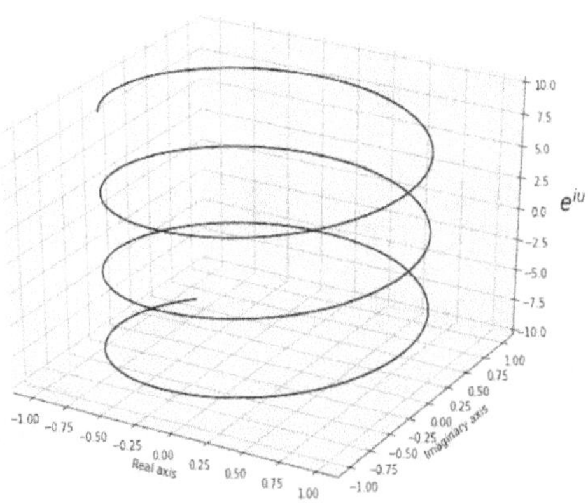

Notice that because the function is periodic its limit as u approaches infinity & negative infinity is not well defined. This is an important point that will come up as we try to formulate our options pricing problem in terms of a characteristic function or Fourier transform.

11.3.2 European Option Pricing via Transform

While the previous technique of using quadrature rules to compute an option price given a known, risk-neutral PDF is natural and intuitive, in practice the number of applications of this technique is limited by the number of processes with unknown or intractable PDFs. In these cases, however, we can rely on methods that leverage a stochastic processes characteristic function rather than its PDF. It turns out this leads to many more practical applications, as many stochastic volatility and jump process models have known, analytically evaluated characteristic functions, as we saw in chapter 13.

The pioneering work in this field was published by Carr and Madan [56] who established the relationship between a stochastic processes characteristic function and European options prices. In this section we derive the standard FFT approach to pricing European options using a characteristic function. In doing this, we follow Hirsa [133] closely[2].

Consider again our standard pricing formula for a European Call:

$$C_T(K) = \tilde{\mathbb{E}}\left[e^{-\int_0^T r_u\,du}(S_T - K)^+\right] \tag{11.18}$$

$$= e^{-\int_0^T r_u\,du}\int_{-\infty}^{+\infty}(x-K)^+\phi(x)\,dx \tag{11.19}$$

In the last section, we saw that if we knew $\phi(x)$ then we could use a quadrature approximation to compute the price of a call option. In the absence of knowledge of this PDF, we will instead transform our problem to work with the characteristic function, which we will find to be known for a plethora of stochastic processes. We begin by changing variables to formulate the call price as a function of the log of the asset price:

$$s = \ln(x) \tag{11.20}$$

$$k = \ln(K) \tag{11.21}$$

This change of variables becomes important and will be analytically convenient when working with the characteristic function $\mathbb{E}\left[e^{iuX}\right]$. After the change of variables, our call pricing formula becomes:

$$C_T(k) = e^{-\int_0^T r_u\,du}\int_k^{+\infty}(e^s - e^k)q(s)\,ds \tag{11.22}$$

where $q(s)$ is now the density of the log of the asset price. It should be emphasized that we are assuming that $q(s)$ is unknown, or intractable, otherwise we would integrate (11.22) directly. Further we assume that the characteristic function of $q(s)$ is known exogenously.

The readers natural first thought might be to try to leverage the relationship between the characteristic function and the PDF defined in (11.17) [122]. In particular,

[2]Readers interested in a more rigorous treatment, or details on other transform based options pricing methods should consult Hirsa's book.

an inefficient, brute force approach would be to evaluate (11.22) by using the Fourier inversion formula defined in (11.16). In doing so, we would extract the probability density function $q(s)$ from its characteristic function directly using Fourier Inversion. This approach however would entail calculating a two-dimensional integral. Instead we will try to formulate our problem such that one of the integrals can be computed analytically. In fact, if a two-dimensional integral was required to price each option, this method would not have the desired efficiency. Luckily, as we will see next, this is not the case.

Instead of the naive approach described above, we do the following:

- We begin by taking the **Fourier Transform** of the option prices $C_T(k)$ viewed as a function of k, and expressing the transform in terms of the **Characteristic Function of the density** of the log of the asset price, $q(s)$.
- Once we have obtained the **Fourier Transform** of the option prices, we can then apply **Fourier Inversion** in order to recover the option price.

To do this, let's start by trying to calculate the Fourier transform of the call price:

$$\Psi(\nu) = \int_{-\infty}^{+\infty} e^{i\nu k} C_T(k) dk \tag{11.23}$$

$$\Psi(\nu) = e^{-\int_0^T r_u \, du} \int_{-\infty}^{+\infty} e^{i\nu k} \left\{ \int_k^{+\infty} (e^s - e^k) q(s) \, ds \right\} dk \tag{11.24}$$

where in the second line we have simply substituted the risk-neutral call pricing formula. Next, we will employ **Fubini's theorem** to switch the order of the integrals, leaving us with:

$$\Psi(\nu) = e^{-\int_0^T r_u \, du} \int_{-\infty}^{+\infty} q(s) \left\{ \int_{-\infty}^{s} e^{i\nu k} (e^s - e^k) \, dk \right\} ds \tag{11.25}$$

Notice that when switching the order of the integrals the limits of integration needed to be modified. Looking at the inner integral, at first glance it appears as though we would be able to calculate it analytically. Further examination, however, reveals a problem that appears because of the periodic nature of the characteristic function. To see this, let's look at the inner integral:

$$\int_{-\infty}^{s} e^{i\nu k} (e^s - e^k) \, dk = e^s \int_{-\infty}^{s} e^{i\nu k} \, dk - \int_{-\infty}^{s} e^{(i\nu+1)k} \, dk \tag{11.26}$$

Focusing on the first term, we have:

$$\int_{-\infty}^{s} e^{i\nu k} dk = \frac{1}{i\nu} e^{i\nu s} - \frac{1}{i\nu} e^{i\nu(-\infty)} \tag{11.27}$$

The problem is revealed by looking at the second term at $-\infty$ which is not well-defined, and certainly does not disappear. Fundamentally, this is a result of the periodic nature of the function $f(u) = e^{iu}$ that we saw in 11.3.1. Therefore, the integral does not converge and we cannot continue. Fortunately, as we will see next, incorporation of a damping factor α will force the integral to converge.

In order to continue, let's define a damping factor of the form $e^{\alpha k}$ where α is a damping factor that we will get to specify and k as before is the log of the strike price. This leads to a damped call price, which has been multiplied by the defined damping factor, of:

$$\tilde{C}_T(k) = e^{\alpha k} C_T(k) \tag{11.28}$$

Next, we continue as before, by calculating the Fourier Transform of the damped option price:

$$\begin{aligned}
\Psi(\nu) &= \int_{-\infty}^{+\infty} e^{i\nu k} \tilde{C}_T(k)\, dk \\
&= e^{-\int_0^T r_u\, du} \int_{-\infty}^{+\infty} e^{(i\nu+\alpha)k} \left\{ \int_k^{+\infty} (e^s - e^k) q(s)\, ds \right\} dk \\
&= e^{-\int_0^T r_u\, du} \int_{-\infty}^{+\infty} q(s) \left\{ \int_{-\infty}^{s} e^{(i\nu+\alpha)k} (e^s - e^k)\, dk \right\} ds
\end{aligned}$$

where in the second line we simply substituted the definition of the call price and in the last line invoked Fubini's theorem again to switch the order of integration and updated the limits of integration accordingly. Examining the inner integral again we now have:

$$\int_{-\infty}^{s} e^{(i\nu+\alpha)k} (e^s - e^k)\, dk = \int_{-\infty}^{s} e^{(i\nu+\alpha)k} e^s\, dk - \int_{-\infty}^{s} e^{(i\nu+\alpha+1)k}\, dk$$

Encouragingly, we can see that if we set $\alpha > 0$, both anti-derivative terms now disappear at $-\infty$ and the integral converges:

$$\begin{aligned}
\int_{-\infty}^{s} e^{(i\nu+\alpha)k} (e^s - e^k)\, dk &= e^s \frac{e^{(\alpha+i\nu)s}}{(\alpha+i\nu)} - \frac{e^{(\alpha+i\nu+1)s}}{(\alpha+i\nu+1)} \\
&= \frac{e^{(\alpha+i\nu+1)s}}{(\alpha+i\nu)(\alpha+i\nu+1)}
\end{aligned} \tag{11.29}$$

Next, Plugging the solution (11.29) of our inner integral back into our equation for $\Psi(\nu)$ leaves us with:

$$\begin{aligned}
\Psi(\nu) &= e^{-\int_0^T r_u\, du} \int_{-\infty}^{+\infty} q(s) \left\{ \frac{e^{(\alpha+i\nu+1)s}}{(\alpha+i\nu)(\alpha+i\nu+1)} \right\} ds \\
&= \frac{e^{-\int_0^T r_u\, du}}{(\alpha+i\nu)(\alpha+i\nu+1)} \int_{-\infty}^{+\infty} q(s) e^{(\alpha+i\nu+1)s} ds
\end{aligned} \tag{11.30}$$

Notice that we are now left with a single integral, and also that this remaining integral is closely related to the characteristic function of $q(s)$. In fact, if we factor our an i we will see that this integral simplifies to a call to the characteristic function:

$$\Psi(\nu) = \frac{e^{-\int_0^T r_u\, du}}{(\alpha+i\nu)(\alpha+i\nu+1)} \int_{-\infty}^{+\infty} q(s) e^{i(\nu-(\alpha+1)i)s} ds \tag{11.31}$$

It should be noted that when making this substitution the identity $i^2 = -1$ was used. We can now see that the remaining integral is just the characteristic function of $q(s)$ evaluated at $\nu - (\alpha+1)i$.

Importantly, if we substitute this back into (11.31) we have now derived the Fourier transform of the call price in terms of the characteristic function $\Phi(\nu)$ of the density $q(s)$:

$$\Psi(\nu) = \frac{e^{-\int_0^T r_u\,du}}{(\alpha+i\nu)(\alpha+i\nu+1)}\Phi(\nu-(\alpha+1)i) \tag{11.32}$$

In particular, if we know the characteristic function of $q(s)$, as we assumed, and as we saw in chapter 13 was often the case, then we can compute the Fourier transform of the call price, $\Psi(\nu)$.

Of course, we want to calculate the call price $C_T(k)$, not the Fourier transform of the call price. But we also learned that we can use Fourier Inversion in order to recover a function from its Fourier transform. We know that the damped call price can be written as:

$$\begin{aligned}\Psi(\nu) &= \int_{-\infty}^{+\infty} e^{i\nu k}\tilde{C}_T(k)\,dk \\ &= \int_{-\infty}^{+\infty} e^{\alpha k}e^{i\nu k}C_T(k)\,dk\end{aligned} \tag{11.33}$$

Solving for the call price, $C_T(k)$ we are left with:

$$C_T(k) = \frac{e^{-\alpha k}}{2\pi}\int_{-\infty}^{+\infty} e^{-i\nu k}\Psi(\nu)\,d\nu, \tag{11.34}$$

where $\Psi(\nu)$ has been computed to be:

$$\Psi(\nu) = \frac{e^{-\int_0^T r_u\,du}}{(\alpha+i\nu)(\alpha+i\nu+1)}\Phi(\nu-(\alpha+1)i) \tag{11.35}$$

Examining (11.16), we can see that this remaining integral in (11.34) is an inverse Fourier transform of the damped call price, which we have formulated in terms of the characteristic function of the log of the asset price.

Finally, as we are interested in extracting an option price, which is a real-valued function and we also know that for a real-valued function, its Fourier Transform is even in its real part and odd in its imaginary part. As we are only concerned with the real part of the function, because we are looking for an option price, we can treat the function as even. This means that the area above and below zero will be the same, and we can replace (11.34) with:

$$C_T(k) = \frac{e^{-\alpha k}}{\pi}\int_{0}^{+\infty} e^{-i\nu k}\Psi(\nu)\,d\nu, \tag{11.36}$$

where $\Psi(\nu)$ is:

$$\Psi(\nu) = \frac{e^{-\int_0^T r_u\,du}}{(\alpha+i\nu)(\alpha+i\nu+1)}\Phi(\nu-(\alpha+1)i) \tag{11.37}$$

Remember that $\alpha > 0$ is a damping factor that is a parameter of the FFT Option Pricing Method. A similar derivation exists for a European put option. In fact, we find that in order to price a put we simply need to specify $\alpha < 0$.

In this section we derived the relationship between the call price and the characteristic function of the density of the log of the asset. If we wish to derive the corresponding formula for additional payoffs we would need to re-derive the analogous relationship between the payoff and the underlying characteristic function. In some cases this may be doable, e.g., for European put and digital options; however in other cases, such as path-dependent options, it will be impossible. In the next section, we consider the case of digital options pricing using characteristic functions and FFT.

11.3.3 Digital Option Pricing via Transform

In the last section, we showed how to derive a pricing formula for a European call in terms of the Characteristic Function using Transform Techniques. One feature of these techniques is that new pricing formulas need to be derived for each option payoff. Along these lines, let's have a look at a slightly modified payoff: European Digital Put Options.

A European Digital Pay pays 1 if the asset is less than the strike at expiry, and 0 otherwise Mathematically, the payoff of a European Digital Put can be expressed as:

$$D_T(K) = \tilde{\mathbb{E}}\left[e^{-\int_0^T r_u\,du}1_{\{S_T<K\}}\right] \tag{11.38}$$

$$= \int_{-\infty}^{K} f(S)\,dS \tag{11.39}$$

$$= \int_{-\infty}^{k} q(s)\,ds \tag{11.40}$$

Where in the final step we applied the same change of variables as in the previous attempt, such that

$$s = \log(S_T) \tag{11.41}$$

$$k = \log(K) \tag{11.42}$$

An astute observer might notice that the pricing equation for a digital option closely resembles the CDF of the underlying risk-neutral distribution, and it turns out this is true. As we can see and as we will explore further in chapter 17 a digital option is a way to buy or sell the CDF of the underlying risk-neutral distribution.

Further, let's include a damping factor, $e^{\alpha k}$ to ensure that the inner pricing integral converges. The price of the damped digital can then be written as:

$$\tilde{D}_T(k) = e^{\alpha k} D_T(k) \tag{11.43}$$

$$= e^{\alpha k}\int_{-\infty}^{k} q(s)\,ds \tag{11.44}$$

If we were to omit this damping factor we would find that the terms at ∞ and $-\infty$ do not disappear, and as a result the integral does not converge, as we saw in the prior example with a European option.

Consider the Fourier Transform of the damped Digital Option Price:

$$\begin{aligned}
\Psi(\nu) &= \int_{-\infty}^{+\infty} e^{i\nu k}\tilde{D}_T(k)dk \\
&= e^{-\int_0^T r_u\,du} \int_{-\infty}^{+\infty} e^{(i\nu+\alpha)k} \left\{ \int_{-\infty}^{k} q(s)\,ds \right\} dk \\
&= e^{-\int_0^T r_u\,du} \int_{-\infty}^{+\infty} q(s) \left\{ \int_{s}^{\infty} e^{(i\nu+\alpha)k}\,dk \right\} ds
\end{aligned}$$

where in the last line we are again employing Fubini's theorem [260] to switch the order of integration. Again, notice the change to the limits of integration when doing so. If we try to take the inner integral, we can see that if our damping factor, $\alpha < 0$, then the term at ∞ disappears and the integral converges.

Taking this inner integral, we are left with:

$$\Psi(\nu) = -\frac{e^{-\int_0^T r_u\,du}}{(i\nu+\alpha)} \int_{-\infty}^{+\infty} q(s)e^{(i\nu+\alpha)s}ds \tag{11.45}$$

Now, as before, all that remains is to express the quantity inside the integral in a way that resembles a call to the characteristic function. In particular, factoring out an i and remembering that $i^2 = -1$ we see the remaining integral is just the characteristic function of $q(s)$ evaluated at $\nu + \alpha i$:

$$\Psi(\nu) = -\frac{e^{-\int_0^T r_u\,du}}{(i\nu+\alpha)} \int_{-\infty}^{+\infty} q(s)e^{i(\nu-\alpha i)s}ds \tag{11.46}$$

$$= -\frac{e^{-\int_0^T r_u\,du}}{(i\nu+\alpha)} \Phi(\nu-\alpha i) \tag{11.47}$$

It should be emphasized that the characteristic function is assumed to be known, in this technique. In fact, in chapter 13, we saw this to be the case for a plethora of models. This includes many stochastic volatility and jump models, where the characteristic function is known and can be evaluated analytically.

Therefore, extracting the digital option price from its Fourier Transform that we just derived amounts to an inverse Fourier Transform operation:

$$D_T(k) = \frac{e^{-\alpha k}}{\pi} \int_0^{\infty} e^{-i\nu k}\Psi(\nu)\,d\nu \tag{11.48}$$

$$\Psi(\nu) = -\frac{e^{-\int_0^T r_u\,du}}{(i\nu+\alpha)} \Phi(\nu-\alpha i) \tag{11.49}$$

where again, Φ, the characteristic function of the model is assumed to be known exogenously. Here, we again used the fact that we are only interested in the real part of $D_T(K)$ and the fact that $D_T(k)$ is an even function in its real part [47] [138].

Importantly, as in the case of the European Call price via Fourier Transform, we were able to simplify the pricing problem into calculation of a single integral. The remaining integral in (11.48) that we are approximating is an **Inverse Fourier Transform**. We could then approximate this remaining integral via quadrature techniques, or use a pre-built FFT algorithm.

However, we can see that, even in the case of simple payoffs, a considerable amount of overhead goes into deriving the Fourier Transform Pricing Formula. This makes employing this technique for non-standard payoffs more challenging as the formula is payoff dependent, and in some cases the payoff will prevent us from calculating the inner integral and using this approach.

11.3.4 Calculating Outer Pricing Integral via Quadrature

So far in this chapter, we have been able to reduce the problem of pricing a European option via its characteristic function to calculating a single integral that was closely related to the inverse Fourier transform of the option price. To see this, consider the case of a European call option. The pricing formula (11.36) still requires us to compute an integral over ν for each option we are pricing. In theory, we can compute this integral numerically using the quadrature techniques discussed earlier in the chapter. Alternatively, it turns out that applying quadrature methods is not the most efficient means of computing this integral. Instead, we will employ a technique called **Fast Fourier Transform** (FFT). Before we introduce the FFT technique, however, let's first see how this works in the context of quadrature.

In this case, in order to compute the remaining integral in (11.36), we first need to choose an upper bound B for our integral. The pricing formula then becomes:

$$C_T(k) \approx \frac{e^{-\alpha k}}{\pi} \int_0^B e^{-i\nu k} \Psi(\nu)\, d\nu \tag{11.50}$$

Next we need to choose a quadrature rule and define our nodes and node weights. In particular, let's divide our interval 0 to B into N equal intervals, that is:

$$\begin{aligned} \nu_i &= (j-1)\Delta\nu & (11.51) \\ \Delta\nu &= \frac{B}{N} & (11.52) \end{aligned}$$

To approximate this integral we can use any of the quadrature rules discussed earlier, such as the mid-point rule or trapezoidal rule[3]. If we were to use the trapezoidal rule, our quadrature approximation would become:

$$C_T(k) \approx \frac{e^{-\alpha k}}{\pi} \sum_{j=1}^{N} \frac{1}{2} \left[e^{-i\nu_j k} \Psi(\nu_j) + e^{-i\nu_{j+1} k} \Psi(\nu_{j+1}) \right] \Delta\nu \tag{11.53}$$

[3]Although for efficiency reasons we would prefer not to use the left or right point quadrature rules.

Expanding this sum, we can see that if $j \neq 1$ and $j \neq (N+1)$ the term will appear twice, and will appear once for $j = 1$ and $j = N+1$. In other words, the function we are integrating numerically is:

$$f(x) = e^{-i\nu k}\Psi(\nu) \tag{11.54}$$

The nodes that we are using in our approximation are:

$$\nu_i = (j-1)\Delta\nu \tag{11.55}$$

$$\Delta\nu = \frac{B}{N} \tag{11.56}$$

Finally, the weight function is:

$$w_j = \begin{cases} \Delta\nu & 1 < j < N+1 \\ \frac{\Delta\nu}{2} & j = 1, N \end{cases} \tag{11.57}$$

We could also apply a different quadrature rule, such as the mid-point rule or Simpson's rule. Alternatively, it turns out that a more efficient approach is to use FFT to take this outer pricing integral, as we explore in the next section. In particular, we will see that FFT techniques lead to a significant improvement in efficiency when pricing options at multiple strikes.

11.3.5 Summary of FFT Algorithm

Fast Fourier Transform (FFT) [67] is an algorithm commonly used in mathematical and quant finance applications where a function is converted from an original domain to a representation in the frequency domain [47]. FFT will be of direct relevance to us when using transform techniques to price options as it allows us to evaluate the outer integral much more efficiently when we are evaluating multiple strikes.

If we use quadrature to approximate the remaining outer integral as detailed above, N operations are required per option. This means N^2 operations will be required to price N options with different strikes. If we leverage the power of FFT, we are able to accomplish this in $N \log N$ total operations, or $\log N$ operations per strike. This is clearly a significant speedup at N increases.

FFT, however, requires these same $N \log N$ operations to price a single strike. In the single strike case, quadrature would only require N operations, conversely making it more efficient than FFT in this situation. FFT achieves this impressive level of efficiency by using a divide and conquer approach and exploiting the periodicity of the Fourier transform when evaluating the option price for multiple strikes. The inner workings of the FFT algorithm are beyond the scope of this book, however, interested readers should consult [67] or [262].

Instead, we take the algorithm as given and focus on aligning our problem so that it is suitable for pre-built FFT algorithms. If we are able to do this, then we can take advantage of the superior efficiency of FFT algorithms to obtain prices for entire volatility surfaces with high accuracy. FFT algorithms allow us to solve problems of the following form:

$$\omega(m) = \sum_{j=1}^{N} e^{-i\frac{2\pi}{N}(j-1)(m-1)} x(j) \tag{11.58}$$

where $m = 1, \ldots, N$

In the next section, we show how we are able to reformulate our outer pricing integral in the form of (11.58) enabling us to benefit from the increased efficiency of the underlying FFT algorithm.

11.3.6 Calculating Outer Pricing Integral via FFT

As we saw in the last section, the FFT algorithm developed by Cooley and Tukey can solve problems in the form of (11.58). Further, remember that the remaining integral that we needed to solve is of the form of (11.36). If we then apply the trapezoidal rule to (11.36), as we did on (11.53) we are left with:

$$\begin{aligned} C_T(k_m) &\approx \frac{e^{-\alpha k_m}}{\pi} \sum_{j=1}^{N} \frac{\Delta\nu}{2} \left[e^{-i\nu_j k_m} \Psi(\nu_j) + e^{-i\nu_{j+1} k_m} \Psi(\nu_{j+1}) \right] \\ &\approx \frac{e^{-\alpha k_m}}{\pi} \frac{\Delta\nu}{2} \left[F(\nu_1) + 2F(\nu_2) \ldots 2F(\nu_N) + F(\nu_{N+1}) \right], \end{aligned}$$

where in the last equation we expanded the sum and used the fact that the first and last term each appear once whereas all other terms appear twice. Further, $F(\nu_j) = e^{-i\nu_j k_m} \Psi(\nu_j)$. The last term, $e^{-i\nu_{N+1}k} \Psi(\nu_{N+1})$ approaches 0 as our upper bound B introduces little approximation error. Said differently, we have the ability to choose B, and choose it such that there is little distributional mass above it. We can just as easily then choose B such that there is no distributional mass on the last node or above B. As a result, we can omit this term from our approximation.

If we omit the last term, then we can create a weight function:

$$w_j = \frac{\Delta\nu}{2}(2 - \delta_{j-1}) \tag{11.59}$$

where δ_{j-1} is an indicator function equal to 1 if $j = 1$ and 0 otherwise. Substituting in the definition of the weight function, we are left with:

$$C_T(k_m) \approx \frac{e^{-\alpha k_m}}{\pi} \sum_{j=1}^{N} e^{-i\nu_j k_m} \Psi(\nu_j) w_j \tag{11.60}$$

Also recall that $\nu_j = (j-1)\Delta\nu$. This means that (11.60) shares the $j-1$ term of the FFT function in (11.58). It does not, however, share the $m-1$ term yet. For this we will need to introduce a grid for the range of strikes we would like to price.

Along these lines, let's define a grid of strikes that we will use in our Fourier Inversion:

$$k_m = \beta + (m-1)\Delta k \tag{11.61}$$

$$\beta = \ln S_0 - \frac{\Delta k N}{2} \tag{11.62}$$

This strike grid may seem somewhat arbitrary, and in fact it is one of a number of strike grids that we could choose. Alternate strike grids are discussed in 11.3.8. A convenient property of this grid however is that the at-the-money strike will fall exactly on the strike grid. To see this, we can set the left hand side of (11.61) equal to $\log S_0$, the at-the-money strike, and solve for the corresponding index, m. Plugging this strike spacing function, k_m into our call price approximation, (11.60), we are left with:

$$C_T(k_m) \approx \frac{e^{-\alpha k_m}}{\pi} \sum_{j=1}^{N} e^{-i((j-1)\Delta\nu)(\beta+(m-1)\Delta k)} \Psi(\nu_j) w_j$$

where here we have also substituted for ν_j. Specifically, we previously defined our grid as:

$$\nu_j = (j-1)\Delta\nu \tag{11.63}$$

$$\Delta\nu = \frac{B}{N} \tag{11.64}$$

Simplifying and rearranging terms, we are left with:

$$C_T(k_m) \approx \frac{e^{-\alpha k_m}}{\pi} \sum_{j=1}^{N} e^{-i(j-1)(m-1)\Delta\nu\Delta k} e^{-i\nu_j\beta} \Psi(\nu_j) w_j$$

This now looks very similar to the function in (11.58). In particular, we can see that both equations have a $-i(j-1)(m-1)$ in the first exponent. Further, if the following condition holds then the exponential terms will be equal:

$$\frac{2\pi}{N} = \Delta\nu\Delta k \tag{11.65}$$

This will then become a necessary condition for using this FFT pricing technique[4]. Lastly, to ensure our call pricing equation is of the same form as (11.58) we can set $x(j) = e^{-i\nu_j\beta}\Psi(\nu_j)w_j$.

Now that we have formulated our pricing problem in terms that FFT algorithms are able to solve, we can now use these algorithms to price N options in $O(N \log N)$. In particular, we have been able to setup our pricing integral, which computed the **Inverse Fourier Transform** as a call to a pre-built FFT package. This will turn out to be an attractive alternative to using quadrature to calculate the outer integral, especially when pricing an entire volatility surface of options for many strikes.

11.3.7 Summary: Option Pricing via FFT

At this point, we have derived all the results that we need in order to use FFT techniques to price European Options. Let's quickly go through the steps that should be taken to implement this technique:

[4]This condition is relaxed when using the Fractional FFT technique which is discussed in [133].

First, we choose a set of parameters corresponding to our FFT technique. These parameters are $\vec{p} = \{\alpha, N, B, \Delta\nu, \Delta k\}$. Note that of these parameters we only have three free parameters. Also remember that the sign of α should depend on whether a call or put is being priced. Specifically:

- **Call Option**: $\alpha > 0$
- **Put Option**: $\alpha < 0$

We also need to choose a corresponding stochastic process that has a known characteristic function, and choose the model parameters that correspond to this stochastic process.

Next, we create a vector of inputs, x_i to provide to our FFT algorithm:

$$x_j = \frac{[(2-\delta_{j-1})\Delta\nu]\, e^{-\int_0^T r_u\, du}}{2(\alpha + i\nu_j)(\alpha + i\nu_j + 1)} e^{-i\left(\ln S_0 - \frac{\Delta k N}{2}\right)\nu_j} \Phi(\nu_j - (\alpha+1)i) \tag{11.66}$$

We then call FFT with our input vector x to obtain an output vector y. From this output vector, y we can recover the option price using the following equation:

$$C_T(k_j) = \frac{e^{-\alpha\left[\ln S_0 - \Delta k\left(\frac{N}{2} - (j-1)\right)\right]}}{\pi} Re(y_j) \tag{11.67}$$

This formula gives us a vector of option prices, $C_T(k)$ that correspond to the strikes on our strike grid.

11.3.8 Strike Spacing Functions

So far, in the derivation of the FFT options pricing framework provided, we have focused on the following strike grid:

$$k_m = \beta + (m-1)\Delta k \tag{11.68}$$

$$\beta = \ln S_0 - \frac{\Delta k N}{2} \tag{11.69}$$

One attractive property of this specification is that it ensures that the at-the-money strike falls exactly on the strike grid. Note that we can find its place on the strike grid by setting $k_m = \ln S_0$ and solving for m. Doing this, we can see that the strike appears at the mid-point of the grid. As at-the-money options are the most common and liquidly traded options, this is certainly a sensible baseline for our strike grid.

Under this grid, however, we can't guarantee that any other strikes will fall exactly on our grid. To extract prices for these strikes, then would require interpolation between the nearest strikes on the grid.

Alternatively, we could simply adjust our strike spacing grid. As an example, if we prefer a different strike is in the center of the strike grid, we could using the following:

$$\beta = \ln K^* - \frac{\Delta k N}{2}, \tag{11.70}$$

where K^* is the strike that we would like to appear on our grid.

More generally, we do have some control of how we setup the strike spacing grid but not complete control. Recall from (11.58) that the FFT algorithm has certain requirements. This means that the $(m-1)\Delta k$ term in our strike spacing grid will be required so that our grid is suitable for the FFT algorithm that we use. Our specification of β, however, is within our control and can be chosen so that it is most convenient for the problem we are trying to solve.

11.3.9 Interpolation of Option Prices

Regardless of the strike grid that we choose, and how we define β, we will only be able to guarantee a single strike appears exactly on the grid. This creates a bit of a conundrum as the main benefit of FFT is its ability to price many strikes very efficiently. If the majority of these strikes, however, are not the strikes that we are looking for, then this is not as big of a selling point.

In practice this leaves us with two choices:

- Invoke a separate call to FFT for each strike so that each strike appears exactly on the grid.
- Invoke a single call to FFT using a dense grid characterized by a high N and high B and rely on interpolation between the strikes on the grid. Generally linear interpolation is used, but other interpolation schemes could be used as well.

Clearly, this presents us with a trade-off between accuracy and efficiency. The first approach will result in less error, as there will be no error introduced by the interpolation scheme, however will be far more computationally intensive. In fact, this method is less efficient than using quadrature to compute our outer pricing integral. The second approach, conversely, will be less accurate but enable us to benefit from the efficiency of FFT when calculating options across strikes.

11.3.10 Technique Parameters

Option pricing via FFT is characterized by the following set of technique parameters:

- α: damping factor introduced to ensure convergence of inner integral.
- N: number of nodes in algorithm, must be of form: $N = 2^n$.
- B: upper bound of pricing integral.
- $\Delta\nu$: spacing between nodes in pricing integral.
- Δk: spacing between nodes on strike spacing grid.

These five parameters are not independent parameters that we can choose freely, however. In particular, as we saw earlier Δk and $\Delta\nu$ must be chosen to make the problem suitable for our FFT algorithm. Additionally, N, B and $\Delta\nu$ together define the spacing grid and are therefore not unconstrained. For example, choosing N and

B defines the value for $\Delta\nu$. As a result, when choosing the technique parameters the following conditions must be satisfied:

$$\begin{aligned} \Delta\nu &= \frac{B}{N} \\ \frac{2\pi}{N} &= \Delta\nu\Delta k \end{aligned}$$

Note that this second conditions means that the strike spacing, Δk and $\Delta\nu$ are inherently related. This leaves us with the ability to choose α and two additional free parameters. Once those are chosen, these relationships will implicitly define the values for the other parameters. Typically, these two remaining free parameters are viewed as either $\vec{p} = \{N, B\}$ or $\vec{p} = \{N, \Delta\nu\}$.

11.3.11 Dependence on Technique Parameters

A desirable property of a pricing technique is a lack of sensitivity to the underlying parameters. Unfortunately, we find this not to be the case with FFT. In this section, we describe the sensitivity to each technique parameter and provide some sense of the trade-off when setting these parameters.

The first parameter, α, was introduced as a damping factor to ensure our inner integral converged and was then removed to obtain an option price. Ideally, we would want our choice of α to not have a significant impact on our model price, and sense it is just a damping factor that we introduced and removed it is reasonable to expect this would be the case. In reality, however, this is not what we observe, instead we observe that different values of α can lead to significantly different prices. In particular we observe that the "wrong" value for alpha will lead to significant pricing errors. This dependence on α is counterintuitive given it is just a damping factor, but nevertheless it is something that readers should be aware of when implementing this technique.

In practice, this creates a challenge as we are generally working with complex stochastic processes with unknown model prices and we do not want our choice of α to bias our results. To safeguard against α biasing our results in this context, we can analyze our option price as a function of α and see what values of α tend to converge to the same solution. Additionally we can look to simplify the model that we are working with such that an analytical solution exists that can be used in comparison. As we gain experience with FFT techniques, we develop some intuition around reasonable values of α in different contexts. A good general rule is to use an $\alpha \approx 1$ in absolute terms.

In addition to α, choice of the additional technique parameters may lead to different option prices when using FFT. As previously mentioned, the remaining set of parameters is defined by two free parameters, and two other parameters which are set implicitly when the free parameters are specified. N is often a parameter that is set as one of these two free parameters. Choosing a higher value for N will mean we are using a tighter strike grid, and make the calculation more accurate for an entire volatility surface. It will also ensure our grid of integration is tight leading to less error. A small value of N, however, may be suitable if we are only interested in a single strike, and we construct our strike grid such that the desired strike is on the grid.

Once N is chosen, we will want to adapt our values of B or $\Delta\nu$ to ensure that they are appropriate. For example, for a small N, a small value of B would lead to potential error from truncating the integral prematurely. Conversely, for large N, a large value of B might result in wasted computational effort going to a region with little to no distributional mass.

The most accurate but also most computationally intensive results will come from high values of N and a large B, or equivalently, a large value of N and a small $\Delta\nu$. This choice of parameters will, however, calculate values for many strikes that are not realistic. It will also leads to a situation where much of the calculation time may potentially focus on parts of the distribution where there is little to no mass. In practice, choice of the best set of FFT parameters end up being somewhat subjective and depends on the problem you are trying to solve.

11.3.12 Strengths and Weaknesses

In this chapter we saw that we could express the price of an option in terms of the Characteristic Function of the density of the log of the asset price. In practice this is quite useful as there are many stochastic processes with known Characteristic Functions. An additional feature of the FFT pricing technique is that it is very general and can be used for many models and asset classes. In particular, we saw in chapter 13 that there are many models with known Characteristic Functions that fit seamlessly into this framework. The general nature of this technique makes it easy to implement via abstract code. In particular, we can implement an FFT Pricing technique that uses an inheritance hierarchy to take different Characteristic Functions for different models dynamically.

Use of this FFT technique also came with considerable potential efficiency benefits. In particular, this method exploits the periodicity of the complex exponential to reduce the computational effort necessary to invert the transform from $O(N^2)$ (naive quadrature) to $O(N \log N)$ (FFT).

The following table summarizes the primary strengths and challenges of using FFT as a pricing tool for European options:

Strengths

- Provides a flexible, efficient method for pricing European options with a wide array of models.
- Prices entire volatility skews with the same effort required to produce single option price.
- Naturally lends itself to a general pricing framework with models easily substituted for each other.

Weaknesses:

- Difficult to extend to non-standard payoffs.

- Impossible to extend to path dependent options.
- Model prices are sensitive to parameters, α, N and B.
- Unable to accurately price highly out-of-the-money options

11.3.13 Variants of FFT Pricing Technique

In this book we detail only the simplest options pricing method that relies on transform techniques. In practice, this is one of a suite of available transform techniques, that can, for example, allow us to disentangle our ν grid from our strike spacing grid. A brief summary of these more advanced methods is presented below:

- **Fractional FFT**: A modification of the standard FFT technique where we decouple the strike spacing grid Δk from the grid related to the integral of the characteristic function $\Delta\nu$.
- **COS Method**[101]: Relies on a Fourier Cosine expansion to using the characteristic function to compute the coefficients in the expansion.
- **Saddlepoint Method** [57]: Treats option prices as tail probabilities and approximates these via a model's Cumulant Generating Function.

These techniques are beyond the scope of this book. Interested readers should consult [133] for a robust, detailed treatment.

11.3.14 FFT Pricing in Practice: Sensitivity to Technique Parameters

In this section we analyze the sensitivity of the underlying FFT algorithm to the input technique parameters. To begin, we start by varying α, the damping factor and find that different values of α can lead to significantly different prices, especially if the value of too low or too high. This phenomenon is depicted in the following chart:

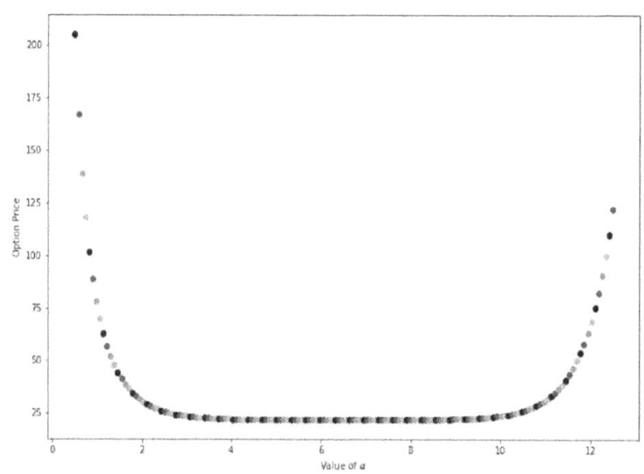

Next, we consider how the FFT price evolves as we vary the number of nodes, N, as well as the strike spacing grid, Δk as is shown in the following charts:

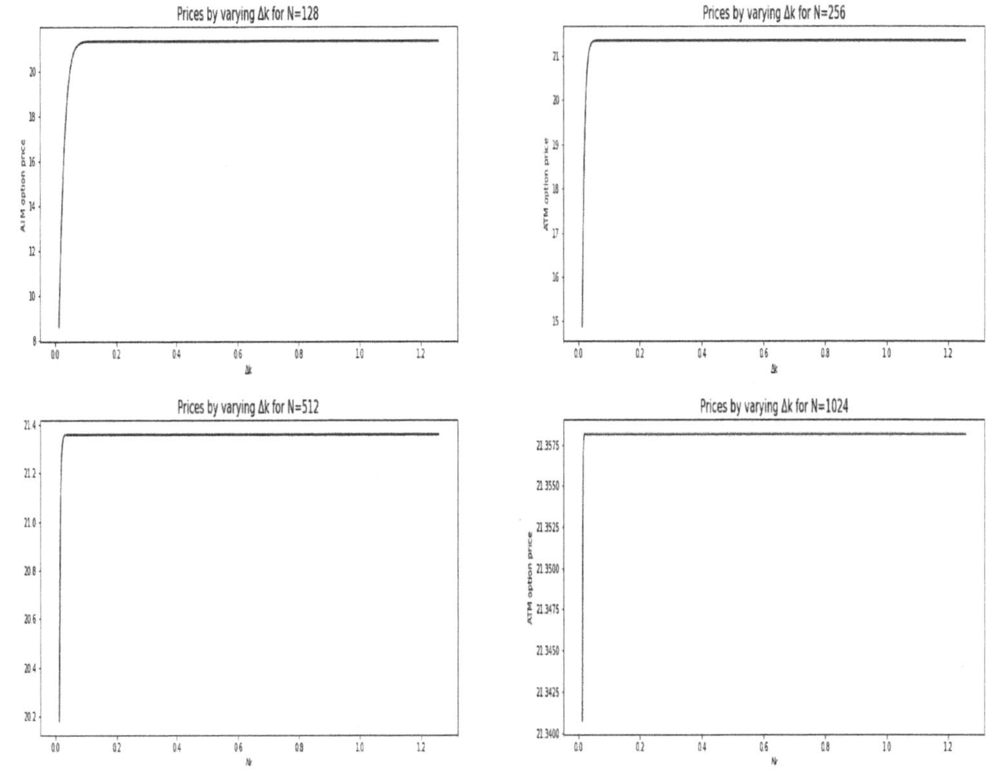

Most importantly, here again we can see that different values of these parameters can lead to different model valuations, which creates a challenge for quants implementing these techniques.

11.4 ROOT FINDING

11.4.1 Setup

A common form of optimization problem is to find the place where a function crosses 0:

$$g(\theta) = 0 \tag{11.71}$$

A solution to (11.71) is called a **root** of g. For purposes of this section we assume we are dealing with functions where the roots cannot be computed analytically, otherwise we would just solve the problem analytically.

In the absence of an analytical solution to the root, an iterative algorithm will be required. This means we will need to start at some initial guess, evaluate the function, and use the information about its value to iterate toward a new, improved guess. We can then do this sequentially until we have hopefully gotten within a certain tolerance of the root. A brief summary of a root finding algorithm for a function g is described below:

(i) Start with some initial guess: θ

(ii) Evaluate the function g at the guess θ and follow the appropriate next step:

   (a) If $g(\theta) > 0$: the value is too high, choose a new θ' such that $g(\theta') < g(\theta)$
   (b) If $g(\theta) < 0$: the value is too low, choose a new θ' so $g(\theta') > g(\theta)$
   (c) If $g(\theta) \approxeq 0$: we have reached the root, no further iterations are necessary

(iii) Update our guess by setting $\theta = \theta'$; repeat step 2

A few questions remain, however, about how to implement this type of iterative algorithm. First, we have to decide how to set the new parameter values, θ', at each step such that we incrementally move closer to the root. Additionally, we have to identify when the root has been found. This means identifying an acceptable threshold ϵ of distance from the root that we are willing to allow.

Root finding problems are commonly encountered in quant finance, for example when extracting an implied volatility from an option price or when extracting a yield-to-maturity from a coupon paying bond price. Many algorithms exists for solving root finding problems. One of the more popular algorithms is Newton's method, which we discuss next.

11.4.2 Newton's Method

Newton's method [262] is an iterative root finding algorithm that uses information from the function and its derivatives to determine the next step in the algorithm, θ'. Newton's method is based on a Taylor series expansion [19] of the function around the current guess $g(\theta)$. This Taylor series expansion can be written as:

$$g(\theta') \approx g(\theta) + g'(\theta)(\theta' - \theta) \tag{11.72}$$

where g' is the Jacobian of g and we are ignoring terms in the expansion beyond first order.

The idea behind Newton's method is that, since finding the root of g itself is impossible, instead we use a linear, first order, approximation to g and find the root of that. We then use that value to refine our guess and determine θ'. To do this, we want to have: $g(\theta') = 0$, therefore we set the left-hand side of (11.72) equal to zero and solve for θ':

$$\theta' = \theta - (g'(\theta))^{-1} g(\theta) \tag{11.73}$$

where $(g'(\theta))^{-1}$ is the inverse of the matrix of derivatives. In the one dimensional case, this equation will simplify to: $\theta' = \theta - \frac{g(\theta)}{g'(\theta)}$

This gives us a first order approximation for the root of the function, $g(x)$. Setting θ' according to this equation at each step is known as **Newton's method**. If we examine the function that we use to generate our next step we can see that it is the previous step, θ, and then a second term which adjusts that step based on the value of the function and its derivatives. We can also see that the second term is the ratio of the function to the derivative. In other words, in the one-dimensional case it is the function value divided by the slope of the function.[5]

One noteworthy feature of Newton's method is that as derivatives get smaller and tend toward zero the algorithms step sizes get larger. This can lead to some instability in functions whose derivatives are not well behaved.

It should be emphasize that we use information about the function $g(\theta)$ as well as its first derivative in each iteration. This makes each iteration more computationally intensive, especially as the number of parameters increases and with it the number of derivatives that need to be estimated, however, in most situations it will require far fewer iterations than competing root finding methods. This is a trade-off that we will see with optimization methods in general. Algorithms that do not require calculation of derivatives evaluate each step faster but require more iterations. Algorithms that incorporate derivative information tend to be slower to evaluate each step but choose their step sizes more efficiently resulting in far less steps.

11.4.3 First Calibration: Implied Volatility

The simplest calibration problem we face in quant finance is an implied volatility calculation. As we saw in earlier chapters, if we know the parameters of the Black-Scholes model r and σ, then we can apply the Black-Scholes formula to obtain the price of a European call or put. Of course, markets don't provide us with the σ but instead provides us with a bid and ask price. This bid-offer two way pricing is generally replaced with a mid-price:

$$\tilde{c_0} = \frac{c_0(\text{bid}) + c_0(\text{ask})}{2} \tag{11.74}$$

We then typically take this mid-price $\tilde{c_0}$ and find the level of σ that enables us to recover that price. Specifically, for each strike we look for the $\sigma_{\text{impl}}(K)$ that solves the following equation:

$$\left(\hat{c}(\tau, K, \sigma) - c_{\tau,K}\right)^2 < \epsilon \tag{11.75}$$

where c is the market price, $\hat{c}$ is the model price and ϵ is the tolerance for our solution.

To extract an **implied volatility**, we need to solve the above one-dimensional optimization problem (11.75). In fact, this is just a root finding exercise so we can apply the techniques learned in this section. Alternatively, these types of root finding exercises can easily be solved in Python using the scipy.optimize function, as we show in the corresponding example in the coding repository. This is the simplest **calibration** problem that we face as quants. Generally, we will have multiple calibration parameters that we need to optimize, as we show later in the chapter.

Note that although an implied volatility surface is a useful visual tool, and helps us compare options across strike and expiry, *it does not provide a consistent set of dynamics for the underlying asset.* In practice we find that this Black-Scholes model which is used to get an implied volatility does not successfully explain the dynamics implied by actual options prices. This means that we will observe different levels of implied volatility for options at different strikes and different expiries. This is a failing

[5]This is also how we would find a y-intercept in a one-dimensional regression.

of the Black-Scholes model as it means there is no single, consistent set of dynamics, or distribution, that represents the volatility surface. Instead, we would have to use a different set of dynamics, or a different underlying distribution, to value options of various strikes and expiries. This is clearly unintuitive and an unsatisfying solution to quants as each option references the same underlying asset.

11.4.4 Implied Volatility in Practice: Volatility Skew for VIX Options

In this section we leverage the code displayed in the previous section to build an implied volatility skew for the VIX index. The resulting skew of three-month VIX options is shown in the following chart:

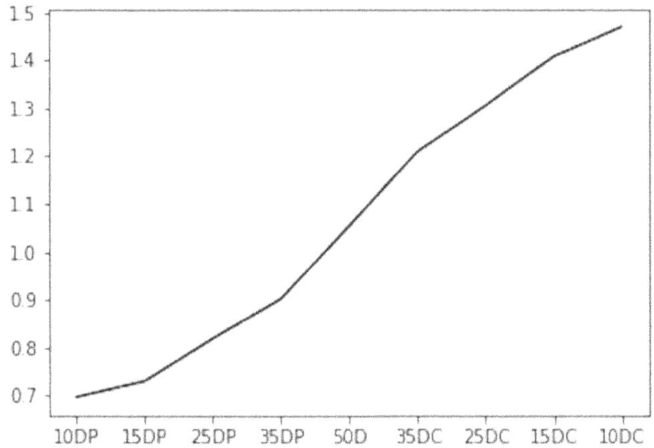

In this chart we can see that skew toward the call side, or, said differently higher implied volatility on higher strike options, is a notable feature of this VIX volatility skew. This also turns out to be a consistent feature in VIX options, both across historical time period and option expiry, and is closely related to put skew in equity options. This phenomenon is driven by risk premium, and the fact that the VIX index measures volatility, a quantity that rises sharply in times of stress. As such, high strike VIX call options have a significant amount of embedded hedging value, and therefore have a potential risk premia associated with them, which is evidenced here by the call skew.

11.5 OPTIMIZATION TECHNIQUES

Optimization is a foundational concept that is ubiquitous in finance both on the buy-side and sell-side[6]. In the majority of quant finance applications of optimization techniques we will need a more general framework than we covered in the last section on root finding. Generally speaking optimization problems in finance are characterized by multiple parameters. A few examples of where optimization techniques may be applied in finance are:

[6] And both in risk-neutral modeling and physical measure modeling

- **Calibration**: Finding an optimal set of parameters to fit a volatility surface
- **Estimation**: Finding the best fit line to forecast an asset price[7] (e.g., via regression)
- **Asset Allocation:** Constructing an optimal portfolio given a set of constraints[8].
- **Yield Curve Construction**: Finding the model rates that best match traded market rate products.[9]
- **Risk Neutral Density Extraction**: Extracting a set of probabilities that best fit market data directly.[10]

As we can see, applications of optimization are wide and span asset classes and modeling approaches. This is intuitive as in many settings we are interested in finding a set of model parameters that provide the best fit to observed market data. In this chapter we focus on applications related to calibration. Other examples are provided later in the book.

In the context of calibration, one way to find the optimal set of parameters would be to find those that minimize the squared pricing error from the market data:

$$\vec{p}_{\min} = \underset{\vec{p}}{\operatorname{argmin}} \left\{ \sum_{\tau,K} \left(\hat{c}(\tau, K, \vec{p}) - c_{\tau,K} \right)^2 \right\} \tag{11.76}$$

where c and $\hat{c}$ are the market and model price for a given option. For example, if we were using the Variance Gamma model, then the parameters would be: $\{\sigma, \theta, \nu\}$. The number of parameters, and the corresponding dimensionality of our optimization depends on the stochastic process that we choose. In the case of Variance Gamma, we have a 3D optimization problem. When working with other models, the space may be larger.

In order to solve equations of the form of (11.76) requires us to be able to solve **optimization problems**. It should be noted that (11.76) is only one possible formulation of a calibration problem. We could choose many different functions to find the parameters that best fit the data. In the following section we show how to formulate and solve these types of optimization problems.

11.5.1 Background & Terminology

Optimization problems have the following form:

$$\hat{x} = \max_{x} \{ f(x) \,|\, g(x) \leq c \} \tag{11.77}$$

---

[7]See chapter 25 for more examples

[8]See chapter 24 for more examples

[9]See chapter 19 for more examples

[10]See chapter 17 for more examples

We refer to $f(x)$ as the **objective function** in our optimization. We refer to $g(x) \leq c$ as the **constraints** in our optimization. In the case of equality constraints, the functions become: $g(x) = c$. Problems with inequality constraints are generally harder to solve than problems with equality constraints. As a result, we generally differentiate between *equality* and *inequality* constraints. We refer to $\hat{x}$ as a maximizer of the function $f(x)$ if it is greater than equal to $f(x)$ at all other points, that is: $f(x) \leq f(x^*) \quad \forall x$.

Many optimization routines in Python only allow us to find minimums. This is not a problem however, as maximizing a function $f(x)$ is the same as minimizing the function $-f(x)$. Therefore, in these situations we can simply use $-f(x)$ as our objective function in the optimization package. It turns out that our ability to transform the objective function is actually a bit more general. For example, we could also apply a constant multiplier to our objective function should it be more convenient. In fact, this type of transformation of an objective function will work for any monotonic function [188]. Mathematically, for a monotonic function h we can re-write our optimization as:

$$\hat{x} = \max_{x} \{h(f(x)) \,|\, g(x) \leq c\} \tag{11.78}$$

To see this intuitively remember that the function is monotonic, meaning that the higher (or lower) the value that is passed in, the higher (or lower) value it will return. That means that maximizing or minimizing the value passed into the function is equivalent to maximizing or minimizing the function itself. This can be a useful trick for making problems more analytically tractable if the right monotonic transformation is applied.

11.5.2 Global vs. Local Minima & Maxima

When solving optimization problems we need to distinguish between local and global minima and maxima. A function may have multiple local minima (maxima) but is characterized by a single global minima (maxima). The following chart shows an example of a function with multiple local maxima:

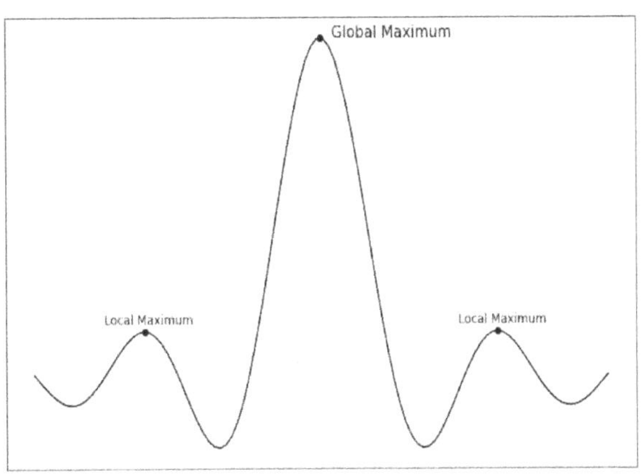

Of course, in calibration exercises, as well as other applications of optimization, we will be looking to find global maxima (minima) rather than local maxima (minima). This will prove to be a more challenging task, however as it inherently much more difficult to find global maxima and minima. Roughly speaking, doing so involves finding all local minima or maxima and selecting the one with the highest value[11]. As a result, when possible it will be crucial to try to formulate our optimizations such that they have a unique local maxima or minima, making our problem more tractable. Because of the increased difficulty finding global extrema, in the remainder of this section we focus on the techniques for find local optimal points.

11.5.3 First- & Second-Order Conditions

We know from calculus that for a differentiable function, first order derivatives tell us whether a function is increasing or decreasing as the variable changes. Stationary points, as a result are of particular interest as they can informally be thought of as where the functions derivative turns, or changes signs. There are three such types of stationary points [237].

- **Local Maxima**: Points where the function stops increasing; characterized by the sign of the derivative changing from positive to negative.
- **Local Minima**: Points where the function stops decreasing; characterized by the sign of the derivative changing from negative to positive.
- **Inflection Point**: Points where the functions changes convexity or concavity but continues to increase or decrease. First derivatives will maintain the same sign on both sides of an inflection point.

Importantly, a property of stationary points is that the functions derivative equals zero. Therefore, when looking for a local maximum or minimum, our approach will be to look for stationary points and confirm whether they are maxima, minima or inflection points. Information about the second derivative can be used to judge this.

Consequently, when searching for a maxima or minima we can do so by finding points that satisfy the so-called first- and second-order conditions. A first-order condition (F.O.C.) is one in which the partial derivative with respect to a given variable is equal to zero:

$$\frac{\partial f(x^*)}{\partial x^*} = 0 \tag{11.79}$$

This is a necessary, but not sufficient conditions for x^* to be a global maximum of $f(x)$. Similarly, a second-order condition is one in which we verify the sign of the second derivative.

[11]Boundary values may also need to be checked as well

The second-order condition (S.O.C.) for a maximum requires that the second derivative be less than zero, meaning that the function switches from increasing to decreasing:

$$\frac{\partial^2 f(x^*)}{\partial x^{*2}} < 0 \tag{11.80}$$

Similarly, the second-order condition for a minimum requires a positive second derivative:

$$\frac{\partial f(x^*)}{\partial x^*} = 0 \tag{11.81}$$

$$\frac{\partial^2 f(x^*)}{\partial x^{*2}} > 0 \tag{11.82}$$

These second-order conditions are necessary but not sufficient conditions for a given point, x^* to be a global maximum or minimum of $f(x)$, respectively.

In the case of a function with many variables, N, the first-order conditions will become a vector, which is referred to as the gradient. That is, the first-order conditions will become: $\nabla f(x^*) = 0$. Similarly, the second-order conditions will become an $N \times N$ matrix of second-order derivatives and cross partial derivatives. The S.O.C. is then that the Hessian is *negative definite* for a maximum and that it is *positive definite* for a minimum. This is analogous to the requirement that they be negative or positive, respectively in the single variable case.

11.5.4 Unconstrained Optimization

Let's start by walking through a very simple unconstrained minimization example:

$$\min_x \quad f(x) \tag{11.83}$$

$$f(x) = x^2 + 3x + 12 \tag{11.84}$$

where $f(x)$ is our objective function and x is our single optimization parameter. This simple unconstrained optimization problem can be solved by finding the point that satisfies the first and second-order conditions:

$$\frac{\partial f(x)}{\partial x} = 0 \tag{11.85}$$

$$\frac{\partial^2 f(x)}{\partial x^2} > 0 \tag{11.86}$$

In this case, our objective function $f(x)$ is easily differentiable leading to the following conditions:

$$\frac{\partial f(x)}{\partial x} = 2x + 3 = 0 \tag{11.87}$$

$$\frac{\partial^2 f(x)}{\partial x^2} = 2 > 0 \tag{11.88}$$

Solving the first-order condition we can see the optimal value of x is $x = -1.5$

which is a minimum for the function $f(x)$. This is validated by the second-order condition being greater than zero at the optimal point. In fact, in this example, we see that the second derivative is positive at all points. That is because the function $f(x)$ is a convex function and means that there is a unique local minimum that is also a global minimum. In cases where the objective function is convex[12], the first-order condition is sufficient to show a point is a minimum. This will turn out to be a very convenient property for objective functions, and wherever possible we will look to formulate our objective functions as convex[13].

11.5.5 Lagrange Multipliers

Constrained optimization problems with equality constraints require that we introduce another piece of theory: Lagrange Multipliers [38]. The key insight behind Lagrange Multipliers is that the function $f(x, y)$ is **tangent** to the constraint $g(x, y) = c$ at optimal points.

Recall that two functions are tangent when their gradients are parallel. Therefore, we have:

$$\nabla f(x, y) = \lambda \nabla g(x, y) \tag{11.89}$$

when the constraint is $g(x, y) = c$.

Lagrange re-wrote this as a single function, referred to as a Lagrangian, which incorporates the constraint. Note the extra variable λ:

$$\begin{aligned} \mathcal{L}(x, y, \lambda) &= f(x, y) - \lambda(g(x, y) - c) \\ \nabla \mathcal{L}(x, y, \lambda) &= \nabla f(x, y) - \lambda \nabla g(x, y) \\ \partial_\lambda \mathcal{L}(x, y, \lambda) &= -(g(x, y) - c) \end{aligned} \tag{11.90}$$

Importantly, it should be emphasized that we have re-formulated our optimization as an **unconstrained** optimization with an additional variable λ. Also notice that if we differentiate the Lagrangian with respect to this new variable, λ we recover the equality constraint. This means that if the first order condition with respect to λ holds, then the constraint will be enforced.

11.5.6 Optimization with Equality Constraints

A natural example of an optimization problem with a single equality constraint is to find the minimum variance portfolio of a fully invested portfolio. To see this, let's consider the case of two assets, leading to the following optimization problem:

$$\min_{w_1, w_2} \{f(w_1, w_2) \,|\, g(w_1, w_2) = c\} \tag{11.91}$$

where

$$\begin{aligned} f(w_1, w_2) &= w_1^2 \sigma_1^2 + w_2^2 \sigma_2^2 - 2 w_1 w_2 \sigma_1 \sigma_2 \rho \\ g(w_1, w_2) &= w_1 + w_2 = 1 \end{aligned} \tag{11.92}$$

[12]Or cases where the objective function is concave for a maximization problem

[13]Or concave if we are solving a maximization problem

where $f(w_1, w_2)$ is the well-known formula for variance of a portfolio with two assets[14] and $g(w_1, w_2)$ is the fully invested constraint. Further, w_1 and w_2 are the portfolio weights, σ_1 and σ_2 are the volatilities of each asset and ρ is the correlation between the two assets.

To solve this problem, we can use **Lagrange Multipliers**. That is, we can rewrite (11.92) in terms of the following Lagrangian:

$$\max_{w_1, w_2, \lambda} \quad \mathcal{L}(w_1, w_2, \lambda)$$
$$\mathcal{L}(w_1, w_2, \lambda) = \left(w_1^2\sigma_1^2 + w_2^2\sigma_2^2 - 2w_1w_2\sigma_1\sigma_2\rho\right) - \lambda\,(w_1 + w_2 - 1) \quad (11.93)$$

Notice that through the use of Lagrange Multipliers we have transformed our problem into an unconstrained optimization. As a result, we can use the same procedure for finding a local maximum as we did in the unconstrained case. That is, we can do so by checking the first and second-order conditions. The First-Order Conditions are:

$$\frac{\partial \mathcal{L}(w_1, w_2, \lambda)}{\partial w_1} = 2w_1\sigma_1^2 - 2w_2\sigma_1\sigma_2\rho - \lambda = 0 \quad (11.94)$$
$$\frac{\partial \mathcal{L}(w_1, w_2, \lambda)}{\partial w_2} = 2w_2\sigma_2^2 - 2w_1\sigma_1\sigma_2\rho - \lambda = 0 \quad (11.95)$$
$$\frac{\partial \mathcal{L}(w_1, w_2, \lambda)}{\partial \lambda} = w_1 + w_2 - 1 = 0 \quad (11.96)$$

This is a system of three equations with three unknowns that can be solved analytically. Doing so yields:

$$w_1 = \frac{\sigma_2^2 - \sigma_1\sigma_2\rho}{\sigma_1^2 + \sigma_2^2 - 2\sigma_1\sigma_2\rho} \quad (11.97)$$
$$w_2 = 1 - w_1 \quad (11.98)$$

We could similarly make the Hessian from the derivatives computed above. However in this case as the objective function can be shown to be convex this step is unnecessary. In chapter 24 we will solve more general version of this problem with various additional constraints.

11.5.7 Minimum Variance Portfolios in Practice: Stock & Bond Minimum Variance Portfolio Weights

In this section we apply the two asset minimum variance portfolio concepts to two assets: the S&P 500 and an overarching US Bond Index (AGG). Doing so yields the following portfolio weights:

We can see that the minimum variance weights are heavily skewed toward the bond index, with only 4% weight in the S&P. Intuitively, we can think of this as because the volatility in the stock index is significantly higher than that of the bond index, meaning that we can lower our variance by holding a portfolio of almost

[14] See chapters 5, 24 and 25 for more information

SPY	AGG
4%	96%

exclusively bonds. It is interesting and noteworthy, however, that the allocation isn't 100% in the bond index. This is an indication the variance in a bond portfolio can be lowered by adding a small amount of equities, which is an example of the benefits of diversification that we have explored in other sections of the book, in action.

11.5.8 Convex Functions

Under certain conditions, we can guarantee that a function will have one and only one local minimum or maximum. These functions will be particularly relevant in optimization because they will significantly reduce the complexity of finding a global minima. In particular this is true for **convex functions**, which have a positive second derivative at all points. This means that for convex functions, if we find a local minimum then we know it is also a global minimum[15]. This reduces the problem of finding a global minimum to the significantly easier problem of finding a local minima, which we can do using the techniques discussed earlier. As this greatly simplifies our optimization problem, it is recommended that we use **convex objective functions** in our minimization problems whenever possible.

11.5.9 Optimization Methods in Practice

When solving any realistic quant finance based optimization problem, we will find that this analytical toolset falls short and instead we need to rely on iterative optimization algorithms. This will in particular be true as we continue to introduce more inequality constraints, and create objective functions that are complex transformations of the parameters. For example, we may be working with objective functions with unknown derivatives with respect to several parameters, making the first- and second-order conditions difficult to evaluate.

In most use cases, quants will (and should) rely on pre-built optimization packages. Nonetheless mastery of the underlying techniques presented in this section are still of the utmost importance as they ensure that the quant is able to react when the optimization doesn't return the expected result. It can also help validate the results returned from the optimizer. Optimization problems are notoriously hard to debug, and knowledge of how the underlying algorithms are working can help isolate potential vulnerabilities. For example, we can see from this section that optimization techniques may struggle in cases where the derivatives are not well-defined. This is also true in cases where the derivatives may have discontinuities. Knowledge of this, based on conceptual understanding of the techniques, can make solving any resulting issues significantly more efficient. In the next section, we explore how optimization techniques can be applied to solve calibration problems.

[15]The same is true of concave functions and global maxima.

11.6 CALIBRATION OF VOLATILITY SURFACES

There are two general methods for calibrating a volatility surface to a set of market data:

- **Parametric**: Begin with a model or stochastic process and find the parameters that lead to the best fit to option prices
- **Non-Parametric**: Attempt to extract a set of probabilities directly

In this chapter we focus on the first approach, where we begin by specifying an SDE and then seek the model parameters that best fit a given set of market data. The second approach, where we extract the probabilities directly is detailed in chapter 17. Fundamentally, however, both cases rely on the same set of optimization tools in order to solve the underlying problem. In one case, we are using an optimizer to find the best set of parameters from an SDE for a given set of data, whereas in another case we are finding a set of probabilities that best fit the data. Either way, we need to formulate and solve and optimization problem to calibrate our model to market data.

In the following sections, we explore the process of calibrating a volatility surface given a specified set of market dynamics. A desirable property of these calibrations would be that the optimal parameters found would be robust to small changes in inputs and different formulations of the problem. We will see, however, that calibration of a stochastic process to a set of market data is often a complicated, ill-posed optimization problem that is highly sensitive. As a result, depending on how we formulate our optimization we might get substantially different results. Against this backdrop, calibration becomes a hybrid of art and science and quants solving these problems need to know how to formulate the problem as robustly as possible, and be aware of potential challenges.

11.6.1 Optimization Formulation

Generally speaking, the optimization problem that we formulate will be to minimize the distance between market and model data. A natural way to do this would be to minimize the squared distance between market and model data:

$$\vec{p}_{\min} = \underset{\vec{p}}{\operatorname{argmin}} \left\{ \sum_{\tau,K} \omega_{\tau,K} \left(\hat{c}(\tau, K, \vec{p}) - c_{\tau,K} \right)^2 \right\} \tag{11.99}$$

where $\vec{p}$ is the set of parameters that correspond to the chosen SDE, $\omega_{\tau,K}$ is the weight we place on each calibration instrument and $c_{\tau,K}$ and $\hat{c}(\tau, K, \vec{p})$ are the market and model prices of each calibration instrument, respectively. As an example, in the case of Heston, $\vec{p} = \{\kappa, \theta, \rho, \psi, \sigma_0\}$ as we saw in chapter 13.

The optimization techniques that we learned in the last section provide us with the tools for solving these types of calibration problems. However, we know from theory that to solve this type of problem we would look for a solution that matches the first and second-order conditions, and would ideally like the objective function to be formulated as a convex function. It should be emphasized, however, that the model prices, $\hat{c}(\tau, K, \vec{p})$ are complex functions of the underlying model parameters. In the case of the Heston model, $\hat{c}(\tau, K, \vec{p})$ will involve a fast Fourier transform and many calls to the complicated characteristic function defined in chapter 13. As a result, solving this problem analytically will be impossible and we will instead have to rely on iterative techniques using pre-built optimization packages. Python has multiple such packages, including scipy and cvxpy.

Depending on the calibration problem we are solving, we may choose to place equal weight in each of the underlying instruments. In other cases, we may choose to place additional weight on options that are closer to the at-the-money strike. Or maybe we are very concerned about calibrating the tail behavior and actually want to place more weight on out-of-the-money options. In other cases, we may choose to determine the weight based on our confidence in the underlying quote, for example by examining the bid-offer spread or weighting the options by their trading volume. In the formulation above we can accomplish this by modifying $\omega_{\tau,K}$.

In many cases we may also choose to place zero weight in options that are in-the-money, as they tend to trade like the underlying and have less information on the volatility surface, and also place zero weight on instruments with one-sided quotes[16].

11.6.2 Objective Functions

Of course, least squares is not the only way that we could formulate (11.99). We could alternatively minimize the sum of the absolute distance between market and model data or use some other power. Additionally, equation (11.99) calculates the absolute pricing error for each instrument, but we could just as easily replace with relative error: $\frac{(\hat{c}(\tau,K,\vec{p})-c_{\tau,K})}{c_{\tau,K}}$.

We could also choose to calculate our error as a function of a different quantity rather than price. For example, we could choose to minimize the distance between the market and model implied volatilities. If we pursue this approach, we would implicitly be placing more weight in out-of-the-money options as their implied volatilities are comparable to those of at-the-money options but their prices are inherently lower.

The optimal method for formulating our optimization problem in (11.99) and choice of a weighting function $\omega_{\tau,K}$ will largely depend on the details of the calibration that we are solving. We will also find that specifying a different weighting scheme and different versions of our objective function can lead to significantly diverging results. As a result it is important to fully understand the implications of these choices and quants should think about the implications of their optimization setup. Are they placing more or less weight on shorter or longer expiry options? Options that are at vs. out-of-the-money? Are these decisions in line with the practical applications of the calibrated model?

[16]That is, where either the bid or offer price is equal to zero

11.6.3 Constraints

Another way to specify our optimization problem would be through the use of pricing constraints. In this case, instead of including an instrument in the squared error term, for example, we could force the optimization to match the price exactly. If we do this, however, there is no guarantee that the resulting optimization would converge as it is possible these prices cannot be matched exactly. This will be especially true if we specify multiple options prices as constraints as we tend to be working with only a few, potentially correlated parameters as in these cases we cannot fit more than a few points exactly. Another related approach would be to define a series of inequality constraints specified using the bid-offer spread of each option.

Additionally, although we have formulated our problem in terms of pricing constraints, there are often **linear constraints** related to the values that the model parameters can take. We can write these linear constraints as:

$$l \leq \vec{p} \leq u \tag{11.100}$$

As an example, in the case of a Heston model, we know that $-1 \leq \rho \leq 1$. For other parameters one of the boundaries will be intuitive whereas the other will be more of a judgment call, such as σ_0 in the Heston model, where we know it must have a lower bound of zero but the upper bound is much less clear. We will see that in addition to being sensitive to the formulation of the optimization and weights on the underlying data, calibrations tend to be sensitive to the chosen upper and lower bounds, as well as the initial starting guess. This is clearly not a desirable feature of a calibration, but is nonetheless something we must learn how to handle robustly. This becomes especially challenging when the underlying pricing technique is itself quite complex, such as in FFT.applications.

11.6.4 Regularization

We can use a **regularization** term in our calibration to try to stabilize the calibration process and make it more robust to many of the different inputs discussed earlier. In particular, it might help in situations where there are multiple local minima by directing the optimizer toward the preferred or global minima. It can also help enforce stability of the parameters in time. This is an important feature as we are often working with potentially correlated parameters that may offset each other, leading to large movements back and forth of these parameters over time.

To add a regularization component, we simply add a term to our objective function that provides some structure to the problem. For example, we could add a regularization term that minimizes distance to some prior set of parameters, or minimizes distance to the optimization's initial guess. As we are able to define the regularization term ourselves, it is sensible to choose it in such a manner that makes the optimization problem as well-posed as possible. In practice, this generally involves formulating the regularization term as a convex function of the parameters when possible[17]. As we

[17]Or a concave function for a maximum

saw earlier, convex functions are characterized by a unique, global minimum, making the optimization problem more tractable.

The following equation shows an example of an augmented objective function taken from (11.99) that includes a regularization term based on distance from a prior set of parameters:

$$\vec{p}_{\min} = \underset{\vec{p}}{\operatorname{argmin}} \left\{ \sum_{\tau,K} \omega_{\tau,K} \left(\hat{c}(\tau, K, \vec{p}) - c_{\tau,K} \right)^2 \right\} + \lambda \sum_{i} (p_i - \hat{p}_i) \tag{11.101}$$

where λ is a constant that specifies the relative weight on the regularization term vs. the least squared distance term. λ can be set to balance the trade-off between fit and stability of the calibration.

11.6.5 Gradient-Based vs. Gradient-Free Optimizers

Optimization algorithms can be classified into two main groups: gradient-based and gradient-free.

- Gradient-based methods utilize information about the function and its derivative in order to determine each iterative step.
- Gradient-free methods utilize only information about the function in order to determine the next step.

In general each step will be more expensive in a gradient-based method because each step requires evaluation of the function and its derivatives. However, gradient-based methods will generally require fewer iterations as they are able to determine an optimal step size at each iteration. Gradient-Free methods are generally preferable in calibration applications because of the complex relationship between objective functions and model parameters. If we were to use a gradient-based approach, we would need to be able to differentiate the objective function with respect to each model parameter, which will generally be intractable. For example, consider the case of an FFT-based pricing technique to calculate $\hat{c}(\tau, K, \vec{p})$. Analytical differentiation will prove to be quite onerous. Instead, we could rely on numerically estimated derivatives using the finite difference techniques we learn in chapter 15, however, the noise in the underlying FFT algorithm will make this a challenge. Consequently, in these types of optimizations we will want to employ gradient-free optimizers.

A notable exception to this will be the Weighted Monte Carlo techniques that we present in chapter 17. Here, the objective functions and constraints will actually be simple functions of the model parameters making gradient-based optimization routines convenient and more efficient.

11.6.6 Gradient-Based Methods with Linear Constraints

Gradient-based optimization requires evaluation of the **objective function**, the **constraints**, as well as the **gradient of the objective function and of the constraints** at each step. For large dimension problems, this means many evaluations of many derivatives are being computed per step in the optimization routine. In this context, linear constraints are of particular interest because they have a constant derivative. This means they only need to be evaluated once rather than once per step. Importantly, this leads to a lower computational burden at each step, and a more efficient optimization procedure.

11.6.7 Practicalities of Calibrating Volatility Surfaces

As we detailed in this section, calibration algorithms tend to be very sensitive to input data. This is clearly not ideal as we would like small changes in market data not to cause significant changes in our calibration results; however, this is not always the case. We also observe that calibrations tend to be very sensitive to the precise details of their formulation, such as how instruments are weighted, what upper and lower bounds are provided on the parameters and what objective function is used. For example, how much weight is given to strikes that are close to near at-the-money strikes vs. out-of-the-money strikes, which is inherently a subjective choice, can have a significant impact on the calibration results.

Generally speaking, the models that we use assume that the market is arbitrage free. If the data violates this assumption then the calibration may not yield meaningful results. This means that the first step in any calibration is to vet the data and perform arbitrage checks based on the relationships defined in 7.5.3. Even in the absence of arbitrage, options data is notoriously dirty, even for the most liquid underlyings. This means that how we clean the underlying data may have a non-trivial impact on our results.

Further complicating any calibration process is the fact that in many cases, the parameters in the SDEs that we are trying to solve are highly correlated. This is in conjunction with complex pricing techniques that make the objective function a complex function of the model parameters. This leads to calibration problems that are sometimes poorly posed. For example, even when we are able to guarantee the objective function is convex, we may find that derivatives are not well behaved. We also often find that these models have a flat bottom with respect to multiple parameters creating a challenge for the optimizer. The end result is that our results are very sensitive to inputs which means building a robust calibration algorithm often requires inclusion of a regularization term.

11.6.8 Calibration in Practice: BRLJPY Currency Options

In this example we leverage the CEV calibration example presented in the supplementary coding repository to calibrate the BRLJPY volatility surface. The calibrated CEV parameters, β and σ, are shown by expiry in the following chart:

The reader may notice that the β parameter decreases as the expiry increases. Recall from 10.1.6 that a lower value of β in the CEV corresponding to a steeper volatility skew. As a result, this pattern of β for BRLJPY indicates that the volatility skew steepens as expiry increases, and in fact this is what we observe in the implied volatility data. For the longer expiries, we find the calibration chooses a β coefficient of 0 which is the value that corresponds to the maximum skew allowed in the model.

Expiry	β	σ
1M	0.62	1.06
2M	0.12	10.70
3M	0.0	18.91
6M	0.0	18.97
1Y	0.0	19.49
2Y	0.0	21.55

This is a reflection of the extreme levels of volatility skew in BRLJPY options, due to the safe-haven nature of JPY and the Emerging Market status of BRL. This phenomenon, as well as other characteristics of volatility surfaces in the currency market, are discussed in more detail in chapter 21. Additionally, we observe that the volatility parameter, σ increases by expiry. The reader is encouraged to keep in mind the difference in the interpretation of σ as β changes when evaluating these results with respect to σ. In particular, as we saw in chapter 13, when $\beta = 0$ the Bachelier model is recovered, which is defined by a σ that is an absolute level of volatility. At the other extreme, when $\beta = 1$ the Black-Scholes model is recovered, leading to a σ that is relative, or percentage terms.

CHAPTER 12

Options Pricing Techniques for Exotic Options

12.1 INTRODUCTION

In the previous chapter we detailed the most common techniques for modeling European options. European options are characterized by a relatively simple payoff structure. Most critically, they only depend on the value of the asset at the option's expiry and not the evolution of the path along the way. This makes them inherently more tractable as we can reduce the entire path of an SDE into a single dimension: the value at expiry. To model options that have more complex payoffs, a new set of techniques will be required. These options, often referred to as exotic options, are characterized by more complex, path dependent payoffs. For example, an exotic option might depend on the minimum or maximum value an asset takes along a given path[1], or its average[2]. In this chapter we focus on two main approaches to handling exotic options: simulation and numerical solutions to partial differential equations. Both of these techniques are important foundational tools that have applications well beyond the scope of exotic options pricing. Simulation in particular is a core technique that quants will continue to use in many contexts.

Exotic options are characterized by more customized structures, less liquidity and less transparent pricing. Many exotics are traded over-the-counter, meaning that, in contrast to European options, market prices for a full set are not directly observed. Instead they are generally only observed when a direct pricing request is made to a dealer or market maker. Additionally, the pricing techniques required in exotic options pricing, such as the ones presented in this chapter, are less efficient than those for European options and too computationally intensive for the purposes of calibration.

A byproduct of this is that exotic options are generally not suitable calibration instruments. Instead, with few exceptions, we focus on calibrating to the European options surface using the techniques discussed in chapter 14 and using the parameters we obtain from this calibration to provide insight into the exotic options of interest.

[1]These are called Lookback options and are detailed in chapter 22.

[2]Which are Asian options and are detailed in chapter 21.

DOI: 10.1201/9781003530626-15

12.2 SIMULATION

12.2.1 Overview

Simulation is the process of using randomly generated numbers in order to approximate an expectation. In contrast to the previous methods that we have studied, which use deterministic estimators, simulation methods rely on random estimators. The idea behind simulation is that we **randomly sample** from some distribution **repeatedly** in order to get an estimate of some function of that distribution. There are some theoretical results, which are summarized in section 12.2.2 that tells us that if this is done properly, the distribution of our sample will converge to the true distribution as we increase the number of draws. Simulation is the most popular technique for dealing with problems with high dimensionality as it does not suffer from the **curse of dimensionality**.

In this chapter we focus on simulation in an option pricing context, however, that is by no means the only place where simulation arises. To the contrary, simulation permeates buy-side and sell-side applications and across the risk neutral measure and physical measure modeling divide. Many of these other applications of simulation are discussed in later chapters, however, the theoretical framework, coupled with options pricing applications is introduced here.

A brief list of other applications of simulation which are discussed in the remainder of the book are:

- **Risk Management**: Computing risk metrics of highly dimensioned complex portfolios. Examples of this are detailed in chapter 26.
- **Portfolio Management**: Simulating asset returns to aid the generation of optimal portfolios. Examples of this are covered in chapter 24.
- **Fixed Income**: Modeling bond prices and interest rate curves which are inherently multi-dimensional. These applications are considered in chapter 19.
- **Credit**: Simulating default events of underlying entities. These applications are discussed in chapter 20.
- **Statistics**: Generating synthetic data from an empirical distribution (e.g., Bootstrapping). Bootstrapping is discussed in chapters 5 and 9.

In the context of options pricing, one of the main benefits of simulation is its ability to handle path dependent payoffs. Generally, simulation techniques are not required for pricing European options, as techniques like FFT and quadrature are inherently more efficient and consequently preferred.

Nonetheless, to warm up let's consider a simplified example of a European option with a known pdf. Clearly, simulation methods are unnecessary for solving these types of problems, instead we could use the techniques discussed in chapter 14. But, this simplified setting will help give us some intuition around simulation before moving to more complex problems.

To proceed, let's revisit our standard risk-neutral valuation formula for a European call:

$$c_0 = \tilde{\mathbb{E}}\left[e^{-rT}(S_T - K)^+\right] \quad (12.1)$$

$$c_0 = e^{-\int_0^T r_u\,du} \int_{-\infty}^{+\infty} (S_T - K)^+ \phi(S_T)\, dS_T \quad (12.2)$$

We have seen how to approximate integrals of the above form using quadrature and FFT. Alternatively, assuming $\phi(S_T)$ is known and we can generate samples from it, another approach to solving this problem is **exact simulation**. To see this, let's define Z_i as a set of i.i.d. random numbers that are sampled from $\phi(S_T)$. The fact that they are i.i.d. means that each draw from $\phi(S_T)$ is *independent*. Additionally, this means that the draws are *identically distributed*, or come from the same distribution, namely $\phi(S_T)$. As we will see later when we discuss variance reduction techniques, we don't necessarily need or want the samples to be independent, but we clearly will always want them to be identically distributed.

Essentially, to create a simulation based approximation, we would generate a sequence of random samples from $\phi(S_T)$, and estimate the option value by calculating its average payoff along the paths. Mathematically, this can be expressed as:

$$c_0 \approx \frac{1}{N}\sum_{i=1}^{N} e^{-rT}(Z_i - K)^+$$

where Z_i is the value of the asset at expiry on the ith draw from $\phi(S_T)$ and N is the total number of simulation paths. The simulation is said to be **exact** because we can sample from the distribution directly. In most simulation applications this will not be possible. There are a few practical applications of exact simulation however, such as Bermudan Swaptions and Forward Starting Options. For Bermudan Swaptions, for example, if possible we might want to simply simulate random numbers at the discrete set of expiry dates rather than form an entire path. As we will see later, when exact simulation is not possible, we can also simulate from the underlying SDE. This type of simulation is far more common, and can also be extended to multiple dimensions.

Of course, to apply these approximations, we need to know that our estimate converges to the true solution. Fortunately it does, which we show in the next section. That means that if we use a large enough sample, we can use the average of the Z_i's in order to estimate our expectation. As a result, a legitimate, albeit inefficient pricing technique for pricing this European option is to repeatedly sample from the density, $\phi(x)$ and calculate the average payoff across all simulated paths.

It should be emphasized that in our applications of simulation we are using Monte Carlo to compute some expectation or integral. This expectation may depend on the asset at maturity, or may depend on the entire path of an asset. Further, the expectation may be based on a single asset or many. The power of simulation algorithms is their ability to handle complex payoffs and multi-dimensional problems seamlessly.

In this chapter we provide an overview of building simulation algorithms within an option pricing context. Those interested in a more detailed treatment should consult [114], [133] or [146].

Generally speaking, a simulation algorithm can be broken down into a few key components. The first component is generating random numbers. We further break this down into two sub-steps, the first of which is to generate random samples from a uniform distribution and the second of which is transforming these random samples from a uniform distribution to an arbitrary and more realistic distribution of our choice. Once we have obtained the correct set of random numbers, the second step is to use these random numbers to generate a path for the underlying asset or assets. In the case above, which we refer to as exact simulation, we sampled from the density at expiry directly, meaning the path would consist of a single random number. More commonly, we will not know the underlying PDF but instead will only know the SDE. As a result, we will need to generate paths by evolving the SDE at small time-steps. Next, once these paths have been created, the final step is to apply the payoff function for our option along each path. Finally, we would then calculate the expectation via Monte Carlo Integration by averaging the payoffs over the full set of draws.

In the remainder of this section, we detail the tools needed to setup each step in the process of building simulation algorithms. Additionally, we begin by examining the convergence properties and highlight the unbiased nature of simulation estimators.

12.2.2 Central Limit Theorem & Law of Large Numbers

The law of large numbers [187] can be used to show that simulation approximations, such as the one defined above in (12.2.1) are unbiased estimators. Said differently, the law of large numbers tells us that the sample mean, the average of the Z_i's, converges as $N \to \infty$ in expectation to the mean of the distribution we are sampling from:

$$\bar{Z}_N = \frac{1}{N}\sum_{i=1}^{N} Z_i \to \tilde{\mathbb{E}}\left[S_T\right] = \mu \tag{12.3}$$

where each Z_i is an independent, identically distributed (i.i.d.) random sample from $\phi(S_T)$ and μ is the expected value of S_T. At a high level, the law of large numbers tells us that if we generate a sufficiently large set of random samples, then our estimator converges to the true, population value, as desired.

Of course, we want to know not only that our estimator converges as $N \to \infty$ but also want to know the speed of convergence. This rate of convergence can be obtained using the central limit theorem [39], which tells us not only that Z_N converges in expectation, but also how the variance changes as we increase N. In particular, if $\phi(S_T)$ has mean μ and variance σ^2, the central limit theorem says:

$$\bar{Z}_N \to N\left(\mu, \frac{\sigma^2}{N}\right) \tag{12.4}$$

In other words, as we increase N, the standard deviation decreases at a rate of $\frac{1}{\sqrt{N}}$. This is equivalent to saying that the rate of convergence of the standard deviation is $\frac{1}{\sqrt{N}}$, or that it is $\mathcal{O}(N^{-\frac{1}{2}})$. This rate of convergence is slower than the other techniques we have considered so far[3], however, importantly, this rate of convergence does not depend on the dimensionality of the simulation algorithm. This is a critical feature of simulation. It is often said that simulation does not suffer from the so-called curse of dimensionality, and this is because its rate of convergence is independent of dimensionality. This will make simulation significantly more efficient in higher dimensions, and as a result it will become our go-to technique in these situations.

In order to see how we arrive at the rate of convergence in (12.4), let's look at the error in our estimate of the mean, defined as ϵ:

$$\epsilon_N = \frac{1}{N}\sum_{i=1}^{N} Z_i - \mu$$

We know from the law of large numbers that as $N \to \infty$ this quantity, ϵ becomes zero. But we also want to know the standard deviation of ϵ_N for a given N, which can be written as:

$$\hat{\sigma}_N = \sqrt{Var(\epsilon_N)} \tag{12.5}$$

$$= \frac{1}{N}\sqrt{Var\left(\sum_{i=1}^{N} Z_i\right)} \tag{12.6}$$

where in the second line we are substituting the definition ϵ and noting that μ is a constant. The remaining term inside the square root is the variance of the sum of random variables. Additionally, as these samples are independent by construction their covariance will be equal to zero. If we then substitute the definition of variance inside the sum, omitting the covariance terms, we are left with:

$$\hat{\sigma}_N = \frac{1}{N}\sqrt{\mathbb{E}\left[\sum_{i=1}^{N} (Z_i - \mu)^2\right]} \tag{12.7}$$

We can then employ Fubini's theorem to switch the order of the sum and expectation:

$$\hat{\sigma}_N = \frac{1}{N}\sqrt{\sum_{i=1}^{N} \mathbb{E}\left[(Z_i - \mu)^2\right]} \tag{12.8}$$

Finally, we know that each Z_i is identically distributed by construction, making each expectation in the sum equal to σ^2. Simplifying, this yields:

$$\hat{\sigma}_N = \frac{\sigma}{\sqrt{N}} \tag{12.9}$$

which provides the well-known convergence rate of simulation algorithms.

[3]For example, in one-dimension standard quadrature techniques have a rate of convergence of $\mathcal{O}(N^{-2})$.

12.2.3 Random Number Generators

In practice, it is not possible for a computer to generate numbers that are truly random. Intuitively, this makes sense as computers are designed to produce predictable outputs as a response to a series of commands. That creates a natural challenge in implementing simulation algorithms, and much work has been done to figure out how to generate numbers that are as close to random as possible. These numbers are referred to as pseudorandom. In this text, we often use the terms random and pseudorandom numbers interchangeably, however, the reader should note the fundamental distinction. We will then refer to the algorithm which generates these pseudorandom numbers for us as a random number generator. These random number generators will be a core building block of our simulation algorithm. They will enable us to generate paths of random samples that we can then use to simulate underlying assets, or portfolios.

Generating pseudorandom uniform numbers is then the most fundamental piece of many Monte Carlo simulations. Clearly, we will need to be able to generate samples from other distributions as well, most notably from normal and exponential[4] distributions. As we will see we can either attempt to do this directly, via a set of convenient built-in Python functions, or by first generating uniform samples and then transforming them to the desired distribution. An alternate approach to pseudorandom numbers is to instead rely on **quasi-random low-discrepancy sequences**. This is discussed in section 12.2.9.

More precisely, a uniform random number generator (RNG) is a method for generating pseudorandom numbers. It should produce a series of numbers $u^1, \dots, u^K \in [0, 1]$, such that:

- Each u^k is uniformly distributed on $[0, 1]$
- The u^k's are independent of each other

It is relatively easy to generate random numbers that are uniformly distributed, but it is harder to generate random samples that are truly independent (i.e. random). Fortunately, Many such random number generators exists for the uniform distribution.

Although much of the work on RNG's has already been done and can be taken as given, and quants in practice should not feel the need to implement customized generators, there are certain desirable properties that are noteworthy. These properties, which are described below, may be beneficial for helping choose between pre-built RNGs:

- **Efficiency:** Many applications require generating a large number of random uniform numbers, therefore, we want our RNG to be as fast as possible.
- **Reproducibility:** Many times, we need to regenerate a sequence of random numbers, e.g., when debugging code and checking for errors and being able to work with the same underlying random numbers will be vital.

[4]Exponential Samples will often come up when modeling jump processes via simulation.

- **Unpredictability:** Because we want the RNG to be as close to random as possible, we do not want the numbers that it generates to be predictable.
- **Robustness:** The output of an RNG should be robust to choice of operating system, programming language, etc. Otherwise, we will have difficulty creating robust simulations.

It should be noted that there is a natural tradeoff between reproducibility and unpredictability. Clearly, we want our numbers to be as unpredictable as possible, as that will be what makes them random. However, unpredictability can create certain practical challenges, for example when we are debugging our algorithm. As an example, if we encounter an error in our code on a specific path, if we are not able to reproduce this path tracing the underlying error will become increasingly challenging. Setting the seed in an RNG will be a critical tool for achieving reproducibility, and will make simulation algorithms much easier to validate and debug.

12.2.4 Generating Random Variables

In this text, the process of generating random numbers is broken into two steps. The first is generating random numbers from the simplest possible distribution, that is a uniform distribution. Uniformly distributed numbers do have some direct practical relevance in finance applications, such as when doing statistical bootstrapping. The vast majority of applications, however, will require us to generate samples from other distributions. Our second step, then, will be to learn how to transform these uniform random variables to the distribution of our choice. The most common distribution of interest might naturally be a normal distribution, as it will arise in every simulation algorithm that includes a Brownian motion component.

In this book, we take this first step of generating random numbers from a uniform distribution as given. For those interested, algorithms for building these uniform(0,1) random number generators are described in [262]. For our purposes, it is sufficient to simply call numpy's random.rand function as outlined in chapter 8. This function will return a set of independent, identically distributed (i.i.d.) uniformal random samples.

Most programming languages, including Python also have functions to generate random variables from other common distributions. This includes samples from a single and multi-variate normal distribution which will be the foundation for simulation from the majority of stochastic processes. As a result, the two-step process presented here for generating random numbers can often be condensed into a single step that simply involves calling a built-in random number generator. Nonetheless, we briefly present the multi-step approach and in particular highlight two techniques for transforming random numbers in the event that the reader is required to sample from a less commonly relied on distribution and needs to leverage these methods.

12.2.5 Transforming Random Numbers

The two most common techniques for transforming uniform random numbers into random numbers from a desired distribution are:

- **Inverse Transform Method**: Enables us to transform our random numbers by inverting the Cumulative Density Function of our desired distribution.
- **Acceptance-Rejection Method**: Enables us to transform our random numbers by first generating samples from a known distribution and then generating an additional sample that enables us to accept or reject whether the sample is in our desired distribution.

In the following sections, we detail each technique and provide examples of using them to generate samples that are exponentially and normally distributed, respectively.

12.2.6 Transforming Random Numbers: Inverse Transform Technique

Recall that we are starting with a uniformly distributed number, U, between 0 and 1. That is, the process for generating U is taken as given and instead we focus on transforming U to a distribution of our choice. The key insight behind the inverse transform technique [79] is to interpret U as the point at which the cumulative probability of our desired distribution matches the random number. Clearly, just like U, the CDF for any distribution is also defined only on the interval between 0 and 1. It is also uniformly distributed over that interval. Put another way, we can think of U as a *random percentile* and notice that this percentile is uniformly distributed.

To be more precise, let X be a random sample from the distribution of our choice. The inverse transform method chooses X such that:

$$X = F^{-1}(U) \tag{12.10}$$

where $U \sim \mathcal{U}(0, 1)$ and F is the CDF of our desired distribution. Clearly, this technique will only be useful in cases where the CDF, $F(x)$ is known and can be inverted. This turns out to be the case in many practical applications, such as for the exponential distribution, as we will see shortly.

To check that (12.10) is a valid sampling technique, let's verify that our random sample in fact comes from the desired distribution [133] [114]:

$$P(F^{-1}(U) < x) = P(U < F(x)) = F(x) \tag{12.11}$$

It should be noted that in the second step the identity $\{U < F(x)\} = x$ is used. So, (12.11) shows that the inverse transform technique does indeed create samples from our desired distribution and is a valid sampling technique.

The inverse transform technique can be applied to any distribution with a known CDF. In many cases, if the PDF is known, we can get the CDF by simply integrating the PDF either analytically or via quadrature methods. In other cases, when we know the **characteristic function** but not the PDF, then we can recover the CDF using FFT techniques[5]. In some cases inverting the CDF might prove challenging, and in these cases we might prefer other methods.

[5]For an example of how to do this, see the Digital option pricing example in our FFT lecture, and recall the natural connection between a Digital option and the CDF of its distribution.

One of the most common uses of the inverse transform technique is for generating samples from an exponential distribution. The exponential distribution is a distribution that corresponds to times between jumps in a **poisson process** and has a single model parameter, λ, which controls the frequency of these jumps. We also saw, in chapter 13, that exponential distributions and poisson processes arise in many stochastic models that contain a jump component. Consequently, generating exponential samples is a directly relevant task for quants. Additionally, the exponential distribution is characterized by an analytically tractable PDF and CDF making the inverse transform technique relatively easy to implement.

The pdf of the exponential distribution is:

$$f(x;\lambda) = \lambda \exp^{-\lambda x}$$

if $x \geq 0$ and 0 otherwise. This pdf has a known anti-derivative which can use to solve for the following cdf:

$$F(x;\lambda) = 1 - e^{-\lambda x} \tag{12.12}$$

The CDF can then be inverted by solving for x in (12.12), that is:

$$\begin{aligned} U &= F(X) \\ &= 1 - e^{-\lambda X} \\ 1 - U &= e^{-\lambda X} \\ X &= -\frac{\log(1-U)}{\lambda} \end{aligned}$$

12.2.7 Transforming Random Numbers: Acceptance Rejection Method

Another common method for generating random samples is the acceptance-rejection method [110] which is ideal when we are trying to sample from some distribution $f(x)$ for which it is difficult to generate random samples directly[6]. Given that we cannot sample directly from our distribution of choice, $f(x)$, we instead introduce another function or distribution $g(x)$ that is more tractable and we can sample from directly. We then apply an acceptance rejection criteria to each sample that maps our sample from $g(x)$ to $f(x)$. Said differently, instead of sampling from $f(x)$ directly, because it is unknown, we sample from a distribution that we do know, $g(x)$ and use its relationship to $f(x)$ to translate our sample.

In this method we begin by specifying some function, $g(x)$, that is referred to as an **envelope function**. This function should be tractable and easy to generate samples from. We choose $g(x)$ such the following property holds:

$$f(x) \leq cg(x) \quad \forall x \tag{12.13}$$

for some constant c. That is, we scale our function $g(x)$ using a constant c, which we get to define, so that the scaled function $cg(x)$ is above our distribution at all points.

[6]However, we do assume that we can evaluate $f(x)$ for a given x.

That is, $cg(x)$ envelopes $f(x)$. The ratio of $f(x)$ to our envelope function, $cg(x)$, will then define our acceptance-rejection criteria.

More precisely, to apply this method, we generate a sample from $g(x)$, which we know to be a tractable function, and accept this as a random sample from $f(x)$ with probability $\frac{f(x)}{cg(x)}$. To determine whether to accept or reject the sample, we then use a $\mathcal{U}(0,1)$ and accept if $u \leq \frac{f(x)}{cg(x)}$.

The following pseudocode shows a sketch of how an acceptance-rejection sampling algorithm would be written:

```
N = 10000
samples = rep(0, N)

j = 1
for i = 1,...,N
  X = GenerateSampleFromGx()
  U = GenerateRandomUniform()

  f = EvaluateFx(X)
  g = EvaluateGx(X)
  p = (f / c g)

  if (U < p)
    samples(j) = X
    j = j + 1
  end if
end for loop
```

As we can see clearly from the coding template, to generate a random sample using this method we need to generate two random numbers: one from our envelope function and another that is $\mathcal{U}(0,1)$, which is used for the accept-reject decision. Further, each draw we reject is completely discarded and not included in our simulation. This means that we are generating two random numbers for the chance to generate a single sample from $f(x)$. In practice, if samples are rejected, then more than two samples on average will be required to generate a single sample from $f(x)$.

A consequence of this is that when using this method we should choose our envelope function such that as many samples as possible are accepted. If a large portion of the samples are rejected, then this method will be inefficient. This means choosing c in (12.13) such that we maximize the portion of accepted samples. It also means choosing envelope functions, $g(x)$ that closely resemble the desired distribution $f(x)$.

To see how acceptance-rejection works in action, let's consider the common task of generating a standard normal random sample. This is clearly of direct relevance as normal random samples will need to be generated when simulating from any model that includes a Brownian motion component in its SDE. As detailed in the supplementary materials, the PDF for the standard normal distribution is:

$$f(x) = \frac{1}{\sqrt{2\pi}} e^{-\frac{x^2}{2}} \tag{12.14}$$

However, instead of working directly with $f(x)$ let's consider sampling from the absolute value of the normal distribution. By symmetry, the probability of standard normal random variable Z and $-Z$ are equal for all Z. As a result, we can choose to work with the PDF of $|Z|$, which is:

$$f(x) = \frac{2}{\sqrt{2\pi}} e^{-\frac{x^2}{2}} \tag{12.15}$$

Next, we need to define our envelope function. This decision is a bit subjective as there are many possible functions that we could choose as $g(x)$ that would be tractable enough to simulate from and could envelope $f(x)$ with the proper choice of c. However, we also want to choose our envelope function such that our algorithm is as efficient as possible. This means having the maximum percentage of samples accepted. For example, a uniform distribution could theoretically be used as an envelope function for a standard normal, however, that would be highly inefficient. It turns out that an exponential density with mean 1 better matches the characteristics of (12.15). This means defining our envelope function, $g(x)$, as:

$$g(x) = e^{-x} \tag{12.16}$$

Recall that in the last section we saw that we can easily sample from an exponential distribution using the inverse transform technique. The next step is to choose c such that (12.13) is satisfied. To do this, we want to find the point at which the ratio between $f(x)$ and $g(x)$ is maximized. We know the ratio can be written as:

$$\frac{f(x)}{g(x)} = \frac{2}{\sqrt{2\pi}} e^{x-\frac{x^2}{2}} \tag{12.17}$$

Again, we want to find the upper bound for this ratio. Once we've done that, we can choose c such that $f(x^*) = cg(x^*)$ at this upper bound. Said differently, we can choose c such that 100% of the samples are accepted at the upper bound.

To do this, we maximize $\frac{f(x)}{g(x)}$ and set c such that:

$$\frac{f(x^*)}{cg(x^*)} = 1 \tag{12.18}$$

$$\frac{f(x^*)}{g(x^*)} = c \tag{12.19}$$

As the first term is a constant and the exponential function is monotonic, we can determine the maximum by focusing on the value in the exponent. Differentiating the exponent in (12.17), we can see that it reaches a maximum at $x^* = 1$. Therefore, in accordance with (12.19), we can set $c = \sqrt{\frac{2e}{\pi}}$, which then makes our acceptance rejection criteria equal to:

$$\frac{f(x)}{cg(x)} = e^{-\frac{(x-1)^2}{2}} \tag{12.20}$$

As this equation can then be used as the acceptance-rejection criteria, we have now developed a framework for sampling $|Z|$ based on the acceptance-rejection method and an exponential envelope function. The final step then, is to determine the sign of Z which can be done via symmetry. In particular, we can generate another uniform random sample and assign a positive of negative sign based on if it is above or below 0.5.

To summarize, the following list shows the steps needed to generate a standard normal via acceptance-rejection:

(i) Generate X from an exponential distribution via the inverse transform technique.

(ii) Generate u_1 from $\mathcal{U}(0,1)$

(iii) Accept the sample if $u_1 < e^{-\frac{(X-1)^2}{2}}$. Set $|Z| = X$. Otherwise reject and go back to the beginning.

(iv) Generate u_2 from $\mathcal{U}(0,1)$. If $u_2 < 0.5$ then $Z = |Z|$. Otherwise set $Z = -|Z|$.

Note that in this case we needed to generate 3 random samples in order to hopefully generate a single standard normal:

- An exponential random sample
- A $\mathcal{U}(0,1)$ to determine whether to accept or reject the exponential sample.
- A $\mathcal{U}(0,1)$ to determine the sign of the standard normal sample.

As this approach requires generating three random samples in order to potentially generate a standard normal, it is generally not seen as the most efficient option. Nonetheless, it is an effective use of the acceptance-rejection method on a distribution of direct relevance. Information on other methods for sampling standard normals is provided in the next section.

12.2.8 Generating Normal Random Variables

Earlier we saw that we can use the acceptance-rejection method to generate samples from a standard normal distribution. This was not the most efficient approach, however, as it required multiple random samples to be created, one from an exponential distribution and two from a uniform distribution in order to generate a single sample from a normal distribution[7]. The acceptance-rejection method is only one of many methods for accomplishing this, and this lack of efficiency makes it an unattractive method in practice. Other methods for generating standard normals include:

- Rational Approximation Method
- Box-Muller Method
- Marsaglia's Polar Method

[7]Further denigrating the efficiency is the fact that some samples will be rejected leading to the three samples on these path being unused.

We omit the details of these methods in this text, and in practice these algorithms can be taken as given, however, they are documented in [133] for those interested.

Once we have standard normal random numbers, we can transform them to be normally distributed with an arbitrary mean, μ and variance, σ^2 using the following well-known equation:

$$X = \mu + \sigma Z \tag{12.21}$$

where $Z \sim \mathcal{N}(0, 1)$ and therefore, $X \sim \mathcal{N}(\mu, \sigma)$

At this point we know how to generate uncorrelated standard normals. This helps us to generate a full path of dW's as each increment is independent. It does not help us simulate the path of multiple correlated assets. This naturally arises in stochastic volatility models that are defined by two correlated Brownian motions for the asset and its volatility. It will also arise for any simulation based on more than one underlying asset. As a result, in addition to being able to generate random normal samples with a given mean and variance, we also will need to be able to generate samples from a multi-variate normal distribution. That is, we need to be able to generate correlated random normals. In order to do this, we need to know how to transform a series of independent normal random variables, which can be generated using (12.21), into a series of correlated random variables that follow a joint normal distribution.

To see how this can be accomplished, let's consider the two dimensional case. That is, let's start with the case of drawing two correlated normals with means μ_1 and μ_2, variances σ_1^2 and σ_2^2 and correlation ρ. The first step in doing this, is to generate two uncorrelated random normals, Z_1 and Z_2 using (12.21). We then need to transform these two samples so that they are correlated with correlation coefficient ρ.

For the first sample, we can simply generate as we did before, that is:

$$X_1 = \mu_1 + \sigma_1 Z_1 \tag{12.22}$$

where $Z_1 \sim \mathcal{N}(0, 1)$.

For the second sample, we cannot simply use Z_2 otherwise the normals would be uncorrelated. Instead, we can create our second sample as part of Z_1 and part of Z_2. In particular, if we select X_2 such that [114]:

$$X_2 = \mu_2 + \sigma_2 \left(\rho Z_1 + \sqrt{1 - \rho^2} Z_2 \right) \tag{12.23}$$

then we can see that X_2 has the desired mean and variance properties, as well as the appropriate correlation to X_1. Consequently, bi-variate normal random numbers with arbitrary means, variances and correlation can be created using (12.22) and (12.23). This approach can be generalized to N correlated random numbers using the following formulas:

$$X = \mu + \sqrt{\Sigma} Z \tag{12.24}$$

where μ is a vector of means and Σ is the covariance matrix of the underlying correlated normals [146].

The reader might notice that creating multi-variate normals requires calculating the square root of a covariance matrix, which can be done using either Cholesky decomposition or Eigenvalue decomposition. Cholesky decomposition, however, requires the underlying covariance matrix to be positive definite whereas Eigenvalue decomposition only requires a positive semi-definite covariance matrix. This makes Eigenvalue decomposition preferable in most applications. This will especially become a concern in higher dimensional simulations where the covariance matrix may not be full rank.

12.2.9 Quasi Random Numbers

The standard approach to generating the random numbers that are used in a simulation algorithm is via pseudorandom numbers. Quasi-random numbers are an alternative to pseudorandom numbers where the premise is that they attempt to sample more efficiently than identically distributed, uncorrelated random points. The intuition here is similar to that of some of the variance reduction techniques, where we may be able to take advantage of knowledge of our function to create more efficient samples. To see this, recall that simulation methods randomly sample points from some space in order to calculate some expectation or integral. But if all we are doing is trying to compute some underlying integral or expectation, then why choose random nodes instead of placing the nodes in some optimal spaces? For example, if we are simulating a $\mathcal{U}(0, 1)$, a quasi-random sequence might observe that most of the previous numbers had been below one-half, for example, and give more weight to numbers above one-half for the rest of the sample.

Sobol Sequences are perhaps the most common quasi-random number/low-discrepancy sequence. Intuitively, we can think of them doing this efficiently by choosing nodes intelligently. Most programming languages, including Python, have packages that include the capability to generate a Sobol sequence [233]. We won't get into the internals of these algorithms in this book, however as they can often lead to increased efficiency interested readers are encouraged to consult [114].

12.2.10 Euler Discretization of SDEs

So far, we have learned how to generate random samples from various distributions, including samples that are jointly normal and exponential. We also saw that, in cases where we the risk-neutral PDF for a given model is known, then European options problems could be solved by sampling directly from this PDF. This method has many limitations, however, as it would only be of use in situations with European style payoffs and for simple models, where more efficient pricing techniques tend to be available[8]. To solve more realistic problems via simulation, such as those involving path dependent payoffs or stochastic volatility models, we instead need to simulate each time step in the underlying SDE. To do this, we need to translate these random samples into paths of the SDE.

[8]Such as the quadrature and FFT techniques described in chapter 14.

SDEs are continuous and describe the evolution of an asset over infinitely small time-steps. In practice, we cannot simulate an SDE exactly but instead need to apply a discretization scheme to approximate the evolution of the SDE. This means choosing a time grid $\Delta t = T/M$ where M is the total number of timesteps and T is the time to expiry. It also means defining how to approximate the integrals over each sub-interval analogous to the decisions we made in our various quadrature rules.

Consider the following generic SDE:

$$\begin{aligned} dS_t &= \mu(S_t)dt + \sigma(S_t)dW_t && (12.25)\\ S(t_0) &= S_0 && (12.26) \end{aligned}$$

In integral form, this is

$$S_t = S_{t_0} + \int_{t_0}^{t} \mu(S_t)\, ds + \int_{t_0}^{t} \sigma(S_t)\, dW(s) \tag{12.27}$$

If we apply the left-point rule to each integral in (12.27) we get the most common discretization scheme, the Euler scheme [161]. It is clear that we can use the left-point rule on the ds integral in (12.27) as it is a deterministic integral. It is less obvious that we can do this to the $dW(s)$ integral [146]. The intuition behind this approximation is that we are specifying a fine grid and the approximation can be expected to be accurate locally. Obviously, the left-point rule is attractive in our simulation as the left-point is the value at the beginning of our sub-interval, which is observed.

In order to apply the left-point rule to the ds integral in (12.27), we can do the following:

$$\int_{t_0}^{T} \mu(S_t)\, ds \approx \mu(S_{t_0})(T - t_0)$$

Essentially, we are estimating the drift term by multiplying the change in time, $T - t_0$ by the drift function evaluated at the asset price at the beginning of the interval.

Similarly, for the $dW(s)$ integral, we can use:

$$\int_{t_0}^{T} \sigma(S_t)\, dW(s) \approx \sigma(S_{t_0})(B_T - B_0) = \sigma(S_{t_0})B_T$$

It should be noted that in both integral approximations we are assuming that $T - t_0 \approx 0$ in order for the approximation to be accurate. That is, we rely on small time-steps. Intuitively, we can see we are approximating the second integral in (12.27) by multiplying a Brownian motion, which we will obtain via a random normal sample by the value of the σ function evaluated with the asset price at the beginning of the period.

Putting this together, we are left with the following approximation for $\hat{S}_T$ using Euler's discretization scheme:

$$\hat{S}_T = S_0 + \mu(S_0)(T - t_0) + \sigma(S_0)B_T$$

where B_T is a standard Brownian Motion with $\mu = 0$ and $\sigma^2 = (T - T_0)$. This can equivalently be written as:

$$\hat{S}_{t_{j+1}} = \hat{S}_{t_j} + \mu(\hat{S}_{t_j})\Delta t + \sigma(\hat{S}_{t_j})\sqrt{\Delta t} Z_j \tag{12.28}$$

where Z_j is a standard normal random sample. Equation (12.28) is called the **Euler discretization** of S_T. It can be proven that if Δt is very small, then the distribution of $\hat{S}_T$ is very similar to the distribution of S_T. That is, Euler's discretization scheme is accurate as long as we specify a sufficiently fine grid of time-steps in our algorithm. Fundamentally, equation (12.28) defines how to take steps along the path of a given SDE. This can then be used to transform a sequence of relevant random numbers to paths of underlying asset prices[9].

The Euler scheme is the most commonly used discretization scheme, however, it is by no means the only approach. Intuitively, we can think of these techniques as trying to achieve higher levels of accuracy by performing an Ito-Taylor series expansion of (12.25) and (12.27) and trying to have our discretization schemes account for additional terms in the expansion.

To see this, consider the case of simulating from a standard geometric Brownian motion process. We know from Ito's Lemma in chapter 4 that the solution to the SDE includes a correction term in the form of $\frac{\sigma^2 t}{2}$. This term is noticeably absent from our discretization scheme and it is intuitive to then think that we might want to incorporate it in our simulation algorithm.

Two of the most common alternative schemes that use this approach are the **Milstein scheme** and the **Runge-Kutta scheme**. These more advanced techniques are omitted here, however, those interested should see [133] for more details.

12.2.11 Simulating from Geometric Brownian Motion

A natural first example of an SDE to simulate from is Geometric Brownian motion, or the dynamics behind the Black-Scholes model. Recall from chapter 13 that this model is defined by the following SDE:

$$dS_t = \mu S_t dt + \sigma S_t dW_t \tag{12.29}$$

where μ and σ are constants.

If we discretize this SDE according to Euler's scheme, then at each step our simulation approximation becomes:

$$\hat{S}_{t_{j+1}} = \hat{S}_{t_j} + \mu \hat{S}_{t_j} \Delta t + \sigma \hat{S}_{t_j} \sqrt{\Delta t} Z_j \tag{12.30}$$

where Z_j is a standard normal distribution and $dW_t \sim \mathcal{N}(0, \Delta t)$.

Equation (12.30) could then be applied at each time increment in a simulation algorithm to generate path's of the asset price. Alternatively, as we saw in chapter 4, it just so happens that we know the solution for this SDE as well, which is:

$$S_T = S_0 \exp\left\{\left(\mu - \frac{\sigma^2}{2}\right) T + \sigma dW_T\right\} \tag{12.31}$$

This means that instead of simulating from the SDE we can simulate the process

[9] Or more than one asset.

directly using its solution. This is sometimes referred to as exact simulation and in this context could be done using the following formula:

$$S_T = S_0 \exp\left\{\left(\mu - \frac{\sigma^2}{2}\right)T + \sigma\sqrt{T}Z\right\} \tag{12.32}$$

where Z is a standard normal random variable. For most SDEs that we simulate from, the solution to the SDE will be unknown and simulation from the discretized SDE as in (12.30) will be required. Where possible, however, simulating from the solution to the SDE reduces the error in the simulation approximation. As a result, it is a preferred approach when possible, but its applications are limited in practice due to the subset of models with tractable known solutions.

12.2.12 Simulating from the Heston Model

In the case of simulating from a Heston model, or any other stochastic volatility model, we need to simulate a system of two SDEs. Additionally, unlike in the above example the solution to the SDE will be unknown and we will have no choice but to resort to discretizing the SDE. Further, along each step in the simulation we will need to evolve two underlying stochastic processes, one for the asset price and the other for the latent volatility process. To do this, we need to apply the techniques previously covered to generate correlated random normals that can be used to simulate the asset and its volatility consistently with its model correlation, ρ.

Recall from chapter 13 that the Heston model is defined by the following system of SDE's:

$$\begin{aligned} dS_t &= rS_t\,dt + \sqrt{\nu_t}S_t\,dW_t^1 \\ d\nu_t &= \kappa(\theta - \nu_t)\,dt + \sigma\sqrt{\nu_t}\,dW_t^2 \\ Cov(dW_t^1, dW_t^2) &= \rho\,dt \end{aligned} \tag{12.33}$$

Applying Euler's discretization scheme to this SDE leaves us with:

$$\hat{S}_{t_{j+1}} = \hat{S}_{t_j} + \mu\hat{S}_{t_j}\Delta t + \sqrt{\nu_{t_j}}\hat{S}_{t_j}\sqrt{\Delta t}Z_j^1 \tag{12.34}$$

$$\hat{\nu}_{t_{j+1}} = \hat{\nu}_{t_j} + \kappa(\theta - \nu_{t_j})\Delta t + \sigma\sqrt{\nu_{t_j}}\sqrt{\Delta t}Z_j^2 \tag{12.35}$$

where Z_j^1 and Z_j^2 are standard normal random variables with correlation ρ.

Note that again we are using the distribution of each Brownian increment to express the discretized SDE in terms of standard normals, namely $dW_t \sim \mathcal{N}(0, \Delta t)$. As we are working with a system of two stochastic processes, we can directly use the bi-variate sampling equations (12.22) and (12.23) to generate the correlated normals Z_j^1 and Z_j^2.[10]

One of the practical issues with applying simulation techniques to the Heston model is that there is some positive probability that the volatility process will become negative. Obviously negative volatilities do not make sense conceptually and will also not work well in our dynamics. Therefore, we will need to adjust our model accordingly to prevent the volatility from becoming negative.

[10]Or alternatively, we could apply equation (12.24).

It should be emphasized that this possibility of a negative volatility is a manifestation of our simulation approximation. That is, the continuous version of the SDE does preclude negative volatilities. This is because of the $\sqrt{\nu_t}$ term in the Brownian component, which will decrease the magnitude of the change as the process ν_t falls, preventing a negative value. Once we discretize, however, this is no longer the case and we will see that with non-zero probability we can generate a Z_j^2 that is sufficiently large to push the process into negative territory [10].

There are three main approaches for handling the possibility of negative volatilities in the Heston model:

- Treat 0 as an absorbing state for volatility
- Truncate the value at 0 by applying a max function to the volatility process.
- Reflect the volatility back above 0 if it goes negative.

When choosing between these three approaches, one should keep in mind the impact to the underlying volatility. As an example, if we choose to make zero an absorbing boundary for volatility, then we are implicitly damping the overall volatility as we are removing uncertainty on certain paths. Conversely, if we choose to make zero a reflecting boundary, this will create an upward bias on volatility. These impacts may be minimal if the probability of obtaining a path with negative volatility is low, however, in practice this can have an non-trivial effect on our price estimates.

12.2.13 Simulating from the Variance Gamma Model

In our final simulation example, we consider simulating from a model with a jump process, the Variance Gamma model. Recall from chapter 13 that Variance Gamma can be viewed as an arithmetic Brownian motion process evaluated with a stochastic clock. This underlying stochastic clock is driven by a jump process defined by the gamma distribution. As a result, simulating from the Variance Gamma model will require generating two random samples, one that is from a gamma distribution, and another that is normally distributed.

Specifically, recall that the Variance Gamma model can be written as:

$$X(t;\sigma,\nu,\theta) = \theta\gamma(t;1,\nu) + \sigma W(\gamma(t;1,v)) \tag{12.36}$$

$$\ln(S_t) = \ln(S_0) + (r - q + \omega)t + X(t;\sigma,\nu,\theta) \tag{12.37}$$

At the heart of simulating from the Variance Gamma model is simulating $X(t;\sigma,\nu,\theta)$. This requires a random sample from a gamma distribution with mean 1 and variance ν, $\Delta(\gamma)$ and a standard normal random sample Z_i. In this text we do not discuss the procedure for generating random samples from a gamma distribution, however, the Python function numpy.random.gamma can be used to do so [176]. Paths for $X(t_i;\sigma,\nu,\theta)$ can then be generated using this discretized SDE:

$$X(t_i;\sigma,\nu,\theta) = X(t_{i-1};\sigma,\nu,\theta) + \theta\Delta(\gamma) + \sigma\sqrt{\Delta(\gamma)}Z_i \tag{12.38}$$

The values of $X(t_i;\sigma,\nu,\theta)$ obtained in the above formula can then be plugged into (12.37) to obtain paths for the asset price, as all of the other terms are known. In the coding repository, we present an example using this technique to simulate from a variance gamma model. It should also be noted that there are additional approaches to simulating from the Variance Gamma model. Interested readers should consult [133] for more information.

12.2.14 Variance Reduction Techniques

The most commonly cited challenge associated with employing simulation techniques is its speed. In spite of this slower convergence relative to other techniques, at least in one-dimensional problems, simulation is a popular and broadly used technique. As a result, any efficiency gains that can be made in simulation algorithms are of the utmost importance. Along those lines, variance reduction techniques, which can potentially improve on this rate of convergence, are a hugely important field of study.

In this text we detail four of the most common variance reduction techniques:

- Antithetic Variables
- Control Variates
- Importance Sampling
- Common Random Numbers

Variance Reduction Techniques attempt to use information about the problem that we are working on (e.g. Control Variates) or information about the samples (e.g. Antithetic Variables) in order to reduce the variance of the samples. In the remainder of this section, each of this techniques is discussed in further detail. Auxiliary treatment of each of these techniques can also be found in [133] or [114]. There are also other variance reduction techniques, such as stratified sampling and conditioning which are omitted here but covered in those texts.

- **Antithetic Variables**:

  Our first variance reduction technique is the use of antithetic variables [114]. The idea behind this technique is to use correlated pairs of samples rather than independent samples. The antithetic variables method uses the correlation between these pairs in order to reduce the variance of the estimator.

  To see this, consider a pair of samples X_1 and X_2 from $\phi(x)$ which has mean and variance μ and σ^2, respectively. The mean of the pair of samples, $\hat{\mu}$ is:

$$\hat{\mu} = \frac{1}{2}(X_1 + X_2) \tag{12.39}$$

  Clearly, $\hat{\mu}$ is an unbiased estimator of μ. Further, using the definition of the variance of the sum of random variables, the variance of μ can be written as:

$$Var\left(\hat{\mu}\right) = \frac{1}{4}\text{var}(X_1) + \frac{1}{4}\text{var}(X_2) + \frac{1}{2}\text{cov}(X_1, X_2) \tag{12.40}$$

In the case of i.i.d. samples, the final covariance term disappears, we are left with only the two variance terms. This leads to the familiar convergence rate of $\sqrt{N}$. Equation (12.40) also shows that the variance of our estimator, μ is smallest when the pairs of samples are perfectly negatively correlated. Further, if the correlation between the pair is negative, then the variance of our estimator will be smaller than when using i.i.d. samples. As a result, the use of such pairs is attractive when we are able to create them such that they are negatively correlated.

In order to apply the antithetic variable technique, we need to leverage our knowledge of the underlying distribution in order to generate samples that are significantly negatively correlated. For most distributions, this process of generating negatively correlated pairs is fairly easy. For example, for a standard normal distribution the pair $(Z, -Z)$ is perfectly negatively correlated and is a natural choice for antithetic variables. Another example would be $(U, 1-U)$ in the case of uniform random samples between 0 and 1. This pair would again be perfectly negatively correlated.

- **Control Variates**

The Control Variate method [133] uses information about a related quantity that has known expectation in order to give us information about a harder problem. In practice this could mean that we are pricing an exotic option with unknown expectation, and using a European call option with known expectation as a control variate. At a high level, we can think of this as adjusting our estimator based on the deviation of our control variate from its known expected value.

When using a control variate, we use the adjusted estimator:

$$\tilde{\theta} = \mathbb{E}\left(F(X)\right) + c(Z - \mathbb{E}\left(Z\right)) \tag{12.41}$$

where $F(X)$ is the payoff function, Z is the value of the control variate, $\mathbb{E}\left(Z\right)$ is the expected value of the control variate and c is a constant that determines the relative weight between the first and second terms. This is in contrast to our standard simulation estimator: $\theta = \mathbb{E}\left(F(X)\right)$. The adjusted estimator defined in (12.41) is still unbiased, as clearly:

$$\mathbb{E}\left[\tilde{\theta}\right] = \theta \tag{12.42}$$

Further, we can write the variance of our new estimator is:

$$\begin{aligned}
\text{var}(\tilde{\theta}) &= \text{var}(\mathbb{E}\left(F(X)\right) + c(Z - \mathbb{E}\left(Z\right))) && (12.43)\\
&= \text{var}(\theta) + (c^2)\text{var}(Z) + (2c)\text{cov}(\theta, Z) && (12.44)
\end{aligned}$$

where in the second step we are again substituting the definition of the variance of the sum of random variables. Recall that c is a constant that can be chosen to minimize the variance of our estimator. To do this, we can use the techniques covered in chapter 14. In particular, we can find the first-order condition by differentiating (12.44) with respect to c. Doing this, we see that c is minimized at:

$$(2c)\text{var}(Z) + 2\text{cov}(\theta, Z) = 0 \tag{12.45}$$

$$c^* = -\frac{\text{cov}(\theta, Z)}{\text{var}(Z)} \tag{12.46}$$

If we plug our solution c^* back into (12.44), we get:

$$\text{var}(\tilde{\theta}) = \text{var}(\theta) - \frac{\text{cov}(\theta, Z)^2}{\text{var}(Z)}$$

Therefore, the variance is lower than $\text{var}(\theta)$ as long as the covariance is not equal to zero. In particular, we can see that control variates will be effective tools for reducing variance if they are highly positively or negatively correlated with the quantity we are trying to calculate. This of course creates a bit of a conundrum as we by construction know little about the payoff we are looking to estimate, $F(X)$, so gauging its correlation to a control variate can present a challenge. Nonetheless, in the context of options pricing, we can often look for similarities in the payoff structure in order to define effective control variates.

- **Importance Sampling**

  Importance sampling [146] is another variance reduction technique that is very closely related to Girsanov's theorem[11]. Essentially, importance sampling works by changing the pricing measure such that the samples contain the most information possible. This variance reduction technique is most commonly applied when simulating rare events. In fact, a canonical example of importance sampling is trying to estimate the probability of 5+ standard deviation moves in a Normal Distribution. Doing this directly would require an unreasonably large number of paths as the probability of obtaining a relevant sample is so low. Instead, when we are attempting to simulate events with very low probability, we can change measure to increase the probability of our desired event, calculate its probability, and then change measure back to adjust the probability back to our original measure. If implemented properly, an estimator with importance sampling can be constructed to have lower variance while still being unbiased.

  To see how importance sampling works, suppose we are computing the expectation of some payoff function $F(x)$ and let's define θ as our estimator: $\theta = E = \mathbb{E}_\phi[F(X)]$.

  Further suppose that X follows the pdf $\phi(x)$. The expected value of our payoff

[11]See chapter 4 for more information on Girsanov's theorem.

can then be written as:

$$\begin{aligned}\theta &= \int F(x)\phi(x)dx \\ &= \int F(x)\frac{\phi(x)}{\phi_g(x)}\phi_g(x)dx \\ &= \mathbb{E}_g\left(F(X)\frac{\phi(X)}{\phi_g(X)}\right)\end{aligned}$$

where in the second line we are changing our probability measure from $\phi(x)$ to $\phi_g(x)$ and in the last step we are writing our payoff as an expectation against this probability measure $\phi_g(x)$.

Therefore, we can compute our estimator in a different probability measure, $g(X)$, and apply an adjusted payoff function $F(X)\frac{\phi(X)}{\phi_g(X)}$. That leaves us with the following importance sampling estimator:

$$\tilde{\theta} = \frac{1}{N}\sum_{i=1}^{N} F(X_i)\frac{\phi(X_i)}{\phi_g(X_i)} \tag{12.47}$$

where the ratio $\frac{\phi(X_i)}{\phi_g(X_i)}$ can be thought of as the term that adjusts the probabilities back to the original measure.

- **Common Random Numbers**:

  The final variance reduction technique that we cover is the use of common random numbers. The idea behind common random numbers [133] is to reuse the same random numbers instead of creating new samples for different parts of related calculations. If done properly, doing so reduces a source of noise in the underlying calculations, leading to smaller sampling errors. The canonical example of common random numbers is for calculation of Greeks[12]. In the context of a simulation algorithm, Greeks are traditionally computed via finite differences. As an example, to calculate Δ via finite differences we would use the following formula:

$$\Delta = \frac{P(S+\Delta S,\vec{p}) - P(S-\Delta S,\vec{p})}{2\Delta S} \tag{12.48}$$

  where ΔS is the shift amount, $P(S,\vec{p})$ is the option price as a function of the asset price, S, and $\vec{p}$ is a vector of model parameters. Similar finite difference equations could be obtain for other first and second-order Greeks. More information on finite differences, including the derivation of the above first-order central difference calculation, is presented in 12.3.3.

  Importantly, when using finite differences a derivative is calculated by evaluating the function at a set of nearby points with slightly modified parameters.

[12]Greeks are the sensitivity of an option with respect to a change in a given variable, and are discussed in more detail in chapter 16.

In the case of delta, these evaluations are of the option price at two different levels of the current asset price, namely $S + \Delta S$ and $S - \Delta S$. In this context we assume that the option pricing function, $P(S, \vec{p})$ is estimated via a simulation algorithm, that is via a set of generated paths. A natural question then is whether we should use the same set of paths when computing $P(S + \Delta S, \vec{p})$ and $P(S - \Delta S, \vec{p})$. The Common Random Numbers technique suggests that we should indeed use the same random samples for each calculation, as it will reduce the variance in our estimate of delta. In contrast, if we were to generate new samples for $P(S+\Delta S, \vec{p})$ and $P(S-\Delta S, \vec{p})$, part of the difference would be due to differences in the random samples and part would be due to the legitimate change in option value as the asset price changes. If we can neutralize the first source of deviation, we obtain a more precise estimate of the sensitivity.

12.2.15 Strengths and Weaknesses

In this section we provided a theoretical framework for using simulation in order to solve options pricing problems. We also saw that it was a technique that had wide-ranging applications across the buy-side, for example in portfolio optimization problems, and the sell-side, most notably in exotic options pricing and risk management. The main strength of simulation as a computational tool is its intuitive nature. Of the options pricing techniques that we study in this chapter as well as chapter 14, simulation is generally viewed as the easiest for quants to understand and implement. This is a key advantage, and leads to broad usage. More fundamentally, we saw that simulation does not suffer from the so-called curse of dimensionality. This will make it our go-to technique, and often our only choice in highly dimensioned problems. These types of problems arise naturally in fixed income, both in interest rate and credit markets, as well as when working with multi-asset portfolios.

Another significant advantage of simulation as a technique, in an options pricing context, is its ability to handle arbitrarily complex payoff structures, including path dependent payoffs. This is because simulation provides the entire path of each asset along each path, which we can then use to translate the path into the payoff of our choice. In some cases, such as a European option, we would only look at the last value in the path. In other cases, such as a Lookback or Asian options, we may find the minimum or average value along the path. In either case, once we have found our simulated paths, these payoffs become trivial to calculate. Simulation is also a technique that naturally lends itself to a generic pricing framework using the object-oriented programming concepts discussed in chapter 8. If done correctly, we can build a simulation engine using inheritance where payoffs and stochastic processes can be swapped seamlessly.

The main drawback to simulation is its slow rate of convergence. As we saw earlier, the rate of convergence for simulation is $\mathcal{O}(N^{-\frac{1}{2}})$. For a one-dimensional problem, this is notably slower than the other techniques that we have studied. This makes **variance reduction techniques** an important field of study as any boost in efficiency can make a big difference in the algorithms practicality. One particularly noteworthy side effect of this slower convergence is that it prohibits us from using simulation as a calibration tool. This is because calibration requires us to search a multi-dimensional parameter space to find the best fit, and doing so will require many evaluations. Therefore, if each evaluation requires many paths, and is computationally intensive as a result, then the calibration problem will become intractable.

A second drawback of simulation algorithms is the difficulty these techniques have with the exercise decisions embedded in American options. This is because simulation inherently works by moving forward in time, whereas exercise decisions are typically done via backward induction. Fortunately, Longstaff and Schwartz [168] have created a methodology for obviating this issue by estimating the exercise decision as the conditional expectation over the entire set of paths.

Understanding these strengths and weaknesses and being able to assess their relevance in different modeling situations is critical as a quant. As a result, readers are strongly encouraged to think through these decisions when completing the end of chapter exercises, and also when applying these techniques in their careers.

12.2.16 Simulation in Practice: Impact of Skew on Lookback Options Values in the Heston Model

In this section we leverage the function provided in the coding repository to calculate the payoff for fixed strike lookback options. We then examine the impact of volatility skew on the prices of these lookback options. In order to do this, we assume a Heston model and vary the ρ parameter to modify the slope of the volatility skew. We then examine the price of lookback options, which give the holder the right to choose their optimal exercise decision at expiry, and are therefore a function of the maximum or minimum of the asset price rather than the price at expiry. In the following plots, we show how the price of both lookback calls and puts at different strikes vary as the skew changes:

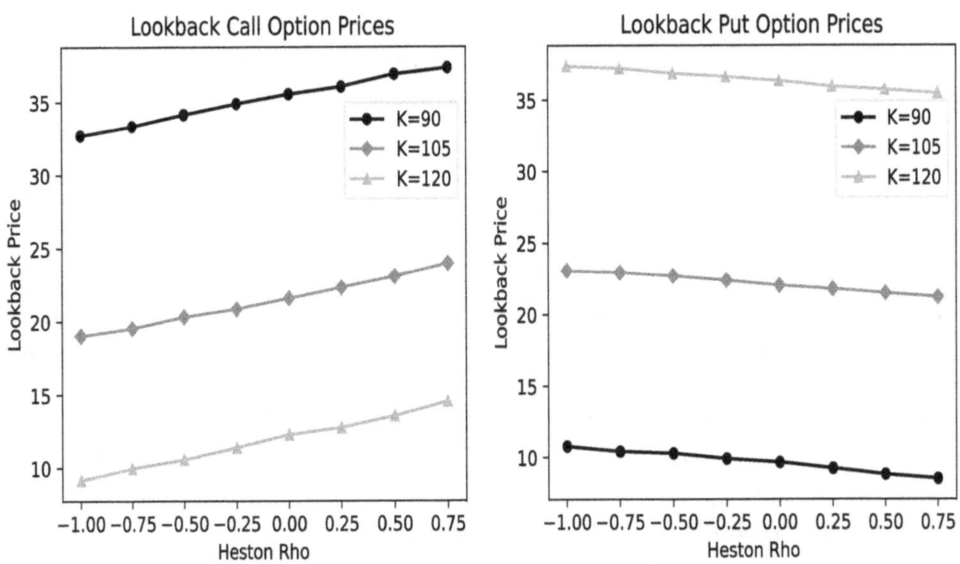

An interesting feature of this chart is that the price of lookback call options increases with, ρ, or as skew increases toward the call side. Conversely, for puts, we find the opposite. That is, the price decreases as we increase ρ. Therefore, we could conclude that lookbacks are sensitive to the slope of the volatility smile with lookback calls wanting implied volatilities to be highest on higher strike options and lookback puts wanting the reverse. Intuitively, this makes sense as this increase in skew increases the distributional mass in the left or right tail of the distribution, leading to larger payoffs for the lookback.

To isolate this relationship between lookbacks and skew even further, we can compare not just the absolute level of a lookback price, but instead look at its premium to a European option with the same characteristics[13]. This is done in the following plot by plotting the ratio of a lookback option price to the analogous European option:

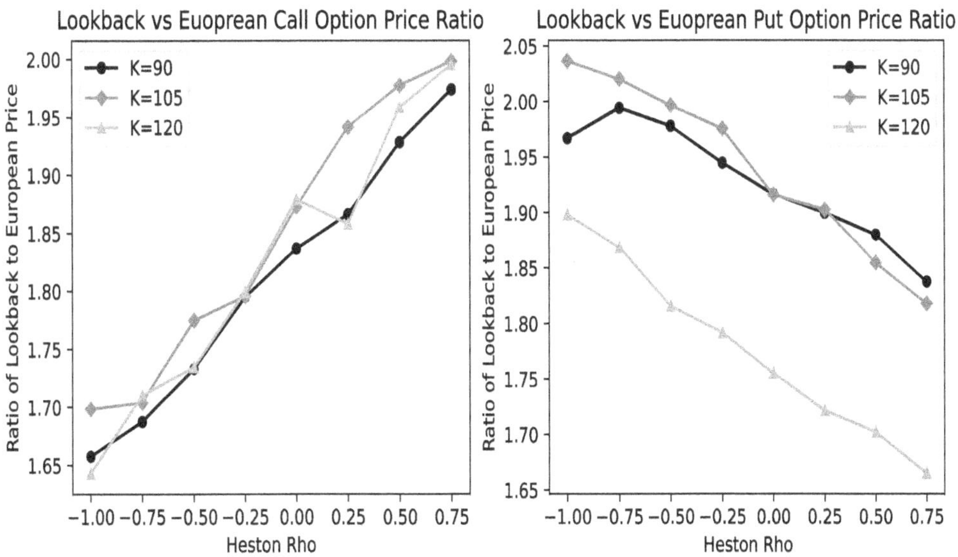

Here, we see a high dependence on ρ, or skew, for the value of a lookback relative to a European option. Readers are encouraged to try to extend this analysis to different types of movements of a volatility surface, and to try to understand the relationship between skew and lookback option valuations. This topic is also revisited in chapter 22 from an investment perspective, where we consider a strategy that is long a lookback option, and short the analogous European option.

12.3 NUMERICAL SOLUTIONS TO PDEs

12.3.1 Overview

As we saw in the last section, simulation techniques are a legitimate way to model exotic, path dependent options. In particular, they are able to handle high dimensions and arbitrary payoffs seamlessly. An alternative to simulation for solving path

[13]Such as strike and expiry.

dependent payoffs is via numerical solutions to partial differential equations (PDEs). PDEs will be especially attractive in the presence of American style exercise decisions. This is significant because the forward evolving nature of simulation techniques made applying simulation in this context more challenging. In this section we detail how a PDE approach might be used to model exotic options. We focus on numerical solutions to PDEs as in most realistic cases there will be no analytical solution. Additionally, many practical challenges of implementing PDEs for more advanced models are omitted from this text, such as the inclusion of an extra dimension for stochastic volatility models or the presence of jumps.

In order to build PDE algorithms, we first need to know how to write down the PDE for our model of choice. This requires understanding the connection between stochastic processes, SDEs and PDEs. In chapter 4 we saw how the Feynam-Kac formula provided this link in a generic manner. We also saw how we could use a hedging argument to motivate the Black-Scholes PDE. Once the PDE has been specified, we then need to specify a grid of the asset price and time over which we will implement our algorithm. It should be noted that inclusion of a stochastic volatility model would introduce a third dimension to this grid corresponding to the level of the stochastic volatility process. We then need to specify the boundary conditions for our instrument of choice. Remember that the PDE that we specify is independent of option payoff, rather, it is only a function of the model that we choose. The payoff then enters our algorithm PDE algorithm via boundary conditions. In general we need to specify boundary conditions at expiry, at the end of our grid, and also at the top and bottom of the grid. Finally, we need to approximate the derivatives in our PDE by discretizing them. To do this, we employ a technique called finite differences.

In the following sections, we discuss each of these steps in greater detail, beginning with a review of how a PDE can be derived for a stochastic process, and then discussing boundary conditions, setting up our time and space grid, and how to approximate the derivatives in the respective PDE numerically. In doing so, we show the implications of using different derivative approximations on the stability of the underlying algorithm. We then show how to apply these concepts in an algorithm designed to model American options, which is a structure that PDEs have a natural comparative advantage when modeling.

12.3.2 PDE Representations of Stochastic Processes

In chapter 4 we established the link between stochastic processes, stochastic differential equations and partial differential equations (PDEs). In particular, we saw that the Feynman-Kac formula enabled us to derive the PDE for a discounted payoff function from a given SDE [86] [144]. The Feynman-Kac formula provides a general link for deriving a PDE for an arbitrary, generic stochastic process albeit under somewhat limiting assumptions[14].

In this chapter we focus on the most well-known stochastic process and its corresponding partial differential equation, the Black-Scholes model. This Black-Scholes PDE can be motivated more narrowly using a hedging & no-arbitrage argument

[14]For example, in the absence of stochastic volatility or a jump process.

where we apply Ito's Lemma to the Black-Scholes SDE and notice that if we are to hedge the delta in this portfolio instantaneously there is no remaining source of randomness. As a result, we conclude that in the absence of arbitrage this portfolio must earn the risk-free rate, leading to the famous Black-Scholes PDE:

$$\frac{\partial C}{\partial t} + \frac{1}{2}\sigma^2 s^2 \frac{\partial^2 C}{\partial s^2} + rs\frac{\partial C}{\partial s} - rC = 0 \tag{12.49}$$

This hedging argument motivating (12.49) was presented in detail in chapter 4. This PDE will be the basis for the numerical solutions provided in the remainder of the chapter. This technique, of course, can be extended to other SDEs and PDEs. However, certain complications arise when doing so. For example, as we mention later in this section, inclusion of stochastic volatility adds another dimension to our PDE, which will unsurprisingly make the problem harder to solve. Similarly, addition of a jump component introduces challenges as they will introduce an integral into our PDE. Additionally, the Black-Scholes PDE is convenient because the parameters σ and r are constant rather than time-varying. This has certain computational advantages and in practice means that, as we will see, the matrix that we form to create our numerical approximation will be the same at every step.

A noteworthy feature of the Black-Scholes PDE is that a natural trade-off can be seen between theta ($\frac{\partial C}{\partial t}$) and gamma ($\frac{\partial^2 C}{\partial s^2}$). To see this, consider an investor who purchases an option and continuously hedges its delta. For this investor, there will be two remaining terms in the PDE: theta and gamma. This is a foundational concept that we will keep returning to throughout the book. Options traders often say that we are long gamma at the expense of theta, and the PDE gives us a clear lens to view this.

Generally speaking, options traders enter into options positions not to obtain access to delta. They can access this exposure directly via long or short positions in the underlying asset. Instead, options traders enter into options positions to access the more subtle aspects of an options (e.g. its convexity). Critically, this means that options traders who purchase options are implicitly making a relative value judgment between the options gamma and its theta. If gamma is expensive, that means that the time decay, or theta is too steep to be warranted by realized volatility. In this case, options traders should sell options and collect the theta. Conversely, if gamma is cheap, then the trader should buy options and continuously delta hedge, collecting the gamma. This concept is revisited in chapter 16, where we explore it in the context of options trading.

In order to solve a PDE numerically we need to define a set of approximations for the derivatives in (12.49). That is, we need to choose a discretization method for $\frac{\partial C}{\partial t}$, $\frac{\partial^2 C}{\partial s^2}$ and $\frac{\partial C}{\partial s}$, respectively. To do this, we can use the finite difference method, which we discuss in the next section. It should be noted that other approaches to this discretization exists, most notably finite elements and finite volumes. More information on these techniques can be found in [234] and [180].

12.3.3 Finite Differences

In order to create a numerical approximation to a partial differential equation, we first need to develop a methodology for approximating the partial derivatives that appear in the underlying PDE. To create these approximations, we can rely on Taylor series expansions of our functions. Recall that the Taylor series expansion of a function, $f(x)$ near $x = x^*$ can be written as:

$$f(x) = f(x^*) + (x - x^*)f'(x^*) + (x - x^*)^2 \frac{f''(x^*)}{2} + (x - x^*)^3 \frac{f'''(x^*)}{6} + \cdots \tag{12.50}$$

Fundamentally, this means that we can calculate the value of a differentiable function at x using information about the function evaluated at a nearby point x^* as well as information about its derivatives. Generally when doing this we will approximate the function using only the first few terms in the expansion. For example, we might consider a second-order Taylor series approximation that includes only the first two function derivatives and omits the remaining terms:

$$f(x) = f(x^*) + (x - x^*)f'(x^*) + (x - x^*)^2 \frac{f''(x^*)}{2} + \mathcal{O}((x - x^*)^3) \tag{12.51}$$

The order of the error in our approximation will then be the order of magnitude of the next term in the Taylor series, $\mathcal{O}((x - x^*)^3)$. As we see next, a truncated version of the Taylor series approximation can then be used to approximate the derivative terms in our PDE.

Forward First Order Difference: To calculate a forward first order difference for our function, $f(x)$, let's begin with a Taylor series approximation of the form of (12.51) at $x + h$ around x, including up to second-order terms:

$$f(x+h) \approx f(x) + hf'(x) + h^2 \frac{f''(x)}{2} \tag{12.52}$$

$$f'(x) \approx \frac{(f(x+h) - f(x))}{h} - \frac{hf''(x)}{2} \tag{12.53}$$

$$f'(x) \approx \frac{(f(x+h) - f(x))}{h} \tag{12.54}$$

where in the second line we solved for $f'(x)$ which is the derivative we are looking to approximate and in the last line we omitted the term $\frac{hf''(x)}{2}$. Therefore, our first order forward difference formula becomes (12.54). Its error will be $\mathcal{O}((h))$ as that is the order of magnitude of the first term in the Taylor series that we omit. As a result, for sufficiently small h, this error will be small and our first order forward difference will provide a reasonable approximation to the first derivative.

Backward First-Order Difference: We can use a similar approach to calculate a backward first order difference to our function, $f(x)$. To do this, let's begin with a

Taylor series approximation at $x - h$ around x:

$$f(x-h) \approx f(x) - hf'(x) + h^2\frac{f''(x)}{2} \quad (12.55)$$

$$f'(x) \approx \frac{(f(x) - f(x-h))}{h} - \frac{hf''(x)}{2} \quad (12.56)$$

$$f'(x) \approx \frac{(f(x) - f(x-h))}{h} \quad (12.57)$$

where we again solved for the first derivative we are looking to approximate and omitted the term involving $f''(x)$. This leaves us with (12.57) as our backward first order difference approximation. Like the forward first order difference estimate, it has error $\mathcal{O}((h))$.

Central First-Order Difference: It turns out, we can get a more accurate first-order difference approximation by combining the Taylor series approximations at $f(x-h)$ and $f(x+h)$, that is, we combine the forward and backward difference estimates. To see this, recall that these Taylor series expansions can be written as:

$$f(x+h) \approx f(x) + hf'(x) + h^2\frac{f''(x)}{2} + h^3\frac{f'''(x)}{6} \quad (12.58)$$

$$f(x-h) \approx f(x) - hf'(x) + h^2\frac{f''(x)}{2} - h^3\frac{f'''(x)}{6} \quad (12.59)$$

If we subtract (12.59) from (12.58) we can see that the $h^2\frac{f''(x)}{2}$ term cancels, improving our error:

$$f(x+h) - f(x-h) \approx 2hf'(x) + 2h^3\frac{f'''(x)}{6} \quad (12.60)$$

$$f'(x) \approx \frac{f(x+h) - f(x-h)}{2h} - h^2\frac{f'''(x)}{6} \quad (12.61)$$

$$f'(x) \approx \frac{f(x+h) - f(x-h)}{2h} \quad (12.62)$$

Importantly, the order of the error in our approximation in (12.62) becomes $\mathcal{O}((h^2))$. Therefore, central first-order difference is preferable to their forward or backward counterparts as they provide substantially less error for a given small h.

Central Second-Order Difference: A similar approach can be used to calculate a central second-order finite difference approximation. To start, we again consider Taylor series expansion at $x + h$ and $x - h$ and consider the first four derivative terms:

$$f(x+h) \approx f(x) + hf'(x) + h^2\frac{f''(x)}{2} + h^3\frac{f'''(x)}{6} + h^4\frac{f''''(x)}{24} \quad (12.63)$$

$$f(x-h) \approx f(x) - hf'(x) + h^2\frac{f''(x)}{2} - h^3\frac{f'''(x)}{6} + h^4\frac{f''''(x)}{24} \quad (12.64)$$

If we add equations (12.63) and (12.64) we are left with:

$$f(x+h)+f(x-h) \approx 2f(x)+h^2f''(x)+h^4\frac{f''''(x)}{12} \quad (12.65)$$

$$f''(x) \approx \frac{f(x+h)-2f(x)+f(x-h)}{h^2}+h^2\frac{f''''(x)}{12} \quad (12.66)$$

$$f''(x) \approx \frac{f(x+h)-2f(x)+f(x-h)}{h^2} \quad (12.67)$$

where again we are solving for the desired derivative, $f''(x)$ in the second line and in our final approximation we omit the final term $h^2\frac{f''''(x)}{12}$ leading to an error of order $\mathcal{O}((h^2))$.

Solving PDEs numerically is not the only application of these finite difference estimates. For example, in chapter 16 we will leverage these estimates in order to approximate Greeks. This analysis can also be extended to multiple dimensions using a multi-variate Taylor series expansion [84].

12.3.4 Time & Space Grid

In order to solve a PDE numerically we need to define a mesh in time & asset price. Generally we do this via a uniform grid in both dimensions. Let's break up our grid into N time intervals and M asset price intervals. This leads to:

$$h_t = \frac{T}{N} \quad (12.68)$$

$$h_s = \frac{S_{\max}-S_{\min}}{M} \quad (12.69)$$

$$S_i = S_{\min}+ih_s \quad (12.70)$$

$$t_j = jh_t \quad (12.71)$$

$$i = \{0,\ldots,M\}$$

$$j = \{0,\ldots,N\}$$

0 is a natural lower bound for most assets, with interest rates being a notable exception, and is thus often a convenient choice for $S_{\min}$. An upper bound $S_{\max}$, would still need to be defined and has no obvious natural value. It should be chosen, however, such that the value along the upper boundary is clear and undisputed. The price of our structure of interest will then be defined at every point on the time and asset price grid, j and i respectively. Specifically, we refer to $P(S_i,t_j)$ as the price of our instrument at the jth timestep with the ith asset price.

12.3.5 Boundary Conditions

If we attempt to implement our finite difference approximations on the edges of our time and space grid, we will find that our approximations involve terms that are beyond the limits of our grid. This is true at the lower and upper bound, and also at the last time step. Fortunately, this doesn't pose too much of a problem as instead, we can specify boundary conditions for these points. To solve an options pricing problem via PDEs will require the following set of boundary conditions:

- At the end of the time grid (e.g. at expiry)
- Along the lower bound of the Asset Price Grid, at $S_{\min}$
- Along the upper bound of the Asset Price Grid, $S_{\max}$

For the vast majority of options pricing problems, the boundary condition at the expiry of the instrument will be fairly easy to compute. In particular, it is just the payoff function of the option. Similarly, for many option structures either the top or bottom condition will be the case of a very out-of-the-money option, which will have zero value. Meanwhile, the other condition will be for a very in-the-money option whose value will be the appropriately discounted payoff value. In other cases, where the value at the boundary condition is less well-defined and harder to construct, we can instead make assumptions about the partial derivative at the boundary[15].

Nonetheless, setting the correct boundary conditions can be one of the more challenging aspects of solving a PDE numerically, and if a slightly wrong boundary condition is specified it can lead to unintuitive results. As a result, to provide the reader some intuition for how to formulate boundary conditions, we consider how boundary conditions would be specified for European Call options and American Digital Put options, below:

European Call Option:

When modeling a European call numerically via PDEs, we can specify the following boundary conditions:

(i) At maturity condition is: $C(S_i, t_N) = (S_i - K)^+ \quad \forall i$

(ii) The lower boundary condition is: $C(0, t_j) = 0, \quad \forall j$

(iii) The upper boundary condition is: $C(S_M, t_j) = S_M - Ke^{-r(T-t_j)}, \quad \forall j.$

Readers should notice and think through the discounting in the upper boundary condition. Fundamentally this is because our option is deeply in-the-money, but we have to wait until expiry to receive payment. Additionally, we expect S_M to accrue at the risk free rate until expiry. This leads to the following future value of the option along the upper boundary: $S_M e^{r(T-t_j)} - K$ which when discounted to time t_j yields the condition above.

American Digital Put Option:

In the case of an American Digital Put option we would specify the following boundary conditions:

(i) At maturity condition is: $D(S_i, t_N) = 1_{\{S_i < K\}}, \quad \forall i$

[15]This approach is known as specifying Neumann boundary conditions whereas specifying a value directly is referred to as a Dirichlet boundary condition.

(ii) The lower boundary condition is: $D(0, t_j) = 1, \quad \forall j$.

(iii) The upper boundary condition is: $D(S_M, t_j) = 0, \quad \forall j$.

Note that this example assumes that the American digital option pays out immediately rather than at expiry, otherwise a discounting term would need to be incorporated in the lower boundary.

12.3.6 Explicit Scheme

The explicit scheme relies on a backward difference for $\frac{\partial C}{\partial t}$ and central differences for $\frac{\partial C}{\partial s}$ and $\frac{\partial^2 C}{\partial s^2}$. We then begin at the end, where the boundary condition specifies the values for the derivative, and iterate backward in time until we have reached the beginning of the grid. Plugging these finite difference approximations for $\frac{\partial C}{\partial t}$, $\frac{\partial C}{\partial s}$ and $\frac{\partial^2 C}{\partial s^2}$ into the Black-Scholes PDE in (12.49), we have:

$$\begin{aligned} & \frac{C(S_i, t_j) - C(S_i, t_{j-1})}{h_t} \\ & + \frac{1}{2}\sigma^2 S_i^2 \frac{C(s_{i+1}, t_j) - 2C(s_i, t_j) + C(s_{i-1}, t_j)}{h_s^2} \\ & + r s_i \left(\frac{C(S_{i+1}, t_j) - C(S_{i-1}, t_j)}{2h_s} \right) - rC(S_i, t_j) = 0 \end{aligned} \tag{12.72}$$

As we are iterating backward in time, the values at time t_j will be known and only those at time t_{j-1} will be unknown. This means that the only unknown in (12.72) is $C(S_i, t_{j-1})$ and we have a single linear equation with a single unknown. Its value will depend on the model parameters, r and σ, as well as three values at the next point in the grid, which are also known: $C(S_{i-1}, t_j)$, $C(S_i, t_j)$, $C(S_{i+1}, t_j)$.

So essentially in the explicit scheme, as we iterate backward we are calculating the value at the previous timestep as a function of three values at the current timestep, as the following chart depicts:

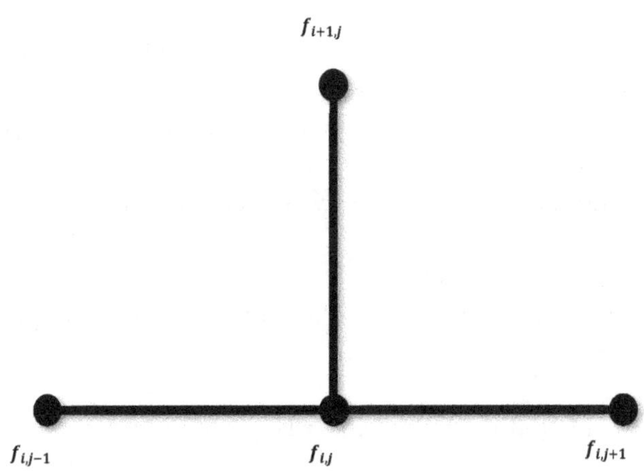

Rearranging the discretized PDE in (12.72) to solve for the sole unknown, $C(S_i, t_{j-1})$, we have:

$$\begin{aligned} C(s_i, t_{j-1}) &= \left(\frac{\sigma^2 s_i^2}{2}\frac{h_t}{h_s^2} - \frac{r s_i h_t}{2h_s}\right) C(s_{i-1}, t_j) \\ &\quad + \left(1 - \sigma^2 s_i^2 \frac{h_t}{h_s^2} - r h_t\right) C(s_i, t_j) \\ &\quad + \left(\frac{\sigma^2 s_i^2}{2}\frac{h_t}{h_s^2} + \frac{r s_i h_t}{2h_s}\right) C(s_{i+1}, t_j) \end{aligned}$$

We can further simplify this equation by defining the constants l_i, d_i and u_i as follows:

$$l_i = \left(\frac{\sigma^2 s_i^2}{2}\frac{h_t}{h_s^2} - \frac{r s_i h_t}{2h_s}\right) \tag{12.73}$$

$$d_i = \left(1 - \sigma^2 s_i^2 \frac{h_t}{h_s^2} - r h_t\right) \tag{12.74}$$

$$u_i = \left(\frac{\sigma^2 s_i^2}{2}\frac{h_t}{h_s^2} + \frac{r s_i h_t}{2h_s}\right) \tag{12.75}$$

Making these substitutions, we are left with the following, discretized, simplified PDE:

$$C(s_i, t_{j-1}) = l_i C(s_{i-1}, t_j) + d_i C(s_i, t_j) + u_i C(s_{i+1}, t_j)$$

Note that the boundary conditions will require special treatment at points $C(S_1, t_{j-1})$ and $C(S_{M-1}, t_{j-1})$[16]. Assuming the lower and upper boundary conditions are L_{t_j} and U_{t_j}, respectively, these points we simplify to:

$$\begin{aligned} C(s_1, t_{j-1}) &= l_1 C(s_0, t_j) + d_1 C(s_1, t_j) + u_1 C(s_2, t_j) \\ &= l_1 L_{t_j} + d_1 C(s_1, t_j) + u_1 C(s_2, t_j) && (12.76) \\ C(s_{M-1}, t_{j-1}) &= l_{M-1} C(s_{M-2}, t_j) + d_{M-1} C(s_{M-1}, t_j) + u_{M-1} C(s_M, t_j) \\ &= l_{M-1} C(s_{M-2}, t_j) + d_{M-1} C(s_{M-1}, t_j) + u_{M-1} U_{t_j} && (12.77) \end{aligned}$$

Put together, this is a set of linear equations that we can solve using standard methods. In particular, we can solve for the vector of prices at time t_{j-1} of length M by collecting (12.76), (12.77) and (12.76) in matrix form:

$$\begin{bmatrix} d_1 & u_1 & & & \\ l_2 & d_2 & u_2 & & \\ & \ddots & \ddots & \ddots & \\ & & l_{M-2} & d_{M-2} & u_{M-2} \\ & & & l_{M-1} & d_{M-1} \end{bmatrix} \begin{bmatrix} C(s_1, t_j) \\ C(s_2, t_j) \\ \vdots \\ C(s_{M-2}, t_j) \\ C(s_{M-1}, t_j) \end{bmatrix} + \begin{bmatrix} l_1 L_{t_j} \\ 0 \\ \vdots \\ 0 \\ u_{M-1} U_{t_j} \end{bmatrix} = \begin{bmatrix} C(s_1, t_{j-1}) \\ C(s_2, t_{j-1}) \\ \vdots \\ C(s_{M-2}, t_{j-1}) \\ C(s_{M-1}, t_{j-1}) \end{bmatrix}$$

[16]Recall that we are assuming that the boundary conditions specified values for $C(S_0, t_{j-1})$ and $C(S_M, t_{j-1})$.

which has the form:

$$C(j-1) = AC(j) + b \tag{12.78}$$

This matrix shows us that at each step, we can compute the values at the previous step by solving a linear system of equations. This means that once we have specified our time and space grids, as well as the appropriate boundary conditions we can iterate backward according to our PDE by solving this system of linear equations. An important feature of the matrix in (12.78) is that is sparse. In fact, it is tri-diagonal, which is an appealing property as it simplifies many matrix calculations and operations.

12.3.7 Implicit Scheme

If we replace the backward difference for the partial derivative $\frac{\partial C}{\partial t}$ with a forward difference we get the implicit scheme. In this scheme we still rely on central differences for the two remaining derivative terms, $\frac{\partial C}{\partial s}$ and $\frac{\partial^2 C}{\partial s^2}$. Substituting these forward and central finite difference approximations into the Black-Scholes PDE in (12.49), we are left with:

$$\begin{aligned}
&\frac{C(S_i,t_{j+1}) - C(S_i,t_j)}{h_t} \\
&\quad + \frac{1}{2}\sigma^2 S_i^2 \frac{C(s_{i+1},t_j) - 2C(s_i,t_j) + C(s_{i-1},t_j)}{h_s^2} \\
&\quad + r s_i \left(\frac{C(S_{i+1},t_j) - C(S_{i-1},t_j)}{2h_s} \right) - rC(S_i,t_j) = 0
\end{aligned} \tag{12.79}$$

Notice that in this case we have three terms at time t_j and a single term at time t_{j+1}, as the following figure highlights:

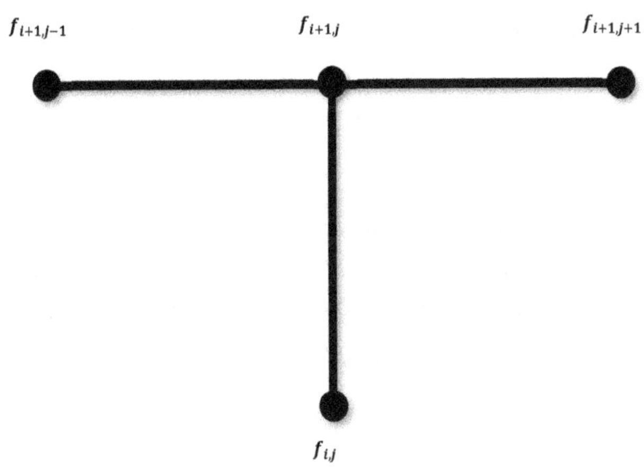

This would lead to a single equation with three unknowns, which we would not be able to solve in isolation. However, as multiple points use the same three unknowns it can be solved if we put it in matrix form. To see this, let's begin by solving (12.79) for the single point at time t_{j+1}, $C(S_i, t_{j+1})$:

$$\begin{aligned} C(s_i, t_{j+1}) &= -\frac{1}{2}\left(\sigma^2 s_i^2 \frac{h_t}{h_s^2} - \frac{r s_i h_t}{h_s}\right) C(s_{i-1}, t_j) \\ &+ \left(1 + \sigma^2 s_i^2 \frac{h_t}{h_s^2} + r h_t\right) C(s_i, t_j) \\ &- \frac{1}{2}\left(\sigma^2 s_i^2 \frac{h_t}{h_s^2} + \frac{r s_i h_t}{h_s}\right) C(s_{i+1}, t_j) \end{aligned}$$

As we did before, let's define the constants l_i, d_i and u_i as follows:

$$l_i = \frac{1}{2}\left(\sigma^2 s_i^2 \frac{h_t}{h_s^2} - \frac{r s_i h_t}{h_s}\right) \tag{12.80}$$

$$d_i = \left(1 + \sigma^2 s_i^2 \frac{h_t}{h_s^2} + r h_t\right) \tag{12.81}$$

$$u_i = \frac{1}{2}\left(\sigma^2 s_i^2 \frac{h_t}{h_s^2} + \frac{r s_i h_t}{h_s}\right) \tag{12.82}$$

Substituting these value back into our original discretized PDE in (12.80), we now have:

$$C(s_i, t_{j+1}) = -l_i C(s_{i-1}, t_j) + d_i C(s_i, t_j) - u_i C(s_{i+1}, t_j) \tag{12.83}$$

Additionally, as before we will need to handle the boundary conditions separately. In particular, for $C(s_1, t_j)$ and $C(s_{M-1}, t_j)$ we will need to use the specified values, L_{t_j} and U_{t_j}, for the upper and lower boundary to avoid stepping outside our predefined grid. For these points we will have:

$$\begin{aligned} C(s_1, t_{j-1}) &= -l_i C(s_{i-1}, t_j) + d_i C(s_i, t_j) - u_i C(s_{i+1}, t_j) \\ &= -l_1 L_{t_j} + d_1 C(s_1, t_j) - u_1 C(s_2, t_j) && (12.84) \\ C(s_{M-1}, t_{j-1}) &= -l_i C(s_{i-1}, t_j) + d_i C(s_i, t_j) - u_i C(s_{i+1}, t_j) \\ &= -l_{M-1} C(s_{M-2}, t_j) + d_{M-1} C(s_{M-1}, t_j) - u_{M-1} U_{t_j} && (12.85) \end{aligned}$$

We can equivalently write this in matrix form, where it will become:

$$\begin{bmatrix} d_1 & -u_1 & & & \\ -l_2 & d_2 & -u_2 & & \\ & \ddots & \ddots & \ddots & \\ & & -l_{M-2} & d_{M-2} & -u_{M-2} \\ & & & -l_{M-1} & d_{M-1} \end{bmatrix} \begin{bmatrix} C(s_1, t_j) \\ C(s_2, t_j) \\ \vdots \\ C(s_{M-2}, t_j) \\ C(s_{M-1}, t_j) \end{bmatrix} - \begin{bmatrix} l_1 L_{t_j} \\ 0 \\ \vdots \\ 0 \\ u_{M-1} U_{t_j} \end{bmatrix}$$

$$= \begin{bmatrix} C(s_1, t_{j+1}) \\ C(s_2, t_{j+1}) \\ \vdots \\ C(s_{M-2}, t_{j+1}) \\ C(s_{M-1}, t_{j+1}) \end{bmatrix}$$

Solving for the values at the previous timestep that we are looking to extract, we have:

$$\begin{bmatrix} d_1 & -u_1 & & & \\ -l_2 & d_2 & -u_2 & & \\ & \ddots & \ddots & \ddots & \\ & & -l_{M-2} & d_{M-2} & -u_{M-2} \\ & & & -l_{M-1} & d_{M-1} \end{bmatrix}^{-1} \begin{bmatrix} C(s_1, t_{j+1}) \\ C(s_2, t_{j+1}) \\ \vdots \\ C(s_{M-2}, t_{j+1}) \\ C(s_{M-1}, t_{j+1}) \end{bmatrix} + \begin{bmatrix} l_1 L_{t_j} \\ 0 \\ \vdots \\ 0 \\ u_{M-1} U_{t_j} \end{bmatrix}$$

$$= \begin{bmatrix} C(s_1, t_j) \\ C(s_2, t_j) \\ \vdots \\ C(s_{M-2}, t_j) \\ C(s_{M-1}, t_j) \end{bmatrix}$$

which will have the form:

$$C(j) = A^{-1}C(j+1) + b \tag{12.86}$$

Fundamentally, this means that when we apply the implicit scheme we can recursively move backward by inverting a matrix at each timestep. Relative to the explicit scheme this will make each step more computationally intensive, however, it can be shown that the implicit scheme is unconditionally stable, meaning we can generally more than offset this added complexity due to the relaxed demands of the time and space grid we will see later. Further, as with the explicit matrix, notice that the matrix in (12.86) is tri-diagonal. This makes matrix inverse operations more tractable, and leads the implicit scheme to be highly efficient.

12.3.8 Crank-Nicolson

Crank-Nicolson [70] tries to improve on the efficiency of the explicit and implicit scheme by using a central first order difference on the partial derivative with respect to time, $\frac{\partial C}{\partial t}$. This central difference approximation can be written as:

$$\frac{\partial C}{\partial t} \approx \frac{C(S_i, t_{j+1}) - C(S_i, t_j)}{h_t} \tag{12.87}$$

This approach introduces another complication, however, as this derivative is centered on a point that is not on the grid, the mid-point of time t_j and time t_{j+1}. This would create challenges with computing the other derivatives, $\frac{\partial C}{\partial s}$ and $\frac{\partial^2 C}{\partial s^2}$. Instead, we can employ the trapezoidal rule, that is for these derivatives we can use the average of the central differences evaluated at time t_j and t_{j+1}:

$$\frac{\partial C}{\partial s} \approx \left(\frac{C(S_{i+1}, t_j) - C(S_{i-1}, t_j) + C(S_{i+1}, t_{j+1}) - C(S_{i-1}, t_{j+1})}{4h_s} \right)$$

$$\frac{\partial^2 C}{\partial s^2} \approx \frac{C(S_{i+1}, t_j) - 2C(S_i, t_j) - C(S_{i-1}, t_j) + C(S_{i+1}, t_{j+1}) - 2C(S_i, t_{j+1}) - C(S_{i-1}, t_{j+1})}{2h_s^2}$$

As we did with the implicit and explicit schemes, we can then collect these terms and put them in matrix form in order to solve the PDE numerically. Doing this, we will see that as in the implicit scheme inversion of a matrix is required as we iterate backward. Fortunately, however, these matrices will generally be tri-diagonal, lessening the computational load and increasing the stability of the inverse operation.

12.3.9 Stability

Robust stability analysis of PDE solutions is beyond the scope of this text. Instead, we will focus on providing readers with intuition and suggestions on how to formulate their time and space grids so that they can ensure that their solutions are stable in each scheme. Stability analysis is, however, a deeply researched field with many established texts that cover the details far more rigorously. Readers interested in this more detailed treatment should consult [133] or [241]. Of particular importance in this field are the well-known CFL conditions [218] that define the required conditions with respect to the time and space grid for which an explicit scheme is stable. These are the fundamental conditions that we must follow in order to ensure that our explicit scheme remains stable.

In the context of the Black-Scholes PDE, remember that the matrix, A that we computed in 12.3.6 was the same at all time-steps. This is because the parameters that enter the PDE, σ and r are constant. That means that, loosely speaking, as we iterate backward in time we are essentially raising that matrix A to an additional power. If we have N timesteps, then solution to the PDE numerically will involve computing A^N. It can be shown that, eigenvalues less than 1 in the matrix A are required for this to be stable [133]. It can then be shown that, the CFL condition required for the Black-Scholes PDE to be stable when using the explicit scheme are:

$$0 < h_t < \frac{h_s}{\sigma^2 S_{\max}^2} \tag{12.88}$$

If this condition is violated, we will encounter numerical issues in solving our PDE. The result of this stability condition is that a very fine time grid will be required to employ the explicit scheme. Both the fact that a tightly spaced time grid is required and that the problem is unstable for certain specifications are both clearly disadvantages of the explicit scheme. Conversely, an attractive feature of the explicit scheme is that matrix inversion is not required at every timestep.

Fortunately, it can be shown that both the implicit and Crank-Nicolson scheme are unconditionally stable, meaning that we are fully free to choose our time and space grids and the solutions stability will be independent of what we choose. This is a notable advantage for these two schemes relative to the explicit scheme. However, both the implicit and Crank-Nicolson schemes do add some level of complexity because of the embedded matrix inversion operations.

12.3.10 Multi-Dimension PDEs

In the case of a stochastic volatility model, we are working with a two-dimension system of SDEs. This in turn increases the dimensionality of the PDE we are trying to solve numerically. In the previous example, we need to specify a two-dimensional grid over the asset price and time. In this case, the presence of a separate stochastic volatility process will result in another dimension to our grid for the volatility. In other cases, such as in credit or interest rate markets, the dimensionality may increase even further, as we will see later in chapter 19 when we discuss term structure modeling. As we introduce these additional dimensions, numerically solving a PDE becomes increasingly challenging. In the case of a two dimensional SDE, such as a stochastic volatility model, PDE solutions are still relatively tractable. As we increase the dimensionality beyond that however, PDE methods become less and less efficient and a simulation approach will be preferred.

Adding another dimension to our PDE in the case of stochastic volatility will unsurprisingly present some computational challenges. First, the PDE that we are solving will include some terms that are cross partial derivatives of the form $\frac{\partial C}{\partial s \partial \nu}$, where ν is the volatility process. To handle these terms, we need a finite difference estimate based on a multi-variate Taylor series expansion. This is of course doable, however, the matrix that we are working with will also increase in size considerably as a result of the additional dimension. Even more problematically these matrices will no longer be tri-diagonal, as the scheme will be a function of neighboring points with respect to asset price and volatility, increasing the computational complexity of the problem and making operations like the matrix inversion required in an implicit scheme much harder.

Fortunately, all of these challenges can be overcome in a reasonably robust way. They are not detailed in this text, however, a more detailed treatment can be found in [133]. Recently, machine learning techniques have been applied to help solve PDE problems in higher dimensions, potentially making them more of a realistic alternative to simulation in these cases [90].

12.3.11 Partial Integro Differential Equations

Another complication arises when the PDE we are working with emanates from a model with a jump process. Specifically, it can be shown that inclusion of a jump component introduces an integral into our PDE [133]. The result is a partial-integro differential equation which needs to be derived for each jump process and then requires approximating an integral to solve numerically. Intuitively, we can think of this as an integral over the expected jumps at each discretized step. Approximating this integral will require leveraging the quadrature techniques that we learned in chapter 14 and adds a fair amount of complexity to our solution. Numerical solutions to Partial Integro Differential Equations are beyond the scope of this text, however, interested readers should see [133].

12.3.12 Strengths & Weaknesses

Numerical solutions to PDEs provide a reasonably efficient approach to solving exotic, path dependent structures in low dimensions. As noted, as dimensionality starts to rise, PDE solutions become less and less tractable, making simulation the technique of choice. In low dimensions, however, PDEs provide a more efficient alternative to simulation. A PDE solution is especially attractive in the presence of an American or Bermudan style exercise decision, as the backward recursion in PDE solutions fits naturally with the backward induction procedure required to model an exercise decision [11]. Recent advances by Longstaff-Schwartz and others, however, have made simulation a more compelling alternative even for American and Bermudan options [168]. Even in cases where PDEs may be more computationally efficient, there is a bit of a trade-off between techniques and many quants may still lean toward simulation due to its intuitive nature, lower upfront cost of implementation and ease of incorporating arbitrarily complex payoffs.

12.3.13 American vs. European Digital Options in Practice

Finally, let's consider an American digital option. This option can be modeled using either the PDE or simulation techniques discussed in this chapter. American digital options have an interesting rule of thumb for pricing that can be used to validate our model. Deviation from this rule of thumb could also be used as a relative value indicator of sorts.

An American Digital option is an option that pays \$1 if an asset price touches a certain barrier level *at any point* in its path. This is similar to a European Digital option which pays \$1 if an asset price touches a barrier at expiry. In FX, American Digital options are referred to as **one-touch** and **no-touch** options. There are also **double no-touch** options which pay if the exchange rate never leaves a given range. In chapter 21 we explore one-touch and no-touch options in more detail, including an example that applies an investment perspective to the rule of thumb relationship that follows. Clearly, European and American digital options are somehow related as they share a binary payoff structure, with the difference between them being that American touch options observe the barrier at every tick whereas European touch options only observe the barrier at expiry.

Let's consider a simple up-and-in one-touch option with barrier K. For simplicity let's assume that the payout is at maturity regardless of when the barrier is triggered. This will be an easy assumption to relax as we move forward. The price of this option can be computed via the following expectation:

$$A_T(K) = \tilde{\mathbb{E}}\left[e^{-\int_0^T r_u\,du} 1_{\{M_T > K\}}\right] \tag{12.89}$$

where M_T is the maximum value S takes over the interval $[0, T]$. The European equivalent would be:

$$E_T(K) = \tilde{\mathbb{E}}\left[e^{-\int_0^T r_u\,du} 1_{\{S_T > K\}}\right] \tag{12.90}$$

In chapter 14 we saw that European Digital options can be priced via FFT techniques. We will also see in chapter 17 that they are closely related to the CDF of the risk-neutral distribution and can be estimated in a non-parametric way. Importantly, European digital options are simpler because they only depend on the value of the asset price at expiry, and are independent of the intermediate path. This means that if the density, or the characteristic function of the asset price at expiry is known, than quadrature and FFT techniques can be applied.

Conversely, if we think about modeling American digital options, it should be clear that they are path dependent and the quadrature and FFT pricing techniques we have cover in previous chapters will not help us. An exception to this would be if we know the distribution of the maximum (or minimum) of the asset value, then the problem reduces to a 1D integral which we can compute analytically or via quadrature. This will be true in some cases, such as under the Black-Scholes model but will generally speaking not be the case. Of course, in those cases we can instead apply simulation techniques or solve the PDE numerically in order to estimate its proper value.

In the absence of those techniques, we can try to use the relationship between the easier to model European and more challenging American digital options to get a general idea of the proper price. In particular, we can use the **reflection principle** in order to give us a rough estimate for the price of the American digital option using a European Digital Option. To see this, let's think about whether paths that pay off for the European digital also payoff for the American digital, and vice versa. First, it is clear that the American Digital pays out on all paths that the European Digital pays out by construction, as the expiry time is part of the monitoring period of the American. This places a lower boundary on the price of the American digital option, in the absence of arbitrage.

Next let's consider a path where the American Digital is triggered prior to expiry. What is the probability that the European will also trigger? If we know this probability, then we can use it to establish a link between the European and American digitals. To answer this, we will leverage the reflection principle [232] conditional on the point that the American digital has hit the barrier. Generally speaking, the reflection principle tells us that for every path that finishes above (or below) the barrier at expiry there is a path of equal probability that is reflected across the barrier and will finish below (or above). Said differently, the reflection principle tells us that for each realization that finishes above the barrier there is an inverted path with equal probability that finishes below the barrier. As a result, the probability that the European will trigger is $\approx 50\%$, and the cost of the American should be approximately twice the cost of the European.

This reflection principle argument is exact in certain circumstances, for example for a stochastic process that is governed by arithmetic Brownian motion. In other cases, such as when we introduce drift of skewness into our underlying stochastic model, this approximation will become less accurate. One can imagine trading the American against the European and using deviation from this equilibrium condition of:

$$A_T(K) \approx 2E_T(K) \tag{12.91}$$

as a filter or signal in our trade construction. It turns out, that this particular structure is a fairly direct way to trade market skew and is a concept that we return to in chapter 21 from an investment perspective. An example of digital option pricing is presented in the coding repository so that the reader can explore this phenomenon in more detail.

12.4 MODELING EXOTIC OPTIONS IN PRACTICE

In this chapter we have introduced the two main techniques for modeling exotic options: simulation and numerical solutions to partial differential equations. In subsequent chapters we will explore how exotic options are used in different asset classes, from interest rates, to FX to equity markets. Prior to moving on, however, we should emphasize a few defining characteristics and features of the universe of exotic options. In particular, a first defining feature is whether exotics are linked to a single asset or multiple assets. This has significant modeling implications. If the former is true, then we might need to model a complex path dependent structure, but the problem is fundamentally simplified because of its low dimensionality. In this case, either a PDE or simulation may be entirely appropriate to value our exotic, and the choice may be somewhat subjective. Conversely, if we are modeling a multi-asset exotic, we may be dealing with a simpler European payoff structure, but may still have a complex model because of the increased dimensionality. In these cases, simulation will become our go-to technique as the increased dimensionality will make solving a PDE intractable.

In the following list, we detail a few of the most commonly traded single asset and multi-asset exotics:

- **Single Asset Exotics**: Single Asset Exotics have complex payoffs but those payoffs are only tied to the behavior of a single underlying. These exotics may be a function of the entire path of the underlying asset, however, the presence of a single asset is a simplifying assumption. The most common single-asset exotics include:
  - **European Digital Options**: A European digital option with strike K pays a predefined amount if an asset is above (or below) K at expiry. European digital calls pay out if the asset is above K at expiry whereas European digi puts pay out if the asset is below K at expiry.
  - **American Digital Options**: American digital options with strike K pay a predefined amount if an asset is above (or below) K at any point prior to expiry. The fact that these options pay if the asset crosses the strike K at any point, rather than at expiry is a notable differentiating feature between American Digital Options and their European counterparts.
  - **Barrier Options**: Barrier options are standard vanilla options that also embed a digital barrier which causes the option to either **knock-in** or **knock-out**. Just as we saw with digital options, these barriers can be observed continuously or only at a specific time.
  - **Lookback Options**: Lookback options give the buyer the right to choose the optimal exercise point at expiry. This means that we get to exercise at the maximum for a lookback call and at the minimum for a lookback put.
  - **Asian Options**: Asian options define their payoff based on the average value attained by the asset over a given period.

- **Volatility/Variance Swaps**: Volatility & variance swaps have payoffs linked to realized volatility. They are fixed for floating swaps, meaning require they buyer to exchange realized volatility/variance for some set strike. However, because they are functions of realized volatility, they are path-dependent and hard-to-value derivatives. Volatility/Variance Swaps provide investors a direct way to express view on the cheapness or richness of realized volatility.

- **Multi-Asset Exotics**: While single asset exotics comprise the lions-share of the exotic options market, there are also many multi-asset exotic options structures that frequently trade over-the-counter. As their name implies, these structures are characterized by payoffs that depend on multiple assets. These payoffs may be linked to only the assets values at expiry, which would make them easier to value, or may be a function of the entire path of each asset. Modeling these types of payoffs requires knowledge of the correlation structure of the underlying assets. Importantly, when working with multi-asset exotics we must calibrate this correlation structure so that we can then build it into our exotic option models.

  Multi-asset options are of direct interest to market participants for a few reasons. First, in some cases, these hybrid structures may be a more precise mechanism for investors to hedge specific risks relative to their more vanilla equivalents. Investors may also use these types of options to isolate cheapness or richness in the implied correlations of different assets. For example, if we believe there to be a correlation premium then it would be natural to want to trade these types of correlation dependent exotic structures in order to harvest this. Finally, in practice, hedge funds also use multi-asset options as a means of cheaply instrumenting their macro views.

  A few of the most commonly traded multi-asset exotics include:

  - **Dual Digitals**: A Dual Digital Option pays a predefined amount if two assets are above/below a certain level at expiry. As such, they are a two asset extension of the single-asset Digital Options described in the previous section.
  - **Basket Options**: Basket options have their payoff based on a linear combination of asset prices, or returns.
  - **Multi-Asset (Hybrid) Knockout Options**: These options are similar to the single asset barrier options discussed previously, however, in this case the barrier in a knock-in or knock-out option is on a different asset. As before, the barrier may still be European or American, that is the barrier may be monitored continuously or only a single date.

CHAPTER 13

Greeks and Options Trading

13.1 INTRODUCTION

In the last two chapters, we built the foundation for modeling options markets. In these chapters, we focused solely on valuation, and saw that this alone was quite complex. In this chapter, we take the next step and try to introduce some intuition on how different options structures behave and describe how some of the more common options structures are traded. Pricing options is inherently an exercise in trying to understand the market's volatility surface. Once we have a model that helps us understand this, then valuation of an arbitrarily complex instrument becomes mechanical. Finding opportunities in the options market, however, is a much more challenging endeavor that requires grappling not only with the valuation piece but also performing some sort of relative value analysis of these implied prices. In other words, we need to juxtapose what the market is implying with what is likely to happen. This will tell us if the market prices are over or underpriced.

In this chapter we also introduce a fundamental concept in managing derivatives portfolios: Greeks. Greeks are an important tool for understanding and managing derivatives books. These Greeks, or sensitivities to various parameters, help us understand how our options portfolios are likely to behave in the presence of different market conditions. For example, they can help us gauge whether we are likely to make or lose money when markets trend significantly. Greeks are also a critical component of building a hedging portfolio. The market standard set of Greeks, which are discussed in the next section, emanate from the Black-Scholes model. These set of Greeks are so ingrained within trading floors and options desks that much of the conversation around options portfolios centers on these Greeks. As such, they are a fundamental concept that quants need to be fluent in when engaging with PMs.

One of the main benefits of options is the ability it gives us to design structures that isolate parts of an asset's distribution. With options, rather than simply buying or selling the mean of an asset[1], we can isolate pieces of the distribution that we think may be overpriced or underpriced. A canonical example of this is the tail of a distribution, which may be overpriced due to an embedded risk premia. Later in the chapter, we describe the most commonly traded structures that options traders

[1]As is the case in delta-one instruments.

DOI: 10.1201/9781003530626-16

engage in. We pay special attention here to the properties of each structure. For example, is it long or short volatility? When doing this, we also connect the structures back to the underlying Greeks to provide intuition to the reader about how these concepts interact.

13.2 BLACK-SCHOLES GREEKS

In chapter 13, we introduced the Black-Scholes model and corresponding SDE. We have also seen that, while this model has serious limitations, including its inability to match market implied volatility skews, it is also a baseline of sorts in many markets. The concept of implied volatility, which is itself a manifestation of the Black-Scholes model, leads to the standard visual representation of the options market, a volatility surface. This model leads to a set of sensitivities, often referred to as a set of Greeks that are pervasive throughout the field of quant finance. As a result, Greeks are another main reason for adoption of the Black-Scholes model.

The Black-Scholes model defines a set of Greeks that help us understand how an option will behave locally with respect to small changes in certain parameters, notably:

- **Delta**: $\frac{\partial C}{\partial S}$: A move in the underlying asset, S_0
- **Gamma**: $\frac{\partial^2 C}{\partial S^2}$: The change in delta corresponding to a move in the underlying asset, S_0.
- **Vega**: $\frac{\partial C}{\partial \sigma}$: A move in the implied volatility of the option, σ
- **Theta**: $\frac{\partial C}{\partial T}$: A change in time to expiry, τ

These are the most dominant Greeks and as such are the ones that the reader will work with most frequently. Importantly, recall that the relationship between an option price and the underlying asset is not linear, leading to a second-order Greek, which is referred to as **Gamma**. This non-linearity is an important part of what makes options trading interesting, and as a result, understanding gamma is central to options modeling and trading.

While these are the main Greeks, and tend to account for most of the risk in options structures, there are others, including Rho ($\frac{\partial C}{\partial r}$), Vanna ($\frac{\partial^2 C}{\partial S \partial \sigma}$) and Volga ($\frac{\partial^2 C}{\partial \sigma^2}$). In fact, as the Black-Scholes formula is infinitely differentiable there are an infinite number of Greeks, but generally we don't look at anything beyond a second-order term as their impact diminishes. In the Black-Scholes model, each of these Greeks has an analytic solution. This will not be the case in more realistic models.

13.2.1 Delta

The delta of an option tells us the first order sensitivity of an option to a move in the underlying asset. It is the partial derivative of an option's price with respect to the underlying asset. Said differently, delta tells us about the linear part of an option's behavior with respect to the asset price. It also tells us how much the option price can be expected to move under small shifts to the underlying asset. As the shift gets bigger, delta will begin to tell less and less of the story. Instead, the non-linear behavior of the option will play a larger role.

Under the Black-Scholes model, a closed form solution for the delta of a call or a put option can be found analytically by differentiating the Black-Scholes formula in (10.2) with respect to the asset price. It should be noted that the asset price, S_0 appears both in the outer equation as well as in the d_1 and d_2 terms. This makes calculation of the derivative more complex, however doing so leads to the following formulas for delta for European call and put options:

$$\Delta_c = \frac{\partial C}{\partial S_0} = \Phi(d_1) \tag{13.1}$$

$$\Delta_p = \frac{\partial P}{\partial S_0} = -\Phi(-d_1) \tag{13.2}$$

where d_1 is as defined previous in (10.2).

Notice that, as $\Phi(d_1)$ is a positive number, the delta for a call is always positive and the delta for a put is always negative. This is consistent with the payoff functions for calls (and puts), respectively, whose value increases when the asset price increases (or decreases). We can also see from this that the magnitude of delta is contained between 0 and 1. Again this is intuitive because an option that is sufficiently out-of-the-money may not increase when the asset price moves in our favor, but a move in our favor should never lower the option price. Similarly, once an option is sufficiently in-the-money, it starts to behave like the underlying asset, hence the delta of 1, but its delta cannot exceed this.

Delta is perhaps the most commonly used Greek and in many markets options are quoted in terms of their deltas instead of their strikes. It is also the most commonly hedged Greek, as we explore later in the chapter. Traders tend to focus on out-of-the-money options, which have deltas whose magnitude is less than 0.5.

13.2.2 Gamma

One of the fundamental reasons that options are interesting to traders is the non-linearity that is embedded within them. That is, options payoffs are convex, when we are long an option. This means that, when the market moves in our favor, we will make more than is implied by our delta. Similarly, when the market moves against us, we will lose less than is implied by our delta. This is a naturally attractive feature for investors, and in practice if we can find cheap methods to buy this convexity it is incredibly valuable. Conversely, we may be able to find pockets of the market where this convexity is overpriced, and extract a premium for selling it. Gamma is the Greek that measures this non-linearity, or convexity with respect to the underlying asset price. It is the second partial derivative with respect to the underlying asset price.

Just as we could with delta, the gamma of a European option can be computed in closed form within the context of the Black-Scholes model. This can be accomplished by differentiating the Black-Scholes formula in (10.2), and leads to:

$$\Gamma = \frac{\partial^2 C}{\partial S_0^2} = \frac{\partial^2 P}{\partial S_0^2} = \frac{\phi(d_1)}{S_0 \sigma \sqrt{T}} \tag{13.3}$$

Notice that gamma is always positive, and is the same for a call and a put with the same strike. An option's gamma relative to its price, however, will be significantly higher for out-of-the-money options.

Gamma is of critical importance to quants modeling options for many reasons. First, traders often delta-hedge their options portfolios. When doing this, gamma becomes a paramount source of risk and return for the resulting trade. This relationship, and the resulting trade-off with theta, are described in 13.3. Secondly, the relative pricing of gamma, whether an investor chooses to hedge the delta or not, is generally a deciding factor when choosing between a position in the underlying and a position in an option's structure. Said differently, investors don't trade options to access the delta of an asset. They can do that via futures, forwards, or cash positions in the underlying. Instead they choose options because of the embedded non-linearity, or gamma.

13.2.3 Delta and Gamma in Practice: Delta and Gamma by Strike

In this example we leverage the code in the repository for computing delta and gamma to illustrate how both the delta and gamma of options with the same expiry, vary by strike. In the following chart we focus on delta, and show how the delta of an option changes as it becomes further in or out-of-the-money:

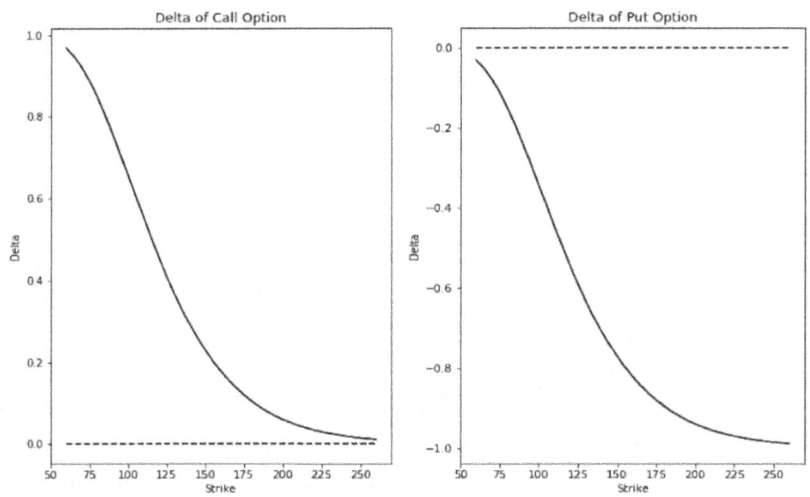

The reader is encouraged to notice that the sign for the delta of a call is indeed always positive, and is bounded by zero and one. Similarly, for a put option we observe consistently negative values bounded between 0 and −1. Notice also that, for a call option, delta decreases monotonically as we increase the strike, as it becomes increasingly out-of-the-money. Conversely, for a put option, in absolute terms, we see the opposite, That is, its absolute value increases monotonically as we increase the option's strike. Finally, the reader is encouraged to notice that, in both the case of call and put options, delta is a non-linear function of strike. This is an important property of options. In fact, as we will show in the next chapter, options are by necessity convex functions of the strike, in the absence of arbitrage.

Next, we consider the gamma of a call and put option, again focusing on options whose characteristics are otherwise equal. Recall from earlier in this section that for a given strike the gamma of a call and put option are equal. Therefore, we display this gamma for both calls and puts, as a function of strike, in the following chart:

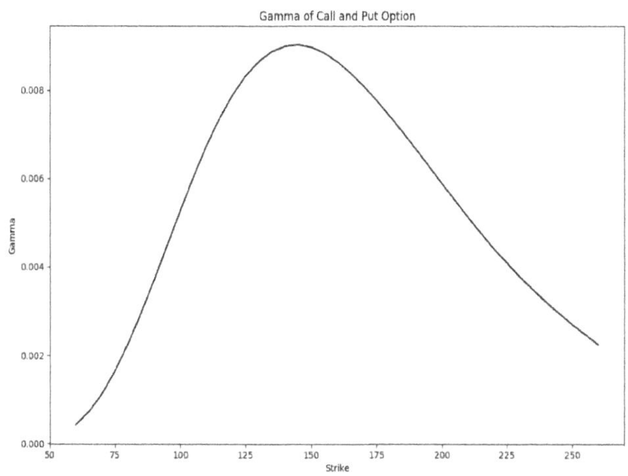

Notice that the absolute level of gamma which is displayed here is highest near the at-the-money strike and decays as we approach either extreme. This picture changes, however, if we think about gamma in terms of the option price. That is, if we instead focus on relative gamma then we find that out-of-the-money options have significantly more gamma per unit of cost of the option, than at-the-money or in-the-money options. A theoretical example of this phenomenon is displayed in the following chart for call options:

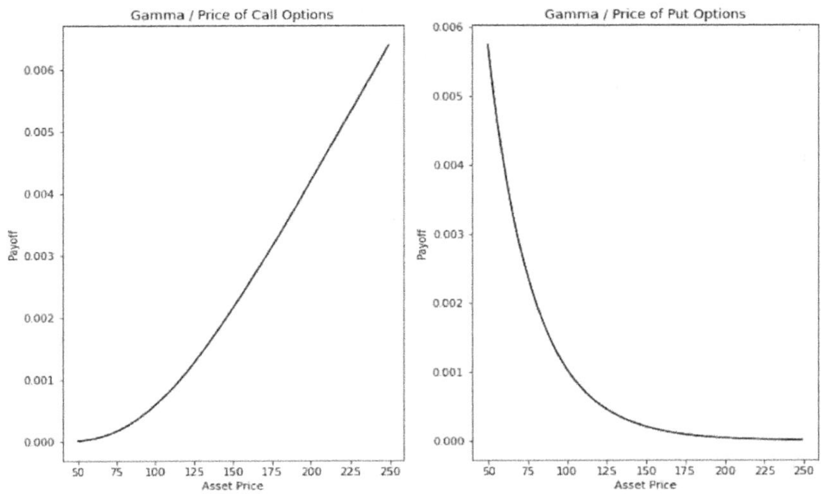

13.2.4 Theta

Theta, or the sensitivity of an option to the passage of time, is another critical Greek. As we saw when discussing gamma, the presence of convexity is an attractive feature of option's structures. Another interesting feature of option is that they decay in value as we approach expiry. Prior to expiry, options are worth more than their payoff value, and as expiry approach, they converge to this payoff, as they lose their optionality. Measuring this decay is therefore a critical component to understanding a book of options. If the theta decay is prohibitive it can have a problematic drag on the portfolio's performance. Benefiting from this theta decay, however, in practice generally requires that we take on other unappealing properties.

Theta is defined as the first order partial derivative with respect to option expiry, T, and, like other Greeks can be calculated in the Black-Scholes model by differentiating the Black-Scholes formula:

$$\theta_c = \frac{\partial C}{\partial T} = -\frac{S_0\phi(d_1)\sigma}{2\sqrt{T}} - rKe^{-rT}\Phi(d_2) \tag{13.4}$$

$$\theta_p = \frac{\partial P}{\partial T} = -\frac{S_0\phi(d_1)\sigma}{2\sqrt{T}} + rKe^{-rT}\Phi(-d_2) \tag{13.5}$$

Notice that theta takes on a negative value for both calls and puts. This reflects this nature of option's to converge to their payoff value over time as the embedded optionality becomes worth less. How this theta decay occurs is itself non-linear, and as a result is of great interest to market participants. This phenomenon is explored in more detail in the next section.

13.2.5 Theta in Practice: How Does Theta Change by Option Expiry?

In this example we leveraged the code provided in the repository to explore how the theta of an at-the-money evolves with option expiry. That is, we explore if theta is constant and independent of expiry or if there are certain patterns. In doing this we uncover a commonly known fact about theta. That is, the behavior of theta is highly non-linear, and when options are close to expiry their theta is particularly onerous.

This phenomenon is depicted in the following chart, where we plot the theta of both a call and put options as a function of expiry:

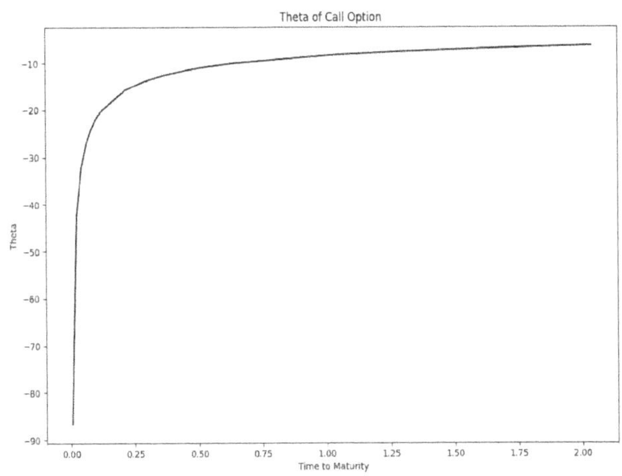

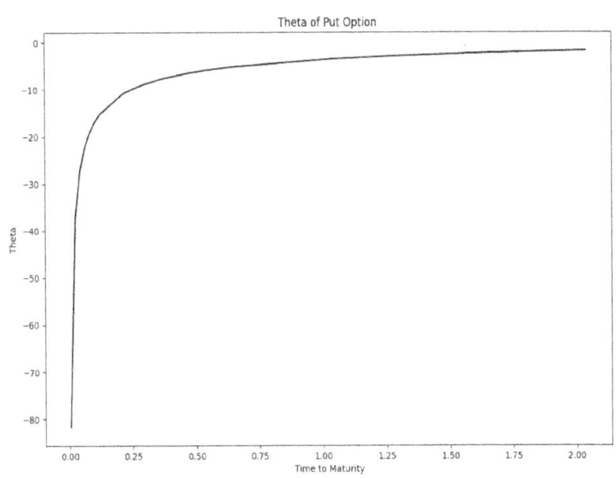

In both charts we can see this similar phenomenon emerge. That is, the value of theta when the remaining time to expiry is small is significantly more negative than when the remaining expiry is longer. This non-linear behavior of theta has important implications for options traders. In particular, to hold an option to expiry requires that we pay a sizable theta bill as expiry approaches. Alternatively, we could attempt to manage our theta decay by rolling our trades before we have reached the steepest point on the above theta curve.

13.2.6 Vega

Vega is defined as the sensitivity of an option to changes in the level of implied volatility, σ. Of course, in the Black-Scholes model, this σ is assumed to be constant. However, in practice it varies with market conditions as well as strike and expiry. In particular, we tend to observe significant increases in levels of implied volatility as market stresses occur. As such, options trader will want to be keenly aware of their exposure to these sharp increases, which can be approximated using vega. Vega measures the first-order change of the option price to a change in implied volatility. Another derivative, volga, measures the second-order derivative of an option's price with respect to implied volatility, capturing the non-linear component.

Differentiating the Black-Scholes formula with respect to σ, the implied volatility parameter, we can see that vega can be written as:

$$\nu = \frac{\partial C}{\partial \sigma} = \frac{\partial P}{\partial \sigma} = S_0 \sqrt{T} \phi(d_1) \tag{13.6}$$

Notice that vega is always positive for long options positions. Somewhat relatedly, options traders tend to say that we are long volatility when we buy options and are short volatility when we sell options. Long volatility trades are characterized by increases in value when implied volatility increases and large market moves occur. Short volatility trades, conversely, tend to suffer as volatility increases and benefit from markets remaining calm and not moving much. There is a strong correlation between trades that are long vega and long gamma, as these two Greeks tend to be tightly intertwined.

13.2.7 Practical Uses of Greeks

In practice, quants and portfolio managers should not think of Greeks as abstract modeling quantities, but instead as a critical piece of management of a book of derivatives. As any options trader or desk quant surely knows, Greeks are a central part of any options based investment process, whether it is a quantitative or fundamental approach. In this context, the Black-Scholes model is still the market standard for Greek calculations due to the intuitive nature of each parameter and ensuing sensitivity.

Greeks are a critical component of hedging, both at the trade and portfolio level. This is a concept that we will explore in more detail later in the chapter, however, intuitively, it makes sense that creating a portfolio with little exposure to movements in the underlying, for example, will be closely coupled with delta calculations.

Greeks can also be used to set position limits and approximate portfolio risk. For example to compute VaR (Value at Risk) of an option's position, we might use our portfolio's delta and gamma and then use that to determine the value after a 2 or 3 standard deviation move in the underlying. Relatedly, Greeks can also be used to estimate P&L in real-time. While larger models may require a new calibration and may take more time to run, we can always multiply our respective portfolio Greeks by the market change to get a good sense of the P&L of our options books.

Greeks also play an important role in portfolio attribution for options structures. That is, if we have an option position we will most certainly want to know how to estimate how much of our P&L has come from movements to the vol. surface (vega), how much has come from movements in the underlying (delta and gamma), and how much is due to lost optionality as time passes (theta).

Greeks also can help to increase our understanding of the general properties of our options positions. As a result, traders also often communicate in the language of Greeks, and may even use Greeks in their trading decision-making process. We explore one aspect of this relationship is the next section, that is, the fundamental relationship between theta and gamma. In this, and many other cases, knowledge of an options Greeks tell us how our structure is likely to perform in various situations. If a structure is short vol, or short gamma, this means something very fundamental to an options trader. Specifically, it means that the position will suffer in large movements of the underlying, or when volatility spikes. Similarly, Greeks can help us to understand the theta decay profile of an option, which is non-linear but instead decays very quickly as the expiry shrinks toward zero[2]. As such, the reader is encouraged to think through how Greeks can be leveraged as part of an investment process, and to understand the different characteristic Greeks of different options structures.

13.3 THETA VS. GAMMA

Recall from chapters 4 and 15 that the Black-Scholes model is defined by the following PDE:

$$\frac{\partial c_t}{\partial t} + \frac{1}{2}\sigma^2 S_t^2 \frac{\partial^2 c_t}{\partial S_t^2} + r S_t \frac{\partial c_t}{\partial S_t} - r c_t = 0, \tag{13.7}$$

This can be seen by applying Ito's Lemma to the Black-Scholes SDE and relying on a hedging argument. This PDE can be solved with the appropriate boundary conditions in order to derive the Black-Scholes formula for vanilla options. Alternatively, as we saw in chapter 15, this PDE can also be solved numerically. To do this, we need to know how to **discretize the PDE** using **finite differences**, which was covered in detail in chapter 15.

However, the PDE also tells us how option prices evolve over time, and as a result we can get a great deal of insight from it. For example, let's consider a Black-Scholes world and a delta-hedged portfolio. That is, a portfolio that is long one call option and short Δ shares of the underlying such that the portfolio delta is 0. If we use equation (13.7) to analyze the remaining risks in this structure, we can see that a delta-hedged portfolio will only have exposure to the first two terms. These two terms are tied to the theta and gamma of an option. It is noteworthy that the sign on both terms is positive, however, we also know that gamma itself is always positive and the theta for an option is always negative. Therefore, the sign on the gamma term is

---

[2]As we explored in 13.2.5.

always positive, and the sign of the theta term is always negative, meaning that they have an offsetting effect on each other.

Therefore, when we buy an option, we have a tradeoff between the theta that we pay, and the gamma that we receive. This tradeoff is a common refrain on trading desks. Relatedly, it is common to hear discussions on whether there will be enough gamma in the underlying to justify the decay of the option price, or theta.

13.4 MODEL DEPENDENCE OF GREEKS

An astute reader may notice that the set of sensitivities that we care about is **model dependent**. For the Black-Scholes model, we care about a certain set of Greeks. Some of the Greeks, such as delta and theta are fairly universal, as these concepts will be embedded in any stochastic model that we choose. Vega, by contrast, is completely dependent on the SDE that we chose. It is entirely model dependent. Importantly, for the Black-Scholes model these Greeks are intuitive and have been universally accepted by the market as the standard in spite of the many limitations of the underlying model. In particular, as we saw in earlier chapters, the Black-Scholes model does not provide us with a consistent set of dynamics for an underlying asset that fit the volatility surface. Instead, to apply the Black-Scholes model, we need to apply a different implied volatility, σ for each strike, K, and expiry, T. As a quant this is not satisfying as it means a different set of dynamics, or different distribution, for different strikes on the same underlying asset. The fact that the Black-Scholes model is the standard for Greeks while simultaneously is insufficient for modeling the volatility surface means that quants must resort to more realistic models to understand the dynamics and then map their Greeks back to the Black-Scholes model to communicate their results to other market participants. This is a problem quants often face as traders generally think in terms of Black-Scholes Greeks but as we move to more realistic models there is no single parameter corresponding to each Black-Scholes model input.

Instead, when a different model is applied, it is less clear what the meaning of some of these Black-Scholes model Greeks become. For example, if we apply the Heston model, how do we calculate the equivalent of a Black-Scholes vega? The answer is not trivial because the Heston model does not have a parameter with a one to one mapping to a Black-Scholes σ. The Heston model, in contrast to the Black-Scholes model, has more than one parameter that deals with volatility. For example, as we saw in chapter 13, there is a parameter for the equilibrium level of volatility, θ and another parameter for the initial level of volatility, σ_0. This gives us the ability to create different, more nuanced shifts to volatility that are not possible in the one parameter Black-Scholes model. For example, we could define a parallel shift to the level of volatility, or steepen or flatten the term structure of volatility. In reality, this makes creating a set of Black-Scholes style Greeks in the context of another model conceptually challenging.

Note that in the Black-Scholes model a change in the underlying asset is assumed to be independent of a change in its implied volatility. This adds another level of complexity due to the fact that the asset price and its volatility depend on each other by definition in a stochastic volatility model. This means that if we shift one, we might be implicitly shifting the other. How do we make sure our Greeks incorporate this? This is a concept that we explore in more detail in 13.7.

In practice, due to a confluence of reasons, such as the ones brought forth in this section, it is common to find that the Greeks across models diverge. This is true both for Greeks like delta where the interpretation is unambiguous, and for Greeks like vega, where it is harder to identify an analogous parameter shifts. This creates a challenge for quants identify which set of Greeks they should trust, and use to construct their hedging portfolios.

13.5 GREEKS FOR EXOTIC OPTIONS

Unsurprisingly, as the complexity in our instruments, or in our underlying models increases, our ability to calculate quantities in closed form is significantly challenged. In the last chapter, we saw this in a valuation context, as most exotic option structures require the use of simulation or numerical solutions to PDEs. For example, in the case of a Lookback option, we needed to use *simulation* to get an option price. It will be the same story for Greek calculations. In some rare cases, such as when employing the Black-Scholes model, we may find certain exotic option Greeks for which we are able to derive a closed-form-solution. In most realistic applications, however, we will need to employ a numerical approximation for these derivatives.

It should be noted that there will be a common set of Greeks that apply to all our exotics. These Greeks may indeed be model dependent, as we saw in the last section. However, once we have selected a model we will be interested in the same set of Greeks for European options, as we would be for Lookback options, or other exotic structures.

The standard way to handle Greeks, and other sensitivity calculations in the absence of a closed-form solution is to utilize finite differences, as we detail in the next section. The uniform nature of Greek calculations, and the lack of dependence on the instrument's payoff are a convenient feature from a coding perspective. In particular, it means that we can write code for Greeks in a highly generic way, leveraging the techniques learned in chapter 8, such as inheritance and the template method.

13.6 ESTIMATION OF GREEKS VIA FINITE DIFFERENCES

As our stochastic process becomes more complex, or the option we are pricing becomes more complex, we should not expect analytical formulas for prices or Greeks. In these situations, the most common method for calculating Greeks is via finite differences. This approach will be required for exotic options, as noted in the last section, and also for more advanced stochastic processes such as stochastic volatility models and jump-diffusion models.

Intuitively, we can do this by **shifting** the input parameter of interest and re-calculating the model price. For example, for Δ we would shift the spot price of the asset, S_0, and would have:

$$\Delta \approx \frac{c_0(S_0+\epsilon) - c_0(S_0-\epsilon)}{2\epsilon} \tag{13.8}$$

where ϵ is some pre-defined shift amount. The motivation for (13.8) is the technique of finite differences, which we introduced in chapter 15. In this chapter, we saw that finite differences could be used to approximate derivatives within the context of a numerical solution to a PDE. We also saw that (13.8) was a first order central difference approximation for the pricing function, c_0 with respect to S_0.

To review, we begin by looking at a Taylor series expansion of the option price around S_0. A central difference in particular takes advantage of increased efficiency by approximating the function at both $S_0 - \epsilon$ and $S_0 + \epsilon$:

$$c_0(S_0+\epsilon) = c_0(S_0) + \epsilon dc_0 + \frac{\epsilon^2}{2} dc_0^2 + O(\epsilon^3) \tag{13.9}$$

$$c_0(S_0-\epsilon) = c_0(S_0) - \epsilon dc_0 + \frac{\epsilon^2}{2} dc_0^2 + O(\epsilon^3) \tag{13.10}$$

Notice that if we subtract 13.10 from 13.9 the dc_0^2 term is eliminated. This leads to increased accuracy in our Greek estimate. This leads to the following equation:

$$c_0(S_0+\epsilon) - c_0(S_0-\epsilon) = 2\epsilon dc_0 + O(\epsilon^3) \tag{13.11}$$

which can be rearranged to obtain (13.8). This approach will enable us to compute a first order derivative with respect to any input parameter, not just the option delta. For example, an option's vega would analogously be analogous:

$$\nu \approx \frac{c_0(S_0, \sigma+\epsilon) - c_0(S_0, \sigma-\epsilon)}{2\epsilon} \tag{13.12}$$

Similar techniques can be used to approximate second-order derivatives as well as cross partial derivatives. For example, for a second-order finite difference estimate with respect to the spot price, commonly known as gamma, we would have:

$$\Gamma = \frac{c_0(S_0-\epsilon) - 2c_0(S_0) + c_0(S_0+\epsilon)}{\epsilon^2} \tag{13.13}$$

13.7 SMILE ADJUSTED GREEKS

As we have detailed in this chapter, Greeks are model dependent and sometimes it is challenging to map from a set of Greeks in one model to a comparable set of Greeks in another. Δ is an exception to this, however, as the definition of a shift in the underlying is unambiguous and comparable regardless of model. Yet even in this case, the Δ's in different models can be significantly different. One fundamental reason for this is the interaction between an asset's price and its volatility in a stochastic volatility model. To be precise, let's define Δ as the sensitivity to changes in the underlying asset leaving all other parameters, including volatility, constant.

That is, we perform the following transformation:

$$S_0 = S_0 + \delta S \tag{13.14}$$

$$\sigma_{\text{impl}} = \sigma_{\text{impl}} \tag{13.15}$$

where δ is the shift amount. This is a sensible thing to do in the context of the Black-Scholes model, where σ_{impl} is a constant. However, in practice we observe that implied volatilities are not constant but rather move with market conditions. This means that, especially in equities, we expect the implied volatility to increase as the spot price, S_0 decreases. Said differently, we expect negative correlation between the asset price and the implied volatility. Stochastic volatility models formalize this connection by explicitly defining the correlation between an asset and its volatility as a model parameter, ρ. For example, we saw this in the Heston and SABR models introduced in chapter 13.

In this setting, if we are calculating the Δ of a call option, we need to incorporate the potential change in volatility that is likely to correspond to a change in the asset. The chain rule tells us that our sensitivity, or smile-adjusted delta will be:

$$\Delta = \frac{\partial C}{\partial S} + \frac{\partial C}{\partial \sigma_{\text{impl}}} \frac{\partial \sigma_{\text{impl}}}{\partial S} \tag{13.16}$$

In the Black-Scholes model, the second term is zero as implied volatility is assumed to be constant. Therefore, the second term can be ignored since the implied volatility does not depend on the asset price. This will not be true in other models, however, in particular those with stochastic volatility if $\rho \neq 0$. This leads to the concept of multiple Δs, those that are standard and those that are smile-adjusted. By smile-adjusted, we mean those that account for the correlation between spot and volatility when adjusting the spot and volatility parameters. In particular, a smile-adjusted delta would create a scenario consistent with the stochastic volatility process, such as:

$$S_0 = S_0 + \delta S \tag{13.17}$$

$$\sigma_{\text{impl}} = \sigma_{\text{impl}} + \delta_S \sigma_{\text{impl}} \tag{13.18}$$

where δ_S is defined by ρ and the particular SDE we are working with.

For other Greeks, such vega, we need to first agree on what a shift in volatility means. In Black-Scholes, it is clearly defined as the sensitivity to the option's implied volatility. In the case of Heston, there is no single volatility parameter. Instead, we have:

- θ can be thought of as a long-run equilibrium volatility.
- ν_0 can be thought of as a short-term volatility.

Consequently, we can think of moving both together as a **parallel shift** of the volatility surface. The alternative would be to bump the input volatility surface in parallel and re-calibrate our Heston model parameters. Once we agree upon the definition of a shift in volatility, the same argument as we used for delta applies, meaning we would need to be thoughtful about whether the underlying asset is likely to move in conjunction with the volatility parameters.

13.7.1 Smile Adjusted Greeks in Practice: USDBRL Options

In this section we leverage the code in the repository for computing smile adjusted deltas and compare the deltas obtained using the Black-Scholes model, standard Heston model deltas, and smile-adjusted Heston Model deltas. We apply these calculations to the FX market where a large amount of skew is prevalent, USDBRL. We begin by calibrating a Heston model to the USDBRL volatility surface, using the techniques introduced in chapter 14, and then leverage the code with the appropriate parameters to produce delta estimates for at-the-money, one-year options. The comparison of these computed deltas is presented in the following table:

Type of Delta	Delta Value
Black-Scholes	56%
Heston	55.3%
Heston Smile Adjusted	55.6%

Here we can see that the deltas are similar, however, the inclusion of the smile adjustment does have an impact. As the Heston model parameters change, along with the slope of the volatility smile, the magnitude of this correction may vary in importance as well. In particular we can see that, relative to the Black-Scholes delta, the smile adjusted lowers the delta, albeit marginally. The fundamental reason for this is because USDBRL is characterized by upside skew, meaning that volatility increases in the delta calculation as the exchange rate increases. How far these deltas diverge will depend on the dynamics of the volatility surface, and the reader is encouraged to experiment with the code under different Heston model parameterizations to analyze the magnitude of the difference in the deltas as market conditions evolve.

13.8 HEDGING IN PRACTICE

A natural use for Greeks is constructing portfolios that are immunized from a source of uncertainty. **Delta**, is most common, but not the only Greek to hedge. To construct a hedging portfolio, we need to identify the underlying sensitivity, as well as the instrument or set of instruments that will be used to construct the hedge. In some cases the choice of a hedging instrument may be trivial, such as when we use the underlying asset to hedge our delta. In other cases, such as hedging ones vega risk, it becomes less clear what the hedging portfolio should entail. Further, when we construct these hedging portfolios, the assumptions that we make about the model that we use, and how the parameters interact with each other, can have a significant impact on the resulting hedging portfolio. Most notably, when delta-hedging, whether we assume that the volatility surface stays constant, or moves in line with the change in spot, may have a substantial impact on the effectiveness of our hedging portfolio.

Theoretical hedging arguments assume continuous trading with no transactions costs or market frictions. In reality, we will choose to update our hedging portfolio at discrete times and will, generally speaking, incur costs for doing so. This presents us with a tradeoff when hedging. Hedging too frequently will result in prohibitively high transactions costs. Conversely, hedging too sparsely will result in larger portfolio volatility.

Remember that, when hedging an option portfolio, a dynamic hedge is required for us to stay immunized to a given risk. Once we construct our hedged portfolio, we have theoretically eliminated all risk associated with a given exposure[3]. This hedged portfolio will then start to accumulate exposure over time as the market moves. For example, if we construct a delta-hedged portfolio consisting of a long call option and a short position in the underlying assets, this portfolio will have no delta at first. But as the stock moves up or down, this portfolio will accumulate positive or negative delta until it is re-balanced again. This creates the need to re-balance to become delta-neutral again.

In the next section we explore the two main approaches to constructing a re-balancing strategy: calendar and move-based. We then proceed to look at delta and vega hedging more specifically, focusing on how we can implement these hedges in practice.

13.8.1 Re-Balancing Strategies

A key component of any practical hedging strategy requires a **re-balance scheme**. This re-balance scheme provides us with a mechanism for deciding when the residual exposure we may have accumulated in our portfolio warrants re-balancing. Re-balancing schemes can be entirely rule based, or can involve a subjective component. For example, a Portfolio Manager may alter their re-balancing decision based on how attractive they feel their residual exposure is given the market climate. That said, many re-balancing schemes are systematic, and are usually based on one of the following two criteria:

- **Calendar Based Re-Balance**: Re-balance the hedging portfolio after a predetermined period of time elapses regardless of the drift that has occurred in the portfolio.

- **Move Based Re-Balance**: Re-balance after some percentage movement threshold in the underlying has been reached, regardless of how quickly or slowly the move occurred.

13.8.2 Delta Hedging in Practice

Delta-hedging is done by combining an options portfolio with long or short positions in the underlying asset. If we are long a call option, then the option's delta is positive,

[3]Assuming our model is 100% correct.

and therefore we will want to sell shares to neutralize our delta. Conversely, if we are long a put option, then our delta is negative, and we will need to buy shares to neutralize the delta.

In the context of a Black-Scholes model, the delta of a European call can be calculated as:

$$\Delta = N(d_1) \tag{13.19}$$

In other, more complex models, the delta may differ from this Black-Scholes delta and may require the use of finite differences as discussed in section 13.6.

Clearly, an important factor in delta hedging will be how successful we are at calculating the options delta to begin with. This is a model dependent concept and will depend on our beliefs about the interaction between the underlying asset and its volatility. In a stochastic volatility model, for example, correlation between the two is embedded in the model. In contrast, in the Black-Scholes model volatility is a constant and thus is unaffected by the level of the underlying asset. Empirical analysis of this relationship in most asset classes shows that the correlation between these two quantities tends to be strong, making it an attractive feature to include in our model and delta calculations. If not, we may over or underestimate our Δ which may lead to increases P&L swings.

Once we have created a delta-hedged portfolio for an options position, it will not remain delta-neutral forever. Instead, as time passes and market conditions changes, the hedged portfolio will accumulate delta until we choose to re-balance. Prior to re-balance then, a phenomenon emerges where we become long (or short) continuation moves vs. short (or long) reversals. This means that, if we have beliefs about whether the underlying asset is mean reverting, or prone to large drifts, it may impact our optimal re-balance strategy.

As an example, suppose we are long an at-the-money straddle. As the straddle is struck at-the-money, initially, its delta will be zero. However, suppose the underlying asset moves by +5%. The delta of this straddle will no longer be zero. Instead, it will be positive. This means that if we do not choose to re-balance we are implicitly betting on a continuation of the initial 5% move and are short a reversal. If we re-balance however, we reset our delta to zero, making us indifferent to the direction of a future move. If we re-balance and the market subsequently reverses, then it is said that we have captured gamma and will have been better off by re-balancing.

13.8.3 Vega Hedging in Practice

In some cases, we may want to hedge an options exposure not to the underlying asset but to a movement to the implied volatility of the option. This is known as vega hedging. To do this, we need to start with a hedging portfolio that has isolated exposure to vega with no other exposures. Unlike in the case of delta, where the hedging portfolio was self-evident, the proper hedging instruments for vega hedging are more ambiguous. A common hedging portfolio is an at-the-money straddle. As we will see in 13.9.4, a straddle is an options structure that is long a call and put option with the same strike and other characteristics. In the case of an at-the-money straddle, this structure is characterized by an absence of delta, and a significant exposure to volatility, or vega. It is also characterized by a potentially large amount of positive gamma, and a negative theta.

An ideal hedging instrument is clearly one with a large exposure to our sensitivity of choice, in this case implied volatility. Additionally, it is of equal importance that the hedging portfolio be neutralized from other key risks, such as delta. This is what makes an at-the-money straddle position an appealing candidate as a vega hedging tool. That is, we are able to introduce an offsetting amount of volatility exposure without introducing any delta into the portfolio. In order to choose the correct number of units, we need to calculate the vega of the portfolio to hedge, as well as of the straddle, which will give us the hedge ratio:

$$H = \frac{\nu_{\text{port}}}{\nu_{\text{hedge}}} \tag{13.20}$$

where ν_{port} and ν_{hedge} are the vega exposures of the options portfolio, and the hedging instrument, respectively. In the Black-Scholes model, these individual vegas for the portfolio and hedging instrument could be computed analytically. In other cases, finite differences would be required.

This type of vega hedging is most common when we have a portfolio that consists of a set of exotic options that contain additional interesting characteristics. The trader may then look to isolate these more subtle characteristics by hedging the delta and vega of the exotic. In these cases, ν_{port} would be the vega of one or more exotics, and ν_{hedge} would be the vega of a straddle, for example. Vega hedging decisions, like their delta-hedging counterparts, may also be dependent on whether we think the underlying asset and volatility are likely to move together.

13.8.4 Validation of Greeks Out-of-Sample

As has been mentioned throughout the chapter, Greeks are a model-dependent concepts and the Greek that we obtain, even for an unambiguous quantity like delta may depend on the model specification. This leads to the fundamental question of which Greeks should be trusted and why. There are many ways to attempt to answer this question. Part of the answer may come from theory. We may conceptually find certain Greeks that incorporate more realistic market features more attractive. Alternatively, a more empirical approach can be employed.

To validate the choice of a model for Greek calculations, one can perform an out-of-sample test to see how well the model's hedge ratios performed when looking at historical data. Said differently, considering the case of delta, we can see how much the value of the delta hedged portfolio moved as the underlying asset moved. The best hedge, would then be the one with the minimum P&L due to variations in spot over the historical dataset.

13.9 COMMON OPTIONS TRADING STRUCTURES

13.9.1 Benefits of Trading Options

Investors who buy options in lieu of a cash position in the underlying asset are given access to **leverage** via a high payout in relation to their premium outlay should the option expire in-the-money. Relative to a position in a forward or futures contract, options provide us with access to a conditional payout if the trade goes against them. Therefore, relative to futures, in cases we are less certain about our payoff, options may be an attractive way to mitigate our losses in adverse scenarios. They also give access to the **interesting non-linear properties** in the option's price as a function of its parameters. Generally speaking, we do not buy options to get long or short access to the **delta** of an underlying asset. We could do that by buying or selling the underlying or a forward/futures contract.

Instead, when we buy options we are generally acutely interested in the options gamma and theta. We may choose to look at this from a risk premia perspective, and believe that selling gamma should reward us with a so-called volatility risk premium. Alternatively, we may choose to take a relative value perspective. Perhaps we have a forecast of realized volatility that we would like to use to express views on its deviation from implied volatility. Or perhaps we have another means of judging whether gamma is cheap or rich. A benefit of options trading is that each of these perspectives are easily implementable via liquid structures. Most notably, these types of trades are often associated with long, or short, positions in straddles or strangles.

More generally, options provide a precise way for quants to express views on parts of a distribution. When we buy or sell a stock, we are implicitly making a bet on the mean of the future distribution of this asset. We cannot, however, do anything more nuanced than bet on this mean. Options, however, provide us the tools for instrumenting more precise views of a distribution. For example, options could be used to wager that the tail of a distribution is over- or under-priced. We could also express a view on a specific slice of a distribution.

In the rest of this section, we detail some of the most commonly traded options structure, and describe their properties in more detail.

13.9.2 Covered Calls

A covered call strategy is one in which we buy the underlying asset and then sell a call option on the underlying. This is a commonly executed strategy in equities, as it enables us to benefit from a perceived volatility premium. Retail investors with existing stock positions will often take advantage of this strategy as a way to receive additional income in their portfolios.

The following table and payoff diagram show the detailed legs of a covered call trade, as well as the payoffs at expiry:

Long/Short	Units	Instrument
Long	1	Underlying Asset
Short	-1	Call Option

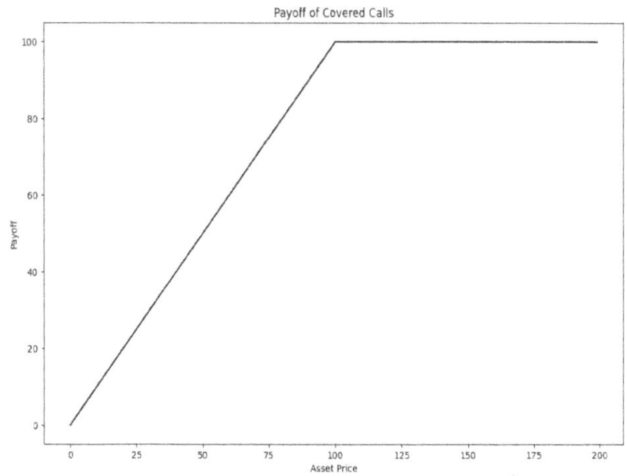

Here we can see that, in contrast to a position in the underlying asset, we have sold some upside, meaning we have foregone payoffs in states of the world where the asset increases substantially in exchange for an upfront option premium that leads to higher payoffs in states where the asset doesn't rally considerably. As we can see from the payoff diagram above, the best case scenario for this structure is that the sold call option expires at the strike, so that the sold call is worthless, but the values of the long position in the underlying shares is maximized. Because of this, this structure is characterized by a short volatility profile. That is, the trade will be short gamma, and short vega by nature. Conversely, the trade will benefit from the passage of time, and will as a result be long theta and be defined by positive carry.

13.9.3 Call & Put Spreads

Spreads are another commonly traded options structure that enables us to sell back some volatility relative to an outright position in a European call or put. This feature of selling back some volatility can be useful as it helps to manage the underlying volatility profile of the structure, and can be particularly attractive in situations with extreme skew. Spreads can be executed on calls or puts, and are then referred to as call and put spreads, respectively. Long spreads are generally characterized by long volatility exposure in that an initial premium outlay is exchanged for a convex payoff profile.

Buyers of call and put spreads, as shown in the following tables, are combinations of long positions in a call (or a put), coupled with a short position at a higher (or lower) strike with the same expiry. The sold option positions are by construction further out-of-the-money than they long option positions. As we will see in the next chapter, call and put spreads whose strikes are close to each other approximate the CDF of the risk neutral distribution. This makes call and put spreads structures that are naturally of interest to quants and other market participants.

Further, the following payoff diagram shows the payoff profile for a call and put spread, respectively:

Long/Short	Units	Instrument
Long	1	Call Option with Strike K
Short	−1	Call Option with Strike K + h

Long/Short	Units	Instrument
Long	1	Put Option with Strike K
Short	−1	Put Option with Strike K - h

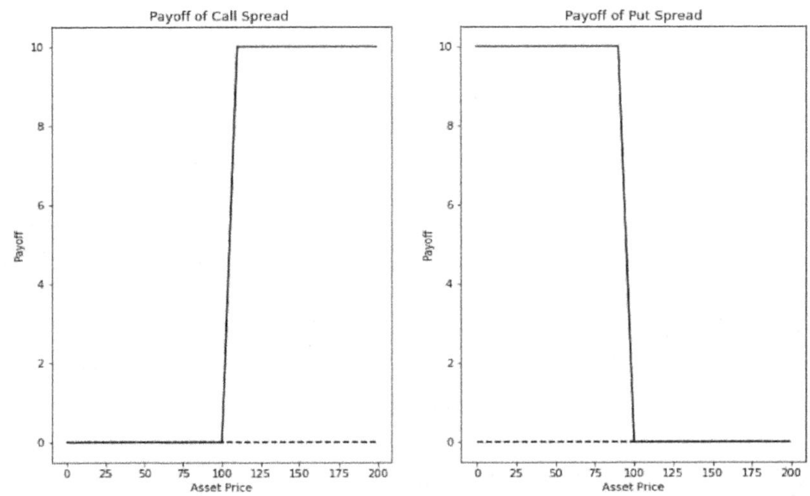

Put and call spreads will be most attractive when the shape of the volatility smile is structured in such a way that the implied volatility of the sold option is highest and the implied volatility on the bought leg is minimized. This means that in cases where skew is most pronounced on low strike puts, then put spreads will naturally be appealing. Long positions in out-of-the-money Put spreads, like puts will be characterized by negative delta, positive gamma and negative theta. Unlike a put option in isolation, however, a put spread is characterized by a changing gamma profile. When the put spread is out-of-the-money, it is defined by long gamma and negative theta, similar to a put option. As the put spread becomes in-the-money, however, the option can become short gamma and accrue positive theta. Analogously, a long position in an out-of-the-money Call spread is defined by positive delta and gamma, and likewise a negative theta, but just as for a put this gamma and theta behavior can flip as the call spread becomes further in-the-money.

13.9.4 Straddles & Strangles

Straddles and strangles are commonly traded options structures when a trader is looking to isolate exposure to volatility. As we will see later in the chapter, delta-hedged straddles, or strangles, are an effective means of capturing a persistent premium between implied and realized volatility. Both straddles and strangles combine long positions in both call and put options with the same expiry. As such, they are by definition long volatility structures with large amounts of gamma and large amounts of negative theta. Straddles involve buying a call and put option at the same strike[4]. Straddles are most commonly traded at the at-the-money strike, where they are then delta-neutral. Strangles, in contrast, combine out-of-the-money options positions in calls and puts at different strikes. The trade is generally constructed such that the call and put have the same absolute level of delta, leading to a delta-neutral structure at trade initiation.

The following tables, and payoff diagram shows the legs of straddle and strangle trades, as well as their payoff characteristics:

Straddle		
Long/Short	**Units**	**Instrument**
Long	1	Put Option with Strike K
Long	1	Call Option with Strike K

Strangle		
Long/Short	**Units**	**Instrument**
Long	1	Put Option with Strike K - h
Long	1	Call Option with Strike K + h

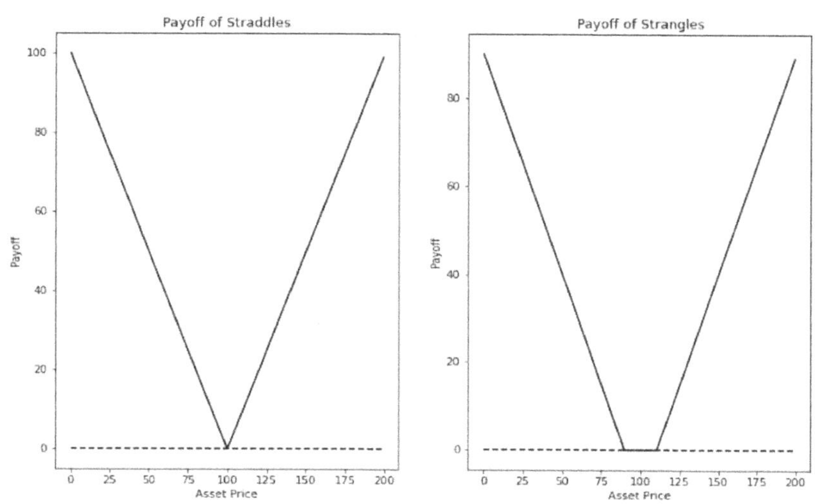

When constructed as described above, the delta-neutral nature of straddles and strangles makes them inherently volatility instruments. Long positions in straddles and strangles are long volatility, and long gamma, instruments, that make significant amounts of profit when a large move in the underlying asset occurs in either direction.

[4]And expiry.

This positive exposure to large moves is clearly a desirable feature. Investors are naturally inclined to structures they are able to hold and wait for the market to move in either direction in order to profit. The cost, or trade-off of these structures then are the significant amounts of theta decay that we must pay to obtain this convexity to large movements in either direction. In fact, this preference for convex, straddle-like asymmetric payoffs is a commonly cited justification for a volatility risk premium. That is, investors do not like taking risks that have asymmetric payoffs not in their favor, and as such require a premium for engaging in such trades.

13.9.5 Butterflies

Butterflies are another commonly traded options structure. Unlike straddles and strangles, butterflies are characterized by long and short positions in call or put options. In particular, the following table shows the three legs of a butterfly structure:

Long/Short	Units	Instrument
Long	1	Call Option with Strike K + h
Short	−2	Call Option with Strike K
Long	1	Call Option with Strike K − h

The following payoff diagram shows the payoff profile for this structure:

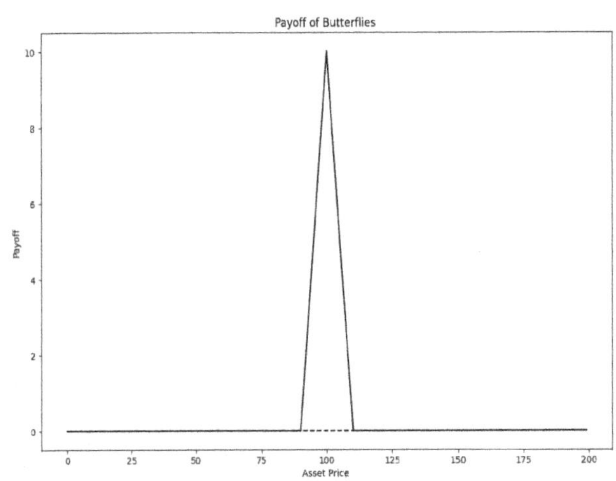

Butterflies are characterized by an initial premium outlay and a strictly positive payoff. The structure of their payoff gives them an interesting volatility and gamma profile. When the asset price and the central strike of the butterfly are equal, or close to each other, then the butterfly has a short volatility profile. That is, it is short gamma, or big moves, and collects the theta decay. For a butterfly option whose central strike, K is significantly far away from the current asset price, this profile changes. In this case the gamma of the option will be positive, as a big move will be required for the option to move toward the payoff zone, and the theta may be significantly negative.

Butterflies, like spreads, are also of direct interest because of their connection to the risk neutral distribution of the underlying asset. In particular, as we explore in chapter 17, a butterfly option with closely spaced strikes approximates the PDF of the risk neutral distribution.

13.9.6 Condors

Condor options are conceptually very similar to butterflies except that the two units of short option positions are disentangled and replaced with short positions at two distinct strikes. The following table and payoff diagram show the updated structure:

Long/Short	Units	Instrument
Long	1	Call Option with Strike K_1
Short	-1	Call Option with Strike K_2
Short	-1	Call Option with Strike K_3
Long	1	Call Option with Strike K_4

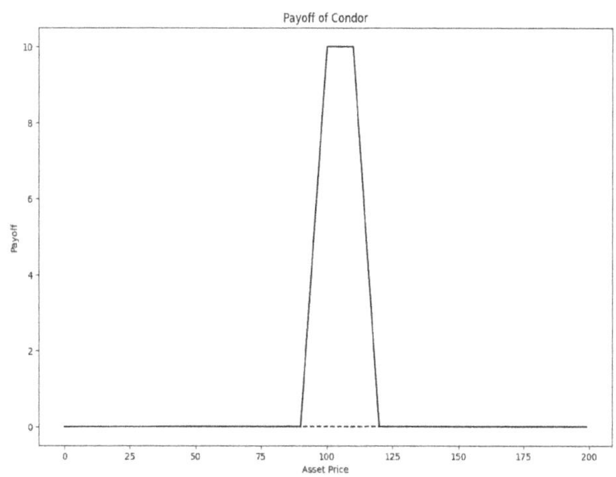

Due to their similar nature, condors have similar properties to butterflies. In particular, depending on where the strikes K_2 and K_3 are relative to the spot asset price, the structure may have positive or negative gamma (and consequently negative or positive theta).

13.9.7 Calendar Spreads

Calendar spreads are a fundamental options structure that enables traders to isolate the term structure of volatility. Unlike the previously considered structures, which focused on options trades expiring at the same time, a calendar spread isolates the term structure, or slope of the volatility surface with respect to expiry, by combining a long and short position at different expiries. The following table shows the legs of a calendar spread trade:

Long/Short	Units	Instrument
Short	-1	Call Option with Strike K_1 and Expiry τ_1
Long	1	Call Option with Strike K_2 and Expiry τ_2

where $\tau_1 < \tau_2$.

The presence of multiple expiries in a calendar spread structure make visualizing the payoffs of a calendar spread more challenging than in the previous examples. However, one can gain intuition on the payoff profile by considering the potential outcomes at the first expiry date, τ_1, as is shown below:

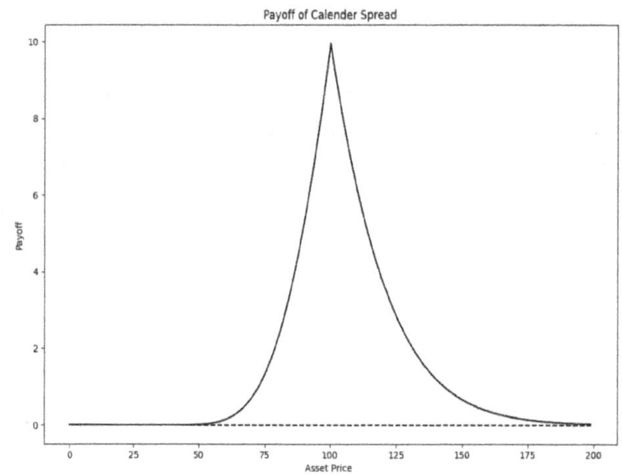

To understand this payoff profile, it helps to consider the extremes. In the absence of arbitrage, a calendar spread costs money upfront, because we are buying an option with a longer expiry. Therefore, a payoff of zero is equivalent to a loss on the trade, and this is exactly what happens at either extreme. At one extreme, our sold option expires deep in-the-money, which is bad. Our long option will be deep in-the-money as well, offsetting the loss, but will be worth little more than that. Put together, this results in a value for the calendar spread that approaches zero. At the other extreme, our sold option will expire deeply out-of-the-money, which at first glance appears to be a good thing. But, if it is deeply out-of-the-money, then our remaining long option position will likewise be deeply out-of-the-money, making it also worth close to zero despite the remaining time on the option. Conversely, if we consider the case that we are at the strike at the first expiry date, the sold option is worth zero, but our long option still has the most remaining value. As a result, this point defines the peak in the chart above.

Of course, just as with the other structures, we could just as easily reverse the legs of a calendar spread, instead buying the shorter expiry and selling the longer expiry. The desirability of this trade is closely connected to the term structure of the volatility surface. In cases where the volatility surface is inverted[5], selling the shorter

[5]With respect to expiry.

expiry is likely to seem most attractive, as we are selling the most expensive option. Conversely, a steeply upward sloping volatility surface may suggest we should sell the longer expiry, higher volatility option and buy the shorter expiry, lower volatility option. An example of a calendar spread is discussed in the context of credit markets in chapter 20.

13.9.8 Risk Reversals

Risk reversals are a commonly traded structure in the options market that help traders isolate skew. This is done by buying (or selling) an out-of-the-money call option and selling (or buying) an out-of-the-money put option. The strikes of the call and put are generally chosen such that their absolute levels of delta are equal. Risk Reversals are often then overlaid with a delta-hedging component in order to make them delta-neutral at trade initiation. This helps isolate the skew component, as, in the absence of this hedge, they can have significant amounts of delta that can dominate the option's profile. This delta-hedge may subsequently need to be re-balanced over time, as positive or negative delta accumulates. Additionally, as we are long one option and short another option, this structure is also characterized by being roughly vega-neutral. This makes the remaining exposure to the volatility differential between calls and puts, or skew.

The following table, and ensuing chart, show the legs of a risk reversal and its payoff profile:

Long/Short	Units	Instrument
Short	-1	Put Option with Strike K_1
Long	1	Call Option with Strike K_2

where $K_1 < K_2$ and both are often chosen such the magnitude of the delta of the put at K_1 and the call at K_2 are equal.

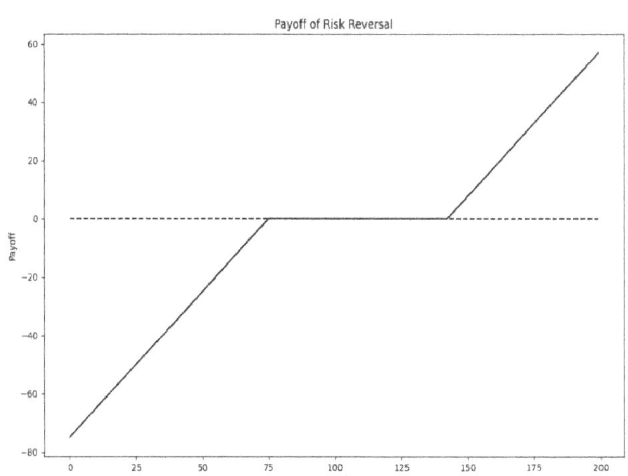

It should be noted here that in the payoff profile we neglect to include a delta-hedge, and, from the profile, can see the significant delta embedded in the trade if it is left unhedged.

Risk reversals can be structured so that we benefit from skew flattening, or steepening. Said differently, we may use this structure in equity markets where we feel put skew is too extreme. In that case, we would like to sell the overpriced put option and buy the underpriced call option. In other cases, we may want to do the reverse.

13.9.9 1x2s

Another commonly traded options structure is a so-called one by two, where we buy one unit of a put (or call) option and then sell two units at a corresponding lower (or higher) strike, with all other characteristics such as expiry remaining the same. The reader might notice that this structure is familiar and it is in fact a close variant of the Put & Call Spread, as well as Butterfly structures introduced earlier in the section. A 1x2 involving puts, for example, consists of the following legs:

Long/Short	Units	Instrument
Long	1	Put Option with Strike K
Short	-2	Put Option with Strike K - h

We can see that this structure is a put spread with an additional sold put at the lower strike, $K - h$. Relative to a butterfly, it is a butterfly without the long position in the lowest strike option. Said differently, a 1x2 can be thought of as a butterfly plus an additional short option at the lowest strike. Similarly, the legs of 1x2 involving calls are:

Therefore we can see the same relationships relative to call spreads and butterflies emerge. This leads to 1x2s having a shorter volatility profile relative to the previously discussed structures, however, this volatility profile is dynamic. As with a butterfly, if we are sufficiently far from our payoff region, then we may have a long volatility profile as we need the asset to move in order to receive a payout. If we are near this second strike, however, then we will observe a sharp short volatility profile. In the following chart, we display the payoff diagram of 1x2's constructed using both call and puts:

Long/Short	Units	Instrument
Long	1	Call Option with Strike K
Short	-2	Call Option with Strike K + h

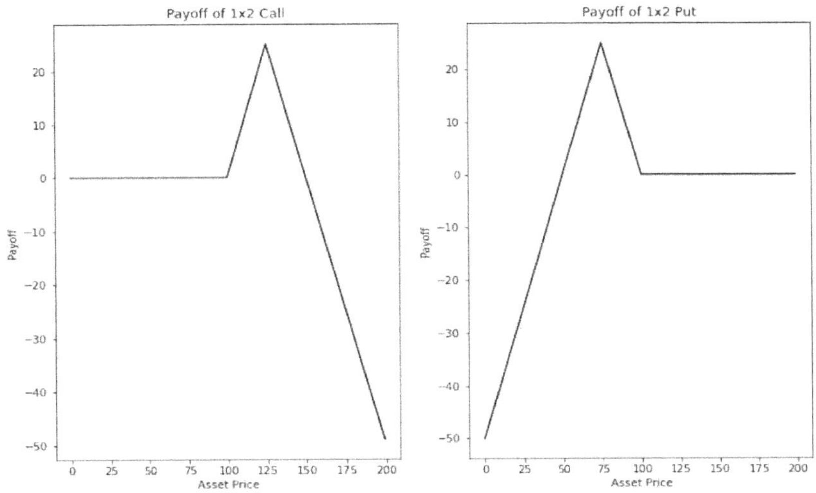

We can also see in this payoff diagram that the delta of a 1x2 will also be dynamic. In some cases, we will want the value of the underlying asset to increase. In others, we will want it to decrease, depending on where the asset is relative to our strikes.

One way to setup a one by two is to choose the most out-of-the-money strike such that the trade requires zero upfront premium. In this case, the trade becomes conditional on the asset being above (or, for a call 1x2 below) the first strike, K. We will then profit if this is the case provided the move has not been too large, in which case we will experience losses. Another way to view this structure is that we are short a tail event but are long the part of the distribution closer to the mean. Therefore, we are short kurtosis but might benefit from some local convexity. We could, of course, invert this structure, that is, sell a one-by-two, in which case we would have a structure that is long the tail, but short convexity locally.

13.10 VOLATILITY AS AN ASSET CLASS

Recently, many market participants have suggested that volatility should be treated as its own asset class with its own properties and risk/reward profile. The use of volatility instruments, such as vanilla and exotic options, provides investors with the ability to implement views on a particular part of a distribution far more precisely, as we explore in chapter 17. It also enables us to benefit from the unique, non-linear behavior of option structures. Perhaps just as importantly, it has been argued the use of options opens up additional return sources and risk premia [127]. The canonical example of this is a so-called volatility premium, where investors take advantage of a persistent premium between implied and realized volatility [143]. This is a phenomenon we explore in more detail in the next section. This however, should not be regarded as the only potential premia in the options space. Instead, as we have an entire volatility surface to mine for premia related to the term structure of volatility, or skew, for example.

In order to understand how this potential asset class works, we need to first understand how to value the underlying instruments, starting with simple vanillas and including the most common multi-asset derivatives. Further, we need to understand how these products will react to different market moves. The behavior of derivatives is non-linear with respect to its underlying parameters, and we need to understand this non-linearity well in order to model, trade or risk manage these products. The foundation for this process was detailed in chapters 14 and 15. Later in the book[6] we connect these fundamental concepts to an investment and trading perspective. At the crux of this venture is understanding the difference between the risk-neutral world, defined by market implied prices, and the physical distribution, of what is like to actually happen. Differences between these two worlds are influenced by the presence of risk premia in various states, and truly understanding this relationship can be a key component to successfully trading options.

As the importance of volatility trading and risk management grows, so too does the need for quants with a deep understanding of these markets. This makes volatility markets a fertile place for quants to explore, given the significant features in volatility skews, volatility term structures and the cheapness or richness in the level of volatility.

13.11 RISK PREMIA IN THE OPTIONS MARKET: IMPLIED VS. REALIZED VOLATILITY

13.11.1 Delta-Hedged Straddles

Consider a portfolio that is long one unit of a call and one unit of a put with the same strike. As we saw earlier in the chapter, this is known as a **straddle**. If it is struck at-the-money, then this portfolio will have no exposure to the underlying asset initially, that is it will be delta-neutral. This is because the delta of the call and put at the at-the-money strike are equal in absolute terms, and the call delta is positive while the put delta is negative. Thus, they offset when the trade is put on. It should be emphasized that this condition of delta-neutrality for a straddle is not true generally, but is instead only true for the at-the-money strike. For another strike, one option will be in-the-money whereas the other will be out-of-the-money, and their deltas will clearly diverge. Because of this feature, at-the-money straddles are of most interest to practitioners trading options, as it doesn't require an additional delta-hedge when the trade is put on.

Although this trade construction is delta-neutral, it will still have other Greeks, most notably **gamma**, **theta** and **vega**. Straddles in particular are characterized by large positive amounts of gamma and vega, at the expense of being short theta. This means that a long straddle position will gain a significant profit when the asset makes a large trending move in either direction, but will lose value as time passes if the asset doesn't move.

Additionally, although the trade is initially delta-neutral, as the asset moves, the trade will accumulate delta. To eliminate this, traders often delta-hedge the straddle to eliminate this exposure. As we saw earlier in the chapter, this need to delta-hedge

[6]For example in chapters 17 and 27.

introduces a corresponding need to construct a re-balancing strategy. When a straddle is hedged, it changes the complexion of its payoff structure, which we explore in the next section.

13.11.2 Implied vs. Realized Volatility

Let's consider a portfolio that buys an at-the-money straddle and delta-hedges it continuously until expiry. As we saw in 13.3, within the context of the Black-Scholes model, in a delta-hedged portfolio the primary remaining risks are theta, the time decay and gamma, the convexity of the option to changes in the spot price.

Another way to view this is that the cost of the portfolio is a function of the implied volatility when we put on the trade. This is equivalent to the total theta that we pay from when the trade is put on until option expiry. Therefore, in an important sense when we buy a straddle we pay the implied volatility.

The payoff of that portfolio, assuming it is delta-hedged continuously, is a function of the realized volatility that we experience during the life of the trade. This realized volatility is very closely connected to the concept of gamma. So, put together, when we buy a straddle and continuously delta-hedge, we are paying implied volatility and receiving realized volatility. Therefore, the profits of this strategy will be highly correlated with the spread between implied and realized vol. When implied volatility is higher than the corresponding realized volatility, a straddle will lose money. On the other hand, if realized volatility exceeds the initial implied volatility then the straddle will generate a profit. This relationship between realized and implied volatility is important for options traders, and tends to play a large role in how options trades are analyzed.

Importantly, the realized volatility that we are receiving is the future realized volatility over the life of the trade. This payoff is not linked to historical volatility, which is a backwards looking measure that we can observe when we enter our trade. At this point this should be a familiar concept to the reader. On the one hand we have a forward looking, market implied quantity, and on the other hand we have to forecast a realized counterpart using backward looking, historical data. We may choose our forecast to be the trailing realized volatility, but we must understand conceptually that we are treating this as a forecast. That is, when applying implied vs. realized volatility we must be aware of the assumptions. There are many additional techniques for trying to forecast realized volatility for a comparison with implied volatility, which are detailed in chapter 25.

More generally, many market participants have observed a persistent premium in implied volatility relative to realized volatility [93]. To take advantage of this, investors can engage in the strategy discussed in this section. That is, they can sell straddles and delta-hedge, collecting the gap between implied and realized. There are of course many other formulations of such a strategy, and may inherent risks in this formulation that the reader is encouraged to think through.

A natural question the reader might ask is why this premium exists, and why we should (or shouldn't) expect it to continue going forward. At heart, this is a risk premia story. Investors who are long straddles are rewarded with convex payoffs that are significantly asymmetric. Conversely, investors are weary of receiving small payoffs in some states with the potential for outsized losses in others. Perhaps as importantly, they are rewarded with large payoffs in times of market stress, or crisis periods. The combination of the asymmetric payoffs of straddles, as well as their potential for outsized gains during market turbulence, have been postulated as reasons for an embedded risk premia in implied volatility relative to realized volatility. Said differently, the owner of a straddle can just sit and wait for the market to move, without a care for the direction it moves. If they didn't have to pay a premium to do so, this is a trade that any investor would rush to do. Similarly, the seller of a straddle has a proverbial penny in front of a steamroller proposition, and will likely require a premium in order to be enticed into the trade.

13.11.3 Implied Volatility Premium in Practice: S&P 500

In this section we highlight the premium between implied and realized volatility in practice by looking at the difference between VIX[7], a measure of one-month implied volatility for the S&P 500, against realized volatility of the S&P 500:

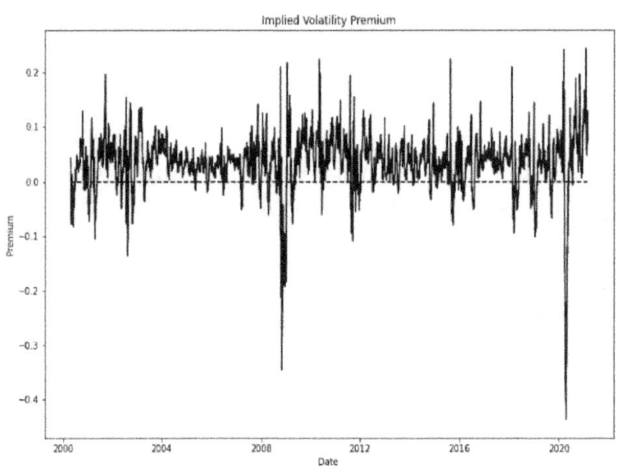

In this chart we can clearly see that the implied volatility premium tends to be positive. In fact, on average we observe a premium of approximately 3.5%. This indicates a potential profit for investors willing to sell straddles or volatility in some other form. However, the large negative spikes in the chart provide somewhat of a cautionary counterpoint as they correspond to periods with substantive losses for those pursuing these types of strategies. In practice, this is indeed what we observe in these types of strategies. We tend to make small profits most of the time, but suffer outsized losses periodically that can wipe out our capital if we are not careful.

[7]VIX is used a proxy for implied volatility in the S&P for simplicity, as it does not change the overarching conclusions.

13.12 CASE STUDY: GAMESTOP REDDIT MANIA

One of the more interesting developments that took place during the aftermath of the Covid crisis and the ensuing so-called work-from-home rotation trade was the proliferation of retail day traders who used the change in work environment to monitor markets closely and trade frequently. The most emblematic example of this is what we saw with Gamestop in early 2021. Leading up to the Covid crisis, Gamestop's fundamentals became increasingly challenged as more and more video game rentals and purchases began to take place online and the need for brick and mortar stores diminished. These challenged fundamentals negatively impacted the stock price in the preceding years, as we can see in the left side of the following chart.

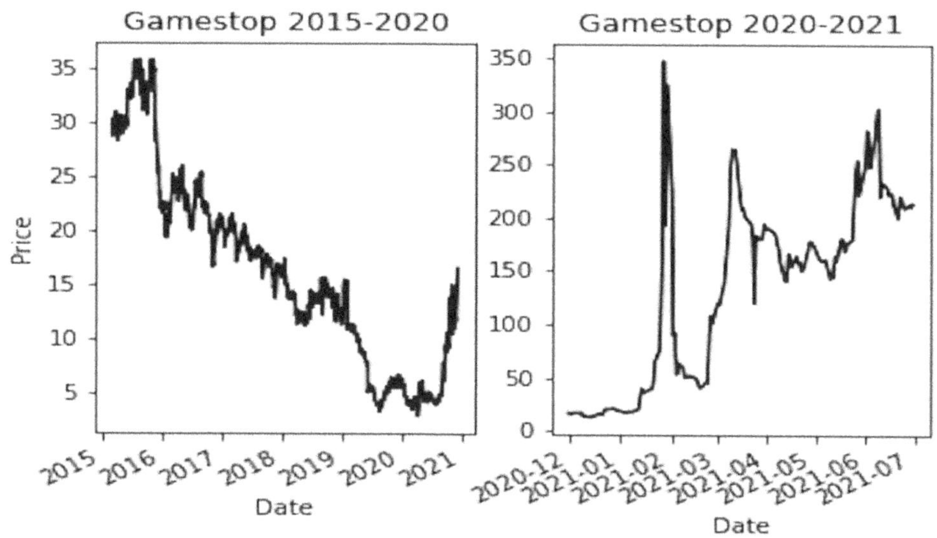

Additionally, these challenged fundamentals made Gamestop a popular short-selling target for hedge funds and other professional investors. This led to a massive amount of short interest in Gamestop. In early 2021, however, as we can see on the right side of the chart below, the stock price exploded as retail investors took the other side of these hedge fund shorts and created a massive short squeeze. This set of retail investors represented a new brand of investors that were working from home and day-trading. Further, these retail investors were acting in unison through social media platforms such as reddit, where they encouraged others to buy shares or call options and coordinated their trades. In doing this, they were able to put enough pressure on the short positions to trigger a short squeeze, where losses on short positions led to margin calls that required positions to be terminated abruptly, which in turn lifted the stock further leading to more painful margin calls for short sellers.

A second factor that contributed to the dramatic rise of Gamestop over the period of interest was the fact that the band of retail investors relied on deeply out-of-the-money call options to gain access to significant leverage on their trades. Recall that earlier in the chapter we saw that gamma per unit of price of the option is highest for deeply out-of-the-money options. This means that, these out-of-the-money options being purchased by retail investors had significant gamma, leading to large profits as the stock began to rise. Conversely, on the other side of these option trades were often dealers, who instead of holding their short option position would delta-hedge them to avoid the risk associated with this type of price increase. These delta-hedges consist of long positions in the underlying stock, because a short call option has a negative delta profile. This means that, as the options that the dealer is short become increasingly closer to at or in-the-money, their delta increases and they must enter the market and purchase shares, further propping up the price. This impact was exacerbated by the fact that the options initially required very small positions in the stock, because they were deeply out-of-the-money, but over time these hedging obligations became substantial, and created a feedback loop.

CHAPTER 14

Extraction of Risk Neutral Densities

14.1 MOTIVATION

In earlier chapters, we discussed several methods for solving options pricing problems, including those based on Quadrature and Fourier Transform techniques. When applying these techniques, we traditionally start by specifying a stochastic differential equation that governs the underlying asset. In this chapter, we will explore an alternative approach where we look to find the market implied density directly without first presupposing an SDE.

To see this, recall that the standard risk neutral pricing formula for a European call option can be written as:

$$c_0 = e^{-\int_0^T r_u\,du} \int_{-\infty}^{+\infty} (S_T - K)_+ \phi(S_T)\, dS_T \tag{14.1}$$

where $\phi(S_T)$ is the density of the underlying asset, S at expiry. If this density, $\phi(x)$ is known, then we can evaluate this integral against the density to obtain an option price. At the crux of solving this options pricing problem, however, is where do we obtain the density $\phi(S_T)$? In traditional methods, this density is defined by our choice of a stochastic process and its parameters. In particular, the choice of an SDE and its parameters will define the risk neutral distribution in all states.[1]

Inherently, the success of this approach is determined by how accurately the SDE and the parameters we choose represent the dynamics of the underlying market. In practice, however, we are not given the proper SDE or the proper set of parameters to use. Instead, we are given a set of option prices for different strikes and expiries and need to find the optimal model. The process of choosing the optimal SDE for a given options pricing problem is often subjective, and requires some knowledge of a particular market's volatility surface behavior.

[1]It should also be noted that in some cases this distribution will be analytically tractable, such as in the case of a Black-Scholes or Bachelier model, whereas in other cases the distribution may not be known and we may need to resort to other methods to proceed.

DOI: 10.1201/9781003530626-17

Additionally, SDEs have at most only a handful of parameters for us to fit to the entire volatility surface. As a result, we generally have far fewer model parameters than market prices. This creates an over-determined problem in which we are likely not able to fit the entire volatility surface exactly.

In this chapter we explore an alternative methodology that attempts to overcome some of these challenges associated with the traditional approach of specifying an SDE and optimizing its parameters. In particular, we consider methods that instead look to solve for the risk neutral density directly using a set of market traded European options.

To see how these methods will work, consider the discretized version of the pricing formula for a European call:

$$\hat{c} \approx e^{-\int_0^T r_u\,du} \sum_1^N (S_{T_i} - K)_+ \phi(S_{T_i})\,\Delta S_T$$

Previously, our approach to solving this problem has involved making assumptions about $\phi(S_{T_i})$ and then applying computational techniques to obtain a solution. Alternatively, in this chapter we explore techniques that begin by parameterizing the problem by the underlying probabilities of the terminal asset price, S_{T_i}, and then find the probabilities that best fit the quoted market prices directly. We explore two approaches for this, Breeden-Litzenberger and Weighted Monte Carlo.

14.2 BREDEN-LITZENBERGER

14.2.1 Derivation

It turns out, that for European call and put options, the risk neutral density can be accessed in a completely non-parametric manner that relies solely on the structure of the payoff function. This was first discovered by Breeden and Litzenberger [48], who found that the underlying risk neutral density can be extracted by differentiating the payoff function for European options.

To see this, let's again start with the payoff formula for a European call:

$$c = e^{-\int_0^T r_u\,du} \int_K^{+\infty} (S_T - K)\phi(S_T)\,dS_T \tag{14.2}$$

If we differentiate the payoff function with respect to strike we obtain:

$$\frac{\partial c}{\partial K} = e^{-\int_0^T r_u\,du} \int_K^{+\infty} -\phi(S_T)\,dS_T = e^{-\int_0^T r_u\,du}(\Phi(K) - 1) \tag{14.3}$$

Encouragingly, we can see that the first derivative with respect to strike is already a function of the risk neutral distribution, in particular the CDF. It should be noted that in order to take this derivative we need to apply the first fundamental theorem of calculus to account for the strike K appearing in the lower limit of the integral.

Differentiating the call payoff formula again with respect to strike we are left with:

$$\frac{\partial^2 c}{\partial K^2} = e^{-\int_0^T r_u\,du}\phi(K) \tag{14.4}$$

Here we can see that $\frac{\partial^2 c}{\partial K^2}$ is directly connected to the risk neutral distribution evaluated at K. In fact, it is just this density discounted at the risk free rate.

Importantly, this means that if we can approximate the second derivative using market traded instruments, then we can use that approximation to extract the risk neutral distribution. Further, if we assume that unique option prices are available for all strikes, then we can in fact use finite differences to estimate this derivative, $\frac{\partial^2 c}{\partial K^2}$. In particular, we saw in chapter 15 that the finite difference approximation for a second derivative can be written as:

$$\frac{\partial^2 f}{\partial x^2} \approx \frac{f(x-h) - 2f(x) + f(x+h)}{h^2} \tag{14.5}$$

Applying this to $\frac{\partial^2 c}{\partial K^2}$ and solving for the risk neutral density, $\phi(S_T)$, we get:

$$\frac{\partial^2 c}{\partial K^2} \approx \frac{c(K-h) - 2c(K) + c(K+h)}{h^2} \tag{14.6}$$

$$\phi(K) \approx e^{\int_0^T r_u\,du} \frac{c(K-h) - 2c(K) + c(K+h)}{h^2} \tag{14.7}$$

Each of the three terms in the numerator is simply a European call option with a different strike. Therefore, we have been able to express the risk neutral distribution as a function of observable, market traded, call prices.[2] As a result, the Breeden-Litzenberger technique provides an elegant, non-parametric approach for extracting the underlying density, assuming that unique call prices are indeed available for all strikes.

14.2.2 Breeden-Litzenberger in the Presence of Imprecise Data

As highlighted, a critical assumption in the Breeden Litzenberger methodology is the presence of exact option prices available for all strikes. Of course, in reality, prices are only available for a finite set of strikes. Additionally, even for the strikes where pricing is available, we observe a range (via the bid-offer spread) rather than a single price. This creates challenges implementing the Breeden-Litzenberger method in practice.

One natural way to overcome the lack of data for all strikes would be to employ an interpolation methodology between publicly traded strikes. This approach requires careful consideration, however, as we must to ensure that the interpolation method that we choose does not introduce arbitrage into our model. For example, applying linear interpolation to options prices would clearly be an inappropriate assumption, as we know from theory that options prices have non-linear behavior. Applying linear or quadratic interpolation to a set of market implied volatilities, however, is likely to be more reasonable and less likely to lead to arbitrage. Further complicating this task is that, even in the absence of arbitrage, interpolation may introduce unrealistic assumptions about the market implied density that are magnified when taking the second derivative.

---

[2]It should be noted that, while this derivation focused on European call options, a similar derivation exists for European put options, as you will see in the end of chapter exercises.

Another approach to potentially obviate this problem would be to first calibrate a stochastic process, based on some predefined SDE, and then extract the risk neutral density after the calibration process. This would help overcome the implementation challenges related to limited data, however, it would require us to lose some of the non-parametric nature of our solution and extracted risk neutral density.

14.2.3 Strengths and Weaknesses

Overall, the Breeden-Litzenberger methodology has intuitive appeal due to its non-parametric nature and its ability to overcome many of the challenges with the traditional approach of choosing and applying a specific SDE. In particular, it helps us transform the problem from an over-determined one, where we only have a few parameters to fit a large number of option prices and will not be able to do so exactly, to a problem where we are able to, under certain assumptions, find a single, unique solution.

Despite this intuitive appeal, the Breeden-Litzenberger technique is not without challenges or limitations. In particular, while the method is elegant theoretically, the data that we observe in the market is often insufficient to use the model in practice without imposing additional structure. This additional structure can be imposed either by specifying an SDE, or by the use of interpolation methods, as described above.

Additionally, this methodology only helps us to uncover a single risk neutral density for a corresponding expiry. It does not, however, help us connect the risk neutral densities at different expires, as a stochastic process would. This limits the applications of the technique, as in many cases, such as pricing path dependent exotic options, we will need a single consistent distribution across all expiries to obtain a price in line with the volatility surface.

14.2.4 Applying Breden-Litzenberger in Practice

Applying the corresponding piece of Python code in the coding repository to a set of S&P options with one-year expiry, we arrive at the following risk neutral density:

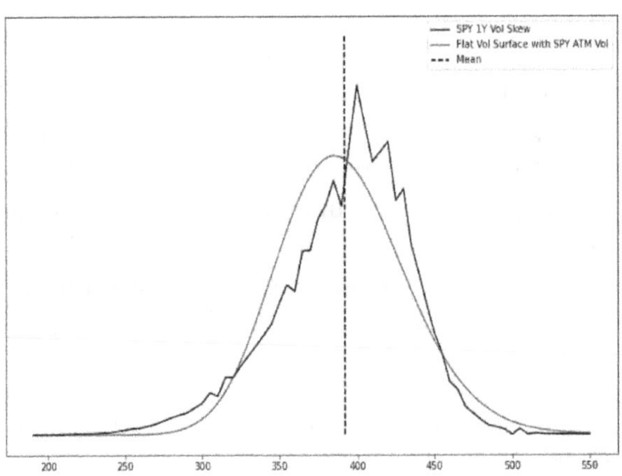

This example underscores the challenges of using this model in practice, even in one of the most liquid options markets. Generally speaking, the data that we observe in the market does not have the level of precision required for such a technique, unless a significant amount of work is done to the data before the technique is applied.

In particular, we can see that there are several jagged points in the extracted density. Given bid-offer spreads, it is far more likely that the jumps are caused by imprecision in the data rather than a real artifact of the density. As we will see later in the chapter, the use of Weighted Monte Carlo is on way to overcome these obstacles and provide a happy medium between parametric and non-parametric approaches.

Having said that, this method can still help us to gain insight into the market implied density. In this example, we compared the extracted risk neutral density to the Log-Normal distribution implied by the Black-Scholes model.[3] Comparison of these two distributions is of direct practical interest as it enables us to see the impact of the volatility skew on the underlying density. In this case, we can see that a fatter left tail, relative to a log-normal distribution, seems to be a key feature of the market implied distribution. Later in the chapter we examine the impact of shifting the volatility skew on the underlying density.

14.3 CONNECTION BETWEEN RISK NEUTRAL DISTRIBUTIONS AND MARKET INSTRUMENTS

Regardless of the challenges of implementing the Breeden-Litzenberger technique, the method exposes a key relationship between the market implied distribution and certain market traded structures. Importantly, application of this technique enabled us to express the PDF in terms of purely European call and put options. Similarly, we can express the CDF of the risk neutral distribution in terms of the same commonly traded instruments. This is a critical relationship that gives quants and other market participants a theoretically precise way to invest directly in parts of a distribution that they believe are mispriced. In this section we explore this link between the risk neutral distribution and market instruments.

14.3.1 Butterflies

Recall from (14.6) that the PDF of the risk neutral density can be found by differentiating the call price formula with respect to strike twice. In particular, we found that the PDF is a function of call options with three different strikes. Focusing on the numerator in this equation, we have the following underlying option positions that allow us to estimate the value of the PDF:

- A long position of one unit of a call option with strike $K - h$
- A short position of two units of a call option with strike K
- A long position of one unit of a call option with strike $K + h$

[3]Note that this can be done by calling the function in the coding repository with a constant volatility.

Where we are assuming that all characteristics aside from strike are the same for all three positions. This portfolio is a commonly traded options strategy that is generally referred to as a butterfly, and we can now see that this strategy approximates the underlying market implied density. Fundamentally, this implies that butterfly options play a critical role in that they enable an investor to bet on an underlying risk neutral PDF.

The following chart shows the payoff function for a butterfly option:

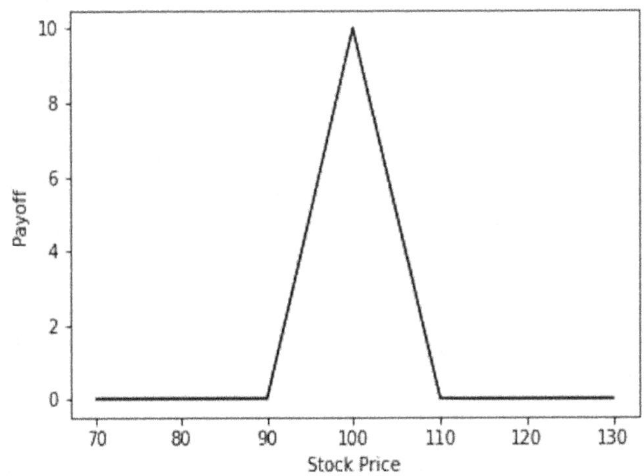

We can see that the payoff for this structure is always non-negative. This means that, in the absence of arbitrage, the price of this structure should be greater than or equal to zero. If this is not the case, and the traded price of the butterfly is negative, that would mean that we could receive payment upfront in order to gain exposure to a structure with only non-negative payoffs. Through another lens, because of the relationship between a PDF and a butterfly option, we can now see that a negative butterfly price implies a negative probability in our risk neutral density, which implies arbitrage.

14.3.2 Digital Options

It can also be seen that a natural relationship exists between the CDF of the risk neutral density and call spreads or digital options.

To see this relationship, recall that a European digital call option is defined as an option that pays \$1 if the asset is above K at expiry. The risk neutral pricing formula for a European Digital Call can then be written as:

$$d = e^{-\int_0^T r_u\, du} \int_K^{+\infty} \phi(S_T)\, dS_T \quad (14.8)$$

$$= e^{-\int_0^T r_u\, du}(1 - \Phi(K)) \quad (14.9)$$

From this we can see that the price of digital options is directly related to the CDF of the market implied distribution.[4] This means that when investors trade digital options, they are implicitly expressing a view on the CDF of the risk neutral distribution. Breeden-Litzenberger exposes a similar relationship between a call spread with a narrow gap in strikes and the CDF, as you will find in the end of chapter exercises.

14.4 OPTIMIZATION FRAMEWORK FOR NON-PARAMETRIC DENSITY EXTRACTION

A second and more general approach to extracting a risk neutral density directly from a set of observed option prices would be to formulate an optimization problem that calibrates the probabilities directly to a set of market data. In particular, to calibrate to a set of European call option prices, we would look for a set of probabilities, p_i that minimizes the distance to market data, for example using the following equations:

$$\vec{p}_{\min} = \min_{\vec{p}} \left\{ \sum_j \left(\hat{c}(\tau_j, K_j) - c_{\tau_j, K_j} \right)^2 \right\} \tag{14.10}$$

where c_j is the market price of the jth instrument and $\hat{c}$ satisfies:

$$\hat{c}(\tau_j, K_j) \approx e^{-\int_0^T r_u \, du} \sum_1^N (S_{\tau_i} - K)_+ p_i(S_{\tau_i})$$

Additionally, in order to ensure we have a valid density we would also need to include the following constraints:

$$p_i > 0 \qquad \forall p_i \tag{14.11}$$

$$\sum_i p_i = 1 \tag{14.12}$$

These constraints ensure positive probabilities in each state and also that the sum of the probabilities equals one. Putting (14.10), (14.11) and (14.12) together we have the necessary components of an optimization problem that is based on the underlying terminal state probabilities rather than a set of SDE parameters. It should be noted that this approach, like the first approach we discussed earlier in the chapter, helps us find the density from the current state until option expiry. It does not naturally define conditional probabilities from one intermediate state to another (as an SDE would).

Much like the Breeden-Litzenberger technique, this approach has the advantage of not requiring any assumptions about the underlying stochastic process in our model. In fact, generally, there will be far more p_i's than option prices, that is, the problem is under-determined. This is in contrast with traditional methods based on a few parameters that are naturally are over-determined. This means that, in the absence of arbitrage in the data, this technique should enable us to match options prices exactly.

---

[4] It can easily be shown that this relationship holds for both European Call and Put digitals.

This is a significant feature compared to traditional SDE based approaches that will be unable to exactly recover market prices for more than a handful of strikes. Further, there will actually potentially be multiple solutions, or multiple densities that fit a set of options prices. As a result, we can use additional criteria, such as smoothness, in order to choose between the solutions that match market prices.

This under-determined nature of the problem is important and enables us to use the extra dimensionality to provide structure to the problem and avoid some of the issues with implementing Breeden-Litzenberger in practice. It also enables us to formulate our optimization differently. Instead of treating the options prices in the objective function, we can treat them as constraints. These may be equality constraints, or we may choose to formulate them as inequality constraints to take into account each option's bid-offer spread.

The natural question then, is what would we choose to be the objective function if we were to pursue this approach. In this approach, with equality options pricing constraints, we have:

$$\vec{p}_{\min} = \min_{\vec{p}} \left\{ \sum_i g(p_i(S_{\tau_i})) \right\} \tag{14.13}$$

with the following constraints:

$$p_i > 0 \qquad \forall p_i \tag{14.14}$$

$$\sum_i p_i = 1 \tag{14.15}$$

$$\hat{c}(\tau_j, K_j) = c_{\tau_j, K_j} \qquad \forall j \tag{14.16}$$

Of course, the remaining question would be which function $g(p_i(S_{\tau_i}))$ should we use as our objective function, given that the constraints ensure that we match market prices and have a valid probability density function. In particular, $g(p_i(S_{\tau_i}))$ should help us choose among the multiple potential solutions to our under-determined problem. The choice of which of the potential solutions is best is somewhat subjective, as they all perfectly fit the supplied market data. It may be defined as the smoothest density, in which case we would incorporate a term that penalizes curvature, or could be defined as a term that measures distance to a prior, economically appealing distribution. In practice, whichever approach we use, the $g(p_i)$ term imposes some preferred structure to the density, although we allow that structure to be overridden to match market prices.

While the main purpose of the function $g(p_i)$ is to provide structure and help us choose between many possible solutions to our problem, it also makes sense to choose a function that has appealing characteristics for optimization problems. Most fundamentally, it would be desirable to choose a convex function for $g(p_i)$, as this would ensure our optimization problem has a unique minimum.[5]

[5] Also notice, that when we do this, the pricing constraints are linear with respect to the probabilities. This is an another important property and makes the optimization problem more tractable.

14.5 WEIGTHED MONTE CARLO

One such approach, that leverages these optimization methods, is Weighted Monte Carlo, a technique developed by Avellenada and others [172]. This approach directly uncovers a set of probabilities of terminal values of the asset price by solving an optimization problem. The optimization procedure generally relies on maximum entropy as the objective function, $g(p_i)$. We can also, however, define this objective function as relative entropy, which enables us to introduce a prior distribution and measure the relative entropy to that prior. That is, we can use this approach to aim for a density that is smooth but also to have minimal distance from some prior or economically motivated distribution.

Lastly, and perhaps most importantly, this technique can be utilized to weight entire simulation paths instead of weighting terminal probabilities. This particular realization is a significant innovation in that it enables a robust method for incorporating additional structure, such as correlation information, into the problem. Additionally, when using this approach we have access to the entire simulated paths which means that we are able to use the technique to connect the densities at different points, allowing us to price path dependent, and not just European option payoffs. In subsequent sections, we explore each of these approaches for constructing and solving a Weighted Monte Carlo based problem.

14.5.1 Optimization Directly on Terminal Probabilities

The first approach to implementing Weighted Monte Carlo is to leverage the optimization framework that we discussed above, with option prices as constraints, and to use the entropy function as the objective function, $g(p_i)$.

To see this, let's begin by reviewing the properties of the entropy function:

$$H(\vec{p}) = -\sum p_i \log p_i \tag{14.17}$$

where we are looking to maximize the entropy function, $H(\vec{p})$.

This function has some attractive properties, in particular it is unbiased and unprejudiced in the absence of data. That is, it only makes the assumptions that are present in the data without imposing any additional structure. In particular, we can see that as probabilities get too high or too low, the function will place a higher penalty on them, through large value of $\log p_i$ or p_i, respectively. Therefore, in the absence of any constraints, entropy will place equal weight in all of the probabilities and result in a uniform distribution.

The following set of equations show how such a maximum entropy optimization would be formulated:

$$\vec{p}_{\max} = \max_{\vec{p}} \left\{ -\sum p_i \log p_i \right\} \tag{14.18}$$

$$p_i > 0 \qquad \forall p_i \tag{14.19}$$

$$\sum_i p_i = 1 \tag{14.20}$$

$$\hat{c}(\tau_j, K_j) = c_{\tau_j, K_j} \qquad \forall j \tag{14.21}$$

An alternate and valid formulation would be to minimize the sum of squared distance from uniformity as the objective function, instead of maximizing entropy.

In the coding repository, we show an example in Python that leverages this optimization formulation. Importantly, the probabilities, p_i are the parameters in our optimization and correspond to the probability of a certain terminal value of the asset price at expiry, $S_{T,i}$. To solve this optimization, we must specify a grid of potential asset price values at expiry, $S_{T,i}$. This is typically done via a uniform grid.

We can see that in the optimization we are looking to maximize entropy subject to the constraints that the probabilities are non-negative, as is implied by lack of arbitrage, and sum to one. Further, we require all option prices, in this case assumed to be call option prices, to be matched exactly. Alternatively, we could formulate these option pricing constraints as inequalities that reflect the bid-offer spread in the market.

There are a few noteworthy features of this approach that make it particularly appealing. First, it is able to overcome many of the data related issues that we encountered with Breeden-Litzenberger by providing a more unbiased view on the data and enabling us to robustly incorporate bid-offer spreads on different options. Secondly, as with the previous approach, it makes minimal assumptions on the underlying distribution. Finally, the optimization itself contains many properties that make the problem more tractable and easier to solve. In particular, the objective function that we choose, the entropy function, is a convex function, which, as we saw in an earlier chapter on optimization techniques, ensures that the optimization has a unique, global minimum. Further, the constraints that we specified in the problem are all linear with the respect to the probabilities, including the option pricing constraints, also making the optimization more tractable and well behaved.

While this is a useful technique with many appealing properties, when implementing it in practice one can find that it gives too much freedom to the optimizer and doesn't give the problem enough structure. In that case, incorporation of a prior distribution that has economic justification can be used to help the optimization make the proper additional assumptions.

14.5.2 Inclusion of a Prior Distribution

Inclusion of a prior distribution in a Weighted Monte Carlo framework can be incorporated by replacing the entropy function above with a relative entropy function.

The relative entropy function that we will minimize is defined as:

$$H(\vec{p}) = \sum p_i \log \frac{p_i}{q_i} \tag{14.22}$$

As we can see, this function will ensure that the probabilities are as close as possible to the probabilities in the prior distribution, q_i.

Using relative entropy, our problem formulation becomes:

$$\vec{p}_{\min} = \min_{\vec{p}} \left\{ \sum p_i \log \frac{p_i}{q_i} \right\} \tag{14.23}$$

$$p_i > 0 \quad \forall p_i \tag{14.24}$$

$$\sum_i p_i = 1 \tag{14.25}$$

$$\hat{c}(\tau_j, K_j) = c_{\tau_j, K_j} \quad \forall j \tag{14.26}$$

This type of approach clearly has intuitive appeal, as it is able to benefit from the same strengths of the previous approach but is likely to make more reasonable assumptions in the presence of sparse data. A Gaussian prior is probably the most commonly used in practice, in which case we would ensure that all option prices are matched, while also making the risk neutral density as close to normal as possible.

14.5.3 Weighting Simulated Paths Instead of Probabilities

Another way to incorporate a prior distribution into a Weighted Monte Carlo framework is via the use of scenarios, or simulated paths as the input nodes for which we optimize the probabilities. Instead of choosing a set of probability nodes so that they are uniformly spaced over a grid, we can define the nodes as Monte Carlo paths from a particular SDE, or as bootstrapped paths of historical data. The optimization then will vary the probabilities of these scenarios or simulated paths until it is able to match the available market data.

This weighting of the Monte Carlo simulation paths is intuitive and quite a powerful insight. It is important in that it helps connect the distributions at various options expiries, a challenge that we had in previous implementations of density extraction techniques. It also enables us to incorporate complex stochastic processes as a prior, and find solutions that have minimal distance to these models but still match the traded options prices.

14.5.4 Strengths and Weaknesses

The non-parametric and unbiased nature of Weighted Monte Carlo makes it quite appealing in practical applications. Additionally, Weighted Monte Carlo enables us to directly overcome many of the issues associated with the Breeden-Litzenberger technique, without compromising on the underlying flexibility. For example, Weighted Monte Carlo provides a more robust approach for handling uncertain and missing data. It also provides a way for us to connect the densities at different expiries, through the use of simulated paths rather than terminal probabilities in the optimization.

These strengths make Weighted Monte Carlo a compelling alternative to the traditional method of specifying an SDE in practical applications. Having said that, its implementation does come with challenges. In particular, while the non-parametric nature of the technique is undoubtedly appealing, it also leads to parameters that have less intuition. This makes Greeks, or sensitivity calculations more challenging.

As an example, in the Black Scholes model that we introduced earlier, we were able to define vega as a sensitivity to a model parameter. Clearly, the Weighted Monte Carlo approach will not have this parameter, and calculation of a similar sensitivity will require more thoughtful research.

14.5.5 Implementation of Weighted Monte Carlo in Practice: S&P Options

Applying the relevant piece of Python code in the coding repository to a set of S&P options with three-month expiry, we arrive at the following risk neutral density:

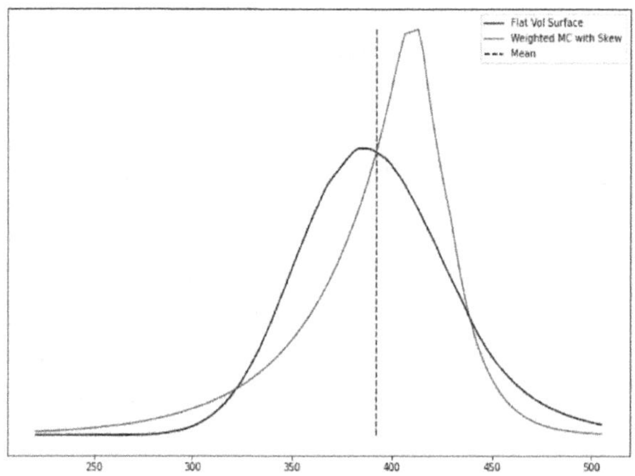

We can see that the density is much smoother than the density produced by the Breeden-Litzenberger technique in practice. We can also see similar properties in the density, in particular, apparent negative skewness and a fat left tail. This is a common structure in equity options, where we tend to see significantly higher implied volatilities on lower strike options. This feature of the distribution is often associated with an underlying risk premium. That is, investors are willing to pay a premium to receive payoffs in states of the world characterized by equity drawdowns.

14.6 RELATIONSHIP BETWEEN VOLATILITY SKEW AND RISK NEUTRAL DENSITIES

As we indicated in the previous example, in addition to the implications of these methods for providing a non-parametric pricing methodology, these extraction techniques can help us learn about the underlying properties of asset distributions. They can also help us learn about the relationship between volatility surfaces and risk neutral distributions. In particular, it gives us a tractable way to see the impact of different shapes within a volatility surface, such as smirks, smiles and skews sloping in either direction. In this section we examine the connection between volatility surface shapes and the underlying market implied distribution.

To provide a baseline, we start with a flat volatility surface. In this case, we know from theory, and earlier chapters, that the risk-neutral distribution will be log-normal in a world with a flat volatility surface. The density extraction techniques above help us to confirm that, as the following chart shows:

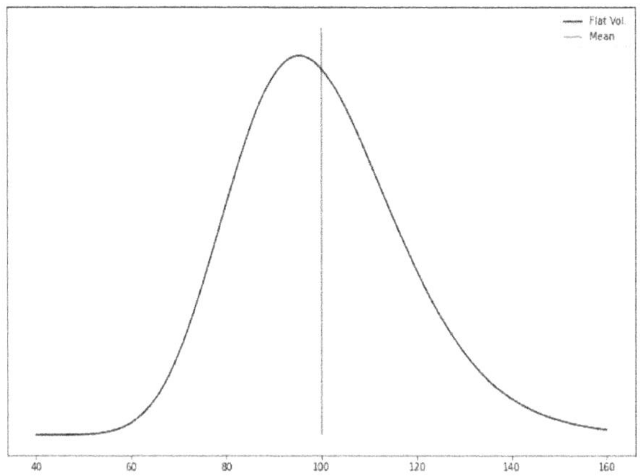

Next we consider a volatility skew with higher implied volatilities on lower strike options (generally referred to as put skew in industry). The following chart shows a comparison of this risk neutral density against a log-normal distribution or flat volatility surface:

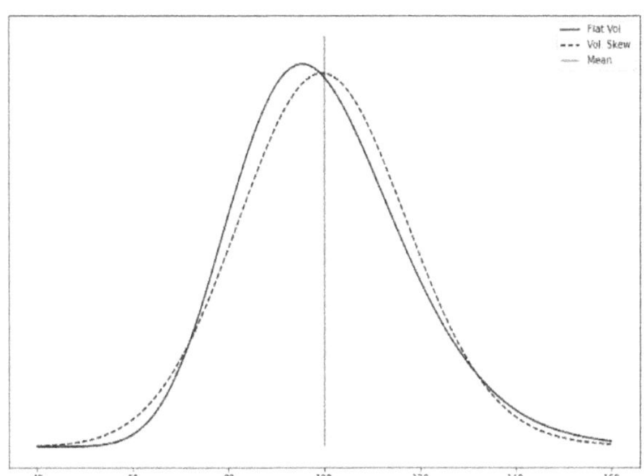

Here we can directly see the impact on the distribution of the slope of the volatility skew. Perhaps most noticeable is the fatter left tail corresponding to low strike puts with higher implied volatilities. This is one defining characteristic of this type of skew. Also notice that the mean of the distribution is unchanged as this is set by an arbitrage argument in the forward and is thus independent on the volatility surface. Because of this constraint on the mean, the additional distributional mass on the left tail must be offset elsewhere in the distribution, and in fact we do see that phenomenon. In particular, we see less probability mass below, but closer to the mean.

Finally, we look at a volatility smile, which is defined by higher implied volatilities on both high and low strike options, and lower implied volatilities closer to at-the-money. The following chart compares the density of a flat volatility surface to that of a volatility surface defined by a smile:

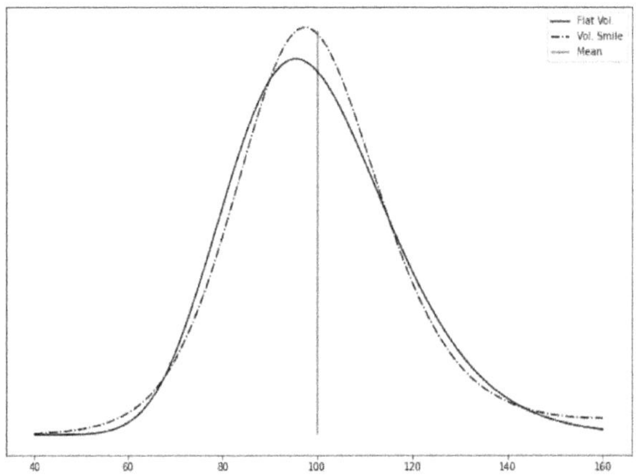

Here we can see that the smile shape of the implied volatilities lead to fatter tails on both sides of the distribution. As we saw last time, this extra mass in the tails is somewhat offset by lower mass closer to the mean of the distribution.

14.7 RISK PREMIA IN THE OPTIONS MARKET: COMPARISON OF RISK NEUTRAL VS. PHYSICAL MEASURES

Up until this point, we have discussed many applications for density extraction techniques from pricing options with arbitrary payoff structures to providing a deeper understanding of the underlying asset's distribution. However, portfolio managers and buy-side institutions will be looking for more. In particular, they will want tools that will help them perform relative value analysis of different parts of the distribution.

As an example, earlier we saw that for equity indices risk neutral distributions tend to have fat left tails. This observation alone, albeit informative, does not lend itself naturally to an investment strategy. To create an investment strategy based on our observation we would need to know if that fat tail in the distribution is warranted or not. Conversely, if we were to say that risk neutral distributions had fatter left tails then we tend to observe in actual equity indices price movements, then we would potentially have the basis for an investment strategy.

This is at its heart a juxtaposition of the risk neutral measure and a forecast of the physical measure. This type of comparison is inherently quite complex, and much of its subtlety is beyond the scope of this book, however is of direct relevance to hedge funds and other buy-side institutions. In the following sections we walk through a simplified example of this type of analysis and explain its connection to market implied risk premia.

14.7.1 Comparison of Risk Neutral vs. Physical Measure: Example

This chapter has provided the tools for estimating the risk neutral measure in a relatively robust manner. In order to complete the juxtaposition, we also need to create a model for estimating the distribution in the physical measure. This itself is a challenging task for which we generally rely on historical data and an econometric model. In the following example, we use Weighted Monte Carlo to estimate the risk neutral density for three-month options and use a simple bootstrapping technique of S&P returns to generate the three month forward distribution of prices in the physical measure:

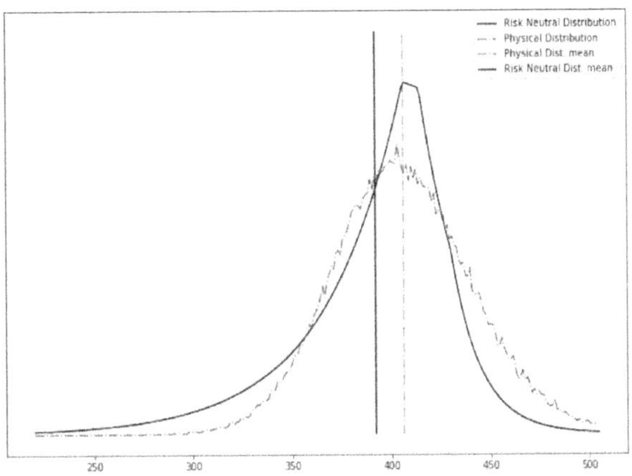

It should be noted that these assumptions in construction of the physical measure distribution are unrealistic and more serious time-series and econometric techniques would need to be applied in practice. Interested readers should think through what the assumptions and ensuing implications of a bootstrapping approach are, and how more realistic features could be introduced.

With that said, the comparison of densities does show some interesting differences in features. In particular, the fatter tail in the risk neutral distribution does not seem to be as prevalent in the physical measure density. This implies that an investor should be able to profit from a strategy where that part of the distribution is sold consistently. In fact, this type of strategy is closely related to selling put options, and is a common options trading strategy implemented by practitioners.[6] Additionally,

[6]Although not without serious flaws if not implemented judiciously.

we can see that the mean of the physical measure density is significantly higher than that of the risk neutral measure.

14.7.2 Connection to Market Implied Risk Premia

Setting aside the many potential issues and challenges with forecasting a physical measure density, deviations between the two distributions still do not in isolation indicate a mispricing. Instead, they may merely be a manifestation of a risk premium, or a preference of investors for returns in certain states.[7]

As we discussed earlier in the book, the physical and risk neutral measures inherently come from different worlds with different risk preferences. The risk neutral measure tell us the probabilities that the market is implying under the assumption investors are risk neutral and is based on hedging and replication arguments. The physical measure, instead tells us the distribution of actual prices in future states where investors are likely to be risk averse and each may have different preferences. In this measure, returns may be discounted at lower or higher rates based on when these future states tend to occur and when investors prefer higher returns. Specifically, in our S&P example we saw a higher distributional mass in the left tail of the risk neutral measure than the physical measure. A natural and intuitive explanation for this is that investors value payoffs in these states more, and may even place a premium on them, and place larger discounts on payoffs in other states.

As a result, one interpretation of this methodology would be as a tool to identify areas of risk premia in options implied distributions and to analyze the level of risk premia in different parts of a distribution. One could then attempt to analyze periods where the extracted levels of risk premia were high or low and use that as an input into an investment process.

14.7.3 Taking Advantage of Deviations between the Risk Neutral & Physical Measure

Once we have identified what we believe to be a persistent deviation between the two measures, and convinced ourselves that this deviation is a valuable premia that we want to harvest, the final question becomes how to instrument trades to isolate deviations in the PDF for each measure.

To do this, we rely on the relationships between distributions and market instruments that we established earlier in the chapter. Specifically, we saw that butterflies can be used to approximate the PDF of a distribution and that call spreads of digital options to approximate the corresponding CDF. In particular, investors should be long butterflies in areas of the distribution where there is a higher mass in the physical measure to the risk neutral measure and be short butterflies when the mass in the risk neutral distribution exceeds that of the physical measure.

[7]Of course just because something is identified as a risk premium rather than a mispricing also does not mean that investors would not want to harvest it.

14.8 CONCLUSIONS & ASSESSMENT OF PARAMETRIC VS. NON-PARAMETRIC METHODS

In this chapter we explored an alternative to the traditional approach of solving options pricing problems by specifying and calibrating a stochastic process. In particular, we saw that there are two approaches that we can use to extract the risk neutral density directly from a set of options prices in a non-parametric manner. These methods enabled us to transform our problem from an over-determined one, where we were unable to match all option prices, to an under-determined one that is able to calibrate exactly to option prices in the absence of arbitrage. The density that we extract can then be used to price arbitrary derivative structures.

Perhaps as importantly, these techniques allowed us to better understand the way that options markets trade and the densities that underlie them. As an example, these methods enable us to measure the properties of the density (e.g. skewness/kurtosis) without imposing model assumptions. Investors can then use this information to make better investment decisions. This might be achieved by examining market implied densities over time, and applying time series techniques. This could also be accomplished by comparing the market implied distribution with some forecast of the physical measure, and attempting to trade on the divergence.

Although risk neutral density extraction techniques have many appealing characteristics, they are not without limitations. Most challenging is the ability to apply these techniques in markets with sparse set of options data. In these cases, where the market doesn't provide enough information to accurately represent the density itself, we need to use interpolation methods, additional constraints, or a prior distribution in order to impose that structure.

Additionally, because these models are parameterized by probabilities instead of traditional SDE parameters, they may be less intuitive. We saw in an earlier chapter that Greek calculations are inherently model dependent. Because of this, Greeks and other sensitivity calculations tend to be less intuitive in these non-parametric models as we don't have parameters like implied volatility to define Greeks with respect to. This is a challenge that we must confront should we want to implement these methods in practice. Although these are legitimate limitations, they each can be overcome and these methods end up being very useful in practice. In particular, the flexibility and lack of assumptions about the underlying process is particularly appealing.

IV

Quant Modeling in Different Markets

CHAPTER 15

Interest Rate Markets

15.1 MARKET SETTING

IN chapter 2 we provided an overview for a few common instruments in interest rate markets: debt instruments (bonds) and swaps. In this chapter we build off of that to provide the reader with a more detailed, practical examination of rate markets, and how buy-side and sell-side rates desks operate in this space. Rates and credit markets comprise a surprisingly large, and often underrepresented section of the overall investment landscape. To that end, they are a natural place for quants as they provide an environment with many, interconnected pieces that move together as part of a larger system.

When modeling equities, commodities or a foreign exchange rate, we are modeling a single stochastic variable. In the case of interest rate markets we are instead modeling an entire yield curve, which can be thought of as modeling a string instead of a single particle. Said differently, we are modeling a system of many underlying pieces, and there are restrictions on how they move together. This creates a more challenging, highly dimensional background where quants can, hopefully, add significant value.

The majority of traded products in interest rate markets are traded over-the-counter[1], meaning they rely on a dealer or intermediary to match buyers and sellers. As a result, they are highly customizable tools that investors can use to hedge their interest rate risk, instrument economic views precisely, or attempt to harvest related risk premium. Each of these potential applications is explored later in the chapter in section 15.13.

The main players in interest rate markets are banks, who serve as intermediaries, hedgers looking to neutralize their interest rate exposures, and hedge funds looking to express their macroeconomic views or pursue relative value opportunities. Hedgers may be corporations or mortgage issuers who hedge their inherent interest rate risk. Holders of mortgages, for example, are exposed to prepayment risk, which refers to the phenomenon that if rates decrease substantially, borrowers are likely to refinance their mortgages to obtain a lower mortgage rate and lower payment. This can impact the cashflows the mortgage holder receives and have an adverse economic impact. As a result, someone who holds a large portfolio of these mortgage-backed-securities may

[1]With notable exceptions such as certain futures contracts.

DOI: 10.1201/9781003530626-19

have substantial embedded prepayment risk and may choose to hedge this risk via an interest rate product. Additionally, the Federal Reserve plays a key role in rates markets by setting policy rates. We explore this relationship later in the chapter.

The following list shows the primary linear products that trade in interest rate markets, and provides a brief description of each:

- **Treasuries**: Coupon paying bonds backed by the US Government.
- **Treasury Futures**: Futures contracts linked to underlying treasury bonds.
- **Libor**: Libor is the instantaneous rate at which banks offer to lend to each other. Until recently, Libor was the standard floating rate on vanilla interest rate swaps and interest rate derivatives. In recent years, this rate has been decommisioned and replaced with SOFR, which is introduced next. Some context on the challenges of using Libor as a risk-free and some of the reasons for the push to retire the rate are discussed in 15.3.4.
- **SOFR**: A secured overnight financing rate (SOFR) that represents the average rate on treasury repo transactions. Unlike Libor, the secured nature of SOFR, as well as its link to actual transaction, make it a more appropriate discount rate.
- **OIS**: An overnight index swap rate (OIS) is a type of swap where an overnight rate is exchanged as the floating leg in exchange for a predetermined fix rate. This overnight rate is reset daily and compounded to calculate the relevant floating rate to be exchanged. In US markets, SOFR and Fed Funds effective rates could be used in an OIS contract.
- **FRAs**: agreements to pay or receive a forward rate, such as SOFR [2] at a future date in exchange for a fixed rate set at trade initiation.
- **SOFR / Eurodollar Futures**: Futures on a forward SOFR [3] rate. Futures are similar to FRAs, in that they are an agreement to exchange the specified forward rate in exchange for a fixed rate set when entering the trade.
- **Swaps**: A swap is an agreement to exchange cash-flows at a set of predetermined dates. The most common type of swap is a fixed for floating rate swap.

Later in the chapter, we show a similar list for so-called interest rate derivatives.

15.2 BOND PRICING CONCEPTS

15.2.1 Present Value & Discounting Cashflows

At the heart of modeling fixed income securities is valuing different cashflows received at many future dates. To do this consistently, we need to know how much each of

[2] Or previously Libor.

[3] Or Libor prior to the transition to SOFR.

these cashflows are worth today. Basic present value concepts suggest that a dollar received today should be worth more than a dollar received in the future[4]. As a result, we must discount each future cashflow at the appropriate rate to know its value today. Once we have applied the appropriate discounting, the present value of a bond or other fixed income instrument can be found by summing over the universe of cash flows. Throughout the chapter, we will see examples of different types of interest rate securities, and how we use discounting principles to calculate their present value. In some cases we will assume the discount process is deterministic, whereas in other cases we will need to assume that it is stochastic.

It should be noted that in interest rate modeling, we assume that the underlying cashflows are guaranteed. That is, we assume that default is not possible and as a result do not need to model default risk. That will change in the next chapter, when we model credit based instruments where default is the primary risk factor.

15.2.2 Pricing a Zero Coupon Bond

The simplest instrument in interest rate markets is a **zero coupon bond**, which pays its principal balance at maturity with no intermediate cashflows. Buying a zero coupon bond is equivalent to lending money today and being re-paid, with interest at maturity. Present value concepts dictate that the buyer of a zero-coupon bond will require interest to compensate them for forgoing their capital upfront. As a result, a natural equilibrium condition for the price of a zero-coupon bond should be less than its principal value[5].

Mathematically, the price of a zero coupon bond today can be written as:

$$Z(0,T) = exp\left(-\int_0^T r(s)ds\right) \tag{15.1}$$

In order to calculate the price of the zero-coupon bond, we will need to model $r(s)$, the underlying interest rate process. Depending on the context, we may make varying assumptions about $r(s)$, ranging from assuming that it is constant to assuming that it evolves according to a multi-dimensional SDE. Throughout the chapter we discuss these different approaches and when we tend to apply each.

Zero coupon bonds play an important role in modeling fixed income securities as they tell us the value today of a specified cashflow at a predetermined later date. This will become a critical component when calculating the present value of a set of cashflows, as we will need to discount them all back to the current time. Zero-coupon bonds, are also referred to as discount factors, and will be the primary tool for accomplishing this.

15.2.3 Pricing a Coupon Bond

In practice, many bonds pay periodic coupons, C in addition to re-paying their principal, P, at maturity. Like zero-coupon bonds, valuation of these instruments will

[4]Although recently negative interest rates have called this basic principle into question.

[5]Although this will not hold in the case of negative rates.

involve calculating the present value of the underlying cashflows, and will require us to discount each cashflow at the appropriate rate. Mathematically, the price of a coupon paying bond today can be written as:

$$B(0,T) = \left\{ \sum_i C \exp\left(-\int_0^{T_i} r(s)ds \right) \right\} + P \exp\left(-\int_0^{T} r(s)ds \right) \tag{15.2}$$

Coupon bonds can be thought of as weighted sums of Zero Coupon Bonds, where the weights are determined by C and P, respectively. Valuation of coupon bonds will require modeling the same interest rate process $r(s)$. As before, in some cases, we will assume this is a deterministic function, whereas in other cases we will treat it as a stochastic process. It should be noted that in this case the coupon for the bond was chosen to be a constant, fixed rate, C in practice bonds also exist where the coupon itself is floating. In this case, the floating rate is often set as a spread to the current SOFR rate.

15.2.4 Daycount Conventions

A subtle but important consideration for pricing both zero-coupon and coupon paying bonds, as well as other interest rate based securities, is that of daycount conventions. Daycount conventions are used to help measure differences between dates/times. We will find that these times, or differences in between two times show up repeatedly in our modeling efforts. For example, to value the cashflow today of a payment that is received in three-months, we need to calculate the appropriate time of that three-month interval. This may sound fairly unambiguous, but there are a set of conventions used in different markets and it is important to use the right one.

The most commonly used daycount conventions are: [179]

- **Actual/Actual**: Number of actual days in period divided by number of actual days in year.[6]
- **Actual/360**: Number of actual days in period divided by 360.
- **Actual/365**: Number of actual days in period divided by 365.
- **30/360**: Number of days in period based on every month having 30 days divided by 360,
- **30/365**: Number of days in period based on every month having 30 days divided by 365.

Daycount conventions are ubiquitous not just in interest rate markets but in any derivative valuation exercise but are of particular interest when modeling securities that are defined by a stream of cashflows. We will find that in some cases, such as swaps, different legs of a trade have different defined daycounts. As a result, understanding of the seemingly esoteric differences between them is pivotal in rates modeling.

[6]Accounting for leap-years.

15.2.5 Yield to Maturity

In practice we are not given information on the discount process $r(s)$ in the above bond pricing equations, directly. Instead we observe a set of bond prices and need to extract information about relevant discount rates.

A natural first assumption would then be to assume a constant discount rate, y, for all cashflows for a given bond. This is a commonly employed method and is known as a yield-to-maturity (YTM) approach. Employing this technique on a zero-coupon bond yields the following pricing formula:

$$Z(0,T) = \exp(-yT) \tag{15.3}$$

$$y = \frac{-\log(Z(0,T))}{T} \tag{15.4}$$

where $Z(0,T)$ is the market price of the zero-coupon bond and T is the time to maturity. Clearly, for a zero-coupon bond the yield-to-maturity is extractable analytically via a simple formula.

In the case of a coupon bond, the pricing equation becomes:

$$B(0,T) = \left\{\sum_i C\exp(-yT_i)\right\} + P\exp(-yT) \tag{15.5}$$

where $B(0,T)$ is the market price of the coupon paying bond. In this case, no analytical closed form solution will exist for the yield-to-maturity, y, however a simple one-dimensional solver can be used to find the yield that matches the market price.

15.2.6 Duration & Convexity

As we can see from the pricing formulas, the risk to holders of zero-coupon or coupon paying bonds is that interest rates will change. When interest rates change, the yield-to-maturity and discount rate process will change with it, leading to potential losses for the bondholder. Two primary measures of a bond's sensitivity to these types of changes in interest rates are **duration** and **convexity**.

- **Duration**: $\frac{1}{B}\frac{\partial B}{\partial y}$: measures the percentage change in bond price for a corresponding change in the bond's yield-to-maturity.
- **Convexity**: $\frac{1}{B}\frac{\partial^2 B}{\partial y^2}$: measures the second-order change in bond price with respect to a change in yield to maturity, usually viewed as a percentage of the bond's price.

Duration and convexity for coupon paying bonds can be found by differentiating (15.5) with respect to yield:

$$\frac{1}{B(0,T)}\frac{\partial B(0,T)}{\partial y} = \frac{-1}{B(0,T)}\left[\{\sum_i CT_i\exp(-yT_i)\} + PT\exp(-yT)\right] \tag{15.6}$$

$$\frac{1}{B(0,T)}\frac{\partial^2 B(0,T)}{\partial y^2} = \frac{1}{B(0,T)}\left[\{\sum_i CT_i^2\exp(-yT_i)\} + PT^2\exp(-yT)\right] \tag{15.7}$$

Later in this chapter we will generalize these concepts so that they apply to a greater breadth of instruments and refer to sensitivity not only with respect to changes in a bond's yield-to-maturity but also shifts of an underlying yield curve.

15.2.7 Bond Pricing in Practice: Duration and Convexity vs. Maturity

In this section we build off the code in the repository to highlight an important feature of bonds, that is, their duration and convexity profile as their maturity increases. To do this, we isolate a particular set of bond characteristics, such as its principal balance, yield and coupon, and vary its time to maturity to see how both the duration and convexity evolve. This is depicted in the following chart:

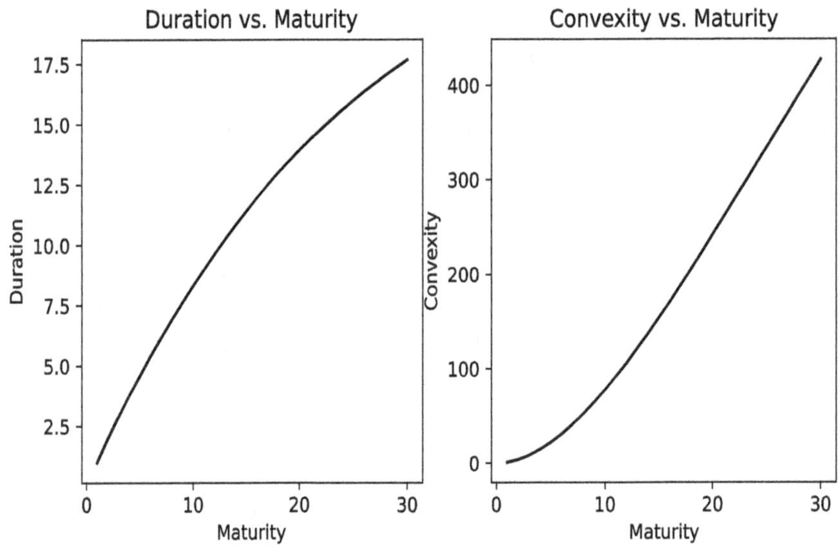

Here we can see that as we increase the maturity of a bond, both the duration and convexity of the bond increase as well. Additionally, we can see that while the increase in duration tends to slow at longer maturities, this is not the case for convexity. In fact, this is something that we observe in markets, that is, bonds with long maturities are characterized by large amounts of convexity, which is an interesting and attractive feature of these types of bonds. Investors may rationally desire this type of convexity, and be willing to accept a lower yield on these longer maturity bonds because of the convexity it provides them.

15.2.8 From Yield to Maturity to a Yield Curve

So far, we have tried to model the underlying interest rate process for different bonds as a constant yield. The primary benefit of this approach is its simplicity. Importantly, however, the yields that we extract from a set of market data will be different for bonds with varying characteristics. This is a challenging feature of a YTM driven approach, as many bonds are likely to have cashflows paid at the same time which will then be discounted at different yields. This is clearly not an appealing feature as the cashflows are economically equivalent but are discounted differently depending on the bond that they are attached to. Said differently, we are working with instruments that have overlapping cashflows, and are modeling these cashflows separately in each bond. Instead, we might want to try to model the discount rate process via non-overlapping segments that can be treated consistently across the cross section of market instruments.

Along these lines, an alternative approach, as we will see in the next section, is to use a set of so-called forward rates to break the curve into distinct, non-overlapping segments. Each segment would then have a different rate that could be applied to all bonds with a cashflow linked to this rate. Later in the chapter we will see how we can use this set of forward rates to price bonds consistently via a unified discount process or yield curve.

Because of these properties of the YTM approach, it is often used for working with individual bonds or other fixed income securities in isolation, but will not be used to build an overarching single, consistent representation of an interest rate process, or yield curve.

15.3 MAIN COMPONENTS OF A YIELD CURVE

15.3.1 Overview

When constructing a yield curve from a set of market data the main sources of market information are forward rate agreements, SOFR futures and swaps. Additionally, prior to the transition from Libor to SOFR, if we wanted to properly incorporate the embedded credit risk in Libor , we would need to model a basis between Libor and OIS into our yield curve,. The interest rate market also trades treasury products, including treasury bonds and treasury futures. Treasury instruments, however, are typically not included in yield curve construction exercises as there tends to be an idiosyncratic component to treasury yields that makes them hard to model in conjunction with other instruments. Further, differences in financing between cash and synthetic instruments lead to a basis between treasuries, which are cash instruments, and swaps, which are synthetic. This relationship is explored further later in the chapter in 15.13.3.

15.3.2 FRAs & SOFR Futures

Forward Rate Agreements (FRAs) and SOFR futures are both agreements to exchange SOFR, or Libor, for a fixed rate at a specified future date. The difference between a FRA and a future is that a FRA is an OTC contract and a future is exchange traded. As forwards and futures are fundamentally similar economically, so to are FRAs and SOFR futures. This means that SOFR futures have more standardized terms whereas FRAs are negotiated via a counterparty and as a result have more customizable terms.

More fundamentally, exchange traded products require adjustments to their posted collateral daily, whereas this is not the case for OTC instruments. Because of the negative correlation between the P&L of a futures contract and discount rates, a convexity correction emerges in SOFR futures. Specifically, if we are required to post collateral when rates are higher, we are posting collateral when the value we could earn on our deposits is highest. This convexity correction that arises in SOFR futures is the most substantive difference between FRAs and futures. It is one of many convexity corrections that arises in interest rate modeling. To fully understand and model these convexity corrections requires incorporating a stochastic model.

FRAs, on the other hand, are simpler as this daily mark-to-market feature is not present. As a result, the pricing equation can be expressed as:

$$V(t) = D(0,t)\delta_t \left(L_t - C\right) \tag{15.8}$$

where $D(0,t)$ is the discount factor, L_t is the floating rate and C is the predetermined fixed coupon.

15.3.3 Swaps

As we saw in chapter 2, a swap is an agreement to exchange cashflows periodically at a set of predetermined future dates. Swaps exist in multiple asset classes, however, they are perhaps most prominent in interest rate markets where investors may use them to hedge floating rate exposures. For example, consider a corporation who has issued a bond with a floating rate coupon. If interest rates rise, then the coupon that the corporation owes its bondholders will increase. Should the corporation not want to bear this risk, they can enter into a swap where they receive a floating rate, which will offset their debt obligation, and then pay fixed. Consequently, we can see that swaps are an effective means of transforming floating rate debt to fixed rate debt and thus are a critical hedging tools for market participants. They also provide an effective tool for speculators to implement their views about changes in future interest rates.

A vanilla swap is a *fixed* for *floating rate* swap. This is the most common type of swap, and typically has its floating legs linked to a SOFR [7] rate. Each payment in a swap is calculated by multiplying the fixed or floating reference rate by the appropriate time period, using the appropriate daycount convention. Swaps also have a specified **reference notional** which is used as a multiplier in these reference rate calculations to determine their present value. Swaps can either be spot starting, meaning that they start immediately or start at a specified date in the future based on terms locked in immediately.

Swaps consist of both fixed and floating legs, as described above. As a result, to model them will require us to value each leg, where we will need to apply the discounting concepts discussed above to calculate their present value.

For a swap's fixed leg, the present value can be written as:

$$\text{PV(fixed)} = \sum_{i=1}^{N} \delta_{t_i} C D(0, t_i) \tag{15.9}$$

[7] Or Libor.

where C is the set fixed coupon and $D(0,t_i)$ is the discount factor from today until time t_i[8] and δ_{t_i} is the time between payments.

Similarly, the present value of the floating leg of a swap can be expressed as:

$$\text{PV(float)} = \sum_{i=1}^{N} \delta_{t_i} L_{t_i} D(0,t_i) \tag{15.10}$$

where L_{t_i} is the floating rate of interest and $D(0,t_i)$ is the discount factor from today until time t_i and δ_{t_i} is the time between payments.

Clearly, in order to compute the present value of a swap at any time, we will need to know how to model discount factors and future forward [9] rates. This will require knowledge of the underlying discount rate process, and will be explored further when we discuss yield curve construction. Once those quantities have been represented and modeled effectively, however, we can see that valuing a swap becomes mechanical.

Of particular interest will be the fixed coupon that sets the present value of a swap contract equal to zero. This is referred to as the break-even, par or fair swap rate and is the fixed coupon that swaps are traded at in practice. To calculate this fair swap rate we simply take the difference between the present value of the fixed and floating legs and set it equal to zero:

$$\begin{aligned} \text{PV(swap)} &= \text{PV(float)} - \text{PV(fixed)} && (15.11) \\ &= 0 && (15.12) \end{aligned}$$

Solving this equation leads to the following formula for the fair swap rate:

$$\hat{C} = \frac{\sum_{i=1}^{N} \delta_{t_i} L_{t_i} D(0,t_i)}{\sum_{i=1}^{N} \delta_{t_i} D(0,t_i)} \tag{15.13}$$

This par swap rate, $\hat{C}$ tells us the fixed coupon at which we can enter into a swap without requiring an initial cash outlay. Clearly, examination of the formula shows that this fixed rate is a function of the expected floating rates during the contract, and we can expect a higher par swap rate as the underlying floating rates increase. It should be noted that the discount factor terms will also vary as we move the underlying discount rate process, however, they move in the same direction so this tends to be a second-order impact. As a result, we can think of the par swap rate loosely as a weighted average of the expected future floating rates.

The denominator in (15.13) is the value of a constant stream of payments and is often referred to as the annuity function:

$$A(t,T) = \sum_{i=1}^{N} \delta_{t_i} D(0,t_i) \tag{15.14}$$

This annuity function, or PV01, tells us the present value of a one basis point

[8]Or the price of a zero-coupon bond that pays 1 at time t_i.

[9]SOFR or Libor.

stream of payments between two dates and will appear repeatedly when working with swaps and options on swaps, swaptions. In fact, it can be proven that the swap rate is a martingale under the annuity numeraire. This will make the annuity measure a convention numeraire for swaption pricing and make calculation of today's annuity value a common task. Although we choose the swap rate at trade initiation such that the present value of the contract is equal to zero, eliminating the need for an upfront exchange of capital, as interest rates move the present value of the swap will change as well. In order to measure the mark-to-market valuation of a swap, we need to reevaluate the difference in the present value of the two legs with an up to date discount rate process, as shown below:

$$V(t) = \sum_{i=1}^{N} \delta_{t_i} L_{t_i} D(0, t_i) - C \sum_{i=1}^{N} \delta_{t_i} D(0, t_i) \tag{15.15}$$

where the valuation, $V(t)$ is from the perspective of the party paying a fixed rate and receiving a floating rate.

We also know however, that there is a par swap rate S, where the present value of the swap contract is zero and the present value of fixed and floating legs are equal:

$$\sum_{i=1}^{N} \delta_{t_i} L_{t_i} D(0, t_i) = S \sum_{i=1}^{N} \delta_{t_i} D(0, t_i) \tag{15.16}$$

An astute observer might notice that the left-hand side of (15.16) is the same as the first term on the right-hand side of (15.15). The remaining two terms also have a similar structure in that they are a swap rate multiplied by the value of the annuity function. Therefore, if we substitute the right-hand side of (15.16) for the first term on the right side in (15.15), the mark-to-market simplifies to:

$$V(t) = (S - C) \sum_{i=1}^{N} \delta_{t_i} D(0, t_i) \tag{15.17}$$

Intuitively, this means that we can express our swap's present value, or mark-to-market value, as the difference in spread between our contract and the par rate, multiplied by the present value of a one-basis point annuity.

15.3.4 Libor vs. SOFR & The Decommissioning of Libor

Prior to the Global Financial Crisis (GFC), LIBOR was the sole rate used as the risk-free discount rate in interest rate markets. It was also the predominant rate used as the floating leg in any vanilla fixed for floating interest rate swap. Coming out of the financial crisis, however, many issues emerged with Libor as a discount rate. First, as Libor is based on an offer rate, participants were able to manipulate the rate by adjusting their offers in a way that might be advantageous to their books. Secondly, Libor is at heart an unsecured rate, meaning that it embeds within it the credit risk of the underlying bank.

As a result of the credit risk embedded in Libor, and the potential for manipulation caused by it not being linked to a traded rate, one of the major shifts in interest rate markets over the last few years has been decommissioning of Libor for discounting [10] and as a floating rate linked to derivative transactions. Its replacement rate, SOFR aims to overcome these issues by used a traded, secured rate that can be used for discounting and as a floating rate in derivative transactions. While this transition is largely complete, and Libor rates are no longer published, many OTC derivatives contracts that were created prior to the transition still exist, creating an added complexity when attempting to fully phase out the rate.

15.4 MARKET RATES

The rates market, even in the absence of non-linear derivative instruments is characterized by a wide range of interconnected instruments, each of which provides a different potential representation of the market. For example, we saw zero-coupon bonds earlier in the chapter, and saw that they are often defined in terms of the constant yield that enables us to match market price. Swaps, conversely are characterized by their par swap rate, that is, the fixed coupon that sets the present value of a swap equal to zero at trade initiation.

The following table describes the main types of rates in fixed income modeling:

- **Spot Rates**: The rate at which we can instantaneously borrow or lend today for a specified amount of time.
- **Forward Rates**: The rate at which we can lock in to borrow or lend at some future time, for some predefined period.
- **Zero Rates**: The constant discount rate that matches the price of zero coupon bonds: $z = \frac{-\log P}{t}$
- **Swap Rates**: Breakeven/Par rates on vanilla fixed for floating swaps.

When we are modeling the yield curve, or the discount rate process, we have the flexibility to choose among these types of rates that are most convenient to model. As we saw earlier, one convenient property will be that they are non-overlapping. This will make zero rates and swap rates less appealing in representing a yield curve. It should also be noted that forward rates are a generalization of spot rates. That is, they are the amount we can borrow or lend at some time in the future. If we set that future time equal to today, then we recover the spot rate. As a result, forward rates are a natural choice for our yield curve construction process, as they provide us with general, non-overlapping rates. Once we have calibrated these forward rates however, we may still choose to visualize the curve by any of the market rates above.

15.5 YIELD CURVE CONSTRUCTION

15.5.1 Motivation

Our goal in building a yield curve is to extract a consistent, coherent discount rate process from a set of quoted products, including swaps, FRAs and futures contracts.

[10] This actually happened earlier, replaced by OIS.

We refer to this as a yield curve, or term structure of interest rates. It is a representation of the interest rate surface that enables us to compute any instantaneous, forward or swap rate. To do this in a unified manner means overcoming the issues of using yield to maturity and having the same cashflow valued using a different rate for different instruments. The first step in doing this is to specify a representation of the yield curve. It could theoretically be any of the rates discussed in the previous section, however, here we will focus on instantaneous forward rate representations of the yield curve. An instantaneous forward rate is the rate at which we can borrow or lend at some point in the future for some period of time, δt, where $\delta t \approx 0$.

Constructing the yield curve will involve taking all market data, including swaps, FRAs and SOFR futures[11] and finding a parameterization that best fits the given market prices. Clearly, a requirement of the representation that we choose, however, is that we can easily and efficiently convert from our internal representation to a set of traded products. When doing this, we must be able to handle all different products that underlie the market. In the case of an instantaneous forward rate representation of the yield curve, zero coupon bond prices, or discount factors, can be written as:

$$Z(S,T) = exp\left(-\int_S^T f(s)ds\right) \tag{15.18}$$

Note that this is the same equation we saw earlier in the chapter in 15.2.2. We can then make assumptions about the functional form of the instantaneous forward rate process. For example, we could assume that it is piecewise constant, or utilize an interpolation technique such as cubic splines that will lead to a more intuitive, smoother yield curve less prone to discontinuities between segments. Valuation of any interest rate based security will require modeling of discount factors or zero-coupon bonds.

We find that discounting terms appear in the pricing equations for swaps, FRAs as well as coupon paying bonds. Additionally, we will find many payments linked to SOFR [12] in our universe of instruments, for example in a swap. These forward rates are traded, simply compounded rates that can also be computed from the instantaneous, continuously compounded forward rate process that defines our yield curve. In particular, a SOFR or other forward rate can be expressed in the following terms of $f(s)$:

$$L(S,T) = \frac{1}{\delta}\left[\exp\left(\int_S^T f(s)ds\right) - 1\right] \tag{15.19}$$

Once we have defined our yield curve representation, as well as our parameterization, the next step is to choose a method to calibrate the parameters to the relevant market data. This is another type of calibration exercise that is conceptually similar to those that we explored in chapter 14. In this setting, there are two main techniques that we can rely on, **bootstrapping** and **optimization**, each of which are briefly summarized below:

[11] With the caveat that proper handling of futures requires a convexity correction that is beyond the scope of this book. For more info see [206] or [51].

[12] Or other floating forward rates, such as Libor.

- **Bootstrapping**: Bootstrapping works iteratively beginning with the nearest expiry instrument. For this instrument, we find the constant rate that enables us to match the market observed price. The algorithm then moves onto the next instrument, fixes the part of the term structure that was used to match the previous set of instruments, and solves for the rate over the remaining period that matches the next market price.

- **Optimization**: Alternatively, we can use optimization to minimize the squared distance from our market data.

In the following sections, we provide more context on how to formulate a yield curve construction problem via bootstrapping, and then via optimization.

15.5.2 Bootstrapping

Bootstrapping is an iterative algorithm that enables us to match the market prices/rates of all instruments exactly under most circumstances. It accomplishes this by treating the instantaneous forward rate process as a piecewise function and working through the instruments sequentially by expiry to calibrate pieces of the function in order. To do this, it begins with the shortest maturity instrument, which is usually a future or FRA, and finds the constant forward rate that enables us to recover the market rate. This is done by inverting formula (15.19). The algorithm then moves onto the next instrument, fixes the forward rate(s) previously fit to market instrument and repeats the operation.

The following is a summary of the procedure that is required to construct a bootstrapping algorithm:

- Order the securities that you have market prices for by maturity.

- Begin with the closest maturity FRA, future, or Swap Rate. Invert the formula to find the **constant forward rate** that allows you to match the market rate. In the case of a swap, this may require a one-dimensional root solving algorithm.

- Fix this **constant forward rate** for this time interval.

- Move to the next traded instrument. Holding the forward rate constant for the cashflows expiring prior to the last instrument, find the **constant forward rate** of the remaining dates that enable us to match market price.

- Continue to do this until all market instruments have been matched.

It should be emphasized that at each point in our bootstrapping exercise there is only one unknown constant forward rate that we are trying to solve for, and we are doing by inverting formulas (15.13) and (15.19), respectively. In practice, futures/FRAs tend to have shorter maturity, and as a result are used to calibrate the front-end of the yield curve. The curve then cuts over to swap around the two-year point, at which point swaps should be processed sequentially in order of their maturity.

Bootstrapping is a conceptually simple approach to fitting a yield curve that enables us to recover market rates exactly. As such, it is a very appealing technique and is commonly used in industry applications. Conversely, yield curves that are constructed via bootstrapping tend to have more jaggedness, especially at the front of the curve, and as a result are less economically intuitive. This will be a particular challenge at the intersection of the traded set of swaps and futures on the front-end of the curve. To overcome this, we can use an optimization based framework to construct a yield curve, as we detail in the next section.

15.5.3 Optimization

Optimization is a natural alternative to bootstrapping as it provides us more flexibility to balance the fit of market data with other desirable properties, such as obtaining a smooth, economically sensible curve. A brief introduction to the theory of optimization, such as first and second-order conditions, was covered in chapter 14. In the context of yield curve construction, our goal is to apply these optimization techniques to find the best representation of the yield curve for a given set of market data. This could be done by trying to match all the traded quantities exactly as constraints in the optimization, or alternatively could be formulated as a least squares fit to the data without requiring all data matches exactly. To formulate an optimization problem in this context, we begin by specifying a parameterization of the yield curve. As an example, we could break the term structure into distinct, non-overlapping segments and assume that forward rates are piecewise constant in each interval.

To proceed, let's define an optimization function $Q(f)$ that will minimize the distance between market and model rates.

$$Q(f) = \sum_{j=1}^{m} (R_j - \hat{R}_j)^2 \tag{15.20}$$

where m is the number of benchmark instruments, R_j is the market rate of instrument j and $\hat{R}_j$ is the model rate of the same instrument. Remember that these benchmark instruments may be a swaps, future, or FRAs, but that in each case we were able to formulate the pricing equation in terms of the instantaneous forward rate process. Clearly, we will want to minimize the objective function we have defined, $Q(f)$.

A first approach might be to assume that the instantaneous forward rate process, $f(t)$, is a non-overlapping, piecewise constant function. While this would work, it would have potential unintended side effects such as discontinuities in the forward rate processes along the edges of each interval. Alternatively, we might prefer a method that assures us that the function $f(s)$, and certain derivatives are continuous. It turns out this is possible via spline and b-spline functions, which have some attractive properties that will ensure a smooth curve with continuous derivatives. More information on using splines and b-splines in yield curve construction can be found in [13].

In addition to fitting to a set of market data, we also may want to add a term to our objective function ensuring smoothness of our curve (eliminating the jaggedness of forward rates we might see otherwise). This will help us balance the trade-off between fit to market data and an intuitive, sensible shape of the yield curve. A common regularization term is a **Tikhonov Regularizer**, which minimizes the curvature of the function:

$$R(\lambda, f) = \lambda \int_0^T \frac{\partial^2 f(t)}{\partial t^2} dt \tag{15.21}$$

where λ is a constant denoted the weight on the smoothness term and Q is the value of the regularization term. This function, $R(\lambda, f)$ would then be an auxiliary term that is added to the objective function $Q(f)$ defined above.

15.5.4 Comparison of Methodologies

In this chapter we explored two distinct approaches to modeling the yield curve. Bootstrapping, in particular, is a conceptually simple, easy to understand technique. It will also generally fit all of the available market data exactly. An unintended consequence of this technique, however, was the potential for unrealistic and jagged forward rates. Optimization techniques provided us with a more flexible alternative to bootstrapping that provides greater control over the yield curve construction process. As an example, an optimization framework allows us to specify more precisely whether to fit rates exactly, or within a bid-offer range. More generally, optimization techniques allow us to balance fit against our prior, normative beliefs about yield curve shapes enabling us to achieve a solution that is economically intuitive while still fitting the data. The optimization methodology, however, adds complexity to the problem. As a result, there is a familiar, natural trade-off between realistic assumptions and practical considerations.

In the next section, we explore an example of how bootstrapping can be used to extract an interest rate term structure, and apply it to the US rates market.

15.5.5 Bootstrapping in Practice: US Swap Rates

In this section we leverage the swap rate bootstrapping coding example in the repository to calibrate a set of instantaneous forward rates to a recent US yield curve[13]. The set of extracted instantaneous forwards, as well as the input swap rates are presented in the following chart:

[13] As represented by a set of swap rates.

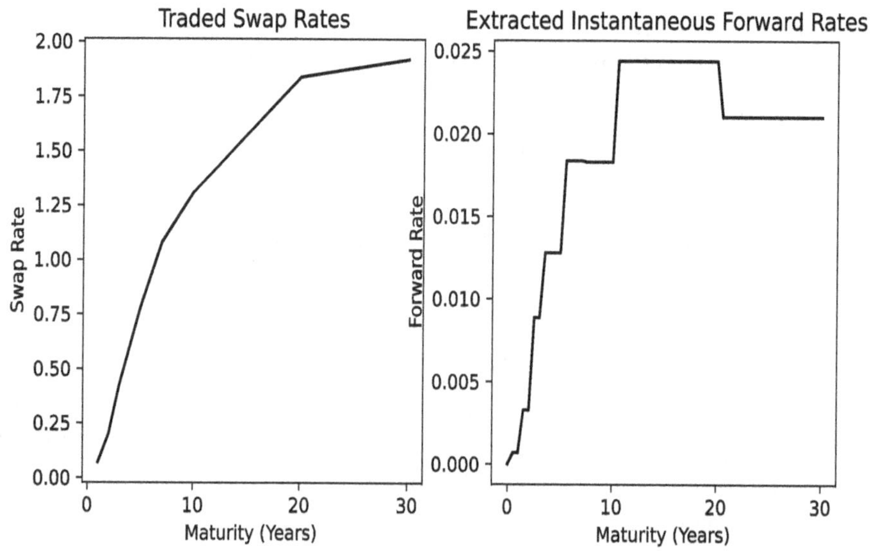

Here we can see that the traded swap rates are upward sloping, which is perhaps evidence of expected rate increases in the future. We see a more subtle shape in the instantaneous forward rates, and find that it peaks prior to the end of the curve, which corresponds to the point at which the slope of the increase in the traded swap rates begins to level off. One reason that yield curves tend to increase more slowly at longer maturities is the inherent convexity in longer maturity securities. This convexity is a desirable feature, meaning investors may demand lower yields to gain exposure to it. As such, it is possible that this is one reason that we observe this phenomenon in the extracted forward rates.

15.5.6 Empirical Observations of the Yield Curve

Although the yield curve is a highly dimensioned object with many interconnected, correlated segments, there is some commonality in the movements that we tend to observe. In fact, a PCA analysis[14] shows that the vast majority of the movements in interest rate curves can be explained by the first few principal components. As a result, when necessary, we can reduce the dimensionality of our problem significantly and work with a handful of underlying principal components.

Empirical analysis also tends to show that most of the movements in the yield curve are characterized by changes in the level of rates, changes in the slope of the curve and changes in its curvature/convexity. Consequently, when modeling a yield curve we must ensure we make realistic assumptions about how the level, slope and curvature can evolve over time. The slope of the curve is of particular interest to market participants as they look for clues regarding expected future Fed policy decisions.

[14]See chapter 5 for more details on PCA.

In economic terms, the yield curve can be decomposed into a few main drivers:

- **Fed Policy Rate(s)**: Short term rate set by the Fed aimed at managing economic growth and inflation.
- **Inflation**: (Nominal) Bonds lose value in the presence of inflation and thus investors will require higher levels of interest when inflation is higher.
- **Term Premia**: Premium investors are compensated for locking up their capital for longer periods.
- **Policy Uncertainty**: Expected future Fed policy decisions and uncertainty around these decisions

Of particular interest is the term premium as this helps develop an appreciation for the fair slope of the yield curve. This is directly relevant to the Fed as well as other market participants. Multiple models of the empirical dynamics of the term premia have been created, such as the Adrian-Crump-Moench term premia model [244].

15.5.7 Fed Policy and the Yield Curve

The Federal reserve plays a large role in the shape of the yield curve as well as its movements over time. Notably, the Fed is responsible for setting short term interest rates via the Fed Funds Rate, which is the rate that banks can borrow or lend excess reserves. When this rate is high, it discourages borrowing because it is more expensive, making banks and other institutions less likely to take on projects because of the higher opportunity cost. Conversely, when the rate is low, borrowing is cheaper which tends to have a stimulating impact on the economy.

In the United States, the Fed makes it policy decisions in line with a dual mandate of maximum employment and stable inflation[15]. There is a natural trade-off between these two objectives. If the economy is allowed to run too hot, it is likely to generate high levels of inflation. Conversely, if policy is tightened to rein in inflation, it is likely to led to lower levels of employment. Increases in the Fed Funds rate are used to tighten monetary conditions and are usually done during periods of high growth to ensure an expansion is sustainable. Decreases in the Fed Funds rate are used to loosen monetary conditions and are usually done during recessions when the economy needs to be stimulated and jobs need to be created.

Additionally, the slope of the yield curve says a lot about market participants view of future policy decisions. If the curve is upward sloping, it may mean that future rate increases are expected. This is interpreted as a signal that investors have confidence in the underlying economy, because the Fed is likely to increase rates only when growth and employment numbers are strong. Conversely, when the curve is downward sloping, it indicates that the Fed is expected to cut rates in the future. This is in turn a potentially bearish indicator, as the Fed steps in to stimulate the economy when growth stagnates and the unemployment rate is high. In fact, research has shown that an inverted, or downward sloping yield curve is a reliable predictor of an ensuing recession [34].

[15]With a 2% long-run inflation target.

15.6 INFLATION LINKED ASSETS

15.6.1 Importance of Inflation Linked Assets

Interest in inflation linked assets has risen dramatically over the last few years as inflation has gone from relatively subdued to one of the primary macroeconomic drivers across markets. This has led inflation to be one of the most topical markets themes and led investors to place an increased emphasis on isolating and hedging inflation risk. At the heart of this are inflation linked assets which enable market participants to hedge inflation risk and to extract forward looking inflation expectations from market data. In the remainder of the section, we discuss two of the most commonly traded inflation linked assets: inflation linked bonds and inflation swaps. Finally, we end this section by showing how a combination of nominal and inflation linked bonds can be used to extract inflation expectations embedded in the market. Later in the chapter, we also explore the theme of inflation post Covid and its impact on fixed income portfolios.

15.6.2 Inflation Linked Bonds

Perhaps the simplest inflation linked assets are inflation linked bonds, which are liquidly traded in many markets. For example, in the US these inflation linked bonds are known as Treasury Inflation Protected Securities, or TIPS [1]. The appeal of these inflation linked bonds, then, is that they do not lose value in the presence of inflation, instead retaining their value. Like nominal bonds, inflation linked bonds are still sensitive to changes in interest rates, however, when this change in rates is not driven by a corresponding change in inflation. This leads to the concept of a real interest rate, that is, the rate in excess of inflation, and inflation linked bonds still have exposure to changes in real interest rates.

The mechanism for these bonds to account for inflation, and adjust their payments upward in the presence of inflation to retain their value, is an introduced **index factor** that measures the change in the relevant inflation index during the life of the bond. As this index factor increases, so too do the coupon and principal payments of the bond, offsetting what would have been a loss due to an increase in interest rates.

This index factor mathematically can be expressed as:

$$I(0,t) = \frac{\text{index}_t}{\text{index}_0} \tag{15.22}$$

where index_t represents the value of a CPI or other inflation index at time t [16]. This index factor is then used to determine the value of the coupon or principal payments received by an investor in an inflation linked bond.

$$L(0,T) = \left\{ \sum_i I(0,T_i) C \exp\left(-\int_0^{T_i} r(s)ds \right) \right\} + I(0,T) P \exp\left(-\int_0^{T} r(s)ds \right) \tag{15.23}$$

[16] And index_0 represents the value of that same inflation index at time 0.

where I_t is the value of the index ratio at time t. As you can see from the pricing formula, in the case of an unexpected change in inflation, we would expect it to be reflected in the discount rate process, $r(s)$, however, we would expect the index factor to increase neutralizing this change.

15.6.3 Inflation Swaps

Interest rate markets also trade swaps that are linked to inflation rates, such as CPI[17]. Like the swaps introduced earlier in the chapter, these swaps are fixed for floating in nature, where in this case the floating leg will be linked to an inflation rate rather than a SOFR or other forward rate. Because they are similar in structure to other vanilla fixed for floating swaps, similar valuation techniques can be used.

For example, the basic structure of a typical fixed leg might look like:

$$\text{PV(fixed)} = \sum_{i=1}^{N} \delta_{t_i} C D(0, t_i) \tag{15.24}$$

Whereas the corresponding floating inflation linked leg of a swap might be written as:

$$\text{PV(inflation)} = \sum_{i=1}^{N} \delta_{t_i} \left(\frac{\text{CPI}_{t_i}}{\text{CPI}_{t_{i-1}}} - 1 \right) D(0, t_i) \tag{15.25}$$

where $\left(\frac{\text{CPI}_{t_i}}{\text{CPI}_{t_{i-1}}} - 1 \right)$ represents the change in CPI over a given period. In this contract, we are assuming that the inflation swap is linked to CPI, however, we could easily substitute this for another inflation index.

The present value of an inflation swap could then simply be expressed as the difference between the present value of the two legs, as expressed below:

$$\text{PV(swap)} = \text{PV(inflation)} - \text{PV(fixed)} \tag{15.26}$$

At initiation, the fixed rate C on the inflation swap would then be chosen such that the present value of the contract is equal to zero.

In practice, many traded inflation swaps are so-called zero-coupon inflation swaps meaning that, unlike the formulation above, a single cash flow is exchanged at maturity with no intermediate cashflows. In this case, we present value of the fixed leg of an inflation swap that matures at time t_M can be written as:

$$\text{PV(zero-coupon-fixed)} = D(0, t_M) \left((1 + C)^{t_M} - 1 \right) \tag{15.27}$$

where C is a fixed annualized coupon agreed at trade initiation that is then compounded over the maturity of the contract, t_M.

Similarly, the PV of the inflation linked leg can be expressed as:

$$\text{PV(zero-coupon-inflation)} = D(0, t_M) \left(\frac{\text{CPI}_{t_M}}{\text{CPI}_{t_0}} - 1 \right) \tag{15.28}$$

[17]Consumer Price Index.

where $\frac{\text{CPI}_{t_M}}{\text{CPI}_{t_0}}$ is the ratio of the inflation index at maturity of the swap and at trade initiation. Importantly, we can see the same concepts emerge in either formulation of an inflation swap, and the same concepts for modeling an inflation and discount rate will be the key factors in valuing these types of securities.

15.6.4 Breakeven Inflation Rates

The presence of both nominal bonds [18] and inflation adjusted bonds, which as discussed above account for inflation through an index factor, allow us to extract a forward looking measure of expected inflation from market pricing. These rates are typically referred to as breakeven inflation rates, and represent the difference in bonds that are nominal, and not inflation adjusted, versus those that are inflation adjusted. Assuming the bonds are otherwise equal, then any difference in yield between these two bonds must be due to expected inflation.

Mathematically, these breakeven inflation rates can be written as:

$$y_{\text{BE}} = y_n - y_r \tag{15.29}$$

where y_n is the yield on a nominal bond, such as a treasury, and y_r is the yield on an inflation adjusted bond that only has exposure to real interest rates.

Because of the information contained about expected future inflation, breakeven inflation rates are of utmost importance to market participants and are monitored closely by fixed income investors. These breakeven rates can be compared to realized inflation [19] to try to gauge potential changes in coming inflation numbers.

15.7 MODELING INTEREST RATE DERIVATIVES

15.7.1 Linear vs. Non-Linear Payoffs

Up until this point in the chapter, we have focused on modeling instruments that make up the yield curve, such as FRAs and swaps. These instruments are characterized by simple, linear payoffs. Fundamentally, the linearity in the payoffs means that these instruments can be replicated statically. This means that we can create our replicating portfolio at the beginning of the trade and will not be required to re-balance until the trade expires, regardless of market conditions. As a direct result of this, these products won't have sensitivity to volatility but instead will only be a function of the expected future value[20]. In this context, a stochastic model is not necessary, instead we rely on these static replication arguments.

In the remainder of the chapter, we turn our attention to more complex derivative structures with non-linear payoffs. As we do this, we will see that it is no longer possible to create a static replication portfolio. Instead, we will need to rely on dynamic replication arguments, such as the one we saw in chapter 4 to motivate risk-neutral pricing in the Black-Scholes model. Importantly, because these replication schemes

[18] Such as treasuries.

[19] Measured by CPI or a similar metric.

[20] A notable exception to this are futures, or other linear products that require a stochastic model because of the inherent convexity correction.

are dynamic, they will require periodic re-balancing and the underlying hedging costs will vary. As a result, we will need to employ a stochastic model in the context of these types of derivatives. Phrased differently, we no longer only care about the expected value of the underlying at expiry. Instead we care about the entire distribution of outcomes at, and potential before, expiry. This is a fundamental difference that explains why we invoke stochastic models to solve non-linear derivative pricing problems but do not need them for simpler, linear payoffs.

15.7.2 Vanilla vs. Exotic Options

A second differentiating feature in derivatives modeling is whether we are working with a so-called vanilla, market standard options or an exotic option. In rates markets, caps and swaptions are considered vanilla options as they are both European call and put options on SOFR [21] or a swap rate, respectively. Many exotic options, such as spread options and Bermudan swaptions trade as well. For these types of derivatives, market pricing is less standardized and they are characterized by lower levels of liquidity.

Generally speaking, valuing a vanilla option can be done via a simple, market standard model such as the Bachelier or Black's model, both of which we saw in chapter 13. This does not mean that larger, term structure models cannot add value in the context of vanilla options, however, we will find them far less critical in as the market is characterized by high liquidity and price discovery.

Exotic options, however, will depend on the interconnection between different rates and thus cannot be modeled using standard one-dimensional market models. Instead, they will need to rely solely on larger, term structure models that model the entire curve. Traditionally, it is desired to price exotic options in a manner that is consistent with the set of traded vanilla options. Consequently, the standard procedure is to calibrate a term structure model to a set of relevant vanilla options and then use this calibration to model the desired exotics. This procedure mimics the more general calibration setup that we detailed in chapter 14.

15.7.3 Most Common Interest Rate Derivatives

A brief description of the most commonly traded vanilla and exotic options in rates market is listed below:

- Vanilla Options
  - **Caps & Floors**: A basket of options each of which is linked to an underlying Forward rate, such as SOFR. Caps are call options on the set of forward rates, whereas Floors are puts options. The underlying component options are referred to as Caplets and Floorlets, respectively.
  - **Swaptions**: An option to enter a swap contract at a specified strike, K. Payer swaptions are call options on the swap rate and provide the owner

[21] Or Libor.

the option to pay fixed on a swap. Receiver swaptions, conversely, are put options on the swap rate and give the owner the option to receive fixed.

  - **SOFR Future Options**: An option linked to a SOFR future. Aside from complications due to convexity corrections, these will be fundamentally similar to caplets & floorlets.
  - **Treasury Options**: An option to buy or sell a treasury at a specified strike price, K.

- Exotic Options
  - **Bermudan Swaptions**: An American style option to enter into a swap with a prespecified set of exercise dates. Further complicating these derivatives it the fact that the underlying swap that the owner would exercise into changes at each exercise date.
  - **Spread Options**: An option on the spread, or difference between two rates (most commonly swap rates). For example, a spread option might have its payoff linked to the difference between the 10 year swap rate and the 2 year swap rate.
  - **CMS Options**: An option linked to a constant maturity swap. A constant maturity swap pays a specified swap rate at each cashflow (i.e. the 10-year swap) instead of using the traditional Libor rate for the floating leg.

In the following sections, we detail the most common approaches to modeling each of these structures, and also detail the properties inherent in each type of derivative. We will find that some of these derivatives are fairly simple to model (i.e. caps), whereas other are some of the more challenging problems that we will find in the field of quant finance (i.e. Bermudan swaptions).

As a note to the reader, when introducing stochastic and derivative pricing models, we provide a general methodology that works with the more general concept of a forward rate, rather than a specific forward rate such as SOFR or Libor. These techniques can then be applied with whatever the prevailing forward and swap rates and conventions exist in the market of interest.

15.7.4 Modeling the Curve vs. Modeling a Single Rate

A key defining feature of models in interest rate markets is whether they attempt to model the entire yield curve or attempt to model a single representation of the curve, such as a forward or swap rate, in isolation. Models that attempt to model the entire curve are referred to as term structure models. As noted earlier, one of the main challenges with rates modeling is the presence of an entire curve that contains correlated, interconnected pieces. Knowledge of how part of the curve moves implies certain behavior about how the rest of the curve moves, and our term structure models need to account for this in a robust manner. Modeling the yield curve requires treating the entire curve as stochastic and modeling its evolution in a realistic manner. This includes making sure that as the yield curve evolves the representation is reasonable, that is, that the pieces move in a realistic way. In some cases, however, this unified curve representation may not be necessary. In those cases we can instead model the underlying swap or forward rate as its own stochastic process and ignore that it is part of a larger, more complex term structure.

The choice between a term structure model and modeling a single underlying rate is the first fundamental problem that a quant faces in rates modeling. In some cases, such as when working with exotic options that are highly sensitive to intra-rate correlations, there will be no choice but to leverage a term structure model. In other cases, such as modeling a single swaption in isolation, we may choose to model the underlying swap rate and ignore the broader, term structure. In the following sections we provide an overview for each modeling approach, beginning with the simpler approach of modeling individual forward or swap rates, and then proceeding to attempting to model the entire curve via short rate and forward rate models.

15.8 MODELING VOLATILITY FOR A SINGLE RATE: CAPS/FLOORS

The simplest interest rate derivative is a caplet or floorlet, which is an option on a single forward rate. As these are the simplest derivatives, they are a natural choice for our first exploration into modeling individual rates in isolation. Because they are options on a forward rate, we can model caplets by specifying a stochastic process for this underlying forward. For example we might assume that the forward follows a *Normal* or *Lognormal* model. The properties of both of these models were detailed in chapter 13. In order to incorporate caplet volatility skew, however, we may need a more realistic model, such as a stochastic volatility model or a jump process.

Caplets and floorlets themselves are not traded directly. Instead baskets of them are traded in the form of caps and floors. A buyer of a cap or floor has the right to exchange a forward rate, such as SOFR, for a specified strike rate, K, at multiple dates. Importantly, these options can be viewed independently as individual options that expire at different times with independent exercise decisions. This means that pricing a cap involves simply pricing each caplet separately and the summing over the set of caplets. Caps are OTC instruments, and are a function of strike, starting expiry and length.

15.8.1 T-Forward Numeraire

In chapter 4 we saw how we could use Girsanov's theorem to change measure and that we can work in an equivalent martingale measure should we choose. Fundamentally, this means that the choice of numeraire is a matter of convenience. In most asset classes, the traditional risk neutral measure that we have seen throughout the book is most convenient. In rates, however, discount rates are part of the underlying stochastic nature of the process. This adds another level of complexity and often means that the risk neutral measure will not be the most convenient. In this book we don't delve too deeply into the derivations of numeraires but it is important to understand the basic concept/intuition[22].

[22]For more details see [50].

When solving options pricing problems, we compute the relative price of an asset or derivative discounted by a given numeraire, N:

$$S^N(t) = \frac{S(t)}{N(t)} \tag{15.30}$$

In standard pricing problems in other asset classes, the numeraire is a zero-coupon bond, that is: $N(t) = \exp(rt)$.

However, arbitrage free pricing dictates that the following more general condition holds:

$$\frac{V(s)}{N(s)} = \mathbb{E}^Q\left[\frac{V(T)}{N(T)}\right] \tag{15.31}$$

where $N(s)$ is today's value of the chosen numeraire and $N(T)$ is the value of the chosen numeraire at expiry. It will be convenient to choose numeraires where this term is equal to 1, such as will be the case for a Zero-Coupon Bond. If we do this, we are left with:

$$V(s) = N(s)\mathbb{E}^Q\left[V(T)\right] \tag{15.32}$$

When working with caplets, the T Forward Numeraire will be of particular interest. It is the numeraire that is defined by the price of a zero coupon bond at time t that matures at time T. This numeraire arises naturally when pricing instruments with maturity T, such as a forward rate. In fact, it can be proven that forward rates, such as SOFR, are martingales under the T-Forward measure. This will make the T-Forward numeraire convenient for pricing derivatives whose payoffs are linked to forward rates, such as caplets.

15.8.2 Caplets/Floorlets via Black's Model

As Libor and SOFR rates are martingales in the T-Forward measure, the price of a caplet, which is a call option on an underlying forward rate, can be written as:

$$C(K) = \delta D(0, T+\delta)\mathbb{E}\left[(F_T - K)^+\right] \tag{15.33}$$

where $D(0, T+\delta)$ is the discount factor from time $T+\delta$ to today, δ is the year-fraction spanned by the forward rate[23] and F_T is the forward rate at option expiry. The reader should notice that this discounting term is similar to that of risk neutral measure, except that we are now discounting from the end of the forward rate accrual period until today rather than the expiry of the option. This is why we see the discount factor range from 0 to $T+\delta$ instead of the standard T in the risk neutral measure.

Pricing of caplets & floorlets can then be done by specifying a set of dynamics for the underlying forward rate. For example, if we are to use the dynamics of the

[23]Using the appropriate daycount conventions.

Black-Scholes model[24], then the pricing equation for a caplet would become:

$$C = \delta P(0, T+\delta)\left[F_0 N(d_1) - K N(d_2)\right] \quad (15.34)$$

$$d_1 = \frac{\log\left(\frac{F_0}{K}\right) + \frac{1}{2}\sigma^2 T}{\sigma\sqrt{T}} \quad (15.35)$$

$$d_2 = \frac{\log\left(\frac{F_0}{K}\right) - \frac{1}{2}\sigma^2 T}{\sigma\sqrt{T}} \quad (15.36)$$

A similar pricing formula exists for floorlets, which are put options on a forward rate. In fact, we can see that these formulas are very similar to the Black-Scholes pricing formulas presented in (10.2). The fundamental difference between the formulas is a slightly different numeraire term. Instead of using the common risk neutral discounting we see in other asset classes we are working in the T-Forward measure, resulting in a different zero-coupon bond, or discount factor being used as the numeraire term. The price of this zero-coupon bond in the numeraire term can easily be calculated from today's yield curve using (15.18). A coding example of caplet pricing is provided in the corresponding repository.

15.8.3 Stripping Cap/Floor Volatilities

Caps are baskets of caplets consisting of multiple options each linked to a different forward rate. As caps are baskets of caplets, their pricing equation can be written as:

$$C(K) = \sum_i \left\{\delta_i P(0, T_i + \delta)\mathbb{E}\left[(F_{T_i} - K)^+\right]\right\} \quad (15.37)$$

where i indicates that we are looping over all individual caplets. We can see that the cap pricing formula is just a sum of the individual caplets which are defined in equation (15.33).

In practice, the market trades caps rather than caplets and caps are quoted in terms of a single (Black) implied volatility. This single implied volatility is interpreted as the *constant volatility* that allows us to match the price of the entire string of caplets. In order to obtain volatilities from individual caplets we need to extract them from the constant cap vols. This process is referred to as **stripping caplet volatilities**.

There are two main approaches to stripping caplet volatilities from a series of cap prices or volatilities:

- **Bootstrapping**: iterate through the universe of caps starting with the shortest expiry. Calculate the caplet vols at each step that enable you to recover the market price.

- **Optimization**: start with a parameterized function for the caplet volatilities and perform an optimization that minimizes the pricing error over the set of parameters in the function.

[24]It should be noted that we employ a slightly modified version of the Black-Scholes model discussed in chapter 13 that models a forward rate.

This procedure mimics the process we saw when constructing a yield curve with a different objective function and underlying pricing formulas.

15.8.4 Fitting the Volatility Skew

In addition to extracting the set of caplet & floorlet volatilities from a set of caps and floors, ideally we will also want to make reasonable assumptions about the skew of different caplets. If done correctly, this would ensure that we are able to match caps at multiple strikes with a consistent stochastic model for each underlying forward rate. Said differently, we'd like all the caplets & floorlets of a given expiry, which correspond to the same underlying forward rate, to have the same dynamics. We can accomplish this using the some of the models discussed in chapter 13. In rates markets, SABR is the most commonly applied model to match the underlying skews.

15.9 MODELING VOLATILITY FOR A SINGLE RATE: SWAPTIONS

Our second case study of modeling a single underlying rate will be swaptions. As detailed in 15.7.3, Swaptions are vanilla options on vanilla fixed for floating swaps, where the floating leg is generally linked to SOFR. Swaptions are referred to by their expiry and tenor. For example, a 1y2y swaption has a 1 year expiry and the underlying is a 2 year swap that starts at option expiry. Just as in the last section, where we wanted to make the underlying forward rate the stochastic variable in a caplet/futures option, here the most convenient approach will be to model the underlying swap rate as its own stochastic process. This will require us to work in a different pricing measure, the annuity numeraire.

15.9.1 Annuity Function & Numeraire

Recall from (15.13) that the par swap rate is the fixed coupon that sets the present value of a swap to zero at trade initiation. The denominator in the swap rate formula, which is often referred to as PV01 or the annuity function, is of particular interest in modeling swaptions. This annuity function can be written as:

$$A(t,T) = \sum_{i=1}^{N} \delta_{t_i} D(t,t_i) \tag{15.38}$$

This annuity function, or PV01, tells us the present value of a one basis point annuity between two dates. Even more fundamentally, however, it can be proven that the swap rate is a martingale under the annuity numeraire. In this text we omit the details of this derivation. Readers looking for a more detailed treatment should consult [52], [13] or [50]. The result of this, however, is that when we price a swaption, today's annuity value plays the role that a discount factor plays in standard risk neutral valuation for other asset classes.

15.9.2 Pricing via the Bachelier Model

In the last section, we saw that the swap rate is a numeraire in the annuity measure. This is a key insight as it enables us to price swaptions using today's value of the annuity function and specifying a set of dynamics for the swap rate. We can theoretically choose any SDE we want to represent the swap rate dynamics. A common market convention is to use the Bachelier SDE[25] to model individual swaptions.

Because the swap rate is a martingale in the annuity measure, in this numeraire the price of a payer swaption can be written as:

$$P = A_0(t,T)\mathbb{E}\left[(F-K)^+\right] \tag{15.39}$$

where in this case F refers to the current par swap rate. To price a swaption, we will need to specify some dynamics for the underlying swap rate F. If we specify the Bachelier or Normal model, as detailed in chapter 13, the SDE will be:

$$dF_t = \sigma dW_t \tag{15.40}$$

As we saw in chapter 13, this models leads to a closed form solution for European options prices. This means that swaption prices will be available analytically in the Bachelier model, which is clearly an attractive feature. The pricing equations will mimic those defined in (10.8) with the only difference coming from the different numeraire. As an example, the pricing equation for a payer swaption under the Bachelier model can be written as:

$$\begin{aligned} P &= A_0(t,T)\sigma\sqrt{t}\left[d_1 N(d_1) + N'(d_1)\right] && (15.41)\\ d_1 &= \frac{F_0 - K}{\sigma\sqrt{t}} && (15.42)\\ d_2 &= -d_1 && (15.43)\end{aligned}$$

where t is the option expiry and T is the end time of the swap.

The Bachelier model relies on a single constant volatility parameter, σ. As a result, while it provides a simple, tractable pricing framework that is a useful reference for market participants it will not match the behavior in markets, including the volatility skew.

15.9.3 Fitting the Volatility Skew with the SABR Model

If we want to incorporate the swaption volatility skew for a given underlying swap rate, we need to rely on a more realistic SDE. As we learned in chapter 13, there are three main ways of trying to fit a volatility smile or surface: local volatility, stochastic volatility or via a jump process. In rates the standard way of incorporating skew is via a stochastic volatility model, namely the SABR model. Recall from chapter 13 that the dynamics for the SABR model can be expressed as:

$$\begin{aligned} dF_t &= \sigma_t S_t^{\beta}\, dW_t^1\\ d\sigma_t &= \alpha\sigma\, dW_t^2\\ Cov(dW_t^1, dW_t^2) &= \rho\, dt && (15.44)\end{aligned}$$

[25]Or alternatively Black's model.

Further recall that an approximate corresponding normal volatility for an option given a set of SABR parameters is identified in (10.26). This formula provides a normal implied volatility for a given combination of SABR parameters (α, β, ρ and σ_0), strike and expiry.

The SABR model allows for a fairly good fit of the volatility surface for a **single expiry** but relies on separate calibration results for each expiry. When calibrating a SABR model, we need to find the best parameters (α, β, ρ, σ_0) for our data. Although we have four distinct parameters, they are correlated. This presents challenges in calibration. As a result, we often fix one parameter, generally β, and calibrate the three remaining parameters.[26]. To calibrate these SABR parameters, we can leverage the techniques that we covered in chapter 14. In the coding repository, we provide an example of calibrating a SABR model to market data.

15.9.4 Swaption Volatility Cube

In earlier chapters, we saw that in equity markets the volatility surface was a two dimensional object (across strike and expiry). In rates, we have another dimension that arises from the tenor or length of a swap. This three dimensional object is referred to as the **volatility cube** and has dimensions:

- Expiry
- Tenor
- Strike

Clearly the volatilities for swaptions with different tenors but the same strike and expiry should be somewhat related. As a result, when we model the volatility cube in rates we need to model them as interconnected underlying pieces of a broader yield curve. In the preceding sections, we have focused exclusively on modeling these underlying swap rates in isolation without awareness of the remainder of the curve. Proper modeling of the swaption volatility cube, however, will require a stochastic representation of the entire yield curve that embeds realistic correlation assumptions of different rates. Because of this extra dimension, fitting an interest rate volatility model is a more complex endeavor than it is for other asset classes. In practice we may have close to 1000 underlying swaptions to fit presenting a tremendous calibration challenge. In the next section, we show how one might try to accomplish this via short rate or forward rate models of the yield curve.

15.10 MODELING THE TERM STRUCTURE: SHORT RATE MODELS

15.10.1 Short Rate Models: Overview

In previous sections, we have attempted to model interest rate derivatives by modeling a single underlying rate as a stochastic process. We saw this was possible for swap

[26] Alternatively, β can be estimated via a time-series regression of at-the-money implied volatility against the forward rate.

rates and forward rates [27] alike. However, what this approach did not provide was a unified framework for modeling the entire curve (and connecting different rates). Instead, we ignored the fact that a given rate was part of a broader yield curve, and instead modeled it in isolation. The first approach to building a unified stochastic model for interest rate derivatives is called a short rate model.

In a short rate model, we choose to model the instantaneous spot rate, that is the risk free rate from time t until time $t+\Delta t$ (with $\Delta t \to 0$). This can be thought of as an instantaneous borrowing and lending rate. Clearly, this short rate is a mathematical abstraction rather than a traded rate. Tradable rates however, can be expressed as functions of this mathematical abstraction. In fact such transformations are required to make short rate models useful in practice. In the simplest case, we can express a zero-coupon bond or discount factor in terms of the short rate we are modeling:

$$P(0,t) = \exp\left(-\int_0^t r_t\, dt\right) \tag{15.45}$$

In some cases, this transformation from our mathematical abstraction to a traded rate may be relatively simple. In other cases, such as for a swap rate, we will see that the swap rate is a complex function of the short rate process, making calculations involving them fairly involved.

The price of a generic payoff F_t can be written as:

$$P_0 = \mathbb{E}\left[\exp\left(-\int_0^T r_t\, dt\right) F_T\right] \tag{15.46}$$

Modeling r_t provides a model for the evolution of the entire yield curve, and gives us a way to price any instrument where the payoff F_T can be expressed as a function of the yield curve at time T.

When working with short rate models, there are certain desirable features that we'd like to make sure are incorporated. For example, a basic prerequisite for a short rate model would be that it is able to **match the current, observed yield curve**. While this may seem trivial, it is more complex than in other asset classes as it involves matching an entire curve rather than a single current asset value as would be the case in FX, equities or commodities. Additionally, we ideally want our short rate model to be able to match the prices, or implied volatilities of **at-the-money options** of different expiries and tenors. Lastly, we would ideally want to fit all **out-of-the-money options** for all expiries/tenors. This is the most daunting task and we will find even fitting the at-the-money surface to be a challenging process.

In this book we cover the following four common one-factor short rate models:

- Ho-Lee
- Vasicek
- Cox-Ingersol-Ross

[27] such as SOFR.

- Hull-White

Additionally, we cover two multi-factor models that incorporate an additional source of randomness, which we will see is an important feature for building a realistic model.

15.10.2 Ho-Lee

The simplest short rate model is the Ho-Lee model [139], which inherits the dynamics from the Bachelier or Normal model[28] and applies it to the instantaneous short rate process, r. As such, the short rate dynamics can be expressed as:

$$dr_t = \theta_t dt + \sigma dW_t \tag{15.47}$$

As we can see from the equation, this is the familiar Bachelier model, with model parameters θ_t and σ. Additionally, because the short rate is a hidden variable, its initial value, r_0, must be specified as well. Notice that the drift is time-varying. This is an important feature of the model as it enables the Ho-Lee model to match the current yield curve. As we saw in the previous section, this ability to fit the current yield curve is one of the core features that we look for in a short rate model. In particular, the following functional form of θ_t will enable us to match the yield curve:

$$\begin{aligned} dr_t &= \theta_t dt + \sigma dW_t && (15.48) \\ \theta_t &= F(0,t) + \sigma^2 t && (15.49) \\ F(0,t) &= -\frac{\partial Z(0,t)}{\partial t} && (15.50) \end{aligned}$$

where $Z(0,t)$ is the price today of a zero-coupon bond maturing at time t. A brief summary of the model parameters is found below:

- r_0: initial value of the (unobserved) short rate
- σ: volatility of the short rate
- θ_t: time-varying drift

Under the dynamics of the Ho-Lee model, the short rate can be shown to be Gaussian with known mean and variance:

$$\begin{aligned} r(t) &= r(s) + \int_s^t \theta(u)du + \sigma W_t \\ \mathbb{E}\left[r(t)\right] &= r(s) + \int_s^t \theta(u)du \\ \mathrm{var}\left(r(t)\right) &= \sigma^2(t-s) \end{aligned}$$

The Ho-Lee model is characterized by a single volatility parameter, σ, which can

[28]See chapter 13 for more details.

be used to match certain points on the volatility surface. In addition, the time-varying θ_t process will be utilized to match the current yield curve.

In addition to knowing the distribution of the short rate, the distribution of sums and integrals of the short rate will also be of direct relevance. Even more saliently, we will need to be able to model exponential functions of the sum or integral of the short rate. To see this, consider modeling a zero-coupon bond under a short rate model. The pricing equation for a zero-coupon bond can be expressed as:

$$Z(0,T) = exp\left(-\int_0^T r(s)ds\right) \tag{15.51}$$

Consequently, to price a zero-coupon bond, or more realistically options on a zero-coupon bond, we will need to model this exponential term. It can be shown that if the distribution of the short rate is normal, then the distribution of a zero-coupon bond is log-normal. To see this, recall that the sum of normally distributed random variables is also normal. As integration is at heart a sum, we should expect integrals of the short rate to be normal as well if the short rate distribution is normal. This leaves us with something of the form $Z(0,T) = e^x$ where x is normally distributed. Therefore, we can expect $Z(0,T)$ to be log-normally distributed.

The main strengths of the Ho-Lee model are its simplicity and analytical tractability as well as its ability to match the yield curve. Additionally, as the distribution of the short rate is Gaussian, negative short rates are permitted within the model. Conversely, the single constant volatility parameter will restrict it from calibrating multiple points on the volatility surface. Further, empirical analysis tends to show that **interest rates show mean-reverting properties**. This has led other models, such as Vasicek and Hull-White to try to incorporate this mean-reversion into their SDE's for the short rate, as we will see next.

15.10.3 Vasicek

In chapter 13 we discussed Ornstein Uhlenbeck processes and saw that they were first order mean-reverting processes. The Vasicek model [254] applies this set of mean-reverting dynamics to the short rate, as the following SDE specifies:

$$dr_t = \kappa(\theta - r_t)dt + \sigma dW_t$$

When we apply an OU process to the short rate, as we do in the Vasicek model, we have four underlying parameters which are described in the following table:

- r_0: the initial value of the latent short rate
- κ: speed of mean-reversion
- θ: level of mean-reversion
- σ: constant volatility of the short rate

Notice that in this model θ is a constant rather than time-varying function. As a result, this parameterization allows for only two parameters, r_0 and θ to match the yield curve, which we will generally not be able to do.

Recall from 10.1.7 that the SDE for an OU process has an analytical, closed form solution and can be shown to follow a Gaussian distribution. As we saw in the previous section, this means that the distribution of integrals of the short rate will also have a Gaussian distribution, and that zero-coupon bond prices will be log-normally distributed.

The main strengths of the Vasicek model are its simplicity and incorporation of mean-reverting dynamics. Mean-reverting dynamics are something that we observe empirically in rates, as evidenced by the traditional hump we see in yield curves, and are therefore an attractive feature. Like the Ho-Lee model, the Vasicek model permits negative rates due to the normal distribution of the short rate. This used to be viewed as a weakness of the model, however, recent occurrences of negative rates in many economies has caused this to be rebranded as a strength. On the other hand, the Vasicek model cannot fit the current yield curve because of the constant level of mean-reversion, θ, which is a major drawback. Additionally, the Vasicek model has only a single volatility parameter, σ, and as a result will also struggle to fit much of the at-the-money volatility surface. In the coding repository, we provide the reader an example of simulating from the Vasicek model.

15.10.4 Cox Ingersol Ross

As detailed in chapter 13, the **CIR Model**, like the Vasicek model, assumes that the short rate follows a mean reverting process. Unlike the Vasicek model, however, there is a square root in the volatility term of the SDE. The dynamics of the CIR model for the short rate can be written as:

$$dr_t = \kappa(\theta - r_t)dt + \sigma\sqrt{r_t}dW_t \tag{15.52}$$

We can see that the CIR model is very similar to the Vasicek model with the exception of the $\sqrt{r_t}$ in the dW_t term. More generally, the dW_t term follows a CEV model with $\beta = 0.5$. Importantly, the CIR model has the same set of model parameters and interpretation as described above, with a subtle difference in the interpretation of the σ parameter caused by the $\sqrt{r_t}$ component. Also notice that the level of mean-reversion parameter, θ is a constant as in the Vasicek model. This means that in this model we will not be able to match the entire yield curve.

The presence of the square root in the volatility term prevents the short rate process from going negative, which is a distinguishing feature of the model. This square root term also increases the computational complexity of the model and means that the underlying distribution of the short rate will no longer be normal. This leads to more involved pricing formulas both for the short rate, and also for the integrals and exponential terms we tend to encounter.

One desirable feature of the CIR model is its incorporation of a mean-reverting short rate process. At one point, another selling point for the CIR model was that it prevented the short rate process from becoming negative, however, as negative rates have become more common this has evolved and is now perceived as a weakness. Additionally the inability to fit the current yield curve is another significant drawback of the model. Finally, in spite of the added complexity caused by the square root, the CIR model still won't capture much of the volatility.

15.10.5 Hull-White

The Hull-White model [139] [50] inherits the mean-reverting nature of the Vasicek model (and its underlying OU process). In the Hull-White model, however, instead of a single long term equilibrium rate θ, we use a time varying equilibrium rate, θ_t. This turns out to be a significant feature as it enables us to recover the current yield curve in the model, which was seen as a considerable weakness in the Vasicek model.

The dynamics for the Hull-White short rate model are:

$$dr_t = \kappa(\frac{\theta_t}{\kappa} - r_t)dt + \sigma dW_t$$

As in the Vasicek model, r_0, κ and σ are still (constant) model parameters. Additionally, we now have a function, θ_t to parameterize. This freedom is used to fit the Hull-White model to the entire currently observed yield curve. In particular, we can express the values of θ_t required to match the yield curve as a function of discount factors, or zero-coupon bond prices, as shown [139]:

$$\begin{aligned} \theta_t &= \frac{\partial F(0,t)}{\partial t} + \kappa F(0,t) + \frac{\sigma^2}{2\kappa}\left(1 - e^{-2\kappa t}\right) \\ F(0,t) &= -\frac{\partial Z(0,t)}{\partial t} \end{aligned}$$

Intuitively, we can think of the time-varying drift, θ_t as a function of the current forward rate, $F(0,t)$ and its change.

Aside from this improved ability to match the curve, the Hull-White model inherits most of its properties from the Vasicek model. This includes a Gaussian distribution for the short rate and the possibility of negative rates. This leads the model to be analytically tractable with closed from solutions to zero-coupon bond prices. Inclusion of negative rates, ability to match the yield curve and incorporation of mean-reverting dynamics are all seen as attractive properties of the Hull-White model. Overall, this makes the Hull-White model an improvement over the other short rate models that we have seen. However, due to the **single volatility parameter**, σ, it will still have trouble matching much of the at-the-money option surface. Additionally, the Hull-White model, like other one-factor short rate models, is known to lead to **unrealistic correlations between traded rates**. This can be overcome by creating multi-factor short rate models, which we explore later in this section.

15.10.6 Multi-Factor Short Rate Models

Among the biggest challenges associated with short rate models is modeling correlation of different traded rates, such as the correlation between movements of two different forward or swap rates. In a single factor model, the short rate is driven by a single source of randomness, which is then used to model all subsequent underlying rates. This leads to an unrealistically rigid correlation structure between rates.

To combat these overly restrictive correlation assumptions, we can introduce additional factors, or sources of randomness into our short rate model. Multi-factor models are those where we incorporate multiple sources of noise. These models provide a more flexible correlation structure, and are conceptually the same as the one-factor models that we discussed earlier in the chapter. Multi-factor models, however, tend to lead to longer, more tedious formulas and pricing equations.

Principal Component Analysis of yield curve movements shows that 90% of the variance in the curve can be explained by the first 2 or 3 factors. This implies that a parsimonious, two or three factor short rate model might be optimal to model the correlation structure that we observe. This will enable us to balance the trade-off between realistic comovement of different rates and computational complexity.

Multi-factor short rate models can rely on any underlying dynamics that we choose, such as those presented in the Ho-Lee, Vasicek, CIR and Hull-White sections. Having said that, an important feature in model selection will be its tractability. Models that lead to simple distributions of the short rate and market traded concepts should naturally be preferred to more complex models that may not have efficient pricing applications. This model selection problem is also a trade-off, and is similar to those faced in other calibration and pricing exercises in chapters 14 and 15.

15.10.7 Two Factor Gaussian Short Rate Model

The simplest multi-factor model is a two-factor model that assumes each factor has Gaussian dynamics. This model is defined by:

$$\begin{aligned} dr_t &= \theta_t dt + dx_t + dy_t \\ dx_t &= \sigma_x dW_t \\ dy_t &= \sigma_y dZ_t \\ Cov(dW_t, dZ_t) &= \rho \end{aligned}$$

As both of the underlying factors dx_t and dy_t are normally distributed, the short rate process will also follow a Gaussian distribution. Notice that the two sources of randomness in this model may be correlated. This correlation parameter, ρ can be calibrated to help model the correlation structure between different rates. Additionally, as θ_t is time-varying, we can expect the model to match the current yield curve. Importantly, a two-factor Gaussian short rate models contains two distinct volatility parameters, σ_x and σ_y. The presence of multiple volatility parameters in our model can be used to help match multiple points on the at-the-money volatility surface. One drawback of this two-factor model is that it will not incorporate mean-reversion in the short rate process. Conversely, its Gaussian nature will lead to computational simplicity.

15.10.8 Two Factor Hull-White Model

If we want to incorporate mean-reversion of rates into a multi-factor model, we can instead use a **Two Factor Hull-White Model**:

$$
\begin{aligned}
r_t &= \theta_t + x_t + y_t \\
dx_t &= -\kappa_x x_t dt + \sigma_x dW_t \\
dy_t &= -\kappa_y y_t dt + \sigma_y dZ_t \\
Cov(dW_t, dZ_t) &= \rho
\end{aligned}
$$

As with the single-factor Hull-White model, θ_t is time-varying and can be used to match the yield curve. Additionally, the underlying components x_t and y_t are both mean-reverting around zero, resulting in a mean-reverting short rate process. As in the previous two factor model, we also have a correlation parameter, ρ, which can be used to help model the correlation between different rates. Further, we now have multiple speeds and levels of mean-reversion that can be calibrated, one for each factor.

The distribution of the short rate in a two-factor Hull-White model will be Gaussian, just as in the one factor case. This will make calculations in the model more efficient and analytically tractable. Similarly, we could choose to add additional factors to our Hull-White model, and as before the underlying concepts would be the same. Ultimately the choice of factors should be made in a manner that manages the trade-off between realistic correlations and parsimony.

15.10.9 Short Rate Models: Conclusions

In this section we detailed the first approach to modeling an entire term structure. We did this by relying on a convenient mathematical abstraction to define the stochastic process underlying the yield curve: an instantaneous short rate. Within the context of the models that we discussed, we saw that finding a short rate model that fit the current yield curve, allowed for mean-reversion of rates, permitted negative rates and was analytically tractable was a challenge. Further, we saw that even if we were able to accomplish these tasks, fitting an at-the-money volatility surface was a far greater challenge. We also saw that single-factor short rate models lead to unrealistically rigid correlation structures between rates.

Multi-factor Short Rate models, on the other hand, enabled us to embed more realistic intra-rate correlation structure into our models. This is a significant improvement, and in practice multi-factor models are preferred to single factor models. In practice we can use PCA to help us determine how many factors we might need in our models. There is a trade-off, however, as the more factors we include the more complex the pricing formulas become. Additionally, adding factors doesn't help us match the volatility skew, for that, we would need to replace the SDE's that govern the short rate with those that incorporate skew. To accomplish that task, we need a different, more advanced set of tools, such as stochastic volatility models.

15.11 MODELING THE TERM STRUCTURE: FORWARD RATE MODELS

15.11.1 Libor Market Models: Introduction

In a forward rate or SOFR market model (also known as a Libor Market Model or LMM) [50] [217], we choose to model forward rates in the form of (3-month) traded SOFR contracts[29]. Modeling the underlying forward rates as stochastic variables is an important differentiating feature of such market models. Whereas in short rate models we chose to model a mathematical abstraction, albeit a convenient one, in this case we choose to model traded rates directly. We then look to model this collection of traded forward rates as a set of correlated stochastic processes. This provides a unified representation of the yield curve with a robust correlation structure. LMM models have become increasingly popular as the appeal of modeling traded rates has led to widespread model adoption. Additionally, solutions to LMM for important SDEs, such as Black-Scholes and SABR has lowered the overhead for implementing a realistic model considerably.

In an LMM implementation, forward rates, such as SOFR rates, are each governed by distinct but correlated stochastic processes. To do this, we generally split up the curve into uniform length, non-overlapping chunks. A term structure must be at least 30 years in maturity, as various 30y swaptions are commonly traded. Each underlying forward rate that we split the curve into would then be its own stochastic variable. For example, if we are modeling the yield curve out to 30 years, we might break the curve into 3 month or 1 year increments. This leads to a problem with huge dimensions. In our example, we would have 120 (or 30) potentially correlated stochastic variables.

When using an LMM framework, each forward rate has its own dynamics. As a result, the model is defined by the following system of SDEs:

$$dF_t^i = \mu_i(t)dt + \sigma_i(F_t^i, t)dW_i(t) \tag{15.53}$$

Here, $\sigma_i(F_t, t)$ is a volatility function for the ith forward rate and will depend on the chosen underlying dynamics. For example, if a Log-Normal model is chosen then $C_i(F_t, t) = \sigma F_t^i$. Importantly, a separate process then governs each forward rate F_t^i and will have its own set of parameters. Additionally, as all forward rates on the yield curve are interconnected, we must also model the **correlations between all the forward rates**, which can be done by controlling the correlation between the $dW_i(t)$ terms.

Lastly, there is a noteworthy subtlety in LMMs related to numeraires. Recall from earlier in the chapter that a forward rate is a martingale under the t-Forward measure. This means that we can conveniently choose our numeraire such that a single forward rate is a martingale. As these forward rates are our underlying stochastic processes, clearly formulating them in a way that they are martingales will be a desirable feature. The subtlety of numeraires in an LMM framework is that only one forward rate can be a martingale at a time because we can only work in a single T-Forward measure. We cannot work in multiple T-Forward measures simultaneously, which would be required for the entire set of forward rates to be martingales. This creates modeling

[29]Or previously Libor rates.

challenges and introduces drift terms into the forward rates that are not martingales. In practice this means we will choose a measure such that one forward rate is a martingale and then will employ Girsanov's theorem to calculate the drift terms of the additional forward rates.

The LMM setup is generic enough to support various underlying SDEs in (15.53). Having said that, we must choose a set of dynamics such that the problem is tractable as a unified model. In the following sections, we provide an overview for how a Libor Market Model, or forward rate model, would be implemented in the case of log-normal and SABR dynamics, respectively. Many of the details and practicalities of LMM implementations are beyond the scope of this book, instead the goal here is to provide a foundation and some intuition. Interested readers should consult [217] for more details.

15.11.2 Log-Normal Libor Market Model

Specifying log-normal dynamics leads to the following dynamics in LMM:

$$dF_t^i = \mu_i(t)dt + \sigma_t^i F_t^i dW_i(t) \tag{15.54}$$

where each forward rate would have a separate volatility, σ_t^i. Therefore, when working with a log-normal LMM a set of σ_i's need to be specified, one for each forward rate. Additionally, the correlations between each forward rate will also need to be specified:

$$\rho_{i,j} = Cov(dW^i(t), dW^j(t)) \tag{15.55}$$

In practice, we will need to fit these parameters to a set of market data, as we saw in the calibration section of 14.

This choice of dynamics has many attractive properties, most notably analytical tractability. Additionally, use of a log-normal LMM will lead to Black's formula in (15.34) for pricing caplets & floorlets. Conversely, although a log-normal implementation will be useful for modeling the at-the-money volatility surface, it will not enable us to recover realistic market skews which will prevent us from fitting the entire volatility cube. To do this, a stochastic volatility model, such as SABR, can be used.

15.11.3 SABR Libor Market Model

The market standard for modeling the interest rate volatility cube is a Libor Market Model or SOFR Market Model where each underlying forward rate is governed by a SABR model. In comparison to the log-normal LMM mentioned in the previous section, the advantage of SABR underlying dynamics is in their ability to match volatility smiles for the underlying forward rates. In order to use SABR in the context of a unified Libor [30] Market Model, we not only need to be able to solve one-dimensional SABR problems, we also need to be able to connect the SABR processes for the different forwards.

[30] or SOFR.

Fortunately, Rebonato and others [217] have created an enhanced SABR model that does just this: it connects the individual SABR models of the underlying forwards into a multi-dimensional model for all underlying rates. The SABR/LMM model is defined by the following SDE's:

$$\begin{aligned} dF_t^i &= \sigma_t F_t^{i\beta^i} dW_t^i \\ d\sigma_t &= \alpha^i \sigma \, dZ_t^i \\ Cov(dW_t^i, dZ_t^i) &= \rho^i \, dt \\ Cov(dW_t^i, dW_t^j) &= \Omega_{i,j} \\ Cov(dZ_t^i, dZ_t^j) &= \Theta_{i,j} \end{aligned}$$

We know that the classic SABR model is defined by the model parameters α, β, ρ and σ_0. In the case of a SABR/LMM we have a much richer parameterization. In particular, we have the classic SABR parameters for each forward rate, α^i, β^i, ρ^i and σ_0^i. We also have a correlation structure of different forward rates and different volatilities. Lastly, we have the cross correlation between forward rates and volatilities. In some cases this will consist of a traditional SABR correlation parameter (ρ), but not in all cases.

To make the problem more tractable, we generally specify a set of parsimonious functions that reduce the number of required parameters. For example, as intuitively it makes sense for correlations to be highest for rates closest together on the curve, we might specify a one or two parameter decay function that defines the entire correlation matrix. Lastly, recall that each forward rate will potentially have a different drift term depending on the T-Forward measure we are working in.

15.11.4 Valuation of Swaptions in an LMM Framework

Because the stochastic variables in our LMM framework are the same underlying forward rates that caps/floors are based, pricing of caps & floors in an LMM will be seamless. To see this, recall that the pricing equation for caps can be written as:

$$C(K) = \sum_i \{\delta_i P(0, T_i + \delta)\mathbb{E}\left[(F_{T_i} - K)^+\right]\} \tag{15.56}$$

Pricing of caps requires modeling the underlying F_T's as stochastic variables, which is exactly what we do in LMM. Further, because caps are a basket of options, we can price each caplet separately. Importantly this means we can work in a different numeraire for each caplet and do not have to worry about drift terms when pricing caps via LMM.

Swaptions, however, turn out to be more challenging as the annuity numeraire is not accessible to us given the way we defined the dynamics. Instead we will have to price swaptions in a single T-Forward measure.[31] Recall that the swap rate can be

[31] So-called swap market models, conversely, innately price swaptions with ease but make cap pricing a more challenging problem.

written as:

$$S(t,T) = \frac{\sum_{i=1}^{N} \delta_{t_i} L_{t_i} D(t,t_i)}{\sum_{i=1}^{N} \delta_{t_i} D(t,t_i)} \tag{15.57}$$

We can see that a swap rate is essentially an option on a basket of forward rates, such as SOFR rates, but the discount factor terms are also stochastic and will move with the underlying stochastic forward rate processes. As a result, swaptions are considered exotic options in the context of a LMM. To price them we can either resort to simulation methods or rely on an approximation. A simulation approach poses significant challenges in the context of calibration to market data and is therefore often not efficient enough. Instead, we can freeze the discount factor terms and try to create a more efficient approximation. If we freeze the discount factor terms the equation above becomes:

$$S(t,T) = \sum_{i=1}^{N} \omega_{t_i} L_{t_i} \tag{15.58}$$

where $\omega_{t_i} = \frac{\delta_{t_i} D(t,t_i)}{\sum_{i=1}^{N} \delta_{t_i} D(t,t_i)}$ is set to an initial value using the current yield curve and then assumed to be constant.

Here we can see that under this assumption the swap rate is just a weighted combination of forward rates, which we have already modeled as part of our LMM. To proceed, we can then create an approximation for what the dynamics of this weighted combination of forward rates would be. Rebonato [217] show that the dynamics will still follow a SABR model and provide estimates for β, α, ρ and σ_0 for each swap rate.

15.12 EXOTIC OPTIONS

15.12.1 Spread Options

Spread options are options on the difference between two rates. These types of options are a key component of the exotic options market and are traded exclusively over-the-counter. Spread options provide investors with a direct mechanism to bet on the slope of the yield curve, which is of direct interest to participants in terms of hedging and also expressing their macroeconomic views. Call spread options are called caps, whereas put spread options are referred to as floors. As an example, a common spread option might be a 1y 2s30s option. This would indicate that the option expires in one-year, and the payout would be based on the 30 year swap rate less the two year swap rate.

As spread options are options on the difference of two random variables, their valuation will clearly depend on the correlation between these rates. To see the intuition behind pricing spread options, recall that for two random variables X and Y, the volatility of their difference, $X - Y$, can be written as:

$$\sigma_{x-y} = \sigma_x + \sigma_y - 2\rho\sigma_x\sigma_y \tag{15.59}$$

This relationship is exact under certain assumptions, but does not account for the volatility smile (among other things). Regardless of whether or not the formula is an accurate approximation, which will depend on market behavior, it does provide intuition on how spread options operate. In particular, we can see that the higher the correlation between the underlying swap rates, the lower the volatility and consequently the cheaper the spread option. So, as a result, spread options are naturally correlation products, and all else equal a buyer of a spread option will benefit from more dispersion, and lower correlation. Spread options can either be priced via a term-structure model, such as SABR/LMM, or can be priced using a bi-variate copula.

15.12.2 Bermudan Swaptions

Bermudan swaptions (Berms) are one of the most complicated interest rate derivatives and present a huge modeling challenge for quants. They tend to be illiquid instruments that are traded over-the-counter. In many ways, Bermudan swaptions are an outstanding issue in rate modeling today, where different models may lead to different prices. As such, there is no single market standard pricing model. Berms are American style options that allow exercise on a pre-determined set of dates. Further complication these instruments is the fact that at each exercise date the underlying swap that is exercised into, is different.

Properly modeling the underlying exercise decision, in addition to modeling the multiple possible underlyings, is one of the harder problems quants face in rates. To value a Bermudan swaption accurately, we need a model that calibrates the entire volatility cube and understands the connection between the different potential underlyings. This can be done via a short rate model or alternatively via a Libor Market Model. Even still, most models have difficulty matching market traded Bermudan swaption prices. Model prices for Berms tend to be persistently higher than market prices, implying more optionality than is priced into the market. As a result, a common technique employed by market participants to match Berm prices is to remove certain exercise decisions, therefore reducing the optionality from the model. While this is conceptually unsatisfying, it does enable participants to match market prices for Berms.

A common Berm structure is a 5 yr no-call 30 yr Berm, which means that the first potential exercise opportunity is 5 years out, at which point additional exercise decisions would occur periodically, for example on a semi-annual or annual basis. Importantly, the length of the swap that is exercised into in this example is 30 years less the time to exercise. This means that the longer that the investors waits to exercise their Berm, the shorter the tenor of the swap that they will receive. This creates an incentive to exercise the Berm sooner, otherwise the annuity of the swap will decrease making the swap less valuable. Conversely, the residual optionality in the Berm will incentivize investors to delay exercise, creating an interesting trade-off.

As valuing Bermudan Swaptions requires valuation of an American style exercise decision, it is most natural to use a tree or PDE approach. This is because, traditionally, we calculate an optimal exercise policy for an American option involves

backward induction. In this case we work backward and compare the **continuation value** of the option to the **exercise value**. However, given the dimensionality of an interest rate curve, this type of approach is not tractable and we must almost always resort to simulation. For example, in a SABR/LMM framework, our only choice for valuation of an exotic option, will be via simulation. The forward looking nature of Monte Carlo simulations, combined with the backward looking nature of optimal exercise decisions, creates an additional modeling challenge for Berms. Fortunately, Longstaff and Schwartz created an accurate approximation to the exercise decision that can be incorporated into a simulation algorithm [168].

15.13 INVESTMENT PERSPECTIVE: TRADED STRUCTURES

15.13.1 Hedging Interest Rate Risk in Practice

Perhaps the most common reason for participation in rates markets is to hedge interest rate risk. Many types of investors may have interest rate exposure. Mortgage issuers, and holders of mortgage backed securities, for example, may have embedded interest rate risk because of the prepayment risk inherent in mortgages. Pension funds also have significant amounts of interest rate risk as they tend to have nominal debt obligations that are fixed, meaning that changes in interest rates can modify the present value of these obligations. Banks may have interest rate risk that is related to the slope, rather than the level of the curve as they tend to lend over longer horizons and borrow over shorter periods. When the curve is upward sloping, they are then able to harvest the spread between longer and shorter horizon rates. Corporations who issue debt instruments also have ensuing interest rate risk.

For a given investor who wants to hedge this interest rate risk, the first step would be to identify their exposure, which is usually done by calculating a portfolio's duration or DV01[32]. The next step would be to identify a hedging instrument, which would be a bond, swap, swaption or other derivative. A hedge ratio for the portfolio with respect to the underlying hedge instrument can then be calculated using the following formula:

$$h = \frac{\Delta P}{\Delta B} \tag{15.60}$$

where h is the hedge ratio or units of the bond or other hedging instrument that should be bought or sold to hedge the interest rate risk in the portfolio. Further, ΔP and ΔB are the changes to the value of the portfolio and hedging instrument, respectively.

Most importantly, if we augment our existing portfolio with h units of the hedge instruments, then we know that the augmented portfolio will have zero duration, or DV01. Said differently, this augmented portfolio will have no expected change in value[33], making it agnostic to interest rate movements.

[32]DV01 is defined as the dollar value change in a portfolio for a one basis point change in rates.
[33]To first order.

Choice of the underlying hedging instrument can provide different profiles for the hedger. If a linear instrument, such as a bond or swap, is used, then the hedge will be unconditional. If swaptions are used, however we can implement a conditional hedge. For example, if a pension fund is exposed to an increase in rates, then a payer swaption would neutralize this risk should rates rise while allowing them to potentially benefit if rates decrease. The trade-off for this implementation, however, would be the upfront premium of the swaption.

It should also be noted that here we are thinking in terms of moving rates, or yields, up or down in parallel. In practice different portfolios, and hedging instruments may have very different exposures to different parts of the curve, and to changes in curve slope. In practice this means that in addition to parallel shifts hedging interest rate shifts requires analysis of sensitivities to different points on the curve, and different types of shocks. More information on risk management and hedging interest rate risk is presented in chapter 26.

15.13.2 Harvesting Carry in Rates Markets: Swaps

A second natural lens that investors may choose to apply to interest rate markets is that of risk premia. Risk premia is a deep and foundational concept in quant modeling that underpins many investment decisions. It is also a concept that permeates all asset classes and can be an important lens for buy-side investors. In chapter 27 we discuss risk premia based strategies in different asset classes in more detail. In this section, however, we explore a specific case study of a carry risk premia instrumented via vanilla swaps.

Carry is defined as the return of a security in the absence of price appreciation or depreciation. It can be thought of as the return of an investment if nothing happens except the passage of time. More formally, Pedersen [214] disaggregated the return of any security into three components: carry (θ_t), expected price appreciation ($\mathbb{E}\left[\delta P_t\right]$) and a random component (X_t)):

$$r_t = \theta_t + \mathbb{E}\left[\delta P_t\right] + X_t \tag{15.61}$$

A significant amount of evidence exists that shows that the carry component, θ is a well-rewarded risk premium [142] [214].

In the context of an interest rate swap carry can be thought of as having two components:

- The present value of any income or intermediate cashflows and
- The change in present value of the swap when revaluing it after the passage of time.

As an example, if we are calculating the one-year carry of a 5y swap, the first component would consist of the cashflows for the fixed and floating legs over the first year. The second component would involve calculating the value of the remaining 4y swap. An ambiguity with this component is the yield curve or par swap rate we would use when valuing the remaining 4y swap. Specifically, we must decide whether we want to assume that the curve and swap rate roll down the curve to spot, meaning the par swap rate for a 4y swap in one-year would equal today's 4y swap rate. This part of the carry calculation is referred to as roll down. Alternatively, we could assume that forwards are realized, and instead use today's 1y forward 4y par swap rate as the par swap rate in our re-valuation.

In order to turn this idea into an actionable portfolio, an investor would need to measure the carry of various swaps, including both components, and construct a portfolio with long positions in swaps with high amounts of carry and short positions in swaps with low amounts of carry.

15.13.3 Swaps vs. Treasuries Basis Trade

Swaps and treasuries provide similar exposure, therefore one might naturally think they would have the same yields. However there is an important distinction between them: treasuries are cash based instruments whereas swaps are synthetic derivatives. Remember that treasuries have principled re-paid at maturity and as a result require an upfront payment when the treasury is purchased. Swaps, conversely, do not have this principal payment at maturity and instead are linked to a reference notional, meaning that at trade initiation no cash flow is required to change hands[34]. The fact that treasuries require upfront capital whereas swaps do not leads them to trade at different yields due to financing requirements.

The spread between these two yields is referred to as a **swap spread**. Generally speaking, the level of the swap spread is a function of the following:

- Value of Collateral
- Liquidity
- Idiosyncratic Supply and Demand Dynamics for Swaps and Treasuries

Swap spreads, and other such bases, tend to have a risk-on feel to them, as collateral is most valuable in times of stress when the value of holding collateral tends to be the highest. Many investors, such as hedge funds try to harvest this spread.

As an example, suppose 30-year swaps and treasuries were trading at the levels in the following table:

Market Quantity	Swap	Treasury	Swap Spread
30y Yield	2.05	2.35	−0.30

An astute investor might notice the 30 basis spread between swaps and treasury and construct a portfolio that is long treasuries[35] and pays fixed on a vanilla swap. If financing for this trade was able to be obtained cheaply, then the investor would be able to harvest the 30 basis point spread between the two products. In order to make this trade more attractive in absolute terms, funds employ leverage in order

[34]Recall that swaps are structured to have zero present value at their onset.

[35]Where the investor may borrow to obtain the capital to purchase the desired treasuries.

to enhance the 30 basis point return. The presence of leverage, however, introduces certain complications and the potential for margin calls to stop an investor out of the trade should the spread widen. In the next chapter, we will see a similar basis in the credit markets between corporate bonds which are cash based instruments, and credit default swaps, which are synthetic derivatives.

15.13.4 Conditional Flattener/Steepeners

A very common trade is one in which investors bet on the slope of the yield curve increasing, which is referred to as a steepener, or decreasing, which is known as a flattener. In the sections on exotic options we saw that one precise way to instrument such a view would be via a spread option. This provides direct exposure through a payoff directly linked to the slope of the yield curve. In some cases, however, the investor may have confidence that the curve will steepen or flatten if rates moves in a certain direction. For that investor, a more precise trade would be one in which they bet on the slope of the curve conditional on a certain rate level.

To see how an investor might do this, remember that a 2s10s spread option had a payoff linked to the difference between the 10-year swap rate and the 2-year swap rate. In the absence of exotic options, another naive approach to try to do this would be via outright swaps. An investor would pay fixed on a 10-year swap and receive fixed on a 2-year swap. This would provide a steepening profile as the investor would benefit from rate increases in the 10-year swap, and rate decreases in the two-year swap. As a result, this naive swap portfolio is another way to express a view on the steepness of the curve without trading options[36].

Instrumenting this trade via options, however, gives us far more control of the underlying payoff. For example, if we wanted a payoff conditional on higher rates, we could consider the following trade legs:

Long/Short	Instrument
Short	1y2y Payer Swaption with strike K
Long	1y10y Payer Swaption with strike K

As we can see, if the rate is below K at expiry, then both options will expire worthless. If the trades are sized such that they required equal premium up front, then there would also be zero cash outlay, giving the trade its conditional nature. The trade is also still clearly exposed to the slope, as increases in 10y rate, or decreases in the 2y rate, would lead to positive P&L.

Alternatively, if we wanted a payoff profile that looked more like a straddle[37] we could substitute the short position in the payer swaption above for a long position in a receiver swaption, as shown:

In this example, we would still benefit from a curve steepening, but this trade would also be characterized by long volatility component. This leads to a trade-off

[36]Interested readers are encouraged to think about how the 10-year and 2-year legs of this trade should be sized, for example to hedge duration, or to be of equal notional.

[37]Refer to chapter 16 for more details.

Long/Short	Instrument
Long	1y2y Receiver Swaption with strike K
Long	1y10y Payer Swaption with strike K

relative to the conditional approach discussed earlier. This structure will require a significant upfront premium but will overlay a straddle with a steepening component. Conversely, the earlier structure had a more neutralized volatility profile and would not require a premium outlay.

15.13.5 Triangles: Swaptions vs. Mid-Curves

Another commonly traded relative value structure in interest rate markets is an interest rate triangle. Interest rate markets provide a large array of opportunities to disaggregate rates into their different components, and that is at heart what this structure does. Interest rate triangles involve trading swaptions and mid-curves. Swaptions were discussed earlier in the chapter and are options on spot starting, vanilla swaps.

Mid-curves, conversely, are options on a forward starting swap. Because they are options on a forward starting swap, there is another variable that defines a mid-curve, namely how far forward the swap starts. It is often referred to as forwardness. As a result, mid-curves defined by their expiry, forwardness & tenor. For example, a 1y2y3y mid-curve refers to a one year expiry on a three year swap which starts 2 years after the option expiry. Modeling mid-curves is fundamentally similar to modeling swaptions, as they are both options on swaps, and is generally done in the relevant annuity measure.

The relationship between swaptions and mid-curves is of particular interest to market participants as they are clearly interconnected pieces. For example, the volatility between mid-curves and swaptions with the same or similar expiries must be related as they will reference similar, correlated rates. In fact, it turns out that a portfolio of swaptions and mid-curves can be put together to form interesting correlation or dispersion trades. In the case of swaps this decomposition is generally not interesting. However, the non-linearity of swaptions and mid-curves makes this decomposition more nuanced.

The main underlying premise is that we can decompose a longer swap into a shorter swap and a mid-curve. Investors can then perform relative value of these two packages to identify trading opportunities. This distills a triangle structure to a trade that compares options on the underlying pieces to an option on the aggregate. This is a common theme in rates derivatives market and is analogous to equity markets where investors might trade index options vs. the sectors or underlying companies[38].

An example of a swaption vs. mid-curve triangle is depicted in the table below:

Notice that all three options have the same expiry date. Additionally, each leg spans the same set of underlying swap dates[39]. Consequently, this portfolio is essentially trading a basket of options vs. an option on a basket. The difference in valuation

[38]This analogous equity structure is detailed in chapter 22.

[39]Of a three-year period in total.

Leg	Description	Instrument
1	Longer Swaption	1y3y Swaption
2	Shorter Swaption	1y2y Swaption
2	Mid-Curve	1y2y1y Mid-Curve

between these two quantities will depend on correlation, and investors might use this structure as a vehicle to trade on the premium between implied and realized correlation, or to express their views on expected future correlations.

15.13.6 Wedges: Caps vs. Swaptions

Trading a cap vs. a swaption is another commonly traded rates correlation structure. As we will see, it is another opportunity to disaggregate a rate into its underlying components, and is another natural example of an option on a basket vs. a basket of options.

To understand the connection between the structure, let's review their pricing equations, beginning with a cap:

$$C(K) = \sum_i \{\delta_i P(0, T_i + \delta_i)\mathbb{E}\left[(F_{T_i} - K)^+\right]\} \tag{15.62}$$

Similarly, the pricing equation for a swaption can be written as:

$$P = A_0(t, T)\mathbb{E}\left[(S - K)^+\right] \tag{15.63}$$

$$\widehat{S(t,T)} = \frac{\sum_{i=1}^N \delta_{t_i} L_{t_i} D(0, t_i)}{\sum_{i=1}^N \delta_{t_i} D(0, t_i)} \tag{15.64}$$

If we treat the discount factor terms in the swap rate, $\widehat{S(t,T)}$ in (15.63) as constants[40], then we can see that a swap rate is fundamentally a weighted average of the underlying forward rates in the swap. The reader may recall we employed a similar argument to value swaptions in the context of a Libor Market Model. Therefore, a swaption can be intuitively thought of as an option on a basket of Forward rates. Likewise, a cap can be thought of as a basket of options linked to a set of forward rates. Consequently, we can use these structures with matching dates to execute a dispersion or correlation trade.

In practice this is a commonly traded structure as it give investors a precise way to bet on forward rate correlations and also harvest any dislocations between the two markets. Specifically, consider the trade details in the following table:

Long/Short	Instrument
Long	1y3y Swaption
Short	1y3y Cap

[40]This is an approximation but is not unreasonable given that changes in the numerator tend to be offset by changes in the denominator.

An investor who engages in this trade will benefit from high realized correlations between forward rates. Conversely, an investor who sells the swaption and buys the underlying cap benefits from increased dispersion and lower correlation in forward rates.

15.13.7 Berm vs. Most Expensive European

Another interesting relationship and ensuing traded structure in rates is a Bermudan vs vanilla Swaption. As mentioned, vanilla swaptions are European instruments whereas Bermudans can be exercised at several potential exercise dates. The difference in valuation between the two isolates the additional exercise premium that an investor receives for the additional optionality in the exercise choice. As a result, investors may choose to trade one against the other, as in the example below:

Long/Short	Instrument
Long	5y no call 30 Bermudan with strike K
Short	5y25y Swaption with strike K

Here, a 5 yr no call 30 year Berm refers to a Berm whose first exercise date is in 5 years and the swap that is exercised into is 30 years less the time to exercise[41]. Let's also assume that exercise is possible on an annual bases, i.e. after 5 years, 6 years, and so on. It should be emphasized that the first exercise date of the Berm has been chosen to match the expiry of the sold vanilla swaption. This structure will require a cash outlay, as a Berm is by definition more expensive than its corresponding European because of the added optionality.

If this Berm is purchased and a corresponding European swaption is sold, then the resulting exposure has interesting properties and it turns out this structure is an effective way of isolating the relative value of the optionality embedded in the exercise decision. To see this, consider the position at the expiry of the swaption that was sold. If that option expires deeply in-the-money, then the Bermudan can be exercised to cover this obligation. This would be the worst case scenario for the trade, however, as the investor would lose his or her initial premium. Conversely, if the first swaption expires out-of-the-money, then the swaption that we sold will expire worthless. Our Berm, however, will still have value as it has exercise dates further out in the future. A natural next step in our investment strategy would be to then sell another vanilla option corresponding to the next exercise date. We can then repeat this exercise until we are forced to exercise to cover our sold option obligation.

It should be noted that the precise remaining value of the Berm in this scenario will be highly dependent on where the par swap rate is relative to the strike of the option. If the Berm has become deeply out-of-the-money, then the value of future exercise dates is likely to be negligible. As a result, this structure leads to a payoff that is dependent on the par swap rate at expiry. Said differently, there is a short gamma profile[42] of this structure as constructed. To see this, consider rate levels at

[41]Making it a 25 year swap if exercised at the first chance.

[42]See chapter 16 for more information on gamma profiles of option structures.

high and low extremes. In each case, our trade becomes close to worthless, resulting in us simply losing our premium.

The argument presented here for the first European can be easily generalized to use the most expensive European as the sold vanilla option. Intuitively, this is because we can choose our Berm exercise policy, meaning we can choose to exercise against the most expensive Berm instead of the first.

As we have detailed in this section, this structure is an effective way to implement a view of the underlying embedded exercise premium. If we feel it is too low, we can engage in a trade that is synonymous with the trade presented here. If we feel the premium is too high, however, we might decide to sell the Bermudan and buy a set of corresponding Europeans.

15.14 CASE STUDY: INTRODUCTION OF NEGATIVE RATES

Perhaps the most noteworthy development in interest rate markets over the last decade has been the proliferation of negative interest rates, where borrowers are charged to store their deposits at a bank instead of receiving interest. Prior to this, 0 was seen as a natural lower bound for interest rates. A side effect of this was that, when interest rates were close to zero, there was a perception that central banks ability to stimulate the economy, which is traditionally done through cuts to benchmark interest rates, was limited. As rates began to decline secularly around the globe, this created an increasing challenge and motivated central banks to invent other means of stimulating economies. Some such economies, such as the European Union and the Swiss National Bank decided to obviate this problem by simply setting benchmark rates to negative values. The European Central Bank (ECB), which had interest rates at the zero lower bound for years, for example, first did so in 2014. Deposit rates for the ECB over time are depicted in the following chart:

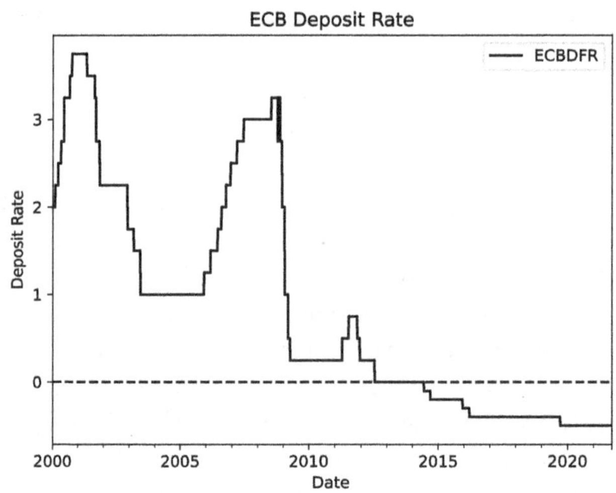

From a financial modeling perspective, negative rates are at odds with basic opportunity cost concepts, i.e. that a dollar today should be worth more than a dollar in the future. Said differently, negative rates imply that borrowers can be paid to take on debt. Perhaps even more saliently for quants, negative rates are not permitted in many standard options pricing models, such as the Black-Scholes model. As a result, to the extent that these models were used for pricing and hedging of interest rate options, the incorporation of negative rates could have significant consequences. For example, consider a bank who relied on a model that assumed that interest rates could not go negative. Theoretically, this means that, in the absence of a bid-ask spread, the bank would charge 0 for a digital put option struck at 0%. This could then leave banks exposed, and often not appropriately hedged for a move toward negative rates, as their underlying model didn't properly account for this probability.

The Black-Scholes model is not the only model to suffer from such a limitation, and in fact, it used to be considered a desirable feature in interest rate models that they did not allow negative rates. The SABR model is another example of this, where in general negative rates are not allowed unless certain parameter values are used. This was perceived as a model strength until rates became negative, at which point the model needed to be adjusted, and the shifted SABR model was created. The ramifications of negative rates go far beyond the world of options trading and interest rate markets. In fact, in many ways the incorporation of negative rates has been an experiment with implications on the entire global economy that we still don't fully understand, given its recent adoption.

15.15 CASE STUDY: POST-COVID INFLATION SHOCK

Stunningly, on the heels of negative rates, one of the most consequential themes in markets in recent years has been the potential of a change in inflationary regime in the aftermath of the Covid pandemic. This development has permeated not just rates markets but also had a huge impact on the broader financial landscape. This picture of increased inflation and inflation volatility was in sharp contrast to what had been observed in the previous decade post the global financial crisis and leading up to Covid. During this period of secularly low inflation, interest rates remained at or near all-time lows. As we explored in the last section, in some cases these rates even went negative. At the onset of Covid, however, this picture began to change as distortions to the supply chain, stimulus and other macroeconomic developments led to a generational spike in inflation and inflation volatility. As we show in the following chart, this paradigm shift could be seen in changes in consumer prices [43]:

[43] As measured via CPI.

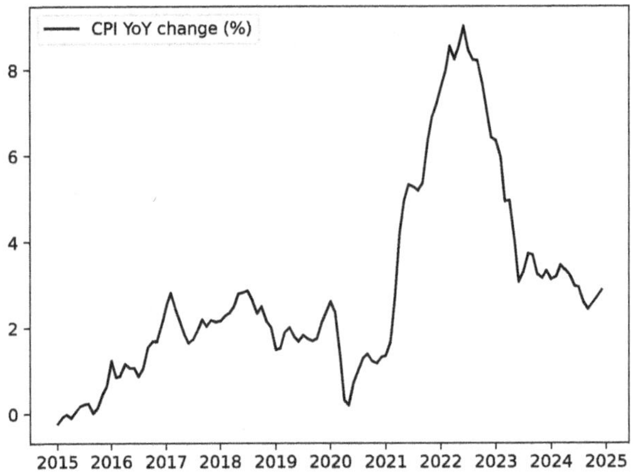

In the chart we can see inflation prior to 2022 was well contained below 2% but that it soon spiked reaching 8% at its peak.

A consequence to this prior period of lower inflation was that interest rates were able to be held anomalously low. Another side effect of this was that because rates were pinned at these lower levels that the magnitude of the changes in rates and the corresponding the volatility of rates were relatively muted. As the spike in inflation transpired, however, the Fed was forced to react by raising rates. These dramatic increases in rates presented themselves not only at the front end of the curve, determined by the Fed, but also to longer term rates. In the following chart, we highlight this by plotting a time series of the Fed Funds target rate as well as the ten-year yield on US treasuries:

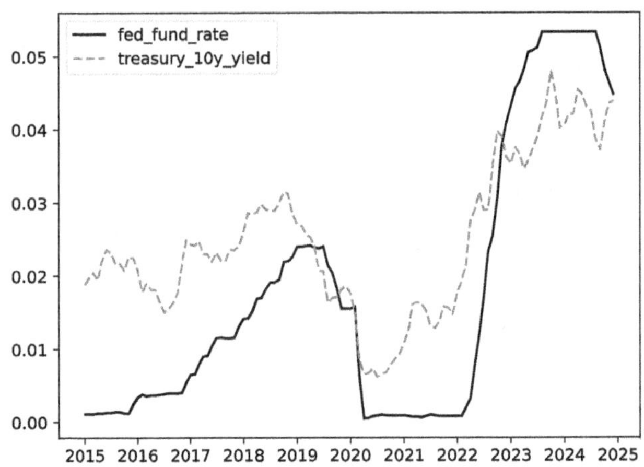

This mechanism of quickly increasing interest rates to curb inflation, potentially tightening the labor market while doing so should be familiar to readers from sections 15.5.6 and 15.5.7. One implication of this policy reversal is that borrowing costs increase. Another implication is increased volatility in interest rate markets as the market digests these policy changes and tries to identify equilibrium rates in the presence of increased inflation uncertainty. This phenomenon is evidenced by the heightened volatility in the time series of ten-year treasury yields. A third byproduct of this large movement in rates is that nominal bonds, among other interest rate products, may suffer unforeseen losses.

In the following chart, we show one such vehicle that sustained heavy losses, a long term treasury bond ETF [44]:

From the chart one can see the huge decline in the ETF as the inflation picture began to change. It is important to emphasize that this chart represents an instrument that is generally agreed upon as a risk-free asset. However, in the presence these massive shifts in rates, we can see that investors in this supposedly risk free product suffered significant losses. It should be noted that, as the astute reader will observe, these investments are in fact not risk-free because they are subject to duration risk as detailed earlier in the chapter. Nevertheless, the magnitude of the losses experienced stressed many parts of the financial system and underscored the importance of risk management even when trading seemingly innocuous instruments.

This dynamic had reverberations throughout the financial markets. In chapter 24 we detail the impact on risk parity portfolios of this generational move in yields. Banks, especially smaller banks, were also adversely impacted by this change in regime resulting in what was known a small bank crisis. At heart, this was due to banks tendency to borrow for longer horizons, and lend for shorter horizons, allowing them to profit from the yield spread between shorter and longer term rates. This exposed these investors to significant mark-to-market losses as yields rose because of the inherent interest rate sensitivity of these positions. In some cases, such as with Silicon Valley Bank, these banks reached solvency issues as these mark-to-market losses mounted and investors rushed to withdraw their balances.

[44]With treasury maturities of at least 20 years.

CHAPTER 16

Credit Markets

16.1 MARKET SETTING

IN the last chapter, we introduced some of the most common instruments and modeling approaches in interest rate markets, ranging from simple bond-like structures to exotic options that required modeling the entire term structure. A defining feature of this set of instruments, however, was the absence of default risk. For example, when we considered bond pricing, the assumption was that the coupons and principal payments would be made with certainty. For many developed market economies, such as US treasuries, this assumption feels perfectly reasonable. For emerging market governments, however, default risk is a major concern with a significant historical precedent [194]. Similarly, corporations are clearly subject to default risk as they may be forced into bankruptcy and unable to re-pay their debts.

As a result, in this chapter we introduce a new class of models and techniques that will help us model potential defaults. Of particular interest will be Poisson processes, and exponential distributions, which will provide us with the mathematical tools for modeling what we often refer to as jumps into default. Merton, for example postulated that firms should exercise their option to jump into default when the value of their assets falls below its liabilities [183]. Viewed from this lens, the equity in a firm can be seen as a call option on the underlying assets, with the liabilities defining the strike.[1]

Much like the rates markets, the landscape of credit derivatives provides fertile grounds for quants to apply their skills due to the large number of interconnected instruments.

The following list briefly summarizes the main products traded in credit markets:

- **Corporate & Sovereign Bonds**: Debt instruments issued by corporations or emerging market governments. This debt could be structured in many ways, with different coupon structures (fixed, floating or zero coupon) and different subordination levels.

- **Credit Default Swaps on Corporates or Sovereigns**: Credit Default Swaps (CDS) are analogous to life insurance on a corporation or sovereign

[1] A concept we revisit later in the chapter.

DOI: 10.1201/9781003530626-20

government. In a CDS a periodic premium is paid in exchange for a lump-sum payment at the time of default of the entity.

- **Index Credit Default Swaps**: CDS indices, or index credit default swaps, are baskets of CDS and are usually comprised of 100 or 125 equally weighted single-name CDS.
- **Options on CDS**: An option to enter into a CDS contract is commonly referred to as a CDS swaption. The option holder is entitled, but not obligated, to pay (or receive) a premium for insurance against default. A *Payer option* is one in which the option holder may enter a contract paying an agreed upon premium, whereas a *Receiver option* is one in which the option holder may enter into a contract receiving the premium.
- **Options on Indices of CDS**: An option to enter into a CDS contract on an index.
- **Collateralized Debt Obligations**: An investment where cashflows from a set of underlying bonds, loans or CDS are re-packaged into different tranches or classes of an index loss distribution. For example, an equity tranche absorbs the first losses within an index, whereas senior tranches have a significant buffer prior to experiencing any losses.
- **Credit ETF's**: Credit ETF's, such as HYG, JNK and LQD, usually consist of baskets of underlying cash bonds. Comparisons of these ETF's against similar indices of CDS are often of interest to market participants because of their natural relationship.
- **Mortgage Backed Securities**: Bonds, CDS or CDO's that are linked to an underlying pool of residential or commercial mortgages.

Bonds, CDS and ETF products are **linear** instruments and are therefore the simplest to model. The vast majority of the products that trade in credit markets trade over-the-counter, just as in rates markets.[2] This means that the structures are highly customizable, but conversely in some cases are characterized by poorer liquidity and price transparency. This relative lack of liquidity, however, may create an opportunity for investors to harvest some sort of liquidity premia. That is, for investing in less liquid credit assets, and bearing the risk that they will be unable to unwind during times of stress, they may collect a premium. These types of liquidity or related risk premia are common in credit markets, and throughout the chapter we will highlight several examples.

While credit markets are often treated as a distinct asset class, their interconnection with equities has been of particular interest to researchers. As mentioned in chapter 2, the connected nature of these asset classes stems from the fact that they both share exposure to success or failure (default) of an underlying firm. Their payoff structures vary greatly, as does there claim on the underlying assets, however, at

[2]Although there has been a recent push to move toward more exchanged traded instruments.

heart they are both exposed to the same underlying source of risk. Many investment firms have tried to use this underlying link between the two asset classes to try to build relative value models between them. For example, one might look to credit markets for a signal about future equity prices, or vice versa.

The main participants in credit markets are corporations and government entities, who use the credit market as a source to raise capital via debt issuance, as well as banks, hedge funds and asset managers. Hedge funds, in particular, may use credit markets as a venue for expressing their macroeconomic views, instrumenting relative value structures or harvesting what they deem to be well rewarded risk premia.

In this text our goal is to provide the reader with an overview of the main concepts relevant to credit models, detailing the most common instruments traded, along with the most popular modeling techniques. Additionally, in section 16.13 we attempt to view the credit markets through the lens of an investor, and show how the various components and instruments can be used to construct risk premia and relative value structures. Many of the idiosyncratic details specific to credit market, including the conventions and settlement procedures are intentionally skipped over in our treatment. Instead, we focus on building the readers underlying intuition and tool set. O'Kane [192] has an entire book dedicated to these details, in addition to many of the models introduced in this text for those interested.

16.2 MODELING DEFAULT RISK: HAZARD RATE MODELS

Pricing instruments subject to default risk, such as defaultable bonds, and CDS, requires modeling default events for the underlying assets. This will also be true for the derivatives based on underlying assets with default risk that we discuss in this chapter, such as options on CDS. To incorporate defaults into our models requires a new piece of mathematical machinery.

The standard approach to incorporating default risk is via **reduced form models** that assume default events occur when a company jumps into default. That is, we treat the transition to default as a jump process and model the probability that this transition occurs over time. This approach provides some mathematical convenience as it enables us to model default using exponential distributions and Poisson processes. In practice, default occurs when a company can no longer meet its financial obligations. In that case, they may be required to file for bankruptcy. Additionally, they may have debt obligations that they are unable to meet, requiring them to miss a coupon payment or work with their debtors to restructure their debt. Traditionally, when we model default risk we use a single overarching definition of default, and do not account for the subtle technical details of default.[3]

Merton postulated that firms will default when the value of their debt exceeds the value of the firm's assets [183]. This creates a situation where default happens once a certain default barrier is crossed. This approach to default modeling is explored in more detail in 16.11.

[3]That said, this is an important issue when trading these contracts as in some cases missing a coupon may constitute default whereas in others a bankruptcy declaration may be required.

When modeling default risk, we will rely on what is called a hazard rate approach where we assume times to default follow an exponential distribution. The theory behind the exponential distribution was introduced in the text's supplementary materials, where we saw that it can be used to measure the time between events in a Poisson process. In the context of default risk, we assume default can only happen once. That is, once a company defaults it no longer exists and thus cannot default again. As we detail in the supplementary materials, the PDF and CDF of an exponential distribution can be expressed in terms of a firms time to default:

$$f(t;\lambda) = \lambda e^{-\lambda t} \tag{16.1}$$

$$F(t;\lambda) = 1 - e^{-\lambda t} \tag{16.2}$$

where t is a default time, and λ is a parameter of the exponential distribution.

In this framework we can see that the probability of default in a given interval is defined by $f(t;\lambda)$ and is a function of λ, which we refer to as the default intensity or hazard rate.

As we increase the hazard rate, defaults will occur more frequently. Similarly, $F(t;\lambda)$ tells us the probability of a default happening prior to a given time, t. While this quantity is of interest, in practice we will be more interested in finding the probability of default in a given interval, which can be calculating using (16.1), conditional on the entity not having defaulted previously. As such, of particular interest will be the probability default does not occur prior to time t. We will refer to this as the entities survival probability, and can express it as one minus the CDF defined in (16.2):

$$Q(0,t) = 1 - F(t;\lambda) \tag{16.3}$$

$$Q(0,t) = e^{-\lambda t} \tag{16.4}$$

where $Q(0,t)$ is the survival probability of a given entity until time t.

This survival probability will be a fundamental tool in credit modeling, as we will use this quantity to discount all cashflows that are conditional on the firm not defaulting.

As the reader may have noticed, when formulated in this way, the structure of a default model is similar in structure to how discount factors and forward rates were modeled when constructing a yield curve. Recall that a discount factor can be written as:

$$D(0,T) = \exp(-\int_0^T f_\tau d\tau) \tag{16.5}$$

Analogously, we define a hazard rate, or default intensity, λ and define an entities survival probability to be:

$$Q(0,T) = \exp(-\int_0^T \lambda_\tau d\tau) \tag{16.6}$$

where it should be noted that in (16.6) we have generalized to the case where λ_τ is no longer constant but instead is a function of time. A commonly used assumption in real-world applications is that the hazard rate, λ, is piecewise constant. The instantaneous forward rate, f_{τ} above is analogous to the instantaneous default intensity, or hazard rate, λ_{τ}. Mathematically speaking, we treat defaults as Poisson processes. The time between defaults then follows an exponential distribution, which will give us probability of default in some period equal to: $\lambda_{\tau}Q(0,T)$.

In subsequent sections, we show how to apply a hazard rate model to pricing bonds and CDS in the presence of default risk, and then proceed to show how we can extract a set of default intensities, or hazard rates, from a set of traded securities.

16.3 RISKY BOND

16.3.1 Modeling Risky Bonds

In the last chapter, we covered different types of bonds that are present in rates markets. A defining characteristic of these bonds was that their payments were guaranteed. That is, we assumed that the issuer of the bonds would not default, and therefore used risk-free discount rates in order to value the bonds. This assumption is certainly reasonable for US Treasuries, but is far less appropriate for Emerging market government entities or corporate bonds. For these bonds, lack of a willingness or ability of the issuer to repay meet its obligations, also known as default risk, will be a dominant risk factor in the instrument.

When pricing a risky or defaultable bond, we need to account for this possibility that the issuer will be unable or unwilling to re-pay a coupon or principal amount[4]. For these bonds, there is a non-trivial chance that the holder will not receive the promised coupons or principal payments. As a result, it is natural to expect investors to require a premium to invest in these securities and bear this risk. As one might expect, this premium might depend on many factors, most notably credit rating. Consequently, pricing a defaultable bond will require knowledge of the probability of an investor receiving each payment, which will depend on the firm's ability to meet its debt obligations. As mentioned in the previous section, these probabilities are generally referred to as survival probabilities.

Should default occur, bondholders may not lose their entire investment. Instead, they may be entitled to some residual claim on the firm's remaining assets. This quantity, known as the recovery amount for the bond, R is an important component of modeling credit securities, especially for firms with relatively large default probabilities. The amount that is recovered will depend on where the bond sits in the firm's capitalization structure, as well as a firm's debt profile. In some pockets of the markets, standard market conventions are used to determine the recovery when pricing credit instruments. As issuers get closer and closer to default, however, this quantity will be of greater interest to investors, and is a closely monitored value.

The presence of default events makes the distribution of underlying bond price bi-modal and we must handle the:

[4]In practice it turns out that even the definition of default can be quite complex. For example, does it include a firm who restructures its debt.

- Default Case
- No-Default Case

This is a defining feature of modeling credit risk and credit securities: we always need to think in terms of the default case and the no-default leg. In the case of a bond, the no-default leg provides a stream of payments which must be discounted appropriately to account for a firm's survival probability. The no-default case is the same as we saw in interest rate modeling, with the exception that we now must include a survival probability component. The default case, on the other hand, is new. Conditional on default, investors will receive the residual value of the firm's assets, or recovery value. This case needs to embed within it the appropriate issuer default probabilities, as well as the proper recovery assumptions. Putting these two legs together, the price of a defaultable bond can be written as:

$$\begin{aligned} B(t,T) =& \sum_{i=1}^{N} \delta_{t_i} C D(0,t_i) Q(0,t_i) + P D(0,T) Q(0,T) \\ &+ C \int_0^T \lambda D(0,\tau) Q(0,\tau) d\tau \\ &+ \int_0^T \lambda D(0,\tau) Q(0,\tau) R d\tau \end{aligned}$$

Here, R is the recovery value, $Q(0, t_i)$ is the survival probability until time t_i and λ is the default intensity, or hazard rate. The reader should notice that the first two terms are familiar. In fact, they are the same as the payment of coupons and principal in the interest rate case with an additional term for survival probability. These terms both emanate from the no-default case, and, as such, are contingent on survival.

The final two terms, which contain integrals, are how we model the payment that occurs after a default. The first of these terms handles the accrued interest that accumulates prior to a default. The second term handles the payment conditional on a default. To calculate this we integrate over all possible default times, each of which is conditional on survival until that point, and calculate the value of the recovery amount paid at that time.

In order to value this term in (16.7), an approximation for the integral must be computed. To do so, we will need to rely on the quadrature methods that we learned in chapter 14. In particular, if we assume that the recovery rate R is a constant, as is done above, and that the hazard rate, λ is piecewise constant between payment dates, then we can use the mid-point rule to obtain the following quadrature approximation of the integral:

$$\int_0^T R D(0,\tau) Q(0,\tau) \lambda d\tau \approx R \sum_{i=1}^{N} \frac{1}{2} \delta_{t_i} D(0,\hat{t_i}) \left(Q(0,\hat{t_{i-1}}) - Q(0,\hat{t_i}) \right) \tag{16.7}$$

where $i = 1, \ldots, N$ corresponds to the payments dates of the bond and $\hat{t_i} = \frac{t_i + t_{i+1}}{2}$ is the middle of each subinterval and we are using the fact that $\int_{t_i - 1}^{t_i} \lambda Q(0,\tau) =$

$Q(0, \hat{t_{i-1}}) - Q(0, \hat{t_i})$ over each sub interval where λ is constant. This approximation is equivalent to assuming that default occurs half-way through each period. In practice, this approximation is quite reasonable as λ is extracted from market data at specified, traded points, and in between is generally assumed to be constant. This is not the only way to approximate the integral, of course, however, its simplicity has led it to become the market standard approach.

16.3.2 Bonds in Practice: Comparison of Risky & Risk-Free Bond Duration

In this section we leverage the risk-free and risky-bond pricing functions in the coding repository in order to compare the duration of an otherwise equivalent risk-free and risky bond. That is, we compare the duration of two bonds with similar characteristics, such as their coupon and maturity, with the only difference being that one includes default risk. As before, we then compare this across a set of bond maturities, which is shown in the following chart:

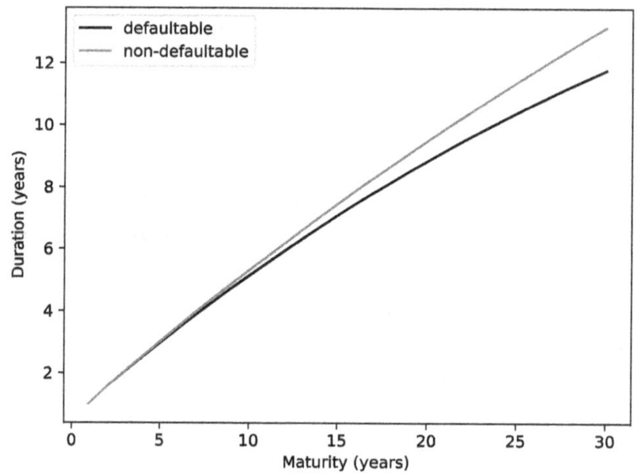

As we can see, the duration of a risky bond is lower than its riskless equivalent. This means that these bonds will have lower amounts of interest rate sensitivity, especially at longer maturities. This is fairly intuitive as, for these bonds, default risk will now be the primary risk factor. That is, when we buy a defaultable bond we are buying a bond with less interest rate sensitivity, at the expense of an added risk of default.

16.4 CREDIT DEFAULT SWAPS

16.4.1 Overview

Credit default swaps are one of the most commonly traded instruments within credit markets. A credit default swap is analogous to buying life insurance on a corporation or government. That is, in a CDS, a periodic fixed coupon payment is exchanged for a payment conditional on default of the underlying issuer. As previously mentioned, there is a precise definition of what exactly constitutes a default, which may impact the CDS's valuation. Like an interest rate swap, a CDS consists of two legs:

- **Premium or Fee Leg**: consists of a fixed, contractual spread that is paid periodically (usually quarterly or semi-annually).
- **Default Leg**: consists of a lump sum payment that is paid upon default.

The reader may notice that the premium or fee leg is similar to the fixed leg of an interest rate swap, whereas the traditional floating leg has been replaced with a payment conditional on default. In order to properly value a CDS, modeling both streams of cashflows will be required. The reader may also observe similarities between bonds and CDS, a connection we explore in more detail later in the chapter. Specifically, the premium leg of a CDS is analogous to the coupon payment stream in a bond. Similarly, the default leg is similar to the loss a bondholder would absorb in the event of a default on their underlying issuer.

By convention, an investor who pays the premium leg and is entitled to receive a payment conditional on default is said to be long protection. Conversely, an investor who receives the premium leg and is obligated to payout on default is said to be short protection. Through this lens, we can view a long position in a bond as a short protection position, that is, we receive coupons, but experience a significant loss in the event of a default. This convention is used throughout the chapter to clarify the type of exposure in a CDS trade.

CDS contracts are themselves complex documents that are negotiated between investors and dealers. Full details of these conventions are beyond the scope of this book, however, a brief summary of all CDS conventions is provided below:

- **Settlement Type**: Either physical or cash settlement. Physical settlement requires delivery of a bond in order to trigger the CDS contract.[5]
- **Daycount**: CDS contracts require a daycount convention to determine exactly how interest on the premium leg is accrued and the exact amount of each payment.
- **Payment Frequency**: CDS contracts are usually paid quarterly.
- **Documentation Clause**: Determines the precise definition of what characterizes default, including restructuring, missed coupon payments and declaration of bankruptcy.

16.4.2 Valuation of CDS

To value a CDS we need to calculate the present value of both the premium and default legs, which can be done by leveraging the hazard rate modeling concepts

[5]This physical settlement feature can create challenges when defaults occur and the underlying bond becomes scarce as investors look to obtain it to settle their CDS contracts.

introduced earlier in the chapter. Specifically, the value of the premium and default legs can be written as:

$$\begin{aligned}
\text{PV(premium)} &= \sum_{i=1}^{N} \delta_{t_i} S D(0,t_i) Q(0,t_i) + \int_0^T \lambda_\tau S D(0,\tau) Q(0,\tau) \\
&\approx \sum_{i=1}^{N} \delta_{t_i} S D(0,t_i) Q(0,t_i) + S \sum_{i=1}^{N} \frac{1}{2} \delta_{t_i} D(0,\hat{t_i}) \left(Q(0,\hat{t_{i-1}}) - Q(0,\hat{t_i}) \right) \\
\text{PV(default)} &= \int_0^T D(0,\tau) Q(0,\tau)(1-R)\lambda d\tau \\
&\approx (1-R) \sum_{i=1}^{N} \frac{1}{2} \delta_{t_i} D(0,\hat{t_i}) \left(Q(0,\hat{t_{i-1}}) - Q(0,\hat{t_i}) \right)
\end{aligned}$$

$$\begin{aligned}
\text{PV(CDS)} &= \text{PV(premium)} - \text{PV(default)} \\
&= \sum_{i=1}^{N} \delta_{t_i} S D(0,t_i) Q(0,t_i) + S \sum_{i=1}^{N} D(0,\hat{t_i}) \left(Q(0,\hat{t_{i-1}}) - Q(0,\hat{t_i}) \right) \\
&\quad -(1-R) \sum_{i=1}^{N} \frac{1}{2} \delta_{t_i} D(0,\hat{t_i}) \left(Q(0,\hat{t_{i-1}}) - Q(0,\hat{t_i}) \right)
\end{aligned}$$

where in the formula for the default leg of a CDS we are using the same approach for approximating the default integral, that is, we are assuming that default happens half-way through the period using the mid-point rule. Notice that there is no exchange of a principal payment, only a periodic exchange of coupons. This is an important and differentiating feature of a CDS relative to a bond, as we will see later.

Alternatively, the above integral in the present value calculation of the default leg can be estimated using the following equation, where we still rely on the mid-point rule however no longer rely on the integral form of $\lambda Q(0,\tau)$:

$$\int_0^T D(0,\tau) Q(0,\tau)(1-R)\lambda d\tau \approx \sum_{i=1}^{N} \lambda (1-R) D(0,\hat{t_i}) Q(0,\hat{t_i}) \delta_t \tag{16.8}$$

where again $i = 1, \ldots, N$ corresponds to the payments dates of the bond and $\hat{t_i} = \frac{t_i + t_{i+1}}{2}$ is the middle of each payment period.

In the absence of accrued interest, the break-even, or fair CDS spread is the spread that equates the values of the two legs, or makes the present value of the CDS equal to zero:

$$\hat{S} = \frac{\int_0^T D(0,\tau) Q(0,\tau)(1-R)\lambda d\tau}{\sum_{i=1}^{N} \delta_{t_i} S D(0,t_i) Q(0,t_i)} \tag{16.9}$$

This is analogous to the par swap rate that we saw within the context of interest rate swaps. Unlike an interest rate swap, however, CDS generally do not trade at their fair CDS spread, instead trading at contractual spreads of 100 or 500 basis points. This means that, an upfront payment, as defined by the initial present value of the CDS, is required to enter into a CDS at the standard contractual spread. This upfront payment, can be found by solving for the left-hand side of the CDS pricing equation in (16.8) using the contractual spread, C:

$$\begin{aligned} U &= \text{PV}(\text{CDS})_C && (16.10)\\ &= \text{PV}(\text{premium})_C - \text{PV}(\text{default})_C && (16.11)\\ &= \sum_{i=1}^{N} \delta_{t_i} C D(0,t_i) Q(0,t_i) + \int_0^T \lambda_\tau C D(0,\tau) Q(0,\tau) && (16.12)\\ &- \int_0^T D(0,\tau) Q(0,\tau)(1-R)\lambda d\tau && (16.13) \end{aligned}$$

Even though it is not the convention to trade at its spread, the par or fair CDS rate is still of direct interest to market participants as it provides the best measure of cost of protection, and is comparable across contractual spreads, maturities and credits. Credit default swaps contain many technical details, conventions and standard features that are omitted from this text. Interested readers should see [59] for more details.

16.4.3 Risk Annuity vs. IR Annuity

Examination of the CDS par spread formula in (16.9) we can see a familiar quantity appears in the denominator:

$$\text{RiskyAnnuity}(0,T) = \sum_{i=1}^{N} \delta_{t_i} D(0,t_i) Q(0,t_i) \qquad (16.14)$$

Recall from chapter 19 that we defined the annuity function as the present value of a one-basis point annuity over a given set of dates. In CDS markets, the same concept appears, however, the annuity is now subject to default risk, as we can see from the survival probability terms $Q(0,t_i)$ in (16.14). This gives rise to a new, related concept which is referred to as risky annuity, or risky PV01. Risky annuity is defined as the present value of a one-basis point stream of payments that are conditional on a credit surviving to fulfill its obligation.

Just as with the annuity function in rates markets, this quantity is important to many CDS modeling calculations, including marking a CDS to market, and will also be our choice of numeraire when we consider options on CDS. As a result, calculations of risky annuities, are fundamental to credit valuation and modeling.

16.4.4 Credit Triangle

If we make some simplifying assumptions to the CDS pricing formula in (16.9) we can get a simple heuristic formula that approximates the relationship between the key quantities in a CDS. In particular, if we ignore the accrual term, assume a constant hazard rate, λ over the entire period, and assume that the recovery rate, R is also constant, the formula for a CDS at the par spread[6] can be written as:

$$S\sum_{i=1}^{N}\delta_{t_i}D(0,t_i)Q(0,t_i) = (1-R)\lambda\int_0^T D(0,\tau)Q(0,\tau)d\tau \tag{16.15}$$
$$S\sum_{i=1}^{N}\delta_{t_i}D(0,t_i)Q(0,t_i) \approx (1-R)\lambda\sum_{i=1}^{N}\delta_{t_i}D(0,t_i)Q(0,t_i) \tag{16.16}$$
$$S \approx \lambda(1-R) \tag{16.17}$$

where in the second line we are approximating the integral in the default leg by the value at each payment date. As we have assumed that the hazard rate, λ is constant, this is certainly a reasonable assumption and amounts to using the left-point quadrature rule. Importantly, this assumption makes the terms inside the summations equal so that they can be cancelled. Equation (16.17) is referred to as the credit triangle and is important in that it establishes a simple, approximate link between credit spreads, recovery rates and hazard rates.

16.4.5 Mark to Market of a CDS

Once a CDS contract is entered into, both participants will have exposure to movements in the underlying CDS spread curve. As this happens, the contracts will need to be marked-to-market.

To see how this is done, suppose a CDS contract was entered into at the prevailing par CDS spread, C. In this section we will examine the value of this CDS contract after a shift in the par CDS spread from C to S. To see how this works, recall that the present value of a CDS can be written as:

$$V(t) = \sum_{i=1}^{N}\delta_{t_i}CD(0,t_i)Q(0,t_i) - \int_0^T D(0,\tau)Q(0,\tau)(1-R)\lambda d\tau \tag{16.18}$$

where the valuation, $V(t)$ is from the perspective of the party that is short protection.

We also know however, that a contract at the par CDS spread, S, will have a present value of zero. That is, the value of the premium and default legs will be equal:

$$\sum_{i=1}^{N}\delta_{t_i}SD(0,t_i)Q(0,t_i) = \int_0^T D(0,\tau)Q(0,\tau)(1-R)\lambda d\tau \tag{16.19}$$

Following the same logic that we followed in the case of marking-to-market interest rate swaps, we notice that the right-hand side of (16.19) is the same as the second term on the right hand side of (16.18). Therefore, we can substitute the right-hand side of (16.19) into our equation and are left with:

$$V(t) = \sum_{i=1}^{N}\delta_{t_i}CD(0,t_i)Q(0,t_i) - \sum_{i=1}^{N}\delta_{t_i}SD(0,t_i)Q(0,t_i) \tag{16.20}$$
$$= (C-S)\sum_{i=1}^{N}\delta_{t_i}D(0,t_i)Q(0,t_i) \tag{16.21}$$

[6] That is, where the present value of the premium and default legs are equal.

Intuitively, this means that we can express our CDS's present value, or its mark-to-market value, as the difference in spread between our contract and the par rate, multiplied by the present value of a one-basis point risky annuity. This is analogous to what we found for interest rate swaps, with the distinction that we are now using a risky annuity in lieu of an ordinary annuity. From this equation we can also observe that an investor who is short protection will lose money should S rise, and will profit when S decreases. Conversely, an investor who is long protection will benefit as spreads rise.

16.4.6 Market Risks of CDS

Credit default swaps are subject to many different market risks. As a result, holders of CDS in their portfolios will want to have a corresponding set of analytical tools that helps them measure the sensitivity of their portfolio along each dimension. The main market risks embedded in a CDS contract are:

- **Jump to Default Risk**: Risk that the underlying issuer will default. This may be a positive or negative event for the investor depending on whether they are long or short protection.
- **Credit Spread Risk**: Risk that the underlying issuer's par CDS spread curve will widen (increase) or tighten (decrease). An investor who is long protection will benefit from an increase of credit spreads, whereas and investor who is short protection will have their position increase in value should spreads tighten.
- **Interest Rate Risk**: Risk that interest rates will rise or fall impacting the discounted value of the underlying cashflows. This risk is generally small for CDS when compared to credit spread risk.
- **Changes to the Recovery Rate**: Risk that the expected recovery value upon default will change.

In addition to the above market risks, due to their over-the-counter nature CDS are subject to counterparty risk which arises from the fact that the counterparty they have engaged in the contract with may be unable or unwilling to fulfill the contract. The risk of this is clearly largest for the default leg, where a counterparty may not be able to meet the obligation of the required lump sum payment. This phenomenon is explored in more detail in chapter 4.

16.5 CDS VS. CORPORATE BONDS

16.5.1 CDS Bond Basis

Astute readers may notice that a **long position in a defaultable bond** and a **short protection position in a CDS** have similar risk profiles. In fact, they are both exposed to **default risk** of the underlying issuer. They also both have sensitivity to changes in interest rates. Intuitively, these two instruments are clearly related as one provides periodic cashflows in the absence of default and the other provides insurance against default in exchange for periodic payments.

As a result, one might naturally think that a long position in a risky bond, coupled with an appropriately matched long protection position in a CDS is riskless, and that the bond and the CDS should therefore trade at the same spread in the absence of arbitrage. In practice, this however is not what we observe and there is instead a basis between CDS and bonds.

To understand this, consider that a fundamental difference between the two instruments, however, is that a defaultable bond is a ***cash*** instrument that requires us to post collateral and a CDS is a ***synthetic*** instrument that requires no cashflows at origination. This funding difference causes bonds and CDS to trade at different spread levels. Said differently, it is natural to expect an investor to demand a higher return or premium on an asset that will tie up relatively more capital, and that is a significant part of this phenomenon. The difference in spread between risky bonds and the corresponding CDS is generally referred to as the CDS-bond basis. As we discuss in the next section, this basis may be a function of several factors, including credit quality and general market sentiment that may increase or decrease the value of collateral. Because of this difference in spreads on the bond and the CDS that arise due to funding, when we model a defaultable cash bond we need to include what is referred to as a basis between the two spreads.

In chapter 19, we saw a similar basis arise between swaps and treasuries for a similar reason: swaps are synthetic instruments that do not require financing whereas treasuries are cash instruments that will require a significant upfront payment. This basis that arises in credit between bonds and CDS is conceptually similar.

From an investment perspective, this CDS-bond basis is of interest to many market participants. Investors may analyze this basis and try to spot anomalies that can be the basis of relative value or risk premia trades. Many hedge funds in particular have been known to engage in CDS-bond basis trades, and even at a high level, it does appear to have attractive characteristics: we are able to harvest the spread differential and have theoretically perfectly offsetting risks. In practice, however, there are many complications of trading these structures. One that is particularly noteworthy is that these trades are typically implemented with a significant amount of leverage in order to make the spread differential attractive in absolute returns terms. Once leverage is included in the trade, than investors run the risk of being forced out of a seeming near arbitrage opportunity by a significant margin call when the basis moves against them. Later in the chapter we discuss a few of the details of implementing CDS-bond basis trades in practice, and some of the properties and dangers of these structures.

16.5.2 What Drives the CDS-Bond Basis?

In practice, the basis between bonds and CDS varies greatly by issuer and market conditions. Those who observe markets and monitor the basis closely will see that it can evolve significantly over time. This evolution can be explained by many potential factors, including:

- **Overall Market Sentiment**: CDS-bond basis has a natural risk-on / risk-off component as in times of stress the value of collateral tends to be the highest.
- **Credit Quality**: Lower rated credits will have significantly different properties of their basis than their higher rated counterparts.
- **Liquidity**: Supply and demand imbalances between the bond and the CDS can lead to disparate behavior in the CDS-bond basis.
- **Contractual Details**: Differences in contracts of the bonds and CDS can lead to a natural, persistent basis, such as if restructuring events are considered default events.
- Other idiosyncratic components such as firm's capitalization structure and debt profile.

The risk-on risk-off nature of CDS-bond basis trades, as well as the liquidity driven component imply that investors who are engaging in such trades may be harvesting an underlying risk premia. That is, they are being compensated for bearing the risk that the basis will widen considerably in a period of market stress characterized by lower liquidity and a higher value on collateral.

16.6 BOOTSTRAPPING A SURVIVAL CURVE

16.6.1 Term Structure of Hazard Rates

So far, we have seen how to model CDS and corporate bonds when we are given a set of discount factors and hazard rates. In practice, however, we do not observe this data but instead need to infer the set of hazard rates from a set of market data, most notably quoted credit spreads. This process is analogous to constructing a yield curve in rates markets. In particular, CDS for a given corporation or government entity may trade for multiple maturities, just as a corporation of sovereignty may issue bonds with different maturities. In practice, the most liquid CDS contract has a five year maturity, however CDS of 3, 7 and 10 year maturities are also commonly traded. Our goal will then be to calibrate, or extract a consistent set of hazard rates that enable us to match the traded credit spreads for all CDS for a given government or corporate entity. This set of hazard rates is often referred to as a term structure of hazard rates. This process is a **calibration** of our hazard rate model to market traded credit spreads, and is conceptually similar to the bootstrapping and optimization approaches that we applied in the context of constructing a yield curve.

In addition to CDS that trade on a particular entity, the entity may have a set of corporate bonds that theoretically could also be used to extract an underlying term structure of hazard rates. However, as we saw in the last section, there is often a basis between bond and CDS spreads which would need to be incorporated into our calibration process. Said differently, if we were to use bonds as inputs into a bootstrapping procedure, we would obtain different hazard rates for bonds and for CDS. Instead, we could try to calibrate them together using a single set of underlying hazard rates but also calibrating the CDS-bond basis in the calibration. In practice, it is far more common to calibrate hazard rates directly to CDS, ignoring bond data, and then if needed separately calculating the basis between bonds and CDS.

16.6.2 CDS Curve: Bootstrapping Procedure

The bootstrapping procedure for a CDS curve is analogous to the bootstrapping procedure we used to construct a yield curve. Specifically, we begin with the shortest maturity CDS and find the hazard rate that enables us to match the traded market spread. When we do this, we need to solve the CDS breakeven, or fair spread formula in (16.9) for λ.

In particular, the quantity on the left-hand side of the equation, the par spread is assumed to be known, as it is provided by the market. We also assume that other parameters, such as the recovery rate and interest rate curve are known and can be fixed in our bootstrapping algorithm.

We then proceed by making assumptions about the hazard rate process. For example, we might assume it is constant. This enables us to recover the hazard rate λ that allows us to match this fair CDS spread by inverting the CDS pricing formula. Note that in order to invert this formula may require use of a one-dimensional root finding algorithm. We then fix this hazard rate for the relevant time period, move to the next maturing CDS and find the hazard rate that matches this CDS's traded spread while still matching the previous CDS spreads.

It should be emphasized that to do this, we need to make λ a function of time to match more than one CDS spread. The standard specification is to assume $\lambda(t)$ is piecewise constant, although this is certainly not required. Instead we could fit a quadratic or cubic polynomial that would enable us to recover a smoother term structure of hazard rates. The details of implementing a bootstrapping based CDS curve construction algorithm are left to the reader, however, we do provide an example using an alternative optimization approach detailed in the next section.

16.6.3 Alternate Approach: Optimization

Just as was the case in the context of yield curve construction, a natural alternative to bootstrapping a term structure of hazard rates is optimization. To do this, we would formulate an optimization problem that, for example, minimizes the squared distance between model and market credit spreads:

$$Q(\vec{\lambda}) = \sum_{j=1}^{m} (S_j - \hat{S}_j)^2 \tag{16.22}$$

where m is the number of benchmark CDS, S_j is the market traded spread of instrument j and $\hat{S}_j$ is the model spread of the same instrument.

We could then specify a parameterization for the set of hazard rates, $\vec{\lambda}$, such as assuming they are piecewise constant. For interested readers, an example implementation of this simplified optimization framework can be found in the book's coding repository.

16.7 INDICES OF CREDIT DEFAULT SWAPS

16.7.1 Credit Indices

Credit default swaps trade not only on single entities but also on credit indices [173] [174]. These products are essentially baskets of CDS, and because of the linear nature of their payoff are relatively easy to model. Most importantly, because the payoff is linear we do not need to introduce a stochastic model to value CDS indices, and instead can value them as the weighted average of the underlying CDS contracts.

When we enter into a CDS contract on an index, we obtain protection on a set of names with a predetermined set of weights. Within the scope of corporate CDS indices, the most common structure is equally weighted protection on either 100 or 125 underlying names. Because of the high number of names in each index, positions in a CDX index provide broad exposure to a particular credit market. Additionally, to the extent that defaults correlations are low, the use of many names in the index theoretically lessens the idiosyncratic risks for investors and provides a significant diversification benefit. Conversely, there is evidence to show that in times of substantial market stress, these correlations tend to increase, lessening any diversification benefit.

As defaults occur, the investor who is long protection in the index would then receive $w_i(1 - R_i)$. In the case of an equally weighted 100 name index, the weight, $w_i = 0.01$ for all i. If an issuer defaults whose recovery is 40%, the payment received by the long protection investor would then be 60 basis points. As was the case for individual CDS, this protection would be provided in exchange for a periodic fixed payment, S_I, which we will refer to as the spread of the index[7]. The value of this protection will clearly depend on the default probability of the entities in the index. All else equal, one should expect that the protection would be more expensive if defaults in the index are expected to occur with higher probability.

The most common credit indices are:

- **IG**: North America Investment Grade Credits
- **HY**: North America High Yield Credits
- **iTraxx Europe**: European Investment Grade Credits
- **EM**: Emerging market sovereign credits

Each credit index may have its own conventions with respect to payment frequency, number of credits and documentation that defines key terms such as what constitutes default.

Additionally, as CDS are inherently over-the-counter instruments, banks and dealers will also quote and trade baskets on a customized set of names. These baskets are referred to as bespoke, and also appear when trading CDOs as we will soon see.

[7]Alternatively, the premium leg may be defined by a fixed coupon C of either 100 or 500 basis points, and an upfront payment.

16.7.2 Valuing Credit Indices

Pricing CDS indices, often referred to as CDX, can be done by leveraging the concepts that we learned in the previous sections on CDS. In particular, we could first build hazard rates and survival probabilities for the underlying components. Once we have these we have essentially reduced our modeling problem back to the case of a CDS, and can leverage pricing formulas for CDS akin to (16.9).

Consider a CDS on an index of M underlying names. If the names are equally weighted, then the weighting function, $w_i = \frac{1}{M}$. As defaults occur, payments of this weight times $1 - R_i$ will be made to the investor who is long protection, where R_i is the ith names recovery. In order to value the default leg, we need to integrate over all possible default times and sum over-all index constituents. This can be expressed as:

$$\text{PV(default)} = \sum_{i=1}^{M} w_i \int_0^T D(0,\tau)Q_i(0,\tau)(1 - R_i)\lambda_i d\tau \tag{16.23}$$

where $Q_i(0,\tau)$ is the survival probability until time τ of issuer i, λ_i is its hazard rate and R_i is the recovery rate of issuer i. Readers should note the similarity between this and the present value equation for the default leg of a single name CDS. Essentially, to find the present value of a CDS on an index we are leveraging the hazard rate and survival probabilities extracted from single-name CDS curves. Using a similar approach, we can write the present value of the premium leg in terms of the single-name survival probabilities:

$$\text{PV(premium)} = \sum_{i=1}^{M}\sum_{j=1}^{N} \delta_{t_j} S_I D(0,t_j)Q_i(0,t_j) \tag{16.24}$$

Putting this together, we could obtain an equation for the present value of an index. However, this present value would contain an integral in (16.23) that we would need to approximate.

Instead we can remember the CDS identity learned earlier, that is for each component CDS we have:

$$\sum_{i=1}^{N} \delta_{t_i} S_i D(0,t_i)Q(0,t_i) = \int_0^T D(0,\tau)Q_i(0,\tau)(1 - R_i)\lambda_i d\tau \tag{16.25}$$

where S is the par spread for a given CDS. The reader should notice that the right hand side in (16.25) is equal to the value inside the outer sum in (16.23). That is, we can replace the value of default leg for each underlying CDS that appears in this equation with a term that includes the par CDS spread and its risky annuity. Therefore, we can make this substitution to compute a present value for the index:

$$\begin{aligned}\text{PV(Index)} &= \sum_{i=1}^{M} w_i (S_I - S_i) \sum_{j=1}^{N} D(0,t_j)Q_i(0,t_j) \\ &= \sum_{i=1}^{M} w_i (S_I - S_i)\, \text{RiskyAnnuity}_i(0,T)\end{aligned} \tag{16.26}$$

where S_I is the index spread and S_i is the par spread of issuer i. This equation provides us an estimated intrinsic or fair value of the index, S_I when the present value equals zero.

In practice, however, an alternate approach is to extract a set of index survival probabilities and hazard rates directly from the set of traded index CDS. One might naively think these two approaches would recover the same present values and par spreads for the index. However, in a real market setting this turns out to not be the case, as we explore in the next section. In addition, more details on the mechanics of CDS index products, as well as the index vs. single name basis can be found in [192].

16.7.3 Index vs. Single Name Basis

Somewhat interestingly, the spread of credit indices in general does not match the spread that we would obtain using (16.26). Instead, there is a basis between the two. This basis is similar conceptually to other bases we have seen in this book. Unlike the CDS-bond basis, however, both sides of this basis are synthetic in nature. As a result, we know this basis doesn't arise for funding reasons. Instead, the following factors might contribute to this basis:

- **Liquidity**: Indices tend to be more liquid than single-names and differences in liquidity may lead to different spread levels.
- **Contractual Definitions**: Mismatches in the precise definitions of default or settlement type.
- **Bid-Offer Spread**: Part of the basis between indices and single-names may be explaining that the contracts cannot be executed at the mid levels generally used to compute the spreads.

Even after accounting for bid-offer spreads, and differing contract details, in practice we still observe different spreads between the index and basket of single-names. It is perhaps most natural to think of this basis in terms of a liquidity premium. In fact, it is one of cleanest examples of liquidity premia present in markets. As indices tend to be more liquid, it is natural that they would trade at a tighter spread. Investors willing to invest in the less liquid product, a basket of single-names, is rewarded with a higher spread, however, must bear the risk of difficulty unwinding the single-names in certain market conditions.

Hedge funds, and other institutions may try to take advantage of this basis by buying protection on the side of the trade with the tighter spread (usually the index) and selling protection on the side of the trade with the wider spread (generally the single-names). This provides investors with theoretically offsetting risks while harvesting the spread differential. This spread differential, however, is generally not appealing in absolute terms. As a result, leverage is applied, introducing the possibility that the investor could be stopped out of their trade at a significant loss by a large margin call.

16.7.4 Credit Indices in Practice: Extracting IG & HY Index Hazard Rates

In this section we leverage the optimization based CDS coding example in the supplementary repository in order to extract the most recent hazard rates for the current on-the-run High Yield (HY) and Investment Grade (IG) credit indices. In doing this, we first found the following quoted par CDS spreads for the indices, respectively:

Tenor	IG Spread (BPS)	HY Spread (BPS)
3y	30	248
5y	49	308
7y	71	331
10y	92	-

Notice that the 10y spread is not available, and as such we proceed with the remaining points in our optimization. The resulting term structure of these hazard rates, found using our bootstrapping algorithm, for both HY and IG is shown in the chart below:

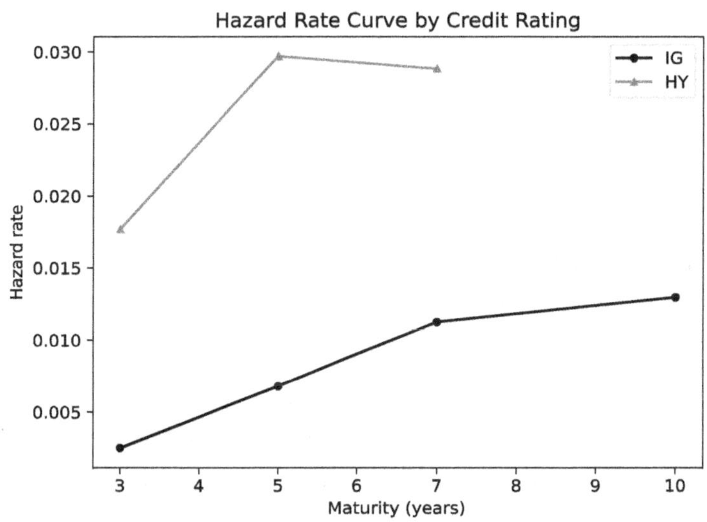

Here we can see that both hazard rate curves are upward sloping, implying that as time progresses the default intensity is likely to increase. This is a commonly observed phenomenon when spreads are low, as they are currently. This low level of spreads is further evidenced by the corresponding low hazard rates, both for HY and IG. Lastly, we find that the hazard rates for HY are considerably higher than their IG counterparts, which is intuitive because of the riskier nature of HY credits.

16.8 MARKET IMPLIED VS. EMPIRICAL DEFAULT PROBABILITIES

So far, in this chapter we have developed a methodology for extracting hazard rates, and the ensuing default probabilities from a set of market instruments such as bonds or CDS. It should be emphasized that the default probabilities that we extract from the market are risk-neutral default probabilities. That is, these probabilities are extracted using risk-free discount rates and are based on hedging and replication arguments. Said differently, they are not a direct forecast of what defaults will actually happen. Instead, they only reveal the probabilities that are currently implied by the market. This is analogous to the difference between risk-neutral and physical probabilities we explored in chapter 17.

Essentially, the market implied default probabilities that we have taught the reader to extract, may, very conceivably, embed a risk premia for investors willing to bear the risk of default. In fact, empirically, this appears to be the case as market implied default rates tend to exceed empirical default rates by a significant margin. Even more striking, we often see market implied default rates in some index products[8] that exceed even the realized default rates that we experienced in the height of the financial crisis. This would seem to provide strong evidence of a well-rewarded risk premia for investors who are willing to bear default risk in credit markets.

While this may be true, and the presence of a premium in default rates has some satisfactory motivations in behavioral finance, a note of caution is also necessary. In particular, defaults tend to occur in times of market stress, and it is therefore plausible that investors would value cashflows in these states more, placing a premium on them relative to cashflows in other states of the world. Although this argument has solid economic footing, one challenge is the difficulty determining the likelihood of rare events. Said differently, because defaults are rare, only a small sample of historical defaults are available to analyze. This creates a great deal of error in our estimates of default probabilities and implies they are quite noisy. To see the intuition behind this, let's assume that default is a three standard deviation event and consider an algorithm that uses Monte Carlo simulation to estimate the probability. In order to estimate this quantity accurately would require many, many draws: far more than the market has provided for investors. This means that, an alternative and more pessimistic view of the gap between market implied and empirical default probabilities is simply that the historical dataset isn't sufficient for the probabilities to have converged, which, if true, would imply that the gap between them is less likely to persist going forward.

16.9 OPTIONS ON CDS & CDX INDICES

Up until now the products that we have covered in credit markets have had linear payoffs. Modeling them required a new set of tools, as we had to incorporate the possibility of a jump into default, however the payoffs have been linear. As we have seen throughout the book, this makes replication arguments significantly less complex and means a stochastic model is not required to value the products. Credit markets also consist of a robust set of option structures, both on CDS and indices. These products are over-the-counter agreements with dealers playing the role of an intermediary, and provide an investor with the right, but not the obligation to buy (or sell) protection on a single-name or index at a specified time, for a specified spread. In this section we examine both types of option structures, starting with options on a single CDS

[8]Such as the indices of CDS discussed in 16.7 that are meant to provide exposure to a broad index.

and the proceeding to those on indices. We will see some similarity between these options and the rate swaptions we detailed in chapter 19, most notably the use of an annuity measure. One unique feature of these options, however, is the possibility of a default occurring between the time the contract is initiated and the expiry of the option.

16.9.1 Options on CDS

An option on a CDS (CDS Swaption) provides the buyer of the option the right, but not the obligation to enter into a CDS contract at a future time. Payer swaptions are options that enable the buyer to enter into a contract where they pay a contractual spread, specified as the strike spread, in exchange for protection. Receiver swaptions, conversely, are options that enable the holder to enter into a contract receiving the spread. This nomenclature matches that of swaptions in interest rate markets, as in both cases the option is defined in terms of whether the fixed leg is being paid or received. In the case of options on CDS, a payer option is equivalent to an option to be long protection, whereas a receiver option is equivalent to an option to be short protection.

As mentioned above, an important feature of CDS options is that default for the entity may occur prior to option expiry. If this happens, then the entity would no longer exist, as default can only happen once. There are two approaches for handling this potential default event:

- **Knockout Options**: Options that cancel should default occur prior to option expiry.
- **Non-Knockout Options**: Options that don't cancel should default occur prior to option expiry. In this case, an option holder who has the right to buy protection on the underlying that has defaulted will exercise and then can immediately receive a lump sum payment.

This feature is only relevant for options that are long protection (payers) as an investor would not exercise their short protection option if default occurs. Therefore, receiver swaptions are unaffected by this potential default prior to expiry. For a payer, we can think of a non-knockout swaption as a knockout option plus additional protection that we receive during the options life. This additional protection is referred to as front-end protection, or FEP. The relationship between the two types of payer swaptions can be written as:

$$\text{Payer(Non-KO)} = \text{Payer(KO)} + \text{FEP} \tag{16.27}$$

$$\text{FEP} = (1-R)D(0,t)\left(1-Q(0,t)\right) \tag{16.28}$$

where the second line defines the value of the front-end protection component. It includes the payment conditional on default, $1-R$, as well as the cumulative default probability within the interval $(1-Q(0,t))$. Additionally, a discount factor is applied to compute the present value of the protection. Because of the relationship between the two types of options, and the relative ease of calculating the front-end protection, we can focus on pricing knockout payer swaptions in lieu of their non-knockout counterparts.

Along these lines, to proceed let's consider a payer knockout swaption with strike spread K. At expiry, the value of the CDS that can be exercised into can be expressed as:

$$(S(t) - K) \times \text{RiskyAnnuity}(t, t, T) \tag{16.29}$$

where $S(t)$ is the market spread of the CDS at the option's expiry, t. Similarly, the payoff for a receiver option, if exercised will be:

$$(K - S(t)) \times \text{RiskyAnnuity}(t, t, T) \tag{16.30}$$

Clearly, an option holder will only exercise if the quantities in (16.29) and (16.30) respectively are positive. Using this we can define the typical risk-neutral pricing formula for payer and receiver swaptions, respectively as:

$$V_{\text{pay}} = \mathbb{E}\left[1_{\tau>t} Z(0,t)\,(S(t) - K)^{+} \times \text{RiskyAnnuity}(t,t,T)\right] \tag{16.31}$$

$$V_{\text{rec}} = \mathbb{E}\left[1_{\tau>t} Z(0,t)\,(K - S(t))^{+} \times \text{RiskyAnnuity}(t,t,T)\right] \tag{16.32}$$

where the term $1_{\tau>t}$ is defined as an indicator function that is equal to 1 if the default time τ is greater than the option expiry, t and 0 otherwise. This term is introduced due to the knockout assumption of our swaption. That is, it ensures that the swaption will not be exercised if default happens prior to expiry.

This will turn out to be an inconvenient pricing formula for pricing CDS options. In particular, we will want to proceed by assuming that the spread process, $S(t)$ evolves according to a stochastic process, such as the Black-Scholes dynamics. The risky annuity quantity, however, in the above equations is also theoretically stochastic which creates a significant challenge. Thankfully, we can simplify the pricing equation by changing our numeraire, as we did with interest rate swaptions. The full details of this change of numeraire are beyond the scope of this book, however, interested readers should consult [192] for a more rigorous treatment. The following equations, however, outline the change of measure required to change from the standard risk-neutral measure to a measure defined by the risky annuity of the underlying credit:

$$\begin{aligned} V_{\text{pay}}(0) &= Z(0,t)\mathbb{E}^{\mathbb{Q}}\left[1_{\{\tau>t\}}(S(t_E) - K)^{+} \times \text{RiskyAnnuity}(t,t,T)\right] \\ &= \text{RiskyAnnuity}(0,t,T)\mathbb{E}^{\mathbb{Q}}\left[\frac{Z(0,t)}{Z(t,t)}\frac{\text{RiskyAnnuity}(t,t,T)}{\text{RiskyAnnuity}(0,t,T)}1_{\{\tau>t\}} \times (S(t) - K)^{+}\right] \\ &= \text{RiskyAnnuity}(0,t,T)\mathbb{E}^{\mathbb{P}}\left[(S(t) - K)^{+}\right] \end{aligned} \tag{16.33}$$

where $\text{RiskyAnnuity}(0, t, T)$ is the value of an annuity subject to default from the option expiry, t to maturity of the CDS, T, viewed as of time 0. Further, the discount factor and risky annuity terms are assumed to be subsumed by the change of measure, or Radon-Nikodym derivative, $\frac{dP}{dQ}$. Importantly, this information is available today once we have bootstrapped the underlying CDS curve. Further, the quantity inside the expectation now has only one stochastic variable that we need to model, $S(t)$. This means that, just as with interest rate swaptions, we can price CDS swaptions as martingales in their natural measure, which is defined by risky annuity. As a result, we can then proceed as we have in other options pricing formulas. That is, we can specify a stochastic model that we believe governs $S(t)$. For example, we could suppose that $S(t)$ follow a log-normal distribution and use the Black-Scholes model. This would lead to the following closed-form solution for swaption prices:

$$V_{\text{pay}} = \text{RiskyAnnuity}(0,t,T)\left(S(t)\Phi(d_1) - K\Phi(d_2)\right) \tag{16.34}$$

$$V_{\text{rec}} = \text{RiskyAnnuity}(0,t,T)\left(K\Phi(-d_2) - S(t)\Phi(-d_1)\right) \tag{16.35}$$

$$d_1 = \frac{\log(\frac{S(t)}{K}) + \frac{1}{2}\sigma^2 t}{\sigma\sqrt{t}} \tag{16.36}$$

$$d_2 = \frac{\log(\frac{S(t)}{K}) - \frac{1}{2}\sigma^2 t}{\sigma\sqrt{t}} \tag{16.37}$$

16.9.2 Options on Indices

Options also trade on CDX indices as well as single-name CDS. In fact, index options are far more liquid that their single-name counterparts and are commonly used as vehicles for hedging credit risk. This is largely due to their broad underlying credit exposure and also the attractive properties of options as hedging instruments. Pricing an index option is conceptually similar to pricing an option on a single-name CDS, however a few additional complications emerge:

- In practice all traded index options are structured as non-knockout options.
- This is further complicated by the fact that the front-end protection we saw previously is no longer a binary outcome in the case of an index. That is, a portion of the index may default.
- Indices trade at a fixed coupon C with a corresponding upfront payment rather than at a par spread as we assumed for CDS options.

Importantly, when we are pricing index options, we can no longer treat them as knockout swaptions and then add the front-end protection. Instead we need to value them as non-knockout options and handle the partial index defaults between option origination and expiry, which is more nuanced. O'Kane developed a model where these challenges were incorporated through an adjustment to the strike and forward CDS spread. Once these adjustments have been made, the problem of pricing an index swaption can be reduced to the principles that we saw in the previous section. That is, we can use the (risky) annuity measure and model the spread as a martingale according to the stochastic model of our choice. The details of these adjustments are omitted in this text, however, readers should refer to [192] for more details.

16.10 MODELING CORRELATION: CDOs

Collateralized debt obligations (CDOs) are another commonly traded derivative in credit markets. Like CDS and index swaptions, they are also non-linear products.

CDOs are perhaps most well-known for causing instability in the financial system during the financial crisis [15]. This caused certain restrictions on these products, however there is still a sizable over-the-counter market where these products trade. The premise behind CDOs is to aggregate the cashflows and risks in a credit index, such as the CDX indices described earlier, and re-package the losses and cashflows to different investors according to a set of rules. This set of rules is referred to as a waterfall. It defines how received cashflows and losses are allocated to different pieces of the CDO. These pieces are known as tranches. Each tranche may be characterized by different loss exposure due to different subordination levels, and may have different rights on any incoming cashflows.

There are two types of CDOs: funded and unfunded. Funded CDOs are linked to underlying cash instruments, such as mortgages or bonds. In contrast, unfunded CDOs are linked to a set of underlying synthetic instruments, most commonly CDS. In this section we focus on unfunded CDOs, which are characterized by simpler waterfall structures. The main feature of these CDOs is that the losses are packaged into different underlying tranches.

16.10.1 CDO Subordination Structure

A CDO tranche is defined by its attach and exhaust point. These points define when the losses of the index begin to be absorbed by the tranche (the attach point) and at what level of index loss the tranche is exhausted. As an example, a common mezzanine tranche structure is a 3–7% loss tranche which would have an attach point of 3% and an exhaust point of 7%. This tranche would begin to experience losses once the index loss exceeds 3% and would be fully exhausted once index losses hit 7%. If index losses reached, say 4%, then the loss in this tranche would be $\frac{4\%-3\%}{7\%-3\%} = 25\%$.

The following list highlight the different types of tranches that trade in a funded CDO:

- **Equity**: Equity tranches absorb the first losses in an index and are therefore the riskiest. They are characterized by attach points equal to zero. These tranches have the least subordination as they have no buffer to protect against losses. A common example of an equity trance is a 0–3% tranche in IG or a 0–10% tranche in HY.

- **Mezzanine**: Mezzanine tranches are next to equity tranches in terms of subordination. Like equity tranches, they are exposed to early losses in the index, however, they have a built in buffer. A common example of a mezzanine tranche is a 3–7% tranche in IG or a 10–15% tranche in HY.

- **Senior**: A senior tranche sits above the mezzanine tranches on the subordination structures but below the final, super senior tranche. These tranches are characterized by a large buffer of losses before their tranche is attached. As an example, a senior tranche might be defined as 15–30%.

- **Super Senior**: Super senior tranches are the safest tranches as they are last to absorb any losses in the index. They are characterized by an exhaust point equal to 100%, which would correspond to all components of the index defaulting with no recovery. A common example of a super senior tranche is a 30–100% tranche, meaning losses don't accumulate until the index has experienced a 30% loss.

16.10.2 Mechanics of CDOs

CDOs can be thought of in a similar vein as CDS. That is, we are buying (or selling) insurance against a certain slice of losses in and index, in exchange for a periodic fixed payment. Tranches are also defined in terms of selling and buying protection, just like as we saw in the case of CDS. Clearly, the required fixed coupons will vary by tranche. For example, as the equity tranche has the highest default risk, it is natural that investors would demand a higher fixed coupon for selling protection on this index slice. At the other end of the spectrum, senior and super senior tranches are likely to pay a much smaller fixed coupon as the inherent default risk is significantly lower.

Aside from a few additional complications, the mechanics of CDOs are somewhat similar to the mechanics of credit default swaps. They consist of two legs: a premium or fee leg where a fixed coupon is paid periodically and a default leg where payments are made conditional on a default that is absorbed by the tranche. For each tranche, we will need to build a loss distribution and a survival probability curve in order to value the tranche.

Mathematically, if we define A as the tranche's attach point and E as the tranche's exhaust point, the losses in the tranche at time t, defined as $L(t)$, can be written as the following function of index losses:

$$L(t, A, E) = \frac{(L(t) - A)^{+} - (L(t) - E)^{+}}{E - A} \tag{16.38}$$

With knowledge of this loss function, we can then define the tranches survival probability as $Q(0, t) = (1 - L(t, A, E))$. Once we have these quantities defined, we can plug them into (16.9) in order to calculate the present value of a tranche. Experts in options might also notice that the tranche loss distribution function in (16.38) is a call spread on index losses with a long position in a call on losses struck at A and a short position in a call on losses struck at E.

The challenge in valuing CDOs is building these underlying tranche loss distributions. It turns out that these distributions are dependent on the correlation structure of default, meaning that when investors trade CDOs they are trading correlation products and expressing an implicit view on correlation.

16.10.3 Default Correlation & the Tranche Loss Distribution

CDOs, due to their non-linear payoff structure are by their nature correlation instruments. This adds complexity to valuing them as it requires in depth understanding of the underlying correlation structure of defaults. Further complicating the problem is the high number of constituents in the underlying indices.

A key feature of the problem, however, is the dependence of tranche valuation on default correlation. Some tranches, such as the equity tranche, may, all else equal, benefit from higher default correlations. Other tranches, such as the senior and super-senior tranches will benefit from lower default correlations. This phenomenon is explored in more detail in 16.10.6. Intuitively, however, to see this consider the case where defaults are perfectly correlated. In this case, with some likely small probability the entire index defaults, otherwise there are no defaults. If the entire index defaults, all tranches will be hit, therefore this increased correlation has the effect of spreading the losses out. Conversely if defaults are uncorrelated, but occur with the same probability, we are much more likely to experience smaller levels of index loss. This mean the equity tranche will bear most of the losses, and the senior and super-senior tranches are unlikely to experience any losses.

16.10.4 A Simple Model for CDOs: One Factor Large Pool Homogeneous Model

One of the most common and simplest approaches to modeling CDOs is via what is called the Large Pool Homogeneous (LHP) model. It is a one factor model with a Gaussian latent variable that defines the correlation structure of defaults. We begin by defining the quantity A_i, the value of the assets of firm i, as:

$$A_i = \beta_i Z + \sqrt{1-\beta_i^2}\epsilon_i \tag{16.39}$$

where both Z and ϵ_i are standard normals with mean 0 and variance 1. Further, we assume that the ϵ_i's are independent. Each A_i therefore is a combination of the firm's idiosyncratic risk and also market risk. A firm with a high coefficient β_i is characterized by a high amount of comovement with the market and a smaller idiosyncratic component. Conversely, a small beta corresponds to a firm that has larger idiosyncratic risk. A fundamental feature of this model is that default correlation is expressible only through shared market exposure. This simplifies the correlation structure greatly, making the problem far more tractable.

In the LHP model we assume that a firm will default if it crosses a default barrier. This approach to default follows the Merton approach [183] which says that firms should default if the value of their debt exceeds the value of the firm's assets. Merton approaches are discussed in more detail in the next section.

Along those lines, we define $C_i(T)$ as a default barrier for issuer i that, once crossed, the issuer will default. We will then calibrate this default barrier to match the survival probability until time T of a CDS contract. To do this, we match the CDF at the default barrier to the default probability up until time T, as shown below:

$$\Phi(C_i(T)) = 1 - Q_i(T) \tag{16.40}$$

$$C_i(T) = \Phi^{-1}(1 - Q_i(T)) \tag{16.41}$$

where in the second line we solve for the value of the default barrier that matches today's survival curve. Returning to (16.39), we know that default will happen if $A_i < C_i(T)$. Making this substitution, we have:

$$\beta_i Z + \sqrt{1-\beta_i^2}\epsilon_i < C_i(T) \tag{16.42}$$

Solving this equation for ϵ_i, we then have:

$$\epsilon_i \quad < \quad \frac{C_i(T) - \beta_i Z}{\sqrt{1-\beta_i^2}} \tag{16.43}$$

If we condition on the market variable, Z, then each ϵ_i will be independent. This is an important trick that has made the problem easier to solve. Conditioning on Z can now express our probability of default for issuer i prior to time T:

$$p_i(T|Z) \quad = \quad 1 - Q_i(T) = \Phi\left(\frac{C_i(T) - \beta_i Z}{\sqrt{1-\beta_i^2}}\right) \tag{16.44}$$

If we assume that all credits are the same, as we do in the homogeneous model, then $p_i(T|Z)$ becomes $p(T|Z)$. The loss distribution of the tranche can then be written as:

$$\mathbb{E}\left[L_i(T|Z)\right] = (1-R)p(T|Z) = (1-R)\Phi\left(\frac{C_i(T) - \beta_i Z}{\sqrt{1-\beta_i^2}}\right) \tag{16.45}$$

$\mathbb{E}\left[L_i(T|Z)\right]$ then provides us with the expected loss of credit i at a given horizon time T conditional on the market value, Z. As we are assuming all credits are homogeneous, and independent conditional on the market variable Z, their losses can be thought of as successive draws from a Bernoulli distribution. This means that the index, or portfolio loss distribution will be binomially distributed with probability of default defined by (16.44):

$$f(L(T|Z)) = \frac{N!}{n!(N-n)!}p(T|Z)^n(1-p(T|Z))^{N-n} \tag{16.46}$$

We can then integrate over Z using quadrature methods to obtain a distribution of the index loss distribution. Finally, once we have the index loss distribution, we can apply (16.38) in order to obtain the tranche loss distribution and a set of tranche survival probabilities.

Of course, in practice, we do not have to assume that the credits are the same. These types of models are referred to as heterogeneous models, as they incorporate dispersion of the underlying credits. These models are more complex, but better reflect the actual dynamics of the loss distribution.

16.10.5 Correlation Skew

In practice we observe that the correlation parameter, β[9] that matches a tranche's price is different for different tranches. This is analogous to how we obtain different levels of implied volatility for different strike options. In the CDO space, this phenomenon is referred to as correlation skew. It is not an ideal feature of the model as

[9]Which it should be noted is not a correlation in the truest sense but is treated by markets as one.

it means that we need to use a different loss distribution to price different tranches. More generally, it is a reflection that our model is oversimplified and doesn't accurately represent the true market implied loss distribution. Just as in the case of options we tried to combat this phenomenon with more realistic dynamics, such as jump processes and stochastic volatility, research has been done to try to fit the correlation skew as well. In particular, some have tried to relax the assumption that the recovery rate is a constant [136], while others have tried to pursue other model extensions such as increasing the number of model factors or using making different assumptions about the correlation of defaults.

16.10.6 CDO Correlation in Practice: Impact of Correlation on Tranche Valuation

In this example we leverage the homogeneous CDO model code provided in the repository in order to show the impact of varying the correlation parameter, ρ on index losses. In the following chart we show this index loss distribution for a select set of values of ρ:

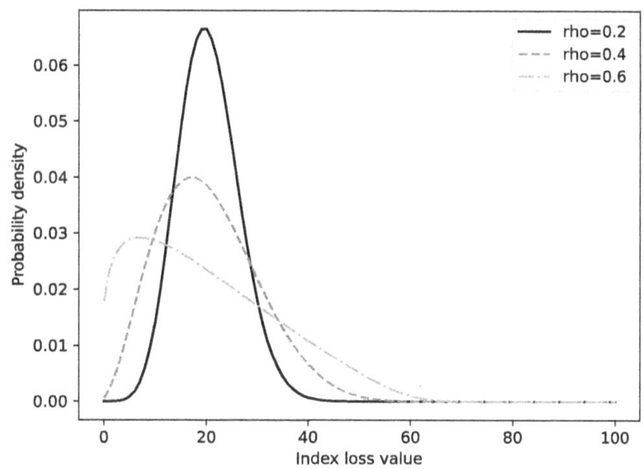

The reader should notice that, as ρ increases, the probability of higher levels of index losses increases with it. The driver behind this is the notion of idiosyncratic vs. systemic defaults mentioned in the chapter. When ρ is high, for example, defaults are highly correlated meaning the presence of a single default implies that many others are prone to default as well. This pushes the losses further to the right on the above chart. This is done at the expense of lower levels of loss, which are now less frequent. This phenomenon is referred to as default contagion. Conversely, when ρ is low, defaults are assumed to be idiosyncratic in nature, or uncorrelated, making high levels of loss less likely, but smaller levels of loss more probably. A significant factor of success when investing in these types of CDOs is being able to discern whether defaults are likely to be systemic, as exemplified here by a high value for ρ, or idiosyncratic.

16.10.7 Alternative Models for CDOs

In this text we only introduce the simplest possible tranche pricing model with the goal of providing the reader some intuition on how these products work and are traded. In practice, the Large Pool Homogeneous model functions as a quoting convention much the same way that the Black-Scholes model does in options. It is leveraged to provide an agreed upon methodology for creating base correlations, which we introduced in section 16.10.3.

This model is by no means the only approach to valuing tranches, and has many weaknesses. In particular, the assumption of a large pool of credits of similar quality is often unrealistic, especially in cases where a few names have diverging spreads. In these cases, other models are likely to be preferable. While these other methods are not described in detail in this text, CDO pricing is a well-developed subject and there are many other references the reader can consult for details on these more advanced models, such as [192] and [227].

- **Heterogeneous Model**: In these models we relax the assumption that all credits have similar characteristics and instead model the underlying single-names directly. This model is better suited to handle cases where there is a lot of dispersion in the index credit quality. This method requires building the full loss distribution, which is computationally challenging due to the dimensionality. FFT techniques, or a recursive algorithm have been developed that can handle this problem, however. These approaches are detailed in [192].
- **Heterogeneous Model with Stochastic Recovery**: Leverages the heterogeneous models ability to incorporate the characteristics of the single-names, allowing it to handle dispersion while also enabling the recovery rate to be a function of market state. Intuitively, we might expcct recoveries to be lower when markets are in a crisis periods, and the inclusion of stochastic recovery enables us to incorporate this in our model. For more information on this approach see [136].
- **Simulation of Default Times via Copula Models**: An alternate approach is to specify a copula [243] that defines the default correlation structure and then use a Monte Carlo algorithm to simulate times to default [192]. This approach will be problematic when employed in the context of a calibration, just as we saw in the analogous options case in chapter 15.

16.11 MODELS CONNECTING EQUITY AND CREDIT

In chapter 2 we alluded to the potential commonality between equities in credit. In particular, both equities and credit are exposed to the same default risk of the underlying firm. That is, if we own an equity or a call option on an equity, it will clearly be negatively impacted by a default. The payoffs of equity and credit instruments, however, are quite different. This interconnected nature and differing payoff structures make these asset classes interesting and has led to a great deal of research trying to connect them. This research has generally focused on trying to look to either the equity market or credit market for potential clues about potential future movements in the other asset class. At heart, this type of model is a relative value analysis where we might look for dislocations where equities appear cheap or expensive relative to credit. In this book, we cover two simple approaches to connecting equity and credit: Merton models and the Hirsa-Madan approach. Much more work has been done on this subject. Interested readers should pursue [134] for more detailed treatment of alternate approaches.

16.11.1 Merton's Model

Merton was one of the first to formally connect equity and credit markets [183]. The premise behind Merton type models, which are referred to as structural rather than reduced form models, is that a firm is incentivized to cease operations when the value of the equity available to shareholders is negative.

Consider that on a firm's balance sheet assets, equities and liabilities are bound by the following relationship:

$$A = E + L \tag{16.47}$$

where A is the value of the firm's assets, E is the value of the firm's equity and L is the value of the firm's liability.

Merton assumed that a firm should default if the value of its equity ever goes below zero. This makes the equity at time T equal to a call option on the firm's assets. If the liabilities, or debt ever exceed the firm's assets, then this means shareholders have negative equity and the firm should default according to Merton. This call option can be written as:

$$E_T = (A_T - L_T)^+ \tag{16.48}$$

If we assume that the liabilities, L_T are fixed, we can then model the firm's assets as a stochastic process in order to better understand the behavior of the firm's equity. For example, if we assume that the asset's follow a log-normal distribution, as Merton did, we are left with a familiar variant of the Black-Scholes pricing formula:

$$\begin{aligned} E_T &= A_T N(d_1) - L_T e^{-rT} N(d_2) && (16.49) \\ d_1 &= \frac{1}{\sigma_A \sqrt{T}} \left(\ln \frac{A_T}{L_T} + \left(r + \frac{\sigma_A^2}{2} \right) T \right), \\ d_2 &= d_1 - \sigma_A \sqrt{T} && (16.50) \end{aligned}$$

To solve this, we need to extract the value of σ_A, which is unobserved. However, we do observe σ_E. It can be shown that under some assumptions the volatility of the assets and the volatility of the equity can be expressed via the following condition:

$$\sigma_A = \sigma_E N(d_1) \frac{E}{A} \tag{16.51}$$

In this context, we are working with a known default barrier, as the volatility of the assets and equity are now known, and the debt is assumed to be constant. We can also compute the firm's distance to default and probability of default:

$$DD = \frac{\log\left(\frac{A_T}{L}\right) + (\mu_A - \frac{1}{2}\sigma^2 T)}{\sigma_A \sqrt{T}} \tag{16.52}$$

$$Q_i(0,T) = N(DD) \tag{16.53}$$

where DD is the distance to default and μ_A is the drift of the firm's assets. This formula follows from the log-normal nature of the asset's stochastic process. The default probability up until time T, $1 - Q_i(0,T)$, can then be juxtaposed with the observed market probabilities in corporate bonds or CDS creating a relative value signal.

The Merton model discussed here only incorporates a single type of debt. More recently, work has been done to try to account for more complex debt structures with different subordination levels. Additionally, research has been done to incorporate an uncertain default barrier into the model [216] as it is hard to pinpoint a single point that the firm's leadership will walk away. As we introduce an uncertain default barrier we will need to integrate over many possible default barriers, with some probabilistic model of the likelihood of each. An example of implementation of a simple Merton model is provided in the text's coding repository.

16.11.2 Hirsa-Madan Approach

Another approach to connecting equity and credit via equity volatility surfaces was proposed by Hirsa and Madan [134]. In this approach, we consider an augmented call and put payoff function that considers both the default and no default cases:

$$C(S_0, K, T) = (1-p)e^{-rT}\mathbb{E}\left[(S-K)^+\right] \tag{16.54}$$

$$P(S_0, K, T) = (1-p)e^{-rT}\mathbb{E}\left[(K-S)^+\right] + pKe^{-rT} \tag{16.55}$$

where p is the default probability between trade initiation and expiry of the option and it is assumed that the value of the equity goes to zero in the event of default. In the no-default case, we are left with are familiar risk-neutral expectation that we can calculate by specifying a set of dynamics for the underlying stock. In the default case, we assume that the call price is worthless, whereas the put has significant value because the equity is worth zero.

In this augmented option pricing setup, the parameter p then becomes another parameter to which we can calibrate our volatility surface. For example, we might postulate that the underlying dynamics follow a Variance Gamma model, or another model of our choice. In the case of the Variance Gamma model we would now have four parameters: $\vec{p} = \{\sigma, \theta, \nu, p\}$. Calibration of this augmented model can then be done using the techniques developed in chapter 14.

16.12 MORTGAGE BACKED SECURITIES

16.12.1 Overview

Mortgage backed securities (MBS) represent a large fraction of credit, and fixed income markets as a whole. MBS products are collateralized by an underlying pool of mortgages. As cashflows come in as homeowners make their mortgage payments, they are then distributed to MBS holders. The products traded in the MBS space may package these cashflows in many different ways. They range from simpler bond like structures, where the received mortgage payments are re-paid to investors in the form of principal, to more complex structures. Tranches are often created, as we saw earlier in the chapter on CDOs, making certain investors exposed to the first losses and giving other investors a buffer. This means that the holders of various MBS products have different claims on the underlying cashflows of the underlying mortgages. Unlike the previous CDOs that we have covered, however, we need to track more than just the losses. For example, we need to define how excess payments (often referred to as prepayments or curtailments) are distributed. Additionally, we need to define which MBS holders will be impacted by missed payments on the underlying mortgages. The set of rules that describe how cashflows are allocated to different investors and different MBS tranches is referred to as a waterfall. In practice, waterfalls are fairly complex documents, however, certain third party programs[10] have conveniently coded these rules enabling quants to leverage this in their modeling. Additionally, in order for investors to achieve better diversification and minimize potential idiosyncratic risk related to individual borrowers, the underlying pools in MBS tend to be quite large, often consisting of thousands of underlying residential mortgages.

MBS are some of the most complex instruments traded in financial markets. To model MBS effectively we need to model the underlying path of each mortgage including when each investor will continue to make payments, when it will default, and when it will pay off the loan. The main risks that a holder of an MBS faces are:

- **Interest Rate Risk**: Risk that interest rates will rise or fall changing the present value of the holders cashflows.
- **Delinquency Risk**: Risk that a borrower fails to make at least one payment. These borrowers are classified as delinquent. These borrowers could then either make a catch up payment, at which would their loan would become current again, or they could continue not to pay and eventually default on the loan.
- **Default Risk**: Risk that a borrower stops paying to the point that the loan is foreclosed on or otherwise defaulted on.
- **Curtailment Risk**: Risk that a borrower pays extra on their mortgage without paying it off in full.
- **Prepayment Risk**: Risk that a borrower pays off their loan early. Most commonly, a loan would be prepaid when a mortgage holder refinances, or sells the underlying property.

[10]Such as Intex.

16.12.2 MBS Waterfall Structure

One of the most challenging features of mortgage backed securities is the uncertainty of when cashflows will be received, and the corresponding challenges in allocating cashflows, or missed cashflows, to different investors. The structure that defines this allocation process is commonly referred to as a waterfall [30]. In contrast to the CDO's covered in section 16.10 where the amount of the cashflows received was unambiguous, and the sole uncertainty was whether the investor would receive the cashflows, in the case of MBS the size of the cash flows can also change in the presence of early payments on the mortgage, known as curtailments or prepayments.

In aggregate, an MBS has the potential for the following four types of cashflows to be embedded:

- Interest Payments
- Regular Principal Payments
- Prepayments
- Curtailments

Importantly, the presence of curtailments and prepayments means that there is a great deal of uncertainty over the timing of cashflows. In other words, the duration profile of an MBS can change significantly. This may be undesirable to some investors and may impact how waterfalls are structured so that this uncertainty can be reduced.

Perhaps the simplest approach to a waterfall is to simply pass through all received cashflows, regardless of the their type, to investors according to some pre-specified subordination schedule. In fact, these type of pass through structures [249] have permeated MBS markets since their origination. The corresponding simplest subordination structure might then be sequential pay [3], where a given a tranche A is specified as having the first right to cashflows up to a threshold, wherein a given tranche B might then begin receiving cashflows [11]. This type of structure is most prone to the uncertainty of the instruments duration discussed above.

Other waterfall structures, such as Planned Amortization Classes (PACs) and Target Amortization Classes (TACs) attempt to reduce the uncertainty in the duration of the MBS by creating a schedule or target of how principal is allocated and creating a suite of different products with different duration profiles.

Another potential waterfall structure for an MBS would be to break the cashflows received into principal only (PO) and interest only (IO) products, known as strips. In that case, the owner of the PO strip would bear the risk of a change in duration due to prepayments or curtailments, whereas the owner of the IO strip would have a relatively more stable duration.

A full treatment of MBS waterfalls and their modeling is beyond the scope of this text, however, interested readers are encouraged to consult [129].

[11]But may receive no cashflows until tranche A is retired.

16.12.3 Modeling Default & Delinquency Risk

Modeling delinquency and default risk will be largely determined by borrower credit quality. Therefore, this can be accomplished by looking at underlying data for each borrower, such as credit score, payment history, and debt to income ratios. If we have access to this dataset, then this becomes one of the most natural machine learning and big data problems in the field of finance. In particular, we are looking to solve a classification problem where the features would be the customer attributes, such as credit score, and the classifiers would correspond to whether loans became delinquent or default, respectively. Many of the machine learning techniques covered in chapter 28, such as Logistic Regression and Support Vector Machines, are well suited to solve such problems.

16.12.4 Modeling Prepayment Risk

Additionally, to model MBS we need to model prepayments and curtailments, which arise when a borrower moves, sells or refinances their home, or makes additional payments beyond schedule. In this case, while borrower quality will matter, a more important driver of prepayments will be mortgage rates. Refinancing behavior in particular will be very interest rate sensitive. Specifically, the lower interest rates are, the greater the incentive for homeowners to refinance their mortgages. This prepayment dynamic creates a negative convexity in MBS bonds. As interest rates decrease, mortgage bonds increase in value by less than standard bonds because of the expected increase in prepayments. When rates increase, conversely, mortgage bonds decrease by more than a standard bond because the corresponding decrease in expected prepayments increases the bonds duration.

In order to model prepayment behavior we need to know the distribution of long term interest rates, which is largely driven by macroeconomic factors and interest rate policy of central banks. One approach to tackle this challenging problem is to leverage a calibrated interest rate model such as SABR/LMM to infer expectations about future rate paths. Alternatively, we could use a macroeconomic term structure model, or insight on the business cycle, to aid our prepayment model. The details of modeling MBS could themselves constitute an entire book. As such, in this book we exclude many details. Readers interested in a more in depth treatment should consult [96] or [129].

16.13 INVESTMENT PERSPECTIVE: TRADED STRUCTURES

16.13.1 Hedging Credit Risk

One of the most common uses of the credit market is for investors to hedge existing credit risk in their portfolios. An asset management firm, for example, might hold a portfolio of corporate bonds and look to hedge part or all of the aggregate embedded credit risk. A natural way to do this would be via broad based exposure to the credit markets. In this chapter we saw that credit indices and options on credit indices in particular seem to fit this criteria.

When constructing a credit risk hedge we need to consider many of the concerns that we discussed in the last chapter with respect to interest rate hedging. That is, we may begin by choosing a set of benchmark instruments, such as index CDS of different maturities, and different index options. Linear (CDS) and non-linear (options) hedges are fundamentally distinct and as such that is the first decision that an investor faces when constructing a hedging portfolio. Option based hedges are attractive in that they allow us to have a hedge only when we need it, however, this comes at a cost in terms of the upfront premium. When deciding whether to use an option based hedge, we need to balance its leveraged payoff vs. the cost of the option. In practice, market participants generally agree that options that hedge tail risks across asset classes tend to be expensive, and credit markets are no exception to this rule. As a result, finding cheap options based hedges is a legitimate challenge and requires a fair amount of skill[12].

Once we have decided on a benchmark hedging instrument, or set of instruments, the next step is to create a hedging portfolio with proper hedge ratios. The challenges and approaches to this will mimic those in the rates example in the previous chapter. In particular, we need to account for different types of movements in the credit spread curve, such as parallel shifts, flattening and steepening. As we detailed in the last chapter, one robust approach to handling this is to use a scenario approach and construct a hedging portfolio that minimizes tracking error over this set of scenarios.

An additional complication arises, however, in the credit markets due to the potential of a default. In particular, when we build a hedge portfolio in theory we want to be hedged to movements in credit spreads. Perhaps even more fundamentally, we want to be hedged in the case of a jump to default of one or more issuers.

That means that in practice we must choose between the following types of hedges:

- **DV01 Based Hedge**: Choose a hedging portfolio such that movements in credit spreads do not impact your portfolio value.
- **Jump to Default Based Hedge**: Choose a hedge such that minimizes exposure should an issuer default. Idiosyncratic default risk will be harder to hedge via a broader index based exposure than systemic default risk.

In practice the hedging portfolios catered to each of these approaches can be quite different, meaning that if we are hedged on a DV01 basis we could still have considerable jump-to-default risk and vice versa. This is an important consideration for investors as they think through what an optimal hedging portfolio looks like.

16.13.2 Harvesting Carry in Credit Markets

Aside from hedging credit risk exposure, a main motivation for investors to participate in credit markets is to harvest the embedded risk premia within the market. Of particular interest to credit investors is a carry risk premia. A significant body of evidence suggests that carry is a well-rewarded risk premium in credit markets [142]. We have seen carry in other contexts, and the framework is the same in credit markets.

[12]If you believe this type of skill exists.

To harvest carry in credits, we would construct a portfolio that is long the assets with the best carry profiles, and short the assets with the worst carry profiles. These trades could be instrumented via corporate bonds, or CDS. At the heart of this will be defining a carry metric that is suitable for each credit product. We can approach this analogously to how we did in rates, that is we can consider two components:

- **Carry Solely Due to Passage of Time**: Amount of appreciation or depreciation due to passage of time holding all other parameters constant.
- **Rolldown**: Amount of appreciation/depreciation due to an implied roll down the credit spread curve. For example, we might expect a 5y CDS to roll in one year to the current 4y CDS spread.

If we aggregate these components with the desired assumptions about the evolution of the curve, we then have a carry metric that can be used across credits and maturities. The final step in creating a carry strategy based on this metric would be to turn these raw signals into a portfolio, characterized by a set of long-short weights. There are many different techniques for approaching this problem. More details on these approaches and how to think about a conversion from a raw signal to a portfolio are provided in chapter 27.

16.13.3 CDS Bond Basis

Earlier in the chapter, we discussed how a basis emerges between bonds and the equivalent CDS due to differences in liquidity and funding of the instruments. A natural interpretation is that this basis is a risk premia that hedge funds, and other entities may want to harvest. They may also want to express some view on the relative value of the basis. That is, they want to model when the basis is extreme, and enter into positions conditional on this relative value judgment.

At heart, this trade is attractive because bonds and CDS have offsetting risks, most notably default and interest rate risks. Consequently, as their spreads diverge, it is natural for investors to want to pair these against each other taking a short protection position in the asset with the higher spread and a long protection position in the asset with the lower spread, thus collecting the spread differential.

The following table shows an example trade that might be entered when the basis is positive, that is when the spread on bonds exceeds that of CDS:

Position Type	Instrument
Long	Corporate Bond
Long Protection	Matching CDS

The natural follow up question is what units of the bond and CDS we should be long and short, respectively. It turns out in practice this is not a trivial question, and many of the challenges faced here mimic the challenges introduced in the section on hedging credit risk.

In particular, if we are short protection on one unit of the bond, we can be long protection in δ units of the CDS. One approach to choosing δ would be to set it such that we had equal notional in the bond and the CDS, that is, $\delta = 1$. This would hedge us in the case of a jump to default, making it an attractive starting point. However, we wouldn't necessarily be hedged in the event that interest rate or credit spread curves moved. That is, we would have neutralized our jump to default risk but our DV01, or credit spread 01 risk may be significant.

As a result, an alternative would be to choose δ such that our DV01 risk is neutralized. This would mean that as the credit spread curve moved we would theoretically not experience P&L shocks. We could then even continually re-balance this δ such that we maintained this DV01 neutral property. A side effect of this approach however, is that, depending on the hedge ratio required to make the trade DV01 neutral, we might now have a significant amount of jump-to-default risk.

16.13.4 Trading Credit Index Calendar Spreads

The presence of swaptions on individual CDS and credit indices means that popular options strategies to harvest a potential volatility premium are also accessible in credit markets. This creates a vast array of potential opportunities as there is theoretically an entire volatility surface to analyze and parse for relative value. Additionally, the techniques for doing so in credit markets will be similar to the techniques for doing so in other asset classes. We could, for example, look for a premium between implied and (predicted future) realized volatility.

In this section, we explore one particular options strategy in more detail: a calendar spread. A calendar spread, as detailed in chapter 16 consists of a long and short position in the same asset with different expiries. In this example, we assume that the strike on the two options is also the same, however, this is by no means required. For example, one might construct the following structure on a single-name CDS or credit index:

Long/Short	**Instrument**
Short	1M At-The-Money Payer
Long	3M At-The-Money Payer

In the absence of arbitrage, this trade is characterized by an initial outflow, as the three-month option should cost more than the one-month option. As a result, if we consider the value of the option in one-month, at the first options expiry, we can consider the value of the package based on whether that first option expires in the money:

- *Out-of-the-money*: the value of the 1M option expires worthless. There will be residual value in the remaining 2M option we are long, however, this residual value will be highest when we are closest to its strike.
- *In-the-money*: the expiring 1M option has intrinsic value and will be exercised. The remaining 2M option also have intrinsic value and can be exercised to cover

the obligation of the short expiring 1M option. If the 1M option is deeply in the money, then its value will converge to the value of the remaining option.

As a result, at the extremes, if the 1M option expires deeply in or out-of-the-money, this package loses money. This is because there is an initial capital outflow and a payoff that approaches zero. The best case for the structure is for the 1M option to expire exactly at the strike. This would make its exercise value zero, but would maximize the value of the remaining long option exposure. This structure has a similar feel to the Bermudan vs. European option structure that we examined in the last chapter. Another feature of this structure is that, if the 1M option expires worthless, we can then proceed to sell another 1M option collecting additional premium, and will now have a 2M remaining long option position against it.

This type of structure is most attractive when short dated options (i.e. 1M) are expensive relative to longer dated options. In particular, when implied volatility surfaces are inverted, the pricing on these packages can be appealing. Another appealing feature of this trade is that even though it has a short volatility profile in some sense, as evidenced by its butterfly like payoff profile, it has defined downside. That is, the worst case loss for the trade is the initial premium outlay.

16.13.5 Correlation Trade: Mezzanine vs. Equity Tranches

A common strategy that hedge funds engage in within the CDO market is to construct relative value trades of tranches against each other. Inherently, these are by nature correlation trades, as we saw tranche loss distributions depend greatly on the structure of default correlations. A fundamental deciding factor in these trades is whether we think defaults are likely to be largely idiosyncratic, in which case they are likely to be mostly absorbed by the equity tranche, or systemic. Recall from earlier that systemic defaults disproportionately impact the more senior tranches as they spread the losses further down the subordination structure.

In this section we consider a specific example of a potential relative value tranche trade that consists of an equity vs. mezzanine tranche. In this example we will choose to be short protection on the equity tranche and long protection on the mezzanine tranche, as shown:

Long/Short Protection	Instrument
Short Protection	0–3% IG Equity Tranche
Long Protection	3–7% IG Mezzanine Tranche

As constructed, this is a relative value trade that is long correlation. That is, this trade will do well should systemic default risk dominate. As a result, we'd expect this trade to do well in a major financial stress event like the global financial crisis. This trade is, conversely, short idiosyncratic defaults.

If this package is constructed in a one-to-one ratio, then it will come with a considerable delta. Said differently, a one-one ratio of these tranches will have significant exposure to changes in credit spreads of the underlying index. This is because the equity tranche is by nature more sensitive to changes in credit spreads than mezzanine tranches. As a result, if implemented in equal units, this particular iteration would experience gains when credits spreads compressed, and losses when spreads widened. A natural alternative to constructing this trade would be to do it so that it is immune to these changes in credit spread, which could mean sizing the as one unit of the equity tranche and Δ units of the mezzanine tranche, where Δ is chosen to neutralize exposure to a small change credit spreads.

CHAPTER 17

Foreign Exchange Markets

17.1 MARKET SETTING

17.1.1 Overview

In this chapter we orient the reader to the foreign exchange market, which is one of the deepest and most liquid asset classes, especially for exotic instruments. Much of this demand for exotics stems from a natural need for many investors to hedge. Companies who generate revenues in a foreign currency, for example, are inherently exposed to the value of these revenues becoming less valuable if the currency devalues. These firms then may choose to hedge their currency risk through forward contracts, or other derivative structures. This same phenomenon exists for individuals who plan to travel to another country in the future. This trip may become more expensive by the time they get there, if the foreign currency has appreciated. Individuals more often than not are willing to bear this risk, however, corporations and other entities instead may choose to hedge this risk. To do this, they would look to lock in the current value of the currency at a specified date in the future. In other words, they would enter into a forward or futures contract. Alternatively, the investor may want to make their hedge conditional. In this case they might choose to use an option structure, ranging from a vanilla option to a barrier. We explore the problem of currency hedging in more detail in section 17.5.1.

The FX market is a predominantly over-the-counter market, where the hedging demand, flows from central banks, and presence of hedge funds and other direct investors, leads to a strong two-way flow and a robust market. This means that our favorite dealer is happy to quote very customized exotic structures, such as Asian options, barrier options and volatility linked derivatives. In section 17.1.5 we provide more color on the types of derivatives traded, and then provide a more detailed treatment of specific structures in section 17.4.

To an investor such as a hedge fund, foreign exchange markets provide another vehicle for expressing their views, and in many cases they may provide a more direct path to do so. Additionally, just like in other markets, we can look to harvest carry and other available premia in both the exchange rates and their volatility surfaces. In contrast to other asset classes, such as investments in stocks or bonds, however, FX is unique in that it is inherently a zero-sum game. Long-term investments in stocks

DOI: 10.1201/9781003530626-21

and bonds should generate positive returns for an investor, due to increased profits for the firm or interest earned for the bond. For currencies, it is less obvious that an investment in say US dollars, relative to say Euros, should appreciate over time. And if it does, it comes at the expense of an investment in Euros. This means that the investment participants in FX tend to have shorter investment horizons and be more tactical in nature.

In the remainder of this section, we highlight the most liquid currencies, as well as the different types of derivatives that encompass the FX market.

17.1.2 G10 Major Currencies

The most liquid currencies within the FX market are commonly referred to as the G10 majors. The following currencies are members of this G10:

- Australian Dollar (AUD)
- Canadian Dollar (CAD)
- Euros (EUR)
- Yen (JPY)
- New Zealand Dollar (NZD)
- Norwegian Krone (NOK)
- British Pounds (GBP)
- Swedish Krona (SEK)
- Swiss Franc (CHF)
- US Dollar (USD)

A more restrictive list, titled the G4 is also sometimes used, referring to Euros (EUR), Yen (JPY), Pounds (GBP) and US Dollars (USD). These currencies have the deepest and most liquid markets, and are often used as reference, or base currencies. Derivative instruments ranging from simple forwards contracts to complex exotic options can be traded within reasonable bid-offer spreads within this set of currency pairs.

17.1.3 EM Currencies

As we move away from the major G10 currencies, liquidity can become a bit more challenging, especially for the more exotic structures we present later in the chapter. Nonetheless, the overall depth of the FX market, and its exotics market in particular means that even as we expand our currency universe a decent OTC market exists with reasonable bid-offer spreads. The following list highlights some of the main EM currencies traded within FX markets:

- South African Rand (ZAR)
- Turkish Lira (TRY)
- Brazilian Real (BRL)
- Mexican Peso (MXN)
- South Korean Won (KRW)
- Russian Rubble (RUB)
- Indian Rupee (INR)
- Hong Kong Dollar (HKD)
- China Renminbi (CNH/CNY)

China's currency has two rates, one which is available for on-shore accounts, and the other which is available to foreign investors via off-shore accounts.

These currencies tend to be characterized by higher available yields, making them attractive carry currency candidates. Conversely, these currencies tend to be more exposed to depreciation in the event of a macroeconomic or geopolitical shock, as the flight to quality usually flows from emerging markets to safe haven currencies like CHF, JPY and USD.

17.1.4 Major Players

An interesting feature of the foreign exchange market is that it is the meeting place of players with different objectives. On the one hand, central banks operate in the FX markets to control their currencies. A currency that is too strong, might discourage foreign investments, whereas a currency that is too weak might erode spending power on a global scale, and might generate undesired levels of inflation. Hedge funds, at the other end of the spectrum operate in FX markets for the same reason they do any other market: their desire to generate alpha. They may look to achieve this via FX specific risk premia, or via expressing their macroeconomic views. Additionally, asset managers may enter the currency market to hedge their equity investment in foreign firms. Similarly, corporations who do business globally will naturally have embedded currency exposure, and may look to hedge some or all of that risk in the FX market.

In the following list, we summarize the main participants in FX markets and the role that they play:

- **Central Banks**: Central banks operate in the currency market to implement monetary policy, changing the money supply as the need arises. A central bank may regulate an exchange rate to avoid the side effects of having a currency that is too weak or too strong. When entering these transactions, central banks have an inherently different objective than most investors in markets. Rather than looking to profit on each transaction, they are looking to maintain financial stability and competitiveness for their local economy. This creates an interesting interplay with investors seeking returns through currency trading.

- **Corporations and Other Hedgers**: Corporations and other hedging entities enter the currency market because they have some sort of exposure linked to a foreign currency, whether it is through an obligation to pay a foreign party, or through receiving profits in a foreign economy. These entities may or may not want to bear the risk that a change in the currency can lead to a material difference in the value of these future assets/obligations. In that vein, they might enter the currency market to lock in an exchange rate today, for these future cashflows.

- **Dealers/Investment Banks**: Like in other OTC markets, dealers play the roles of intermediary matching buyers and sellers who have different hedging needs or different objectives in the FX market. Dealers may create exotic options structures for their clients, and then look to offload this risk either entirely through an offsetting party, or by immunizing the trade's major Greeks.

- **Hedge Funds**: Hedge funds use currency markets primarily as a tool for generating returns, or alpha. This may be through systematic strategies that harvest dislocations in the market, persistently harvesting a set of available risk premia, or using currencies to implement their macroeconomic views in a more precise manner.

- **Asset Managers**: Asset managers, like other hedging entities, may use currency markets to hedge currency risk that emanates from their global equity investments. Like hedge funds, they may also view FX and FX derivatives markets as an additional source of return, or another mechanism to instrument their macroeconomic views.

17.1.5 Derivatives Market Structure

The FX market is a predominantly over-the-counter market with robust depth for simple and complex, exotic structures alike. The natural hedging demands of many market participants, combined with the wide array of investors looking to provide liquidity to the market and harvest dislocations, creates a deep two-way market even for some of the most exotic payoff structures. Generally, speaking, the derivatives market can be broken into four levels of varying complexity. On the simpler end we have forward contracts, followed by vanilla, European style call and put options. The exotic options market is then typically broken down into first and second generation exotics, with first generation exotics covering simpler payoffs such as digital payouts and barriers, and second generation including multi-asset options, more complex path dependent such as Asian options, and volatility derivatives.

In the following table, we provide a brief description of each of these four types of derivatives that appears in FX markets:

- **Forwards**: FX forwards are the simplest derivative instrument. They are linear in nature, meaning they can be priced by static replication arguments and therefore can be valued without the use of a stochastic model. Nonetheless, they are a common hedging vehicle that enables investors to lock in an exchange rate at a future date, today.

- **Vanilla Options**: The vanilla options market in FX consists of European call and put options that are typically traded over-the-counter. These are the simplest non-linear derivatives traded in the market, and make up the lion's-share of the transactions and notional traded.

- **First Generation Exotics**: First Generation Exotics refer to the simpler exotic structures traded in the FX derivatives market. Digital options, one-touch and barrier options on a single currency pair are commonly known as first generation exotics.

- **Second Generation Exotics**: Second Generation Exotics refer to the most complex exotic option structures traded in FX markets. They are usually path dependent, and often rely on more than one currency pair. Some common examples of second generation exotics are Asian options, and multi-currency barrier options.

17.2 MODELING IN A CURRENCY SETTING

17.2.1 FX Quotations

One of the challenges of modeling FX and FX derivative markets is the litany of corresponding market conventions. This permeates the quotations of exchange rates all the way through to how volatility surfaces and exotics are quoted. In this section, we provide a broad overview of these conventions, omitting some details for the sake of brevity. For more detailed treatment of the practicalities of modeling currencies, the interested reader should consult [58] or [264].

Exchange rates are quoted in pairs. The value of a given currency pair tells us the value of currency X in terms of, or per unit of currency Y. To take a more concrete example, we cannot know the value of the dollar (USD) without knowing what units to measure it in. The value of a dollar to an investor holding Euros (EUR) may be very different than the value of a dollar to an investor holding Brazilian Real (BRL). Exchange rates, then, tell us the rate at which we can convert from one currency to another. In exchange rate markets, the currency that we are measuring in terms of, currency Y above, is referred to as the foreign currency, whereas the other currency, currency X above, is known as the domestic currency:

- **Domestic Currency**: Also known as the quote currency. The domestic currency is the numerator in the currency quotation, or the currency we are looking to convert into.

- **Foreign Currency**: Also known as the base or reference currency. The foreign currency is the denominator in the currency quotation, or the currency we are looking to convert from.

FX pairs are quoted using these three letter codes, and the convention is to quote the foreign currencies three letter code followed by the domestic currencies three letter code: FORDOM. Mathematically we can write an exchange rate in terms of the domestic and foreign currencies as:

$$\mathrm{FX}_{\mathrm{FORDOM}} = \frac{\mathrm{Domestic}}{\mathrm{Foreign}} \tag{17.1}$$

As an example, consider an investor who is looking to convert their USD to Swiss Francs (CHF). This investor wants to know how many CHF they will receive **per** dollar, or USD. Thus, they would like to know the FX rate where USD is the foreign, or base currency and CHF is the domestic. This rate is known as the USDCHF rate. A quote rate of 0.9 USDCHF, for example, would mean that for every unit of USD (dollar) that is converted, 0.9 units of CHF are received. Equivalently, investors looking to convert the other way, from CHF to USD would need to know the CHFUSD rate, or the value of a unit of USD per CHF. This exchange rate must simply be the reciprocal of the rate above, as if we convert our money to CHF and back to USD instantaneously, we should end up with the value as we started with. That is:

$$\mathrm{FX}_{\mathrm{XXXYYY}} = \frac{1}{\mathrm{FX}_{\mathrm{YYYXXX}}} \tag{17.2}$$

where XXX and YYY are the currencies in the pair, respectively.

There are certain conventions about how standard FX rates are quoted, that is, which are used as the domestic and foreign currency. For example, USD is quoted as the foreign currency when quoted against most other currencies, including JPY, CAD and ZAR. Conversely, it is quoted as the domestic currency when quoted against EUR and GBP. As an example, as USD is quoted as the foreign currency against CAD, the standard rate would then be USDCAD. The quoted exchange rate of USDCAD then tells us how many units of CAD, the numerator, are required to purchase 1 unit of USD, the denominator. An increase in this USDCAD rate means that USD has appreciated relative to CAD, and CAD has depreciated relative to USD.

Conversely, as USD is quoted as the domestic currency against EUR, the FX quote would then be EURUSD. This means that an EURUSD quote is telling us how many dollars, or USD are required to purchase a Euro, or EUR. With this quoting convention, and increase in the rate, for EURUSD, means a depreciation of USD versus and appreciation of EUR.

This same concept exists in other asset classes as well, although we tend not to even think about these underlying conversions. For example, a share price for Apple of $125 tells us that we can convert from shares of Apple to USD at a rate of $125 per 1 unit (share) of stock. An investor who is looking to sell shares of Apple is interested in doing just that, converting from units of Apple to USD. The rate at which they can do this is can then be thought of as an exchange rate with USD as the domestic currency that we are converting into, and Apple shares as the base or reference currency. An investor who is looking to buy Apple stock, is analogously looking to convert their USD into AAPLE. The relevant rate for this investor then is the number of shares (AAPLE) per unit of USD. In this case the foreign, or base currency is USD and the domestic currency is AAPLE. This rate would then be one over the share price.

These conventions for how different FX pairs, or crosses, are quoted permeates the spot and derivatives market. As such, it is important for aspiring quants to build intuition and orient themselves to the quotes, and how they impact the interpretation of an appreciating or depreciating currency, respectively. This is especially important to master in this simplified context, before adding the complexity of option payoffs and modeling.

17.2.2 FX Forward Valuations

In chapter 2, we learned that forwards and futures contract valuations were determined by an arbitrage argument between the spot and the forward wherein we could statically replicate the forward using positions in the underlying asset, or spot, as well as borrowing or lending at the risk-free rate. In equity markets, we also saw how a dividend impacted this relationship. Specifically, a long position in the forward means foregoing any dividends that are paid in the interim. Thus, we needed to adjust the forward price lower for any intermediate dividend payments, which is often accomplished via a dividend yield term q.

In FX, dividends are not a factor in forward valuations, however, we do have to account for the local interest rate that we are able to obtain. Depending on what currency we hold, we may have access to differing deposit rates on our capital. A deposit in BRL, for example, where rates are high, is likely to earn more interest than a deposit in JPY, where rates are low. Thus, if we choose to hold JPY, we are implicitly foregoing the higher rate that we could earn if we were willing to hold BRL. Along these lines, the interest earned in the foreign currency that we are foregoing is akin to a dividend yield in the equities space. This leads to following foundational FX forward valuation equation [97]:

$$F = S_0 \exp\left((r_d - r_f)\,T\right) \tag{17.3}$$

where T is the time to expiry of the forward contract, S_0 is the current exchange rate, and r_d and r_f are the interest rates of the domestic and foreign currency, respectively. r_f in particular plays the role of a dividend that is foregone for not holding the foreign asset but instead using a forward contract. The term $(r_d - r_f)$ is referred to as the interest rate differential, which in turn defines the forward curve for a currency pair r to spot.

17.2.3 Carry in FX Markets: Do FX Forward Realize?

In the last section, we saw that FX forward valuations were determined by an arbitrage argument that is analogous to the one presented in chapter 2, and that forward curves are ultimately determined by the difference in local interest rates. This forward then fundamentally tells us what the mean, or expected future value of the exchange rate is at some point in the future. Importantly, however, this is the mean in the **risk-neutral measure**. In the physical measure, where investors are no longer risk-neutral but instead risk averse, it could be a very poor estimate of the mean. This difference between the risk-neutral and physical measure is one of the key philosophical emphases of this text and the reader is encouraged to think through the embedded subtlety.

Among practitioners, there is a great deal of debate about whether the forward, the spot, or another value are the appropriate expectation for the future value of the exchange rate. Assuming the forward is the expectation, then from (17.3) we can see that a currency with a high interest rate is expected to depreciate by an offsetting amount. If spot instead is the proper expectation, then we can invest in economies with higher deposit rates, collecting the extra yield, without a corresponding move in the exchange rate. That is, in this case we could earn a premia for investing in higher yield currencies relative to lower yielding currencies. And, in fact, this is a well-documented potential risk premia known as FX carry [214]. It is referred to as a carry premium because it relies on the propensity of exchange rates to experience less price appreciation, or depreciation, than is implied by their forward curve. One often cited justification for the existence of this potential premia is that investors are being compensated for investing in higher risk economies with higher deposit rates because of their tendency to depreciate significantly in periods of market stress. Additionally, these economies may be viewed as having less stable financial systems, leading to higher probabilities of outsized moves going against these carry trades. Ultimately, these rationale are based on behavioral biases and an aversion for large losses.

In the remainder of this section we analyze this phenomenon for two particular currency pairs of interest: USDBRL and AUDJPY. In each case we analyze the effectiveness of carry by examining the interest rate differential and subsequent move in the spot exchange rate to see whether the forwards realize over long periods of time. If carry is indeed a risk premium we would expect the forwards not to realize but instead for spot to appreciate or depreciate by less than was implied in the forward curve.

To see this, we begin with USDBRL which is a natural candidate for currency carry as rates in BRL can be expected to be uniformly higher than corresponding rates in USD by a significant margin as Brazil is an emerging market currency. In the following charts we summarize the performance of USDBRL, its forward curve and the indicative performance of a naive carry strategy in USDBRL:

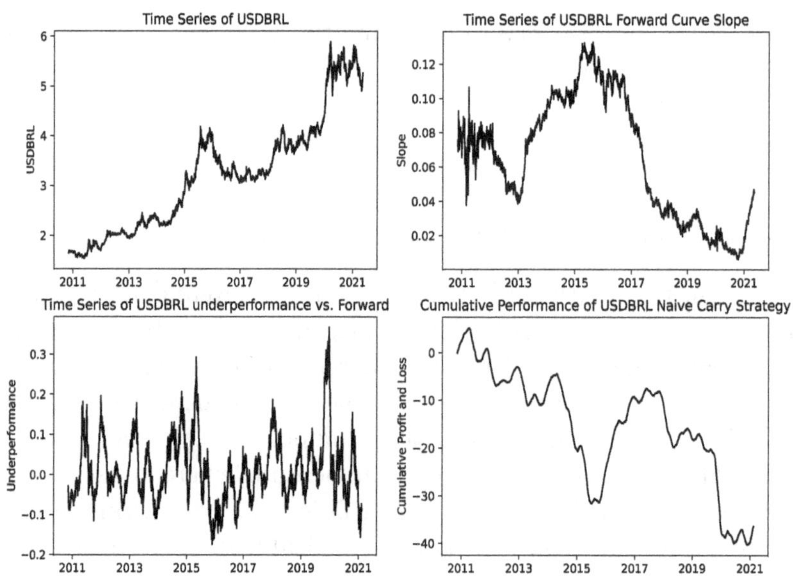

In these charts we can see that the forward curve is persistently upward sloping, with an average slope of approximately 6.5% annualized. However, the USDBRL spot rate has depreciated significantly, approximately 12% per year. Said differently, the USDBRL rate has increased by far more than was implied by the forward curve. This is because, even though the interest rate differential was high throughout the period, making it an attractive carry candidate, the exchange rate actually overshot the forward because of a secular decline in the BRL during the period of analysis. As we can see in the lower right chart, this would have led to substantial losses in a carry strategy relying on USDBRL.

Next we consider another common example of currency carry, AUDJPY. Deposit rates in AUD are also significantly higher than those in JPY. This is largely due to the anomalously low nature of rates in JPY. Here, if we look at the analogous set of charts we find that AUDJPY has indeed been an attractive carry candidate:

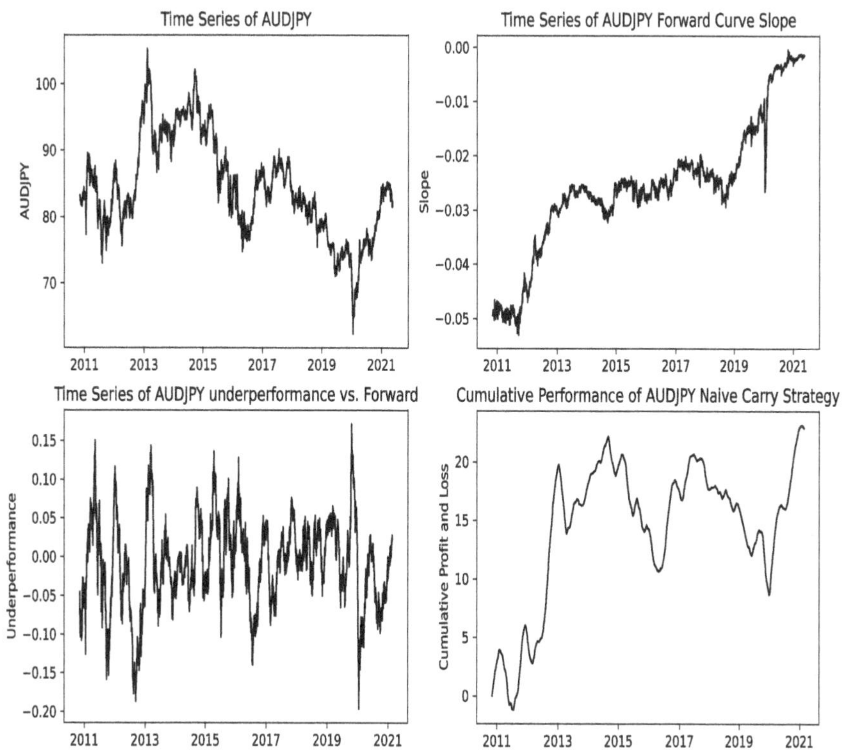

Here, we can see that forward curve is persistently downward sloping, with an average slope of -3.5% annualized, however, the spot remained largely unchanged over the period of interest. Unlike USDBRL, this led to potential gains from carry strategies applied to AUDJPY.

More generally, a substantive body of research exists validating FX carry as a potential risk premia across the cross-section of currency pairs [214]. As we saw in this section, however, harvesting such a premia can be quite nuanced, as there may be long periods where high-yield currencies with higher deposit rates depreciate significantly more than their forward curves imply, or depreciate in line with their forwards. This is something we explore more from an investment perspective in section 17.5.2.

17.2.4 Deliverable vs. Non-Deliverable Forwards

Another challenge that emerges in currency markets is that not all currencies can be held by all investors. Some currencies, instead have certain restrictions on who can take ownership of their currency. Perhaps a bank account at a local institution is required to store the currency, for example. This leads to two types of forward contracts within the FX space:

- **Deliverable Forwards**: Forward contracts that are settled by taking on the specified number of units of the specified currency. These are the most common set of forwards traded in currencies without restrictions.
- **Non-deliverable Forwards (NDFs)**: Applies to currencies with barriers to entry to hold the currency. In these cases, the forward contract describes a pre-determined cash settlement mechanism.

In NDF markets, this can lead to differences in FX rates between those who can access the market, and be delivered into the currency, and those who can't and must instead cash settle. A common example of an NDF is one on CNH and CNY, where one corresponds to an on-shore rate, available to those who are able to take delivery, and the other applies to off-shore investors who cash settle. The basis between these two represents the difference in exchange rate a local vs. a foreign investor can achieve. Additionally, INR and BRL are two other examples of NDFs within the FX market.

17.2.5 FX Triangles

One of the interesting things about currency markets is the presence of so-called currency triangles. A currency triangle is a set of three currencies where we can obtain the exchange rate for a given currency pair within the triangle as the product or quotient of the two other currency pairs in the triangle. To take an example, consider the currencies BRL, CHF and USD. Within this ecosystem there are three underlying currency pairs, CHFBRL, USDBRL and USDCHF. The CHFBRL rate, however, is completely determined by knowledge of the other two rates, in the absence of arbitrage. In particular, it is defined by the following relationship:

$$\frac{\text{BRL}}{\text{CHF}} = \frac{\text{BRL}}{\text{USD}}\frac{\text{USD}}{\text{CHF}} \tag{17.4}$$

In spot trading space, any deviation from this relationship will be quickly identified and corrected by an arbitrageur or automated trading algorithm. As we move into the options space, however, currency triangles become more interesting, and the strict arbitrage argument we have here is replaced with a dependence on correlation, and bounds for option valuation. This concept is analogous to the difference between an option on a basket vs. a basket of options, as we have seen in other parts of the text. The idea behind this is that once we add a non-linear payoff, then dispersion matters. This is something we explore in more detail in section 17.5.3.

17.2.6 Black-Scholes Model in an FX Setting

In a currency setting, the Black-Scholes model, as well as other stochastic models, need to be modified to account for the differences in FX forward valuations relative to other asset classes. In chapter 13 we detailed the standard Black-Scholes model, which has many iterations depending on whether dividends are paid on the underlying asset, etc. In FX, we could leverage this Black-Scholes model and consider foregone interest, r_f as the dividend.

Alternatively, the following variant of the Black-Scholes formula, which is adapted to use the value of the forwards instead of the spot, interest rates and dividends, can be use to value FX options:

$$
\begin{aligned}
C_0 &= e^{-r_d T}\left(\Phi(d_1)F - \Phi(d_2)K\right) && (17.5)\\
P_0 &= e^{-r_d T}\left(\Phi(-d_2)K - \Phi(-d_1)F\right) && (17.6)\\
d_1 &= \frac{1}{\sigma\sqrt{T}}\left(\ln\frac{F}{K} + \left(\frac{\sigma^2}{2}\right)T\right), \\
d_2 &= d_1 - \sigma\sqrt{T}, && (17.7)
\end{aligned}
$$

where F is the value of the FX forward contract as defined in (17.3), K is the strike, and T is the option expiry time. Similar adjustments would need to be made to other stochastic processes, however, these modifications are typically easy to integrate either through the use of r_f as a dividend, or in terms of the forward rather than the spot.

Additional subtleties arrive in FX markets as some options are quoted in terms of spot premium and other are quoted in terms of forward premium. (17.5) and (17.6) are quoted in terms of spot premium, meaning it is the value of the option today rather than at expiry. Quoting in terms of forward premium, or the value at expiry rather today, is also quite common in FX markets. To convert equation (17.6) to forward premium, we would simply remove the discount factor outside the patenthesis, $\exp(-r_d T)$. An example of this is provided in the book's coding repository.

17.2.7 Quoting Conventions in FX Vol. Surfaces

In contrast to most markets, where strikes and prices (or implied volatilities) are the standard building blocks in a volatility surface, FX markets are quoted in terms of three underlying components:

- At-the-money volatility
- Risk Reversals
- Butterfly[1] volatilities

[1]Or strangle.

The presence of these unique quoting conventions increases the complexity of FX volatility surface calibration problems.

One of the challenges with calibrating FX volatility surfaces is that strikes are not provided at all. Instead, the points are quoted in terms of delta. Intuitively, this actually provides a great deal of convenience in a market setting as that means that as spot moves the volatility surface doesn't need to be re-quoted. Additionally, there is a link between the other quoted FX volatility surface points and the moments of a risk-neutral density. The at-the-money volatility, for example, is linked to its variance. A risk-reversal, as we will soon see, it linked to its skewness. A butterfly or strangle, quote, finally is linked to the kurtosis in the distribution.

If the strike is unknown, but the implied volatility is known, this doesn't present much of an analytical problem as we know the formula for the delta of an option under the Black-Scholes model. For example, for a call option, the delta can be written as:

$$\begin{aligned} \Delta_c &= N(d_1) \\ d_1 &= \frac{1}{\sigma\sqrt{T}}\left(\ln\frac{F}{K} + \left(\frac{\sigma^2}{2}\right)T\right), \end{aligned} \tag{17.8}$$

If we are provided with the delta and the implied volatility, the only remaining unknown is K which can be solved by inverting the above equation. This leads to:

$$K = F\exp\left(-\sigma\sqrt{T}N^{-1}(\Delta) + \left(\frac{\sigma^2}{2}\right)T\right) \tag{17.9}$$

A similar derivation exists for a put, or we use put-call parity to infer the delta of a call option at the same strike as our desired put of interest[2].

In the following list, we describe each of the three components of an FX volatility surface quote.

- **At-the-money**: The volatility for the at-the-money strike, σ_{ATM} which is defined as the point where the delta of a call and a put are equal in absolute terms. This point typically corresponds to the 50 delta point. With this volatility, the corresponding at-the-money strike, K_{ATM} can be found using (17.9). This particular piece of an FX quote is the most well-defined and most tractable.

- **Risk Reversals**: In FX, risk reversal quotes are defined as the difference in implied volatilities between an out-of-the-money call and an out-of-the-money put with the same absolute level of delta:

$$\sigma_{\text{RR}} = \sigma_C - \sigma_P \tag{17.10}$$

  Given that we don't know the strikes corresponding to these risk reversals, and also don't know their individual implied volatilities, risk reversals will be problematic for us to deal with in isolation. Instead we will need to leverage additional market information and structure.

[2] A long position in a call plus a short position in a put is equivalent to owning the underlying and must therefore have a delta of 1.

- **Butterlies or Strangles**: The final type of FX volatility surface data is a butterfly or strangle, and measures the difference in implied volatility on out-of-the-money strangles relative to at-the-money straddles[3]. The butterfly or strangle quote is treated as an implied volatility such that:

$$\sigma_{\text{fly}} = \sigma_{\text{ATM}} + \text{offset} \tag{17.11}$$

where the offset quantity is what is quoted by the market as a butterfly quote. With σ_{fly} now defined, we can extract strikes corresponding to the call, K_c^s and put K_p^s such that the call and put delta match the quote, using (17.9). With these strikes, we can then value both the call and the put with this implied volatility, σ_{fly}. The price of this strangle, a call and a put with the specified delta is then what is agreed upon in the market, that is:

$$v_{\text{strangle}} = c(\sigma_{\text{fly}}, K_c^s) + p(\sigma_{\text{fly}}, K_p^s) \tag{17.12}$$

where $c(..)$ is the pricing function for a call and $p(...)$ is the pricing function for a put. This price of this strangle, v_{strangle} is what we want to match in our calibration algorithms.

In practice these points are also available at multiple expiries. Additionally, risk reversal and butterfly points are often available at more than one delta, with 10 and 25 delta being to most commonly quoted points.

When solving calibration problems using this input data, we have too many degrees of freedom to uniquely define a volatility smile without making additional assumptions. For example, risk reversals provide a single point, the spread between their volatilities, but don't provide us with the underlying strikes or individual volatilities. Further, because the volatilities are different on each leg of the risk reversal, we cannot yet apply (17.9) to extract the strike.

Instead, one approach is to use standard calibration techniques, as detailed in chapter 14 to provide additional structure to the model. That is, we specify a stochastic process and find the set of parameters that match the conditions implied by the at-the-money, risk reversal and butterfly points. For a given set of parameters, we know the implied volatility at any strike. Once we know this, we can then find the delta for each strike as well. This means we can extract the strikes for both sides of the risk reversals, although doing so is still not trivial because as we shift the strike the implied volatility changes as well. Thus, it is not as simple as a single call to (17.9). Instead, we may need to perform a one-dimensional root search. With the strikes extracted, we can then measure the at-the-money volatility, the value of the risk reversal, and the value of the butterfly.

The goal of our calibration, then, will be to perform an optimization that enables us to fit $\theta = (\sigma_{\text{ATM}}, \sigma_{\text{RR}}, v_{\text{strangle}})$. There is some flexibility in how we could set this optimization up. One option would be to make each quoted point a constraint in the optimization, and create an objective function that maximizes smoothness of the volatility surface or smile. Alternatively, we could define an objective function

[3]Which, when put together makes a butterfly structure.

which minimizes the distance to each quote without requiring any be matched perfectly. Other approaches exist, such as the Vanna-Volga method or using some form of interpolation or other parametric function for the implied volatility smile. These methods are not necessarily arbitrage free and don't capture different volatility skew behavior effectively. As such, their details are omitted from this text. Readers interested in further background can consult [264] or [58].

17.3 VOLATILITY SMILES IN FOREIGN EXCHANGE MARKETS

17.3.1 Persistent Characteristics of FX Volatility Surfaces

In contrast to other markets, such as equities, FX volatility surfaces can look fairly different depending on the underlying currency pair. In equities, for example, we have seen, and discussed at length, the presence of a persistent put skew, especially for shorter dated options. As we have emphasized throughout the text, this is at heart driven by risk premia. Essentially, the put skew that we observe is a function of the hedging value of low strike put options in risk-off events. This is in stark contrast to FX where the definition of a risk-off event can itself be relative to the currency pair that we are referencing. This means that, in practice, in some cases we observe put skew, or higher implied volatilities on lower strike options. Conversely, in other cases we might observe a call skew, or even a true smile. Even within a single currency pair, there may be periods, or regimes, where the volatility surface characteristics shift. That is, a currency pair may contain call skew during one market regime but then switch to put skew, or a smile during another. This could be determined by many factors, including the stability of the currency and the embedded financial system, and potential concerns regarding inflation. Complicating this picture even further is the underlying conventions. For example, when comparing the skew of EURUSD and USDCAD we need to keep in mind that a depreciation in the dollar will increase the value of one exchange rate and decrease the value of the other. Therefore, in this case we might expect put skew in one pair and call skew in the other.

Taking aside conventions, to understand these volatility surface dynamics recall that the shape of the volatility smile is ultimately driven by risk preferences and supply and demand dynamics. In equities, periods of market stress lead to a flight from stocks into safer assets such as cash or bonds. Investors pay a premium for payoffs in these states, raising the prices and implied volatility of low strike put options. In FX this same flight to quality argument exists but is perhaps a bit more nuanced. Remember that in FX we always need to think about the value of a currency relative to another reference currency. Therefore, when we think about a risk-off event, we need to think about the relative performance of one currency vs. another. We may know, for example, that a currency, for example USD, is likely to appreciate relative to another currency in a risk-off event, say CAD. A totally different story, however, might present itself if we were to instead consider the case of USD against JPY. As a result, we can often in practice observe that a currency, such as USD is a hedge in risk-off events relative to some currencies, but relative to others in may tend to depreciate during these periods. This can lead to substantial differences in the dynamics of the volatility surfaces between currency pairs, even relative to a single common currency.

Further, there may be some currency pairs where the type of risk-off event will have a significant impact on this dynamic. In that case, we might observe a smile, where both tails are priced higher because of the potential for idiosyncratic events in either direction. To take an example, if we consider the case of EUR against USD, the extent to which the currency pair appreciates or depreciates in a risk-off event may be highly dependent on whether the event is European or US centric in nature. Theoretically, this could lead to a volatility smile instead of a skew as a means for incorporating these more idiosyncratic potential risk-off valuation events.

Certain currencies, such as JPY and CHF consistently appreciate relative to most currencies in the presence of broader market stress. This increases the volatility of options on these currency pairs linked to JPY and CHF appreciation. As we explore in the next section, this manifests itself through a volatility skew that is akin to equities. On the other side of the spectrum, emerging market currencies tend to be the hardest hit in periods of stress. This leads to volatility smiles with higher implied volatilities on states with EM currency depreciation relative to G10 currencies. In the next section we explore the dynamics of volatility smiles in different currency pairs, and highlight the differences across the cross section.

17.3.2 FX Volatility Surfaces in Practice: Comparison Across Currency Pairs

In this section, we highlight the heterogeneity of FX volatility smiles and skews by plotting the three-month implied volatility for a cross-section of currency pairs:

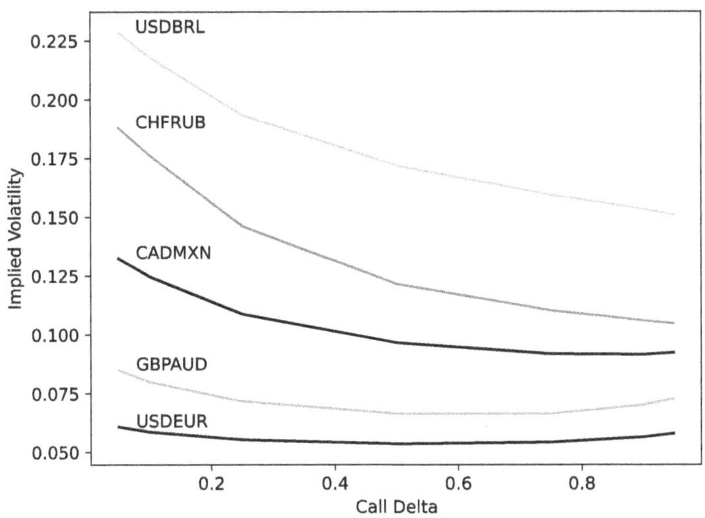

In this visualization we can see a great deal of variation in the level of volatility, and the slope of the volatility skew. Additionally, in some cases we see a smile or smirk emerge with heightened volatilities at both extremes. In the coding repository for the text, we provide an example of an FX volatility surface calibration that incorporates the quoting conventions prevalent in FX markets.

17.4 EXOTIC OPTIONS IN FOREIGN EXCHANGE MARKETS

17.4.1 Digital Options

The simplest so-called exotic option that trades in FX markets is a digital, or binary option. As their name implies, these options have a payoff or \$1 if a condition is met and 0 otherwise. Digital options are by definition European, meaning that they are not path dependent but instead are only concerned with the value of the asset, in this case an exchange rate, at the option's expiry. These types of options have appeared throughout the text. Perhaps most notably, in chapter 17 we established a link between these options and the CDF of the risk-neutral distribution of the asset.

Mathematically, the price of an upside digital call option can be written as:

$$D_T(K) = \tilde{\mathbb{E}}\left[e^{-\int_0^T r_u\,du} 1_{\{S_T>K\}}\right] \tag{17.13}$$

where K is the threshold or strike and S_T is the value of the exchange rate at expiry. This expectation above in (17.13) can be expressed in terms of the distribution of the terminal value of the rate, S_T, as:

$$D_T(K) = e^{-\int_0^T r_u\,du} \int_K^{\infty} \phi(S_T) dS_T \tag{17.14}$$

Here in this integral representation we can see the close connection between the CDF of the risk-neutral distribution and digital option prices. As such, digital options are of direct interest to market participants. Digital options can also be useful vehicles for investors looking to immunize their foreign currency exposure.

17.4.2 One Touch Options

In addition to digital options, FX has a deep market for American digital options. As we saw earlier, an American Digital option is an option that pays \$1 if an asset price touches a certain barrier level *at any point* in its path. This is similar to the European Digital options discussed in the last section, with the difference being that the barrier is continuously monitored rather than only being observed at expiry. In FX markets, American digitals are referred to as **one-touch** and **no-touch** options. One-touches, pay their binary \$1 payout if the barrier is crossed, whereas no-touches pay their binary payout if the barrier is not crossed prior to expiry. There are also **double no-touch** options which pay if the exchange rate never leaves a given range.

Let's consider a simple up-and-in one-touch option with barrier K. For simplicity we assume that the payout is at maturity regardless of when the barrier is triggered. This will be an easy assumption to relax as we move forward. The price of this option can be computed via the following expectation:

$$D_T(K) = \tilde{\mathbb{E}}\left[e^{-\int_0^T r_u\,du} 1_{\{M_T>K\}}\right] \tag{17.15}$$

where M_T is the maximum value S takes over the interval $[0, T]$.

Similarly, an up-and-out no-touch option could be written as:

$$D_T(K) = \tilde{\mathbb{E}}\left[e^{-\int_0^T r_u\,du} 1_{\{M_T < K\}}\right] \tag{17.16}$$

where M_T is the maximum value S takes over the interval $[0, T]$. Because one-touches and no-touches depend on the maximum or minimum value an exchange rate takes over a given period, they are inherently path dependent, exotic structures. This makes their valuation more challenging. In some cases, with primitive stochastic models such as the Black-Scholes model, we may be able to define an analytical solution for the maximum or minimum value an FX rate takes over a given interval. When using more realistic models, however, this will not be possible and instead the techniques for exotics covered in chapter 15 will be required. In particular, either simulation of PDEs can naturally be used to model one and no-touches.

Recall from chapter 15 that we can use the reflection principle to better understand the relationship between European Digital options and their American equivalents, one-touches. In particular, we can consider a path where the one-touch has triggered, and think about the probability that the European digital will be above the threshold at expiry as well. In many models, such as arithmetic Brownian motion, this probability is 50% because for every path that expires above the barrier there is a reflected path below the barrier with equal probability. This means that the one-touch should be twice as expensive as the European. As we move to models that incorporate other distributional features, such as drift and skewness, we will find that this probability diverges from 50%. Even in these cases, the approximate 2x ratio of one-touch vs. European digital price serves as a helpful rule of thumb. Perhaps more fundamentally, it gives us something to isolate as investors as it enables us to hone in on this skewness, and other distributional features by trading one-touches against their European counterparts. This is a topic we explore in more detail in the next section.

17.4.3 One-Touches vs. Digis in Practice: Ratio of Prices in EURJPY

In this section we test the previously developed rule of thumb for one-touches vs. European in the context of the EURJPY currency pair. To do this, we calibrate a SABR model to the volatility skew for each expiry, and use the calibrated model to price a series of digitals and one-touches via simulation. In the following chart, we show the ratio of One touch prices to European Digitals for 25-delta calls on EURJPY:

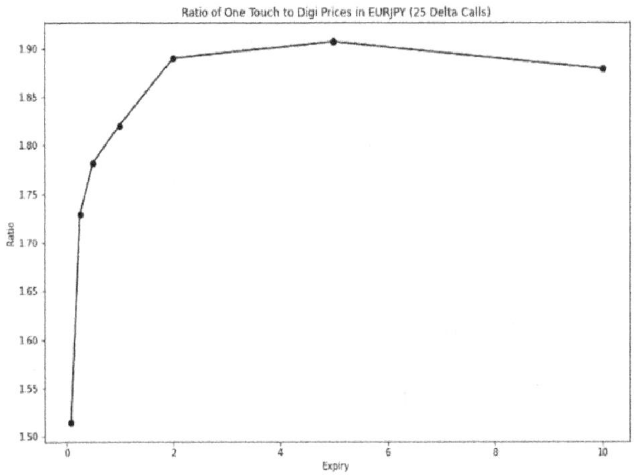

As we can see in the chart, the cost of all such one-touches are below the specified 2x rule of thumb. This is likely to be driven by the volatility surface for EURJPY, in particular, the skew or volatility smile that we observe. Further, we observe this phenomenon to be most pervasive at shorter expiries, when skew is likely to be most extreme. We then see it normalize and converge to slightly below the rule of thumb for longer expiries. The reader is encouraged to think through the driving factors causing this divergence, and to see how persistent these results are in different currency pairs. The reader is also encouraged to consult the example in the coding repository to deepen their understanding of these subtle concepts.

17.4.4 Asian Options

Asian options are another structure that is commonly traded over-the-counter in FX markets due to its robust market for exotics. Asian options define their payoff based on the average value attained by the asset over a given period. Because they depend on the average of the asset, they are also path-dependent, and in this way are similar to lookback options[4]. A differentiating feature, however, is that Asian options are defined as the average value of the asset instead of the maximum or minimum value. As with lookbacks, Asian options can be either fixed or floating strike (which are defined in the same manner). Further, Asian options can be defined to be based on an arithmetic average over a given period or a geometric average.

The pricing equation for a fixed string Asian option can be expressed as:

$$c_0 = \tilde{\mathbb{E}}\left[e^{-rT}(\bar{S_T} - K)^+\right] \quad (17.17)$$

$$p_0 = \tilde{\mathbb{E}}\left[e^{-rT}(K - \bar{S_T})^+\right] \quad (17.18)$$

where $\bar{S_T}$ is defined as the arithmetic or geometric average of the asset price over the

[4]Which were introduced in chapter 15.

interval, depending on the contract. Similarly, the pricing formulas for floating strike Asian options can be written as:

$$c_0 = \tilde{\mathbb{E}}\left[e^{-rT}(S_T - K\bar{S_T})^+\right] \quad (17.19)$$

$$p_0 = \tilde{\mathbb{E}}\left[e^{-rT}(K\bar{S_T} - S_T)^+\right] \quad (17.20)$$

Asian option contracts must also define the frequency of observations in the average calculation. This is true for lookbacks as well, where we needed to define the frequency of observations used to compute the maximum or minimum. Due to their path dependent nature, simulation is an appealing technique for modeling Asian options. This pricing technique would be very similar to that of a lookback, which was detailed in chapter 15, with the only difference coming in the payoff embedded in the simulation algorithm. To do this, we would generally begin by choosing a stochastic process that we believe accurately represents the exchange rates dynamics.

17.4.5 Barrier Options

Another class of exotic options that are frequently traded in FX markets are barrier options. They are options that combine the barrier type of payoff that we saw in digital and touch options, with a European option payoff as well. For example, we might define a barrier option as an option whose payout is a vanilla call but whose value knocks out, or becomes 0, if the value of the asset declines by more than a given amount at any point prior to expiry. These types of barrier structures are fairly prevalent in FX markets and are exclusively traded over-the-counter. The first defining feature of a barrier option is whether the barrier will be a knock-in or a knock-out:

- **Knock-In**: Options that knock-in, or whose payoff becomes active once a condition is reached. For example, an up-and-in call can be written as:

  $$c_0 = \tilde{\mathbb{E}}\left[e^{-\int_0^T r_u\,du}(S_T - K_1)^+ 1_{\{M_T > K_2\}}\right] \quad (17.21)$$

  where M_T is the maximum value of the exchange rate attained over the interval.

- **Knock-Out**: Options that knock-out, or become worthless, once a condition is met. For example, a down-and-out put could be written as:

  $$p_0 = \tilde{\mathbb{E}}\left[e^{-\int_0^T r_u\,du}(K_1 - S_T)^+ 1_{\{M_T > K_2\}}\right] \quad (17.22)$$

  where M_T is now the minimum value of the FX rate.

It should be noted that the up-and-in-call, and down-and-out-put considered above are both American style barriers. That is, like one-touches, they check to see if the barrier is reached at any point rather than only at expiry. More generally, the embedded barriers in these option structures can be European or American. That is, they may be monitored continuously (i.e. American), or may be only observed at expiry (i.e. European). Clearly, just as with digital options vs. one-touches, American barrier structures introduces a path dependency and therefore increase modeling complexity.

In the case of a Black-Scholes model, barrier options often have closed form solutions, which are detailed in many other texts, including [126]. As we start to use more realistic models that better capture the volatility surface and underlying risk neutral distribution, however, this analytical tractability fades. In these cases, the techniques in detailed in chapter 15 can be used to model these barrier structures. Simulation in particular is a convenient, intuitive framework for valuing these derivatives. Additionally, as with other payoff structures, the weighted Monte Carlo techniques discussed in chapter 17 can be applied for a more non-parametric approach.

Barrier options are often intriguing to investors because of their ability to cheapen an underlying options structure. As barrier options include an additional condition, they mechanically lower the price of an options structure. This can be attractive to investors, especially hedgers, who are looking for low-cost hedging vehicles. Along these lines, one approach that many investors take to viewing these options is looking at the **rebate** they receive for adding the barrier or contingency[5]. Of course, just because we have lowered the price of a structure doesn't necessarily mean that we have improved our value proposition, and that is something we should have at the forefront of our mind when pursuing these more exotic structures. Nonetheless, if we can use a barrier to sell a part of the volatility surface with heightened levels of implied volatility, then inclusion of a barrier can lead to a more efficacious hedge. For example, if we are able to sell into the highest part of a volatility smile, a barrier structure might be appealing.

These types of barriers also exist in the multi-asset class space, in which the barrier is based on a different asset. For example, we could buy a call option on the S&P 500 that is contingent on the level of US rates being below a certain level, or the EURUSD exchange rate not crossing a threshold. This class of options are referred to as hybrids.

17.4.6 Volatility & Variance Swaps

In addition to digital, one-touch and barrier options, FX markets have a fairly deep market for volatility linked products, most notably volatility and variance swaps, where we exchange volatility for a predetermined, fixed coupon. As such, they are fixed for floating swaps, in the same spirit as the swaps we saw in chapters 19 and 20. **Volatility & variance swaps** have payoffs linked to realized volatility. They are structured as swaps, meaning they require the buyer to exchange realized volatility or variance for some set strike. However, because they are functions of realized volatility, they are path-dependent and hard-to-value derivatives. These types of swaps show up in multiple markets but are most common in FX, where they are commonly traded OTC structures. Due to their payoff structure, Volatility and Variance Swaps

[5]Where the rebate is defined as the difference in price between the equivalent vanilla and the barrier.

provide investors a direct way to express view on the cheapness or richness of **realized volatility**.

Mathematically, the price of a volatility swap can be written as:

$$\sigma_R = \sqrt{\frac{1}{T}\sum_{i=1}^{N}\left(\frac{S_t - S_{t-1}}{S_{t-1}}\right)^2} \tag{17.23}$$

$$c_0 = \tilde{\mathbb{E}}\left[e^{-rT}(\sigma_R - \sigma_K)\right] \tag{17.24}$$

Similarly, the price of a variance swap would be same, without the outer square root:

$$\sigma_R^2 = \frac{1}{T}\sum_{i=1}^{N}\left(\frac{S_t - S_{t-1}}{S_{t-1}}\right)^2 \tag{17.25}$$

$$c_0 = \tilde{\mathbb{E}}\left[e^{-rT}(\sigma_R^2 - \sigma_K)\right] \tag{17.26}$$

Notice that, unlike many exotic option structures, the payoff at expiry is not conditional on being positive. Nonetheless, the complex, path dependent and non-linear nature of the realized volatility calculation makes volatility and variance hard-to-value exotic options. One common approach for modeling volatility and variance swaps is via simulation, where we would simulate the entire path of the underlying exchange rate and then be able to calculate its realized volatility. To do this, we would start with some stochastic process, such as the processes introduced in chapter 13, and then calibrate the volatility surface using European options data. We would then use the calibrated volatility surface to obtain an estimate for the value of the volatility or variance swap. Alternatively, others have used a non-parametric, static replication approach to modeling volatility derivatives. These replication techniques are not covered in this text, however, interested readers should consult [55].

The presence of the square root in the payoff of a volatility swap leads to some nuanced differentiating properties between volatility and variance swaps. In particular, it can be easily shown that the square root function is concave, meaning that a trade that is long volatility swaps and short variance swaps is short convexity. Conversely, a trade that is short volatility swaps and long variance swaps will be long convexity.

Further, because the payoffs for the two instruments are otherwise identical, this convexity will be the defining feature of such a structure. Throughout the text we have discussed many other ways to buy or sell convexity, such as via buying or selling delta-hedged straddles. Here, we see another approach based on volatility derivatives. Said differently, trading volatility vs. variance swaps provides us a direct mechanism to harvest the embedded convexity when it is over or underpriced. This is conceptually similar to the delta-hedged straddle trades discussed in chapter 16. As such, investors may be inclined to trade volatility swaps against variance swaps, or vice versa, therefore isolating this inherent convexity, and in fact, this is a strategy that is commonly engaged in by hedge funds and other institutions in practice.

17.4.7 Dual Digitals

A Dual Digital Option pays a predefined amount if two assets are above/below a certain level at expiry. It is a natural extension of the digital options we detailed at the beginning of this section, and are another common OTC instrument in FX markets. Dual digitals are similar to European single asset digis in that they only depend on the distribution of the terminal value of the asset. Dual digitals, however, are multi-asset options. That is, they depend on more than one exchange rate at expiry, rather than a single asset price. This means that, unlike single asset digitals they are concerned with the **joint distribution of the two assets**, and the **correlation structure between them**. This makes dual digitals inherently more complex to model as we need to model both underlying exchange rates, as well as their correlation structure, robustly.

Mathematically, the price of a call dual digital can be written as:

$$c_0 = \tilde{\mathbb{E}}\left[e^{-rT} 1_{S_T > K_1} 1_{X_T > K_2}\right] \tag{17.27}$$

where K_1 and K_2 are the barriers respectively and S_T and X_T are the prices of the exchange rates at the expiry of the dual digital. Similarly, a put dual digital can be written as:

$$p_0 = \tilde{\mathbb{E}}\left[e^{-rT} 1_{S_T < K_1} 1_{X_T < K_2}\right] \tag{17.28}$$

Note that the barrier on one asset may be above while the other may be below. The expectation in (17.27) can be written as

$$c_0 = e^{-rT} \int_{K_1}^{\infty} \int_{K_2}^{\infty} \phi(S_T, X_T) dS_T dX_T \tag{17.29}$$

Importantly, we can see that, if the joint distribution is known, however, then quadrature techniques can be used to price dual digitals. It is noteworthy that for a dual digital that this would involve a two-dimensional integral and a bi-variate joint distribution. Modeling of dual digitals can also be done via simulation techniques, where we would pick an underlying stochastic process for each underlying exchange rate, as well as some additional correlation parameters. The choice of these parameters is typically obtained via a calibration of each individual exchange rates to its European volatility surface. The additional correlation parameters, however, can be more challenging to model as there is often no market instrument for which to calibrate this correlation. This, coupled with their OTC nature, is one of the main challenges to valuation of dual digitals. In addition to simulation from a given stochastic process such as Heston or Variance Gamma, we could also look to employ a more non-parametric approach. Along these lines, the weighted Monte Carlo framework introduced in chapter 17 is a convenient choice for doing so. Alternatively, copula techniques, which ensure that the marginal densities match while also embedding a correlation structure, can be used. In this case, we would be able to match the risk neutral densities of the individual exchange rate while also embedding a Gaussian, or other correlation structure assumption. More information on copulas can be found in [184].

More generally, just as single asset digital options are tightly linked to the CDF of the risk-neutral distribution for the asset, a dual digital will likewise be closely related to the CDF of this two dimensional, joint distribution, which we can see clearly in (17.29). This makes dual digitals an interesting instrument from a quants perspective because they provide us with a mechanism for precisely buying or selling parts of a joint distribution.

17.5 INVESTMENT PERSPECTIVE: TRADED STRUCTURES

17.5.1 Hedging Currency Risk

Perhaps the most fundamental reason to enter the currency, or FX derivatives market is to hedge existing currency risk. This currency exposure may arise from a firm that operates globally and generates profits in foreign countries. From an asset managers perspective, it also may emanate from equity investments in foreign companies or indices. Investors may then not want to bear the risk that these foreign currencies will depreciate, lessening the value of their investments, or cashflows in their domestic currency. To obviate this, an investor would use the currency markets to instrument a hedge. As is the case in other asset classes, the first decision in implementing a hedging strategy is deciding on a single or set of hedging instruments.

In some cases, a simple linear hedge will be effective. In this case, the use of FX forwards is most appropriate. In other cases, investors may want to benefit from appreciation of the foreign currency while still hedging the risk of a depreciation. In this case, options structures, ranging from vanilla options to customized exotic structured products, may be preferred. Options structures have the benefit of a conditional payoff, meaning that our return is capped should the value of our hedge go against us. Forward contracts, in contrast, do not have this feature. Conversely, forward contracts do not require an upfront cost, whereas options structures require an initial premium outlay. Thus, there is a trade-off between the two. If we are concerned with the foreign currency appreciating substantially, then the payoff of an options structure may be more appealing than a forward. If we feel the price of this extra optionality is unjustifiably high, however, we would lean toward a forwards contract in lieu of an option.

As an example, consider a US based investor who owns a significant amount of stock in an EUR denominated company. The return to this investor will, of course, be a function of the price appreciation and dividends of the stock. As the investor is US based, however, they will need to convert their profits from EUR to USD in order to utilize them. As such, this investor is exposed to a change in the EURUSD exchange rate. If EUR appreciates relative to USD, meaning the EURUSD rate increases, then the profits will be worth more when converted to dollars. If EUR depreciates, meaning the EURUSD rate decreases, however, the value of the profits converted back to USD will be lower, leading to a worse return for the investor. To avoid this risk of EUR depreciation relative to USD, the investor can lock in the rate at which their future profits are converted at using a forward contract. Alternatively, they can hedge against the EURUSD exchange rate decreasing by buying a put option on EURUSD. If they prefer, they can even make this option a one-touch, barrier, or other exotic structure.

The dynamics of this currency component, along with the equity investment have an interesting interplay that has potential repercussions for such investors making hedging decisions. As we noted above, the return of an equity investment to a foreign investor, r_f is a function of the local return of the equity market, r_l, as well as the currency return, r_c:

$$r_f \approx r_l + r_c \tag{17.30}$$

Should the foreign investor choose to hedge, their return stream can be written as:

$$r_f = r_l + h \tag{17.31}$$

where h represents the costs of hedging the given currency.

If we assume that currencies have zero expected returns over the long-term, as the currency market is inherently a zero sum game, and ignore hedging costs, then from a return perspective we are indifferent between hedging our currency risk or not. These are both potentially dubious assumptions in practice, but nonetheless illustrates a key point about hedging currency risk. We can similarly write the volatility, or variance of returns from the perspective of a foreign investor approximately as the variance of the sum of two random variables, as shown below:

$$\sigma_{r_f}^2 \approx \sigma_{r_l}^2 + \sigma_{r_c}^2 + 2\rho\sigma_{r_l}\sigma_{r_c} \tag{17.32}$$

where $\sigma_{r_f}^2$ is the variance of returns to the foreign investor, $\sigma_{r_l}^2$ is the variance of returns to the local investor, $\sigma_{r_c}^2$ is the variance of returns of the currency, and ρ is the correlation between the return of the currency and return of the local equity market. For the hedged investor, the last two terms would disappear and the variance would simply be a function of the variance of the local equity market.

More generally, we can see that, unsurprisingly the variance of the returns to an unhedged foreign investor depends on the correlation between fx returns and local equity returns. If the two return streams are uncorrelated, then we can see from (17.32) that introducing fx risk increases the variance from a foreign investors perspective relative to a hedged investor. We can also see that the variance to a foreign investor is lowest when the fx and local equity returns are perfectly negatively correlated, and is highest when they are perfectly positively correlated. This means that, ultimately, the attractiveness of hedging currency risk is closely related to the correlation between the FX rate and the equity. If there is positive correlation between FX and equity returns, hedging may seem like an attractive way to reduce volatility, in the absence of a strong view of the currencies expected return. When they are strongly negatively correlated, however, leaving the currency unhedged can actually reduce the volatility from the perspective of a foreign investor.

In FX markets the dynamics of this correlation can depend greatly on the currency pair and macroeconomic landscape. There are certain persistent safe-havens or flight to quality currencies, such as CHF, JPY and USD. These currencies are thus likely to significantly appreciate in the presence of equity market stress. This appreciation in times of stress leads to potential significant negative correlation with equity returns. As such, foreign investors may desire to leave this type of currency risk unhedged. On the other hand, EM currencies like BRL are likely to have a completely different picture, with the tendency to depreciate significantly in risk-off events. This is likely to drive positive correlation between fx and local equities, making hedging a relatively more attractive proposition.

17.5.2 Harvesting Carry in FX Markets

In addition to hedging, many participants enter FX markets in order to generate positive investment returns through systematic or discretionary trading strategies. One such approach is based on harvesting risk premia. Risk premia is a common theme in this text, and one that permeates asset classes and traded structures. The FX markets are no different, with their own set of potentially harvestable premia which hedge funds and other buy-side institutions look to take advantage of. Perhaps the most well-known and widely documented such phenomenon in FX markets is the carry premia. Earlier in the chapter, in section 17.2 we discussed the dynamics of FX forward valuations and the propensity for FX spot exchange rates to under-realize relative to their forwards. This is ultimately the basis for carry as an investment strategy.

Carry is a type of strategy that many institutions pursue in a variety of different forms. Even those who are not directly pursuing carry strategies are acutely aware of the carry properties of their portfolio. Simply stated, carry tells us what happens to our portfolio if nothing happens except the passage of time. As such, it is a quantity that is of direct interest to any portfolio manager, regardless of the type of instrument or asset class they are engaging in. From a quant perspective, it can be thought of as the partial derivative of our portfolio value with respect to time.

In the context of FX rates, carry means that exchange rates remain unchanged. There are many ways to structure carry trades in fx, ranging from spot positions in the currency to forwards to options. In the following table, we highlight how FX carry trades would be constructed in the forward and options space:

Condition	Forward Type	Option Type	Description
$S_0 > F$	Buy at F, Sell at S_T	Long Call, Short Put	S_T expected higher than F
$S_0 < F$	Sell at F, Buy at S_T	Short Call, Long Put	S_T expected lower than F

where S_0 is the current value of the exchange rate and F is the value of the forward contract.

In the case of a forwards implementation, we can see that, if we observe that the current exchange rate is above the forward, then we expect the exchange rate at expiry in the future to be above the current forward rate. Therefore we should buy the forward at F and then at expiry sell at S_T which we expect to be above F, earning a profit of $S_T - F$ if our premise ends up being true. Conversely, if we observe

S_0 to be below F, then we should sell the forward contract at F and buy at expiry for S_T as we expect the value of the exchange rate at expiry to be below the forward contract value, collecting a profit of $F - S_T$ if we are correct. In the case of options, we can structure the carry trade multiple ways depending on whether we want to be long or short volatility.

Just as there are many ways to implement a carry strategy, there are also many ways to build an underlying carry signal and translate it into a portfolio. The canonical carry definition in FX is one that is based exclusively on the interest rate differential between the two economies [214]. If we were to pursue this approach, and an approach that buys and sells the top/bottom N[6], a carry strategy at any given time could be formulated by comparing the magnitude of the deviations in F and S. The strategy would then be long the carry trade structures in 17.5.2 where F is the further from S and being short the same structures above where F and S are the closest.

17.5.3 Trading Dispersion: Currency Triangles

Another commonly traded structure within FX derivatives markets is based on the presence of triangles, which uniquely define the third exchange rate in a triangle as a function of the two other pairs within the triangle. This relationship for the hypothetical currencies XXX, YYY and ZZZ can be written as:

$$S_{\text{XXXYYY}} = S_{\text{XXXZZZ}} S_{\text{ZZZYYY}} \tag{17.33}$$

In delta-one space, such as transactions in the underlying currencies, this relationship is bound by arbitrage. In options, however, the non-linear payoff of an option means this ceases to be the case. Instead it depends on correlation. A vanilla option on XXXYYY, in particular, can be written as:

$$c_{\text{XXXYYY}} = \mathbb{E}\left[(S_{\text{XXXYYY}} - K)^+\right] \tag{17.34}$$

$$c_{\text{XXXYYY}} = \mathbb{E}\left[(S_{\text{XXXZZZ}} S_{\text{ZZZYYY}} - K)^+\right] \tag{17.35}$$

where in the second equation we have substituted for S_{XXXYYY} the triangle relationship in (17.33). From this equation we can see that an option on the currency pair XXXYYY can be thought of as an option on the product of two other currency pairs, XXXZZZ and ZZZYYY[7].

It is natural to then try to compare this option on XXXYYY to a portfolio of options on XXXZZZ and ZZZYYY. For example, consider the structure in following table:
where, typically, the weights w_1 and w_2 in 17.5.3 are chosen such that the two long positions best replicate the short position, which is on their product.

[6] This approach for translating a signal to a portfolio, and others, are discussed in more detail in chapter 27.

[7] In some cases it might be the quotient depending on FX quoting conventions.

Long/Short	Units	Currency Pair	Description
Short	1	XXXYYY	Sell Call (or Put) on XXXYYY
Long	w_1	XXXZZZ	Buy Call (or Put) on XXXZZZ
Long	w_2	ZZZYYY	Buy Call (or Put) on ZZZYYY

If it weren't for the non-linear payoff of the options, then we know this relationship would be defined by arbitrage. In the options space, however, a more nuanced picture emerges. If XXXZZZ and ZZZYYY move in the same direction, then their product does as well, leading to offsetting gains in the options we are long with losses in the option we are short in 17.5.3. If XXXZZZ and ZZZYYY move in opposite directions, however, it leads to a differentiated payoff. In that case, their product, XXXYYY is likely to be little changed, leading to no increase in the payout of the option we are short. One of the options we are long, however, will have a positive payout. Therefore, in this situation, when we observe movements in opposite directions, the structure in 17.5.3 achieves a profit. Because of this, this type of structure is often known as a dispersion trade, which benefits from lower correlation among the pairs within the product.

The construction of the trade in 17.5.3 is setup as long dispersion, or short correlation. That is, when the returns on XXXZZZ and are very different and offset each other, leading XXXYYY to be little changed, then the payoff of this structure will be highest. Of course, we could choose to buy correlation and sell dispersion instead by buying the option on XXXYYY and selling the two component options.

This trade construction is commonly traded in practice as a way to harvest dislocations between realized and implied levels of correlation. For instance, if we believe in a correlation premium due to the tendency of correlations to rise toward one in periods of market stress, then we may subsequently want to buy dispersion and sell correlation, collecting the difference between the two.

17.5.4 Trading Skewness: Digital Options vs. One Touches

As FX has the deepest exotic options market across asset classes, buy-side institutions and hedgers also often leverage these exotic structures to generate more customized payoffs, or to isolate different parts of a volatility surface. One such example, which has appeared in some form a few times in this text, is a one-touch vs. digital option. As we saw earlier in the chapter, one-touch options are American digital options that pay out if a barrier is reached at any point between now and expiry. Digital options, conversely are European-style, meaning they payout only if the barrier is breached at expiry. In chapter 15, we discussed how the reflection principle can be used to provide a baseline for the relative pricing of these structures.

From a trading perspective, when one-touches are traded against digital options the resulting structure has some interesting properties that are closely related to skew and mean-reversion. In particular, consider an investor who purchases the following structure:

This investor has a long position in a one-touch option and a short position in a digital option with the same characteristics. These may be upside or downside

Long/Short	Units	Description
Long	1	One Touch with expiry T and barrier K
Short	1	Digital with expiry T and barrier K

one-touches and digitals. In either case, this investor is long an option that pays out if a barrier is crossed at any point. They are then short an option that pays out if the barrier is crossed at expiry. As a result, the best outcome for this investor is for the barrier to touch prior to expiry, triggering their one-touch, and then for the spot to revert. This reversion would then lower the price of the remaining sold digital option. Therefore, the owner of this option is said to be long mean-reversion. Conversely, the owner of this package is short a continued move once the barrier has been breached.

Because of the extra optionality in one-touches, they are by necessity more expensive than digitals. This means that, we must pay a premium to enter into the trade described in 17.5.4. If neither option is triggered we then lose our premium. Similarly, if both are triggered the payments offset and we lose our premium as well. These are both negative outcomes. But if the one-touch is triggered, and the price reverts such that the digital does not trigger at expiry, then we make a profit. So, from a mean-reversion perspective this trade is long reversions and short trending moves. As this trade is inherently priced in the risk-neutral measure, it is priced in a world without mean-reversion. Therefore, if we believe that mean-reversion exists in the physical measure, then this trade can be an interesting way to isolate and harvest it.

Another lens of viewing the relationship between one-touch and digital prices is through the lens of skewness in the underlying distribution. We know that the rule of thumb is based on a 50% probability of the digital option expiring in-the-money conditional on the one-touch being triggered. One way for this probability to diverge from 50% is the presence of skewness in the distribution. As we introduce skewness into the risk-neutral distribution, then, as learned in chapter 17, the conditional median of the risk-neutral distribution shifts while the conditional mean remains constant. The result of this is that the conditional probability of finishing above or below the barrier at expiry shifts as well. If this probability were to increase significantly, then we would observe one-touches which cost less than the 2x rule of thumb. If this probability were to decrease significantly, conversely, we would see prices of one-touches in excess of the rule of thumb. Therefore, if we believe that the skewness in the risk-neutral distribution is mis-priced, or if we believe there to be an available skewness risk premium, structures like 17.5.4 can help us take advantage of it. As such, one-touch vs. digital structures enable us to isolate some interesting distributional effects, most notably mean-reversion and skewness in the risk-neutral dynamics.

17.6 CASE STUDY: CHF PEG BREAK IN 2015

As mentioned in the chapter, one of the defining features of currency markets is the presence of central banks who buy and sell currencies to manage their money supply.

These central bankers aim to do this in a way that is optimal for their local economy. A currency that is too strong, or weak might have repercussions and might cause a central bank to step in. This type of central bank policy is handled differently by different economies. Some countries let their exchange rate float freely. Others choose to fix their currency in terms of another. This is often referred to as a pegged currency. The Swiss Franc choose to peg the minimum value of its currency in 2011 to the Euro at a value of 1.2 EURCHF[8]. The impact of this peg was capping the level of appreciation in the Franc against the Euro.

From an investment perspective the presence of this type of peg is of direct interest. If a central bank is telling you it won't let the EURCHF rate fall below 1.2 then a downside one-touch at or below this strike is essentially a bet that the peg will break. Perhaps more attractively, if we collect a substantial amount of premium selling this type of structure, then all we need to happen if for the central bank to maintain credibility and not change their policy.

One of the interesting features of CHF is that it is a natural risk-off, flight to quality currency. In periods of market stress, historically we have seen CHF strengthen vs. other currencies, including EUR and USD. This is largely because of investors confidence in the stability of Switzerland and its financial system. This flight to quality aspect added an interesting layer to the peg as it meant in periods of stress there would be more pressure on the Swiss Central bank to maintain the peg. Over the years following 2011 there were periods where the peg was tested and the rate lingered very close to the 1.2 EURCHF limit, as we can see in the following chart:

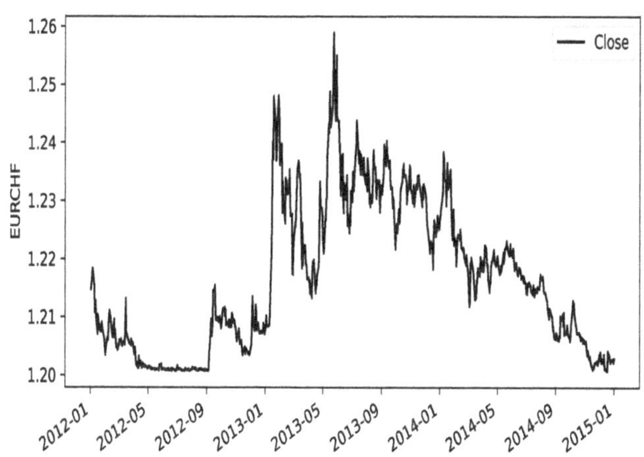

In January of 2015, as pressure continued to build, and the transactions required for the Swiss central bank to maintain the peg increased, it became unsustainable and finally broke. Once the peg broke, and the Swiss central bank made clear they would not defend the 1.2 threshold any longer, an extraordinary move in the currency began. As we can see in the following chart, in the immediate aftermath of the peg

[8]Or 1.2 Swiss Francs per Euro.

break, the Swiss franc appreciated significantly, over the course of one or two days, to the point that it approached parity:

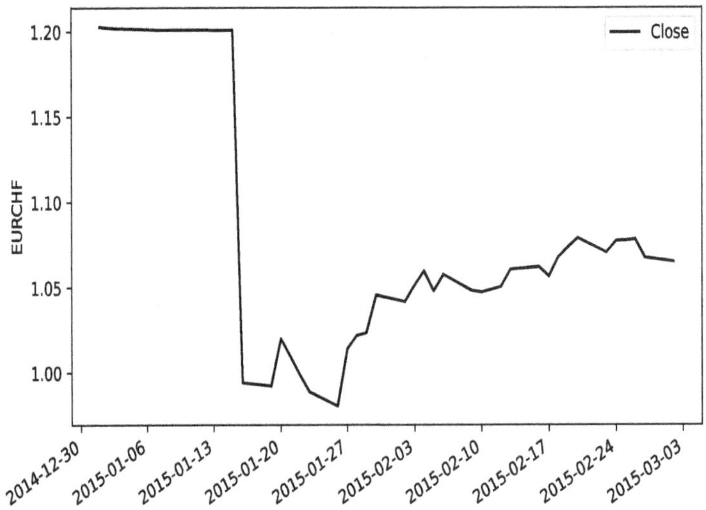

This period of market activity highlights several important points about foreign exchange markets. First, it highlights the tension that can occur between speculators and central banks when engaging in the market. Secondly, it shows the unique feature of a potentially pegged currency. A trade such as selling a digital option below the EURCHF peg may have seemed to be a near arbitrage opportunity, but as we can would have led to massive losses in 2015. Along these lines, this is also a great example of the non-stationarity of the underlying data that we are working with as quants.

CHAPTER 18

Equity & Commodity Markets

18.1 MARKET SETTING

EQUITY markets are perhaps the most familiar market to aspiring quants, and are generally the primary focus in most graduate texts and courses. As we introduced in chapter 2, equities provide us with an opportunity to participate in the profits of a publicly traded firm via purchases of shares[1]. In doing so, they provide a mechanism for investors to harvest an underlying equity risk premium [23], where investors are compensated for bearing the systemic risks facing these firms.

The vast majority of investors, especially retail investors, have investment portfolios that are highly concentrated in equities. In addition to the ability to buy shares in single companies, the equity market provides participants with the ability to obtain broad market exposure through indices, such as the S&P 500 or NASDAQ. Investing in these indices is equivalent to investing in a properly weighted basket of the underlying components.

Additionally, over the last few decades a plethora of exchange traded products have been created that provide exposure to different segments of the broader equity market. These segments could be defined by their sector, their geographic region, or by other defining characteristics, such as their factor exposure. These ETFs can be used as additional building blocks in an overarching portfolio construction algorithm, or can be the basis for a trading strategy. These areas are explored in more detail in chapters 24 and 27, respectively.

The equity market is accompanied by a corresponding liquid derivatives market that allows investors to hedge their overarching equity risk or benefit from the more nuanced properties of options that we have emphasized throughout the book. This natural demand for hedging, coupled with a demand to harvest risk premia in volatility surfaces, particularly the well-known volatility risk premium, creates a strong two-way market for these derivatives, with both natural buyers and sellers of options. Vanilla options are the most liquid equity derivative instruments. By convention, most of the these vanilla options are traded via exchanges, leading to

[1] Or privately traded firms in the case of private equity.

DOI: 10.1201/9781003530626-22

standardized terms and price transparency. For single-name stock options, the convention is for listed options to be American style, creating an extra quantitative challenge when modeling their volatility surfaces. Index options, conversely, are European by convention.

In addition to vanilla options on stocks, ETFs and equity indices, some exotic options structures trade over the counter. These exotics are characterized by significantly lower liquidity and less price transparency, but can be catered to an investors needs, and, as we will see later in the chapter, have some interesting properties relative to their vanilla counterparts. Later in the chapter we consider two of the more commonly traded exotic structures, basket and lookback options. We also provide an example of how a lookback might be traded against a European option to isolate some of the more nuanced properties of the exotic.

Commodity markets, in contrast, comprise a much smaller market than equities with a smaller investor base. Nonetheless, they can provide investors with interesting exposures, and provide another asset class in which we can harvest risk premia. The market consists of futures contracts in underlying commodities, such as oil, natural gas, corn and gold, and also contains liquid European options on these commodities. These futures are augmented by some commodity ETFs, which typically gain their exposure via a basket of futures positions. Exotic option structures are also available in commodities, however, relative to markets like equities and FX they are less liquid and a bit more niche. This is potentially driven by the smaller overall hedging demand relative to these other asset classes.

In the following section, we detail the mechanics of futures curves and volatility surfaces in equity and commodity markets, with a particular emphasis on how the unique features in each market affect futures curve and implied volatilities.

18.2 FUTURES CURVES IN EQUITY & COMMODITY MARKETS

18.2.1 Determinants of Futures Valuations

In chapter 2 we established a link between current, or spot, prices of the underlying asset, and prices of futures and forward contracts. In particular, we saw that an arbitrage argument dictated that the two must be in line, otherwise an investor would borrow (or lend) to buy (or sell) the spot and take an offsetting position in the future. This leads to the following well known futures pricing equation:

$$F = S_0 \exp\left((r - q)\, T\right) \tag{18.1}$$

where r is the risk-free rate, q is the dividend the asset pays and T is the time to expiry of the future. In both equity and commodity markets, this relationship underpins the futures curve.

In the case of equities, an investor must choose between investing via the future, in which case they forego any dividends, or the spot. Thus, higher dividends push futures curve lower as they represent lost income to the investor. In the case of commodities, investors who choose to invest via spot rather than futures must pay to store the asset. This storage cost acts as a negative dividend, pushing up the futures curve relative to spot. In cases where the underlying asset is tradeable and easily storable, (18.1) is a tightly bound relationship. In commodity markets, however, some assets are inherently difficult to store, increasing storage costs, disincentivizing investors from pursuing the arbitrage, and weakening the relationship. In the next section, we discuss the implications of storage challenges on futures curves.

In futures markets, equity and commodity curves are said to be in contango, or backwardation, depending on whether they are upward sloping, or inverted respectively:

- **Contango**: $F > S_0$: An upward sloping futures curve defined by futures that increase with time to expiry.
- **Backwardation**: $F < S_0$ A downward sloping, or inverted, futures curve defined by futures that are declining by expiry.

18.2.2 Futures Curves of Hard to Store Assets

In many cases, even in commodity markets, it is relatively inexpensive to hold the underlying asset, making the relationship between spot and forward trivial and generally uninteresting. Gold, for example, is a commodity that is inherently storable, albeit at some cost. That is, we need to pay for a safe, potentially with a trusted security guard, each of which might be costly, but it is undoubtedly storable. Commodities like oil and natural gas, on the other hand, are much harder, and in some cases dangerous to store. This unsurprisingly significantly increases the cost of storage. Even more extreme, for some assets, the spot is completely untradeable. The VIX index, a frequently monitored proxy for market implied volatility, is a common example of this, where futures and options are liquid, however, the spot is not accessible directly.

In the case of underlying assets that are difficult or impossible to hold and or store, the trivial relationship between spot and forward expressed in (18.1) becomes much more nuanced. In the absence of this iron-clad arbitrage argument, futures prices can be driven by many other factors, including risk premia and the convenience of owning spot in different periods.

In these cases, the concept of an implied dividend is introduced. That is, we look at the value of the dividend that equilibrates the relationship between the spot and forward. This implied dividend incorporates storage costs, as well as supply and demand dynamics of different futures contracts vs. the spot. We could then look to extract a risk premia from the implied dividend using some of the econometric techniques presented in other chapters. The implied dividend can be found by solving (18.1) for q:

$$q = r - \frac{\log\left(\frac{F}{S_0}\right)}{T} \tag{18.2}$$

A negative implied dividend pushes the forward curve up relative to the spot, whereas a positive value would decrease the forward vs. spot. In the case of VIX in particular, this quantity is highly volatile depending on the market regime.

18.2.3 Why Are VIX & Commodity Curves Generally in Contango?

In practice, we tend to observe that, more often than not, VIX and difficult to store commodity curves tend to be in contango. Said differently, contango seems to be the equilibrium condition that occurs the majority of the time. Periods of backwardation, in contrast, tend to be brought on by exogenous shocks. For example, in the case of VIX, these exogenous shocks would correlate with equity market drawdowns. For commodities, these shocks would be far more idiosyncratic, related to supply disruptions for a given commodity. These periods of backwardation then tend to revert somewhat quickly to their equilibrium counterpart.

In the following list we highlight some potential explanations for why contango might be the more persistent, steady state:

- **VIX**:
  - Spot is untradeable and VIX futures tend to be rich.
  - Risk Premium (fear of large spike in VIX & prolonged backwardation).
  - Investors are willing to overpay for insurance against large risk-off events.
- **Commodity**:
  - Storage Costs act as negative dividend.
  - Storage of some commodities is difficult
  - Risk Premium (fear of supply disruptions, inventory running out, etc.)

18.2.4 Futures Curves in Practice: Excess Contango in Natural Gas & VIX

In this section we examine the potential presence of excess levels of contango in VIX futures curve as well as a difficult to store commodity, Natural Gas. As previously mentioned in cases where storage of the asset is difficult or impossible, then the dynamics of the futures curves are determined by, among other things, risk premia. In this section we provide a simple method for analyzing whether this risk premia exists in the context of these two futures curves.

To motivate the analysis, we begin by plotting a time series of the VIX Index as well as a position in a continually rolled front month VIX future, which we see in the following chart:

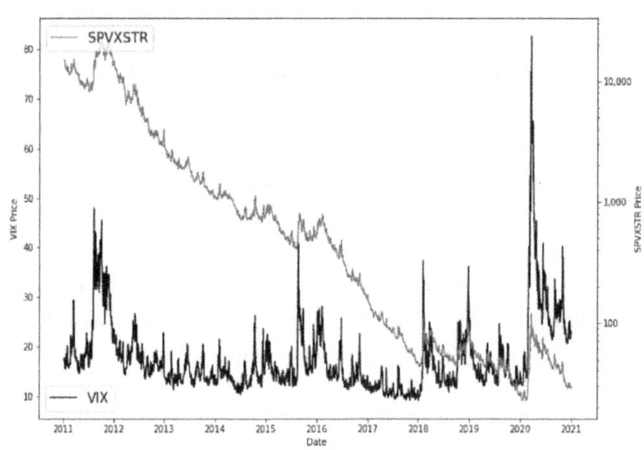

Interestingly, we can see that while the level of the VIX index has maintained a relatively consistent value over the period, a position in the rolling front month future, which is labeled as SPVXSTR in the chart, has declined significantly. A natural potential explanation for this phenomenon is that the curve was upward sloping during the period, but the upward sloping futures didn't realize. Instead, the futures rolled to spot, resulting in lower values and poor returns for holders of the futures contracts. Naturally, we might postulate that this phenomenon is due to a persistent risk premia in the VIX futures curve. Said differently, because of the tendency for VIX futures to increase in periods of market stress, their subsequent hedging value may entice investors to pay a premia to invest in them. This is a canonical example of a risk premia. If we are willing to bear the risk of significant losses during crisis period, we are potentially compensated with a premium over time. Looking at the chart, it appears as though this is exactly what we observe. Positions in the rolling front month future decay over time, evidencing excess contango in the market, but also spike significantly during periods of stress such as the Covid crisis in 2020.

To better understand this potential market dynamic, we next look at a time series of the slope of the front of the VIX future:

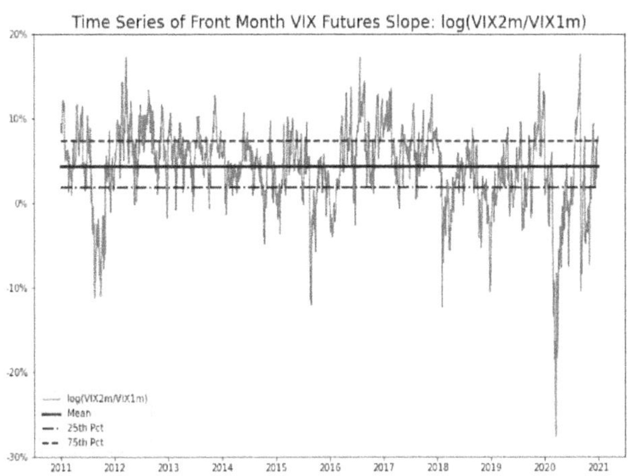

From this chart, we can see that this curve is upward sloping in the vast majority of cases, leading to positive values in the chart. However, we do observe a few cases where they are severely inverted. We can also visualize this as a histogram to try to better understand the distribution of the front month VIX slope:

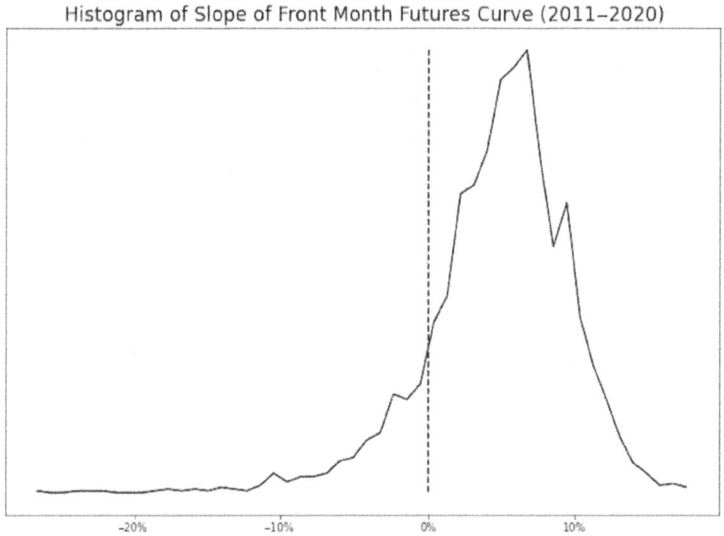

The natural next question, then, is whether these upward sloping futures values leads to subsequently higher realizations of the level of VIX in the future, or whether the futures roll to spot. To isolate this, we plot a histogram of the underperformance of a VIX future, measured as the difference between the futures value at a given point against the subsequent value when that future expires. We show this in the following histogram:

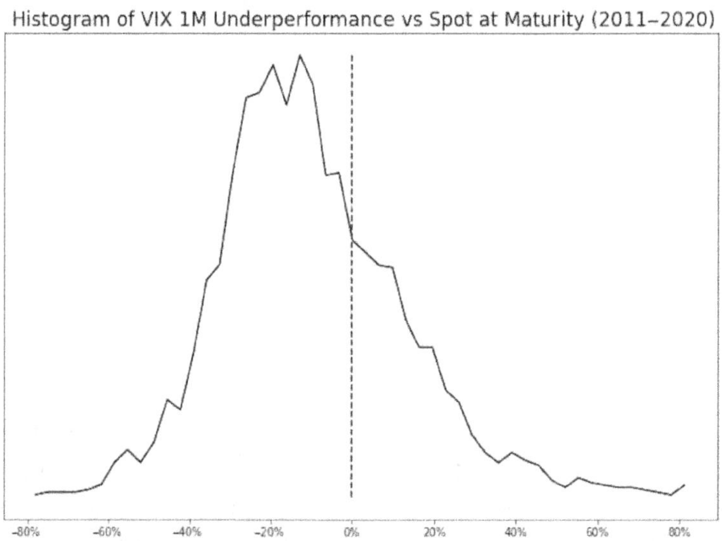

Here we can see an interesting phenomenon emerge. Most of the time, VIX futures underperform, which leads to the decay we observed in our initial chart above. However, when they outperform they tend to outperform significantly, as is evidenced by a significant amount of skewness in the histogram above. Taken collectively, this analysis seems to indicate that there are interesting levels of risk premia in the VIX future curve, however, that there are also significant spikes in the futures that make this a difficult dynamic to harvest.

It turns out that this type of market dynamic isn't specific to VIX, and also presents itself in some commodities that are difficult to store. To see this consider the analogous set of charts described above for Natural Gas:

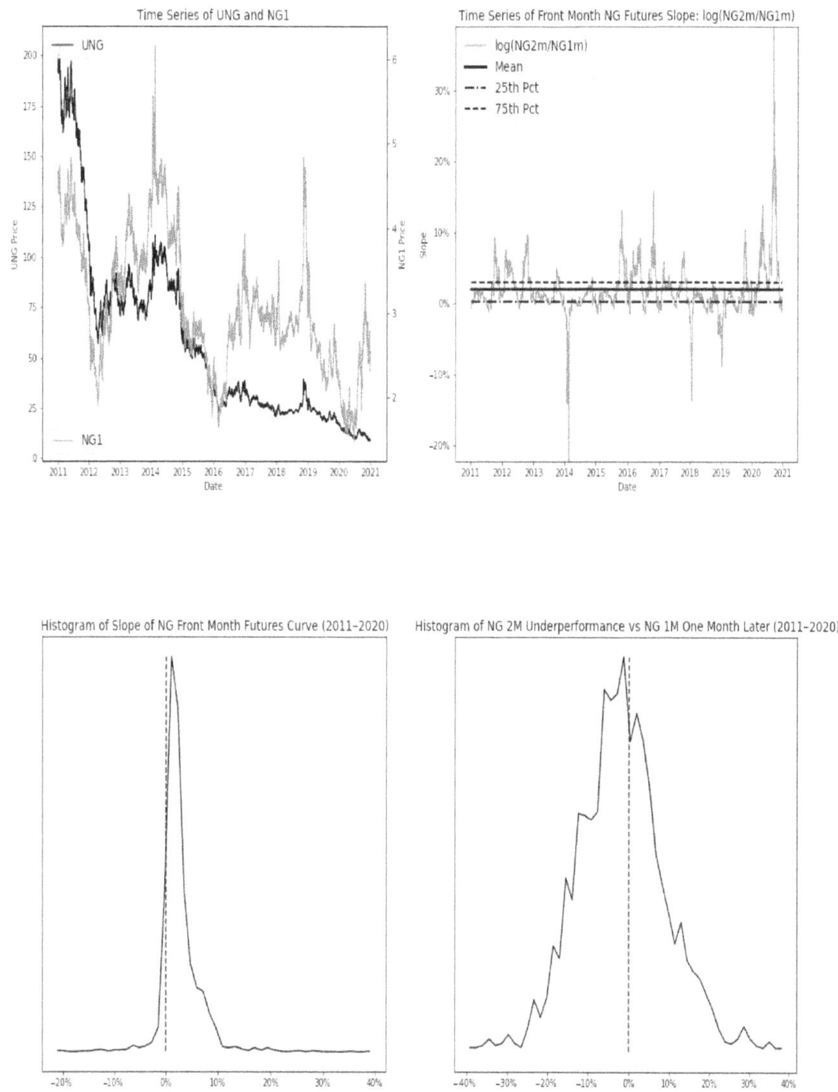

In these charts we see a similar phenomenon, albeit one that is not quite as pronounced. In particular, we find that Natural Gas curves tend to be upward sloping, and these valuations don't tend to be realized in the future. This is also evidence of an inherent risk premia in the futures curve, but, as we saw in the case of VIX, is prone to large spikes and suddenly inverted curves.

Our goal in this section is merely to present this potential market dynamic to the reader, and to provide some intuition on why it might be occurring. Identifying whether this is truly a persistent feature of the market, and how one might try to take advantage of it, are subtle questions that are left as exercises for the reader.

18.3 VOLATILITY SURFACES IN EQUITY & COMMODITY MARKETS

18.3.1 Persistent Characteristics of Equity & Commodity Volatility Surfaces

Throughout this text, a great deal of emphasis has been placed on the properties of the volatility surface. In chapters 13–17 we went to great lengths to build models that enable us to fit the types of volatility surfaces that we observe in real markets. Additionally, we placed a great deal of emphasis on the intuition behind volatility surfaces. In particular, we highlighted how a change in the volatility surface might impact the corresponding risk-neutral distribution. When introducing a set of models meant to explain volatility surface behavior, our goal was to be able to match a wide array of market conditions and types of volatility smiles and smirks.

Ultimately, the actual features of these volatility surfaces are determined by supply and demand dynamics, and investors preferences for payoffs in different states. Investors may rationally prefer payoffs when their other assets, and maybe even their career prospects, are likely to be threatened. As a result, certain patterns do emerge in some volatility surfaces. These more persistent features are noteworthy in that they inform the types of behavior we most need to be able to handle in our stochastic models. From an investment perspective, the persistent nature of these features also gives investors something to try to harvest, or take advantage of deviations from.

In equity markets, the most widely documented feature is so-called put skew. Put skew, or the tendency of low strike options to have higher implied volatilities than higher strike options, is something that we observe consistently across time and across the cross section of equities. A common volatility smile in equities, with respect to strike, for example, might look like:

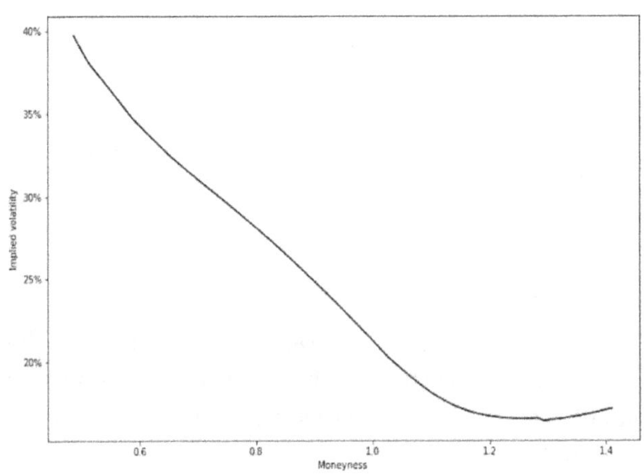

This put skew has a very natural risk premium interpretation. One way to think about this is that investors value payoffs in different states of the world differently, and they place the highest value on payoffs in equity drawdowns. Said differently, these options provide a natural hedging value, leading investors to be will to pay a premium to access their payoff profile. Interestingly, although this put skew is persistent across time, there are certain times when this feature is more or less pronounced. In fact, in times of market stress we observe that the level of implied volatility rises, but the skew flattens.

Another feature in equity volatility surfaces is that the term structure, or slope of the volatility surface as expiry increases [2], tends to be upward sloping. We saw an example of this earlier in the chapter based on VIX, and the concept behind the slope of a volatility surface is analogous. In particular, during calm, normal periods, we find implied volatilities increase as we increase the option expiry. Conversely, when markets are stressed, this behavior reverses. That is, the level of volatility is quite high, but the expectation is (generally) that this will subside after some period of time, leading to lower implied volatilities at longer expiries. In a normal market environment, the following plot shows how an equity volatility term structure might look:

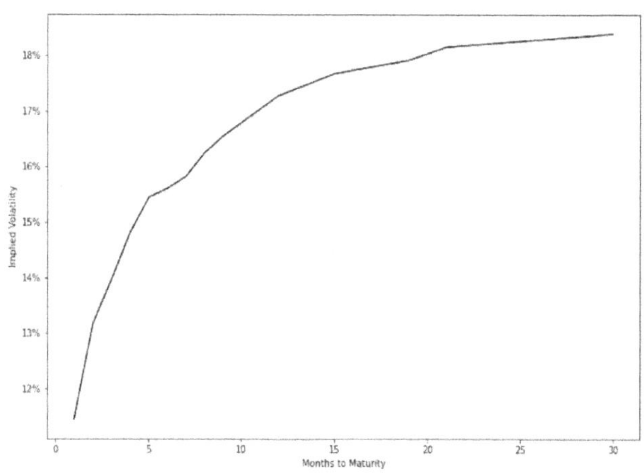

In commodity markets, these features are much less uniform than in equity markets. This is because individual commodities have different underlying properties. Some commodities, such as oil, are inherently pro-cyclical with a significant risk-on/risk-off component. These types of commodities tend to have volatility skew that aligns with their expected movement in a broad based risk-off event, mimicking what we see in equity markets. Other commodities, such as corn and wheat, might be viewed as more idiosyncratic, leading to a flatter volatility surfaces or those that are driven by more specific perceived tail risks.

[2]For a given strike.

18.4 EXOTIC OPTIONS IN EQUITY & COMMODITY MARKETS

18.4.1 Lookback Options

Earlier in the text, in chapter 15, we introduced lookback options and provided an illustration of how market skew impacts their pricing. It turns out that lookback options are frequently used in practice, especially in the equity derivatives markets. Lookback options give the buyer the right to choose the optimal exercise point with the benefit of hindsight. That is, the exercise point is chosen at expiry. This means that we get to exercise at the maximum for a lookback call and at the minimum for a lookback put.

There are two types of lookbacks, **Fixed Strike** and **Floating Strike**. Fixed Strike lookbacks set the strike at trade initiation and determine the payoff of the option based on the optimal point. Mathematically, the price of a fixed strike lookback can be written as:

$$c_0 = \tilde{\mathbb{E}}\left[e^{-rT}(M_T - K)^+\right] \tag{18.3}$$

$$p_0 = \tilde{\mathbb{E}}\left[e^{-rT}(K - N_T)^+\right] \tag{18.4}$$

Where M_T is the maximum attained over the period and N_T is the minimum value over the period, and c_0 and p_0 are lookback call and put options, respectively. The presence of the maximum and minimum values in (18.3) is a reflection of the fact that the holder of the option will choose to exercise at the point that is most advantageous to them, which is the point at which the asset, and subsequently the option payoff are the highest.

Floating Strike lookbacks, on the other hand, set the strike based on a percentage of the optimal point in the asset and calculate the payoff from the terminal asset value. Mathematically, the price of a floating strike lookback option can be written as:

$$c_0 = \tilde{\mathbb{E}}\left[e^{-rT}(S_T - KM_T)^+\right] \tag{18.5}$$

$$p_0 = \tilde{\mathbb{E}}\left[e^{-rT}(KN_T - S_T)^+\right] \tag{18.6}$$

Where M_T and N_T are again the maximum and minimum attained over the period, respectively, and K is a contractually specified percent of the strike that you observe. Lookbacks also define how frequently the price is sampled (daily, weekly) in the calculation of the minimum or maximum value.

As we can see from their payoff structure, Lookback options are path-dependent and depend on the distribution of the maximum or minimum of the asset. This makes them exotic, and increases the computational effort required to model them. In particular, generally speaking we will need to rely on the techniques presented in chapter 15 to value them. This means leveraging either a PDE approach, or a simulation algorithm. Alternatively, depending on the underlying model that we choose, the distribution of the maximum and minimum values of the asset may or may not be known. If it is known, then this simplifies the problem significantly. That is, the problem then reduces to a one-dimensional integral over that distribution, which we may be able to compute analytically or alternatively could estimate via quadrature techniques, instead of requiring the entire path. Of course, with realistic models, it is far more common and should be expected that this distribution of the maximum/ minimum is unavailable, leaving simulation and PDEs as the primary tools for solving these types of pricing problems.

In practice, lookback options have some nuanced properties, especially compared to their vanilla, European counterparts. As we explore later in the chapter, lookback options, when combined with Europeans, are an effective way to isolate mean-reversion and skew in the volatility surface. Because of this, Lookbacks are often traded against European options, a structure we explore in 18.5.4.

18.4.2 Basket Options

Basket options are other commonly traded exotic options within the equity derivative markets. Due to their exotic nature, baskets are almost exclusively traded OTC, allowing for significant customization of terms, including the desired items in the basket. Basket options are multi-asset options, making them quantitatively interesting and challenging. As we will see, they are also inherently correlation, or dispersion based instruments. This means that they are a good way for an investor to place a view on forward looking correlations, or to harvest an embedded correlation risk premia in the market.

Basket options have their payoff based on a linear combination of assets. Mathematically, the price of a basket call option can be written as:

$$c_0 = \tilde{\mathbb{E}}\left[e^{-rT}\left(\sum_{i=1}^{N} w_i \frac{S_{T,i}}{S_{0,i}} - K\right)^+\right] \tag{18.7}$$

Where the w_i's represent the weights of each underlying asset, the $S_{T,i}$'s represent the terminal asset prices. Similarly, best-of and worst-of option are defined on the max or min payoff of a set of underlyings:

$$b_0 = \tilde{\mathbb{E}}\left[e^{-rT}\left(\max_i \frac{S_{T,i}}{S_{0,i}} - K\right)^+\right] \tag{18.8}$$

$$w_0 = \tilde{\mathbb{E}}\left[e^{-rT}\left(\min_i \frac{S_{T,i}}{S_{0,i}} - K\right)^+\right] \tag{18.9}$$

where b_0 is the price of a best-of call and w_0 is the price of a worst of call. As we can see from the formulas, these options are usually normalized by their initial asset values, $S_{0,i}$, to account for potentially different levels in the underlying.

Because of their non-linearity in the payoff function, each of these products depend on the **correlation** between the assets in the basket. If the payoff of the basket was linear, then we could employ Fubini to switch sum and expectation in (18.7) and we would see that the payoff would be equivalent to positions in the underlying assets themselves.

This non-linearity in the payoff, however, makes things more interesting, and in these cases we observe a difference between an option on a basket of assets, and option positions in the underlying assets. Later in the chapter, we explore this relationship between a basket of options and an option on a basket in the context of single stock options vs. index options, and show that correlation, or dispersion is a fundamental driver of the difference between the two.

Because of their multi-asset and multi-dimensional nature, modeling basket options is quite complex and is generally done via simulation. Like any other multi-asset option, use of a standard SDE specification will require us to parameterize and calibrate a multi-dimensional SDE. For example, if we assume that the individual assets follow a Black-Scholes model, then we need to define the correlation of the increments of the asset processes. One approach would be to rely on realized correlation between the assets to value the basket, however, there tends to be a basis between realized and implied correlations. Further, if we use this approach, even if we are able to match the correlation, we are still unable to match the individual volatility smiles because of our overarching Black-Scholes assumption. In a stochastic volatility model, such as Heston, we would then be able to match the individual volatility smiles fairly well, at least potentially, however, the correlation structure becomes more complex because it now includes volatility processes of different assets, which are inherently unobservable and more challenging to calibrate.

Alternatively, the weighted Monte Carlo technique that we introduced in chapter 17 provides a convenient, robust framework for modeling basket and other multi-asset options. To apply this, we would apply weighted Monte Carlo to the underlying assets, with its European options as the problem constraints, and then use the calibrated model to value the basket, best-of or worst-of structure of our choice. This approach is appealing in that it would, at least theoretically, enable us to match the individual volatility smiles of the underlying assets while making reasonable correlation assumptions.

18.5 INVESTMENT PERSPECTIVE: TRADED STRUCTURES

18.5.1 Hedging Equity Risk

One of the main uses of equity markets, of course, is to participate in the profits of underlying firms and in doing so access an associated risk premium. One of the features of this premia, and perhaps one of the reasons that it is well-rewarded, is that it has a pro-cyclical component. In strong economic times, the majority of public companies are able to generate profits, leading to strong equity returns. In times of stress, firms struggle to generate profits, however, and equities often correct. The times of these corrections are often associated with times of broader uncertainty for investors as well. Their jobs may be uncertain, or, from a fund managers perspective, their capital may be more uncertain. As a result, a common use of the equity markets is to hedge overarching equity risk inherit in their portfolios. Almost every investor is naturally inclined to hedge this tail risk associated with equity drawdowns, because of the uncertainty it eliminates.

Hedging equity portfolio risk can be done in a number of different ways. The equity derivatives market, in particular provides several options for this type of hedge. For example, a short position in an equity index future can provide this type of exposure. Alternatively, long positions in put options, or put spreads, on equity indices, provide a leveraged, conditional downside hedge. Relative to other asset classes, hedging in equity markets is comparably straightforward, because of the lack of a yield curve or a jump to default component. The natural demand for these types of hedges, however, mean it can be extremely difficult to find an inexpensive way to hedge broad equity market risk. This is something we should keep in mind when thinking through our hedging decisions. In some cases hedging still makes sense, even if we have to overpay for the hedge[3]. In other cases it might lead us to try to hedge less. It also means that a tremendous amount of value is placed in investors who are able to construct this type of exposure in a cost effective manner.

18.5.2 Momentum in Single Stocks

Momentum is one of the most widely documented potential market anomalies, and is also one that has been around the longest and seemingly persisted [66]. As such, many investors employ momentum strategies, where they take long positions in companies with strong recent performance and offset those long positions with short positions in firms that have recently struggled.

When constructing a momentum strategy, the first step is to identify an explicit momentum signal. Perhaps most commonly, it will be based on the previous period's returns, however, there are many ways to construct such a signal. In particular, we need to choose the lookback window. AQR, for example, proposed using the last twelve month's return, less the last month's[4] [66]. Once we have this momentum signal, the next step is to identify a methodology for translating this signal into a portfolio. Perhaps most simply, we could enter a single long position in the stock with the highest momentum signal and a single short position in the stock with the lowest momentum signal. Or more generally, we could choose an equally weighted portfolio with long positions in the top N momentum signals and short positions in the bottom N momentum signals. In chapter 27 we revisit building this type of momentum signal and discuss more generally how to approach the different phases of building a quant strategy.

18.5.3 Harvesting Roll Yield via Commodity Futures Curves

Earlier in the chapter, we discussed the interesting nature of commodity futures curves, especially when the underlying commodity is difficult to store. Another prime example of this is VIX, where the index itself is completely untradeable. This has led many investors to pursue carry strategies on these assets, which harvest the roll yield, or curve slope, in commodities, as well as VIX.

---

[3] As is the case with homeowner's insurance, for example.

[4] The last one month return has a document reversal component to it and is thus often excluded from a momentum signal.

In order to do this, one could either employ a cross-sectional approach or look at each commodity in isolation and engage in spread trades on individual commodities. In a cross-sectional approach, the framework would closely mimic that of the momentum strategy discussed in the previous section. That is, we would start by building an underlying carry signal. Most naturally, this carry signal might just be a function of the curve slope relative to spot or a shorter dated future, as detailed in the following equations:

$$s_{spot} = \log\left(\frac{f_i}{S_0}\right) \tag{18.10}$$

$$s_{fut} = \log\left(\frac{f_i}{f_j}\right) \tag{18.11}$$

where S_0 is the current spot price of the commodity, f_i is the price of the ith future and f_j is the price of the jth. The slopes, s_{spot} or s_{fut}, respectively, would then be input carry signals. We would then want to be short commodities with the highest, positive slope and long commodities that have the flattest or most inverted curves. In a cross-sectional approach we may choose to take a single position in a single future for each commodity. We could do this at a fixed point on all curves, or choose the point on each curve where the signal is maximized or minimized (i.e. where the curve is steepest / flattest). We might then assume that holding a portfolio of different commodities, with some long and some short positions, neutralizes the overarching commodity exposure. To do this, we could use the same approach discussed in the previous section on momentum. For example, we could choose to be short the commodities with the N steepest curves, and choose to have corresponding long positions in the N flattest curves.

Alternatively, we could construct a set of so-called spread trades on the underlying commodity futures based on the slope of the curve. Spread trades are inherently long-short structures where we long one future and take an offsetting short position in another future along the same curve. The basic premise of this type of spread trade is to go long the point on the curve with more attractive roll properties, and lower roll costs, and then go short another point on the same curve that has the highest roll costs & least attractive carry properties. This is summarized in the table below:

Units	Future
-1	Steepest, Highest Roll Cost Future
β	Flattest, Lowest Roll Cost Future

where β is the number of units we choose to buy to offset our short position. β can be chosen in a few different ways, for example it might be set such that the spread trade is neutralized to parallel shifts in the futures curve. It also might be set based on the relative volatilities, or similarly, the betas to movements in the spot price, of the underlying futures. To do this, one approach is to calculate the β by looking at the realized volatilities of the two points on the futures curve. Alternatively, we might choose to use an equal notional approach, that is, set $\beta = 1$. This could leave us with residual directional exposure to a significant increase or decrease in the futures curve, however. This is because different futures may have significantly different volatilities. In fact, in practice we observe that futures curves are more volatile in their front end than later expiries. Thus, an equal ratio would lead to more exposure on the shorter expiry side of the trade, relative to its longer counterpart. We also observe that the relative volatilities, or betas to the underlying asset are dynamic and change according to market conditions. This makes choosing a proper β for such a spread spread inherently challenging.

Most commonly, curves in contango are characterized by their concave nature. This means that, the short end of the curve is steeper than the longer end. In this framework, this will lead to short positions at the front end of the curve and corresponding long positions further out the curve. If a β of one is chosen, in this case we would retain some residual short exposure to the asset or commodity. If β was chosen to neutralize the exposure to parallel shifts of the curve, then that would lead to a larger long position further out the curve, as it is inherently less volatile. A major risk for this trade then would be a change in the slope of the futures curve. In particular, in this construction a flattening, or inversion of the curve could lead to significant losses. When curves are in backwardation, the story changes slightly however the concepts are the same, with the main difference being that in this case the futures would be expected to roll higher, instead of lower.

A common tenet of these structures is that the steepest parts of a futures curve have the highest roll costs and are therefore likely to have the most challenging future returns. This is a premise that is deeply rooted in theory, and is the basis for many investment strategies [214]. At heart, the assumption that we are making is that the futures values will converge toward their lower expiry counterparts and ultimately to the spot value of the asset. Said differently, we are betting that the futures curve has risk premium in it that leads to persistent steepness in the curve that is not subsequently realized. As such, the success or failure of these strategies is largely determined by whether futures contracts roll toward the spot for the underlying, or whether the spot will drift higher or lower to match the forward. This is a highly philosophical question that is actively debated by quants. Efficient markets theory suggests that spot should mechanically drift toward the forward, however, a considerable amount of evidence suggests that, in some markets, futures have a persistent tendency to roll to spot. Again, this could be a manifestation of an underlying risk premium in the futures curves, perhaps because of a fear of a prolonged backwardation or supply disruption.

18.5.4 Lookback vs. European

Earlier in the chapter we discussed lookback options and learned that when trading them we were able to make our exercise decision with the benefit of hindsight. That means that lookback options are concerned with the distribution of the minimum or maximum value over an interval, in contrast to a European option which is concerned with the distribution of the terminal value. This ability to choose our exercise decision with the benefit of hindsight is clearly an advantageous feature, making lookback options more valuable, and thus more expensive than the equivalent European option. In practice, market participants often try to isolate the unique features of lookbacks by pairing them against European options. A typical lookback vs. European structure would then look like the following:

Long/Short	Description
Long	Fixed Lookback Put Option with strike K and expiry t
Short	European Put Option with strike K and expiry t

In this case we have focused on a fixed strike lookback, and put options, however, floating strike lookbacks or calls could easily be substituted. As a reminder, the price of a fixed strike lookback can be expressed as:

$$p_0 = e^{-rT}\mathbb{E}\left[(K - S_T)^+\right] \tag{18.12}$$

$$l_0 = e^{-rT}\mathbb{E}\left[(K - N_T)^+\right] \tag{18.13}$$

where N_T is the minimum value of the asset over the interval from 0 to T. Assuming that K and T are the same for the European put and the lookback, we can use (18.13) and (18.12) to compute the payoff of the lookback vs. European:

$$l_0 - p_0 = \begin{cases} (N_T - S_T) & S_T \leq K \\ (K - N_T) & S_T > K, N_T \leq K \\ 0 & S_T > K, N_T > K \end{cases} \tag{18.14}$$

Here, we can see an interesting feature of this trade emerge. If the minimum value and the terminal value of the asset are equal, our payoff is zero. As this is a trade that costs an upfront premium, because $l_0 > p_0$[5], this is a negative outcome for the investor. Similarly, if the asset never decreases to the strike K, then our payoff will also be zero as both legs will expire worthless. The best outcome for an investor in a lookback vs. European trade is a large spread between the minimum and the terminal value. That is, we need a significant drawdown in order for the value of our lookback to kick in, but will then want a significant reversal in order for our sold European to have as little value as possible. Said differently, this structure has a mean-reversion component. Our goal is for the asset to have a large move in one direction, but then for mean-reversion to pull the asset back lowering the value of our sold option. This dynamic gives the lookback vs. European an interesting delta profile. At the beginning of the trade, while the lookback is out-of-the-money, the delta will be negative. However, as the trade gets more in-the-money and closer to expiry, the delta becomes positive as we want the reversal to kick in to maximize the payoff of our lookback relative to the European.

18.5.5 Dispersion Trading: Index vs. Single Names

A common structure that is traded by many hedge funds and other major players in equity derivatives markets is to trade index options against their single-name

[5]Because of the added optionality in choosing our exercise time.

counterparts. For example, an investor might look to sell a put option on the S&P 500 index, and subsequently buy put options on the constituents in the S&P with notionals that match the weights in the index, as is depicted in the following table:

Long/Short	Description
Long	Call or Put Options on S&P Constituents
Short	S&P Index Call or Put Option

At heart, this is a dispersion trade. If we are long the single-names, then we benefit from their lack of correlation. In such a scenario, where there is little correlation, the index may remain unchanged, leaving our sold option worthless, however, the dispersion or lack of correlation may lead to non-zero payoffs in some of the individual put options.

One of the main reasons that this trade is attractive, and is commonly targeted by hedge funds and other sophisticated market players, is that implied correlations tend to exceed realized correlations, just as implied volatilities tend to exceed realized volatilities. This correlation premium may lead index options to be expensive relative to the underlying set of single names. There is an intuitive, economically motivated rationale for this correlation premium, and that is that correlations tend to rise in crises or other periods of stress. These states with high correlations are then states that investors place a high premium on returns, leading to an associated risk premium.

Additionally, index options are far more liquid than single-stock options, meaning that part of the premium at work in this trade might be a liquidity premium. If we are willing to bear the risk of holding single-stock options, and risk potential delivery into a less liquid underlying, then perhaps we should be compensated with a premium for doing so. This might be another factor in the pricing dynamics of index vs. single-stock options, and the prevalence of a premium in implied correlations.

Lastly, index options are more commonly used as hedging vehicles, which may inflate their prices, and implied volatilities relative to their single name counterparts. This concept of dispersion trading is not unique to equity markets. We saw similar themes emerge in FX and interest rate markets, for example. We also revisit this theme again in more detail from a trading perspective in chapter 27.

18.5.6 Leveraged ETF Decay

Over the last decade, another recent innovation within equity markets has been the launch of leveraged ETFs. These leveraged products provide investors with access to leverage without having to actually use margin accounts, or trade derivatives. Instead, these leveraged ETFs by mandate return a multiple of the underlying index. For example, a 3x S&P leveraged ETF promises to deliver three times the return of the S&P every day. Additionally, within the leveraged ETF universe many inverse products exist. These products provide a negative multiple of the index return. As an example, a -3x S&P leveraged ETF by mandate would give negative three times the return of S&P each day.

This is a potentially appealing structure, especially to retail investors who may not want to manage their margin account and otherwise may be unable to engage in short positions, however it also comes with some unintended consequences. In particular, over longer periods of time, this daily re-balance mechanism leads to a feature that is commonly known as volatility decay. Essentially these leverage ETFs re-leverage their portfolios daily. This is a really good thing if the underlying asset trends in their favor, however, it makes them naturally prone to reversals.

To see this, in the following equation we write the expression for the value of a leveraged ETF at time t, as a function of the returns of the index, and its leverage ratio, β:

$$L_t = L_0 \prod (1 + \beta r_i) \tag{18.15}$$

$$\log\left(\frac{L_t}{L_0}\right) \approx \sum \left(\beta r_i - \frac{\beta^2}{2} r_i^2\right) \tag{18.16}$$

Note that in the second equation, we have taken the log of both sides, which enables us to turn the product into a sum, and leveraged the Taylor series approximation $\log(1+x) \approx x + \frac{1}{2}x^2$. If we were to do the analogous decomposition of the underlying asset at time t, S_t, we would have:

$$S_t = S_0 \prod (1 + r_i) \tag{18.17}$$

$$\log\left(\frac{S_t}{S_0}\right) \approx \sum \left(r_i - \frac{1}{2} r_i^2\right) \tag{18.18}$$

$$\beta \log\left(\frac{S_t}{S_0}\right) \approx \beta \sum \left(r_i - \frac{1}{2} r_i^2\right) \tag{18.19}$$

Where in the last line we have multiplied by the leverage ratio, β. If we then subtract (18.19) from (18.16), we can isolate the differences in properties of the leverage ETF vs. the underlying asset. Doing this yields:

$$\log\left(\frac{L_t}{L_0}\right) - \beta \log\left(\frac{S_t}{S_0}\right) \approx -\frac{\beta^2 - \beta}{2} \sum \left(r_i^2\right) \tag{18.20}$$

Finally, solving for L_t, the value of the leveraged ETF at time t, we get:

$$L_t \approx L_0 \left(\frac{S_t}{S_0}\right)^{\beta} \exp\left(-\frac{\beta^2 - \beta}{2} V_t\right) \tag{18.21}$$

where $V_t = \sum (r_i^2)$ is the realized variance of the underlying asset S from time 0 to time t. In this derivation we neglected terms such as the cost of borrow of the underlying asset, S and the funds expense ratio. These terms can be added easily within the same framework, however, and interested readers should consult [26] for more details.

Importantly, we can see that the leveraged ETF has a dependence on volatility. Further, the negative sign tells us that the higher the volatility, the lower the future price of the leveraged ETF, hence the term volatility decay. This is quite the unintended side effect of holding leveraged ETFs. Over long periods, as volatility accumulates, we should anticipate our positions in leveraged ETFs to lose money. We can also see from the equation that inverse ETFs suffer the most decay. For example, because of the structure of the $\frac{\beta^2-\beta}{2}$ term an ETF with $\beta = -3$ will decay more than a comparable ETF with $\beta = 3$. At first glance this decay is striking, and means that it is possible for both the long and inverse leveraged ETFs to lose money over the same period. And in fact, historically, this has been the case for many such pairs of leveraged ETFs. This has led many investors to develop strategies that short these leveraged ETFs, both on the long and short side. This is a trade that initially has no market exposure, but over time may accumulate exposure while also collecting the volatility decay. The profitability of such strategies is also highly dependent on an ability to borrow the leveraged ETFs cheaply to maintain short positions. Additionally, this trade is by no means an arbitrage, as leveraged ETFs do outperform in a prolonged trend. This is evidenced by the $\left(\frac{S_t}{S_0}\right)^{\beta}$ term. Put together, however, these features mean that leveraged ETFs provide quants a fertile place to explore.

18.6 CASE STUDY: NAT. GAS SHORT SQUEEZE

In this chapter, as well as throughout the remainder of the text, we postulate that implied volatility exceeds realized volatility over long stretches, an indication of a premium for investors who are willing to sell options. In this case study, we provide a cautionary tale based on real market events in natural gas for those who are looking to harvest this premium.

Natural Gas is a particularly interesting commodity because it has interesting dynamics in its futures curve, as well as its volatility. With respect to its futures curve, we saw in 18.2.4 that Nat. Gas curves tend to be upward sloping, and more poignantly that these forwards in general are not realized. This is an indication of a potential risk premium in the futures curve, made possible by the difficulty of storing Nat. Gas. Similarly, like the volatility surface for other asset classes, historical realized volatility has been lower than the implied volatilities traded in the market. This has led many market participants to focus on harvesting both the futures roll yield, and the volatility carry in Nat. Gas markets. The following time series chart of a rolling position in a Nat. Gas future[6] shows us why investors may find this appealing:

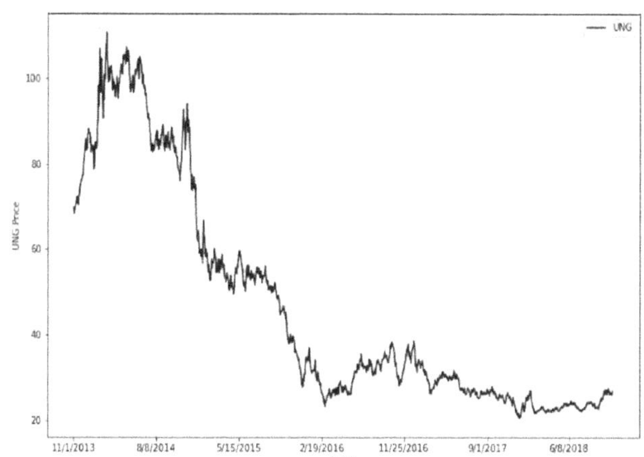

[6]Depicted in this chart as UNG, an ETF that mimics a position in this rolling futures strategy.

These strategies, however, are, of course, susceptible to sudden spikes in the price of Nat. Gas. Because Nat. Gas is primarily used in the winter as a source of heat, and in the summer as a source of cooling, these spikes are inherently dependent on the weather. This means that, a particularly cold winter, can lead to a supply shortage causing a large spike in prices. This in turn leads to losses for investors harvesting the carry strategies mentioned above, which might in turn lead to investors being stopped out, which then in turn could lead to further price increases.

One such episode in took place in November of 2018 when a spike in the Nat. Gas markets notoriously led to severe losses by many fund managers, even forcing some to liquidate. In the following chart we show the price of the first Natural Gas future during the first two weeks of November when this played out:

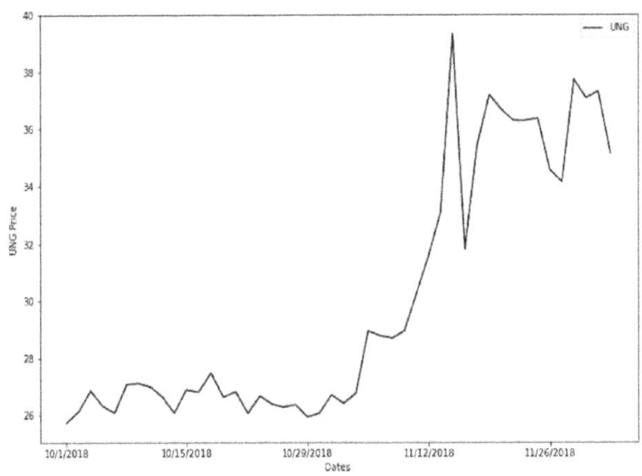

As we can see, there was a significant price appreciation in natural gas over this period. One fund in particular, which unfortunately gained attention because of a YouTube video posted by a co-founder of the firm, experienced losses that were larger than their assets under management and were forced to both close and also ask their investors for additional capital to cover their losses. The co-founder had also written a book related to selling options [68] where he described how they approached harvesting the spread between implied and realized volatility. Along those lines, the fund was engaging in short volatility strategies, and in particular were selling straddle or strangle like structures.

Recall from the payoff diagram below that a short position in a strangle is characterized by small gains when the asset's price movements are small and large losses when they exceed a threshold. That is, the payoffs are asymmetrically against the investor, as we see in the chart:

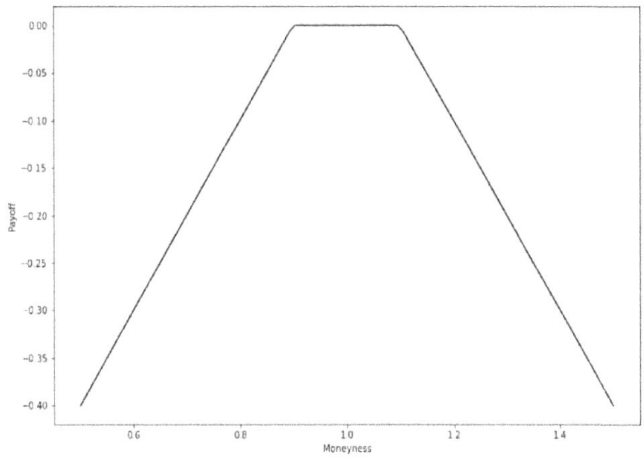

Because of this, as the price of Nat. Gas rose, the fund experienced increasing, asymmetric losses. Another key point here is that, prior to expiry, the profile of the value of the structure can look quite different from this payoff profile, and immense losses can occur well before we reach the traded strikes in our strangle. This is because, as the asset moves in either direction, the distribution of outcomes at expiry changes significantly. This means that, as we get closer to the traded strike in either direction, significantly more mass is placed in the distribution where losses occur, and far less is placed in places where we make money. Greeks, in particular gamma, provide another lens for viewing this same phenomenon. Because we are short gamma, we know that we are short large movements in either direction, and that is exactly what we saw in the market. To see this in practice, the following chart shows how the valuation profile for a short strangle position is different at expiry vs. half-way through the trade:

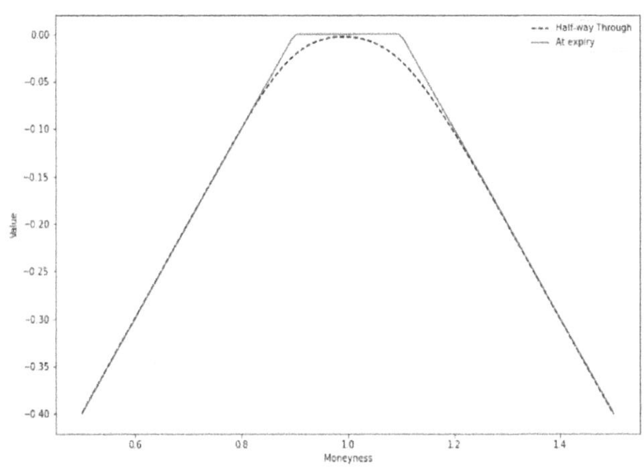

The problem with this type of approach, of course, is that the losses that you may experience are undefined. We make money most of the time when employing these strategies, but when losses occur they are extreme and have the potential to accumulate quickly. This episode highlights the dangers of trying to sell volatility and underscores a critical point with respect to risk management of these strategies. These strategies must be constructed with survival in these types of episodes at the front our minds. Had the fund managers sold put and call spreads, for example, instead of strangles, then their maximum loss would have been defined, and they would have been better able to withstand these price movements. While this period in time contained seemingly idiosyncratic movements, this is a broader lessen to be learned for investors, quants and traders. When harvesting a seemingly attractive risk premia with negative skewness, we need to ensure that we can survive the inevitable drawdowns by managing our tail losses.

18.7 CASE STUDY: VOLATILITY ETP APOCALYPSE OF 2018

Another anomalous market event that occurred in 2018 that is worth exploring was the implosion of the inverse volatility exchange traded products in February 2018. Earlier in the chapter, we argued that VIX futures tend to have significant levels of risk premium in them, and that the contango that we observe in the markets is not accompanied by an ensuing move in the VIX index. Like the last case study, this period of market activity serves as a warning to investors and highlights the subtle difficulties in harvesting even the most documented, widely agreed upon risk premia.

In the years leading up to 2018, trading VIX related products had increased significantly in popularity. As we saw earlier in the chapter, investing directly in the VIX index is not possible, however exposure to VIX can be accessed through positions in futures and options. Additionally, several exchange traded products exist[7], that provided another mechanism for investors, especially retail investors, to participate in the VIX market. A caveat of all of these products, however, is that they do not provide direct exposure to movements in the underlying VIX index. VIX futures, for example, give us the ability to access the spot VIX index only a certain future date. Their valuation on other dates will include many other factors, such as the dynamics of the futures curve. This means that, among other things, just because we observe VIX being anomalously low, does not mean we can take advantage of it. Instead, if the futures are significantly higher than then spot, then we would be buying our exposure to volatility at a higher price than the spot VIX index, significantly altering our payoff profile. Exchange traded products, which hold a rolling basket of future, are then one step further removed. In the following table, we provide a snapshot of the ETF products offered that are linked to VIX:

Two of these products, VXX and UVXY, provide long exposure to volatility, via VIX futures, and as such are potential tail hedge instruments. Inverse products, conversely, are short increases in volatility but have significantly positive carry, making them attractive to investors looking to harvest a volatility risk premium. This leads

[7] Or used to exist.

Ticker	Description
VXX	VIX Short Term Futures ETF
UVXY	Ultra VIX Short Term Futures ETF
XIV*	Inverse VIX Short Term Futures ETF
SVXY	Short VIX Short-Term Futures ETF

to a strong two-way market in these VIX products, with the natural hedging demand driving demand for the long volatility products and the desire to generate income and take advantage of the carry properties driving positions in the inverse. Another important caveat is that three of these products, UVXY, SVXY and XIV[8] are leveraged ETFs and susceptible to the properties discussed in 18.5.6.

As we show in the following time series charts, performance in the long volatility products historically has been particularly onerous, whereas the inverse products tend to exhibit a sharp upward trend with drawdowns associated with volatility spikes:

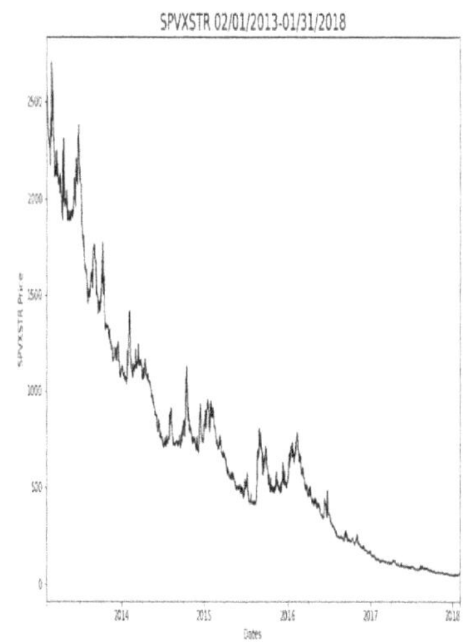

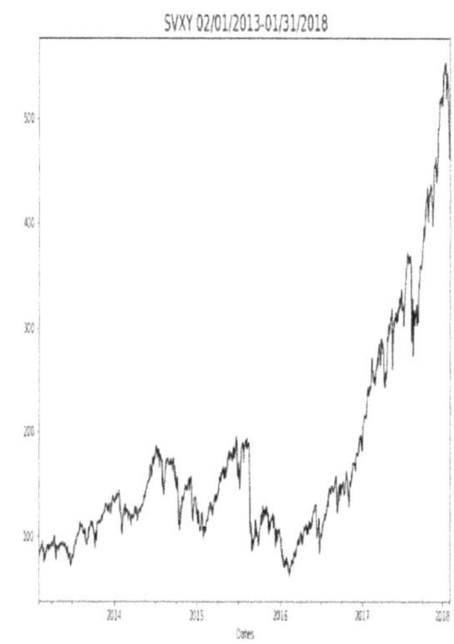

This type of return behavior led many investors to pile into inverse ETPs. Investors in these inverse ETPs enjoyed spectacular returns and many questioned whether buy and hold strategies in these ETPs were guaranteed to make money over sufficiently long periods. This was potentially compounded by the lack of volatility in markets in 2017, lulling investors into a false sense of security. In early 2018, however, as the short side of the trade got more crowded, it also got more fragile. Further, because volatility started at a low level, the potential for a large percentage changes increased. This meant a larger chance of excessive losses in inverse products, who by mandate were required to deliver -1 times the return of the underlying index. This means that when the underlying index returns more than 100%, the NAV of the ETP becomes negative and will naturally liquidate. This is exactly what happened

[8] Which no longer exists.

on February 5, 2018, when a one-day volatility spike wiped out one of the inverse products, XIV and nearly wiped out the other inverse product, SVXY[9]. This striking performance of the ETFs during this period is shown in the following time series charts of the index as well as the inverse ETP SVXY:

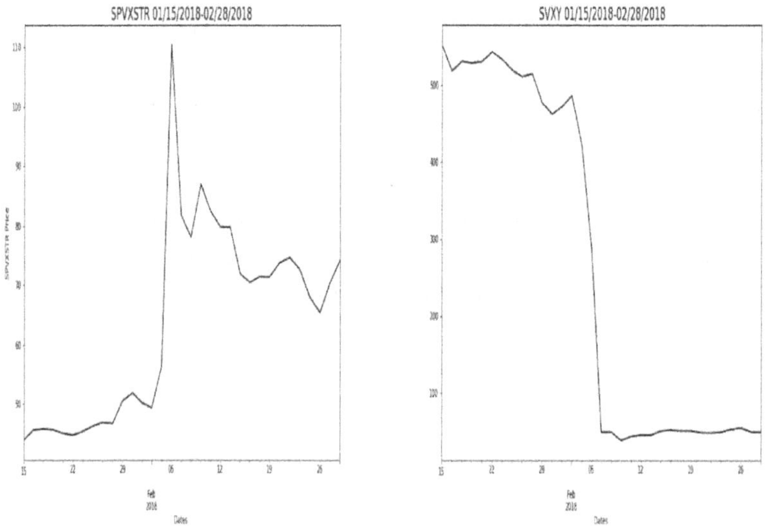

In these charts, we can see that the downward move in SVXY on this day was significantly larger than the corresponding movement in the underlying index or futures. Further, in subsequent days we saw the index equilibrate somewhat, but we don't see the same phenomenon in SVXY, which lost 80% of its value and was unable to recover. Some have speculated that market manipulation was a factor in the magnitude of this particular move. More generally, it brought to the forefront some potential pitfalls of investing in inverse or leveraged ETF products. In particular, it highlighted that the product will close should its NAV turn negative, meaning these inverse products have an embedded knock-out feature. Even if the move had normalized more quickly, the fact that the NAV turned negative led the ETP to become insolvent and not be able to benefit from the recovery. Additionally, this case study serves as another reminder of the perils of trying to harvest these types of premia and show the importance of managing tail risks. Collectively, this case study, as well as the previous one, show how dangerous, and hard it can be to make money being short volatility. This creates a conundrum as it is equally if not more challenging to profit when we are long volatility, given, among other things the persistent implied vol. premium.

[9]SVXY has since changed their mandate to replicate returns equal to $-\frac{1}{2}$ the return of the underlying index.

V

Portfolio Construction & Risk Management

CHAPTER 19

Portfolio Construction & Optimization Techniques

19.1 THEORETICAL BACKGROUND

19.1.1 Physical vs. Risk-Neutral Measure

At this point in the text, the majority of the focus has been on things like derivatives modeling. Said differently, we have spent the majority of the text focusing on the **risk-neutral measure**. When operating in the risk-neutral measure, a certain set of tools are required. In particular, there is an emphasis on hedging and replication arguments and using those arguments to construct market implied distributions under the assumption of risk-neutrality. In practice, while we know that the assumption of risk-neutrality is unrealistic, the presence of these underlying hedging & replication arguments makes this assumption both reasonable and convenient.

This is in sharp contrast to modeling of the physical measure, which will be the primary focus for the remainder of the book. When working with portfolio optimization, risk management and quantitative trading strategies, we are no longer interested in risk-neutral modeling techniques but instead will want to rely on forecasts of the actual distribution (that is, the **physical distribution**). When doing this, replication arguments no longer exist and instead we need to rely on a different part of our quantitative skillset. In the remainder of the text, we build off of this skillset, which was introduced in chapter 5, and discuss some of the main topics of relevance related to the physical measure.

In the physical measure, we know that investors are not risk neutral but are, in fact, risk averse, and need to account for these investor risk preferences in our models. In the next section, we describe an example of this that relies on an investor utility function to construct an optimal portfolio. Additionally, unlike in the risk neutral measure, when we work with the physical measure, we need to estimate expected returns. This creates a significant challenge as forecasting expected returns is unsurprisingly a non-trivial venture, especially in the presence of limited data. When working in the physical measure we need to remember the overarching modeling principles highlighted in chapter 2. In particular, we need to remember that finance is not a hard science and that as quants we are deprived from repeatable experiments.

DOI: 10.1201/9781003530626-24

This creates a great deal of uncertainty in our estimates, and as a result we will need to ensure that our solutions aren't highly sensitive to changes in the inputs. Unfortunately, it turns out that if we are not careful, the opposite is likely to be the case and our optimal portfolios can vary greatly under small shifts to expected return estimates. This phenomenon and suggestions for mitigating it are explored later in the chapter.

The techniques discussed in this section are central to the skillset for buy-side quants, asset allocators and quantitative portfolio managers. It is often said that there are two types of quants: P measure quants, who inhabit the buy-side and focus on econometrics, portfolio optimization and machine learning, and Q measure quants who inhabit sell-side institutions and focus on exotic options, derivatives modeling and structuring. While this is clearly an oversimplification, it highlights an important difference in how each type of problem is approached.

19.1.2 First- & Second-Order Conditions, Lagrange Multipliers

In chapter 14 we introduced optimization techniques and showed how optimization problems could help us solve model calibration problems of volatility surfaces. In the chapter, we mentioned that there are many other applications of optimization techniques. In fact, they are truly ubiquitous and permeate both physical and risk-neutral modeling. In this chapter we explore optimization techniques in a different context, where we are trying to determine the optimal portfolio given a set of expected returns and a covariance matrix.

To do this, we will again need to leverage the same underlying optimization theory. In particular, we found that optimization problems can be solved by finding the relevant first-order and second-order conditions. Assuming an optimization problem where we are looking to maximize[1] the value of a function as shown below:

$$\hat{x} = \max_{x} \{ f(x) \,|\, g(x) \leq c \} \tag{19.1}$$

where $f(x)$ is referred to as the objective functions and $g(x) \leq c$ are the constraints. Recall that we defined a first-order condition, which is a necessary but not sufficient condition for a maximum or minimum point, as:

$$\frac{\partial f(x^*)}{\partial x^*} = 0 \tag{19.2}$$

That is, at maximum or minimum points, the derivative of the function is equal to zero. This is intuitive because if the function were still increasing or decreasing then we know we could obtain a higher (or lower) point by taking a small step in the proper direction. Points that met the criteria in (19.2) are referred to as stationary or critical points. In order to determine whether a stationary point is a minimum, maximum or saddle point, we then need to check the second-order conditions. For a

[1]Recall that we can equivalently write our optimization as a minimization or maximization problem by toggling the sign in the objective function.

maximum, the second-order conditions can be written as:

$$\frac{\partial^2 f(x^*)}{\partial x^{*2}} < 0 \tag{19.3}$$

Similarly, for a minimum we have:

$$\frac{\partial^2 f(x^*)}{\partial x^{*2}} > 0 \tag{19.4}$$

It should also be noted that for certain types of functions, such as convex or concave functions, a unique minimum or maximum is guaranteed as the second derivative has the same sign over the entire domain. This is convenient as it means in practice for convex (concave) functions we do not need to check the second-order conditions for a minimum (maximum), as their will be one and only one solution to the first-order condition and it will be a global minimum (maximum).

This means that, given an objective function $f(x)$, we can apply standard calculus techniques to potentially find minima or maxima. In the next section, we will see an example of this in the context of mean-variance optimization. In practice most optimization problems that we will need to solve will contain at least one constraint. As we saw in chapter 14, solving optimization problems with equality constraints can be done using Lagrange Multipliers. To do this, we re-write our constrained optimization problem as an unconstrained optimization problem with an augmented objective function that includes the constraint, which is referred to as the Lagrangian:

$$\hat{x} = \max_x \{f(x) \,|\, g(x) = c\} \tag{19.5}$$

$$\mathcal{L}(x, \lambda) = f(x) - \lambda(g(x) - c) \tag{19.6}$$

where λ is an additional parameter in the unconstrained Lagrangian. Because it is now an unconstrained optimization problem, (19.6) can now be solved by finding the first and second-order conditions. To do this, we would need to compute the partial derivatives with respect to x and λ for the first-order conditions, for example. An important observation is that the first-order condition of (19.6) with respect to λ recovers the constraint. Therefore, we can see that solving this Lagrangian will enforce the equality constraint.

As we incorporate constraints, especially inequality constraints, solving optimization problems analytically becomes harder and harder, at which point we need to resort to iterative methods.

19.1.3 Interpretation of Lagrange Multipliers

As we saw in the last section, Lagrange Multipliers are an important tool for solving constrained optimization problems. It turns out that, beyond this, Largrange Multipliers are useful because of their practical interpretation. Lagrange Multipliers measure the *rate of change in the objective function as we relax or modify the constraint.* This means that, if the Lagrange Multiplier is large, a small change in the constraint could lead to a significant increase (or decrease) in the objective function.

If we assume the underlying optimization problem is to maximize investor utility, then the Lagrange Multiplier tells us how much additional utility an investor could get from increasing or decreasing a given constraint.

To see this, recall that the first-order condition of the Lagrangian is:

$$\begin{aligned}\nabla f(\vec{x}) - \lambda \nabla g(\vec{x}) &= 0, \quad g(\vec{x}) = c && (19.7)\\ \nabla f(\vec{x}) &= \lambda \nabla g(\vec{x}) && (19.8)\\ \frac{\partial f(\vec{x})}{\partial x_i} &= \lambda \frac{\partial g(\vec{x})}{\partial x_i} \quad \forall i && (19.9)\end{aligned}$$

That is, the Lagrangian is tangent to the constraint at critical points, as discussed in chapter 14. Further, we know that as we move the constraint value, c, the optimal solution $\vec{x}$ changes. Therefore, $\vec{x}$ depends on c. If we write $\vec{x}$ as a function of c and take the derivative of $g(\vec{x}(c)) = c$ with respect to c we get:

$$\sum_{i=1}^{N} \frac{\partial g(\vec{x})}{\partial x_i} \frac{dx_i}{dc} = 1 \tag{19.10}$$

where on the left-hand side we are using the chain rule and calculating the total derivative. On the right-hand side, clearly the derivative of c with respect to a change in c must be 1. Said differently, as we increase the constraint value, c must change by the amount of our increase or decrease in the constraint by construction. Intuitively, this equation tells us that the total change in the constraint function $g(\vec{x}(c))$ must be equal to the total change in c, which, given that it is an equality constraint, makes sense.

Next, let's calculate the change in $f(\vec{x})$ for a corresponding change in c:

$$\begin{aligned}\frac{df(\vec{x})}{dc} &= \sum_{i=1}^{N} \left\{ \frac{\partial f(\vec{x})}{\partial x_i} \frac{\partial x_i}{\partial c} \right\}\\ &= \sum_{i=1}^{N} \lambda \left\{ \frac{dg(\vec{x})}{\partial x_i} \frac{dx_i}{dc} \right\}\\ &= \lambda && (19.11)\end{aligned}$$

where in the second step we substituted the left-hand side of (19.9) for the right-hand side and in final step we used the fact that the total derivative must equal one as shown in (19.10).

Therefore, the Lagrange Multipliers, or the λ's, have an important economic interpretation. In practice, the λ's are often referred to as shadow prices of inventory and in this section we show why. Fundamentally, they represent the extent to which our optimal solution would improve if we were able to modify our constraint. Or said differently, it is a measure of how much the constraint is costing us. In the context of portfolio optimization, the λ's may tell us the additional utility that we could get by relaxing a given constraint. In the context of calibration formulated with option prices as constraints, the λ's will tell us the increase in the objective function that we could get by changing the option price. We saw an example of this in chapter 17, where we minimized the curvature of a density. In this example, the λ would tell us how much curvature is being introduced by each option price constraint.

19.2 MEAN-VARIANCE OPTIMIZATION

19.2.1 Investor Utility

Because we are now working in the physical measure, we need to incorporate investors preference and utility in our models. A portfolio with certain characteristics may be optimal for one investor who is risk averse but highly suboptimal for another with a higher risk tolerance. Typically, this is structured by creating a utility function, $U(\tilde{w})$ of an investors wealth. We know from economic theory that investors should prefer a higher levels of wealth. That is, $U(\tilde{w})$ should be monotonically increasing as we increase $\tilde{w}$. It is also sensible that as investors accumulate wealth, the value of additional wealth will decline. This is referred to as decreasing marginal utility of wealth, and should be similarly incorporated into $U(\tilde{w})$.

$$\tilde{w} = \tilde{w_0}\,(1 + r_p(w)) \tag{19.12}$$

$$U(\tilde{w}) = U(\tilde{w_0}\,(1 + r_p(w))) \tag{19.13}$$

where $\tilde{w}$ is the end of period wealth, $\tilde{w_0}$ is the starting level of wealth and r_p is the return of the investors portfolio, which will be a function of the weights, w.

If we assume that $r_p(w)$ is known with certainty, then our utility function should reflect a higher utility for higher returns. That is, our utility function should be characterized by a positive coefficient on r_p, and again, assuming the returns are constant, would lead investors to the weights, w that lead to the highest return. Of course, in reality, the returns of our portfolio are not known but instead are stochastic. This means that we need to incorporate the entire distribution of $r_p(w)$ into our utility function in order to choose an optimal set of weights, w.

We know that investors should be risk averse, that is, all else equal, they should prefer a portfolio with less volatility to more volatility. This means that, as the variance of $r_p(w)$ increases, utility should decrease.

In the context of stochastic returns for the portfolio, we might look to maximize expected utility:

$$\max_w f(w) = \mathbb{E}\left[U(\tilde{w_0}\,(1 + r_p(w)))\right] \tag{19.14}$$

$$\max_w f(w) = \mathbb{E}\left[U(\tilde{w_0}\left(1 + \sum_i w_i r_i\right))\right] \tag{19.15}$$

where $r_p(w)$ is the portfolio return which can be written as the dot product of the weights and the returns of the individual assets.

This is an optimization problem that can be solved using the techniques reviewed at the start of the chapter. In particular, if we apply the chain rule we get the following first and second-order conditions:

$$\frac{\partial f(w)}{w_i} = U'\left(\tilde{w_0} + \sum_i w_i r_i\right) r_i = 0 \tag{19.16}$$

$$\frac{\partial^2 f(w)}{w_i^2} = U''\left(\tilde{w_0} + \sum_i w_i r_i\right) r_i^2 < 0 \tag{19.17}$$

Economic theory teaches us that we should prefer higher to lower returns, therefore we should expect the first derivative of the utility function, $U'(\tilde{w})$ to be positive. Similarly, we should expect the second derivative, $U''(\tilde{w})$, to be negative to reflect investors risk aversion.

One could make many different assumptions about the form of investor utility functions. A simplistic approach that leads to the intuitive, familiar, mean-variance optimization framework is a quadratic utility function such as:

$$U(\tilde{w}) = \tilde{w} - \alpha\tilde{w}^2 \tag{19.18}$$

If $\alpha > 0$, this form of the utility function is concave, meaning the second order condition will be never be violated, and the first-order condition will represent a unique, global maximum. A positive α, and a corresponding negative sign on the $\tilde{w}$ term, therefore incorporate decreasing marginal utility of wealth.

19.2.2 Unconstrained Mean-Variance Optimization

Markowitz [175] suggested that we formulate optimal portfolio problems in terms of mean and variance. This is often referred to as Modern Portfolio Theory. This formulation, and the exclusion of higher moments from the optimization, can be motivated by making a few different assumptions about investors utility functions. Most notably, an assumption of quadratic utility or of an underlying Gaussian return distribution leads to a mean-variance optimization approach.

As we know that investors prefer higher to lower returns, and lower to higher levels of volatility, Markowitz wrote the investor utility function to be maximized as:

$$\max_{w} f(w) = w^T R - \lambda w^T C w \tag{19.19}$$

where λ is an investor's risk aversion coefficient and is assumed to be a constant in the optimization. The presence of λ in the optimization means that as different investors are likely to have different risk preferences, they may naturally end up with different optimal portfolios. The optimization techniques that we introduced in chapter 14, and then reviewed at the beginning of the chapter can be used to find the solution to (19.19) analytically. Specifically, the first-order condition can be written as:

$$\frac{\partial f(w)}{\partial w} = R - \lambda C w = 0 \tag{19.20}$$

Further, note that the function $f(w)$ in (19.19) is concave and therefore the first order conditions define a unique, global maximum. Therefore, solving for w, the optimal weights, we get:

$$w = \frac{1}{2\lambda} C^{-1} R \tag{19.21}$$

A defining feature of this approach is that, because each investor's risk aversion coefficient will be different, under this model we can't identify a single portfolio or set of weights that all investors should hold. We can however say that some portfolios dominate others. For each risk aversion coefficient, there will be a single set of weights that are optimal for an investor. Aggregating this for all risk aversion coefficients provides us with what we call an **efficient frontier** of portfolios. This efficient frontier is a set of portfolios that are optimal for investors with different risk preferences. In the next section, we develop the concept of an efficient frontier and show how it can be used as a visualization tool.

19.2.3 Mean-Variance Efficient Frontier

We can equivalently rewrite the unconstrained optimization problem in (19.19) as a minimization problem with a single equality constraint. In particular, we can find the portfolio that has the minimum variance for a given expected return:

$$\max_{w} \left(-w^T C w\right) \tag{19.22}$$

$$\text{s.t.} w^T R = \hat{R} \tag{19.23}$$

where C is the covariance matrix of the returns of the underlying assets, R is the vector of expected returns for the underlying assets and $\hat{R}$ is the target expected return, which is a constant.

Notice that a risk aversion parameter no longer shows up in our optimization. This ends up being a useful formulation as the concept of a risk aversion parameter is a bit nebulous and difficult to interpret. Instead, it is much more naturally to say that we are targeting the portfolio with the least volatility for a given return, and then have investors specify their desired target returns.

The optimization problem in (19.22) is now an optimization with a constraint. In particular, we can use the technique of Lagrange Multipliers to solve for the optimal weights analytically. The Lagrangian in particular can be written as:

$$\mathcal{L} = \left(-w^T C w\right) + \lambda \left(w^T R - \hat{R}\right) \tag{19.24}$$

Notice that the Lagrangian above looks similar to the previous formulation in (19.19) with the exception of the placement of λ. As such, it can be solved using the same techniques. In particular, the first-order conditions with respect to w and λ respectively, can be written as:

$$\frac{\partial L}{\partial w} = -2Cw + \lambda R = 0 \tag{19.25}$$

$$\frac{\partial L}{\partial \lambda} = w^T R - \hat{R} = 0 \tag{19.26}$$

We can then use algebra to solve for w and λ, respectively in these two equations. To do this, we begin by solving the first-order condition with respect to w, (19.25), for w:

$$\lambda R = 2Cw \tag{19.27}$$

$$w = \frac{\lambda}{2} C^{-1} R \tag{19.28}$$

We can then use (19.26) to solve for λ by plugging in the above solution for w into the first-order condition and solving for λ. This leads to:

$$w^T R = \hat{R} \tag{19.29}$$

$$\left(\lambda C^{-1} R\right)^T R = 2\hat{R} \tag{19.30}$$

$$\lambda = \frac{2\hat{R}}{R^T C^{-1} R} \tag{19.31}$$

Finally, if we plug the value for λ back into the solution for w above, we get:

$$w = \frac{\hat{R} C^{-1} R}{R^T C^{-1} R} \tag{19.32}$$

In the next section, we revisit the concept of an efficient frontier with an additional constraint: that the portfolio must be fully invested.

19.2.4 Mean-Variance Fully Invested Efficient Frontier

Generally speaking, solving realistic portfolio optimization problems will require many additional constraints. These constraints will not, however, change the approach or formulation of the overall optimization problem. Instead they are tweaks to the framework introduced so far. The most obvious constraint is that we want our portfolio to be fully invested. Incorporating this constraint leads to the following optimization problem:

$$\min_{w} \quad w^T C w \tag{19.33}$$

$$\text{s.t.} \quad w^T R = \hat{R}, \quad w^T 1 = 1 \tag{19.34}$$

Because the constraint we have added is an equality constraint, we can use Lagrange multipliers to compute the solution to this optimization problem analytically. That is, we can use the technique of Lagrange Multipliers, and find the first-order conditions that must be met for a minimum. Further, as the function is convex, we know that the solution to the first-order conditions is a unique, global minimum.

Doing this, we see the Lagrangian is:

$$\mathcal{L} = \left(-w^T C w\right) + \lambda_1 \left(w^T R - \hat{R}\right) + \lambda_2 \left(w^T 1 - 1\right) \tag{19.35}$$

This leads to the following set of first-order conditions:

$$\frac{\partial L}{\partial w} = -2Cw + \lambda_1 R + \lambda_2 1 = 0 \tag{19.36}$$

$$\frac{\partial L}{\partial \lambda_1} = w^T R - \hat{R} = 0 \tag{19.37}$$

$$\frac{\partial L}{\partial \lambda_2} = w^T 1 - 1 = 0 \tag{19.38}$$

This is a system of equations with a well-known solution that can be obtained, albeit somewhat tediously, using linear algebra[2]. Solving the system of equations leads to:

$$\begin{aligned} w &= \frac{1}{d}\left[cC^{-1}R - aC^{-1}1\right]\hat{R} + \frac{1}{d}\left[bC^{-1}1 - aC^{-1}R\right] \\ a &= 1^T C^{-1} R \\ b &= R^T C^{-1} R \\ c &= 1^T C^{-1} 1 \\ d &= bc - a^2 \end{aligned} \tag{19.39}$$

where it should be emphasized that a, b, c and d are constants and c is distinct from C which is an N by N covariance matrix.

If we consider the solutions to (19.33) for many different expected returns, we obtain a set of optimal portfolios that are referred to as an efficient frontier. These efficient frontier portfolios dominate all other portfolios that are below the efficient frontier and represent the portfolio an investor should hold for a given, desired expected return. It is common to plot the efficient frontier with volatility on the x-axis and expected return on the y-axis. An example of this visualization is provided in section 19.2.5.

19.2.5 Mean-Variance Optimization in Practice: Efficient Frontier

In this section we leverage the code developed in the supplemental coding repository to generate a popular visualization of the efficient frontier. In the chart, we plot portfolio returns on the y-axis and portfolio volatility on the x-axis for all portfolios that are on the efficient frontier for different desired returns. Here we use the sector ETF data as the investment universe, which yields the following efficient frontier:

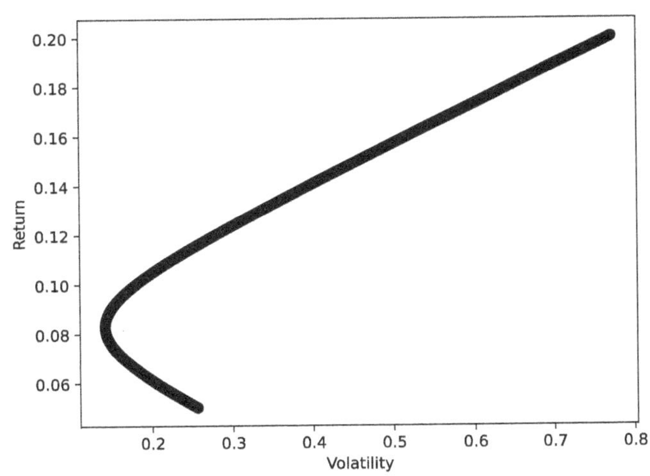

[2]Just as we did in (19.22).

In addition to the analytical approach described in 19.2.4 for generating an efficient frontier, we can also generate an efficient frontier using simulation. In this approach, we simulate random weights for every individual asset, rescaling such that the portfolio are fully invested[3], and compute the return and variance for each random portfolio. This results in the following scatter plot, where we can find points both on and inside the efficient frontier:

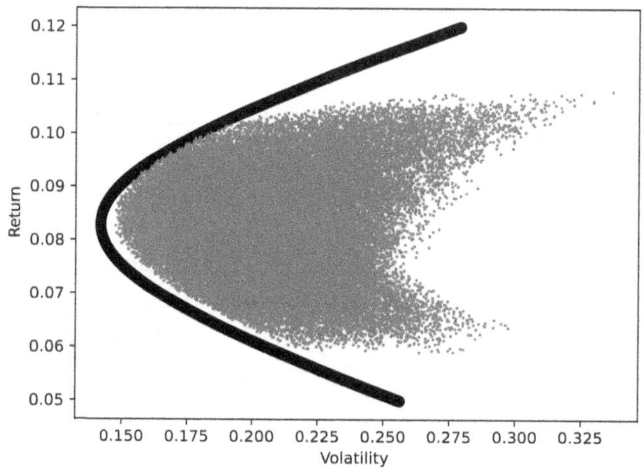

The random portfolios with the best risk/return characteristics would then define the efficient frontier in this methodology. As we can see in the above visualization, these two approaches lead to a similar and consistent efficient frontier.

19.2.6 Fully Invested Minimum Variance Portfolio

Another portfolio of interest is the fully-invested portfolio that has minimum variance. This is known as the **minimum variance portfolio**, and can be found analytically using the optimization techniques discussed in this text. In particular, the minimum variance portfolio is the solution to the following optimization problem:

$$\min_{w} \quad w^T C w \tag{19.40}$$

$$\text{s.t.} \quad w^T 1 = 1 \tag{19.41}$$

If we omitted the constraint that the portfolio was fully invested, that is that the sum of the weights was equal to one, then we could simply set all weights equal to zero and achieve a portfolio with no variance. If we require the assets to be invested, however, the problem becomes more interesting. As the constraint in (19.40) is an equality constraint, we can use Lagrange Multipliers to find the necessary first-order

[3]That is, we ensure that the weights sum to one.

condition for the minimum:

$$\mathcal{L} = -w^T C w - \lambda \left(w^T 1 - 1\right) \tag{19.42}$$

$$\frac{\partial L}{\partial w} = 2Cw - \lambda 1 = 0 \tag{19.43}$$

$$\frac{\partial L}{\partial \lambda} = w^T 1 - 1 = 0 \tag{19.44}$$

It should also be noted that since the objective function is convex, and the constraint is linear, we know that the solution to the first-order condition above will be a unique global minimum. To find this, we can solve (19.43) and (19.44) for w and λ, which yields the following optimal weights:

$$w = \frac{C^{-1}1}{1^T C^{-1} 1} \tag{19.45}$$

This is a multi-dimensional, matrix based extension of the two asset case considered in (11.97). An important feature of the minimum variance portfolio is that it does not depend on expected returns for the underlying assets. We will see later that this is a useful feature. As we will see later in the chapter, the nature of estimation error in expected returns makes this a critical feature.

19.2.7 Mean-Variance Optimization with Inequality Constraints

As more realistic constraints are introduced into a mean-variance optimization framework, we will lose analytical tractability and instead will have to rely on iterative algorithms. When doing so, it is highly recommended that quants rely on pre-built optimization procedures instead of trying to create their own. Python, in particular has a robust set of optimization packages, as we have seen throughout the book. Most notably, the scipy and cvxpy libraries can be used to solve these types of optimization problems. Additionally, there are several professional, commercial optimizers that can be leveraged to solve more complex problems. **IpOpt**, in particular, is an interior point optimization algorithm that is robust and highly recommended if it is required to go beyond the scope of Python's built-in packages. In the next section, we detail some of the most common constraints that are applied when leveraging mean-variance techniques in practice.

19.2.8 Most Common Constraints

So far, the only constraints that we have seen utilized in a mean-variance optimization problem are fully invested constraints. In practice, however, realistic optimization problems require the inclusion of several additional constraints, some of which may be equality constraints and others that might be inequality constraints. The following list details some of the constraints that are most common to embed in a portfolio optimization problem:

- **Long Only Constraint**: For many institutions, short positions are either not allowed or significantly restricted. In these cases, such as mutual funds, they may only hold long positions resulting in $w_i > 0$ constraints for each asset i.
- **Maximum Book Size or Leverage Constraint**: When short positions are allowed, we may want to constrain how much total exposure we have by constraining the sum of the absolute value of our weights: $\sum_i \text{abs}(w_i) \leq B$ where B is the maximum allowable book size. Alternatively, we could relax our fully invested constraint and instead include a leverage constraint that measures the sum of the weights[4], and compares it to a predefined threshold.
- **Market Exposure Constraint**: We may want to define acceptable ranges of overall market exposure. In some cases, we may want to have no market exposure, whereas in other cases we might have a given lower and upper bound.
- **Factor Exposure Constraint**: Similar to a market exposure constraint, we may also want to limit the amount of exposure we have to additional risk factors.
- **Tracking Error Constraint**: Instead of limiting a portfolio's overall level of volatility we may instead want to limit its relative volatility measured against a benchmark. This is referred to as tracking error and is a common task for fund managers.
- **Max Position Size Constraint**: Investors might want to limit how much weight they put in any individual asset. These limits may be based on the market capitalization of the underlying assets, and not wanting to own more than a certain amount of the equity, or may be based on a desire to not have a portfolio that is too concentrated.
- **Turnover Constraint**: To limit the amount the portfolio changes as we rebalance over time, and thus minimize transactions costs, we might want to include a constraint to minimize the deviation of the portfolio from its current weights. This type of constraint is known as a turnover constraint.
- **Daily Volume Constraint**: Investors might want to limit their trading in individual assets based on how much is traded on the asset over the ensuing period. This would look similar to a turnover constraint, however, would vary from asset to asset.

In the next few sections, we provide examples of how one might implement factor exposure, turnover and tracking error constraints, respectively.

19.2.9 Mean-Variance Optimization: Market or Factor Exposure Constraints

One frequently used constraint to include in an optimization would be to limit the β of our portfolio either to the overall market or to another risk factor, such as the

[4]In the absence of cash.

size and value factors defined by Fama-French [98] [99]. To do this, we would begin by measuring the sensitivity of each asset in the optimization to the desired factor or market variable. We could then define a vector, $\hat{\beta}$, that holds the sensitivity, or beta of each asset to the factor. For a given set of weights, we know that the factor exposure for the portfolio would then simply be: $\beta_p = w^T\hat{\beta}$ where β_p is the beta of the portfolio to the risk factor.

Mathematically, this might lead to the following optimization problem if we consider the case where exposure to a factor must be within a given range:

$$\min_{w} \quad w^T C w \tag{19.46}$$

$$\text{s.t.} \quad w^T R = \hat{R} \tag{19.47}$$

$$w^T 1 = 1 \tag{19.48}$$

$$\beta_l \leq w^T\hat{\beta} \leq \beta_u \tag{19.49}$$

where β_l is the minimum allowable beta, β_u is the maximum allowable beta and $\hat{\beta}$ is a vector of market betas for each instrument in the optimization. Importantly, $\hat{\beta}$ is assumed to be determined exogenously and is constant within the optimization.

In this case we have considered the case where β within a range is suitable, meaning that we have formulated the new constraints as inequality constraints. We could, alternatively, choose to have zero exposure to a factor, in which case the inequality constraints above could be replaced by a single equality constraint.

19.2.10 Mean-Variance Optimization: Turnover Constraint

Another frequently used constraint is that of a turnover constraint. This type of constraint discourages portfolio trading activity, and the transactions costs associated with it. A turnover constraint could be formulated in a few different ways. It could be added as an additional term to the objective function, which would discourage but not forbid turnover. Instead, we may choose to forbid turnover in excess of a certain amount, which could be done through its inclusion as an inequality constraint. If this approach is pursued, we can write the new optimization problem mathematically as:

$$\min_{w} \quad w^T C w \tag{19.50}$$

$$\text{s.t.} \quad w^T R = \hat{R} \tag{19.51}$$

$$w^T 1 = 1 \tag{19.52}$$

$$d^T d \leq T \tag{19.53}$$

where T is the maximum allowable turnover and d is the distance between the weights and previous weight $\hat{w}_i$: $d_i = w_i - \hat{w}_i$. This enforces that, if the current weights in the portfolio are $\hat{w}_i$, then the total squared deviation after a re-balance will be less than T.

19.2.11 Minimizing Tracking Error to a Benchmark

Active managers are commonly asked to generate superior returns compared to some benchmark rather than in absolute terms. To do this, they may want to minimize their tracking error for a given level of expected outperformance and will be less concerned with the overall volatility of their fund. Instead they will want to know how much this volatility differs from their benchmark.

Mathematically, we can re-formulate the mean-variance optimization problems that we have seen in terms of tracking error as follows:

$$\min_{w} \quad \sqrt{\text{Var}(r_p - r_b)} \tag{19.54}$$

$$\text{s.t.} \quad w^T R = r_p \tag{19.55}$$

$$w^T 1 = 1 \tag{19.56}$$

where r_p is the return of the portfolio and r_b is the return of the benchmark. Using this framework, fund managers can then make an efficient frontier analogous to what we saw before, except now tracking error would be on the x-axis and expected outperformance, $r_p - r_b$, would be on the y-axis.

While (19.54) is a perfectly valid optimization formulation, we could equivalently substitute the definition of the variance of the difference in two random variables, yielding:

$$\text{Var}(r_p - r_b) = \text{Var}(r_p) - 2\text{Cov}(r_p, r_b) + \text{Var}(r_b) \tag{19.57}$$

If we look through this lens, we can see that the last term is the variance of the benchmark. As mentioned above, this quantity is not of interest to the fund manager as it is exogenous and beyond his or her control. Therefore, he or she can focus on the first two terms in the optimization and omit the final term. This means rewriting the optimization as:

$$\min_{w} \quad w^T C w - 2w^T \text{Cov}(r_p, r_b) \tag{19.58}$$

$$\text{s.t.} \quad w^T R = r_p \tag{19.59}$$

$$w^T 1 = 1 \tag{19.60}$$

19.2.12 Estimation of Portfolio Optimization Inputs

Up to now, in this section, we discussed many different variations of portfolio optimization techniques, and we found there to be a great deal of commonality between them. In particular we saw that as we changed the constraint set, or the desired optimization procedure, it was generally a tweak on the same underlying optimization framework. This is a convenient feature of Modern Portfolio Theory. It also tends to be true for the post-modern approaches to portfolio optimization that look to extend the basic mean-variance framework. In subsequent sections, we discuss a handful of these post-modern approaches, such as risk parity, Black-Litterman and resampling.

One commonality in applying portfolio optimization techniques, is the need to estimate a set of input parameters. We saw in a mean-variance framework that we need to feed the optimization a set of expected returns and a covariance matrix in order to get an optimal portfolio. This means we need to estimate:

- Expected Returns
- Asset Volatilities
- Asset Correlations

The process for developing an estimate of each of these quantities may require different techniques. For example, we may use a factor model, or ARMA approach to estimate expected returns. Conversely, we may choose to use an exponentially weighted moving average or GARCH model for the volatility terms. Techniques for estimating these input parameters, with a particular emphasis on returns and volatility, are discussed in chapter 25.

Importantly, when we are providing these inputs we are attempting to make a **forecast**. We are also trying to do so by looking at limited amounts of historical data. In the next section, we explore the impact of this type of estimation error, solely due to the small sample sizes that we encounter in finance. These challenges mean that expected return estimates will necessarily be noisy and have a significant amount of estimation error. Because of this, it is best practice to make sure that our portfolios are robust to these potential errors. A desirable property of an optimization is that it is robust to small changes in the input parameters, and when leveraging a portfolio optimization framework we should ensure that small changes in return, volatility or correlation estimates don't lead to fundamentally different portfolios.

19.3 CHALLENGES ASSOCIATED WITH MEAN-VARIANCE OPTIMIZATION

While mean-variance optimization techniques are a useful tool for constructing portfolios, there are many issues associated with applying it in practice that quants should be aware of. Much research has been dedicated to overcoming these issues, as we detail later in the chapter, however, in this section we provide a synopsis of some of these main challenges. Awareness of these potential pitfalls is critical in formulating portfolio optimization solutions that are robust, reliable and have the best chance to deliver strong out-of-sample performance.

Most notably, as we saw in the last section, portfolio optimization requires inversion of a covariance matrix, however, in practice this covariance matrix is often not full rank. If this is not handled properly it can lead to unstable and undesirable results when performing the inversion operation. Further, our portfolio optimization algorithms assume that the returns (and covariances) that we pass in are known with certainty, but in reality there is a great deal of error in our estimates. This especially creates challenges when optimizing among highly correlated assets, and can lead to unstable and implausible portfolios.

19.3.1 Estimation Error in Expected Returns

The first potential pitfall of a mean-variance approach is related to estimation error of the input parameters. As mentioned in the previous section, in order to construct a mean-variance optimization, we need to input expected returns and a covariance matrix. To do this, we need to create forecasts. Clearly, principles like the Efficient Market Hypothesis[5] imply that forecasting expected returns will be quite challenging and we should not assume that we will be able to do this accurately. It turns out, however, that even in the absence of this difficulty in forecasting, the relatively small samples that we are working with, and the lack of repeatable experiments to enhance our sample size, creates a significant amount of estimation error.

It can be shown that the error in our estimate of a mean, or expected return is [2]:

$$\epsilon_{\mu} = \frac{\sigma}{\sqrt{T}} \tag{19.61}$$

where ϵ_{μ} is the estimation error, σ is the volatility of the asset and T is the length of the measurement period.

As an example, if we are given 20 years of data, as is often the case in financial datasets, and are estimating the mean for an asset with a volatility of 20%, then the estimation error in the mean is almost 4.5%. To achieve an estimation error of 1% for this asset, we would need 400 years of data. Even more problematic, a 90% confidence interval of the mean estimate would be close to +/− 7%, a range of 14%. The magnitude of this estimation error and the implicit requirements regarding the length of the measurement period are truly striking. Clearly, if we shift our input expected returns by close to 5% we are likely to uncover a fundamentally different optimal portfolio. This could lead us to think that we are investing in a portfolio that is along the efficient frontier, but in reality is sub-optimal.

It should also be emphasized that this estimate of the error assumes perfect forecasting ability, and is still quite large. As such, in practice, our estimates are likely to be significantly less accurate than this. A byproduct of this is that techniques have arisen in recent decades that try to formulate portfolio construction problems without requiring expected return estimates as inputs. Most notably, the minimum variance portfolio detailed in section 19.2.6 does not use expected returns in its optimization. Similarly, in section 19.9 we introduce risk parity as an alternative to mean-variance optimization. As we will see, like the minimum variance portfolio, risk-parity portfolios depend only on estimates of the covariance matrix, and do not require assumptions on the underlying expected returns.

19.3.2 Mean-Variance Optimization in Practice: Impact of Estimation Error

In this example, we highlight the impact of estimation error in expected returns by analyzing how the efficient frontier changes as we shift expected returns within their confidence intervals. Recall that (19.61) defines the standard error of an expected return. We then assume, for illustrative purposes, that the return distribution is Gaussian, allowing us to obtain a confidence interval. We then use the upper and lower boundaries, as well as our estimated expected returns, to construct three efficient frontiers. These efficient frontiers, using 90% confidence intervals and a portfolio of five sector ETFs, are shown in the following chart:

[5] As detailed in chapter 5.

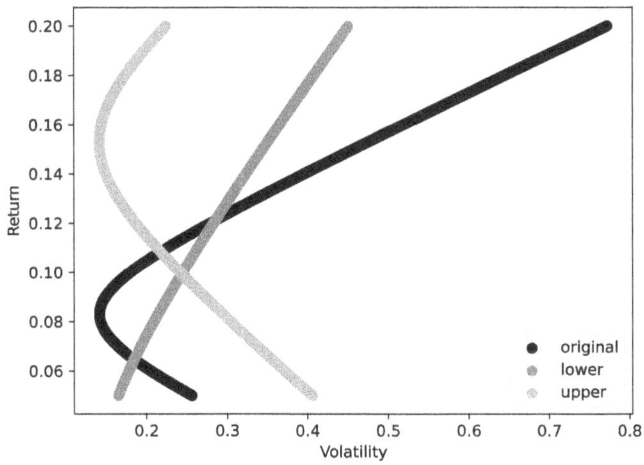

It is interesting to see that the two new efficient frontiers deviate quite a lot from the original one. Further, as we shifted all assets expected returns in the same direction, this is likely to understate the impact of estimation error. For example, increasing some ETFs to the upper bound of their confidence and other to their lower bound would likely lead to larger deviations. This is left as an exercise for the reader.

If we pick a point on the efficient frontier and look at the corresponding weights, we can also see a considerable difference in the weights. An example of this is shown on the following chart, where we plot the weights using the estimated returns, as well as using the upper and lower bounds of our confidence intervals:

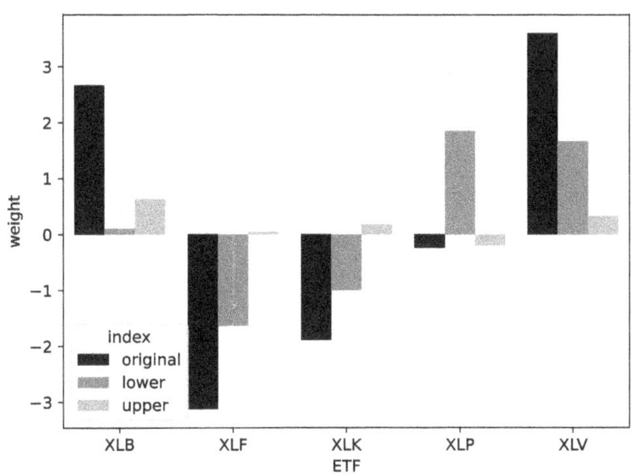

Here again we see significant deviations in the weights due to the imprecision in our estimate. In particular, we see large differences in magnitude as well as a change in the sign of the position for one of the XLP. It should be emphasized that, had we shifted the individual ETF expected returns in different directions, these dramatic changes in weights would have been even more pronounced.

19.3.3 Estimation Error of Variance Estimates

Just as our expected return estimates will contain error, so too will our estimates of the variance and covariances. In this section we detail the uncertainty in a variance estimate, again ignoring any uncertainty that arises due to regime changes, non-stationary or uncertainty about the future state.

Campbell, Lo and MacKinlay showed that the estimation error of a variance estimate can be written as [152]:

$$\epsilon_{\text{var}} = 2\left(\frac{T}{\Delta_t} - 1\right)^{-1} \sigma^2 \tag{19.62}$$

where T is the length of the measurement period, Δ_t is the frequency at which the variance is calculated, and ϵ_{var} is the estimation error of the variance estimate.

An important distinction between the estimation error in a variance estimate and an estimate of the expected return is the presence of Δ_t in (19.62). This implies that to get a more accurate estimate of the variance, we can simply sample the data more frequently. This is in contrast to the error in expected returns, where only the length of the measurement period was involved in the calculation.

19.3.4 Singularity of Covariance Matrices

Another common issue in mean-variance analysis arises due to poorly conditioned covariance matrices. In the last section, we saw that mean-variance optimization problems involve inverting a covariance matrix. If this is not done judiciously, then it can cause numerical issues.

In fact, if the covariance matrix is not full rank, i.e. it is singular, then it will not be invertible. In finance problems, this is very likely to be the case due to the structure of the data we are working with. Financial problems are often defined by high breadth, via a large number of assets, but relatively short histories. In this case, if the number of assets, N, exceeds the historical time periods available, T, then the covariance matrix will be singular by construction [163]. More generally, for N assets, we need far greater than N time periods to have matrix that is full rank. In cases where the matrix we encounter is singular, we need to rely on pseudo-inverse techniques, which will leverage techniques like Eigenvalue or Singularvalue decomposition to invert the low rank matrix in a stable way [248].

Theoretically, there is even some question that the underlying covariance matrix should be full rank. If we consider the case of the S&P 500 index, it is quite reasonable to assume that the index is driven by a set of factors far less than 500 in number. Said differently, do we really believe there are 500 orthogonal risks contributing to the movements in S&P or is it driven by a much more parsimonious set of factors? This has led many participants to rely on dimensionality reduction techniques in order to map the asset universe into a set of factors.

Even more problematically, we may find that the empirical covariance matrix has negative eigenvalues. Negative eigenvalues are not something that should happen naturally, as it implies a negative variance for a portfolio, which is clearly not sensible. It can however, occur in practice depending on how the covariance matrix is created. As an example, negative eigenvalues may occur if we use different windows for calculation of different pairwise covariance terms. As quants, if we encounter negative eigenvalues we need to correct for them prior to inverting the covariance matrix.

Finally, even in the case of a covariance matrix with all positive eigenvalues, inverting the covariance matrix may still lead to undesirable properties. Essentially, if not done carefully we may be amplifying the noise as we invert the covariance matrix. To see this, recall from chapter 5 that an Eigenvalue decomposition of a covariance matrix can be written as:

$$\Sigma = Q\Lambda Q^{\top} \tag{19.63}$$

where Λ is a diagonal matrix with eigenvalues λ_i on the diagonal. If we invert Σ, then we are left with the following:

$$\Sigma^{-1} = Q\Lambda^{-1}Q^{\top} \tag{19.64}$$

Because Λ is diagonal, its inverse will also be diagonal with elements equal to: $\Lambda_{ii}^{-1} = \frac{1}{\lambda_i}$. This has important implications for mean-variance optimization. It should be emphasized that the smallest eigenvalues have the largest weight after we take the inverse. This is problematic because the smallest eigenvalues are most likely to be noise. They are also the eigenvalues that are statistically indistinguishable from zero in many cases. Yet in our optimization approach we have amplified these negligible eigenvalues to have a significant impact on the solution. As a result, it is recommended that when working with covariance matrices in practice, and performing inverse operations, that small eigenvalues be replaced with zero and the resulting low-rank matrix be handled via a pseudo-inverse operation, as discussed above. In the next chapter, we discuss some modern techniques for dealing with these issues with covariance matrix estimation. In particular, we consider shrinkage estimators, random matrix theory and detail how to correct for a covariance matrix with negative eigenvalues.

19.3.5 Mean-Variance Optimization in Practice: Analysis of Covariance Matrices

In this example, we analyze the covariance matrix of 25 stocks. The chart below shows the actual value of the eigenvalues of this covariance matrix:

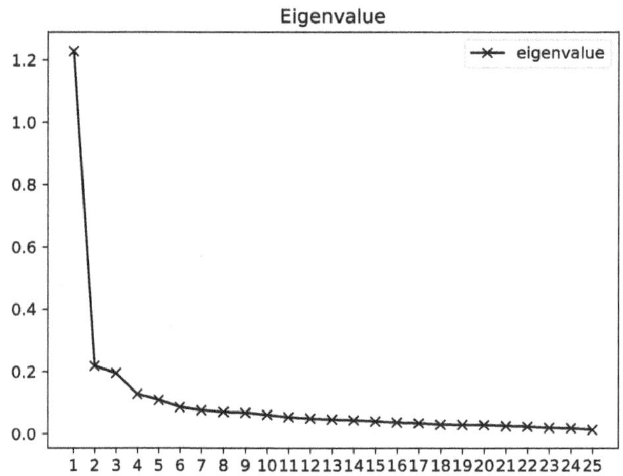

We can see the covariance matrix is characterized by a few large eigenvalues and then get progressively smaller. We then propose a significance threshold of 0.01, meaning we remove the eigenvalues that are below this threshold as we postulate that they are likely to be noise. This leads to five eigenvalues that will be removed as noise. This choice of a significance level for the magnitude of the eigenvalues may seem subjective and somewhat arbitrary, however, in the next chapter we examine how we can apply the insights of random matrix theory in order to provide a more quantitative approach. In the chart below, we show a comparison of the percentage of the variance explained by the eigenvalues both before and after the small eigenvalues have been removed:

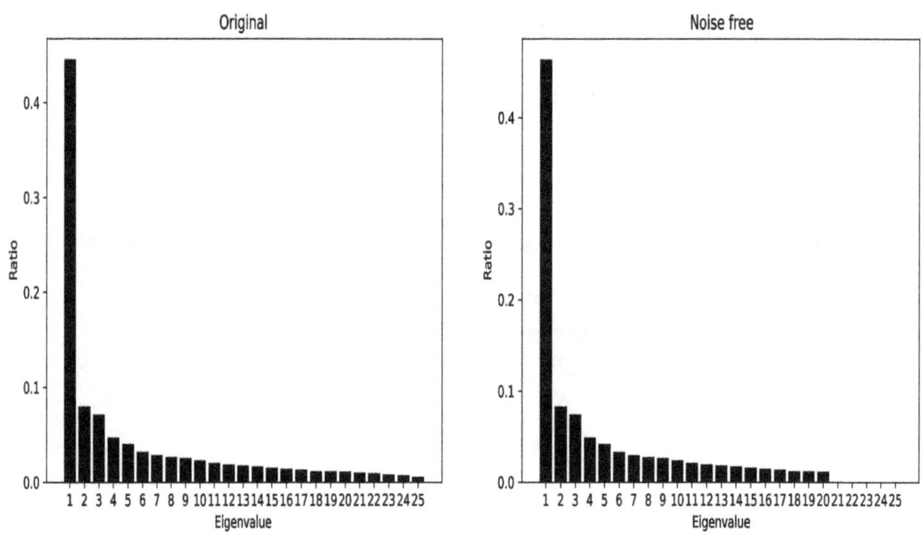

19.3.6 Non-Stationarity of Asset Correlations

Another complicating factor in solving portfolio optimization problems is that the underlying correlations themselves are often non-stationary. Instead, there is a phenomenon in which correlations tend be lower in equilibrium conditions and then spike in times of market stress. When using traditional mean-variance techniques, we input a single overarching covariance matrix and assume it is representative of the future. That is, we assume that it is stationary when in reality it might depend on market conditions.

If correlations are structurally different in different market conditions, then it is natural to want to build that into our optimization procedure instead of using an average correlation over the entire period. There are many potential ways to approach this problem. Chow, Jacquier, Kritzman and Lowry [111], notably suggest creating a covariance matrix of the following form:

$$\Sigma = p\Sigma_s + (1 - p)\,\Sigma_e \tag{19.65}$$

where Σ_s is the covariance matrix during periods of market stress, Σ_e is the covariance matrix during equilibrium periods and p is the probability of being in a period of market stress. p may be estimated from historical data but also more importantly may vary from investor to investor depending on their risk preferences. Investors who are more risk averse, may want to use a larger value for p when constructing their portfolio. This Σ can then be used instead of C in portfolio optimization formulations such as (19.22).

In order to determine Σ_s and Σ_e we can define a distance metric and use a clustering algorithm such as k-means to break the data into stress and equilibrium periods. These techniques are both detailed in chapter 28. We can then calculate the covariance matrix separately over these periods.

19.4 CAPITAL ASSET PRICING MODEL

19.4.1 Leverage & the Tangency Portfolio

If we introduce a risk-free asset into our asset universe, then we can combine positions in any point on the previously constructed efficient frontier with positions in the risk-free asset. A long position in this risk-free asset is equivalent to lending our cash whereas a short position is analogous to borrowing at the risk-free rate. That is, when we introduce leverage into our model, via borrowing and lending at the risk-free rate, the resulting opportunity set expands. A natural question might be how a portfolio along the efficient frontier compares to another point on the same frontier where we borrow or lend to match the expected return of the other point on the frontier.

To take a simple example, if we consider following two portfolios on a hypothetical efficient frontier:

Further, if we suppose the risk-free rate is 1%, then we can apply leverage to portfolio B to achieve a return of 7% with the same volatility as portfolio A. This can be done by having a -100% weight in the risk-free rate, or borrowing 100% of our initial capital, and investing it, plus our original capital in portfolio B. Therefore,

Portfolio	Expected Return	Volatility
A	6%	10%
B	4%	5%

in the presence of the risk-free rate, and the ability to take on leverage, portfolio B dominates portfolio A even though they are both on the efficient frontier. That is, a leveraged portfolio B generates a higher return for the same level of volatility.

More generally, we can draw a line from the risk-free rate on the y-axis of an efficient frontier to each point on the efficient frontier. The point on the y-axis in particular refers to a portfolio of 100% in the risk-free asset and 0% in the frontier portfolio, where a portfolio on the efficient frontier corresponds to a portfolio of 0% cash and 100% in the frontier portfolios. Further, we can move to the right of the point on the efficient frontier by combining a negative position in the risk-free rate with a position larger than 100% in the frontier portfolio. Each of these lines are referred to as Capital Allocation Lines and will result in different risk-adjusted performance for a comparable level of volatility. This means that, in the presence of leverage[6], there is a single dominant portfolio, which is the Capital Allocation Line with the highest slope. Investors can then combine this efficient frontier portfolio with a long or short position in the risk-free to achieve the optimal portfolio. It can be shown that this is achieved at the point at which the Capital Allocation Line is tangent to the efficient frontier. This has led this portfolio to become known as the tangency portfolio. In section 19.4.4 we explore the tangency portfolio in more detail and provide a visualization of the efficient frontier with different Capital Allocation Lines.

19.4.2 CAPM

The Capital Asset Pricing Model (CAPM) [228] was created as an extension of Markowitz's Modern Portfolio Theory by leveraging the risk-free asset and tangency portfolio as discussed in the previous section. It states that, under certain assumptions, the tangency portfolio must be the market portfolio, because adding each asset to the portfolio increases its diversification and reduces the portfolios idiosyncratic risk. The powerful implication of this model is that all investors should simply hold the market portfolio and then use long-short positions in the risk-free rate to achieve the portfolio with the desired risk and return.

Under CAPM, returns for each stock are proportional to a firm's exposure to the market, with the remaining component being uncorrelated noise:

$$R_i = R_f + \beta_i (R_m - R_f) + \epsilon_i \tag{19.66}$$

where R_i is the return of asset i, R_f is the risk-free asset, R_m is the return of the market, β_i is the sensitivity of asset i to a change in the market and ϵ_i is the return of asset i due to its idiosyncratic component. The return from the market, R_m, and

[6]Where financing for borrowing and lending can be executed at the risk-free rate.

the idiosyncratic component, ϵ_i are further assumed to be independent. This distinction between market and idiosyncratic risk is a foundational feature of CAPM. In the model, returns for an asset arise either because of their market exposure or are idiosyncratic in nature. The idiosyncratic term, ϵ_i is assumed to be normally distributed with mean zero, reflecting the fact that idiosyncratic risks are not rewarded. This concept of systemic vs. idiosyncratic risks are explored further in the next section.

19.4.3 Systemic vs. Idiosyncratic Risk

CAPM is based on an important concept regarding two types of risk that we observe in financial markets. One type of risk is idiosyncratic risk. This risk is firm specific and relates to the fact that firm specific events may occur that affect their profitability. This type of risk can be mitigated by investing in many firms and applying the fundamental diversification concepts discussed throughout the book. In short, a portfolio of many assets, like the market portfolio, has little to no idiosyncratic risk, in contrast to a highly concentrated portfolio in a single stock, which would contain significant firm specific, idiosyncratic risk.

Systemic risks, conversely, cannot be eliminated via diversification. They are risks that are more fundamental to financial markets and are therefore common to all firm's. As an example, all firms' profitability may be tied to the economic business cycle and all firms' share the risk of degraded performance should the economy enter a recession. This is a risk that is common to all portfolios, regardless of how well-diversified or highly concentrated. A fundamental piece of CAPM theory is that systemic risk, which cannot be diversified away may be rewarded with a risk premium, however, idiosyncratic risks, which can be removed by diversification techniques, should not provide a premium.

19.4.4 CAPM in Practice: Efficient Frontier, Tangency Portfolio and Leverage

In this example, we leverage the relevant functions in the coding repository to plot the efficient frontier, the Capital Market Line, and the tangency portfolio in a single graph, which is displayed below. Every point on the Capital Market Line will yield the same, highest possible, Sharpe Ratio. The tangency portfolio has no leverage, whereas any other point on the Capital Market Line indicates either positive (left-hand side of the efficient frontier) or negative (right-hand size of the efficient frontier) positions in the risk-free asset.

The graph shows a visualization using the same data as in section 19.2.5:

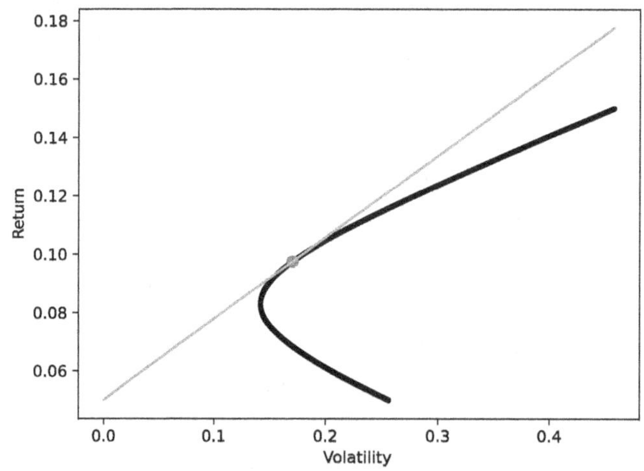

19.4.5 Multi-Factor Models

While CAPM provides a helpful baseline for investment managers, its uses are limited in practice for a number of reasons. In particular, the market portfolio as designed by CAPM consists of all assets available in the market. This includes illiquid assets with no price transparency, such as real estate and art, making the model very challenging to validate or reject empirically. In practice, however, it tends to be more useful to build multi-factor expected return models. In a multi-factor model there are a few ways to identify the underlying factor set:

- **Fundamental & Technical Analysis**: We could rely on either fundamental analysis, where we would leverage information from different firms' balance sheets in order to build a factor, or technical analysis, where we would rely on previous price patterns.

- **Risk Premia**: We could identify a set of factors from a risk premia lens, with the idea that bearing one of these risks should reward us with a premia over time. Commonly cited risk premia that we may want to include as factors are carry, value, momentum, quality and low volatility.

- **Statistical**: Alternatively we could identify the factors statistically, via methods such as PCA. If we leveraged a PCA approach, then each principal component would be a factor in our model.

- **Alternative Data**: Another option would be to try to leverage newer data sources that may rely on NLP methods to parse news articles, press releases or other documents. Other examples of incorporating alternative data include leveraging information on website traffic and satellite images of store parking lots.

In chapter 27 we discuss how both PCA-based factor models and separately a risk premia approach can be leveraged to create an investment strategy. Additionally, in chapter 25 we discuss how to formulate expected return models given a set of data for the underlying factors. These approaches are naturally extendable to cases such as alternative data, once the data has been processed into a signal.

While multi-factor models of this nature can in fact be quite useful, it is worth emphasizing that the presence of these factors that influence future returns is at odds with the concepts of market efficiency introduced in chapter 5. In particular, we learned that weak-form market efficiency meant that we should not be able to generate excess returns using historical price data. To the extent we can use technical analysis, or create a momentum risk premia signal, this is a violation of weak-form market efficiency. Similarly, semi-strong form market efficiency relates to the inability of participants to achieve excess returns from things like fundamental data. This is also at odds with many of the potential factors suggested here. Conversely, the persistent presence of certain behavioral phenomenon suggest that markets, while highly efficient, may not meet the strict standards of the Efficient Market Hypothesis, even its weak form. Nonetheless, a byproduct of this is that we should expect somewhat low explanatory predictive power in our models, and our prior should be that new factors do not generate alpha until we are able to disprove that convincingly.

19.4.6 Fama-French Factors

A canonical example of a multi-factor model was developed by Fama and French [98] and defines three factors that explain market returns:

- **Market (Equity) Risk Premium**: Firm's returns are proportional to their market exposure as in CAPM.
- **Size Premium**: Firms with smaller market capitalizations tend to have higher returns than firms with large market capitalizations.
- **Value Premium**: Firms with high book to market valuations tend to outperform firms with low book to market valuations.

Mathematically, in this model the returns of a given asset, i can be written as:

$$\begin{aligned} R_i = \alpha_i + R_f \quad & + \quad \beta_{i,M}\left(R_m - R_f\right) \\ & + \quad \beta_{i,S}\left(R_S - R_B\right) \\ & + \quad \beta_{i,V}\left(R_H - R_L\right) \end{aligned} \tag{19.67}$$

where R_S is the return on small cap stocks, R_B is the return on large cap stocks, R_H is the return on stocks with high book to market values and R_L is the return on stocks with low book to market valuations. α_i is then the return not attributable to the factors. Variations of the Fama-French model have been created that include additional factors for momentum, quality and low volatility. The βs in (19.67) can be estimated via a time series regression of asset returns on the set of factor returns.

This factor representation is extremely useful and represents a baseline for generating expected returns. Once an investment strategy has been created, a contemporaneous regression is often used to decompose the returns of the strategy into the factor returns, and the remaining alpha. If the remaining alpha is positive, that means the strategy is generating returns in excess of the market, size and value risks that it is taking. Conversely, if it close to zero or negative, it suggests that the profits are primarily due to exposure to these factors. This type of regression analysis is referred to as an attribution analysis, in which we are not required to forecast, but instead simply explain the returns in a given return stream.

19.5 BLACK-LITTERMAN

19.5.1 Market Implied Equilibrium Expected Returns

Earlier in the chapter, we saw how to formulate and solve an unconstrained mean-variance optimization problem. Let's look at the following slightly modified objective function:

$$\max_{w} w^T R - \frac{\lambda}{2} w^T C w \tag{19.68}$$

Notice that the risk aversion coefficient has been multiplied by $\frac{1}{2}$ for convenience. Otherwise it is the same as the problem introduced earlier in the chapter and has the following solution[7]:

$$w = \frac{1}{\lambda} C^{-1} R \tag{19.69}$$

The standard approach to proceeding is to specify a vector of expected returns, R, and a covariance matrix C and compute the optimal weights, w. Alternatively, we could specify a set of weights, and extract the set of returns that would be required for this portfolio to be optimal. These implied returns essentially show us the returns that would justify a given set of weights. Of particular interest might be the implied returns that justify the market capitalization weighted portfolio. This has special relevance in the context of CAPM, where the market portfolio was proposed to be the optimal portfolio. As such, if we extract a set of implied returns, R_{implied} from the market cap weighted portfolio (w_{mkt}), then these represent the required returns to justify the market portfolio being optimal. Solving (19.69) for the implied returns, we see that we can do this using the following formula:

$$R_{\text{implied}} = \lambda C w_{\text{mkt}} \tag{19.70}$$

where λ is an investor's risk aversion coefficient. In this context, λ is usually extracted from the equity risk premium using the following relationship [226]:

$$\lambda = \frac{\left(\mathbb{E}\left[R_{\text{mkt}}\right] - R_f\right)}{\sigma^2_{\text{mkt}}} \tag{19.71}$$

In the numerator we have the expected excess returns of equities, and in the denominator we have the variance of the equity market. The Black-Litterman model, which we will detail in the remainder of the section, relies on this set of implied returns to form a prior distribution to which we can impose additional subjective views. In the next section, we discuss how these views are incorporated.

[7] Which can be found by finding the first and second-order conditions.

19.5.2 Bayes' Rule

The Black-Litterman model uses a Bayesian approach to incorporating both a prior distribution, based on the equilibrium expected returns implied from current market cap weights, and additional information based on a set of market views. The goal of Black-Litterman is to blend these two components together into an overarching posterior distribution. To do this, Black-Litterman relies on application of Bayes' rule [238], which can be written as:

$$P(A|B) = \frac{P(B|A)P(A)}{P(B)} \tag{19.72}$$

where $P(A|B)$ is the probability of A conditioned on B occurring and $P(B|A)$ is the probability of B conditional on A, and $P(A)$ and $P(B)$ are the unconditional probabilities of events A and B respectively. Bayes' rule has many applications not only in finance but also in many other fields. For example, it is often used to compute the likelihood of type I and type II errors.

In the context of likelihood estimation, the events A and B have the interpretation of data and parameters respectively, leading to the following application of Bayes' rule:

$$P(\text{params}|\text{data}) = \frac{P(\text{data}|\text{params})P(\text{params})}{P(\text{data})} \tag{19.73}$$

More generally, $P(A)$ is referred to as the prior distribution and $P(B|A)$ is the conditional probability of the observed data given the prior. This is also known as the sampling distribution.[8]

19.5.3 Incorporating Subjective Views

A key component of the Black-Litterman model is its ability to overlay a set of views on the equilibrium expected returns implied by the current market capitalization weights. Further, we are not only able to specify a set of views we are also able to define confidence levels of them. This is a particular attractive feature as it enables us to seemingly blend two worlds, a more quantitative process of determining optimal weights and expected returns, which we define as the prior, and the more subjective process of incorporating views.

Financial markets consist of different types of portfolio managers. Some PMs are entirely quantitative, with investment processes that are driven by systematic signals and a systematic portfolio construction. At the other end of the spectrum are PMs that are more fundamental in nature. They may rely on quantitative techniques to inform their views, however, they build and size their trades based on their own judgment. A Black-Litterman model has potential appeal to both types of PMs. On the more systematic end, the Black-Litterman model may provide a more robust way to incorporate systematic signals into a coherent systematic portfolio. Doing this

[8] $P(B)$ then is the unconditional or marginal probability of the observations, and is often ignored in Bayesian analysis as it is simply a normalizing constant.

through a Black-Litterman framework, with a prior distribution, instead of a mean-variance approach, may result in more stable weights that are less prone to extreme values[9]. On the more discretionary end, Black-Litterman gives fundamentally oriented PMs a way to continue to leverage their own personalized investment processes to generate views but still benefit from quantitative techniques to ensure that these views lead to an optimal portfolio. Relative to a purely fundamental process, this approach may enable them to better benefit from diversification, and develop risk budgets for each view that are sensible.

Within the context of Black-Litterman, the views are specified as the returns, Q of an asset or portfolio of assets P. These views also have an associated covariance matrix, Ω. In the following list, we detail the components in Black-Litterman that are used to express the underlying views:

- P is a matrix representing the weights of the views, with each row in the matrix representing a view. If a single element in the row is non-zero, the view is expressed on a single asset (i.e. the return on S&P is 5%). Conversely the weights may be a set of long-short weights (i.e. the return of a portfolio that is long Financial stocks and short Technology stocks will have a return of 10%).
- Q: is a vector of returns for the portfolios defined in P. In the example views above the Q vector would be populated with 5 and 10% respectively.
- Ω: A diagonal covariance matrix of the views with the diagonal elements representing the variance of each view.

In the Black-Litterman model, we create a posterior distribution that relies on a matrix, P of view portfolios, a vector, Q of view expected returns, and a matrix Ω that incorporates uncertainty of each view. It also relies on the equilibrium expected returns, R_{implied} derived earlier from a set of market cap weights, and uncertainty in these equilibrium returns.

19.5.4 The Black-Litterman Model

In the Black-Litterman model [42] [259], we combine the view structure discussed in 19.5.3 with a prior distribution defined by the market implied returns discussed 19.5.1. In this approach, we assume that both the views and prior distribution are normally distributed, and we use Bayes' rule to blend these two distributions together into a single, posterior distribution. It should be emphasized that the prior and view in this context are distributions rather than point estimates. This means that, for example, the market cap weighted implied returns will be the mean of the prior distribution, however, there will be uncertainty around that as well. This is true for the subjective views as well, as we saw from the presence of a covariance matrix in their formulation.

---

[9]In chapter 27 we detail other approaches to portfolio construction on a set of underlying quant signals.

Returning to Bayes' rule in (19.72), our prior distribution, which is defined by the implied returns, will be denoted $P(A)$, and our observations, or views given this prior will be $P(B|A)$. $P(B)$ is a normalizing constant. Importantly, it is not a function of our model parameters and therefore has no impact on our solution and can be ignored. Lastly, $P(A|B)$ is the posterior distribution that we are seeking.

The first component, the prior distribution, $P(A)$, can be expressed as:

$$P(A) \sim N(R_{\text{implied}}, \tau\Sigma) \tag{19.74}$$

where τ is a scaling constant of the covariance matrix Σ which is the empirical covariance matrix. R_{implied} are the implied returns extracted in 19.5.1. Said differently, we assume that our prior is normally distributed and has means matching the implied returns. We further assume that its covariance is proportional to the historical covariance matrix, and define the exogenous constant τ to scale up or down this matrix.

The second component, the view distribution of the underlying assets, $P(B|A)$, can be expressed as:

$$P(B|A) \sim N(P^T Q, P^T \Omega P) \tag{19.75}$$

where P, Q and Ω are all as defined in 19.5.3.

The posterior distribution, $P(A|B)$ is then the product of $P(A)$ and $P(B|A)$. Under the assumption that each distribution is normal, it can then be shown that the posterior distribution is equal to:

$$\begin{aligned} P(A|B) &\sim N(\mathbb{E}[R], \text{var}(R)) \\ \mathbb{E}[R] &= \left[(\tau\Sigma)^{-1} + \left(P^T\Omega^{-1}P\right)\right]^{-1}\left[(\tau\Sigma)^{-1} R_{\text{implied}} + P^T\Omega^{-1}Q\right] \\ \text{var}(R) &= \left[(\tau\Sigma)^{-1} + \left(P^T\Omega^{-1}P\right)\right]^{-1} \end{aligned} \tag{19.76}$$

where $P(A|B)$ defines the posterior distribution. It should be emphasized that this posterior distribution is the distribution of returns that incorporates both the equilibrium implied returns and the investor views. The updated set of expected returns, $\mathbb{E}[R]$ can then be leveraged in a mean-variance optimization, in conjunction with an updated covariance matrix that reflects the uncertainty in the distribution of returns. These quantities, and more generally the posterior distribution can then be used to solve the relevant mean-variance optimization problem of our choice. For example, the unconstrained mean-variance problem could be written as:

$$w = \frac{1}{\lambda}\hat{\Sigma}^{-1}\mathbb{E}[R] \tag{19.77}$$

$$\hat{\Sigma} = \Sigma + \text{var}(R) \tag{19.78}$$

where w is the set of optimal weights in a Black-Litterman framework, λ as always is a risk-aversion parameter and $\mathbb{E}[R]$ and $\hat{\Sigma}$ are the updated expected return and covariance matrices, respectively.

The appeal of Black-Litterman is two-fold. First, it bridges the gap between subjective views and a quantitative mean-variance framework. Secondly, the inclusion of implied returns as a prior has a stabilizing effect on the numerical solution to the problem, leading to results that are more stable over time. The Black-Litterman model is only one example of applying Bayesian methods in the context of portfolio optimization and more generally, portfolio management. Readers interested in more detailed treatment of these additional methods should see [154].

19.6 RESAMPLING

19.6.1 Resampling the Efficient Frontier

Michaud [186] [185] proposed a resampling approach to overcome many of the inherent issues in mean-variance optimization, in particular those related to estimation error. The idea is that while mean-variance optimization relies on point estimates for expected returns and covariance terms, they are each estimated with a great deal of error. As we saw, this is particularly problematic for expected return estimates, where our confidence intervals tend to be surprisingly wide. The underlying optimization then uses those point estimates to generate a portfolio that may not be robust to small changes in the inputs. In short, optimization may be too precise a tool for the inputs we are able to provide.

Resampling methods were proposed as a means of overcoming some of these issues. The overarching idea is that the historical data that we observe, which we generally use to create our return and covariance estimates, is simply a single realization from a broader, unobserved distribution. If we treat it as such, then we can try to repeatedly sample from this broader distribution to get a new efficient frontier. In other words, we resample from this distribution, and obtain an efficient frontier along each sample. We then average the weights across the sampled frontiers to obtain a new, resampled efficient frontier.

In the following list, we detail the steps that would need to be included in a resampling algorithm:

- Generate a resampled path that is defined by expected returns $\hat{\mu_i}$ and a covariance matrix $\hat{\Sigma}_i$ for the underlying assets. These resampled paths can be generated using bootstrapping[10] or by making assumptions about the joint distribution of the returns of the assets. For example we could assume that the returns are jointly normal, with means and covariances equal to what we observe historically. In this case, each resampled path would require a simulation from a multi-variate normal distribution[11].

- Generate an efficient frontier for each resampled path i using $\hat{\mu_i}$ and $\hat{\Sigma}_i$. In this step standard mean-variance techniques are used to generate the efficient frontier. For example, we might use (19.22) or (19.33).

[10] Which was detailed in chapter 5.

[11] Which can easily be done in Python via built-in functions and was detailed in chapter 15.

- Calculate the final, resampled efficient frontier as the average of the weights of the efficient frontiers along each resampled path.

There is some evidence that using a resampled efficient frontier leads to a more robust solution that is less sensitive to the input data and has more stable weights over time. This, coupled with its conceptual simplicity, make resampling an attractive algorithm to leverage in practice. In the next section, we detail how the code for a resampling algorithm would work, and then proceed to show how it works in practice.

19.6.2 Resampling in Practice: Comparison to a Mean-Variance Efficient Frontier

In this section we leverage the resampling coding example in the book's repository to generate a resampled efficient frontier for sector ETF portfolios and compare this to the standard efficient frontier as defined in 19.2.5, which is displayed in the following chart:

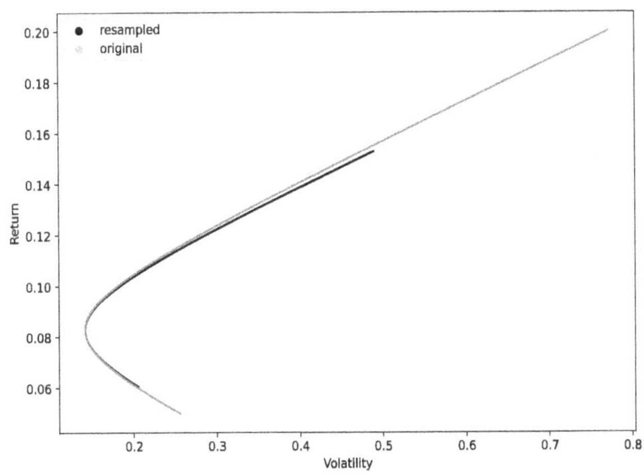

It should be noted that to perform this analysis we began by computing the efficient frontier in the standard manner. We then calculated a corresponding resampled efficient frontier, and then calculated the mean and volatility of the resampled efficient frontier using the actual historical data that we observed. As a result, we observe that the resampled efficient frontier is inside the original efficient frontier, but that it is quite close. One way to think about this is that the original efficient frontier, if properly constructed, defines the best possible portfolios for the historical dataset. As the resampled efficient frontier deviates from these best possible portfolios, it will be inside the original efficient frontier by construction. However, the fact that it is able to stay close to the original efficient frontier is encouraging. It means that, when using the resampling approach we are able to get a portfolio that is only marginally worse should the historical data repeat, however, is more robust to different potential futures outcomes as it was stressed under a set of simulated paths. This makes resampling a more robust approach in practice as a means of incorporating potential estimation error.

19.7 ROBUST PORTFOLIO OPTIMIZATION

It turns out that resampling, as discussed in the last section, is one example of a set of robust portfolio optimization techniques. These techniques, by definition, attempt to provide more robust and stable solutions and be less sensitive to the inherent estimation error in expected returns and covariance matrices [12]. In the case of resampling, this is done by consider not just historical, realized returns, but instead creating return distributions, and then corresponding alternate realizations from that distribution. Covariance matrix shrinkage [13] is another example of a robust portfolio optimization technique, where we work with a more stable covariance matrix in order to improve the robustness of the overall optimal solution.

More generally, the goal of these techniques is to reduce the sensitivity of the optimal weights on the inputs, especially the expected return estimates. Clearly, a valuable property of an optimal portfolio would be that as we make small changes to the inputs, that the solution wouldn't change [14].

Mathematically, we might think of a robust optimization as one that optimizes to not just the input expected returns and covariance, but is also optimal while the input parameters are shifted within a threshold. One way to formulate this is via the following equation:

$$\begin{array}{ll} \max\limits_{w} & \left(\min_{[\mu \in U_\mu]} w^T C w - 2w^T\right) \\ \text{s.t.} & w^T 1 = 1 \end{array}$$

where U_μ is the uncertainty set for μ, the set of expected returns.

In this equation, we consider a series of expected return estimates, and optimize over the worst case solution among the expected returns [15]. It should be noted that using a worst case methodology is only one example of incorporating an uncertainty set, and that here we are only incorporating uncertainty into the expected returns, and that the covariance matrix that is input to the optimization is still static.

Once formulated, this type of optimization can be solved using standard optimization techniques. More details on this type of optimization, and the potential uncertainty sets that can be used to solve a robust portfolio optimization problem, can be found in [106].

19.8 DOWNSIDE RISK BASED OPTIMIZATION

Mean-variance techniques are still the most commonly applied portfolio construction algorithms, and arise naturally under certain sets of assumptions. For example, when the underlying distribution of returns is assumed to be Gaussian, then only the first two moments will be relevant as the normal distribution is described entirely by its

[12] More details on this can be found in 19.3.1.
[13] As discussed in more detail in 20.2.9.
[14] Or would only change by a small amount.
[15] A similar approach is proposed in [225].

mean and variance. In spite of this, recently research has been dedicated to extending these techniques to incorporate higher moments of the distribution of returns[16] [54].

There are many approaches to incorporating higher moments of return distribution, most notably skewness and kurtosis, into our utility and optimization framework. In this section we discuss two such approaches, which are based on the popular risk measures value at risk and conditional value at risk. In reality, we are interested in the entire return distribution and any attempt to simplify this to a set of characteristic numbers is bound to be limiting. With that in mind, these particular optimization frameworks can be relevant for finding portfolios that are sensitive to the left tail of the return distribution.

19.8.1 Value at Risk (VaR)

Value-at-Risk (VaR) is a popular risk measure that is used by many buy-side and sell-side institutions to help determine position sizes, risk limits and estimate worst case outcomes. VaR measures the predicted maximum loss that we can expect to occur within a specific confidence level. In other words, it tells us with probability $1-\epsilon$ that our portfolio won't lose more than X in a given period. For example, a VaR calculation might tell a portfolio manager that with 95% probability he or she will not lose more than one-million dollars in a given day. Because of its intuitive appeal, VaR is perhaps the most commonly used risk measure both for sell-side institutions and hedge funds.

VaR, however, does not come without challenges and in fact has some inconvenient mathematical properties [83]. VaR turns out to not be a coherent risk measure because it is not sub-additive. Sub-additivity would state that a portfolios VaR should not grow as we increase the number of assets in the portfolio. This is an intuitive property as it is very much in line with basic diversification concepts, however it turns out to not be the case for VaR.

Additionally, VaR says nothing about the severity of the loss that we will incur should we exceed our threshold. This can be particularly problematic for distributions that are bi-modal or have fat tails. For example, it is certainly useful to know that with some confidence a portfolio manager will not lose more than a million dollars in a day, however, it tells us nothing about the distribution of losses conditional on exceeding this threshold. Clearly, if this conditional loss was $10 million, then the VaR calculation could be misleading.

Mathematically, VaR for a given level of confidence, ϵ, can be found by inverting the CDF of the distribution of returns, as shown:

$$\mathrm{VaR}_{\epsilon} = F^{-1}(1-\epsilon) \tag{19.79}$$

where $F(x)$ is the CDF of the distribution of returns for the asset or the joint distribution of returns for the portfolio. Said differently, VaR with confidence level ϵ is the distributions $(1-\epsilon)^{\mathrm{th}}$ percentile return.

---

[16]In the physical measure.

19.8.2 Conditional Value at Risk (CVaR)

Conditional Value at Risk (CVaR) [219], also known as expected shortfall was designed to overcome many of the challenges noted above with respect to VaR. In particular, it can be proven to be a coherent risk measure that obeys the property of sub-additivity [220]. It also helps to define the expected loss conditional on a VaR threshold being breached. As such, it is a risk measure that has gained popularity in recent years.

Conditional Value-at-Risk (CVaR) measures the expected loss conditional on exceeding a given VaR threshold. Mathematically, for a given threshold ϵ, CVaR is equal to:

$$\text{CVaR}_\epsilon = \int_{-\infty}^{\text{VaR}_\epsilon} P dP \tag{19.80}$$

where P is the portfolio profit/loss.

As we can see, CVaR and VaR are mathematically related as VaR defines the percentile return corresponding to a given threshold, ϵ and CVaR defines the expected value of returns of the left tail of the distribution, from negative infinity until the calculated VaR at the given threshold.

As CVaR overcomes the main challenges associated with VaR, it is naturally an appealing alternative. In particular, it does not suffer from the same issues lack of coherence and consistency as VaR, and it also estimates the loss in the tail of the distribution. In chapter 26 we explore the strengths and weaknesses of VaR and CVaR as risk measures. Additionally, in the next sections we explore how VaR and CVaR can be incorporated into a portfolio optimization and highlight some of the main challenges to doing so.

19.8.3 Mean-VaR Optimal Portfolio

The standard approach to implement a downside risk based portfolio optimization problem is to replace the volatility term in a mean-variance optimization with another downside risk measure, such as VaR or CVaR. We could then build an efficient frontier of portfolios that are mean-VaR or mean-CVaR optimal instead of mean-variance variance optimal.

In the case of VaR, this would require solving an optimization problem of the following form:

$$\max_{w} \quad \text{VaR}_\epsilon(w) \tag{19.81}$$

$$\text{s.t.} \quad w^T R = \hat{R}, w^T 1 = 1 \tag{19.82}$$

Clearly, just as investors prefer higher to lower returns, and prefer less volatility, they will also prefer portfolios with the highest VaR for a given return[17]. If we choose to employ a mean-VaR approach to portfolio optimization, we have to be mindful

[17] That is, we want VaR to be the smallest negative number possible.

of the impact the mathematical properties of VaR will have on the underlying optimization.

In particular, optimizations work best on smooth, convex functions, and we should not expect a portfolio VaR to have either of these characteristics with respect to changes in weights. Additionally, we should think about the impact of estimation error on our optimization. As we are estimating a feature of the tail of the distribution, we should expect a great deal of uncertainty in our estimates.

Lastly, depending on the properties of the underlying return distribution, calculation of portfolio VaR may be computationally challenging. Recall from (19.79) that we need to invert the CDF of the return distribution. In the case of a portfolio VaR, this mean inverting a joint distribution of returns for many underlying assets, which requires intimate knowledge of their underlying correlation structure. If we assume that the distribution of returns is jointly Gaussian, then this simplifies the problem greatly and we can use standard methods for inverting the CDF, that is, our optimization problem becomes:

$$\max_{w} \quad N^{-1}\left(1-\epsilon, \mu_p(w), \sigma_p(w)\right) \tag{19.83}$$

$$\text{s.t.} \quad w^T R = \hat{R}, w^T 1 = 1 \tag{19.84}$$

where $N^{-1}(\epsilon, \mu_p(w), \sigma_p(w))$ is the inverse of the cumulative normal distribution function evaluated at $1-\epsilon$ of a normal distribution with mean $\mu_p(w) = w^T R$ and standard deviation $\sigma_p(w) = w^T C$, respectively of a portfolio with weights w.

As the complexity of the underlying return distribution increases, so too does the complexity of inverting the CDF of the joint return distribution. In practice this means that, as we incorporate non-normal return distributions, simulation may be required to estimate the portfolio VaR. This simulation algorithm could be based on an underlying assumption of the joint dynamics, most notably we could specify a multi-dimensional SDE or copula that defines the return distribution[18]. Alternatively, we could rely on a bootstrapping algorithm to leverage historical data to generate a set of simulated paths.

Nonetheless, in spite of the many challenges associate with a mean-VaR approach and efficient frontier, it is appealing to formulate our optimization in terms of a downside deviation based metric and the intuitive nature of VaR makes it an ideal candidate.

19.8.4 Mean-CVaR Optimal Portfolio

Instead of replacing the variance term in a mean-variance with a VaR term we could alternatively use a CVaR based approach, creating a mean-CVaR efficient frontier [201]. To do this, we would be solving an optimization problem of the following form:

$$\max_{w} \quad \text{CVaR}_{\epsilon}(w) \tag{19.85}$$

$$\text{s.t.} \quad w^T R = \hat{R}, w^T 1 = 1 \tag{19.86}$$

[18] Where this multi-dimensional SDE would need to be calibrated in the physical measure using estimation techniques.

where $\text{CVaR}_\epsilon(w)$ is as defined in (19.80). Just as investors prefer higher to lower values of VaR, they will also prefer higher to lower values of CVaR. That is, in both cases they will prefer smaller to larger losses in the tail of the return distribution.

A mean-CVaR based optimization is preferable to its VaR based counterpart in many ways, most notably because the coherent nature of CVaR as a risk measure means the underlying optimization is likely to be better behaved. In particular, it can be shown that, for general return distributions, CVaR is both sub-additive and concave. The concave nature of the function, in particular, is a key feature of the ensuing optimization setup, as we saw in chapter 14. Estimation error, however, is still a legitimate challenge in a mean-CVaR based approach as we are trying to estimate the tail characteristics of a return distribution which can be highly sensitive to input data.

Calculation of CVaR within a mean-CVaR framework is analogous to the calculation of VaR we discussed in the last section. In particular, under the assumption of a jointly normal return distribution, the CVaR of a portfolio with mean μ_p and variance σ_p^2, can be computed analytically as:

$$\text{CVaR}_\epsilon = \mu_p + \frac{\sigma_p \phi\left(\Phi^{-1}(\alpha)\right)}{1-\alpha} \tag{19.87}$$

where $\phi(x)$ is the PDF of a standard normal distribution evaluated at x and $\Phi^{-1}(x)$ is the inverse of the CDF of a standard normal evaluated at x.

It can be shown that, if we assume that the underlying return distribution is normal, then the mean-CVaR efficient frontier converges to the mean-variance efficient frontier. As such, this particular application will not be of interest. Instead we will want to leverage models that more accurately reflect the tail of the return distribution. As we introduce more complex distributional assumptions, simulation will be required to estimate the CVaR of a portfolio. We can think of this as creating a large number of (simulated) scenarios and estimating CVaR over that set of scenarios. Just as we discussed in the last section, this simulation algorithm could be based on a multi-dimensional SDE, a copula, or a bootstrapping algorithm.

19.9 RISK PARITY

19.9.1 Introduction

Another approach to portfolio construction is to think in terms of the risk each asset is adding to the portfolio in order to decide the portfolio weights. This approach is referred to as risk parity [222], and has gained notoriety in recent decades with Bridgewater Associates, a large hedge fund, and others famously employing a risk parity approach [46].

The idea behind risk parity is that positions should be sized based on their risk contributions, with the goal of equating the risk contribution of each underlying building block. These underlying building blocks may be individual securities, such as stocks, or they could represent asset classes, factors, or investment strategies.

One particularly appealing feature of risk parity is that its formulation doesn't depend on expected returns. Instead only an overarching covariance matrix is required to estimate a risk parity portfolio. Additionally, if we assume that the underlying building blocks have equal Sharpe ratios, then it can be proven that the risk parity portfolio is on the efficient frontier [212]. This feature makes risk parity a natural procedure to rely on in situations where a quant feels that their underlying strategies are likely to have comparable Sharpe ratios in the long run. Further, given the estimation error associated with expected returns and consequently Sharpe ratios, the underlying assumption of equal Sharpe ratios is a natural baseline from which we should only deviate from if we have substantial empirical and theoretical evidence.

19.9.2 Inverse Volatility Weighting

Intuitively, one might think of risk parity's desire to weight building blocks by their risk contribution as a volatility based weighting. In fact, under simplified assumptions, this is the risk parity portfolio. We would want higher weight in the building blocks with lower volatilities and lower weight in the building blocks with higher volatilities. So, we can simply make a weighting scheme that is inversely proportional to their individual volatilities:

$$w_i = \frac{\frac{1}{\sigma_i}}{\sum_j \frac{1}{\sigma_j}} \tag{19.88}$$

These weights are often then normalized such that they sum to one, that is:

$$\hat{w}_i = \frac{w_i}{\sum_{i=1}^{N} w_i} \tag{19.89}$$

If the building blocks are uncorrelated, then the inverse volatility portfolio is the same as the risk parity portfolio. If the building blocks are correlated, however that might impact how we would like to construct the weights. If a building block is highly correlated with the others, we might want to reduce its weight, and conversely, if the correlation is negative it might lead to a diversification benefit and we may actually want to increase the weight. The risk parity portfolio, unlike the inverse volatility weighted portfolio, leverages this type of information about the correlation structure in addition to the individual volatilities. To do this, we utilize a concept called marginal risk contributions, which we detail in the next section.

19.9.3 Marginal Risk Contributions

Marginal risk contributions are a pivotal concept within a risk parity framework. They represent the change in portfolio volatility as we change the weight in a given asset. To compute a marginal risk contribution, we begin with the definition of portfolio volatility, which, as we have seen, for a portfolio with weights w can be expressed as:

$$\sigma_p(w) = \sqrt{w^T \Sigma w} \tag{19.90}$$

where w is a vector of weights, Σ is the covariance matrix of returns and $\sigma_p(w)$ is the portfolio volatility.

If we differentiate this portfolio volatility with respect to w_i, we get the local change in the portfolio volatility for a given change in the weight of asset i. This quantity, $\frac{\partial \sigma_p}{\partial w_i}$, is called asset i's marginal risk contribution and can be calculated as:

$$\frac{\partial \sigma_p}{\partial w_i} = \frac{(\Sigma w)_i}{\sqrt{w^T \Sigma w}} \tag{19.91}$$

where in the right hand side we have taken the partial derivative of (19.90) with respect to w_i using the chain rule. In the denominator, we have the portfolio volatility, and in the numerator we have the ith element in the vector that we get when multiplying the covariance matrix, Σ by the weights, w.

19.9.4 Risk Parity Optimization Formulation

The goal of risk parity, then, is to choose a set of portfolio weights such that the risk contributions are equal [88]. Instead of equating the marginal risk contributions, we look to equalize the total risk contributions for each building block, which we write as:

$$\begin{aligned} \mathrm{RC}_i(w) &= w_i \frac{\partial \sigma_p}{\partial w_i} \\ &= \frac{w_i (\Sigma w)_i}{\sqrt{w^T \Sigma w}} \end{aligned} \tag{19.92}$$

That is, the total risk contribution of an asset is simply the marginal risk contribution, $\frac{\partial \sigma_p}{\partial w_i}$, multiplied by the weight w_i. If we sum the risk contributions of all the underlying building blocks, we obtain the portfolio volatility:

$$\begin{aligned} \sigma_p(w) &= \sum_{i=1}^{N} \mathrm{RC}_i(w) \\ &= \sum_{i=1}^{N} w_i \frac{\partial \sigma_p}{\partial w_i} \end{aligned} \tag{19.93}$$

In a risk parity portfolio, we want RC_i to be equal for all i. There are multiple approaches to solving for this equilibrium condition of equal risk contributions. For example, we could use an iterative optimization approach and minimize the squared distance from equal risk contributions, as shown here:

$$\min_{w} \sum_{i=1}^{N} \left(\frac{w_i (\Sigma w)_i}{\sqrt{w^T \Sigma w}} - \frac{\sigma_p(w)}{N} \right)^2 \tag{19.94}$$

where N is the number of assets or building blocks in the portfolio and $\mathrm{RC}_i(w)$ is risk contribution for asset i with weight w_i, where the risk contributions are computed using (19.92).

19.9.5 Strengths and Weaknesses of Risk Parity

From a quantitative perspective, Risk Parity's agnostic view of expected returns is undoubtedly appealing. This is especially true in a world with substantial estimation error, making the lack of expected returns in the problems formulation noteworthy. Risk parity also provides a robust framework for pursuing diversification, through a lens of risk contributions rather than weights. In this chapter, we have focused on risk parity that is defined by volatility, however, downside deviation metrics could alternatively be used [22].

One of the most commonly cited challenges with risk parity, however, is the required use of leverage. In many ways, leverage is a double edged sword. On the one hand, there is some theoretical and empirical evidence that suggests that investors are naturally averse to leverage [65]. Along these lines, to the extent that risk parity portfolios tend to leverage lower volatility assets, they may be rewarded with a premium for doing so. On the other hand, leverage introduces extra complexity into an investment process, as we need to decide how to manage that leverage. For example, we could attempt to maintain a constant leverage target, or could scale our portfolio such that it has the same expected future volatility at all times. Additionally, it introduces the possibility of being forced out of certain trades should margin calls occur. Further, if we apply a volatility targeting approach and then apply significant amounts of leverage, small errors in our volatility forecast can lead to large errors in our realized volatility.

19.9.6 Asset Class Risk Parity Portfolio in Practice

In this section we utilize the risk parity coding example in the repository on a set of ETFs: SPY, GOVT, HYG, DBC. These ETFs correspond to different broad asset classes, notably equities, treasuries, high-yield credit and commodities. Doing so yields the following portfolio weights [19]:

SPY	**GOVT**	**HYG**	**DBC**
12.7%	57.3%	19.0%	11.1%

As we can see, this risk parity portfolio is characterized by a large weight in treasuries (GOVT) and small weights in the naturally more volatile asset classes of equities (SPY) and commodities (DBC).

Because of their tendency to have large weights in lower volatility assets, risk parity portfolios are often characterized by small expected returns and a low volatility. This leads them to have potentially compelling levels of risk-adjusted returns, but less appealing absolute levels of return. In practice, this problem is overcome by adding leverage to obtain a more suitable level of return, or volatility. In this case, the optimal portfolio generates an expected return of 3.5%, with a volatility of 5.4%. Readers are encouraged to try other weight allocations, and compare the return, variance, and risk

[19]Using a historical period of 2016-03-01 through 2024-02-29.

contributions, and also experiment with different amounts of leverage and underlying asset universes.

19.10 COMPARISON OF METHODOLOGIES

In this chapter we have detailed many approaches to building portfolios, each of which was based on a different set of assumptions and had its own strengths and weaknesses. Importantly, none of these techniques should be viewed as a panacea. When solving these problems, quants need to remember that they are trying to model a rapidly changing world characterized by non-stationarity and regime changes. This means that no technique will be perfect and means that we will always be starving for more data. Instead, quants should be fully aware of the underlying assumptions embedded in each model and choose the approach that best suits their application.

Perhaps the most striking result in this chapter was the estimation error in expected returns that is present in most realistic quant problems. This is an often under-appreciated phenomenon and, if we are not careful, can lead us to higher confidence than is warranted in our expected return estimates. Optimizing to these expected returns is then even more problematic as the out-of-sample returns will inevitably deviate from the input expected returns. This makes portfolios like the risk parity portfolio, and the global minimum variance portfolio seem particularly appealing as expected returns are nowhere in their formulation. These techniques, however, are not without their own challenges, as both portfolios tend to be associated with leverage which can create its own problems. One problem of note is the potential for variance and covariance terms to be non-stationary. In a risk parity or minimum variance portfolio if these terms are estimated inaccurately due to their non-stationarity, and then leverage is applied, it can lead to unintended consequences. In chapter 28 we discuss some machine learning based approaches, such as clustering techniques, that could be used to extend the concept of risk parity.

Further, while the lack of dependence on expected returns is conceptually appealing to quants, in some practical applications it is unrealistic to assume that the underlying building blocks will have similar Sharpe ratios. In these cases, it is desirable to bring expected returns into the equation, with the caveat that it is the responsibility of the quant to make sure their particular formulation is robust to small changes in the inputs and all of the challenges discussed in this chapter. One technique that we saw that might help provide some structure, or stabilization to the mean-variance process was Black-Litterman, where we were able to use a prior distribution in addition to a set of views. These views, theoretically, could be outputs from a systematic alpha model, or the expected returns we would normally provide to a mean-variance optimization.

19.11 CASE STUDY: RISK PARITY AND THE POST-COVID INFLATION SHOCK

In chapter 19 we introduced one of the key markets themes in the past few years, inflation. In that chapter we detailed how inflation led to a generational rise in interest rates, and highlighted its dramatic impact on bond markets [20]. This change in market regime to one of higher inflation had widespread impacts across markets, and there was perhaps no better example than risk parity products. Earlier in the chapter, we introduced the concept of risk parity in and also provided an in practice example of what a typical asset class based risk parity portfolio might look like. Risk parity strategies have tended to be sold to investors as all weather portfolios that, due to their diversification, perform well across a wide array of market environments, with Bridgewater Associates being perhaps the most well-known example. The underlying premise behind this is tied to the fundamental power of diversification that tends to be most beneficial in periods of market stress.

At heart, however, a defining feature of these risk parity strategies is there tendency to place larger weights, or lever up lower volatility assets. Bonds, due to their historically low volatility are an example of this, and as a result tend to have large weights in risk parity portfolios. This led risk parity products to suffer greatly during the post-Covid inflationary shock period when rates rose, as the magnitude of the decline in bond prices, in conjunction with the ex-ante assumption of bonds being a low volatility asset class proved to be a problematic combination. To see this, in the next chart we show the performance of a hypothetical risk parity fund comprised of stocks, bonds and high yield credit with comparable volatility to the S&P 500 compared to performance of the S&P index itself:

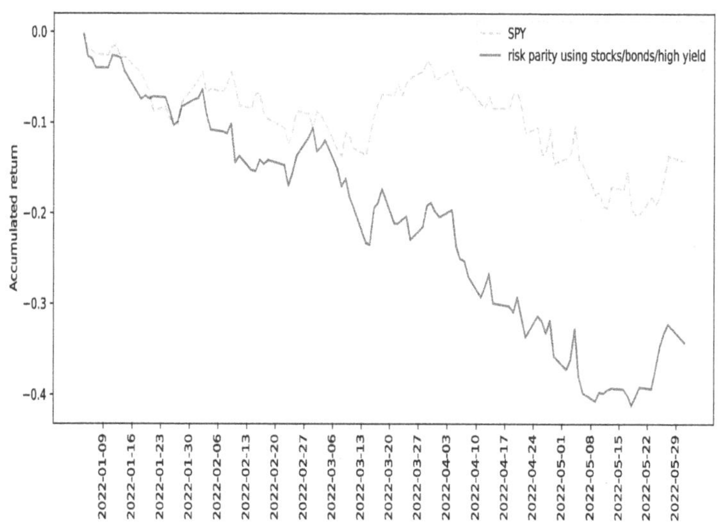

In this chart we can see that the risk parity fund significantly underperformed an investment in equities in both absolute and relative terms, going against the traditional all-weather narrative. This was at heart because of the rapid sell-off in bonds, coupled with the ex-ante estimates of bond volatility being low. This episode was so severe that it in fact caused many risk parity products to shut down, and presented many challenges to those remaining and the industry more broadly. This episode

[20]For example see 15.15.

shined a light on the use of leverage in these funds, which can be highly problematic as realized volatility significantly exceeds ex-ante estimates. It also speaks to an important discussed throughout the chapter, estimation error and showed that forecasting volatilities, just like returns is done with a great amount of uncertainty. When that is done on an asset that is perceived to have little to no volatility, this impact is amplified, as the portfolio levers up. Additionally, this market episode highlighted an important lesson to investors about correlations. Risk parity portfolios often rely on stocks and bonds to be diversifying to achieve better risk adjusted returns, however, in this period the traditionally negative correlation between stocks and bonds became highly positive. This is a reminder to quants and investors that the data that we are working with is non-stationary, and that we cannot expect historical relationships to persist, especially as we enter a new inflationary regime.

In the following chart, however, we show another hypothetical risk parity fund with commodities integrated into the portfolio, along with stocks, bonds and high yield credit and can see the performance improves substantially, albeit while still underperforming:

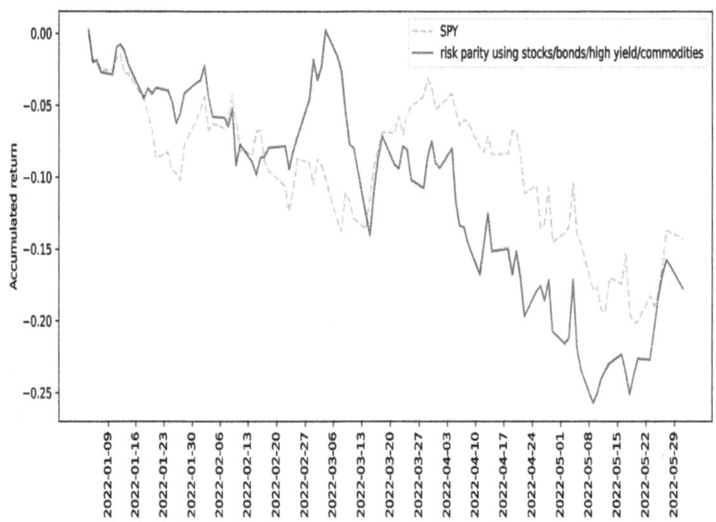

Here we can see that the performance of the fund that diversified across commodities as well was significantly better, and close to in line with the S&P. This illustrates another important lesson related to diversification, that different assets, or asset classes may provide diversification in different market environments. In periods of inflation, it is reasonable to think commodities will provide diversification relative to stocks and bonds. In contrast, in these periods, stocks and bonds are likely to provide little diversification benefit and may instead be positively correlated. Understanding these types of phenomenon and the complex relationships between assets, not just unconditionally but in different market environments is a fundamental part of building optimal portfolios.

CHAPTER 20

Modeling Expected Returns and Covariance Matrices

20.1 SINGLE & MULTI-FACTOR MODELS FOR EXPECTED RETURNS

20.1.1 Building Expected Return Models

IN the last chapter, we detailed the main approaches to building a portfolio given a set of potential investments. Building this portfolio required certain assumptions about the underlying assets in the universe. In some cases, with risk parity and the minimum variance portfolios being notable exceptions, an estimate of the future, expected return of each asset was needed. As we saw in the previous chapter, this can often be problematic as forecasting returns is inherently noisy, and any signal or advantage that we obtain in an expected return model is likely to be small. In fact, according to the efficient market hypothesis, it is impossible to estimate this type of expected return, as all available information should already be incorporated in the price. Further, even in the absence of any forecasting error, the estimation error is surprisingly large.

Putting aside these challenges and caveats, if we are going to leverage any of these portfolio optimization techniques discuss in chapter 24, then we need to build a model for the expected return of each asset. There are of course many ways that we could do this. The simplest approach would be to just use the historical average return over some prior dataset as the expected future return in the next period. Doing so yields the following equation for an expected return:

$$\mathbb{E}\left[r_{t+1}\right] = \frac{1}{N}\sum_{i=1}^{t} r_i \tag{20.1}$$

This assumption is effectively a momentum strategy. That is, we believe the assets who have done the best in some historical window are likely to perform the best in the future. More generally, we may want to place different weights on different returns depending on how recent they are, or how statistically significant they are. We also might want to allow both positive and negative signs on the weights to account for both momentum and mean-reversion. That would lead to the following more general

DOI: 10.1201/9781003530626-25

equation:

$$r_{t+1} = \sum_{i=1}^{t} \beta_i r_i + \epsilon_t \tag{20.2}$$

where β_i represents the weight of r_i in the forecast for r_{t+1}, and ϵ_t is the residual return. Note that this formulation is equivalent to (20.1) if we set $\beta_i = \frac{1}{N}$. The flexibility of (20.2), however, allows for a momentum model, characterized by positive β's, or a mean-reverting model, characterized by negative β's, and varying coefficients at different lags.

Of course, previous returns are not the only datasets that might be relevant for future expected returns. Along those lines, we may want to incorporate additional factors into our return model, such as those based on fundamental data, like book-to-market or price-to-earnings ratio. If we define a set of n additional factors, our updated expected return model then becomes:

$$r_{t+1} = \sum_{i=1}^{t} \beta_i r_i + \sum_{j=1}^{n} \beta_j f_j + \epsilon_t \tag{20.3}$$

This type of model should be familiar to the reader from chapter 5. Essentially, we will want to choose these β's such that we are best able to explain our dependent variable, in this case a future return on a set of observed data. To do this, we want to minimize the unexplained, idiosyncratic component, ϵ_t. In the context of using prior returns as the explanatory variables, this is an ARMA time series model. In the context of fundamental factors such as price-to-earnings, it is a standard regression model.

If we take a step back, we can see that this framework for expected returns that we have built is quite general. In fact, although the previous equations assumed a linear relationship between future returns and the explanatory variables, this is not required. A linear relationship is often assumed for computational ease, however, the model in (20.3) can be replaced with a machine learning or other non-linear state space time series model.

Ultimately, the success or failure of an expected return model such as (20.3) will largely depend on how well formulated the factors are, and how stable their coefficients tend to be over time. In many cases, choosing too many factors can lead to problems related to multi-collinearity, especially when relying on regression techniques. Even when using a machine-learning model that doesn't suffer from these types of issues, inclusion of too many factors often leads to overfitting and poor out-of-sample performance. This creates a natural trade-off between simplicity and interpretable, economically justifiable coefficients, and model complexity. Solving this trade-off in practice is a bit of an art, and requires a diligent research process. In chapter 28 we explore using machine learning techniques in place of regression techniques to estimate expected returns.

20.1.2 Employing Regularization Techniques

The inherent difficulty in estimating expected returns means that in practice reliable, stable expected return models that have any forecasting capability can be difficult to come by. They are at the same time both incredibly valuable, but elusive. Part of this is due to a sparsity of available data, which is further complicated by the underlying non-stationarity, and presence of differing market regimes. Another factor is that models that tend to work, and explain the future well, are pursued by many market participants, causing them to decay over time.

In this context, building an expected return model, or signal is clearly a challenge and requires economic justification, as well as empirical support for our analysis. As a result, when building a regression based expected return model, we may prefer that a certain coefficient have a positive or negative sign that reconciles with economic theory. We also may want to incorporate a prior belief about the coefficients, and shrink toward this prior belief. This has the impact of stabilizing the coefficients over time.

In order to this, we can leverage regularization techniques such as Lasso and Ridge Regression. In both techniques, we apply an auxiliary term to our objective function, a cost function, which discourages the model from increasing the coefficients too much. In the following list, we provide the underlying optimization framework for both Lasso and Ridge regression:

- **Lasso Regression**:

$$\min_{\beta} (Y - X\beta)^{\top} (Y - X\beta) + \lambda \sum_{i=1}^{N} |\beta_i| \tag{20.4}$$

- **Ridge Regression**:

$$\min_{\beta} (Y - X\beta)^{\top} (Y - X\beta) + \lambda \sum_{i=1}^{N} \beta_i^2 \tag{20.5}$$

where in both cases λ is a shrinkage parameter that determines the relative penalty of increasing the coefficients. As these optimizations are both formulated as minimization problems, we can see that an increase in any coefficient increases the value of the function. The relative cost of this increase is then determined by the parameter λ. A high value of λ will place a greater cost on increasing the coefficient estimates, leading to more bias from an ordinary least squares approach. A low value of λ, conversely, will cause the results to converge to an OLS approach. In particular, the case of $\lambda = 0$ simplifies directly to the OLS estimator.

Additionally, we can see the similarity between the two regularization techniques. Lasso regression implements a cost function that is based on the sum of the absolute value of the coefficients, or the L1 norm. Ridge, conversely, uses a cost function that is based on the squared values of the coefficients, or the L2 norm.

The result of both Lasso and Ridge regressions is more parsimonious models with smaller coefficients, making them less susceptible to multi-collinearity issues. At the extreme, as $\lambda \to \infty$, in particular, all coefficient estimates will be zero regardless of fit. Lasso regression in particular, can be interpreted as a feature selection technique due to the nature of the absolute value function, or L1 norm. Because the function is linear, the optimization problem will find it valuable to remove certain features by setting their coefficients to zero. This phenomenon is less true in the case of Ridge regressions, where the square function means there will be diminishing marginal benefit from continuing to lower a particular coefficient to zero.

In addition to the regularization techniques discussed in this section, we could leverage Bayesian regularization techniques, where a prior estimate for each beta could be given, or models that constrain the sign of certain parameters such that they are in line with their economic rationale. These approaches are not discussed in this text, however, interest readers should consult [239].

20.1.3 Regularization Techniques in Practice: Impact on Expected Return Model

In this example we leverage the coding sample in the repository to extend the Fama-French factor exposures of Apple explored in chapter 8. To do this, we regress Apple returns on the same three Fama-French factors, however, we introduce regularization via Ridge and Lasso regression, respectively. In the following chart, we can see how the coefficient estimates with respect to the three factors evolve as we vary our regularization parameter α:

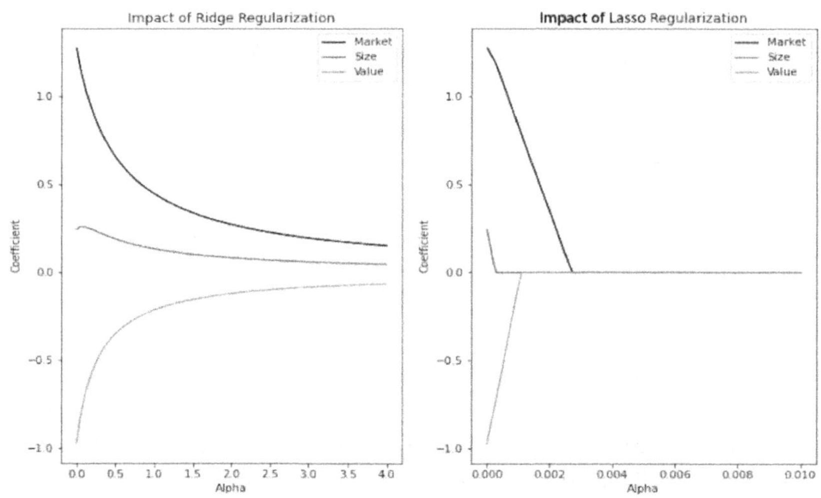

In the chart we can clearly see that Lasso regression serves as an effective means for feature selection, as it is much quicker to set the coefficients with respect to the factors to zero, beginning with the size factor coefficient. Additionally, we can see that as we increase α, the magnitude of the coefficients decreases, and that this impact is magnified for Lasso regression as well. As a result, we can see how these techniques can be used to build more parsimonious models. In chapter 28, we return to these models and describe in more detail how an optimal value for α might be estimated.

20.1.4 Correcting for Serial Correlation

One of the main challenges in estimating expected returns, and in building regression models in a financial setting, is that often the observations are correlated. This correlation between our observations goes against one of the underlying assumptions in a linear regression model, and, as a result, we must identify and correct for it. This correlation could arise from working with a cross-section of assets whose returns, and regression residuals may be correlated, or it could arise from autocorrelation in a time-series. As an example, this type of correlation in a time series arises if we are building a regression model that is based on returns with overlapping periods. For example, if we are building a return model that forecasts the next six month return based on a set of conditioning period, and include monthly returns in our dataset, then by their very nature each new month's observation will have five of the six-month return in common with adjacent observations.

The effect of this serial correlation in our dataset, either cross-sectionally because of cross-asset correlation, or in a time series, is that we will have, if ignored, bias in the estimates of our standard errors. In particular, the standard errors that we estimate in a standard OLS regression model will be too low in the presence of either cross-sectional or time series serial correlation. This means we must adjust our standard errors, and our subsequent t-statistics, to have a meaningful gauge of statistical significance in these situations. The easiest way to see this is to build a model with a long horizon with overlapping observations. These overlapping observations are highly correlated by construction, as they are largely based on the same returns, which leads to significant bias, and significantly inflated t-stats if not corrected.

Fama-Macbeth [100] developed a methodology for correcting the standard errors in a regression for cross-sectional serial correlation. Similarly, Newey-West [190] developed an analogous methodology for adjusting standard errors in the presence of serial correlation in a time series. In the following list, we briefly detail each methodology.

- **Cross Sectional Serial Correlation**: In chapter 5 we introduced some basic techniques for working with panel data, which was defined by having a cross-sectional component as well as a time series component. In particular, we saw that we could use a fixed-effects model, or random-effects model in these situations. Working with panel data is a well-developed field and common area of focus for statisticians in finance and other areas. Another approach to working with cross-sectional data, which is commonly employed in finance, is the approach pioneered by Fama and Macbeth.

  Fama-Macbeth proposed a two-step approach to estimate the model parameters for a set of panel data. In the first step, a separate time-series regression is run for each asset in the cross-section. The β's from these regressions are then used in a cross-sectional regression in each period, where we regress the previously obtained β's against the same return series. The coefficients in this second step are then averaged across time. Mathematically, the first step in the

Fama-Macbeth regression can be written as:

$$R_{i,t} = \alpha_i + \sum_{k=1}^{N} \beta_{i,k} F_{k,t} + \epsilon_{i,t} \tag{20.6}$$

where R_i, t is the return of asset i at time t, $F_{k,t}$ is the return of factor k at time t and $\beta_{i,k}$ is the beta of asset i to the kth factor. Again, it should be emphasized that this first step is equivalent to ignoring the panel component of the data and treating them as distinct, unrelated entities.

Similarly, we can write the second step as:

$$R_{i,t_j} = \gamma_{t_j} + \sum_{k=1}^{N} \lambda_{k,t_j} \beta_{i,k} \tag{20.7}$$

where again $R_{i,t}$ is the return of asset i at time t and $\beta_{i,k}$ is the previously obtained beta of asset i to factor k. It should be further emphasized that in this step we are building a regression model for each period, and doing so using the entire cross-section of data. The reader should notice that these β's, and subsequently the right-hand side of the equation are the same for each time period, t_j. The left-hand size, however, which contains the returns, vary by time period, and thus the regressions are different.[1]

The coefficients, λ_{k,t_j} are calculated for each time period and then assumed to be i.i.d. Therefore, we then average them to compute the coefficient for the entire period:

$$\lambda_k = \frac{1}{T} \sum_{j=1}^{N} \lambda_{k,t_j} \tag{20.8}$$

One of the interesting features of the Fama-Macbeth approach is that it provides a measure of attractiveness, or risk premia embedded in each factor. A positive, statistically significant λ_k, for example, implies that positive exposure to the factor has coincided with higher returns. Conversely, a negative λ_k or one that is close to zero implies that no premium is available in a given factor.

Another key result of Fama-Macbeth is that they find that the standard errors in the second step, of the time series of the λ's, should be used to calculate the resulting t-statistics, in lieu of the traditional OLS standard errors. It should be noted that the methodology employed by Fama-Macbeth assumes that the data in subsequent time periods are not correlated, and therefore only cross-sectional serial correlation needs to be accounted for. That is, it assumes there is not autocorrelation in the underlying asset time series. To correct for any such autocorrelation, we also need to leverage the Newey-West estimator.

[1] In Fama-Macbeth's work, they define $R_{i,t}$ as the excess return over the market, however, this is not required and depends on the particular circumstances surrounding the problem.

- **Time Series Autocorrelation**: The Newey-West estimator can be used to correct for auto-correlation in a time-series regression model. This estimator uses the correlation between observations, and observations of prior lags, to adjust our standard error estimates, as shown in the following formula:

$$\hat{\sigma_\beta} = \sigma_\beta \left(1 + \sum_{i=1}^{T} \frac{T-i}{T} \rho_i\right) \tag{20.9}$$

where ρ_i is the correlation between an observation at time T and observation at time $T-i$. We can see that, in the presence of positive correlation between subsequent observations, we need to amplify our standard error. Once the adjusted standard error, $\hat{\sigma_\beta}$ has been calculated, then the analogous t-statistics should be updated to be based on $\hat{\sigma_\beta}$ rather than σ_β. Derivation of this Newey-West estimator is beyond the scope of this text, however, interested readers should consult [190].

20.1.5 Isolating Signal from Noise

When building a forecasted expected return model we are inherently trying to identify some underlying signal in a dataset that is unavoidably noisy. As such, separating the signal from the noise, to the extent possible, is of paramount importance. In some cases, a dominant source of noise, such as that coming from the market factor may hide a subtler signal. In this case, a forecasting model that includes the noisy, market factor returns may not indicate that there is any signal in the data. If we remove the market factor, however, then the signal may start to reveal itself more strongly. In his book, Marcos Lopez de Prado [76] describes this process as detoning a signal, and this concept is a common theme in this book as well.

In chapter 27, the reader may notice that many of the quant trading models are built on residual, or idiosyncratic returns, after the primary systematic risk factors have been removed. The motivation for this is to separate the noisy, and potentially less relevant factor returns from the underlying signal. As an example, in the section on PCA strategies, residual returns are analyzed for mean-reversion. The premise here is that there is mean-reversion present in the market that is not visible when looking at raw returns data for a given asset, but can be seen after isolating for the primary sources of variance in the market. Said differently, we postulated that the top principal components were purely random and, once removed, the remaining signal became amplified and easier to observe. In practice, of course, removing the noise can be far more complicated than simply removing the top principal components. Nonetheless, this is an important concept that aspiring quants should be aware of as they build expected return models.

20.1.6 Information Coefficient

Once we have built an expected return model, or other signal, we will naturally want to gauge its effectiveness, both statistically and practically. Often this is done in practice via a back-test, which is a trading rather than a statistical measure. Alternatively, from a more statistical perspective, we can analyze the relationship between our signal, and forward returns. One way to do this is to analyze the statistical significance of our regression coefficients or the R^2 of the overarching regression equation. Another useful tool is to plot the signal against future expected returns and visually see if a relationship appears to exist, as we do in the following example:

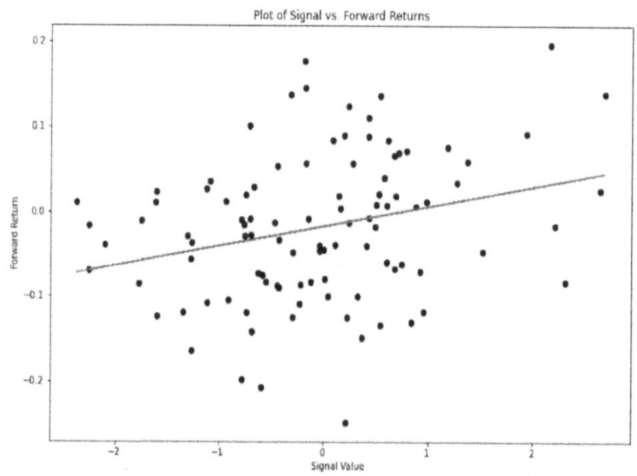

This type of visualization is helpful as it provides a sense of how well our signal explains future returns. It also does not assume a linear relationship between the two. If the results of this plot look like pure noise, with no discernible difference between the future value as our signal increases, then it implies that our signal has little to no explanatory power. Conversely, by eye we may be able to see future returns that tend to increase, or decrease as the signal increases.

Another way to gauge the power of an expected return estimate, or other signal, is to measure the correlation between our signal and future returns. This is in the same spirit as the plotting exercise discussed above. Positive correlation between the value of the signal and the future return means that the signal works. Attractive values of the signal leads to better returns, and less attractive values of the signal lead to comparably worse returns. Mathematically, the information coefficient, or IC, encompasses this correlation, and can be written as:

$$\text{IC} = \text{Corr}\left(\mathbb{E}\left[r_t\right], r_{t+1}\right) \tag{20.10}$$

ICs are often calculated on a rolling basis, in addition to over the entire time period, so that we can identify time periods where the relationship between our forecast and the ensuing returns break down. This also helps us to see if our IC is high due to a consistent positive correlation between our signal and future returns, or perhaps a single anomalous period that is driving the correlation.

20.1.7 Information Coefficient in Practice: Rolling IC of a Short Term FX Reversal Signal

In this example we leverage the IC based coding example in the repository to show how the concept of information coefficient can be applied to judging the validity of an underlying signal. To do this, we consider a simple reversal signal in the currency markets, applied to the pair EURUSD. In particular, we use the prior one-month[2] as a predictor of the next one-month[3]. For simplicity, rather than calibrating a parameter we simply use the negative of this prior one-month return as our signal, giving it an interpretation as a reversal signal. This means that higher prior month returns are likely to correspond to lower subsequent returns. In the following chart, we plot the cumulative correlation between our simple reversal signal and forward returns, also known as a cumulative information coefficient:

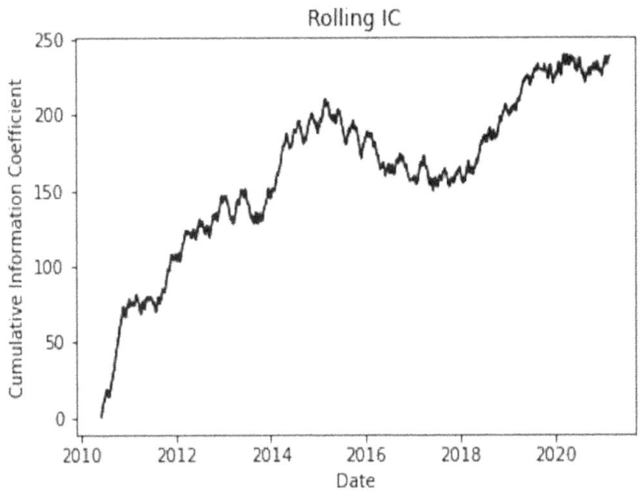

Here we can see a noticeable pattern of positive correlation between our reversal signal and future returns, which is an encouraging indication. It is left as an exercise for the reader to determine whether this is truly an attractive approach after incorporating additional market frictions such as transactions costs. The chart also helps us identify periods where the strategies performance was particularly positive or negative, which is useful for diagnostic purposes.

20.1.8 The Fundamental Law of Active Management: Relationship between Information Ratio & Information Coefficient

Grinold and Kahn [120] developed a relationship between the information coefficient, which measures the correlation of our signal and future returns, and information ratio. This relationship is known as the fundamental law of active management and states the information ratio that a portfolio manager is able to achieve as a function of this information coefficient, and breadth, is:

$$\mathrm{IR} = \mathrm{IC} \times \sqrt{\text{breadth}} \tag{20.11}$$

where breadth is defined as the number of uncorrelated bets that the signal is able to take at each time interval. Of course, in practice, for many strategies these bets are

[2]Or 21 day return.

[3]Or 21 day return.

likely to be at least somewhat correlated, meaning the effective breadth as specified by Grinold and Kahn may be far lower than the number of bets a strategy is actually taking.

Importantly, this foundational relationship shows that, to achieve a high IR we either need to have very strong positive correlation with future returns, as evidenced by a high information coefficient, or have a lower information coefficient with a larger breadth. This equation also implies that we can improve our IR mechanically simply by expanding the breadth of our signals. This relationship can also be quite useful in translating the statistical measure, IC, into a traded, market standard quantity that is near to the heart of any portfolio managers, whether fundamental or quantitatively oriented.

20.2 MODELING VOLATILITY

20.2.1 Estimating Volatility

Modeling realized volatility is of interest to many market participants for different purposes. As we saw in chapter 16, traders buying or selling straddles and delta-hedging them are implicitly making a judgment on expected future realized volatility. As such, if they are able to predict future realized volatility accurately, then they will be able to build a profitable trading strategy that trades this realized volatility, or gamma, against implied volatility. Volatility estimates are also a critical component of building a covariance matrix, which is required to implement all of the portfolio optimization and construction techniques detailed in the last chapter. Additionally, many firms try to maintain a constant ex-ante portfolio volatility and will as a result scale up or down their portfolio risk based on its expected future realized volatility. This might be pursued at the overarching product level, or at the individual strategy level where each strategy is given a risk, or ex-ante volatility budget to determine the size of their trades. Finally, risk managers implicitly make assumptions about expected future realized volatility when computing firm-wide risk-metrics such as VaR, CVaR or other measures.

There are a number of ways that we can obtain these volatility estimates. The simplest approach, is to use a backward looking, historical standard deviation as a prediction for the next period's volatility. We could also look to build a model based on historical volatility patterns. There is much research that shows that volatility exhibits mean-reverting behavior, and this observation is the basis for many volatility models, such as the ARCH and GARCH models that we explore later in the section.

In particular, we often observe that market volatility spikes are brought on by exogenous events and are then followed by subsequent reversions. Some periods, such as the tech bubble or financial crisis in 2008 were inherently more persistent and longer lasting. Other, more recent shocks, such as the Covid crisis, have been characterized by faster reversion speeds. As a result, it is a common place to apply time series techniques to try to capture these post-shock reversions, in order to try generate an improved realized volatility estimate.

Alternatively, we could look to build a volatility estimate that is based on forward-looking, currently observed, implied volatility information. This approach is not without complications and readers pursuing this avenue should remember the fundamental difference in an implied volatility estimate that emanates from the risk-neutral measure and a realized volatility estimate that is inherently rooted in the physical measure.

From the perspective of an option's trader, the underlying goal of this process will be to create an estimate of realized volatility that can be compared to today's known implied vol. For an asset manager leveraging a portfolio optimization framework, or a portfolio manager of a strategy that is sized based on volatility, the goal of this exercise is to ensure that realized, ex-post volatility matches the ex-ante volatility estimate as closely as possible.

In the following sections we explore the use of historical realized volatility estimates, based on rolling and expanding windows, and subsequently an approach based on an exponentially weighted moving average (EWMA). We then proceed to mean-reverting volatility models and estimation challenges of covariance matrices.

20.2.2 Rolling & Expanding Windows Volatility Estimates

The simplest approach to building a volatility estimate is to use some sort of sample standard deviation over a given rolling or expanding data set. Clearly when we calculate our input volatilities in a portfolio optimization, or options trading model, we don't want to look ahead. Instead, we want to rely solely on information available at each period of analysis. As we iterate forward in time, and new data becomes available, we will naturally want to incorporate this data. We may also want to phase out data that is older and potentially less relevant. In an expanding window, we simply keep adding to our dataset as we move forward in time. By contrast, in a rolling window we will remove the oldest period each time a new period becomes available to us. Use of a rolling window volatility calculation gives us an extra parameter, that is, the length of the lookback window. In the following list, we provide a brief description of each:

- **Rolling Window**: Use the last N periods to obtain an estimate of future volatility. As we iterate forward in time, new data replaces the older, potentially less relevant observations. For example, we might use a rolling window with a one-year lookback, meaning that as the data becomes more than a year old, we would no longer include it in the estimate. Data within the rolling window is then equally weighted to form a volatility estimate.

$$\sigma_t = \sqrt{\frac{1}{N} \sum_{j=1}^{N} \mathbb{1}_{(t_j > t - \omega)} \left(r_{t_j} - \bar{r} \right)^2} \tag{20.12}$$

- **Expanding Window**: An expanding set of periods is used to estimate future volatility such that at the time of each calculation all data prior to the given date is used in the estimate, with no data rolling out of the sample. All available data is then equally weighted to estimate volatility.

$$\sigma_t = \sqrt{\frac{1}{N}\sum_{j=1}^{N}\left(r_{t_j} - \bar{r}\right)^2} \tag{20.13}$$

Rolling windows tend to be more dynamic, and also have the attractive property of relying on only more recent, and more relevant data. This of course can be very attractive when forecasting a volatility, which itself is likely to be dynamic. Conversely, they can be sensitive to the chosen lookback window, and in some cases their dynamic nature marks them undesirably unstable. Expanding windows, on the other hand, lead to more stable estimates, but may include some data that comes from another regime and is perceived as less relevant in current market conditions. In the presence of estimation error and the relatively small amount of available data in the majority of finance problems, expanding windows can be an attractive proposition, as their rolling counterpart may have a large standard error associated with its estimate.

One point to emphasize is that in both the rolling and expanding window approaches we relied on an equal weighting of our observations included in the calculation. In the case of an expanding window, this means that significantly older observations which may be less relevant are given the same weight as the most recent observation. In the case of a rolling window, we get around this by excluding observations prior to a given time entirely, that is, we set their weight to zero. One side effect of this is that their weight changes suddenly from having an equal weight in the calculation, to zero, when they are no longer in the lookback window.

More generally, we might want to leverage a different weighting scheme in our volatility calculations, such as the following, which enables different weights on each observation:

$$\sigma_t = \sqrt{\sum_{j=1}^{N}\frac{w_j}{\sum_{j=1}^{N} w_j}\left(r_{t_j} - \bar{r}\right)^2} \tag{20.14}$$

In the case of an expanding window, we define $w_j = 1$ for all i. In the case of a rolling window, we would define $w_j = 1_{\tau > T_{\min}}$. But of course there are many other weighting schemes that we could postulate, such as one that gradually de-weights the observations as they get older. It turns out that, this type of approach is exactly what we do in an Exponentially Weighted Moving Average (EWMA) volatility estimate, which we detail in the next section. The most common weighting schemes used in (20.14) are based on recency of the data, however, we could also use other factors, such as liquidity, or volume, to construct our weight function.

20.2.3 Exponentially Weighted Moving Average Estimates

An Exponential Weighted Moving Average [104] is a natural alternative to a rolling or expanding window, as it places greater weight on more recent observations but is smoother as it doesn't suffer from the same binary nature of weightings as in the rolling window calculation. In an exponentially weighted moving average, observations decay smoothly as they become older, rather than being removed. In an EWMA model, the weights can be written as:

$$\sigma_t = \sqrt{\sum_{j=1}^{N} \frac{w_j}{\sum_{j=1}^{N} w_j} \left(r_{t_j} - \bar{r}\right)^2} \tag{20.15}$$

$$w_j = \exp\left(-\lambda \left(t - t_j\right)\right) \tag{20.16}$$

As we can see from the weight equation, as the distance between the current time and time of the observation increases, the weight, w_j, decreases. w_j is a function of single parameter, λ, which defines how slowly or quickly the weights decrease as we go back in time. As we increase the parameter λ, we lower the weights on older observations relative to newer observations. If we set $\lambda = 0$, then our EWMA model no longer decays and simplifies to an equally weighted, expanding or rolling window[4]. In the following chart, we show how a set of EWMA weights compares to a given rolling window:

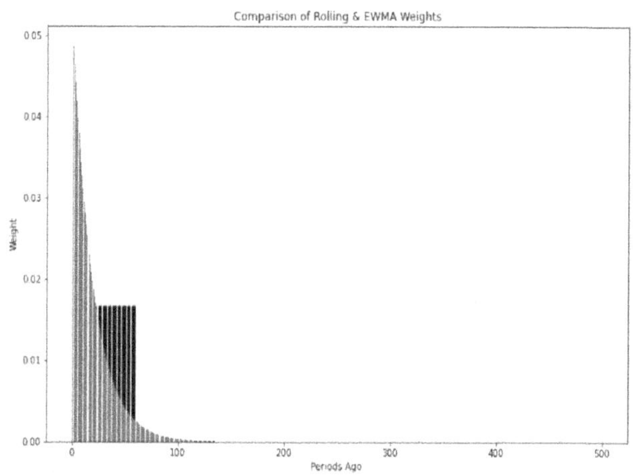

The smooth nature of the weights in an EWMA model is often attractive relative to the more sudden jumps in weights present in a rolling window, and provides a potential nice happy medium between a rolling and expanding window.

An often cited measure of how quickly the weights decay is referred to as the half-life. This half-life defines how much time it takes for the weight to decay to a weight of one-half. For a given value of λ, this half-life can be obtained by setting the left-hand side of (20.16) equal to one-half and solving for t. Doing so yields:

$$t_{\text{half-life}} = -\frac{\log(0.5)}{\lambda} \tag{20.17}$$

[4]Traditionally in an EWMA all available data is used, however, it is theoretically possible to use a rolling window approach and then apply an EWMA weighting as well.

which is the formula for half-life of an EWMA. EWMA models are often described by this half-life parameter. An EWMA with a one-year half-life is seen as a dynamic, fast updating volatility model, whereas an EWMA with a ten-year half-life is more stable, rigid and slower to adapt.

When using an EWMA, traditionally we utilize the entire available dataset and then apply the appropriate weightings. That is, an expanding window is the basis for our EWMA calculation, however, instead of the equal weighted approach described in the last section we will instead rely on an exponentially decaying function of the form of (20.16). We could also choose to use a rolling window as an input to an EWMA algorithm should we choose, however.

20.2.4 High Frequency & Range Based Volatility Estimators

Traditional volatility estimates rely on end-of-day, closing prices in their calculation, that is, they are based on the following equation:

$$\sigma = \sqrt{\sum (r_t - \bar{r})^2} \tag{20.18}$$

where r_t is the close to close period return of an asset at time t, and $\bar{r}$ is the mean return.

Another approach to calculating volatility is to incorporate intraday price data into the underlying volatility calculation. One approach for doing this is to simply rely on higher frequency returns data and then apply the same rolling window or analogous calculation. The benefit of this approach, of course, is that it increases the amount of data that we have to work with. Recall from chapter 24 that the standard error of a variance or volatility estimate was a function of the sampling frequency. As a result, the use of high frequency data increases the available dataset and lowers the estimation error in our volatility estimates. Others have suggested using information on the range of trading in a given period, by leveraging the high, low, open and close, instead of higher frequency data. A series of these so-called range estimators exist, most notably Garmin-Klass and Parkinson's volatility estimators. In the following list, we describe some of the most common approaches to range based, or high frequency, volatility estimates:

- **High Frequency**: Relies on intra-day, up to tick data over a shorter interval in order to calculate volatility. Generally, these approaches account for volatility near the open and close, as well as the volatility over given intervals throughout the trading day.

- **Parkinson**: Parkinson attempted to improve on traditional close-to-close volatility calculations by incorporating information about the High and Low prices obtained over a given period [197]. In doing this, Parkinson incorporated additional information about how volatile the asset was within the period. Periods with a large difference between the High and Low, for example indicate a large amount of intra-day volatility. Periods with similar values for the High and the Low, in contrast, indicate calm, less volatile intra-day trading. The

Parkinson volatility estimator can be written as:

$$\sigma_{\mathrm{P}} = \sqrt{\frac{N}{4T \ln 2} \sum_{i=1}^{T} \ln \left(\frac{H_i}{L_i} \right)^2} \tag{20.19}$$

As we can see, instead of relying on the volatility of close-to-close price movements, in a Parkinson's estimator we instead use the intraday range, as specified by the high and the low, to estimate volatility.

- **Garmin-Klass**: Garmin-Klass suggested using information about the High, Low, Open and Close for an asset to improve its volatility estimate [107]. This is an extension of the Parkinson estimator in that it incorporates the intraday range, via the high and the low, as well as the close-to-open volatility. This is an important extension because markets tend to trade a significant amount of the daily volume around the open and the close. Therefore, it is natural to want to capture information about these points when estimating volatility. The Garmin-Klass estimator for volatility can be written as:

$$\sigma_{\mathrm{GK}} = \sqrt{\frac{N}{n} \sum \left[\frac{1}{2} \left(\log \frac{H_i}{L_i} \right)^2 - (2 \log 2 - 1) \left(\log \frac{C_i}{O_i} \right)^2 \right]} \tag{20.20}$$

Many more such range based volatility models exist, such as Yang-Zhang [265] and Rogers-Satchell [221]. The details of these models are omitted here for brevity, however, interested readers can consult [165] for more information.

20.2.5 Mean-Reverting Volatility Models: GARCH

Two of the more common patterns that market participants observe in volatility data are clustering and mean-reversion. Volatility clustering refers to the fact that that periods of high volatility are usually followed by subsequent periods of high volatility. Mean reversion then refers to the fact that after these high volatility periods volatility tends to normalize. One way to think about this is that volatility tends to spike around exogenous shocks, and these exogenous shocks tend to keep volatility high for some time, but eventually subside, leading to an inevitable mean-reversion. This has led many market participants to want to incorporate these phenomena into their volatility models, and one way to do this is via GARCH models. GARCH models try to incorporate this via a deterministic volatility function. This is a defining feature of the model as we assume that volatility is time varying, but not stochastic. Other models, as we saw earlier in the text, in chapter 13, try to model volatility as its own stochastic process. GARCH instead uses a time-varying function to try to incorporate volatility clustering and mean-reversion. In practice, there are many types of GARCH models with additional features added. In this text, we consider only the standard GARCH model implementation. Readers looking for more in-depth treatment of different GARCH various should refer to [250].

In a GARCH model, we assume that the return process, y, is of the following form:

$$y_t = \sigma_t \epsilon_t \tag{20.21}$$

where $\epsilon_t \sim N(0, 1)$,

Further, in a GARCH model we assume that the volatility, σ_t is a deterministic function of the following form:

$$\sigma_t^2 = \alpha + \beta y_{t-1}^2 + \delta \sigma_{t-1}^2 \tag{20.22}$$

where α, β and δ are model parameters that we can fit to our data, y_t. This model matches the characteristics we observe in markets as we may observe exogenous shocks, which in this process are introduced by the ϵ_t process. These shocks then persist in the process for some time, and then their impact diminishes over time and volatility reverts to its long run equilibrium level. This means that GARCH models allow us to incorporate both the clustering and mean-reverting nature of volatility that we find in many market settings. The estimated α in a GARCH model has a natural interpretation as the long run, equilibrium level of volatility. This is the quantity that we expect volatility to revert to after a shock. The other parameters, β and δ define how persistent these shocks are, or conversely, how quickly they revert. Larger values of β and δ lead to more persistent shocks that are slower to decay, whereas smaller values lead to faster reversions.

It can be shown that, in order for a GARCH(1,1) process to remain stationary, the following conditions must be met:

$$\beta \geq 0 \tag{20.23}$$

$$\delta \geq 0 \tag{20.24}$$

$$\beta + \delta < 1 \tag{20.25}$$

Intuitively, the final condition can be thought of as ensuring that the impact of a volatility shock diminishes over time rather than gets magnified over time. Conversely, if $\beta + \delta \geq 1$, then volatility shocks are magnified over time, hence the process is non-stationary.

More generally, we can have a GARCH process with an arbitrary number of lags on both y_t and σ_t, which leads to the following more general GARCH equation:

$$\sigma_t^2 = \alpha + \sum_{i=1}^{n} \beta_i y_{t-i}^2 + \sum_{j=1}^{p} \delta_j \sigma_{t-j}^2 \tag{20.26}$$

where we now have n β's and p δ's as model parameters in addition to α. This is referred to a GARCH(n, p) model with n lags of the y_t process and p lags on the σ_t process. Readers might notice that this is similar to the ARMA time series models that we introduced in chapter 5.

Importantly, if we condition on previous values in the process for y_t and σ_t, then the process y_t is conditionally Gaussian with mean zero and variance as specified in

(20.26). This proves to be a convenient feature in practice as it makes estimation of GARCH models to real time series more tractable.

In practice, we are given a time series, y_t for a particular asset and must select the appropriate GARCH model, including the number of lags n and p. Additionally, we must choose the model parameters, $\Theta = \{\alpha, \vec{\beta}, \vec{\delta}\}$ that best fit the data. To do this, a common statistical technique is to maximize the likelihood of the data given the parameters.

In a GARCH context, this is a convenient method because of the conditional Gaussian nature of the distribution. For each point, we can calculate the probability of the observed data, y_i given our chosen parameters, Θ, as well as the subsequent value of σ_i which can be calculated from (20.26) using prior values of the observed data series. This probability, $\phi(y_i|\sigma_i, \Theta)$ can be written as:

$$\phi(y_i|\sigma_i, \Theta) = \frac{1}{\sqrt{2\pi\sigma_i^2}} \exp\left(-\frac{y_i^2}{2\sigma_i^2}\right) \tag{20.27}$$

where Θ is the set of model parameters for the GARCH model including α, β_i for all n GARCH y_t lags and δ_i for all p GARCH σ_t lags.

We can then define the joint likelihood function of all of the observed data, $L(\Theta)$ as the product of the likelihood in (20.27)[5]:

$$L(\Theta) = \prod_{i=1}^{T} \phi(y_i|\sigma_i, \Theta) \tag{20.28}$$

$$\log L(\Theta) = \log\left(\sum_{i=1}^{T} \phi\left(y_i|\sigma_i, \Theta\right)\right) \tag{20.29}$$

$$\log L(\Theta) = -T\log\left(\sqrt{2\pi}\right) - \frac{1}{2}\sum_{i=1}^{T}\log\left(\sigma_i^2\right) - \sum_{i=1}^{T}\left(\frac{y_i^2}{2\sigma_i^2}\right) \tag{20.30}$$

Here we can see that working with the log likelihood function will be more convenient, as it enables us to switch the product in (20.28) to a sum. Recall from chapter 14 that we can apply monotonic transformations to the objective function in our optimization. Looking at the final equation, the first term in equation (20.30) is a constant and can therefore be ignored in the resulting optimization. This equation, (20.30), can then be maximized over the set of parameters, Θ using standard optimization techniques in Python, where at each point in the sum we simply need to calculate the value of σ_i given the parameters, as well as the previous values in the process.

It should be noted that the initial value[6] of σ_t will also need to be specified in the optimization in order for us to calculate subsequent values of the process. This can either be treated as an additional calibration parameter, or can be set to the long-run equilibrium volatility, α. Fortunately, in cases where a long enough time series is present, the estimated of a GARCH model tend to not be too sensitive to the choice of an initial value for σ.

[5]This follows from conditional independence of the observations.

[6]Or initial values if $p > 1$.

20.2.6 GARCH in Practice: Estimation of GARCH(1,1) Parameters to Equity Index Returns

In this section we apply the GARCH coding example in the coding repository to analyze the volatility characteristics of the S&P 500 index. When doing this, we are able to gauge whether the structure of volatility has a mean-reverting component to it, which is a commonly observed phenomenon. It can also lend insight into the structure of this mean-reversion, specifically, are the shocks that occur more persistent and slower to revert, or more transitory. A high coefficient for δ in particular coincides with more persistent, slowly decaying volatility shocks, whereas a higher β will be characterized by shorter, less persistent bursts.

In this example we focus on a GARCH(1,1) model, however the interested reader is encourage to attempt to extend this and think through how they would select the optimal number of lags. We also conduct the analysis over two different time periods, 2000–2010 and 2010–2021, with the goal of isolating any changes in the behavior of mean-reversion in different decades. Calibrating a GARCH(1,1) model to each period yields the following coefficients:

Period	GARCH Beta(1)	GARCH Delta(1)
2000–2010	0.1	0.87
2010–2021	0.2	0.78

Notice that in each case the calibrated set of model parameters satisfy the condition in (20.25). We can see from this table that there is significant evidence that volatility is mean-reverting, with significant coefficients for both β and γ. Interestingly, it does also appear as though the dynamics in the most recent decade, from 2010–2021 vary slightly from the prior decade. In particular, we observe more persistent shocks in the 2000's, which was a decade characterized by the global financial crisis, and more transitory recent volatility shocks. The 2020 Covid crisis is an example of a more transitory shock. This is a phenomenon that is well known to market participants, and we see it play out in the data here. Of course, it is hard to predict whether future shocks will continue to have more transitory characteristics, or are more likely to be persistent. Thinking this through is an exercise that is left to the reader.

20.2.7 Estimation of Covariance Matrices

In order to estimate a covariance matrix we need to estimate a series of asset volatilities as well as a series of pairwise correlations. That is, we must compute a matrix containing the following terms:

$$\Sigma = \begin{bmatrix} \sigma_1^2 & \rho_{1,2}\sigma_1\sigma_2 & \dots & \rho_{1,n}\sigma_1\sigma_n \\ \rho_{2,1}\sigma_2\sigma_1 & \sigma_1^2 & \dots & \rho_{1,}\sigma_2\sigma_n \\ \dots & \dots & \sigma_i^2 & \dots \\ \rho_{n,1}\sigma_n\sigma_1 & \rho_{n,2}\sigma_n\sigma_2 & \dots & \sigma_n^2 \end{bmatrix} \tag{20.31}$$

where $\rho_{i,j}$ is the correlation between asset i and asset j and σ_i^2 is the variance of asset i. Note the high number of underlying parameters that we are required to estimate in a covariance matrix of n assets, $\frac{n(n-1)}{2}$. The foundation for estimating these variance and volatility terms was presented at the beginning of this section. In particular, we can choose to use a rolling or expanding window, an EWMA, or rely on a mean-reverting GARCH-type model. We could also look to leverage exogenous data sources, such as implied volatility, in these estimates. Estimation of the correlation terms is based on the same principles and is also traditionally done either via a rolling or expanding window, or an EWMA approach. Theoretically, we can also estimate different parts of a covariance matrix in different ways. For example, we can rely on a more elaborate, dynamic model for volatility, but a simpler, more rigid model for correlations (or vice versa). Alternatively, we could try to leverage a model that builds in the mean-reverting nature of volatility as well as the regime dependent nature of correlations. In the context of an EWMA, this would lead to a smaller λ for correlation terms than volatility terms.

Generally, these calculations will be based on historical, backward-looking data. As such, if volatilities or correlations are non-stationary this cause problems in estimating the matrix. In chapter 24 we discussed some techniques for handling a non-stationary, regime dependent correlation structure. In this chapter we also discussed some of the more common issues that arise with covariance matrices, most notably singularity. So far in this section, we have detailed some additional techniques and practicalities for estimating volatility terms. These are complications that we should have in mind as we build our underlying volatility and correlations models that help us generate a covariance matrix.

20.2.8 Correcting for Negative Eigenvalues

One of the issues that can arise in practice with covariance matrices is the presence of negative eigenvalues. This is not something that should happen naturally, as it implies a negative variance for a given portfolio, which clearly doesn't make sense. Negative eigenvalues can arise, however, in the presence of messy data. For example, it can be shown that use of different historical periods for different pairwise calculations in a covariance matrix can lead to negative eigenvalues.

In addition to their conceptual issues, negative eigenvalues present a practical problem when working with covariance matrices, and that is that the matrix decompositions and transformations that we rely on assume that the covariance matrix is either positive definite or positive semi-definite. Cholesky decomposition, for example, requires a positive definite matrix, meaning that all eigenvalues must be positive. Eigenvalue decomposition, is a bit more flexible, and instead requires that the matrix is positive semi-definite, meaning that the matrices can be low rank as long as no eigenvalues are negative. This means that, in order to use these decompositions to help us invert a covariance matrix, as we need to do in the context of portfolio optimization, or take the square root of a matrix, as we do in the context of simulation, for example, we will first need to correct the matrix for negative eigenvalues.

There are many approaches to correcting a covariance matrix for negative eigenvalues. The simplest approach is to set the negative eigenvalues equal to zero. When doing this, however, we want to preserve the other properties of the covariance matrix.

For example, we do not want to alter the total variance in the covariance matrix when zeroing out the negative eigenvalues[7]. This can be achieved by re-scaling the non-zero eigenvalues by their sum:

$$\hat{\lambda}_i = \max\left(\lambda_i, 0\right) \frac{\sum_i \lambda_i^{\mathrm{O}}}{\sum_i \lambda_i^{\mathrm{Z}}} \tag{20.32}$$

where λ_i^{O} is the ith eigenvalue prior to zeroing out the negative eigenvalues, λ_i^{Z} is the ith after the negative eigenvalues have been zeroed out but before the remaining eigenvalues have been re-scaled and $\hat{\lambda}_i$ is the ith re-scaled eigenvalue.

This is the same approach that we employed in chapter 24 to zero out small positive eigenvalues that we deemed as noise and statistically insignificant. In the remainder of this section we explore two alternative methods for handling negative eigenvalues, singular covariance matrices and statistically insignificant eigenvalues: shrinkage and random matrix theory.

20.2.9 Shrinkage Methods for Covariance Matrices

One approach for improving the stability of a covariance matrix is to combine the empirical covariance matrix, which may be low-rank, have negative eigenvalues or be prone to estimation error, with a prior, structural covariance matrix that is in line with our intuition and improves the statistical properties of the matrix. This technique is referred to as shrinkage, and was created by Ledoit and Wolf [164]. In a shrinkage estimator, we blend the empirical covariance matrix, which itself may be ill-posed, with a simpler covariance matrix that we know to be full-rank. As we incorporate more of the simpler, structural covariance matrix, the properties of the blended, or shrunk covariance matrix will likewise improve. This shrinkage adjusted covariance matrix will then be more stable when inverted. This is of the utmost importance, as know matrix inversion is required for many portfolio optimization techniques.

This shrinkage adjusted covariance matrix can be expressed by the following equation of the empirical covariance matrix, as well as our prior, structural matrix:

$$\tilde{\Sigma} = \delta F + (1 - \delta)\,\Sigma \tag{20.33}$$

where δ is a constant defining the amount of shrinkage applied, Σ is the empirical covariance matrix, F is the normalizing matrix and $\tilde{\Sigma}$ is the resulting, shrinkage adjusted covariance matrix. Ledoit and Wolf suggest using a normalizing matrix, F that has volatilities that match the empirical covariance matrix and a constant correlation between all assets with correlation parameter ρ. For example, we could define ρ as the average pairwise correlation of the assets.

[7] Mathematically this is equivalent to saying we want the trace of the matrix to stay the same.

In practice, using a normalizing matrix, F, that is diagonal can be particularly appealing from the perspective of improving the properties of the covariance matrix. This is because diagonal matrices are full-rank by construction. On the other hand, use of a diagonal normalizing matrix inherently lowers the underlying cross asset correlations. In the case of a diagonal normalizing matrix, we can think of shrinkage as a technique for damping the off-diagonal, correlation terms while preserving the diagonal, volatility terms. This increase in the diagonal elements relative to the off-diagonal elements naturally improves the stability of the matrix but also may cause undesirable shifts to the correlation dynamics as the procedure is agnostic to which components are statistically significant or insignificant.

20.2.10 Shrinkage in Practice: Impact on Structure of Principal Components

In this example we examine the universe of 11 US sector ETFs and see how the structure of their principal components evolves as we shrink toward a diagonal covariance matrix with matching volatilities. This evolution is illustrated in the following chart:

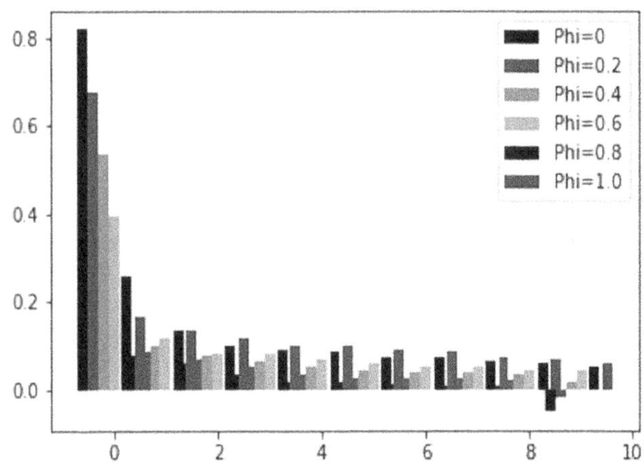

Here we can see that when no shrinkage is applied, that is when $\phi = 0$, the first eigenvalue accounts for the majority of the variance and the lower eigenvalues are quite small, making the matrix potentially low rank. As ϕ increases however, the smaller eigenvalues increase in magnitude, improving the stability of the covariance matrix. We can see this especially at the extreme when $\phi = 1$, where all the eigenvalues are quite similar in magnitude because we are now using a diagonal covariance matrix.

20.2.11 Random Matrix Theory

So far in this text, we have discussed a few methods for improving the stability of a matrix, with the overarching goal of making that matrix more stable under matrix inversion, or similar operations. In the previous section, we employed shrinkage to improve the rank of the matrix, and found this be a simple and effective technique. Shrinkage, however, requires us to specify a normalizing matrix, which is ultimately subjective. Additionally, it modifies the structure of the entire matrix rather than focusing on the components that we can identify as noise. We also saw an example earlier where we simply zeroed out the small eigenvalues in a matrix. This approach has the benefit of focusing on the embedded noise, however, up until now we have not provided the tools for defining a small eigenvalue quantitatively. This can present a problem in practice because for different matrices different thresholds of a small eigenvalue may be appropriate.

Random matrix theory, which we briefly introduce here, provides the tools for quantifying the statistical significance of the eigenvalues in a matrix by comparing them to the distribution of eigenvalues in a random matrix. As such, this provides a natural tool for identifying which eigenvalues are insignificant and should be set to zero. Readers interested in a more rigorous, detailed explanation of the use of random matrix in this theory should refer to [76]. The distribution of eigenvalues in a random matrix was studied by Marchenko and Pasteur [177], who found the limiting distribution of these eigenvalues for different classes of random matrices. In particular, a key result from their work is that the eigenvalues in a random matrix will be contained within the following range:

$$\lambda_{\pm} = \sigma^2 \left(1 \pm \sqrt{\lambda}\right) \tag{20.34}$$

where $\lambda \approx \frac{m}{n}$ with m being the number of assets and n being the number of time periods. As a result, we can conclude that when we encounter eigenvalues that exceed the bounds defined by $\lambda_{\pm}$, that they are statistically significant, and conversely, those that fall within this range can be considered noise.

When working with a correlation matrix, σ is one by construction and therefore the equation simplifies to $1 \pm \sqrt{\lambda}$. When working with a covariance matrix, σ represents the volatility of the random terms in the matrix. This can make the appropriate σ more difficult to estimate in the context of a covariance matrix, however, one approach is to begin by working with the correlation matrix, and making sure that only statistically significant eigenvalues are included. Once we have done this, we can rescale this transformed correlation matrix by the asset volatilities to recover a transformed covariance matrix based on only statistically significant eigenvalues. Importantly, this obviates the need to estimate σ in the context of a (20.34) for a covariance matrix.

In quant finance applications, especially those in the realm of portfolio construction and optimization, the underlying matrix of interest is generally a covariance[8] matrix. Given that, we should be dealing with matrices that are positive semi-definite, meaning we are working with matrices with eigenvalues that are non-negative. Because of this absence of negative eigenvalues, λ_+ is a more relevant threshold than λ_-. This means that, in practice, we can employ the result of Marchenko-Pastuer and use λ_+ as a minimum required eigenvalue for it to be deemed as significant in our portfolio optimization techniques. Once we have chosen this cutoff, we can follow

[8]Or correlation.

the same approach that we did in chapter 24 as well as in the last section. That is, we can zero out the eigenvalues below the chosen threshold, λ_+, and rescale the remaining eigenvalues such that the transformed matrix has the same variance, which can be done by making sure the trace of the matrix, or the sum of the eigenvalues is unchanged.

The primary benefit of using this approach, then, is that it provides a more robust framework for choosing an eigenvalue threshold. Instead of relying on a somewhat subjective process as we did previously, we now have a quantitative measure of statistical significance of an eigenvalue, which we can use to determine a proper cutoff. Additionally, this approach preserves the structure of the larger, more structural principal components that are more likely to contain a signal and less likely to be statistically insignificant. This can be seen as an appealing feature when compared to a technique like shrinkage which improves the statistical properties of the matrix by shrinking both the statistically significant and insignificant eigenvalues without regard for the differences in their structure.

CHAPTER 21

Risk Management

21.1 MOTIVATION & SETTING

21.1.1 Risk Management in Practice

RISK management is one of the most important tasks for buy-side and sell-side institutions alike. It is the process of trying to define the various exposures and risks that these institutions face. These risks could be along many dimensions. A buy-side institution, or hedge fund, will naturally focus on risk in terms of large scale investment losses. Similarly, sell-side institutions will use risk management to monitor their trading desks, setting limits for what trades they can warehouse. Even more fundamentally, risk management at sell-side institutions plays a role in how much capital the institutions need to maintain. This means that at these institutions, robust risk management is directly linked to the institutions profits. That is, it is a source of alpha. If a more efficient risk model leads to a bank being able to allocate their capital more efficiently, it can lead to them being able to structure and warehouse more trades that can lead to significant profits. This makes risk management a desirable location for quants on the sell side.

For buy-side institutions, especially hedge funds, risk management procedures are less standardized. This does not mean, however, that are not of the utmost importance. In fact, many concrete cases exist showing that the difference between successful and failing investors is often risk management. Examples such as Long Term Capital Management [166], who suffered immense losses on convergence spread trades come to mind as cautionary tales of risk management. Funds with disciplined, rigorous risk procedures are able to remain solvent during drawdowns and then reap the potential benefits during subsequent recoveries. On the other hand, funds that take outsized risks may find themselves insolvent, even if their trading premise is ultimately proven correct. In these cases the outsized losses that they suffered would prevent them from benefitting in any subsequent recovery, and ultimately threaten their survival.

At the heart of this exercise is understanding the properties of the tail of an underlying portfolio distribution. Fundamentally, the reader is reminded that in risk management we are interested in what we expect will happen, and not what is implied in market prices. Just because the market is currently pricing in less risk of a certain

DOI: 10.1201/9781003530626-26

event, for example an equity market correction, certainly does not mean that this isn't a risk we would want to guard against[1]. Said differently, we are working with the physical rather than the risk neutral measure, and need to do so leveraging historical data. Inherently, this makes risk a complex issue as we are trying to estimate the tail via limited historical data with all of its caveats.

Several summary risk metrics are commonly used in risk management processes, most notably VaR and CVaR. These techniques were both introduced in the context of portfolio optimization in chapter 24. Estimation of the tails of a distribution is unavoidably sensitive to the historical window, meaning that if we are not careful our risk metrics may be unstable over time and even more problematically may not accurately represent the true risks in the distribution following relatively calm periods.

Of course, historical data is not the only useful input into a risk management process. As we will see in the next section, we can alternatively define a series of adverse shocks to the portfolio, which are called stress tests or scenario analysis, and use performance of our portfolio in these states to gauge our risk. These stress tests may be motivated by shocking a set of underlying variables, or leveraging key features of certain noteworthy historical periods. For example, we might want to define a stress scenario that mimics the Global Financial Crisis, or the Market Crash of 1987. Alternatively, we may want to shock an asset classes, such as commodities, and then use correlation assumptions to develop a coherent scenario for the remaining assets.

21.1.2 Defined vs. Undefined Risks

Throughout this text we have delineated risks along many dimensions. A common delineation was the concept of systemic vs. idiosyncratic risks. In risk management we are also forced to grapple with another important distinction: the risks we know, and can clearly express, and that have presented themselves in the data, and latent risks that we can't quantify until after they have happened. Clearly, the latter are much harder for us to account for in our risk models. For example, in this chapter, we will consider a few approaches for managing risk based on looking at historical data. These methods are inherently susceptible to regime changes, or the appearance of new features in the data. At the time of our analysis, these are unknown and unquantifiable. Ex-post, it is easy to see how we should have incorporated them into our risk calculations.

Exogenous shocks such as the Covid-19 pandemic or the Global Financial Crisis immediately come to mind as these types of events. The magnitude of some of the moves would have seemed inconceivable beforehand. Now, they are standard scenarios that are run by almost all risk teams. Another example is the volatility apocalypse that we covered in a case study in chapter 22. The breaking of the Inverse Volatility ETF very well encapsulates the idea of latent or hidden risk, that few if any could have accurately quantified beforehand.

[1] In fact it might mean the opposite!

Of course, risk management is not the only place we have to reckon with this. As any investor in this inverse volatility ETF will tell you, this clearly shows up in trading, as well as portfolio optimization and other areas. However, its importance is magnified even further when we are focused on the tail of a distribution.

In this text we focus mainly on applying a set of quantitative techniques to help us in the pursuit of estimating defined risks. But we should we also be aware of the overarching latent risks that might be facing a given portfolio and the presence of these phenomenon should have a humbling impact on us as quants.

21.1.3 Types of Risk

In most of this chapter, and in this text, the focus is on what happens to investment returns in different contexts. As such, it is natural to view risk through this same myopic lens. That is, we focus almost exclusively on market risk. However, there are other important risks that firms face, that also may need to be integrated into their risk management systems, such as counterparty risk, model risk, and operational risk:

- **Market Risk**: Risks emanating from potential adverse movements in the portfolio's assets. These risks could be idiosyncratic, such as default of a company, or could be systemic, such as exposure to a broader equity market sell-off. Modeling these risks requires intimate knowledge of the correlation structure of the assets, and the resulting tail of the joint return distribution.

- **Model Risk**: Risks associated with the use of quantitative models. This might include an algorithmic trading algorithm that runs on bad data and generates trading losses, an execution algorithm that executes trades erroneously, or an incorrect derivatives pricing model that leads to poor trading decisions. While it is not possible to eliminate this type of risk, model risk can be greatly mitigated with a strong model validation and production model change process, using the framework discussed in chapter 11.

- **Operational Risk**: Risks associated with failure in a firm's operations especially due to improper processes or lack of oversight. This could be the result of lawsuit facing the company, insider trading or other illegal practices or fraud.

- **Counterparty Risk**: Risk that a counterparty will not meet its contractual payments. This risk tends to be negligible for exchange traded instruments, but can be significant for OTC contracts. As discussed later in the chapter, this is perhaps most noteworthy when trading CDS, as the magnitude of the lump sum payment conditional on default makes the potential of a counterparty default a significant liability.

This list is by no means exhaustive but is meant to give the reader a sense for the main risks that firms face that risk managers need to model. Readers interested in more details on handling related non-market risks should consult [8].

21.2 COMMON RISK MEASURES

Most risk metrics by necessity focus on defined risks. That is, they focus on risks that have presented themselves in the data in some form. When thinking about risk, we are trying to understand a complex multi-dimensional distribution. Therefore, any attempt to summarize it into a single or set of characteristics is bound to be somewhat limiting. Nevertheless, these simplified metrics can help make the problem of risk management more tractable, and can help integrate risk management into an investment process, for example by helping a fund to set a desired overall risk level, or to size its individual trades. These metrics range from simple volatility calculations to more complex VaR and CVaR calculations.

The following list highlights some of the more common risk metrics used in practice:

- **Volatility**: Standard deviation of returns. A higher volatility clearly coincides with higher portfolio risk.
- **Downside Deviation**: Standard deviation of returns only on downside movements. These calculations might be used to account for differences in the distribution of positive and negative returns, and provide further emphasis on losses which are most relevant from a risk perspective.
- **VaR**: Value at Risk provides the maximum loss at a given confidence level.
- **CVaR**: The expectation of returns conditional on breaching a given VaR threshold.
- **Extreme Loss / Stress Test Scenarios**: Pre-defined adverse shocks to the portfolio. These may be motivated by historical periods of market stress, or may be motivated by shocking underlying macroeconomic variables. The creation of stress tests and extreme loss scenarios is often subjective, although banks are subject to certain well-defined stress tests.

In the remainder of the section, we discuss several of these metrics in more detail.

21.2.1 Portfolio Value at Risk

As we detailed in chapter 24, one commonly used tail risk measure is value at risk. It is not without faults as a risk measure, for example, it has some unappealing mathematical properties[2]. Nonetheless it can be a useful measure for monitoring the tail risk in a portfolio and has natural applications to risk management. In particular, funds, and sell-side trading desks often size their portfolios at least in part based on an expected VaR. A 95% daily VaR, of $1 million, for example, would tell us that we should only exceed $ 1 million in losses approximately 5% of the time, or once a month. This type of rough guardrail can be helpful for setting higher level limits, and also making sure individual strategies aren't sized such that their potential losses are too big.

[2]Such as it not being subadditive.

As we saw in chapter 24, Value at Risk for a portfolio can be calculated as:

$$\text{VaR}_{\epsilon} = F^{-1}(1-\epsilon) \tag{21.1}$$

where F^{-1} is the inverse of the CDF of the return, or P&L distribution of the portfolio and ϵ is the specified VaR confidence level. In order to calculate this metric for a portfolio, clearly the first step would be to build the joint multi-dimensional return distribution. Once we have this, VaR calculations become mechanical. Two potential approaches for calculating this underlying return distribution, and an ensuing VaR metric, are discussed in the next section.

21.2.2 Marginal VaR Contribution

In addition to knowing the total portfolio value at risk, which can be useful for helping to set risk limits and ensuring financial stability, it can also be useful to know which assets are contributing the most and least to this VaR. This is analogous to the concept of risk contributions that we introduced in chapter 24 with respect to risk parity. Assets in the portfolio that are contributing disproportionately to the portfolio VaR might be natural candidates to scale down, whereas small VaR contributions might indicate the strategy can be scaled up without significantly impacting the risk profile.

Mathematically, the marginal VaR contribution can be expressed in terms of the partial derivative of VaR with respect to a change in the weights of each asset:

$$\text{MVC}_i = \frac{\partial \text{VaR}_{\epsilon}}{\partial w_i} \tag{21.2}$$

Just as with risk parity, we could then estimate the total VaR contribution by multiplying the marginal VaR contribution, MVC_i by the weight, w_i. Alternatively, we can simply zero out the weight of asset i, recalculate the VaR and then define the difference as the total VaR contribution:

$$\text{TVC}_i = \text{VaR}_{\epsilon,\vec{w}} - \text{VaR}_{\epsilon,\vec{\tilde{w}}} \tag{21.3}$$

where TVC_i is the total CVaR contribution for asset i and $\vec{w}$ is the original weight vector and $\vec{\tilde{w}}$ is the weight vector with the ith assets weight set to zero.

21.2.3 Portfolio Conditional Value at Risk

The concept of conditional value-at-risk, CVaR, was presented within the context of portfolio construction in chapter 24 as well. In that chapter, we saw that CVaR was a conditional expectation, where we calculated the average return contingent on a VaR threshold being breached. Mathematically, CVaR has more attractive mathematical properties than VaR and as such is a frequently used alternative. It also overcomes some other potential drawbacks of VaR. For example, we know that Value at Risk provides us with a confidence level that losses won't exceed a certain amount. However, importantly, VaR tells us nothing about what losses we should expect should we happen to exceed this threshold. For risk purposes, the magnitude of these rare losses are of direct relevance, and CVaR provides a direct way to measure them by calculation of the conditional expectation.

As we saw in chapter 24, CVaR can be expressed mathematically as:

$$\text{CVaR}_\epsilon = \int_{-\infty}^{\text{VaR}_\epsilon} P dP \tag{21.4}$$

where VaR_ϵ is as defined in (21.1) above, ϵ is a specified confidence level and P is the distribution of returns, or P&L. Just as we discussed with portfolio value at risk calculations, the first key component of calculating portfolio CVaR is to generate the joint distribution of returns for the underlying assets in the portfolio. Once this is done, CVaR can be calculated by approximating the integral in (21.4) using standard quadrature or simulation techniques. Two common approaches for doing this are discussed in the next section, historical simulation and Monte Carlo simulation.

21.2.4 Marginal CVaR Contribution

Just as with VaR, we might be interested in not only the overarching absolute number, which we can use for overall risk budgeting purposes, but also how the various underlying components contribute to that CVaR.

The approach for doing this directly mimics the approach presented above for marginal value at risk. That is, we zero each asset out individually and recompute the CVaR without that asset included. Therefore, the total CVaR contribution can be calculated as:

$$\text{TCC}_i = \text{CVaR}_{\epsilon,\vec{w}} - \text{CVaR}_{\epsilon,\vec{\tilde{w}}} \tag{21.5}$$

where TCC_i is the total CVaR contribution for asset i and again $\vec{w}$ is the original weight vector and $\vec{\tilde{w}}$ is the weight vector with the ith assets weight set to zero.

21.2.5 Extreme Loss, Stress Tests & Scenario Analysis

An alternative approach to measuring risk is to create a scenario that defines the returns of the underlying assets in your portfolio. There are a few main ways of creating such a scenario, or stress test:

- **Historical Stress**: In many cases, these scenarios are motivated by past historical events. For example, we could use the equity market movements in the Crash of 1987 as a basis for one scenario. To do this, we would need to embed assumptions about the returns of other asset classes, which we would also get from the same historical period. For assets that were unavailable during that period, we could use correlation to other, available assets to help infer returns for these missing assets.[3]
- **Stress Test Defined By Regulator**: In some cases, most notably at banks, the stress tests that need to be run are created by regulators and completely

[3] For example, we could do so using the missing data techniques presented in chapter 9.

beyond their control. These stress tests would define specific returns for different asset classes, and requires banks losses to be contained during these circumstances.

- **Subjective Adverse Scenario**: Alternatively, a risk manager might use his or her knowledge of the market as well as the funds underlying portfolio to create a set of relevant adverse scenarios. Ideally, to avoid conflicts of interest, these scenarios should not be creating by a PM who may be incentivized to understate their risk.
- **Scenario Defined By Shocking Underlying Macro Variable**: Lastly, we could begin with a desired shock to a key variable, and then use correlation assumptions to infer the returns of other asset classes. For example, we might consider the following shifts to fundamental variables as the driving factors in our scenario:
  - Oil price increases by 50%
  - A +3 standard deviation move in the EURUSD exchange rate
  - The VIX Index increases to 40
  - US Treasury Curve Steepens with Long End Yields Increasing by 100 bps

  Once we have defined this primary shift, we would then choose how we would want to move other, potentially correlated variables.

These types of stress tests and the previous more statistically driven approaches are of course not mutually exclusive and in many cases may be combined to generate a more holistic risk process. These extreme loss scenarios are a natural way to incorporate latent, undefined risks that we fear are not present in our underlying historical dataset. On the other hand, they are by nature less dynamic than their statistical counterparts, and much more subjective.

21.3 CALCULATION OF PORTFOLIO VaR AND CVaR

21.3.1 Overview

In the last section, we introduced the concepts behind VaR and CVaR and discussed their uses in a risk management framework. Of particular and direct relevance is calculating these metrics for an arbitrarily complex portfolio. This complexity might emerge from the portfolio's high dimensionality, or may emerge because of the structure of the underlying components[4]. When doing so, we are examining the tail properties of a multi-dimensional, joint distribution. As such, assumptions about the correlation between different components, may have a significant impact. The two most common approaches to calculating VaR/CVaR for a complex portfolio are:

[4] For example, if the components are options or other derivative structures.

- **Historical Simulation**: Resample historical data to generate simulated paths for the underlying assets. This methodology is simply bootstrapping by another name and embeds the correlation structure from realized price movements.
- **Monte Carlo Simulation**: Assume that the assets are defined by a joint stochastic processes with a specified copula that defines their structure, and then simulate paths from this joint, multi-dimensional process. The most common approach in this context is to assume that the joint distribution is Gaussian, and to estimate the means and covariance matrix from historical data.

Note that both techniques involve simulation, and once the simulation has been properly defined the calculation of VaR or CVaR is the same, just with a slightly different formula. That is, once we have generated the simulated paths, using either simulation from a stochastic model, or historical data, the process for calculating the relevant risk metric given these paths in the same. In the following sections, we detail how to apply these two methods in practice, and highlight some of the strengths and weaknesses of each.

21.3.2 Historical Simulation

Historical Simulation is a non-parametric approach that relies on resampling from our historical dataset repeatedly. Earlier in the text, in chapter 5 we introduced statistical bootstrapping, and historical simulation is just an application of this methodology applied to risk calculations. Leveraging the concepts from chapter 5, we can see that a bootstrapping algorithm applied to a VaR or CVaR calculation can be summarized by the following steps:

- Generate a sequence of uniform random integers between 1 and N, each of which correspond to a distinct historical date.
- At each time-step, look up the return for all assets that corresponds to that randomly chosen date.
- At each time-step, apply these returns to the entire asset universe.
- Continue for many paths, and calculate the change in the portfolio value along each path.
- Using these simulated paths, calculate the relevant risk metric, such as VaR or CVaR. VaR is simply the nth percentile return whereas CVaR is the average of the returns that are less than or equal to the calculated VaR.

Fundamentally, to implement our historical simulation algorithm we begin by generating a series of simulated returns. Each simulated return will be a randomly selected day of historical returns for all assets. The fact that we are using the same historical date for all assets is significant and means that we inherit the realized correlation structure. We then apply these returns to our portfolio positions to compute the P&L of our portfolio. It should be noted that because we are resampling from our historical dataset, we are assuming that our dataset is representative of the future, that is, we are assuming the data is stationary. This means that our VaR and CVaR estimates will only be as good as the underlying data that we have available to us. If we don't have enough data, then we could be missing parts of the tail of the distribution that have not presented themselves yet[5]. Further, if our data is characterized by strong returns for an asset, and that is unlikely to continue going forward, then we might to account for this by detrending the returns.

21.3.3 Monte Carlo Simulation

The second main approach to calculating VaR/CVaR is to use a Gaussian (or some other) Copula and create a simulation algorithm that generates simulated asset paths from a given joint distribution which can then be used to construct risk metrics. Unlike historical simulation, this approach assumes a parametric structure for the return distribution. Notably, in the Gaussian case, we assume that the joint distribution of returns is jointly normal with some mean and covariance matrix. We generally assume that the means are zero for risk purposes, and use the historical covariance matrix. In chapter 25 we discussed some common approaches for generating a covariance matrix from historical data, such as via an EWMA.

In this setting, we can then generate simulated returns that are jointly normal (or specified to have some other structure in the joint distribution) and use these to create our risk metric. In particular, as we learned in chapter 15, the following equation can be used to generate a random set of returns with a given covariance matrix, Σ and vector of means, μ:

$$X = \mu + \sqrt{\Sigma} Z \tag{21.6}$$

where Z is a vector of uncorrelated, normally distributed random normals and μ and Σ are the means and covariance matrix of the assets, respectively. In order to implement this algorithm, we need to leverage the Python functions detailed in chapter 8 to generate normal random variables in order to:

- Estimate a covariance matrix and set of assumed means.
- Generate a series of correlated asset returns according to (21.6). Instead of choosing a return for each time-step, in this framework we can generate a single return spanning the entire period of analysis.
- Apply these returns to the entire asset universe.
- Continue for many paths, and calculate the change in the portfolio value along each path.
- Using these simulated paths, calculate the relevant risk metric, such as VaR or CVaR. Again, VaR is simply the nth percentile return whereas CVaR is the average of the returns that are less than or equal to the calculated VaR.

[5] And are therefore neglecting undefined risks.

It should be emphasized that, in order to leverage this algorithm we need to be able to take the square root of the covariance matrix. This is a common operation on covariance matrices, and can be done either via Eigenvalue decomposition or Cholesky decomposition. Prior to using these decompositions, however, we should be aware of the assumed structure of the underlying matrix. Cholesky decomposition, for example, require a positive definite matrix, meaning all eigenvalues must be positive. Eigenvalue decomposition, instead requires a positive semi definite matrix, meaning the eigenvalues can be greater than or equal to zero. This is an important distinction and makes the use of Eigenvalue decomposition preferred in almost all practical situations. Regardless of which technique is used, we should also ensure that the matrix being given to the algorithm meets the requirements prior to performing the transformation, correcting it if necessary.

Clearly, a Gaussian assumption is not the only one that we could make here. For example, we could try to assume that the dynamics for each asset followed a stochastic volatility model, or a jump process by simulating from a multi-dimensional SDE. This unsurprisingly adds complexity, however, especially because our goal is to generate the joint distribution. This means that we not only need to know how to express the densities of the underlying assets, it also means being able to formulate their joint distribution. In the case of a stochastic vol model, this would mean estimating the correlation of different latent, volatility processes. These parameters are inherently difficult to calibrate, as the market does not provide information on current or historical values of these quantities to base a correlation calculation.

21.3.4 Strengths and Weaknesses of Each Approach

In this section we detailed two main approaches to calculating tail risk measures, such as VaR and CVaR. These methods can of course also be used to compute other preferred tail risk measures, at heart, they are methods for estimating the properties of a distribution via simulation. The two approaches have some important differences, however, and rely on different assumptions, leading to different strengths and weaknesses.

Notably, the historical simulation methodology that we detailed, which is really just statistical bootstrapping, is non-parametric in nature which is an appealing feature. This means that we don't have to make assumptions about the underlying return distribution. This is in contrast to the Monte Carlo method which begins with an assumption of the underlying joint return distribution. As we are estimating a tail when calculating risk, this is an important delineating feature, and is a natural reason to prefer historical simulation. The normal distribution, for example, is known to have tails that are not sufficiently fat to match what we observe in markets. As a result, historical simulation may often lead to more realistic assumptions about these tail characteristics.

Conversely, historical simulation is inherently only as good as the underlying data that is available. If that dataset is biased, or does not match the assumptions in a bootstrapping algorithm, then it might lead to risk calculations that are not robust. For example, remember that in chapter 5 when we introduced bootstrapping, we discussed the assumption that there was no autocorrelation in the underlying data. Monte Carlo simulation, by contrast, provides a general framework that may be easier to integrate additional features into. For example, later in the chapter we consider risk for credit instruments that are characterized by the risk of default. In this context, for these instruments, along each simulated path we must simulate whether or not the firm survives, in addition to how their credit spread and other characteristics evolve. Monte Carlo simulation provides a convenient framework for incorporating this type of additional feature.

21.3.5 Validating Our Risk Calculations Out-of-Sample

Once we have completed the analytical work needed to produce risk estimates, it is pivotal to make sure that the estimates we are providing are tethered to actual, real-world outcomes. If we create a daily VaR metric that is breached every week, it is a clear warning signal that something in our methodology is flawed. For example, perhaps we are not incorporating the fat-tailed nature of our underlying distribution. Conversely, if our VaR is never breached, this could be just as problematic. A 95% VaR, of course, should be breached 5% of the time. This means that, for a daily VaR, we should see the threshold crossed around every 20 days, or once a month. If we go an entire year without breaching this 95% daily VaR, it would be an indication that our methodology is too conservative and needs to be modified.

Clearly, if we underestimate our VaR, and relatedly CVaR, then to the extent that these metrics were used to size the firm's risks, this could create serious issues for the financial institution and could lead to unnecessarily large losses that in extreme cases could lead to solvency issues. On the other hand, if we consistently overestimate VaR and CVaR this is not ideal either. In this case it would lead us to continually take on less risk than we might be able to otherwise, leading to reduced profits.

The best way to safeguard against these issues is to perform an out-of-sample test of our VaR and CVaR metrics, identifying, for example, how often VaR was breached historically in an out-of-sample period. If these occurrences are too common, or too rate, it is a warning sign that our methodology or assumptions should be altered. Just as we look to affirm our portfolio optimization and trading models with out-of-sample results, so too should we when building risk models. As we are estimating tail properties of a distribution, ideally this out-of-sample period should contain a wide array of market conditions, including periods of significant stress. This can be challenging because of the limited available data, especially the limited data on drawdown and stress periods.

21.3.6 VaR in Practice: Out of Sample Test of Rolling VaR

In this example, we show how to compute a rolling historical VaR's for a given set of ETFs, leveraging the VaR coding example provided in the repository. In order to test the reasonableness of our VaR estimates, it is good practice to test them out-of-sample. Said differently, we should compare the estimates for VaR that we estimate to the ensuing, actual returns in the subsequent period. In the following chart, we do this by plotting the 5% VaR estimate for a set of ETFs, as well as an equally weighted portfolio, over rolling periods:

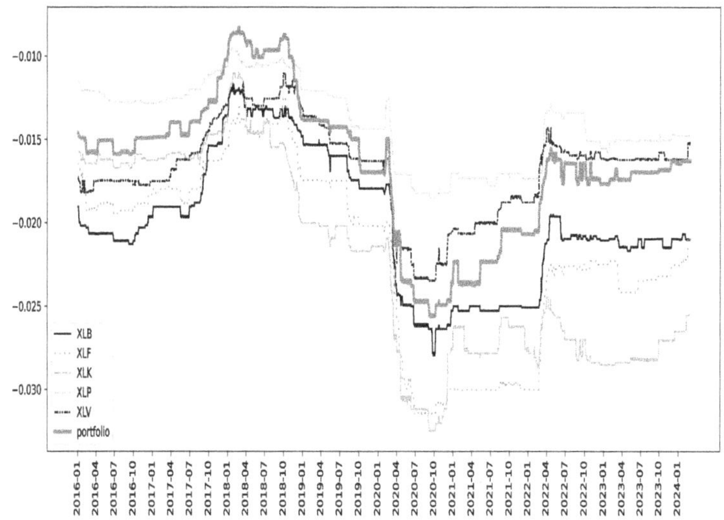

In this example we use a simple historical VaR calculation in order to compute the VaR estimate for each day, however, this could easily be replaced with another technique introduced in the chapter. Because we are focusing on a daily VaR that means comparing the calculated VaR to the following day's returns. In the following chart, we then compare this rolling estimated VaR for the equally weighted portfolio to its one-day forward returns:

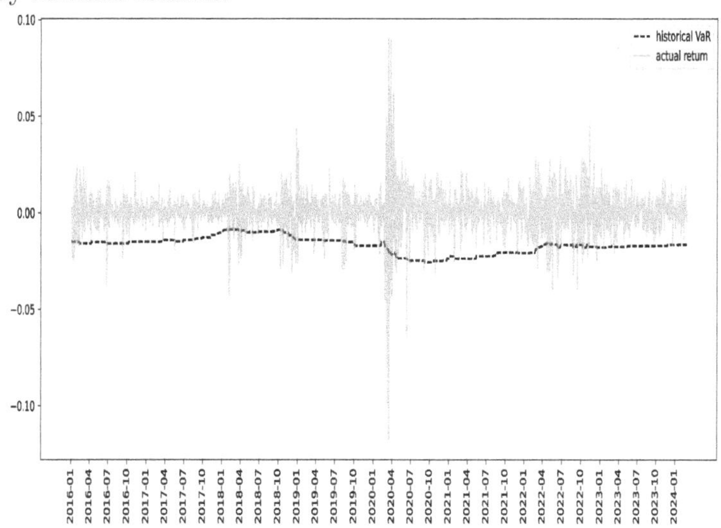

This chart is a strong visualization of portfolio VaR and is extremely useful as a robustness check. The visualization shows that, when the VaR threshold is breached, the violations can be extreme, such as during the Covid crisis in March of 2020. This highlights a limitation of VaR as a metric as it provides no information as to the size of the breaches when they do occur. We can further check the number of days when the actual return is above or below the historical VaR and compare to what we would theoretically expect given our confidence level. During our period of analysis, 2016-01-01 to 2024-02-29, we observed 1,950 returns above VaR, and 101 returns below VaR, meaning our VaR was breached 4.992% of the time. Given that we use a 5% confidence level, the out-of-sample performance of this equal-weighted portfolio is quite good and shows that our measurement of VaR is reasonable by this standard. This type of out-of-sample analysis is something that is second nature when we are conducting back-tests, however, it is also important to perform in our risk management applications as an important validation step. In the coding repository, we also provide an example analyzing the distribution of VaR estimates at a given time. Of perhaps most interest, the reader should consult this to get a sense of the potential uncertainty, or variability in a VaR estimate.

21.4 RISK MANAGEMENT OF NON-LINEAR INSTRUMENTS

21.4.1 Non-Linear Risk

When working with linear instruments, applying the results obtained from a historical or Monte Carlo based simulation is quite simple. We just take the simulated percent change and apply it the asset and multiply by our position size. This worked well because the change to the underlying asset was the only factor that affected our P&L for that position. Further, deep time series of the underlying asset prices is generally available for us to base our risk metrics on.

When working with options and other non-linear instruments, however, this will no longer be the case and other complexities will be introduced. For example, in addition to simulating asset prices, do we also need to simulate implied volatilities? In practice the answer is yes, and even more challengingly we might really need to evolve the entire volatility surface. Further, rather than relying on a deep history of prices to directly base our risk calculations on, we will instead need to rely on changes to multiple variables, including asset price, volatility and time to expiry in order to estimate the value of the instrument. Risk metrics in this context then will be dependent on multiple variables, meaning we have to track significantly more data in our calculations.

There are two distinct approaches to computing risk measures such as VaR and CVaR for options and other non-linear instruments:

- **Full Revaluation**: In this approach we apply our full pricing model in order to calculate the value of the option in every state. This may involve a separate calibration in each state, making this approach computationally prohibitive in many cases. Essentially, in this approach we would evolve, via simulation or a scenario, a set of underlying assets and their volatility surfaces. At the end of each path, we would then re-price each option using terminal values obtained from paths for the underlying asset and volatility surface. Mathematically, this can be expressed as:

$$\mathrm{PL} = c(\hat{S}_0, \hat{\sigma}, K, \hat{\tau}, \dots) - c(S_0, \sigma, K, \tau, \dots) \tag{21.7}$$

where $\hat{S}_0$ is the terminal value of the asset along the path, $\hat{\sigma}$ is the terminal value of the implied volatility for the asset and $\hat{\tau}$ is the remaining length of the option after accounting for the time in the simulated path.

In the context of a Black-Scholes model with a flat implied volatility surface, this approach is reasonably tractable provided we have a historical time series of implied volatilities. In this case, we can simply evolve both the level of implied volatility and the asset price together in our simulation or stress test. This leads to a well-defined set of parameters that can be plugged into the pricing model in (21.7). If we assume a Black-Scholes model but enable the skew, then we will see this approach starts to get more complex. In this case, instead of a single implied volatility to evolve we would have a more complex object, and we would need to keep in mind expected interactions between the asset and its volatility. This would also be true as we introduced more complex stochastic models into our framework. For example if we were to rely on the Heston model, we would begin by calibrating a set of Heston parameters. Next, we would evolve the volatility surface and asset price together. We would then need to identify the set of Heston parameters that would fit the new, evolved volatility surface. Said differently, we would need to re-calibrate. In this case, as we can see, there will be a significant efficiency degradation.

- **Estimation via Greeks**: An alternative and perhaps the most common approach to calculating risk on non-linear instruments is to use a Greeks-based approach to approximate the P&L of each option/derivative. This approach is computationally more efficient as the Greeks are known and do not need to be recomputed along each simulated path. Essentially, to apply this approach we evolve the underlying variables, such as the asset price and its volatility, and then approximate the change in value of the option structure using the changes in these underlying quantities, in addition to their sensitivities. As we saw in chapter 16, options structures have many factors that can lead it to change in value, such as delta, gamma, theta and vega. We also know that using only an option's Δ would lead to a poor approximation of P&L, especially for large moves. This is in part because sensitivity to the underlying asset is not the only dimension on which an option changes in value. Instead, changes in implied volatility, and even the passage of time, can impact this as well. It is also partly due to the non-linear behavior of option prices with respect to the underlying asset, a concept we explored in depth in chapter 16.

  To use Greeks to estimate a new value for the instrument, we would need to include:

  - Changes to the options time to expiry, Θ
  - Changes to the underlying asset price, Δ
  - Changes to the options volatility, ν
  - Changes to other factors such as interest rates

  We would also want to incorporate second-order terms, such as gamma, Γ, which defines the convexity of an option with respect to the asset price. As large moves and tail risk properties are at the heart of risk management exercises, these second-order terms like gamma will be of the utmost importance for risk purposes.

We can then use a Taylor series approximation to estimate the change in the value of the portfolio as a function of the changes to these quantities. This would lead to an approximate risk calculation of the form:

$$\text{PL} = \delta(S)\Delta + \frac{1}{2}\delta(S)^2\Gamma + \delta(\tau)\Theta + \delta(\sigma)\nu + \dots \tag{21.8}$$

It should be noted that we could also choose to include other second-order Greeks in (21.8), such as vanna and volga.

21.4.2 Hedging Portfolios via Scenarios

Another way that we can approach risk management for complex, option payoffs, especially exotic options is to first decompose the more complex structures on a basket of simpler, more tractable instruments and then calculate risk metrics on this basket instead of the complex exotic.

To do this, we can begin by defining a set of scenarios and choose the hedging instruments that best match our payoff over these scenarios. This is conceptually similar to finding a minimum tracking error portfolio as we discussed in chapter 24. The portfolio that we would choose would be the portfolio of benchmark instruments that has the minimum tracking error to our more exotic, less tractable structure over the set of specific scenarios.

These scenarios could be based on historical data, generated via a calibrated stochastic model, or via another approach, such as named scenarios. Ideally, however, the scenarios should be orthogonal. We also need to specify a set of benchmark instruments. We can then mathematically find the hedging portfolio that minimizes our tracking error, which can be written as:

$$\hat{\Delta} = \min \sum_i (\Delta_i dB_i - dP)^2 \tag{21.9}$$

This approach amounts to a regression of the payoffs against the benchmark instruments over the specified set of scenarios.

21.5 RISK MANAGEMENT IN RATES & CREDIT MARKETS

21.5.1 Introduction

In most asset classes, such as equities, FX and commodities, risk management is relatively easy. Rates and credit markets, in contrast introduce additional complications, both for linear and non-linear instruments.

In rates markets, for example, the following complicating factors exist:

- We must account for the stochastic nature of rates and the impact that has on the payoffs and discount terms.

- We must model an entire yield curve when modeling risk.
- Historical data for interest rates is quoted in terms of yields, which we must convert to a change in price of our asset.

Similarly, in credit markets we must contend with the following potential issues:

- We must model an entire credit curve, much like the yield curve we must model in rates.
- We must account for the possibility of jumps to default.
- Historical data for credit is quoted in terms of spreads, which we must then convert to a change in price of our asset.

Additionally, in both cases when working with options we need to handle the normal complexities discussed in the previous section. That is, we must account for volatility surfaces properly when managing option risk.

Even for simple, linear products, risk management is more involved in rates & credit markets because of the presence of an entire yield or credit curve, respectively. In practice certain instruments may have very different sensitivity to different parts of the curve, meaning that how the shape of the curve changes is a pivotal component of risk management. In addition, other factors like the slope of the curve, or the convexity in the curve, may also lead to substantial changes in valuation.

In the remainder of this section, we attempt to address these additional complexities, both in credit and in rates, beginning with the process for working with historical yield data, and converting it to a change in price using the annuity, or risky annuity function[6].

21.5.2 Converting from Change in Yield to Change in Price

The first complication that arises when calculating risk for fixed income instruments is that the relevant time series that we observe tend to be quoted in terms of yields, or credit spreads, instead of price. This means we need another step to value the portfolio in our simulated states.

To see this, consider a bond with a known, quoted yield-to-maturity. In this case, our simulations will evolve this yield, and then the terminal yield of the simulated path would need to be converted to a price shift of the bond. Mathematically, we can express this calculation as:

$$\mathrm{PL} = b(\hat{y}, \hat{\tau}, \dots) - b(y, \tau, \dots) \tag{21.10}$$

where y is the observed yield, $\hat{y}$ is the terminal simulated yield, and $b(y, \tau)$ is the bond pricing function. This methodology would require repricing the bond in all states which can be somewhat computationally intensive. Instead, we can rely on the concepts we learned in chapter 19 to calculate the change in price for a given change

[6]Which were introduced in chapters 19 and 20, respectively.

in yields. That is, we can rely on a Taylor series expansion and use the current values of duration and convexity of the bond to estimate the change in value, as shown below:

$$\text{PL} \approx B(0,T)\text{duration}(B)\Delta_y + \frac{1}{2}B(0,T)\text{convexity}(B)\Delta_y^2 \tag{21.11}$$

$$\text{duration}(B) = \frac{1}{B(0,T)}\frac{\partial B(0,T)}{\partial y} \tag{21.12}$$

$$\text{convexity}(B) = \frac{1}{B(0,T)}\frac{\partial^2 B(0,T)}{\partial y^2} \tag{21.13}$$

As duration and convexity are often known analytically, such as in (15.6) as they are for simple bond structures, then this approach can be highly efficient. Importantly, however, duration and convexity need to be computed only once, rather than along each simulated path, leading to an improvement in efficiency relative to the full revaluation methodology. Further, the second-order term involving convexity tends to be smaller under most shifts and is therefore sometimes omitted in practice. The reader should notice the similarity between the methods presented here for bonds and the two methods discussed in the previous section for non-linear instruments. In particular, in both cases we can use a Taylor Series approximation, based on the most important sensitivities, or we can rely on a full revaluation methodology.

It should be emphasized that in this example we are considering a single, yield to maturity rather than an entire yield curve, which would add additional complexity. In that case, we would instead need to evolve the yield curve, and then efficiently re-price the bond, or other fixed income instrument with the terminal, simulated yield curve. We also considered a bond, however, the same principle can easily be applied to other standard fixed income instruments. For example, for a CDS or an interest rate swap, we could leverage the mark-to-market formulas defined in (15.17) and (16.20), respectively, which show that the annuity function, or its risky annuity analog, can be used to calculate the change in the market value of a CDS or interest rate swap based on a change in yields. In the presence of an entire yield curve that may impact valuation of our instrument, including its slope, how we define curve shifts can significantly impact our risk calculations. In the next sections, we discuss the most common approaches for defining these curve shifts.

21.5.3 DV01 and Credit Spread 01: Risk Management via Parallel Shifts

The most common starting point for interest rate risk management is to consider exposure of an instrument, or portfolio, to parallel shifts of the yield curve. Similarly, in credit markets we usually define credit spread 01, or risky DV01 as the change in value due to a parallel shift of the credit spread curve. The following chart shows what type of curve shift this would entail:

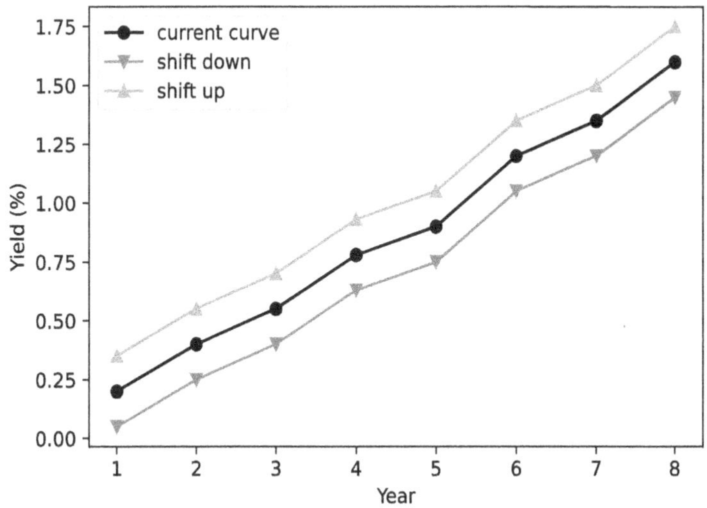

Mathematically, we can express the DV01, or credit spread 01, as the derivative of the price with respect to a parallel shift in yields, as shown below:

$$\Delta_i = \frac{dP}{dr} \tag{21.14}$$

where dr is the change of all underlying benchmark rates, and dP is the change in value of the bond or other instrument.

We could choose to build a hedging portfolio based on parallel shifts by defining a set of benchmark instruments, and then using their relative sensitivities to parallel shifts to construct our hedge as detailed in 21.4.2. This can be expressed as:

$$\Delta_i = \frac{dP}{dB_i} \tag{21.15}$$

where dB_i is the change to the value of the ith benchmark instrument and dP is the change in value of the portfolio after a parallel shift of the underlying rate or spread curve.

In a risk management context, parallel shifts, or moving the level of the yield/credit curve are certainly a key part of interest rate & credit risk management. To incorporate such shifts into a risk management framework, we might start by simulating the level of the rate curve, as proxied by a given key point, and shifting the entire curve based on movements in that key variable. Ideally, of course, although parallel shifts are perhaps the most fundamental feature of changes in curve, it will also be critical to capture other features as well, such as changes to the slope of the curve. To do this, we can either evolve parts of the curve individually, as we do in the calculation of partial DV01's, or look into other techniques such as PCA that help us understand what types of movements in the curve are most likely.

21.5.4 Partial DV01's: Risk Management via Key Rate Shifts

Because many different fixed income instruments have exposure to different parts of the yield curve, it is often the case that parallel shifts don't tell us the full story. For example, some instruments may be much more exposed to movements in long end rates than short end rates. One way to help identify exposures more precisely across the curve, is to use Partial DV01's, or key rate durations.

In this approach, rather than shifting the entire curve in parallel, we shift each key point on the curve in isolation, and measure the instruments sensitivity to a change in that rate. The following chart shows how a set of key-rate shifts would be made on an underlying yield curve at the five-year point:

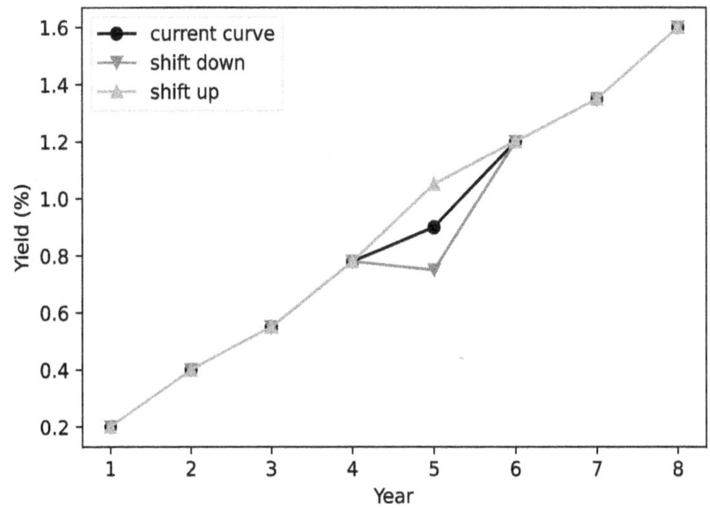

To apply this technique, the first step is to define a set of benchmark rates. For example, we might choose to include the 2y, 5y, 10y and 30y rates. We also need to define the type of rate we are shifting. In some cases we might shift forward rates, whereas in other cases we might prefer to shift zero or swap rates. This leads to the following definition of a key rate shift, or partial DV01:

$$\Delta_i = \frac{dP}{dr_i} \tag{21.16}$$

where dr_i is the change of the ith benchmark rate, and dP is the change in value of the bond or other instrument.

We can also use this approach to find hedging ratios along different parts of the curve. To do this, we would define a set of benchmark instruments, such as swaps, FRAs, etc. We would then define a set of shifts such that the benchmark instruments are moved in isolation, and compute our hedge ratio as the ratio of the change to the benchmark instruments relative to the portfolio:

$$\Delta_i = \frac{dP}{dB_i} \tag{21.17}$$

where dB_i is the change to the value of the ith benchmark instrument and dP is the change in value of the portfolio.

Partial DV01's, or key rate sensitivities are an important tool for isolating where the risk in a portfolio lies on a yield curve. One challenge, however, with key rate sensitivities is that the underlying shifts to each rate are inherently correlated. This means that, while they are useful in isolation, it becomes more challenging to think about combining them in a way that incorporates the markets expected rate and benchmark instrument correlation.

21.5.5 Jump to Default Risk

In credit markets, another critical risk that emerges is default risk, as detailed extensively in chapter 20. Along these lines, our risk models need to account for this risk as well, as it is a potentially significant source of risk for these securities. Because of this, in credit markets, jump to default scenario calculations are a core part of a risk management process, and, more generally, the potential for defaults must be incorporated into broader risk calculations such as VaR and CVaR. Fundamentally, the presence of default risk in credit securities leads to a bi-modal return distribution, with one mode conditional on survival and the other mode centered on the expected value in the event of a default.

This means that, when managing risk on credit securities that are prone to default risk we must use a fundamentally different set of tools. Instead of assuming that the return distribution is normal, we now need to take into account its bi-modal nature, and incorporate this possibility of a jump into default. One approach to incorporating such a feature into our risk models is to rely on a simulation where we first simulate random numbers to determine which mode we are in and then simulate within the mode with a second random number. The following chart illustrates the impact this jump component might have on a distribution:

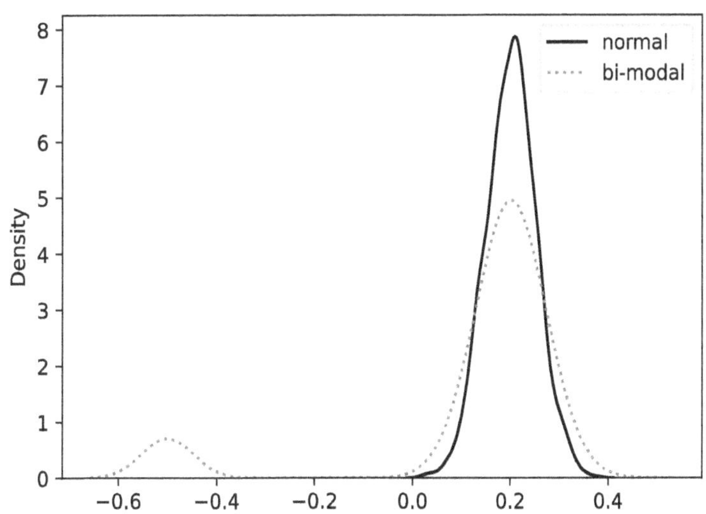

In some cases, inclusion of jump to default risk and jump to default scenarios is relatively straightforward. In these cases, we know what the value of the security will be conditional on default. For example for a corporate bond it would be the recovery value, R, and for a CDS it would be $1 - R$. For more complex derivatives, such as CDOs, jump to default analysis is far more challenging, and may require full revaluation of our CDO pricing model. Further, it requires us to make assumptions about what happens to default correlations as defaults occur. Does it increase, indicating that the default was systemic, or is it unaffected implying the default was more idiosyncratic?

Jump to default risk calculations on index and correlation products lead to an interesting phenomenon in practice. That is, investors in CDOs will have the highest jump to default risk to firms that are regarded as the safest and have the smallest credit spreads. Conversely, the higher spread names actually have smaller jump to default numbers. Intuitively, this is because default is more of a surprise for the safer names, leading to a larger re-pricing shock. For higher spread names, however, the potential for default was much more priced in, leading to much less of a shock should that default actually occur.

21.5.6 Principal Component Based Shifts

When defining risk in the context of a curve, we are implicitly asking what types of interest rate movements are most likely. Said differently, we are asking how rates tend to move together. When rates, or spreads, increase, what happens to the slope of the curve? Do the 5y and 10y tend to move together or are they relatively uncorrelated? Understanding this correlation is a key piece of risk management for rates & credit.

One way to define a set of these most likely curve shifts is to define them statistically based on our knowledge of the empirical distribution of the yield curve. That is, we can identify the shifts via PCA or another equivalent tool. As mentioned in chapter 19, PCA can be used to find a set of orthogonal factors that drive co-movement in the yield curve. In this approach, we use PCA to identify the factors that explain the majority of the variance in the yield curve, and then define these as a set of scenarios. Critically, these PCA scenarios, or curve shifts that we have created are now orthogonal, meaning we no longer have to worry about their interaction or correlation. This is an attractive property when we think about combining shifts.

To see the essence of how this works, recall that PCA is based on the following underlying matrix decomposition:

$$C = VDV^T \tag{21.18}$$

where C is the covariance matrix[7] of movements of different rates on the curve, D is a diagonal matrix of eigenvalues and V is a matrix of eigenvectors. Importantly, we know the diagonal elements of D, the eigenvalues, λ_i tell us how much variance is explained by a given factor:

$$\text{PctExplainedPC}_i = \frac{\lambda_i}{\sum_i \lambda_i} \tag{21.19}$$

[7] Or correlation matrix.

Therefore, the largest eigenvalues explain the most variance in the underlying curve, and we can select these top N eigenvalues in order to capture the majority of the risk in the curve. We could then define risk in terms of movements of these principal components, rather than of the individual, correlated rates.

Importantly, if we look at the eigenvectors in our decomposition, they contain the weights of each rate in the underlying shift. We can write each eigenvector, or row in the matrix, V, as:

$$\nu^j = \left(\nu_1^j, \ldots, \nu_n^j\right) \tag{21.20}$$

where n is the number of assets in our covariance matrix and PCA model. These weights, ν_i^j tell us the weight of each rate, yield or credit spread in each principal component. They are how we map from the concept of individual rate movements to movements in the principal components. Although PCA can be notoriously difficult to interpret, the weights ν_i^j, tell us about the structure of the principal component. For example, a principal component with weights that are all the same sign and similar magnitudes would indicate that the principal component maps to the **level of the curve**. Similarly if the weights on earlier rates are positive, with rates on longer maturities turning negative, then that would be an indication that the underlying principal component is capturing the **slope**.

21.6 MEASURING COUNTERPARTY RISK

21.6.1 Counterparty Risk: Introduction

Earlier in the chapter, we discussed four main types of risk that market participants encounter: market risk, model risk, operational risk and counterparty risk. While market risk is the predominate source of risk and the center of most risk management exercises, these other three areas can be crucial to maintaining a successful risk management framework, and among them.

Counterparty risk [49] refers to the risk that the other party in a contract will be unable or unwilling to meet their obligation. In this case, we as investors would not receive payment. Factoring in this potential to not receive a payment is crucial to understanding the value of our contracts. Because of this, Counterparty risk is an important part of risk management, especially in credit markets in the presence of default.

To take an example that we will revisit in more detail later, if we are long protection on a CDS contract [8], then the potential to not receive our payment on default will play a huge role in how we value the contract. Along these lines, as we will see, counterparty risk depends on the details of the contract, such as whether it is OTC and what the payoff structure of the instrument looks like. In some cases, where the risk of a counterparty not meeting its obligations is sufficiently large, then counterparty risk may lead to adjustments in the valuation of the contract. As we will see, this includes cases like a CDS where the counterparty risk may be asymmetric. These adjustments are referred to as **credit valuation adjustments** [236].

[8]That is, we pay the premium leg in exchange for a payment on default.

21.6.2 Counterparty Risk: Exchange Traded vs. OTC Contracts

A defining feature of the amount of potential counterparty risk in a contract lies in whether the contract is traded over-the-counter or cleared on an exchange. In the case of exchange traded contracts, counterparty risk is significantly lessened as the exchange serves provides a form of insurance. This is because the exchange will typically step in if one of the parties cannot meet their contracts obligations. OTC contracts, conversely, do not have this type of embedded insurance and are thus more dependent on the credit worthiness of the party on the other side of the trade. If this party refuses, or is unable to pay, then the party in an OTC contract has no recourse other than to seek payment from the other party.

In a sense, we can think about the counterparty risk that emanates from exchange traded contracts as being immune from idiosyncratic counterparty defaults as it has protection from the exchange in the event of such an event. OTC contracts, conversely, are exposed to this type of idiosyncratic event, making them inherently more prone to counterparty risk. Exchange traded contracts, however, are still exposed to systemic risks that would affect the liquidity of the underlying exchanges. In some sense, OTC contracts are actually less exposed to this type of risk as they are only concerned with the impact of this on their specific, idiosyncratic counterparty.

21.6.3 Wrong Way Counterparty Risk & CDS

Traditionally, counterparty risk is largely ignored when valuing a derivative instrument. This is perfectly sensible when two parties of equal credit worthiness engage in a trade with symmetric payoffs. In that case, the possibility of the risk of either counterparty not paying is roughly equal and offsetting. One example of this is an interest rate swap. Because both parties receive payments at the same frequency and of similar magnitude, there is every reason to believe that these risks offset and cancel when valuing the contract.

In credit markets, on the other hand, counterparty risk plays a much larger role and may even impact the valuation of an underlying contract. A CDS contract provides a perfect illustration of this and the resulting need for modeling counterparty risk. To see this, consider which of the following two parties is exposed to more counterparty risk in a CDS contract:

- **Buyer of Protection**: Pays a periodic coupon in exchange for a receiving a large lump sum payment in the event of default.
- **Seller of Protection**: Receives a periodic coupon in exchange for being responsible for paying a large lump sum in the event of a default.

The astute reader will notice that a **Buyer of Protection** is more exposed to large losses in a CDS contract due to counterparty default. This is due to the nature of the contract which requires them to pay out a premium periodically in exchange for a single larger cashflow. If this cashflow is not received, it will have a significant impact on the value of the contract. This creates an asymmetry in the counterparty risk for a CDS contract. This potential for *asymmetric* counterparty risk in a CDS contract is known as wrong way risk, or wrong way counterparty risk [81]. Further, the extreme nature of this asymmetry means that it might be something that we want to factor into our valuation model for a CDS.

When we incorporate this wrong way risk into our valuation models, we need a model for the likelihood of default of our *counterparty*. As these models rely on modeling default risk, they will leverage many of the same models and concepts detailed in chapter 20. These adjustments to the value of a contract for the presence of counterparty default risk are known as **Credit Valuation Adjustments (CVAs)** [236] and are detailed in the remainder of this section.

21.6.4 Credit Valuation Adjustments: Definitions & Terminology

Credit Valuation Adjustments are the adjustments that we make to contracts to account for the propensity of one or both parties to default, or not satisfy their obligation. Alternatively, we can think of CVA as the amount required for a party to **hedge** their counterparty risk. CVA is also the difference between the risk-free value of an OTC contract vs. a risky contract.

As we saw in the last section, modeling the CVA of a contract is an exercise in default modeling for the counterparty that we are engaging with. As such, the calculation of CVA is a function of the default probability of the counterparty. It is also inherently connected to the structure and payoff of the underlying contract. The calculation of CVA can be broken into the following three components: [208]

- **Expected Exposure**: The present, expected value that is tied to, or exposed to a default of a given counterparty. This represents our potential loss in the event of a counterparty default.
- **Probability of Default**: Refers to the probability that a counterparty is unable or unwilling to meet the terms of the contract.
- **Loss Given Default**: Refers to the percentage of assets that are lost in a counterparty default.

In the next section, we leverage these three components to model CVA.

21.6.5 Credit Valuation Adjustments: Modeling

If we assume that the counterparty of interests credit worthiness is uncorrelated from our expected exposure, then CVA can be calculated using the following formula: [208]

$$\text{CVA} = LGD \int_0^T \text{EE}^*(t)\; d\text{PD}(0,t) \tag{21.21}$$

where $EE^*(t)$ is the expected exposure, LGD is the loss given default and $PD(0,t)$ is the probability of default. The reader should note that $PD(0,t)$ is the probability of default between time 0 and time t and $dPD(0,t)$ is the probability of an instantaneous default between time t and time $t + \delta t$ that we are integrating over.

To calculate (21.21) we need to model the default risk of the counterparty. This can be done using the default risk modeling techniques covered in chapter 20. In particular, if CDS are available on the counterparty, then we can use a CDS bootstrapping procedure [9] to extract a set of default probabilities from a set of traded credit spreads. If not, we can use a model based on credit ratings to determine estimates for the default probabilities.

Calculation of (21.21) also requires knowledge of the loss given default, or percentage of the counterparties assets that would be lost in a potential default. Said differently, Loss Given Default measures the severity of the loss should the counterparty default, and is dependent on its recovery value, R. If CDS are available on the counterparty, R can be estimated from the quoted CDS data. In the absence of CDS data, which provides an estimate of the market's recovery expectation, approximation of the recovery value can be quite challenging. In fact, in practice there is significant uncertainty in the recovery amount, and we can observe very different actual recovery values should a default occur. Having said that, standard recovery assumptions range between 30% and 40% depending on the credit worthiness of the issuer, with lower credit entities assumed to have lower recovery values.

Finally, in order to calculate CVA using the formula in (21.21) we need to estimate our discounted, expected exposure to the counterparty. This quantity represents the potential loss at a given time, t due to a counterparty default. Mathematically, expected exposure can be written as:

$$EE^*(t) = \mathbb{E}\left[\exp\left(-rt\right) E(t)\right] \tag{21.22}$$

where r is the discount rate, t is the time of exposure and $E(t)$ is the random variable measuring counterparty exposure. In practice, to estimate exposure, $E(t)$ would require us to aggregate all of the contracts we have with a given counterparty and estimate the potential loss.

21.6.6 Credit Valuation Adjustments & Wrong Way CDS Risk

The CVA formula in (21.21) assumes independence between exposure and credit worthiness of the counterparty. In the event of a CDS with wrong way risk, this assumption is violated as increases in exposure may lead to a deterioration in the credit risk of the counterparty. In this case, we would need to use the more general formula [208]:

$$\text{CVA} = \int_0^T E^Q\left[LGD\frac{B_0}{B_t}E(t)\,\middle|\, t=\tau\right] d\text{PD}(0,t) \tag{21.23}$$

21.6.7 Other Valuation Adjustments

In this section we have detailed counterparty risk and seen how it has increased importance in credit markets and when accounting for the possibility of large payments in the event of default [10]. At the heart of this was determining whether an adjustment

[9] For more information see 16.6.

[10] For example, on a CDS contract.

needed to be made to this contract to account for any asymmetry in the counterparty, or wrong-way risk. We then discussed how to model this type of adjustment, known as a Credit Valuation Adjustment.

Credit Valuation Adjustments are one of a set of valuation adjustments that might need to be incorporated into a contract. For example, we might want to incorporate an adjustment related to the presence of margin or funding requirements in a contract. X-Value Adjustment (XVA) [118] refers to the broader collection of adjustments that might be made to contracts to account for credit risk, funding, etc.

In the following list, we highlight a few relevant valuation adjustments that are commonly used, in addition to CVA:

- Funding Valuation Adjustment (FVA) [213]
- Margin Valuation Adjustment (MVA) [32]
- Valuation Adjustment for Regulatory Capital that must be held for transaction (KVA) [12]

The details of these valuation adjustments are beyond the scope of this book, however interested readers are encouraged to consult [119].

21.6.8 Counterparty Risk in Practice: Impact of a Credit Valuation Adjustment on a CDS Contract

In this section we show the impact of CVA on a CDS contract done with a counterparty a different levels of the credit spectrum. Interested readers can find the code for this example in the corresponding repository. As highlighted in the chapter, CDS contains the potential for wrong way risk due to its asymmetric payout and therefore is an example where use of a CVA is of the utmost importance. To calculate the CVA, we rely on equation (21.21), assuming the following credit spread curves for low, medium and high risk counterparties:

Year	Low Spread	Medium Spread	High Spread
1	100bps	300bps	500bps
2	120bps	320bps	520bps
3	150bps	350bps	550bps
4	180bps	380bps	580bps
5	200bps	400bps	600bps

We further assume a 40% recovery, and therefore a 60% loss given default, and a maturity of 5 years on the underlying CDS contract.

This leads to the following CVA table in percentage terms:

Long/Short	Description
Low Risk	7.14%
Medium Risk	15.87%
High Risk	23.76%

Importantly, we can see that even for a low risk counterparty, the valuation of a CDS is significantly impacted by this valuation adjustment. At heart, this is of course due to the nature of wrong way CDS risk. Further, we see this risk increases proportionately with credit spreads, highlighting the importance of choosing a reliable counterparty in credit markets, having proper diversification across counterparties, and undergoing a robust CVA modeling exercise. The reader is encouraged to leverage the code in the repository to see how the magnitude of CVA changes as the different underlying parameters change.

CHAPTER 22

Quantitative Trading Models

22.1 INTRODUCTION TO QUANT TRADING MODELS

So far in this section, we have discussed the mathematical techniques for building risk and return models, and also discussed approaches to assembling portfolios given these inputs. In this chapter we focus more specifically on the field of algorithmic or quant trading. As we will see, this leverages many components from chapters 24 and 25. For example, it is quite common to use regression techniques, time series models or machine learning to build an alpha signal that is then an input into a quant trading application. In this chapter, however, we focus less on the machinery, and more on the signals and signal generation process. Quant trading is a broad field with a robust set of literature detailing many of its uses [61] [62] [157]. In this section we introduce the fundamental concepts central to quant trading, and then proceed to detail several types of examples that are commonly employed in practice.

22.1.1 Quant Strategies

Quantitative investment strategies rely on mathematical methods, or pre-defined rules, in order to deduce patterns in underlying asset prices. Quantitative investment strategies may be based on many different types of underlying inputs, such as price or fundamental data or some hybrid of the two. They may also be based on so-called alternative datasets, such as NLP algorithms parsing news sites or earnings transcripts. Depending on the type of quant strategy, and whether it is based on previous price behavior[1], or more elaborate sources, an underlying quant strategy may look quite different. There is an overarching set of foundational principles, however, that all quant strategies rely on.

Perhaps most fundamentally, Quant investment methods try to take advantage of the law of large numbers in a context where the odds are marginally skewed in your favor. The canonical example of this is the flip of a biased coin (say with 51% probability of heads). That is, they try to find a signal that skews the odds of making a profit slightly in their favor. As markets are highly competitive and efficient a small edge is the best we can hope for. This means that, on any individual observation, we have no idea what will happen. As time progresses, however, and the number

[1] As is the case in momentum and mean-reversion strategies.

DOI: 10.1201/9781003530626-27

of repeated experiments accumulates, other factors tend to cancel out and these strategies are able to take advantage of the bias. The degree to which the odds are skewed in your favor is ultimately determined by our **underlying signal**.

The process of uncovering these signals is referred to as **alpha research**, and is at the heart of quant and algorithmic trading. One such approach would be to build signals based on underlying risk premia, such as carry or liquidity. An alternate approach would be to build a relative value model that attempts to identify an anomaly or price dislocation quantitatively. Alpha research is then the most important component of quant trading, and will ultimately decide the fate of a quant PM. In the next section, we describe the alpha research and signal generation process in more detail and describe how this type of research should be structured.

22.1.2 What is Alpha Research?

Alpha research is broadly defined as research that assists in the pursuit of alpha, or excess market returns. In other words, alpha research is the process of building models that help us generate trading signals. As mentioned above, the goal of this exercise is to find some sort of method that improves our forward looking investment outcomes. For example, perhaps we believe that stocks that did poorly last year should do well this year. Or maybe we think that stocks that have been trending on twitter are likely to outperform. Each of these could be the premise behind an alpha signal.

An alpha research project should begin with some piece of theory or market observation that is turned into a hypothesis to prove or disprove during the research process. In a perfect world, the origin of this observation should be tied to some structural dislocation in the market, piece of economic theory, or behavioral bias. As an example, the volatility risk premium, which as discussed earlier is the tendency for options to be overpriced relative to subsequent realized volatility, is naturally motivated by loss aversion and investors inherent desire for convex payoffs.

In any alpha research project our null hypothesis should be that the signal does not work and we should require the burden of proof to show us that the signal is valid. After the initial hypothesis, the next step is to identify and store the appropriate data sets. These data sets must then be cleaned and should then be split into in and out of sample periods. Later in the chapter we discuss how to split the data, and ensure the integrity of the out-of-sample period. The next step would be to build out the signal and whatever parameter estimation is associated with the signal. This introduces a significant amount of complexity, and as such we need to be careful how many parameters we include in our models, and need to make sure they are added in a robust way. This is a topic we explore later in the chapter as well.

Next, we might take that signal and analyze its correlation with market movements in the in-sample period. This should be familiar to the reader from chapter 25 where we introduced Information Coefficients. Obviously, we want our signal to be positively correlated with future market behavior, indicating the presence of positive returns, and potential alpha. If these results are compelling, we would then subject our signal to a full-blown back-test on the in-sample period, and would tweak the model as necessary. Finally, we would use our out-of-sample data to validate any positive results that we got in-sample. When doing this, however, we need to be cognizant of how many times we look at the out-of-sample period and modify our methodology. If the results are still compelling, then we would continue with a truly live out-of-sample test, usually initially without any capital. The process of back-testing, parameter estimation and in vs. out-of-sample analysis are discussed in more detail in the next section.

22.1.3 Types of Quant Strategies

One defining feature of quant trading models is their trading frequency. Some strategies may hold their positions for long periods of time and re-balance infrequently. These are referred to a low-frequency strategies. Other strategies may trade intra-day, even leveraging tick data [5] to trade seconds apart (or less). A common measure of a strategy, which helps to define its trading frequency is turnover. This is a metric that we formally introduce later in the chapter, however, intuitively tells us how much our positions are changing each period.

Higher frequency models are fundamentally different than their lower frequency counterparts. Higher frequency models may essentially be algorithms for making markets, and trying to profit off of knowledge of market microstructure. These models are characterized by large amounts of available data, making machine learning a natural tool of choice. Conversely, for the highest frequency applications, efficiency and speed of execution is of the utmost importance when trying to take advantage of any imbalances. This leads to the use of simpler, faster executing algorithms and often leads quants to prefer the most efficient computational languages, such as C++.

Lower frequency strategies are fundamentally different. They are strategies where positions may be held for days, weeks, or even months. This limits the amount of available observation periods but also lessens the need for high speed computing. As a result, more complexity can be embedded in these algorithms, as long as the data can support it. As our trading frequency decreases, the need for an economic justification backing our underlying research increases. This makes a robust alpha research process even more critical, as once results have been seen compelling research results it is easy to concoct a story that justifies them. This makes having an economically motivated set of prior beliefs, that are only deviated from in the face of overwhelming evidence, a necessity.

22.2 BACK-TESTING

When we back-test an investment strategy our goal is to simulate as closely as possible what would have happened if we had been engaging in the strategy during the period. This means building a set of hypothetical trades and adjusting our portfolio accordingly based on a set of quantitative rules. When we build a back-test we need to be cognizant of what information is available to us at the time we need to make trading decisions. That is, we should never look ahead. There are certain obvious examples, but sometimes these occurrences are more subtle.

Back-testing is a useful tool for judging the validity of a quant strategy, and is a fairly unavoidable step in the alpha research process. That is, of course we want to know how our proposed strategy worked historically. But we likewise must not be over-reliant on back-test results, as they are only one realization from a vast, complex distribution. Again, we know that one of the challenges in finance is its dynamic and non-stationary nature. This means that just because something has worked in a back-test may have little bearing on it working in the future. As a result, back-testing must be only one component of an alpha research process, and should be augmented with, among other things, scenario analysis, economic justification and analysis of the underlying statistical significance. Additionally, back-tests should be used as a tool for validating a hypothesis, rather than a tool for uncovering profitable strategies. That is, we should begin with an idea to test rather than try to invent an idea that fits our back-tests, which we will no doubt be able to do regardless.

22.2.1 Parameter Estimation

In order to generate our signals, for example by calculating an expected return model, we almost always need to estimate a set of parameters. This is often accomplished via regression, machine learning or other econometric techniques.

Throughout the book, we have provided the mathematical foundation for doing this, most notably in chapters 5 and 25. We have also discussed simplified trading signals, such as momentum or mean-reversion. If we apply momentum or mean-reversion in the context of a single asset we would be considering a model of the following form:

$$r_t = \alpha + \beta r_{t-1} + \epsilon_t \tag{22.1}$$

In this case, a positive β would coincide with the presence of momentum, and a negative sign would signal mean-reversion. More generally, this calibrated coefficient β is a coefficient in our model that we use to forecast the next period's returns. When we estimate this parameter, β we need to do so in a robust manner. For example, one cardinal rule in quant trading is to not look ahead when calculating trading model parameters, otherwise results are likely to be inflated. One way to do this is to estimate the set of parameters dynamically, that is, we may want to perform the parameter estimation periodically in the back-test as we evolve time forward to take into account the dynamic nature of the underlying relationships. Each time we do so in a back-test, we would only include data up until that time in our parameter calibration, thus avoiding looking ahead. In the lieu of dynamic coefficients that are computed at set periods in the back-tests, the importance of in-sample and out-of-sample periods become more critical, and quants should be aware that the performance during the in-sample period may be overly optimistic. In this case we may calibrate a single β for the entire in-sample period, and then validate this calibrated β on the out-of-sample data. More context on setting up this framework is provided later in the chapter, however, it should be emphasized that a dynamic approach which doesn't look ahead is a more honest reflection of expected future results.

In order for our signals to work we need to have signals that are not only statistically significant but also have a good fit to future data that is not yet known. This becomes quite challenging in the presence of market regime changes, and evolving markets. It is further complicated by the fact that finance, unlike hard sciences like Physics is a place where people interact, leading to a significant psychological component. Finance also does not allow for repeatable experiments. This presents a natural trade-off between in-sample fit and minimizing the probability of overfitting. In practice, in this context it is recommended to make trading models as parsimonious as possible and only use parameterizations that have some economic justification. This means that the bar for adding complexity or a new parameter to a trading model should be quite high. Against this backdrop, it is best to focus on parameters with stable coefficients in time and across market conditions. For example, it would be a desirable feature of the model above to have stable, consistent β throughout our entire period of analysis. That said, this process is a bit of an art and different quant PMs apply different approaches depending on the frequency and markets they trade.

22.2.2 Modeling Transactions Costs

Handling of transaction costs is a critical component of a robust back-testing algorithm. Many strategies appear to be appealing until the proper transaction costs are added. This phenomenon is especially true for higher frequency strategies which by nature of their higher turnover are more sensitive to transaction costs.

Modeling transaction costs requires us to estimate the following costs:

- **Commissions**: Fixed costs that traders pay for accessing an exchange or interacting with a broker/dealer. These costs tend to be either a flat rate or dependent on the number of shares/contracts traded. As this component is fixed, and contractually known, it is the easiest to model.

- **Bid-Offer**: The bid-offer spread is the level that a security is available to be sold or bought, respectively. Current and historical information about bid-offer spreads, at least for exchange traded instruments, can be found in the order book. For OTC contracts, two-way pricing including the bid and offer is usually provided, giving us the current level. Historical information for OTC bid-offer spreads, however, is rarely available.

- **Market Impact/Slippage**: An important component of transactions costs is that, as the size of our trades grow, so too do our costs. This is largely because as our orders become too large, they can distort the market. This might manifest itself through driving up the price of an asset that we are buying, or driving down the price of something we are trying to sell, giving us a relatively worse traded level. This piece of a transactions costs is most difficult to model as it is only possible to observe if we actually place the trades. It should be noted that, for OTC transactions, banks may wrap this market impact component into the bid-offer that they quote. In doing so, they bear the risk of adverse market impact as the trade is being processed but are likely to charge a premium for doing so in the form of a widened bid-offer.

- **Borrow Costs**: If we are shorting a security, we tend to incur additional costs in our strategy due to the fact that we have to borrow that security from someone who owns it. Doing so will require us to pay interest on the borrowed shares. These fees are referred to as borrow costs. Historical data on borrow rates can be challenging to find, making this another challenging cost to model. Generally speaking, there are certain hard to borrow shares that may charge exorbitant rates and may not even be available to borrow, and others that are widely available that will charge a nominal spread over SOFR [2]. For long-short strategies, these costs must be incorporated in order to get an accurate portrayal of performance in a back-test.

In back-testing algorithms, transactions costs can be thought of as the expected difference between where the model thinks a trade can be executed and where it will actually be able to execute. Some of these costs (i.e. commissions) are fixed, and do not depend on the size of your trade relative to the overall. This makes them inherently easier to estimate. Others will depend greatly on the size of each trade, especially the market impact component. This piece is more difficult to model, however there is a substantive body of research on market impact models. These models are not presented in detail here, however, interested readers should consult [9] or [124].

The simplest approach to modeling transaction costs is to assume that each time you trade a certain percentage is paid in transaction costs. For liquid stocks and ETFs in a reasonably small account, in practice 5–10 bps can be a fairly accurate assumption, if a little conservative [3]. That is, we assume the price we can execute at can be written as:

$$P_{buy} = P_{mkt} * (1 + \omega(N, i)) \tag{22.2}$$

$$P_{sell} = P_{mkt} * (1 - \omega(N, i)) \tag{22.3}$$

where $\omega_{N,i}$ is a transactions costs function and N is the size of the trade. We may assume that $\omega_{N,i}$ is a constant, or a separately calibrated constant for each asset. More realistically, $\omega_{N,i}$ could be specified as an increasing function of the trade size, where that function incorporates the assets daily volume.

Finally, it is best practice to embed within back-tests, and trading algorithms, a cap on the shares traded, which is set based on the asset's adjusted daily volume. For example we may not want to trade more than 5% of the adjusted daily volume in a given period.

Ultimately, the transactions costs that we must pay depend on the following three underlying factors:

- **Size**: The size of our portfolio, and subsequent size of the component trades plays a key role in our transactions costs. As our size grows, we should expect transactions costs, most notably due to market impact, to rise as well.

[2] Or another risk-free rate.

[3] Depending on the size of the trade, of course.

- **Type of Instrument**: The type of instrument that we are trading also plays a key role in the transactions costs we should expect to pay. Options, for example, inherently have higher transactions costs than single stocks or futures contracts. In the case of options back-tests, 5% is a reasonable estimate for transactions costs of European, vanilla options, in stark contrast to 5bps for a stock or futures trade. If we then look to use more exotic options, we should expect our transactions costs would continue to grow.

- **Liquidity**: The relative liquidity of the asset we are trading will also play a large role in our expected transactions costs. If we are employing an ETF trading strategy that trades SPY, our transactions costs are likely to be negligible due to the liquidity of the ETF. Conversely, if we are looking to trade less liquid, specialty ETFs, we should expect to pay higher costs.

22.2.3 Evaluating Back-Test Performance

Once you have finished your rigorous research process, beginning with a piece of theory that you validated experimentally via historical data, the next step is to decide how enticing the strategy is, and whether it should be allocated capital. This involves evaluating back-tested performance, and making a judgment about the likely persistence of this performance going forward. Due to its nature, this process is somewhat objective and can rely on standard metrics of back-tested performance but also somewhat subjective. Some of the more helpful metrics that can guide us here (i.e. Sharpe & Calmar Ratio) are discussed in 22.2.5. The subjective element arises from the need to decide how likely these results are to persist out-of-sample and how reasonable our research assumptions were. We also need to be sure that we didn't fall for any common quant traps, such as looking ahead or making unrealistic execution assumptions. The most common of such traps, are detailed in the following section.

22.2.4 Most Common Quant Traps

When back-testing quant strategies there are many potential pitfalls that we must overcome. Relying solely on back-tests is itself a quant trap in many ways because we are relying on backward looking results to inform our decisions about expected forward looking results from a complex, non-stationary system prone to feedback loops and regime changes. This, of course, doesn't mean back-tests are without use or merit, but it does mean that back-tests should be augmented with additional robustness tests, scenario analyses and economic justification. Further, we need to keep firmly in mind that outcomes that we observe in a back-test are just a single realization from a broader, latent return distribution. This no doubt creates challenges but also is an essential philosophical anchor.

That said, avoiding certain traps are fully within our control and should be avoided at all costs. A few of the most common traps that quants often find themselves running into are:

- **Looking Ahead**: Using data in the back-tests prior to when it became publicly available. This could happen via an in-sample parameter calibration, or by not incorporating the proper lag to economic data that is released with a delay.
- **Overfitting**: Building a model that fits the in-sample data well but has too many parameters leading to poor performance on out-of-sample data.
- **Unrealistic execution assumptions**: Making inaccurate assumptions about transactions costs, accessibility of assets (especially for short positions) or the speed or quantity that can be traded at a given time.
- **Survivorship Bias**: Inclusion of only firms that have survived until the end of the period and neglecting to include potential long positions in firms that defaulted.

To some extent, fully avoiding some of these traps is not possible. In practice, even parsimonious models may in reality be overfit, given the lack of data we have to work with and non-stationary nature of the data.

More generally, some of these traps are easy to spot but others are more pernicious. Even within each of the items listed some examples are blatant whereas others are very subtle and hard for quants to identify. For example, all quants (hopefully) know not to look ahead in their back-tests, so barring bugs in code it is rare to find examples of back-tests that rely on future data to calibrate a model. A more nuanced version of this however is to assume that a model can use closing price information for an asset while simultaneously trading at its closing price. In a real market setting, we would need time to run our model prior to the close to generate a set of trades to be executed at the close. This means the price available to us when running our model would be prior to the close. The impact of this may be minimal in some cases, but in other higher frequency strategies could lead to more slippage or alpha decay. To handle this robustly, we would need to integrate intra-day data and observe prices prior to the close that give us time to calibrate our model and submit orders. Depending on the strategy, this precision might not be necessary, but it is still important to be aware of. One way to test this is to run the strategy with various execution or implementation lags, and examine the decay as we force a day delay between running the model and trading, for example.

Ultimately, the only way to really avoid these traps is to follow a robust research process. In some cases, something like survivorship bias may significantly inflate our back-test results (i.e. in a long-only strategy). In other cases, the risk of any bias being introduced is much less likely (i.e. in a long-short strategy). Full knowledge and analysis of these types of traps are a crucial piece of alpha research. Only the researcher, or research team will know how legitimate the results are, and a cautious, transparent and honest approach is highly recommended.

22.2.5 Common Performance Metrics

The following is a list of performance metrics that are useful in evaluating the performance for a given strategy:

- **Annualized Return & Excess Return**: For any strategy we want to know the return profile, both in absolute terms and in excess of the risk-free rate. These quantities can be calculated using the following formula:

$$\hat{\mu_i} = \frac{d}{N}\sum_j r_j \tag{22.4}$$

$$\tilde{\mu_i} = \frac{d}{N}\sum_j r_j - r_f \tag{22.5}$$

  where N is the number of periods and d is an annualization factor that translates the returns from their natural frequency to annual returns. For example, if we were starting with monthly returns, then $d = 12$. Conversely, if we started with daily returns, then we would have $d \approx 252$. In this expression, we further assume that r_f is a constant risk-free rate over the period, [4] and $\hat{\mu_i}$ is the absolute annualized average return while $\tilde{\mu_i}$ is the excess return over the risk-free rate.

  These return statistics help us to understand the attractiveness in absolute terms of the return stream of the strategy.

- **Annualized Volatility**: To get a sense of the risk-adjusted return profile for a strategy we need to introduce volatility as well as expected returns. To do so, we simply calculate the standard deviation of historical returns and then annualize then using a scalar as done above:

$$\sigma_i = \sqrt{d}\sqrt{\frac{\sum_j (r_j - \mu_i)^2}{N}} \tag{22.6}$$

  where d is again a constant annualization factor and μ_i is the mean return over the entire period calculated above. Notice that the annualization factor now appears inside a square root. To see this, recall that for a Brownian motion the variance scales linearly with time, meaning the standard deviation, or volatility scales with $\sqrt{T}$ as in the above equation. It should also be stressed that this approach for annualizing volatility is only an approximation. In particular, it will be exact under certain circumstances, such as if the underlying process is Geometric Brownian Motion, however, in cases where there is auto-correlation in the data it will become less accurate. This is closely related to the so-called variance-ratio autocorrelation test. Essentially, if the data is mean-reverting, then we observe higher volatility at higher sampling frequencies, which then tends to cancel, leading to lower volatility at lower sampling frequencies.

  Nonetheless, this type of approach is the most common way to compute annualized volatility for a strategy. It can, however, lead to the concept of different volatilities when using daily vs. monthly returns, for example. If we happen to see lower annualized volatility in the monthly returns, then it is likely due

[4]However, we could easily adjust the formula to make this time varying.

to autocorrelation in our returns, which is a helpful thing to know about. As such, risk-adjusted returns, and Sharpe ratios are often computed and compared based on daily and monthly returns as it provides an estimate of the impact of autocorrelation.

- **Sharpe Ratio**: A measure of risk-adjusted excess return where we adjust the return of a strategy by the risk-free rate, and then divide by its volatility:

  $$\text{SR} = \frac{\mathbb{E}[R] - r}{\sigma}$$

  where r is the risk-free rate and σ is the strategies expected volatility.

  It should be noted that subtracting the risk-free rate in this type of analysis is standard for long-only, fully invested strategies, as this serves as a natural opportunity cost. Long-short strategies, however, often have short positions offsetting their longs, and as a result don't require a capital outlay. In these cases, we would forego the risk-free rate in a Sharpe ratio calculation as an investor pursuing such a strategy would retain their capital and be able to invest it at the risk-free rate.

  In this text we refer to Sharpe ratios and information ratios interchangeably, however, there is a subtle difference between the two. A Sharpe ratio is a measure of risk adjusted performance for an entire portfolio, or entire investment outcome. Conversely, an information ratio focuses on only an active components performance, relative to a benchmark.

- **Market Beta**: A measure of exposure for a given strategy to the broader market. This is relevant because quant strategies tend to charge high fees, and if there returns are highly dependent on the broader market, then this is less appealing to investors who could achieve this dependence on the market for a lower fee. Market betas can be calculated using standard regression techniques using an equation of the following form:

  $$r_t = \alpha + \beta r_M + \epsilon \tag{22.7}$$

  where r_t is the return of a strategy and r_M is the return of the market. This type of beta that relies on realized returns is known as an ex-post beta. An alternative approach would be to calculate a strategy beta that relies on the more stable betas of the underlying securities by multiplying the current weights in the underlying securities times these separately calibrated individual betas. This approach is known as an ex-ante beta. We could also choose to include other variables as well, such as fixed income returns or the Fama-French factors.

  Of course, when we perform this type of ex-post analysis of a back-tested strategy, it is by definition backward looking. This creates some challenges in interpreting whether the market beta obtained was a strategic decision of the underlying strategy, or a persistent exposure that an investor could obtain more cheaply elsewhere. If the market goes up during the period of interest, maybe the strategy was able to predict this and as a result had a significant amount of exposure to the market. Conversely, perhaps in a period where the market declined the beta would look quite different.

One way to check this is to develop a time series chart of the rolling n-period market betas. If the exposure is static or consistent over time, it is less likely to occur due to timing of the market environment. If it is dynamic, however, then it is less fair to use the overall beta as a single summary characteristic.

The α in this regression is also of direct interest to market participants as it computes the ex-post alpha, or excess return not due to the market of the strategy. This alpha is then a better estimate of the portion of the return that was due skill rather than market exposure.

- **Max (peak to trough) Drawdown**: Max drawdown measures the maximum loss from the strategy's high watermark (its peak) to the ensuing trough.

$$\mathrm{MDD} = \min \frac{V_{i,t} - \max_{\tau \leq t} V_{t,t}}{\max_{\tau \leq t} V_{t,t}} \tag{22.8}$$

where $\max_{\tau \leq t} V_{i,t}$ is the maximum value of the strategy, or high watermark prior to time t, and $V_{i,t}$ is the current value of asset or strategy i at time t.

- **VaR/CVaR**: VaR and CVaR, as introduced in chapter 24 and 26, are both tail risk measures that help inform our expectations for worst case outcomes for a given strategy. A VaR of 95% for a strategy, for example, tells us the worst case loss with 95% confidence. As such, it is a relevant metric for quant PMs and managers allocated to many strategies, and may even be part of the criteria for determining how much capital is allocated across strategies.

- **Book Size/Leverage**: For long-short strategies, it is standard to report some characteristics about how much leverage is being taken on over time. It is a common characteristic of many strategies, such as those with higher frequency, that they require the use of leverage in order achieve attractive levels of absolute return. This leverage can be achieved by borrowing money from another financial institution.

Strategies that require large amounts of leverage may be more vulnerable and more dependent on strong relationships and terms with Prime Brokers. Otherwise, a quant PM may not be able to achieve the desired level of leverage, and the absolute risk/return profile of a strategy may become less compelling. For these strategies, leverage is an additional risk factor that should be monitored.

The leverage and books sizes, respectively, for a portfolio of linear assets can be calculated by summing the absolute value of the weights:

$$L_t = \sum_i w_{i,t} \tag{22.9}$$

$$B_t = \sum_i \mathrm{abs}\left(w_{i,t}\right) \tag{22.10}$$

where $w_{i,t}$ is the weight in asset i at time t and the weight in the risk-free rate, or cash is not included in the calculation of leverage, L_t. As we introduce options, especially short options positions, this calculation can be more complex.

When the leverage ratio, L_t exceeds 1, that means we must borrow capital in order to finance our strategy. Conversely, when $L_t < 1$ then we have excess capital on hand that we can lend while still running our strategy. This leverage ratio is usually tuned such that the strategy is running at a tolerable level of volatility.

Relatedly, for long/short strategies, we may want to apply some limits to the total position sizes we are allowed to take, in absolute terms. If our long and short positions offset each other, then theoretically we could place infinite weight in the long and short legs without requiring excess capital. Book size type metrics are used to place a limit on this. In practice, the leverage ratio and book size for a strategy can vary greatly by the type of strategy, and frequency that it rebalances. For long-short strategies operating in the mid to high frequency space, it is not uncommon to see book sizes of around 10–15x the original capital.

- **Turnover**: Another useful measure of a strategy is how often the underlying securities are traded, or turned over. This is a helpful measure as it can help gauge how realistic a strategy is and how punitive the transactions costs are likely to be. Roughly speaking turnover can be written as the difference in weights period over period. An average of this quantity is then expressed on an annualized basis.

- **Downside risk metrics**: While Sharpe ratio is the most commonly used summary statistic for a quant portfolio, others have been introduced that focus on downside risk instead of volatility:

  Sortino Ratio: Replaces volatility in the standard Sharpe ratio calculation with a standard deviation calculated only using downside movements. The justification is that as investors we are not concerned with volatility if it leads to profits, and instead should focus on the standard deviation of losses in our metric, as in the equation below:

$$\text{SR} = \frac{\mathbb{E}[R] - r}{\sigma_{\text{downside}}} \tag{22.11}$$

where σ_{downside} is defined as the standard deviation conditional on the return being negative.

Calmar Ratio: Replaces volatility in the Sharpe ratio calculation with the maximum, peak to trough drawdown as explained above.

$$\text{CR} = \frac{\mathbb{E}[R] - r}{\text{MDD}} \tag{22.12}$$

where MDD is the max drawdown in the period as defined in (22.8).

In aggregate, these metrics help compare performance across strategies. It is common practice for a back-testing engine to output a set of statistics, including those detailed above, for each strategy that is run through a back-test. This saves work for the quant PM as they work through their research process and try different iterations of their signals.

Importantly, none of these metrics in isolation is sufficient to determine the validity of a strategy. Quite the contrary, the entire set of metrics provide a different perspective into the risk, return or downside tail characteristics of a model.

22.2.6 Back-Tested Sharpe Ratios

Although no metric is a perfect judge of a strategy's performance, even on a backward looking basis, the gold standard for measuring the success of a quant strategy is it's sharpe ratio[5]. Sharpe ratio is, however, not without flaws. For example, it does not include the higher moments of the return distribution, such as skewness and kurtosis. This may lead to a situation where a strategy has a high, seemingly attractive sharpe ratio, but a large amount of negative skewness or kurtosis that make the strategy far less appealing in practice. Of course, in reality we want to know about the entire return distribution for our strategy, meaning any attempt to simplify it to a single number will help with interpretability but will by definition oversimplify the problem. Nonetheless, Sharpe Ratio does provide a consistent way to compare strategies while accounting for differences in volatility and expected returns. Said differently, it helps us compare the risk-adjusted return.

Another potential drawback of a sharpe ratio is that it doesn't factor in latent risk factors, such as leverage, that may lead to higher than expected vol. and large drawdowns in certain market conditions. Along these lines, an alternative metric, known as calmar ratio, is to use the strategy's maximum drawdown instead of volatility in the denominator of the calculation. This will help to incorporate the strategy's tail behavior. While it is attractive to incorporate tail behavior into the ratio, one challenge with Calmar ratios is that drawdown calculations are very sensitive to the period of observation. For example, post financial crisis a strategy may have a misleadingly low drawdown, which might be substantially different if the crisis is included. At heart this isn't surprising as estimation of tail behavior, while appealing conceptually, is also inherently noisy.

In practice, for quant strategies, legitimate sharpe ratios above 2 are considered excellent and these types of strategies are highly sought after. Alas, they are also hard to find. In some settings, sharpe ratios of 2 are a required starting point, such as many proprietary trading shops. In other contexts, these types of sharpes are viewed as impossible, with some larger players skeptical of sharpe ratios in excess of a half. The truth in reality may depend largely on the size of the institution, frequency of the strategy and market that they are operating in. Ultimately, however, this is up for the reader to discern should they pursue a career in quant trading.

One of the uses of back-tested sharpe ratios is that they provide some anchor for expected future performance. For example, if we assume that the returns of our

[5]Or information ratio.

underlying strategy are known with certainty and normally distributed, then our strategies Sharpe Ratio has some intuitive rules of thumb. If we have a Sharpe Ratio of X, the it will take an X standard deviation move for the strategy to break even. Said differently, if we have a Sharpe Ratio of X, then the probability that the strategy will lose money on a given period is the CDF evaluated at $-X$. So if our Sharpe Ratio is 1, then it takes at least a one standard deviation move for the strategy to lose money. The probability of losing money can then be found by evaluating the CDF of a normal distribution at -1, and is 16%. Clearly, this sounds like an attractive proposition, assuming the size of the wins and losses are comparable.

The following table shows the implied probability of a loss, assuming an underlying Gaussian return distribution, at various sharpe ratios:

Sharpe Ratio	Probability of Losing Money
0	50%
0.25	40%
0.5	31%
0.75	23%
1	16%
1.25	11%
1.5	7%
1.75	4%
2	2%

These approximation will be less accurate for strategies with high levels of skewness, kurtosis, or asymmetric payoff structures, however, they provide a useful baseline. As we can see in the table, a higher Sharpe leads to a lower loss probability. Further, according to these assumptions a sharpe is 2, has only a 2% chance of losing money. It is recommended that one keep these rules of thumb in mind when judging your sharpe ratios and doing sanity checks on your results. For example, given this, is a sharpe ratio of 5 is realistic, given that it implies that the strategy almost never loses money? Even in the example of a sharpe ratio of 2, this implies a 95% success rate. Do we really think our strategy is that strong? Or maybe, more pessimistically, but also more likely, we have achieved this high success rate by being willing to suffer outsized loses in that 5% of the distribution.

22.2.7 In-Sample and Out-of-Sample Analysis

When back-testing quant strategies, we almost always try many sets of parameters before finally choosing a calibrated set. This inflates our results, as we won't have this luxury when we are live trading! Additionally, we often calibrate our parameters over some fraction of the data and then run the back-test on that same period (so for this period we have looked ahead). This leads to further bias in the results. More generally, the more the data is mangled beforehand, the less likely the results are to persist. Against this context, it is important to follow as robust a research process as possible in order to have a model with any chance to succeed out-of-sample.

One defense against this is to create in-sample and out-of-sample periods, where we are able to try as many iterations as we want in-sample, but only view the out-of-sample data and results once[6]. One general rule of thumb is that you should maintain an out-of-sample period of at least one-third of the data. This out-of-sample period should then only be used after the parameters have been fully optimized in the in-sample period. The degradation of results in the out-of-sample period vs. the in-sample period is then quite informative. When we introduce machine learning techniques in chapter 28 we provide a more detailed treatment of this in-sample and out-of-sample framework.

22.2.8 Out-of-Sample Performance & Slippage

Once we have finished developing a strategy and are convinced that the risk/return profile is compelling, the next step is to conduct a true out of sample test. In this out-of-sample test we generally perform simulated or paper trading. This stage is designed to catch any look ahead or other biases or inconsistencies, and make sure that the model is implemented correctly. This is a critical step prior to allocation of capital, even if we had an out-of-sample period in our research process, as it can help identify some of the quant traps above, such as unrealistic execution assumptions, or looking ahead.

Assuming that the paper trading period goes smoothly, the next step will be to begin live trading (generally with a small amount of capital at first). Once we begin live trading, the performance of the strategy will no doubt deviate from the back-tests. This is due to many potential factors, including:

- **Unanticipated Transactions Costs**: Inaccurate and overly optimistic transaction costs assumptions can lead to poor realized performance relative to a back-test. For example, perhaps the true cost of shorting was not incorporated.
- **Slippage**: More generally, slippage could occur because of poor transactions cost assumptions, assumed availability of shares, or unappreciated market impact.
- **Regime Changes**: Even the most well-formed research project is prone to changes in market regime. The results in a back-test may be entirely legitimate, and we still may do poorly if we enter into a new phase of the business cycle, or an exogenous shock occurs. These regime changes are by nature latent, and therefore hard to definitively pinpoint.
- **Small Sample Sizes**: Out-of-sample periods are generally characterized by small sample sizes leading to relatively large standard errors in their returns[7]. This means that some out-of-sample variation is expected.

[6] Or a handful of times, if the first look isn't attractive.

[7] As we detailed in chapter 24.

- **Capital Constraints & Stop Losses**: In practice, we generally don't have control of our capital, instead it is provided to us via some institution. This creates a short option profile as our capital is more likely to be recalled should we experience a drawdown that may have been totally expected and in-line with our upfront research.

Some of these factors are beyond our control, such as if our risk budget is revoked during a natural period of underperformance. In practice it is also incredibly difficult to determine which of the above factors is the root cause. For example, when we underperform, we don't know for sure whether we have entered into a new regime, indicating maybe we should pull the strategy, or we are just in a normal period that is likely to revert. Slippage, however, is a tractable factor for us to monitor and guard against. In particular, every day that the strategy is live we should compare the strategies actual profit or loss to the profit or loss obtained in the back-test for the same day. If these are in line, it is a sign that we have modeled transactions costs and market impact accurately, and should give us confidence. If these diverge, it is a serious warning signal that we should try to explain immediately. If it is likely to be a recurring divergence, it should give us much less confidence.

Regardless of how robust our research process is, generally speaking, once we start trading a strategy we will often observe a sharpe ratio that is lower than our back-tested sharpe. The degree that it is lower will of course depend on the individual strategy, how robust the research process was, how many parameters were estimated, etc. Practitioners, however, have developed a few rules of thumb for what to expect in terms of sharpe ratio decay:

- Subtract $\frac{1}{2}$
- Divide the Sharpe Ratio by 2
- Subtract $\frac{1}{2}$ and then divide by 2

These rules incorporate the tendency for even the most robust research to overfit the data, the presence of regime shifts and non-stationarity, as well as the propensity for others to engage in similar strategies lessening their profitability. If we remember that some institutions require a sharpe ratio of 2 as a starting point, these rules of thumb might help explain why. Perhaps they are expecting an out-of-sample sharpe that follows the most punitive rule of thumb above. This would lead to a sharpe of 0.75, which is still very respectable, but significantly lower nonetheless.

22.3 COMMON STAT-ARB STRATEGIES

In this section we highlight some of the main approaches to building statistical arbitrage strategies on different assets. In this section we focus on strategies that trade the underlying asset directly, however, in the following section we then consider the relevant options counterparts. We begin with simplified strategies that look at a single price or return series for predictability, and then proceed to introduce complexity by consider multiple assets, cross-asset autocorrelation, factor models and pairs trading.

22.3.1 Single Asset Momentum & Mean-Reversion Strategies

Perhaps the natural first approach to trying to build a quant strategy would be to examine an assets price behavior and look for predictability. In doing this, we are ultimately performing a test of weak-form market efficiency. In this book, we focus on quantitatively oriented approaches for doing this, however, it should be noted that other market participants instead would apply a technical lens to finding this predictability.

In order to find predictability in a time series of an asset's returns, we are looking for stationarity in the underlying price process. Said differently, we want to know if the current return is a function of the previous sequence of returns. In a random walk, we know that this will not be the case. Each periods return is just a function of a white noise, Brownian component. In the physical measure, it is an open question whether this is the case with debate centering on market efficiency vs. behavioral or other anomalies, as discussed in chapter 5.

When looking for this type of single asset predictability we are trying to solve models of the following form:

$$r_t = \alpha + \beta r_{t-1} + \gamma r_{t-2} + \cdots + \omega r_{t-n} + \epsilon_t \tag{22.13}$$

where the parameters β, γ, and ω, among others describe the mean-reverting nature of the process, or conversely the momentum embedded in it. Further, ϵ is assumed to be white noise with zero mean. That is, we are trying to explain the return at time t, r_t with returns from periods $t-n$ to $t-1$. As we learned earlier in the text, this is a standard AR(n) model that can be estimated using standard econometric techniques. If we assume that the current return is only a function of the last periods return, at time $t-1$, then we are left with an AR(1) model of the form:

$$r_t = \alpha + \beta r_{t-1} + \epsilon_t \tag{22.14}$$

Importantly, if we are able to calibrate these models for each individual asset, and we find predictability, as evidenced by a statistically significant coefficient β, then it gives us an estimate for the next periods return r_t for each stock that can be treated as an alpha signal. A positive coefficient estimate for β corresponds to momentum for a given asset, that is positive returns are likely to be followed by positive returns. Similarly, a negative coefficient estimate corresponds to mean reversion, where a negative return is more likely to follow a positive return, and vice versa. It should also be noted that, when building this type of signal we have control over the lookback window in the previous periods return. In this case we have considered the last periods return, however this in practice is a model parameter. It could just as easily be the last 12 periods return, or the last 12 periods less the last 2 periods return. This is part of the art of building these types of signals, that is, choosing an optimal model without overfitting the data. While the above model was formulated as a linear model, we could easily try to replace this with a non-linear model or machine learning based approach. Unsurprisingly, however, in general it is hard to find profitable strategies using solely these techniques. As a result, this has led quants to try to incorporate information about multiple assets in the forecasts, as we examine in the next section.

22.3.2 Cross Asset Autocorrelation Strategies

If we are unable to find predictable behavior within a single asset, a natural next step would be to introduce information about other assets to try to improve our prediction. This type of cross-asset auto-correlation strategy can be used to incorporate lead-lag dynamics such as those found by Lo and MacKinlay [167]. For example, one might postulate that large cap stocks lead small cap stocks, or that US stocks lead Emerging market stocks. Whether or not this is true is an open question, however if these premises are true then cross-asset autocorrelation strategies could be used to profit from them. Like the single asset momentum and mean-reversion strategies we looked at in the last section, tests of cross asset auto-correlation strategies are inherently tests of weak-form market efficiency.

In a cross-asset autocorrelation model we would consider something of the following form:

$$r_t^i = \alpha + \beta_1 r_{t-1}^i + \beta_2 r_{t-1}^j + \epsilon_t \tag{22.15}$$

where again ϵ_t is a white noise term and β_1 represents the familiar single asset momentum/mean-reversion phenomenon discussed in the previous section. The new coefficient that appears, β_2 represents the return in stock i that is dependent on the previous return of stock j. Taking the example of a lead-lag relationship in small vs. large caps, we might expect a positive β_2 in the small cap regression:

$$r_t^{\text{small}} = \alpha + \beta_1 r_{t-1}^{\text{small}} + \beta_2 r_{t-1}^{\text{large}} + \epsilon_t \tag{22.16}$$

When building these types of models, we can choose to separate the cross asset components, from the single asset momentum/mean-reversion components, or instead choose to put them into a single integrated VAR model with a richer, less parsimonious parameter set. If an integrated model is chosen, then this type of model can be generalized and aggregated into matrix form to construct a VAR[8] model. Just as we saw in the last section, this type of model provides a return forecast for each constituent asset, which can then be used to generate an alpha signal.

22.3.3 Pairs Trading

If we cannot find auto-correlation in a single stock, or a lead-lag relationship between two stocks, a natural next place to look is at a pair of closely related stocks. This is the idea behind pairs trading. A stock's price process might not be stationary on its own, but a linear combination of two stocks might be stationary. If this is true, we say that those two stocks are cointegrated. There are multiple ways to implement pairs trading strategies, and to test for cointegration of two assets. In this text we provide an introduction into the concepts and techniques. Interested readers should consult [60] or [256] for further details.

Pairs trading conceptually relies on the idea that correlated or related firms should be characterized by similar performance, and deviations from this are likely to revert,

[8]Vector Autoregression.

creating a potential trading opportunity. For example, if we consider the commonly cited case of Pepsi and Coca Cola, their performance is clearly likely to be driven by similar global factors. If we were to observe Pepsi outperforming Coke by a significant margin, it might lead us to want to buy shares in Coke and sell shares in Pepsi. Pairs trading provides a quantitative framework for this approach. The most common implementation of a pairs trading algorithm is to combine a long position in an asset whose relative performance is anomalously poor, and a short position in a highly correlated asset whose performance has been superior. The long-short nature of the trade eliminates some of the exposure to overarching factors, such as the market, that would be present if we were to focus only on the long or short leg in isolation.

The first step in building a pairs trading algorithm is to identify a cointegrating relationship, or set of cointegrating relationships that we want to take advantage of. A handful of different approaches have been proposed to accomplish this, such as the Engle-Granger two step method [92] and the Johansen test [151]. In the Engle-Granger method, we take a two-step approach, the first step of which is use standard regression techniques to find the optimal hedge ratio between two assets, x and y. We do this via a regression of the following form:

$$y_t = \beta x_t + \epsilon_t \tag{22.17}$$

where y_t and x_t are the price processes at time t for x and y, respectively. The fact that price processes are used here should be emphasized, as it is a deviation from many of the factor model constructions that we have seen earlier in the book. Fundamentally, the hedge ratio is chosen to minimize the idiosyncratic, spread component and maximize the part that is explained by the overarching relationship between the two assets.

Once the proper hedge ratio has been identified, we can then test the residual, or spread, ϵ_t for auto-correlation. Conditional on a given hedge ratio, this can be accomplished using standard stationarity test, such as an augmented Dickey-Fuller test. Calculation of the optimal hedge ratio, however is an important step, as a pair might not be stationary if we use the "wrong" hedge ratio, but still very well could be with the proper hedge ratio. Importantly, pairs trading tells us that even though the individual price processes for x and y are non-stationary, that if we combine them in the right ratio, β, they are collectively stationary. This stationarity can then be exploited by taking positions when the spread, ϵ_t becomes large.

Once we have identified a cointegrating relationship, we can trade it by monitoring the spread between the price processes and entering trades when the deviation exceeds a given tolerance. A common approach, but by no means the only methodology, is to make a trading rule that is based on the rolling z-score of the underlying spread, ϵ_t, which would lead to this type of rule:

- $z > z_{\text{threshold}}$: Enter a position that is short the spread, ϵ by selling one share of asset y and buying β shares of asset x.
- $z < -z_{\text{threshold}}$: Enter a long position in the spread, ϵ that is long one share of asset y and is short β shares of the asset x.

Here, our trade entry criteria is defined by our z-score exceeding some specified threshold. Generally, trades are only put on when this barrier is initially crossed. That is, we usually observe that the z-score has cross the higher or lower threshold and then put the appropriate spread trade on. Should the spread remain above or below the threshold in subsequent days, we would not increase our position until the spread normalizes to below the threshold.

In practice, many other considerations might go into a pairs trading strategy. For example, we need to account for the fact that most of the time our threshold criteria for each pair will not be met, leading to a cash position most of the time, at least for each pair. If we then aggregate this across pairs, the number of invested in pairs can vary greatly. Sometimes there may be none that meet the threshold, in other cases there may be many. This creates challenges figuring out how to weight the pairs in a broader pairs trading portfolio. Cointegrating relationships themselves are also quite complex and can be difficult to uncover. Even for a pair that we believe to be cointegrated, we may observe varied levels of statistical significance in our cointegration tests. Sometimes it may be stationary, sometimes it may not be. We may also observe the hedge ratio changes over time, and would need to reconcile this in our model.

We also need to incorporate additional trading rules in a pairs trading algorithm. For example, we need to develop an exit criteria, where we determine when we exit the trade after the spread has converged. This is often done using a z-score approach just we defined for our trade entry criteria. For example, assuming we entered the trades at ± 2, we could then define a rule to exit trades once the z-score has reverted to ± 1. The choice of this exit z-score is often determined by analyzing the half-life of the spread. Additionally, many traders overlay a stop loss criteria to ensure losses are contained should a pairs cointegrating relationship break down. There are many approaches to such a stop loss, with the simplest being to define a new z-score threshold, $z_{\text{stop-loss}}$.

22.3.4 Pairs Trading in Practice: Gold vs. Gold Miners

In this example we apply the previously developed pairs trading algorithm to a commonly cited pairs trade: Gold vs. Gold Miners. These two entities are intimately connected, and as such are a natural place for us to apply our algorithm. Others, such as Chan [61] have proposed this as a cointegrated pair. Testing this over the period of 2010–2020 we do not find evidence that the pair is stationary using an Augmented Dickey Fuller test, which is clearly a warning signal. In spite of this, in the following charts we show the evolution of the calibrated spread as well as the profit and loss that would emanate from following a pairs trading strategy on this pair with a z-score threshold of ± 2:

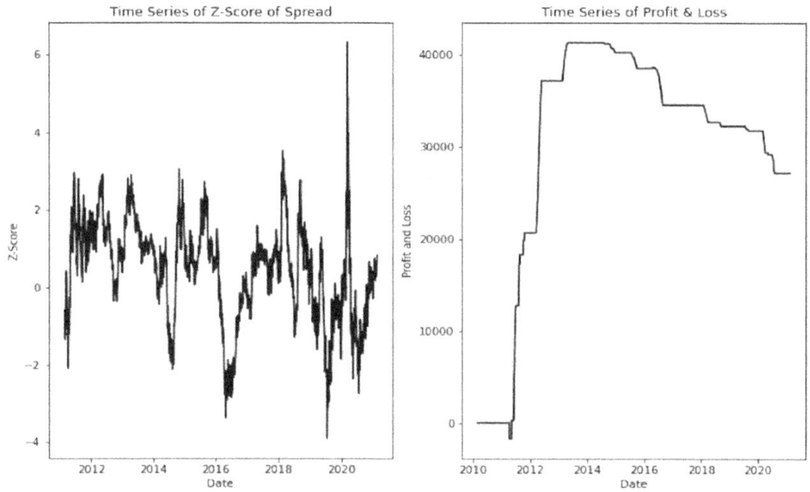

Interestingly, we can see that, in spite of the lack of evidence of stationarity in the pair, we observe a fairly compelling profit and loss profile. The reader should also notice that the many periods of flat performance are due to the fact that the strategy is uninvested at many historical points. It should also be emphasized that the calibration of the hedge ratio was done in sample, using the entire period. This may be a partial reason for the seemingly compelling results. The reader is encouraged to try to extend this example to a more realistic trading strategy.

22.3.5 Factor Models

More generally, we can use factor models that include previous return terms as well as other, non-price related signals in our strategies, and alpha models. When doing this, we can test not only weak-form market efficiency but also its semi-strong form. That is, that news and other publicly available information are incorporated into market pricing instantaneously, meaning investors are unable to profit from strategies based on news and fundamental data.

Factor models, as we have seen throughout the book, tend to have the following form:

$$r_t^i = \alpha + \beta_1 f_{t-1}^1 + \beta_2 f_{t-1}^2 + \epsilon_t \tag{22.18}$$

where f_{t-1}^1 and f_{t-1}^2 are the factor values at the beginning of the period. We could define a factor based on valuation, for example price to earnings ratio. Similarly, as we saw in the last section, we could use previous asset returns, or a portfolio returns as factors as well. For example, we might use the previous market return as a factor. Importantly, it should be emphasized that this regression is formulated as a forecast. That is, we are trying to forecast the next period's return conditional on the current values of various quantities, whether they be news signals, fundamental data based, or price based. Of course, the success or failure of this approach will largely depend on whether we choose the right factors, and may be sensitive to how exactly we define each factor. It should also be noted that as we are trying to forecast, we should expect low levels of explanatory power in our models, meaning it will require some art to build a model that is robust and performs well out of sample.

An alternative is to build a contemporaneous regression model rather than a predictive one, as shown below:

$$r_t^i = \alpha + \beta_1 f_t^1 + \beta_2 f_t^2 + \epsilon_t^i \tag{22.19}$$

where again f_t^1 and f_t^2 are two pre-defined sets of factors. The interpretation of this regression, and the regression coefficients β_1 and β_2 are fundamentally different than in the previous predictive regression in (22.18). Because this regression is contemporaneous, the β's simply tell us the exposure of a given asset to a factor. This means that, at first glance they don't help us predict the next period's return for the asset. However, there are two things that we can do to glean insight from this type of model.

The first is that we could examine the residuals, ϵ_t^i for auto-correlation. If the residuals are mean-reverting, for example, then we can build a signal based on these residuals, and it will inform our return estimates for each asset in the next period.

To test this, we can model the residuals using an ARMA process, as we did before for the asset returns. For example, if we assume that the residuals follow an AR(1) process, we would have something of the form:

$$\epsilon_t^i = \beta^i \epsilon_{t-1}^i + z_{t-1} \tag{22.20}$$

where z_{t-1} is a white noise term with zero mean[9].

Intuitively, assuming that the residuals are mean-reverting, or β^i in (22.20) is negative, this would mean that we would want to buy the assets with the lowest residuals, and sell those with the highest.

This type of model can be conceptually motivated by thinking about the factors as defining the systemic risks for the asset and ϵ as its idiosyncratic risk [27]. In this model, we assume that the structure of these two types are risks are inherently different. The first are broad risk exposures common to all assets, or companies. As such, these risks are likely to be persistent and efficiently priced. The second are asset or company specific and tend to be transitory, indicating that large, idiosyncratic deviations may be more likely to revert. Again, this is the same model that we worked with previously in the single asset momentum & mean-reversion models, however, now we are working with residuals instead of asset returns. The idea behind this approach then is that this transformation helps any existing autocorrelation become more visible because of the distinct nature of the types of risks.

A second approach to leveraging (22.19) to try to build a strategy is to examine if the factors themselves have any auto-correlation. If they do, then we may be able to forecast the factor value, or values, in the next period, which could be used as part of a model for the next period's asset returns. This would mean modeling the

[9]But strictly speaking does not have to be normally distributed for a mean-reverting strategy to work.

factor using an ARMA model, or an alternative time series technique, such as in the equation below:

$$f_t^i = \alpha + \beta^i f_{t-1}^i + \epsilon_t \tag{22.21}$$

Importantly, if we are able to forecast the factors, then we can build a portfolio with high exposure to the factor when it is likely to increase, and vice versa, providing an alternate approach for predicting the asset's return in the next period.

To summarize, we can apply three distinct approaches to building factor models to inform our trading models:

- Predictive Regression
- Model Auto-Correlation of Residuals in a Contemporaneous Regression
- Model Auto-Correlation of Factors and use this to build Signal

Essentially this comes down to which component of a return we want to forecast. Do we want to forecast the entire return, or do we want to isolate a particular component? Regardless of our choice, one defining characteristic of this type of model is that it will be highly sensitive to our choice of factor set. Avelleneda [27], in particular, focused on modeling the residuals, and defined a set of sector ETFs as the factor universe. Another choice would be to use the Fama-French factors and test the residuals and the factors themselves for mean-reversion.

Of course, regression is also not the only manner in which we can construct a factor model, however, the concepts will remain the same if we choose to replace the above regression approach with, say, machine learning[10].

22.3.6 PCA-Based Strategies

In the last section we introduced trading strategies based on factor models and showed three different approaches for building a signal from a given factor model. One defining feature of all of the factor models, however, was that we chose to name the factors. An alternative approach would be to choose these factor statistically, using, for example, PCA techniques [27]. In doing this, we leverage the techniques discussed in chapter 5 to create a set of factors. These factors have some convenient properties, in particular they are orthogonal.

Recall that principal component analysis relies on the following matrix decomposition of the covariance or correlation matrix:

$$C = VDV^T \tag{22.22}$$

where C is the correlation matrix, or covariance matrix, of historical returns, D is a diagonal matrix of eigenvalues and V is a matrix of eigenvectors. Avelleneda proposed that we work with standardized returns, [27], however this is not required. More generally, performing PCA on the correlation matrix vs. covariance matrix helps to

[10] See chapter 28 for examples of this.

normalize each asset, in particular for differences in their levels of volatility. If we use a covariance matrix, conversely, we may find that the more volatile instruments dominate the embedded signal.

If we arrange the diagonal elements in the D matrix in descending order, we know from theory that they top eigenvalue, λ_1 explains the most variance, followed by λ_2, and so on. We further know that the eigenvalues, λ_i tell us precisely how much variance is explained by each principal component.

In practice, we tend to observe that the first principal component, corresponding to λ_1 explains a significant amount of the variation. It also tends to have a consistent structure and natural interpretation. That is, it tends to define the market factor, or risk-on/risk-off element in markets. This observation is fairly robust to the universe of asset classes that we include in our PCA. The remaining PCs, unfortunately, can then be harder to interpret, and more dependent on the universe and details of the problem.

When constructing a PCA-based trading strategy, our goal will be to build a factor model based on statistically chosen factors. These factors will be the principal components defined by the largest eigenvalues. We will then test to see if the residuals, once the top eigenvalues have been removed, are mean-reverting. If so, this can be the basis for a trading strategy analogous to the one we detailed in the last section, with the defining difference being how we built the factors.

A caveat here is that in doing this we have to choose how many PCs to include, and in theory this should depend on the structure of the eigvenvalues. Essentially in this approach we hypothesize that the larger principal components represent the systemic market risks, and the smaller PCs instead represent transitory, idiosyncratic risks that are more likely to mean-revert. The choice of which PCs are systemic vs. idiosyncratic however, is not clear and is in practice a bit of an art form. Generally speaking there are two approaches. First, we could choose enough principal components to explain a given level of variation, say 95%. This would lead to a dynamic number of PCs being used over time. Alternatively, we could instead fix the number of PCs to a reasonable number. This would lead to deviations in how much of the variance our set of PCs explain over time, but would have the benefit of a static number of PCs.

Importantly, as we saw in chapter 5, the matrix V is comprised of a set of eigenvectors. Viewed differently, the elements of each eigenvector are the weights, or loadings on the asset universe:

$$\nu^j = \left(\nu_1^j, \ldots, \nu_n^j\right) \tag{22.23}$$

where n is the number of assets in our correlation matrix and PCA model.

So, the eigenvectors are a linear combination of these underlying assets, or in the other words, they themselves are portfolios. As such, the rows in V are sometimes referred to as eigenportfolios. We can then use the eigenportfolios that correspond to the largest eigenvalues as factors in a factor model. At any time, the return of an eigenportfolio can be written in terms of the underlying asset returns as:

$$f_t^j = \sum_i \frac{\nu_i^j}{\sigma_i} R_{i,t} \tag{22.24}$$

where the σ_i is the volatility of the ith asset and appears in the denominator appears because we are choosing to work with normalized returns. It should be emphasized again that this normalization, and the subsequent σ_i in the denominator is not required. It is, however, recommended by Avelleneda and helps account for differing levels of volatility in the underlying asset universe. Said differently, it prevents the factors from being disproportionately driven by a few higher volatility assets. Similarly, ν_i^j is the loading of asset i in the jth eigenportfolio, $R_{i,t}$ is the return of asset i at time t and f_t^j is the return of the factor defined by the jth eigenportfolio at time t. These factor return streams can then be aggregated across time and across the different eigenportfolios and can then be used in typical factor models such as (22.19).

One of the main challenges associated with this approach, and the use of PCA in many financial applications is the difficulty in interpreting the underlying principal components. As mentioned, the first principal component is unique in that it persistently presents as a risk-on/risk-off factor. Unfortunately, other PCs tend to be harder to interpret economically, and, perhaps more problematically may be unstable over time. This instability over time of the PCs creates significant challenges in our model. If the factors themselves are unstable, how can we be confident that the residuals from this factor model display mean-reverting properties? In practice, this means that a great deal of effort needs to be done to understand the underlying PC structure and make sure that the framework leads to stable results.

An astute reader may have noticed that the creation of the factors via PCA is disconnected from the subsequent estimation of residuals and exposures to those generated factors. This is true and means that in practice the universe of assets used to generate the factors does not need to coincide with the universe that we are trading. Instead, we could use a totally different set of variables to generate the factors, and these variables aren't even required to be tradable quantities. The (somewhat elusive) goal when doing this, would be to choose a set of assets in the PC-based factor generation process that most effectively isolates the set of systemic and idiosyncratic risks, such that the remaining idosyncratic risks are transitory and (hopefully) mean-reverting. Lastly, although we focused in this section on a residual approach to building a PCA strategy, we could similarly use any autocorrelation in the principal components themselves to inform a PCA-based strategy.

22.3.7 PCA Decomposition in Practice: How Many Principal Components Explain the S&P 500?

In this section we leverage the code in the repository to perform a principal component analysis of the components in the S&P 500. To do this, we use the correlation matrix of daily returns and then perform eigenvalue decomposition as is shown in the above coding example. Of particular relevance in this particular example is how many distinct factors we observe in movements of the members of the S&P 500. Are there really 500 underlying factors, or is there a smaller more parsimonious set of dominant factors that drive returns for all constituents?

In the following charts, we answer this by first analyzing the structure of the principal components, as shown in the following chart:

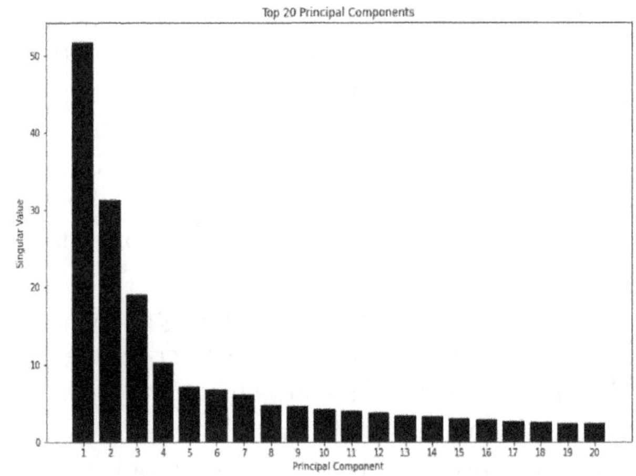

In this chart we can see that there are a few larger, dominant principal components, and that they gradually decline such that they become very close to zero. This would be a strong indication that the members of the S&P 500 are driven by a smaller underlying set of factors. To view this another way, we plot the cumulative variance explained by the top N principal components:

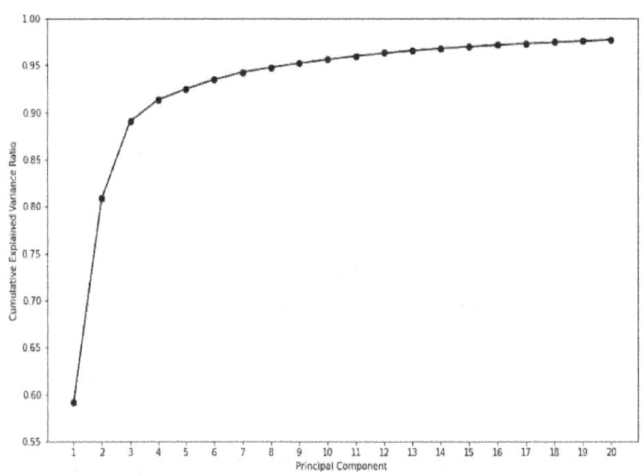

Here, we can see that the first twenty principal components explain more than 95% of the variance. This is another indication that the true set of orthogonal factors driving the S&P components may be far less than 500. It also implies that, when working with these components, dimensionality reduction techniques may be appropriate to isolate the significant principal components.

22.3.8 Risk Premia Strategies

Another approach to building a quant strategy is to leverage the concept of risk premia and attempt to identify a well-rewarded set of risk premia that can then be harvested systematically. This is a common approach in equity markets, where risk factors like carry, momentum and value have been largely commoditized. In other asset classes, such as fixed income and derivatives this approach is less common, but beginning to gain traction.

The idea behind a risk premia strategy is that we find a risk that is compensated over time, and develop a strategy with isolated, persistent exposure to this risk. This premia may be caused by structural dislocations in markets, psychological or behavioral factors, or may simply be caused by preferences for certain types of payoffs, or an unwillingness to bear certain tail risks. Of course, the trick behind this approach is finding the right premia that will be rewarded in the future. Additionally, a significant amount of dispersion in the investment outcome for these strategies can emanate from precisely how the underlying different metrics are defined. For example, in equity markets there are many ways to potentially define value ranging from a metric based on earnings to a metric based on a long term reversal of returns[11]. In the following list, we detail some of the most popular risk premia and provide a simplified, basic framework for how each might be defined:

- **Momentum**: Relies on the simple premise that assets whose recent performance has been strong tend to continue to have strong performance. One potential justification for momentum as a risk premia is the potential for investors to underreact to new information, perhaps because their expectations remain anchored to a prior view. Another potential explanation is a so-called disposition effect where investors are adverse to unwinding losing trades, but quick to try to lock in profits on winning trades, creating an underreaction phenomenon. Regardless of the theoretical justification, it is undeniable that a significant body of empirical evidence exists showing that it is persistent, and further has some appealing payoff characteristics[12].

  Momentum indicators are linked to a previous return for the asset. A standard momentum signal in equities is the previous twelve-month return, less the last one month's return. The last month's return is removed due to its strong reversal characteristics. AQR, for example, has proposed using this twelve month less last one month's return [66]. We could however, more generally choose another lookback window. This lookback period is a model parameter in our momentum signal. Traditionally, momentum models use the previous return directly in their signal, however, this is a specification of the following model:

  $$r_T = \beta r_{\tau,t} + \epsilon_t \tag{22.25}$$

  where $\mathbb{E}\left[\epsilon_t\right] = 0$. Of course, when we use the previous return directly, we are simply assuming that β in the above model is equal to one. More generally, an alternate approach would be to use econometric techniques, such as those detailed in chapters 5 and 25, to estimate β. This is then a specific application of the single stock autocorrelation strategy framework discovered earlier in the chapter.

[11] And, in fact there are upwards of 50 different value definitions in the equity markets today.

[12] Most notably momentum strategies tend to have convex payoffs and positive skewness.

- **Carry**: Carry is another standard risk premia that is based on the concept that securities with high levels of carry tend to outperform their lower carry counterparts. In chapter 19 we discussed an interest rate based example of this, and carry is a premia that is powerful in fixed income and derivatives markets. Within the context of options, carry can be loosely thought of as the profit or loss solely due to the passage of time, or theta. Similarly, in fixed income and credit markets securities with the highest income, in the form of a coupon, and rolldown, have the highest carry, and also tend to have the highest subsequent returns. Moskowitz et al. [214] proposed a methodology for defining carry across asset classes. In equities, the standard measures of carry are based on income in the form of dividend yield.

- **Value**: Another canonical risk premia is the value premium, which states that undervalued and overvalued assets tend to revert to their mean over long horizons. Many market players attempt to define value in terms of a fundamental valuation. For example, if equities it is natural to view value through the lens of something like P/E or P/B ratio. Alternatively, it has been suggested that value can be stated as a longer term reversal factor, using for example the previous 5-year return. This is the approach employed in [66].

  The following highlights a few of the many potential ways to construct a value signal:

  - **Long term Reversal (e.g. 5-year return)**:
  - **Book to Market**
  - **Price to Earnings**

  As was true in the other risk premia, a value signal can be created by simply using the relevant value metric directly to create a factor portfolio, or econometric techniques to estimate the corresponding value coefficients.

- **Volatility**: Another commonly harvested risk premia is a so-called volatility risk premia. More precisely, volatility is a negative risk premium, leading to a premium for the absence, or lack of volatility. This is a premia that exists in the options market, as we explore later. In the options market, the explanation for this type of premium is quite intuitive. Investors prefer convex payoffs and long volatility structures, and are typically willing to pay a premium to obtain them. Additionally, it is premia that may exist in the underlying assets as well. The underlying phenomenon is that lower volatility assets tend to achieve higher risk adjusted returns than higher volatility assets. A natural explanation for this potential premia is leverage aversion. Investors, rightfully prefer not to take on leverage, and in some cases are actively prevented from doing so in certain account types. As a result, it is conceivable that investors would be paid a premium for be willing to leverage lower volatility assets to obtain attractive absolute levels of return.

- **Size**: In the set of Fama-French factors, size was identified as one of the underlying risk factors, and they in particular observed that small firms tend to achieve higher returns than large firms, which they attributed to an underlying premia. That is, small firms may contain larger inherent risks, and therefore investors may require compensation for investing in them relative to their larger counterparts. Size premia is closely connected to a liquidity premia, as smaller firms tend to be associated with lower levels of liquidity, and as a result it is conceivable that at heart this is really a specification application of a liquidity premia. This size factor in equity markets is typically measured by a firm's market capitalization.

These are by no means the only types of risk premia strategies, for example other could be based on quality or liquidity signals, however, the ones detailed in this chapter are some of the most intuitive, common and the ones that tend to extend across asset classes.

22.3.9 Momentum in Practice: Country ETFs

In this section we leverage the code in the repository to apply a simple momentum signal to a list of country ETFs. These country ETFs are by no means the only universe we could apply such a strategy but are convenient in that they have similarly named tickers beginning with EW. The Japanese country equity ETF, for example, is defined by the ticker EWJ. The Canadian country equity ETF, conversely is named EWC.

Using this universe of country ETFs, we develop a strategy that is long the single, highest momentum asset and short the single lowest momentum asset. This leads to the following cumulative profit and loss chart for the strategy from 2014–2021:

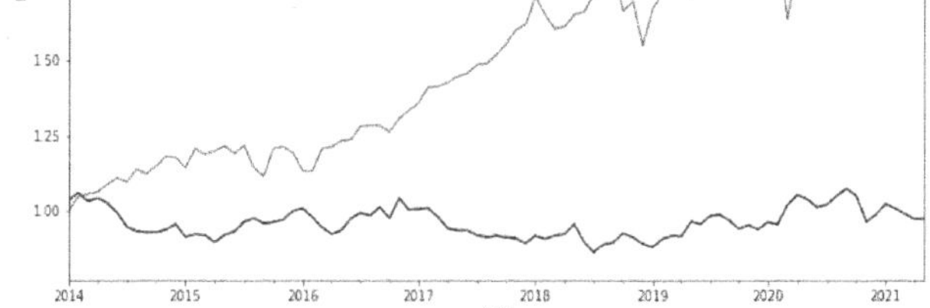

As we can see, this simple momentum strategy is characterized by cumulative returns that are near zero over the entire period. We can also see that the returns from this momentum strategy are significantly lower than a buy and hold investment in the S&P 500. As this is a long-short strategy, the correlation between the momentum strategy and the S&P 500 is likely to be low, meaning that, even though the returns are significantly lower than for the S&P, they may still be compelling to investors. Choice of the proper benchmark, or reference, against which we judge a strategy's returns in these types of situations, can be quite challenging in practice. However in this case, we can pretty clearly see that the return profile is not particularly enticing. The reader is encouraged to think about how this could potentially be extended to a more appealing strategy by analyzing different variations and modifying the number of long short/positions, the lookback window and the asset universes.

22.3.10 Translating Raw Signals to Positions

In this section we have covered many simplified approaches to generating alpha signals, from those based on relative value models (i.e. pairs trading) to those based on risk premia. Once we have created our signals, an important subsequent step is to turn them into a portfolio. This can be broken down into two levels. In the first level, we take a single signal, and translate it into a single set of weights, or position sizes for the signal. That would essentially give us an isolated trading strategy focusing on just that signal. In practice, we are likely to do this for many signals, hoping to benefit from diversification of these signals. This creates a second level, where we weight the individual signals into an overarching portfolio. This second level is discussed in further detail in 22.5. In this section we focus on that first level that is taking a set of raw signals, and turning them into an investible portfolio. There are of course many ways to do this, and depending on the strategy different approaches may be optimal. Experience and intuition are fundamental to choosing a robust approach, however, in this chapter we describe the main considerations and a few of the most common translation types.

If we think of this in the context of a single-asset momentum strategy, then we would have as input the assets with the strongest and weakest momentum. Assuming we are pursuing a long-short strategy, we would still need to decide which assets to be long, of the highest momentum assets, and which of the lowest momentum assets to be short. Further, we need to decide how big each long and short position should be. Should all long positions in the highest momentum stocks be equally weighted? Or, conversely, should we be aware of the assets volatilities or differing levels of momentum. Similarly, for a pairs trading strategy, the weights within each pair are well-defined such that the pair is stationary. However, we need to decide how to weight between the multiple potential pairs we would like to invest in. In this case this process is further complicated by the fact that the number of pairs we are investing in may vary greatly over time leading to a large cash balance or forced re-balances as new pairs are introduced, depending on how this is handled.

In the following list, we detail a few approaches for translating from a set of raw signals to a set of position sizes. It should be noted that the raw signals themselves may come in different forms. In some cases, they may naturally be forecasts of expected returns. In other cases they may be z-scores, percentiles or other metrics.

- **Top N Equally Weighted**: In this approach we rank the signal from highest to lowest and place equally weighted long positions in the top N ranked signals and, if it is a long short strategy, short positions of the same equal weight in the bottom N ranked signals.

- **Top N Inverse Volatility Weighted**: In this approach we again use a ranking to select the assets we want to be long and short, however, in this case instead of using an equal weighting of these we apply an inverse volatility weighting as described in 19.9.2.

- **Map Signals to Expected Returns & Use Portfolio Optimization Techniques**: The ranking based approaches above benefit from simplicity and a lack of assumptions about expected returns. In the presence of estimation error, this agnostic view of expected returns is naturally appealing, however, in some cases we may want to integrate expected returns in our calculation of weights or position sizes. To do this, we would first need to translate our signal to expected returns. In some cases this may be easy, as our signal may be built on an expected return forecast. In other cases, it will require more thought. Once we have done this, we can create a portfolio using our signal using the portfolio optimization techniques discussed in chapter 24. For example, we could use mean-variance optimization with the expected returns generated from our signal. One drawback of this approach is that our underlying signals tend to have a high noise to signal ratio, making an expected return calculation inherently noisy. On the other hand, using this approach enables us to incorporate conviction and avoid a somewhat arbitrary cutoff of assets that warranted inclusion in our portfolio.

- **Multi-Factor Expected Return Models incorporating multiple Signals**: Another approach, which combines the two levels of turning our signals into a portfolio would be to create an expected return model using multiple underlying input signals, and use these overall expected returns in one of the methodologies above to generate a single portfolio.

  For example, if we consider a set of raw signals (s^1, s^2, s^3) then we can formulate expected returns of every asset as a function of the signals using an equation of the following form:

$$r_{i,t+1} = \alpha + \beta_1 s^1_{i,t} + \beta_2 s^2_{i,t} + \beta_3 s^3_{i,t} \tag{22.26}$$

  In the previous iterations, we have attempted to build separate portfolios for s^1, s^2 and s^3 and then use some sort of portfolio optimization technique to aggregate each signal portfolio. Here, in contrast, we build a portfolio by first generating an aggregate expected return, $r_{i,t+1}$ for each asset i and then generate a portfolio using these aggregate expected returns. Using this type of integrated approach gives us more control over the signal weights, defined here by the β's, and enables them to be dynamic.

22.4 SYSTEMATIC OPTIONS BASED STRATEGIES

22.4.1 Back-Testing Strategies Using Options

Back-testing options strategies requires an additional level of complexity because we need additional info to value an option. In the case of a stock model, we simply need to monitor the evolution of that stock's price. However, as we have seen, valuation of an option requires additional information. Notably, if we use the Black-Scholes model, it requires an implied volatility. Even more challenging, this implied volatility is likely to vary as the option's expiry and strike move. In practice, this means that in addition to asset prices, we need to incorporate changes to forward/future curves and volatility surfaces. This means that in an options back-test we need to embed a model for the volatility surface within it, and we need to do so carefully without introducing arbitrage. This can be done using either of the methods described below:

- **Interpolation of Implied Volatility**: Rather than modeling the volatility surface we may choose to simply perform some sort of interpolation between the liquidly traded option strikes and expiries. This must be done carefully, however, to avoid introducing arbitrage, as was discussed in chapter 17. For example, we might have options data quoted at a set of delta points, and may use these to create an interpolated vol for our desired traded strike.
- **Calibration of Stochastic Model**: Another option would be to leverage the calibration techniques that we learned in chapter 14 to fit a model to the volatility surface given the options data provided. We could choose to use a stochastic volatility model, a jump process, or a local volatility model. This approach has the added benefit that it will not permit arbitrage, as the underlying models are arbitrage free. However, these models are likely not to fit the entire surface perfectly leading to error even at the quoted points.

Options also have fixed expiries so monitoring them historically is more challenging because there is no single ticker with a time series for the entire historical period. This means that in our back-tests we will need a roll strategy for our options. Perhaps we will choose to hold them until expiry. Alternatively, we may choose to roll them into new options as they approach expiry or become sufficiently in or out-of-the-money. This creates another set of rules that we need to include in our back-testing algorithm.

It is important to remember that we are trying to simulate the behavior of engaging in this strategy historically, so we need to simulate buying the option at the price that it would have traded at and selling the option at the price that it would have traded at the time of the sale. We should not, then, lock today's option pricing, or volatility and forward curve and only simulate changes to the asset price. While this may be useful analysis, it is not a back-test. Instead we need historical data on the slope of the futures curve and the entire volatility surface. This makes volatility strategy back-tests less precise because there is less certainty about the true market implied volatility that we would be able to trade at. This is further compounded by the larger bid-offer spreads on options relative to stocks, ETFs and liquid futures. That said, this lack of precision can easily be incorporated by increasing the embedded transactions cost assumptions.

22.4.2 Common Options Trading Strategies

In this section we detail simplified versions of some of the most commonly implemented systematic options strategies. Our goal in this section is simply to introduce the concepts behind these strategies and provide some intuition on how they work. There are many subtleties of these systematic options strategies that are omitted in this text, however, we provide the basic framework and leave it to the reader to suite their interests and applications [189].

- **Put Writing Strategies**: A simple systematic options risk premia strategy is to sell put options on the basis that they are likely overpriced due to their value as a hedging instrument. This is especially common in markets like equities, where puts define a clear insurance against a risk-off event. In some cases, a put writing strategy may literally consist of a sold put option that is re-balanced periodically. The payoff of each sold put option in this strategy will be as follows:

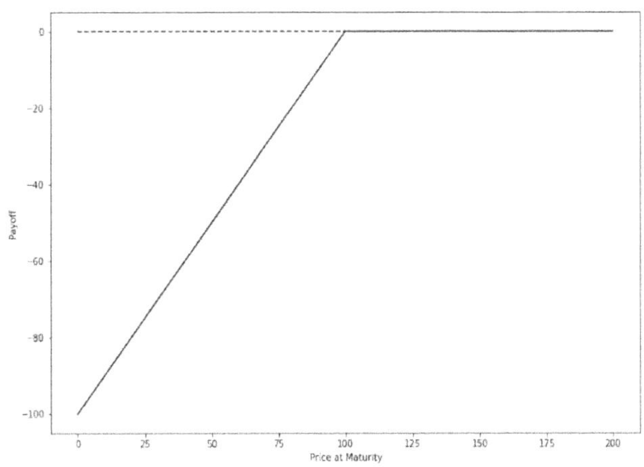

  Clearly, we can see from the payoff diagram that the payoffs are asymmetrically against us. This is by design and part of what we are receiving a premium for. This premium is then heightened by the fact that payoffs in states where put options pay off tend to be valued more than payoffs in other states. To take the equity example, investors universally value payoffs in scenarios characterized by large equity selloffs, because these periods are likely to be correlated with potential losses in most investment losses, and potential job losses for investors. A re-balancing scheme may be needed after the passage of time, as options approach expiry, or may be triggered based on a move based threshold.

A sequence of put options, usually with a consistent delta or moneyness define a put-writing strategy. For example, a simplified put-writing strategy might be to sell at-the-money, or 25 delta one-month S&P put options every day and hold them to expiry. Alternatively, we may choose to sell a single at-the-money put option and choose to re-balance only when we approach expiry, or if the underlying has moved by a given threshold.

Put-writing strategies are an effective way to harvest a volatility premium present in the markets and to harvest overpriced tail hedges. As such, it is a simple yet fairly attractive strategy. However, one feature that is particularly unappealing about these strategies is that the downside is undefined in some cases and in all cases is disproportionately large. To combat this, an astute reader might suggest using put spreads[13] instead of outright put options in this strategy, and in practice this is sometimes a nice alternative. In the following table and chart, we show the updated payoff diagram as well as the legs of a spread based put-writing strategy:

Long/Short	Description
Short	Put at Strike K
Long	Put at Strike $K-h$

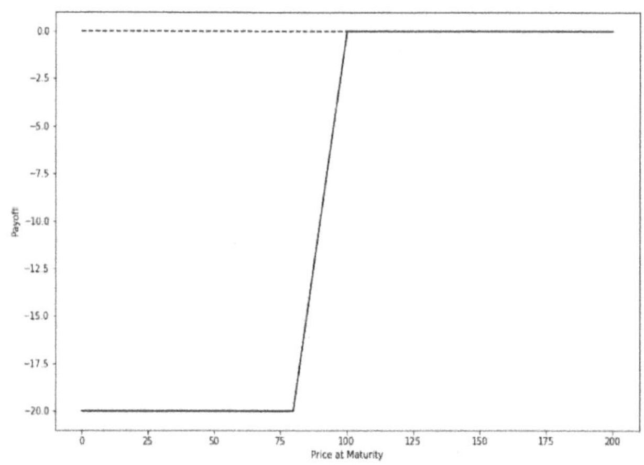

One challenge of this updated implementation is that in markets characterized by high levels of volatility skew, the put that we are buying tends to be very expensive, potentially taking away much or all of the available premia.

- **Short Volatility Strategies**: Another common options risk premia strategy is to systematically sell straddles or strangles and overlay a delta-hedging component. The motivation behind these systematic strategies is similar to that of the

[13]See chapter 16 for more details.

put-writing strategy. That is, we are trying to harvest an underlying volatility risk premia. Volatility, or optionality tends to be overpriced as investors prefer convex to concave payoffs, therefore if we provide this convexity, or gamma, to the market an investor should receive a premium for it. The following payoff diagram shows the payoff for each leg of this strategy, which we can see is again asymmetrically against the investor in a short vol strategy:

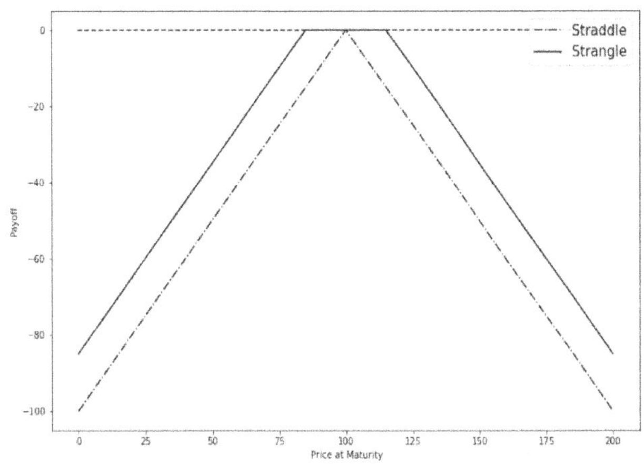

These short volatility strategies will benefit from the passage of time (e.g. they will collect theta), but will suffer in large movements in either direction. Remember that we learned in chapter 16 that delta-hedged straddles provided a way to buy or sell realized volatility vs. implied. In this type of systematic strategy, we rely on a persistent bias in implied volatility.

Many of the questions that we considered with respect to re-balancing in put writing strategies will need to be considered here as well. In this case, we will need to think about re-balancing the short option positions, as well as re-balancing our delta-hedge. Re-balancing for the hedge generally happens on a much faster time horizon. In both components, however, we might consider either a calendar based re-balance or one that is move-based.

- **Implied vs. Realized Volatility Strategies**: If one wanted to try to benefit from the premium in implied volatility relative to realized, while having a more dynamic strategy, then we may be tempted to modify the short volatility strategy above to choose dynamically between being long straddles (long volatility) and short straddles (short volatility). Doing this requires a relative value model for gamma, or realized volatility. To see this, consider the case where we could forecast volatility perfectly. If we are able to delta-hedge continuously, and there are no transactions costs, then the payoff of a straddle is purely a function of the difference between the implied volatility that we pay and the ensuing realized volatility. Therefore, if we can forecast realized volatility, then we should buy straddles when we forecast realized volatility will exceed implied, and sell straddles when the reverse is true. This would lead to the following type of rule:

Position	Condition
Short Delta-Hedged Straddle	$\frac{\partial^2 \hat{C}}{\partial S^2} < \sigma_{\text{implied}}$
Long Delta-Hedged Straddle	$\frac{\partial^2 \hat{C}}{\partial S^2} > \sigma_{\text{implied}}$

where $\frac{\partial^2 \hat{C}}{\partial S^2}$ is the forecasted level of realized volatility, or gamma, that we are able to harvest.

To employ this type of strategy requires an effective forecast of future realized volatility, and it should be emphasized that the payoff of a delta-hedged straddle is linked to **future** realized volatility of the life of the trade. This of course makes the exercise more challenging. The simplest approach, is to simply use a backward looking historical estimate of realized volatility to estimate the condition above. Alternatively, we could leverage the techniques discussed in chapter 25 and leverage a mean-reverting volatility model in our forecast. Other market participants may choose to use higher frequency estimates of realized volatility [63].

Once we have agreed on our underlying relative value model for realized volatility, similar considerations with respect to how we re-balance and hedge positions are required. Additionally, we must choose whether we want to initiate new positions daily, or at some other frequency, or focus on a more parsimonious set of positions that are re-balanced more frequently.

A final note on this type of approach is that, it is easy to apply it cross-sectionally. As an example, we could calculate the condition above for many ETF, indices, or other assets and rank the assets from most to least attractive. We could then translate this into a single portfolio that is long some straddles, and short other straddles.

- **Mean-Reversion Strategies**: Another common options strategy that can be employed systematically is to use the empirical mean-reverting nature of volatility as the basis for trading decisions. A frequent observation about the nature of volatility, which can easily be verified by examining a time series of the VIX index, is that it is characterized by sudden large spikes and then subsequent reversions to the mean. This motivates systematic strategies that sell volatility in some form after volatility spikes (e.g. via short positions in straddles, or butterflies) and buy it when it has decayed to anomalously low levels.

Position	Condition
Short Volatility	$\sigma_{\text{implied}} > \hat{\sigma_u}$
Long Volatility	$\sigma_{\text{implied}} < \hat{\sigma_d}$

where $\hat{\sigma_u}$ and $\hat{\sigma_d}$ are upper and lower thresholds for selling and buying volatility, respectively. These thresholds may be significantly apart from each other, leading to a strategy that is only invested part of the time. As an example, a simple rule would be to set $\hat{\sigma_u}$ and $\hat{\sigma_d}$ based on a z-score calculating using historical implied volatility data.

Clearly, the success or failure of this type of approach will largely be dependent on how well we are able to model the mean-reverting nature of volatility. Additionally, the speed of mean-reversion of volatility is a huge consideration here. If this reversion speed is slow, then it can lead to negative carry costs outweighing the benefit from any desired reversion. Carry costs are a noteworthy challenge in this type of approach as well, especially when buying volatility. This type of trade tends to have particularly onerous carry properties, meaning that the timing has to be very precise in order to profit.

- **Skew Strategies**: In the options strategies that we have considered so far, we have attempted to take advantage of the level of implied volatility. In our first two iterations, we took a more static approach to harvest a risk premia in the level of implied volatility. Next, we considered more dynamic approaches that allowed for relative value comparisons with respect to the level of implied volatility. However, an options volatility surface is a multi-dimensional object, and there is no reason to solely focus on the level of volatility in our strategies. This is an advantage of working with options and our strategies should attempt to take advantage of it by looking in other parts of vol surfaces. Skew in particular is one place that we may want to try to harvest systematically. Perhaps, a motivation for this would be there is a risk premium in skew itself which is separate from the risk premium in the level of volatility.

There are many options pricing structures that enable us to trade the implied volatility skew. One noteworthy structure, which was discussed in chapter 16, is a delta-hedged risk reversal. An attractive feature of these structures is their delta and vega neutral nature. This lowers potential correlation between a skew strategy and strategies that are based on trading the underlying asset, as well as strategies that focus on the level of implied volatility.

Skew strategies may be pursued cross-sectionally. For example, we could choose to sell skew in the assets with the highest skew and buy skew in the assets with the flattest skew. Our input for this type of model might simply be a volatility differential, such as:

$$\sigma_{RR} = \sigma_C - \sigma_P \tag{22.27}$$

where σ_C and σ_P are the implied volatilities of calls and puts with the same absolute level of delta, such as 25 delta calls and puts.

Alternatively, investors may choose to build a more involved, relative value skew model in lieu of the simple metric in (22.27). As these trades can accumulate delta over time, they need to be re-balanced periodically over time. Maintaining a delta-neutral profile helps to isolate the skew piece and eliminate noise, while also lowering correlation to other volatility strategies. As was true in the other potential systematic options strategies that we considered, building a skew strategy requires a re-balancing strategy and a hedging framework, among other things.

- **Calendar Strategies**: Another area of the volatility surface that systematic strategies may naturally want to focus on is their term structure, or slope with respect to expiry. These types of calendar strategies, which we introduced an example of as a standalone trade idea in chapters 16 and 20, involve buying and selling options of different expiries with otherwise similar characteristics. When done systematically, this can be an effective manner in which to harvest any available premia in the slope of the implied volatility surface. The idea behind a calendar strategy would be to isolate cases where the slope appears to be too steep, and vice versa.

  In the case that the volatility surface is inverted, or appears too flat, something like the following structure would be employed:

Long/Short	Description
Short	Call (or put) with strike K and expiry τ_1
Long	Call (or put) with strike K and expiry τ_2

  where it is assumed $\tau_1 < \tau_2$.

  Conversely, in cases where the volatility surface appears too steep with respect to expiry, for a given strike, the following structure might be employed:

Long/Short	Description
Long	Call (or put) with strike K and expiry τ_1
Short	Call (or put) with strike K and expiry τ_2

  where again it is assumed $\tau_1 < \tau_2$.

  This is a strategy that could be harvested cross-sectionally, where we construct trades in the form of the top table above when the volatility surface is flat or inverted and construct trades in the form of the lower table above when the slope is the steepest. Our metric from this could range from something simple, like the volatility differential between the two expiries, or something more complex that tries to make a proper relative value comparison.

- **Dispersion Strategies**: The strategies that we have discussed so far have focused on a single volatility surface. However, it is natural to want to broaden this across multiple surfaces, as we did, for example, in pairs trading in the case of models for the underlying asset. It turns out that a particularly interesting implementation of this approach is to buy (or sell) an index option and then sell (or buy) options on the underlying index components. These types of trades are known as dispersion strategies as the lack of correlation, or dispersion in the underlying constituent returns determines the profitability of the strategy. To see this, consider the risk-neutral payoff equation for each leg, namely an index option and an appropriately weighted basket of options on the underlying constituents:

$$c_{\text{index}} = \mathbb{E}\left[\left(\sum_i w_i S_i - K\right)^+\right] \tag{22.28}$$

$$c_{\text{basket}} = \sum_i w_i \mathbb{E}\left[(S_i - K)^+\right] \tag{22.29}$$

If the non-linear max function were not present in the payoff, then we could employ Fubini's theorem to switch the sum and expectation and the pricing equations would be equal. The presence of the non-linear max function, however, makes things more interesting, and makes the difference between these two structures dependent on correlation.

Intuitively we can see this by considering the case where at expiry half of the index goes up by 50% and half of the index goes down by 50%, in such a way that the index is unchanged. Because the index is unchanged, the payoff of the index option is zero. The payoff of the basket, however, will not be zero, as half of the constituents increased substantially and will thus be in-the-money. Therefore, we can see that if we sell the index option, and buy the constituents, we are long dispersion, and short correlation.

This means that the following structure can be used to instrument a long dispersion trade:

Long/Short	Description
Long	Basket of Constituent Call (or Put) Options
Short	Index Call (or Put) Option

Conversely, if we want to be short dispersion, and long correlation, we would reverse the short and long positions as follows:

Long/Short	Description
Long	Index Call (or Put) Option
Short	Basket of Constituent Call (or Put) Options

Dispersion strategies, like many other systematic options strategies, are generally overlaid with a delta-hedging component. Therefore, a delta-hedging scheme, as well as an options re-balancing strategy must be built into these types of models.

The motivation for this trade, like many others we have discussed, is based on risk premia. It is a common observation that correlations tend to increase toward 1 in times of stress. Market participants have observed that this places a premium on implied correlations relative to their realized counterparts. If true, this could make these type of dispersion strategies in the above table that sell correlation attractive. This argument particularly implies that dispersion trades based on put options are likely to warrant a premia. Another justification for this phenomenon is the liquidity differential between single-name, underlying constituents and indices, or that investors tend to use downside index options rather than single name options to hedge their portfolio.

Clearly, this is the most inherently complex of the trading strategies that we have introduced, and building a systematic approach to leveraging this phenomenon will likewise we quite involved. One approach to take would be to leverage the Weighted Monte Carlo techniques discussed in chapter 17 as Marco Avelleneda suggested [172]. This is a convenient framework as it enables us to handle the highly dimensioned problem of modeling the entire basket thus providing a potential relative value signal for the index volatility. This type of model would help us to not only build a strategy that harvests a correlation premium in the options market but also one that does so in a dynamic manner.

22.4.3 Options Strategy in Practice: Covered Calls on NASDAQ

In this section we apply the covered calls back-test code in the repository, with all of its simplifying assumptions, to the NASDAQ ETF QQQ from 2010 to 2021. The following chart shows the cumulative profit and loss of this strategy that is long the underlying asset, QQQ, and short an at-the-money call option:

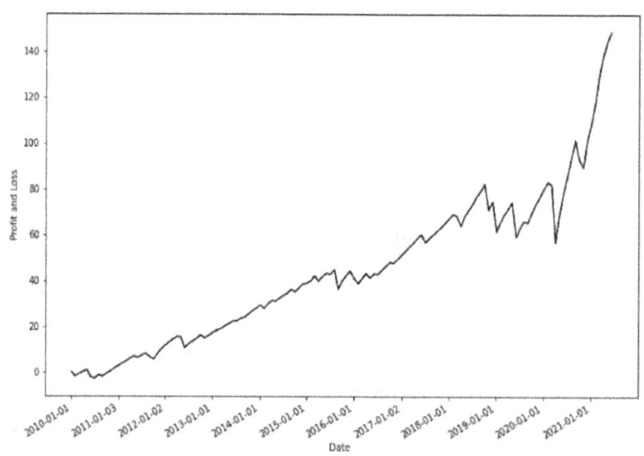

Here, we can see a clear pattern of gains and a significant upward trend in our profit and loss profile over time, albeit with some notable drawdowns that seem to correspond to periods of market stress. It turns out, this is in fact the case, and this strategy is characterized by a large number of relatively small profits and a small number of disproportionately large losses. In the following chart, we examine the histogram of payoffs to our individual covered call positions:

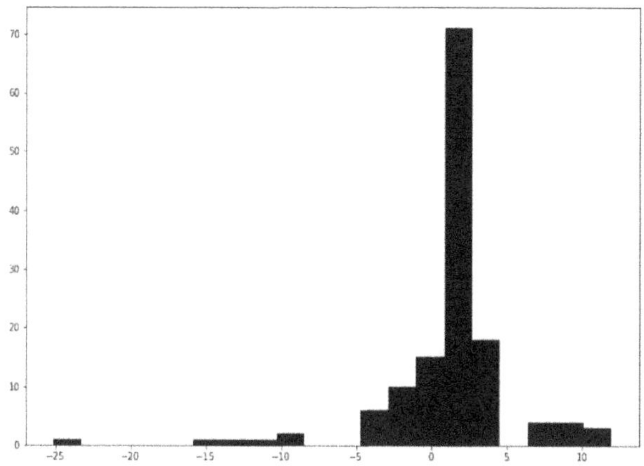

Here we can see this same pattern at work. That is, we see a few large losses and a consequently negatively skewed profit and loss distribution, with a large mass of slightly positive outcomes.

Again it should be emphasized that the profit and loss here only considers the at-expiry outcomes. This is an oversimplification, as drawdowns in each monthly period could have significant mark-to-market calls that aren't represented in this chart. Nonetheless, this chart does display the typical characteristics that we observe in a put writing or covered call strategy. Specifically, we observe a strong trending line up, with mostly positive returns, and a few large, asymmetric drawdowns. The process of translating this phenomenon into a tradeable strategy is then largely about building a strategy that is robust enough to survive these inevitable drawdowns. That is, however, beyond the scope of the text, and is instead left as an exercise for the reader.

22.5 COMBINING QUANT STRATEGIES

Once we have completed the signal generation & strategy design portion of our alpha research, the final step is choosing how to allocate between the one or more strategies that we've created, as well as setting the overall size and leverage characteristics. Of course, this is ultimately a portfolio construction and optimization exercise, and as such, the techniques developed in chapter 24 are of direct relevance, as are the caveats related to estimation error, non-stationarity, etc.

Clearly, the optimal allocation between the strategies will depend on:

- Strategy Expected Returns
- Strategy Volatilities
- Correlation between the strategies

Each of which is estimated in the context of a back-test that is generally based on a relatively short sample period and therefore has potentially substantial estimation error.

Finding the proper allocation of our sub-strategies is a critical part of building a successful portfolio. We could have a set of good signals but if they aren't properly allocated to we may still see large and unnecessary drawdowns driven by a single strategy. That is, just as with the portfolios we built in chapter 24, we need to make sure that our quant strategies benefit from as robust diversification as possible.

To gain some intuition, let's start with the case of two strategies and see how our portfolio varies as we change the sub-strategy weights. In the case of two assets, with returns μ_1 and μ_2, volatilities σ_1 and σ_2 and correlation ρ, the properties of the portfolio become:

$$\mu_p = w\mu_1 + (1-w)\mu_2 \tag{22.30}$$

$$\sigma_p = \sqrt{w_1^2\sigma_1^2 + (1-w)^2\sigma_2^2 + 2w(1-w)\sigma_1\sigma_2\rho} \tag{22.31}$$

$$SR_p = \frac{\mu_p}{\sigma_p} \tag{22.32}$$

Using these equations, let's first consider two uncorrelated strategies with identical risk and return characteristics:

Strategy	$\mathbb{E}[R]$	σ_R
A	10%	15%
B	10%	15%

The following chart shows the sharpe ratio of a portfolio with weight w in strategy A and $1-w$ in strategy B:

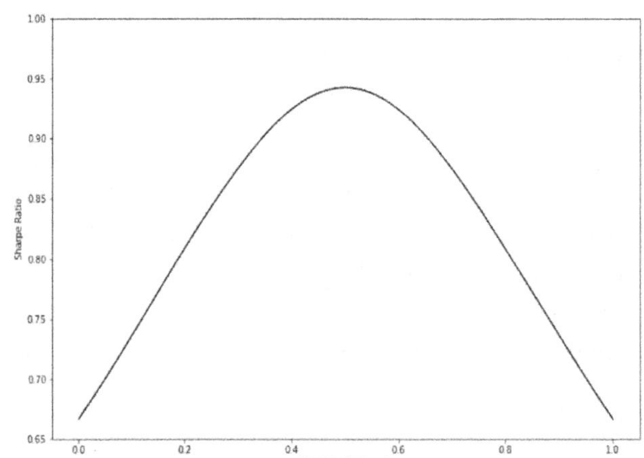

Importantly, we can see that our overall sharpe increases significantly if we combine strategies A and B, and that the optimal point is $w = 50\%$. Next, if we now assume that strategy B has a lower sharpe ratio, but the strategies remain uncorrelated, that is:

Strategy	$\mathbb{E}[R]$	σ_R
A	10%	15%
B	7.5%	15%

We can see from the following plot that the optimal weights in strategies A and B shift toward strategy A, unsurprisingly. However, we can see that we are still able to improve the portfolio sharpe ratio by combining A and B rather than investing solely in A. That is, investing in strategy B improves our sharpe ratio even though its sharpe is lower, simply because of its uncorrelated nature.

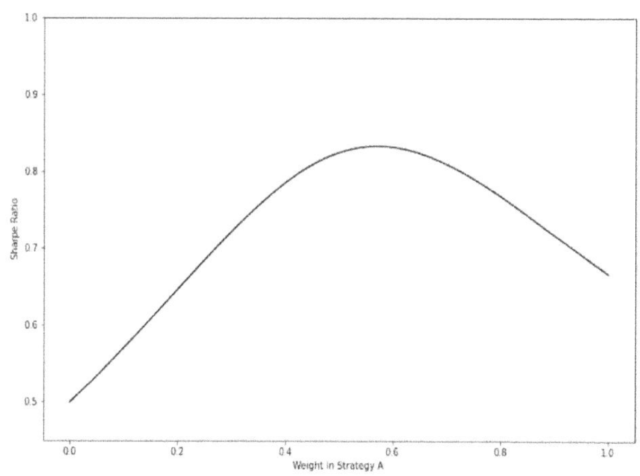

Clearly, this profile will adjust as we change the properties of the strategies. Of critical importance is their correlation. When they are uncorrelated, then we find a substantive diversification benefit from combining them. Unsurprisingly, when the two strategies are perfectly correlated, then this diversification benefit disappears completely. For example, consider again the following two strategies with the same risk and return profile, but now perfect correlation:

Strategy	$\mathbb{E}[R]$	σ_R
A	10%	15%
B	10%	15%

Now we can see that the sharpe ratio is the same regardless of how much we invest in A and B and that there is no benefit to combining the two:

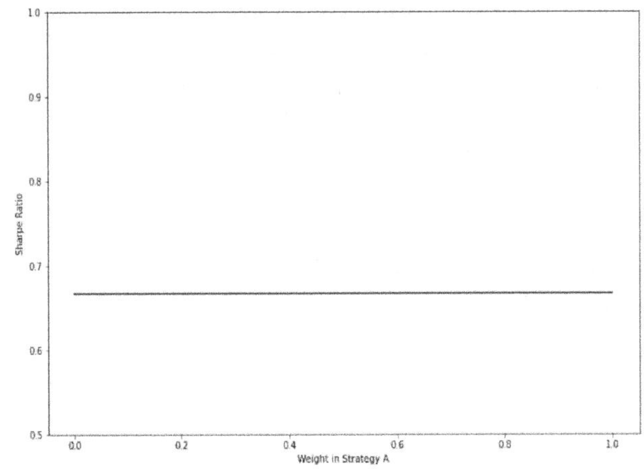

Here we can see that there is no diversification benefit when the strategies are perfectly correlated. This is not true, however, when the two strategies are highly positively correlated, but not perfectly correlated, as shown in the following chart:

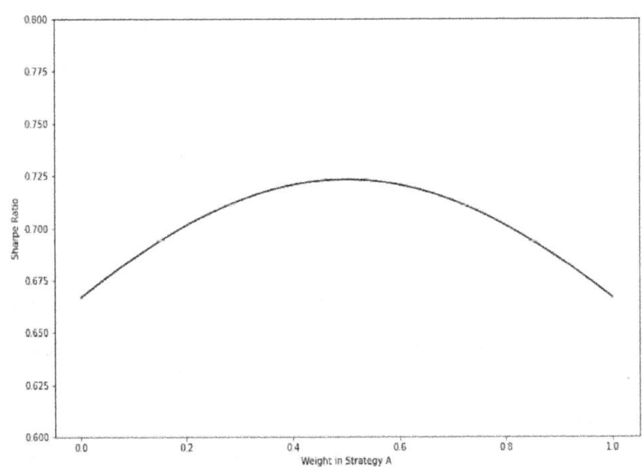

We can see that there is some marginal improvement to the sharpe ratio, but it is not as much as in the case of uncorrelated assets. Finally, if the two strategies are negatively correlated then we get the maximal benefit from diversification:

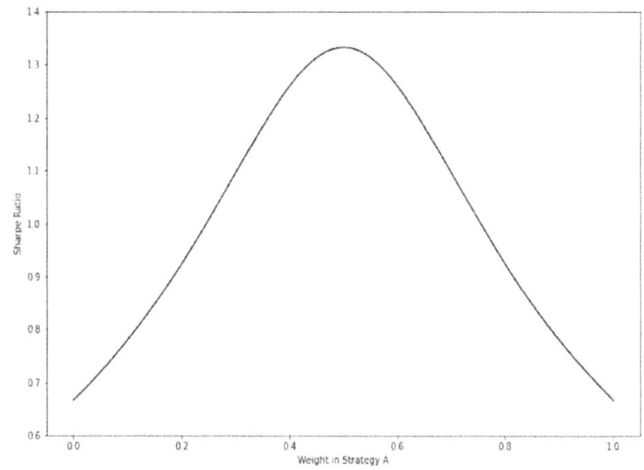

The previous examples illustrate one of the most powerful and important concepts in quantitative investing and portfolio management. In particular, as we know from mean-variance analysis and CAPM, diversification enables us to eliminate idiosyncratic risk without sacrificing expected return. We see this phenomenon play out here. That is, we can improve our portfolio sharpe ratio by adding a strategy that is not perfectly correlated with our existing strategy. It has been said that diversification is the only free lunch in finance, and this concept is illustrated in this example.

Further notice that the benefit of adding a strategy is highest when the new strategy is negatively correlated. This makes finding strategies that are negatively correlated incredibly valuable, and implicitly lowers the required bar in terms of raw performance for the strategy. We might find a strategy has little to no expected return, but whose diversification properties are so appealing that it is still quite worthwhile to allocate to. We would not invest in this strategy in isolation, but in conjunction with our other building blocks it becomes quite appealing.

In practice this principal of diversification is critical and is a key tool for building a successful portfolio. In many cases, we can develop an attractive portfolio by building a set of uncorrelated sub-strategies with relatively low individual sharpe ratios. If enough uncorrelated building blocks can be created, which in and of itself may be a challenge, then the portfolio itself may have a compelling sharpe.

In my experience, this approach is more realistic and more effective than trying to build a single source of return that in isolation is overwhelmingly attractive. This means that, rather than focus on perfecting a strategy which will likely result in overfitting, we may be better served using a simplified version and trying to find more complementary strategies that improve their attractiveness in aggregate. To see this, consider the case of adding N uncorrelated strategies with the same risk and return characteristics. It can be shown that in this simple case, the portfolio sharpe ratio grows at the rate of $\sqrt{N}$, as the following chart shows:

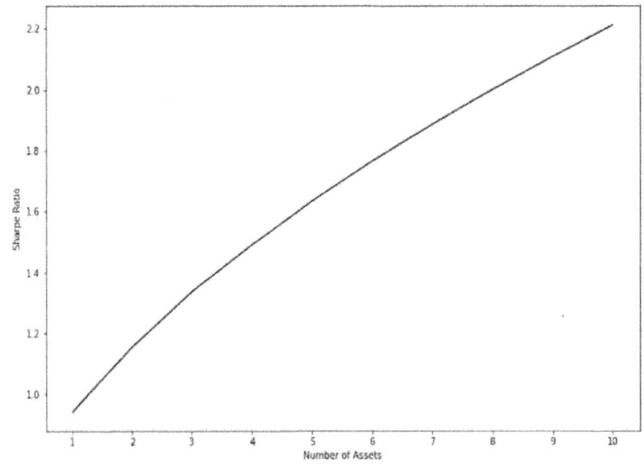

Of course, finding uncorrelated sources of return that are attractive is no small task. But the key point is that diversification can help improve outcomes in the context of quant strategy design. The proverbial whole can exceed the sum of the parts.

In the case of quant models with many sub-strategies, we might consider the following different types of strategy weightings:

- Equally Weighted
- Mean Variance Optimal Portfolio Weights
- Inverse Volatility Weighted
- Risk Parity Weights

Unless we choose equally weighted, allocating between them becomes a non-trivial exercise in portfolio optimization. If we do choose equal weights, then we are not taking advantage of the correlation structure of the sub-strategies. We are also implicitly weighting the higher volatility strategies the most. If we use an inverse vol weighting, then we are incorporating volatility information but still neglecting correlation information as well as information about a strategy's expected returns. Risk parity improves on this by integrating correlation information, but still ignores expected returns.

This may be a feature and not a bug, however, as remember that there is a great deal of uncertainty in the expected returns that we compute. Two strategies with back-tested returns of 10 and 20 percent, respectively, in reality may have very similar true expected returns. Because of this, a natural prior is that the underlying strategies all have similar sharpe ratios. We might deviate from this in cases where the theoretical and empirical evidence is overwhelming, however, an equal sharpe ratio assumption serves as a reasonable baseline. Further, it is highly recommended that some sort of regularization process be applied to the return and volatility estimates, shrinking them toward an equal sharpe ratio assumption. This can be accomplished using the regularization techniques discussed in chapter 25.

22.6 PRINCIPLES OF DISCRETIONARY VS. SYSTEMATIC INVESTING

In this chapter we have provided a high level summary of many systematic approaches to trading and investing. Of course, in the market many participants choose to structure their investment process in a purely quantitative manner. Others, however, prefer to use a discretionary approach[14], for example based on their macroeconomic views. In between, so-called quantamental strategies choose to operate somewhere in the middle with certain discretionary inputs and certain quantitative aspects.

In addition to highlighting some of the approaches to systematic or quantitative investing, we also attempted to emphasize a few philosophical concepts that belie these types of strategies. For example, we saw that, roughly speaking, quantitative techniques try to skew the odds slightly in their favor and then leverage the central limit theorem to increase the power of the investment strategy over time. These strategies will then specialize at finding patterns that are present in the data, and then trying to repeatedly isolate those patterns with attractive conditional returns. In the case of the risk premia strategies detailed in this chapter, these patterns that we find in the data may be augmented by an economic or theoretical argument for why the pattern should emerge. Perhaps we are being compensated for taking on a tail risk that others are disincentivized from doing. Volatility selling strategies are a canonical example of this. To profit from these types of quantitative strategies, we will want persistent exposure to the underlying pattern or premia. On any given trade we have no idea what will happen, but we have confidence that over time the mean of the distribution will converge to an attractive investment. This means that these strategies are characterized by less of a need to time them, or choose our spots to enter and exit. Instead, we harvest the underlying phenomenon persistently. Of course, this doesn't mean there aren't situations where an underlying premia might be higher or lower, and we'd want to adjust our exposure, but it does mean that a natural baseline is consistent exposure.

Discretionary strategies, in contrast, often look to find what is missing, or not properly incorporated in the data, and leverage that to make profitable trading decisions. As an example, as Covid cases began to rise in early 2020 in other parts of the world, traditional datasets didn't reflect its potential impact until late February or early March. A discretionary process may have observed this potential looming risk to the system, and been able to structure a conditional trade customized to that specific set of market conditions. More generally, this is a defining feature of discretionary trading strategies. Instead of looking for persistent exposure to a given phenomenon, we instead look to isolate what the market, standard models, and available data might be missing or not fully appreciating. When doing this, timing will be of the utmost importance for our strategy to be profitable. If we persistently invest in this strategy, it may very well be that the return distribution is unattractive, however, at this particular moment it may be quite compelling. Of course, uncovering this is what make discretionary trading such a sought after art form.

An example of this in action would be the juxtaposition of long volatility vs. short volatility strategies. As mentioned, short volatility strategies are most attractive

[14]Which still may be informed by quantitative tools.

when invested in consistently. If done over a sufficiently long period, they provide us with a volatility risk premium, or the difference between implied and subsequent realized volatility. When we are long volatility, however, we will not want to do this persistently, and in fact, in this case the law of large numbers works against us. If we do this unconditionally in fact, we can infer from theory that it will lead to negative returns. Instead, we will want to do this conditionally, only at a few instances where we can isolate something in the macroeconomic environment that is missing in the data, such as the onset of Covid.

22.7 CASE STUDY: DO TREND FOLLOWING STRATEGIES PRODUCE CONVEX RETURNS?

In this chapter we introduced momentum strategies as one of a number of quantitative based investment approaches that an investor could pursue. These strategies have a long history, with momentum being one of the first documented market anomalies. Interestingly, even as the phenomenon has been written about, evidence has shown that it has persisted. This has made strategies that use momentum based signals, commonly known as trend following strategies quite popular.

In recent years these strategies have become even more popular, as these products were generally able to achieve strong returns during otherwise difficult market conditions in 2022, with stock and bonds both declining significantly. At first glance, the reason for this is that trend following strategies don't have a directional bias. Or, said differently, trend following strategies choose their market direction dynamically in line with market trends. This makes them do well in periods where trends persist, and prone to fast reversals in direction. In the case of 2022, this led most trend products to be long commodities, which had an exceptional year, and short both stocks and bonds.

In the following chart we show the performance of each individual asset class in 2021 and 2022:

As we can see, the poor returns of stocks and bonds, as well as the strong performance of commodities starting towards the end of 2021 turned into persistent trends that these strategies were able to take advantage of, at a time when balanced stock & bond portfolios struggled greatly. To take an example, in the following chart we show the performance of AQR's managed futures product, AQMIX, over the same time:

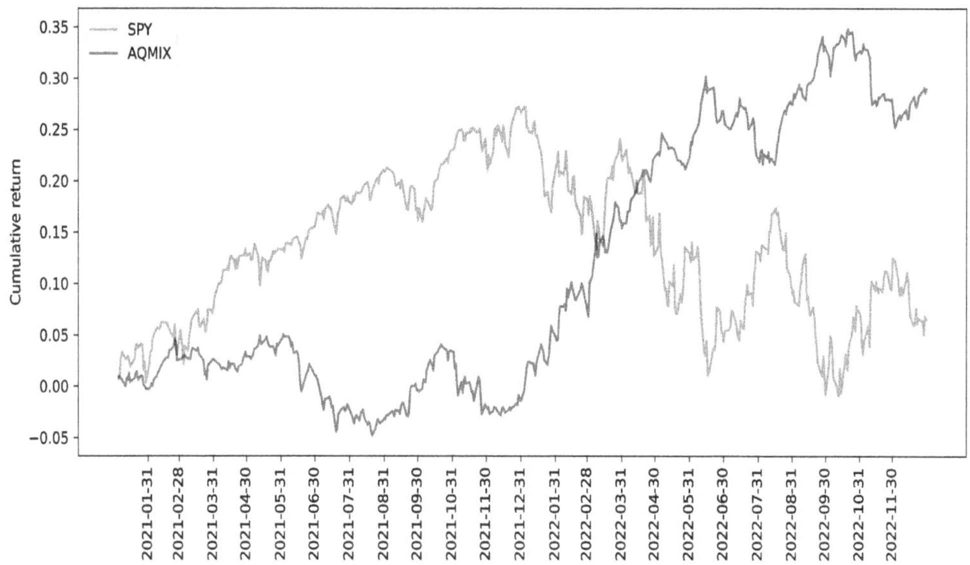

Here, we can see the strong uptick in performance beginning at the end of 2021 as the trends in stocks, bonds and commodities were identified and then continued

Astute readers might notice a similarity between how trend following strategies are presented here and how a certain options structure is presented in chapter 16: a straddle. In each case, an investor is agnostic to the direction the market moves, and in each case a persistent trend in either direction is beneficial. In the case of a straddle the offsetting cost is the theta, or negative carry. In the case of a trend following strategy the cost is more nuanced, but comes in periods where trends do not persist and instead consistently revert.

For a straddle, this leads to a well-known and fundamentally convex relationship between the value of the position and changes in the underlying asset, as shown here:

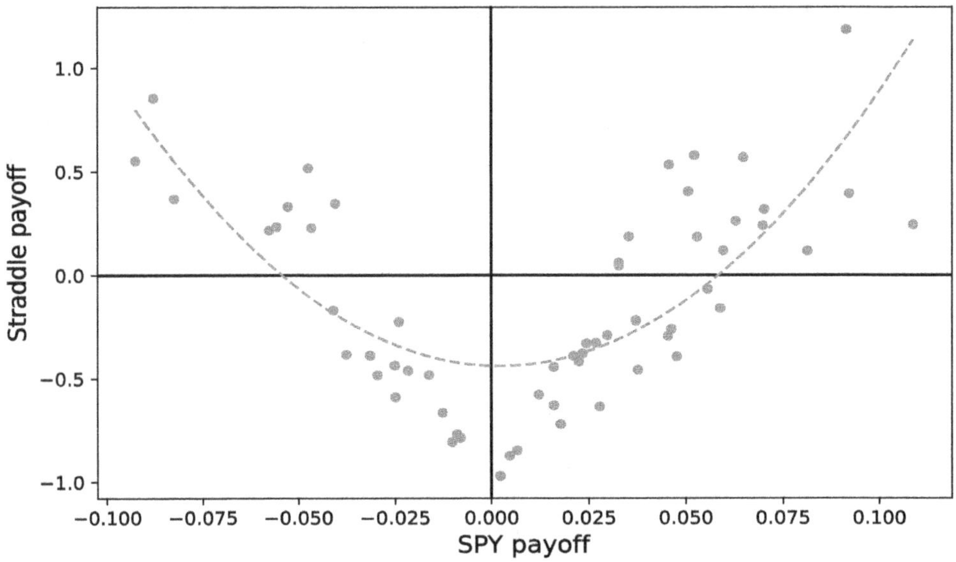

where in this chart we are showing the payoff of a hypothetical straddle with VIX as a proxy for implied volatility, the dots in the chart represent actual P&L's over a given period and the dotted line represents a calibrated parabolic function fit to the data.

Given their strong performance in 2022, investors debated whether trend following strategies had a similarly convex return profile embedded within them, and whether they had comparably better carry relative to a straddle position. Historically, trend following strategies have been shown to have strong returns in crisis periods like 2022, a potential reflection of this convexity. A key difference, however, is that the convexity is not contractual as it is with an options contracts. Instead, for a trend following strategies to behave convexly requires markets to have persistent trends in periods of crisis or market stress. In 2022, we saw these types of trends, and trend strategies benefitted. In other periods of stress, such as the onset of the Covid pandemic, these trends reversed much more rapidly, causing relatively challenging performance for trend following strategies. The reader is encouraged to think through this phenomenon and how the potential for convex returns in trend following strategies works.

CHAPTER 23

Artificial Intelligence: Incorporating Machine Learning Techniques

23.1 ARTIFICAL INTELLIGENCE & MACHINE LEARNING: LANDSCAPE

In recent years, we have seen a proliferation of advances in the fields of artificial intelligence (AI) and machine learning (ML). This has led to increased adoption within the field of finance[1]. It has also led to the promise of continued innovation and evolution. In the next two chapters, we aim to provide a foundation in these concepts and orient the reader to the most important techniques used in these spaces.

Prior to introducing the techniques, we first want to provide context on what constitutes AI and machine learning. Generally, when we think of Artificial Intelligence, we naturally think of sophisticated algorithms like Chat GPT [224]. While this is indeed one example of AI, it is only one of many that fit within a much broader definition. Simpler techniques, like rules based systems [121] and linear regression are also equally valid examples of AI.

As a result, when working with Machine Learning and AI, it is important to understand the context of the problem and also where within the wide ecosystem the techniques being leveraged fit. Similarly, it is also important to choose the right technique for the problem at hand. For example, while more advanced AI/ML algorithms may be preferable in certain circumstances, there are occasions when, perhaps because of a lack of data, linear regression may be more appropriate. Because of this, aspiring quants should be judicious in selecting an approach to solve a problem and not be prone to selecting overly complex methods.

In general, we can break the universe of techniques related to AI and Machine Learning into the following categories:

- **Statistics & Econometrics**: These are the more traditional techniques such as linear regression, PCA and time series models. While we generally do not

[1] And more broadly.

DOI: 10.1201/9781003530626-28

think of these as machine learning or AI tools, they are part of their broad definitions.

- **Classical Machine Learning**: These techniques employ a supervised or unsupervised approach to understanding datasets, identifying patterns and make predictions. This includes clustering techniques that help us identify the structure in a dataset, regression based approaches that might allow us to incorporate non-linear features, [2], as well as classification techniques [3] that allow us to solve problems with binary outcomes. [4] Classical Machine Learning also includes within it the techniques behind statistics and econometrics, such as linear and logistic regression [137].
- **Deep Learning** [266]: A subset of advanced Machine Learning that employs multiple layers within a neural network. Deep learning includes Recurrent Neural Networks (RNNs) [171], Convolutional Neural Networks (CNNs) [255] and Generative Adversarial Networks (GANs). The defining feature of deep learning algorithm relative to a neural network in a classical machine learning application is the use of multiple hidden layers within the network.
- **Large Language Models**: Large Language Models (LLMs) are trained on large amounts of text data in order to repeatedly predict the next word in a sequence. In doing so, LLMs are able to generate readable text responses in paragraph form. LLMs are based on an underlying neural network [5] that trains the model to interpret text input data and guess the most likely next word in a sequence.

Importantly, each of these four classes of techniques fit within the broader definition of AI. Moreover, Machine Learning is a subset of AI, but is in itself quite broad, encompassing the first three classes of techniques, ranging from linear regression to deep learning.

The techniques behind the statistics and time series models were introduced in chapter 5, where we covered introductory econometric models. Applications of these techniques, have then been a point of emphasis throughout the book. In this chapter, we focus on the techniques described in the second bullet, that is, classical machine learning techniques such as decision trees [257], random forest [135], SVM [69] and SVR [125]. In the next chapter, we then provide an introduction to Deep Learning [266] and Large Language Models [21] and describe how alternative data can be leveraged in finance.

---

[2]Such as SVR [125].

[3]Such as SVM [69].

[4]While also incorporating non-linear relationships.

[5]An Artificial Neural Network.

23.2 MACHINE LEARNING FRAMEWORK

23.2.1 Machine Learning vs. Econometrics

Earlier in the text, in chapter 5, we introduced the basic concepts leveraged by quants in the field of econometrics. These econometric tools, such as regression and time series analysis were then our go-to tools for solving many of the problems that we encountered in the buy-side type problems in this section. Specifically, we leveraged regression and time series models to build expected return forecasts and alpha signals, and to calculate underlying factor exposures. Recently, many researchers have shown that machine learning techniques provide a potentially compelling alternative to these previously discussed econometric techniques. Further, as we will see, machine learning techniques can help us to tackle other common tasks as well, as they can help us understand our data better, and can, for example, help us understand high dimensional correlation structures.

In this chapter we provide a brief, high-level introduction into some of the most popular machine learning techniques and applications in quant finance. Along these lines, our intent is to whet the readers' appetite and provide an overview as to how machine learning problems are structured, and how we can try to apply them in a quant finance context rather than focus on the details of the littany of available machine learning techniques.

Prior to introducing these techniques and applications, however, we want to orient the reader to the philosophical differences between machine learning and econometrics. Up until now, we have developed econometric problems, such as expected return models, or alpha models, using a particular framework. This framework was predicated on starting with an economically motivated hypothesis, and using a set of mathematical tools in order to attempt to validate, or disprove this hypothesis. In machine learning, the procedure is often reversed. That is, rather than starting with a hypothesis, we begin with a set of data and allow the algorithm to find the underlying relationship. In the case of so-called unsupervised learning, this relationship is extracted without any guidance from the user. This change in philosophy creates both opportunity and challenges. On the one hand, it may enable us to uncover more complex relationships that may be hidden from our standard econometric approach. On the other hand, this approach can make the relationships harder to understand and visualize. Thus, the model may become less interpretable. It should be noted that, in contrast to so-called unsupervised learning methods, supervised learning methods do enable us to maintain some of the structure that is familiar to econometricians. For example, supervised learning methods enable us to select the features in our model, meaning that we can limit them to those that have some underlying economic justification. Having said that, these algorithms tend to work with a far larger set of features. This can be seen as a positive, as in standard econometric methods larger feature sets often coincide with issues associated with multi-collinearity. Conversely, it can lead to interpretability challenges. These interpretability challenges, for both supervised and unsupervised learning, are detailed later in the chapter.

More generally, this change in perspective, where instead of starting with a hypothesis we tend to allow the algorithm more room to uncover subtle relationships, is a defining feature of machine learning algorithms, and when applying it in practice we need to be cognizant of how appropriate it is. In some cases, such as analysis of a covariance structure or high-frequency trading, it is quite appropriate. In other cases, where the available datasets are small, and non-stationarity in the data presents a large challenge, we may want to lean on more parsimonious models that can be connected back to an economic justification.

23.2.2 Stages of a Machine Learning Project

Just as we saw earlier with respect to generic quant projects, in chapter 2, there are certain steps that occur in machine learning based projects, regardless of which model or technique is being used. The following list highlights these high-level steps:

- **Data Collection**: In this stage, just as with any other quant project, we begin by obtaining our required data from some source whether it be external via a website, or internal via a database.
- **Data Cleaning**: Once we have obtained our data, we need to ensure the quality of our data. This is no different from any other quant project, however, can be of critical importance in a machine learning project as a few mis-classified observations can cause significant harm when training an ML model. This phase of a project might consist of the following substeps:
  - *Visualization*: Visualizing the data can be a key tool for catching outliers or other anomalies and making sure the data quality is sufficiently high.
  - *Outlier Detection*: In addition to visualizing the data, we may want to examine the data for outliers using some of the techniques introduced in chapter 9.
  - *Exploratory Data Analysis*: Exploratory data analysis can, as we explored in chapter 10, help us to understand the data that we are working with better. This can then enable us to make better decisions on the structure of our model. For example, it can help us identify what features should be included. Within this step we may perform, among other things, some unsupervised machine learning methods, such as clustering analysis.
- **Model Implementation**: Of course, the core of implementing a machine learning model lies in the implementation phase, where we identify the appropriate machine learning technique, choose between supervised and unsupervised methods, select the universe of previously collected data to use, and then design the model in Python[6]. At a high-level, this phase of an ML project consists of the following steps:
  - *Model & Feature Selection*: The first step in model implementation is to choose a model. In an ML context, that means first deciding what class of model is appropriate, between supervised, unsupervised and reinforcement

[6]Or another language.

learning. Next, we need to choose an appropriate technique from this class of ML models, and provide whatever additional structure is needed. For unsupervised learning, very little additional structure will be required. Supervised learning, in contrast will require us to create output labels for our data, and define a set of features in the model.

- *Model Design*: Once we have defined the model sufficiently well, including which algorithm we want to employ and what data we want to feed into it, the next step will be to develop a Python implementation of the model.
- Next, we will want to split the dataset into the following subsets:
    * *Training Set*: The in-sample dataset that is used to optimize all dimensions of the model.
    * *Validation Set*: The validation set is used to validate that the technique hyperparameters are chosen optimally. This validation set often evolves, or is re-sampled from the original data set, in order to verify these technique parameters.
    * *Test Set*: The out-of-sample dataset that is used to judge performance on new, foreign observations to measure the degradation in quality of the prediction. A large deterioration in this period signals a high likelihood of overfitting.

  The model will then be fed these different datasets in the appropriate context, beginning with the in-sample, training set, then proceeding to the validation sets, where we use cross-validation to validate the technique parameters, and finally to the test, or out-of-sample set, in order to train the ML model.

- **Model Validation**: Just as with any project, once we have completed the research and model implementation phase, a robust model validation process should be performed to identify any issues or inappropriate assumptions in the implementation of the model.

One noteworthy difference between this setup and an econometric approach, is the separation of the data into three parts. In particular, while in econometrics we often, and should, split the data into two parts, the validation set used to optimize the model's underlying hyperparameters is new and unique to machine learning projects. In the next section, we provide more context on this step.

23.2.3 Parameter Tuning & Cross Validation

As mentioned, one defining feature in machine learning problems is the presence of a cross-validation step. In this step, we tune what we refer to as the model hyperparameters. These hyperparameters are the parameters that are intrinsic to the model. Because they are intrinsic to the model, they can be difficult to choose in practice as they often have little to no intuition behind them. We have seen these types of parameters emerge in many models that we have introduced throughout the text. For example, FFT algorithms are defined by several internal model parameters, such as an α parameter that was interpreted as a damping factor. When solving options pricing problems via FFT this parameter had an impact on the solution, however, as we saw it was very difficult to determine what an optimal setting for α might be. We saw a similar concept when we introduced regularization techniques, notably LASSO and Ridge regression. Both of these techniques were characterized by a λ parameter, which, roughly speaking, defined the relative penalty on increasing the model parameters. Later in the text, when we introduce Support Vector Machines, we will see a similar cost parameter, C, which defines the model penalty when a label crosses the separating hyperplane. These concepts with respect to SVM will be discussed further in 23.5.5, however, more generally, these same internal model parameters tend to be present in most ML algorithms.

Up until now in the text, our approach to selecting these types of internal model has been somewhat subjective. Fortunately, however, Machine learning provides a robust, quantitative approach for choosing these types of parameters. That is, in ML techniques we use cross-validation to optimize these parameters. One way to choose these parameters would be to rely on ourselves, as quants to select them. Of course, there is no reason to expect we would do this well, especially since these hyperparameters may be difficult to interpret and have little intuitive meaning. At the other extreme, we could optimize over this set of parameters over the entire training dataset, such that the model fit is the best. The problem with this approach is that it leads to overfitting, making the model less likely to perform well out-of-sample, in the test dataset. Cross-validation provides a happy medium as it enables us to optimize the parameters while minimizing the chances of overfitting. Cross-validation works by withholding part of the training dataset when optimizing the hyperparameters, and then validating the chosen hyperparameters on the data that was withheld. This validation on data that was not part of the initial optimization, minimizes the risk of overfitting. In many cases, this process is performed repeatedly with different parts of the dataset withheld as the cross-validation set.

One type of cross-validation is referred to as k-fold cross validation, where we separate the training dataset into k pieces, or folds. We then withhold one of the folds, and optimize the hyperparameters on the remaining $k - 1$ folds. After the parameters are chosen, they are then validated on the remaining, withheld fold. This process is done repeatedly in a loop so that each of the k folds is withheld. An important feature of cross-validation is that it enables us to refrain from using the out-of-sample test set when tuning the hyperparameters. This reserves the test set for actual iterations of the model, reducing the likelihood of overfitting.

Like other high-level languages with robust support for ML, Python has built-in functions and tools for tuning an algorithm's hyperparameters, In the coding repository you can find a coding example covering how to perform cross validation and parameter tuning.

23.2.4 Classes of Machine Learning Algorithms

Broadly speaking, there are three main types, or classes, of machine learning algorithms: Supervised Learning, Unsupervised Learning and Reinforcement Learning.

In the remainder of this section, we describe the defining features of each type of algorithm:

- **Supervised Learning**: The algorithm is trained on a set of inputs, or features, along with a corresponding set of desired outputs, or labels.
- **Unsupervised Learning**: The algorithm tries to extract structure from a dataset with no additional structure given. In particular, the outputs, or labels, that are provided with supervised learning, are not provided when performing unsupervised learning.
- **Reinforcement Learning**: The algorithm interacts with a dynamic environment while trying to perform a certain task. Reinforcement learning allows for the specification of one of more intelligent agents whose decisions evolve as the model is trained according to a specified cost function.

In this text we focus on the first two techniques, and how they can be applied in a financial setting, however, Reinforcement Learning has also gained popularity in recent years. Readers interested in more background on Reinforcement learning should consult [141] or [215].

23.2.5 Applications of Machine Learning in Asset Management & Trading

In recent years, machine learning techniques have become a trendy topic in the field of quant finance and has left many believing that we are on the precipice of a paradigm shift in the field [75]. Adoption of machine learning techniques within the field has been broad, encompassing execution, hedging, risk, portfolio optimization and alpha generation.

In the high-frequency space, machine learning tools seem ideally suited to help understand the dynamics of market impact and optimal execution. In this case we have a dataset that is sufficiently large to warrant more sophisticated tools. Further, there tends to be less of an emphasis in this space on an underlying economic rationale, a commonly cited challenge in employing black-box machine learning algorithms.

As we move toward lower frequency alpha research and trading strategies, incorporation of machine learning encounters several challenges that we discuss in the next section. Nonetheless, use of machine learning expands our set of tools, and can help us overcome many of the challenges associated with standard regression models and econometric techniques. In this context, the type of alpha signals that were formulated in chapter 27 based on regression techniques, and principal component analysis, can potentially be upgraded with equivalent ML tools. Particularly appealing is that some of these machine learning techniques can incorporate non-linearity in the relationships to the prediction and be less dependent on understanding the correlation structure of the underlying features. Lack of data, however, becomes a challenge as the investment horizon expands and less truly independent observations become available.

Machine learning algorithms also can be of great use when trying to understand the correlation structure, and covariance matrices of large groups of assets. This means that they are of great use to quants on the buy-side, especially working for asset management firms, whose primary role is to build robust portfolios. An increased understanding of the underlying covariance structure, whether achieved through clustering or another dimensionality reduction technique, can then lead to great insights. These same techniques can then be of great use to risk managers, whose goal is to forecast the tail of a distribution, which is ultimately highly dependent on the underlying correlation structure.

Machine learning techniques have also greatly expanded the set of data at our disposal, making so-called alternative datasets available to many practitioners. These datasets come with challenges, and truly understanding them is a natural machine learning problem. For example, as we explore in the next chapter, machine learning algorithms can help us parse news stories and attempt to identify positive or negative news. They can also help us to transform unstructured data into a more usable format. More generally, machine learning techniques can make certain alternative datasets that were previous intractable, tractable.

23.2.6 Challenges of Using Machine Learning in Finance

While the potential incorporation of machine learning algorithms in finance brings great promise, it also comes with many challenges [223]. Perhaps most problematically, machine learning algorithms require large amounts of data in order to train their models. In finance, this abundance of data is often not available, at least in traditional price based datasets. Instead these datasets are characterized by short histories going back a few decades, and a relatively small cross-section. A notable exception to this is high-frequency strategies, where the number of ticks in a given day allows for far more observations. But, for quants relying on daily, or even lower frequency data, applying machine learning techniques and trying to calibrate large sets of parameters presents a serious challenge. This is in contrast to more canonical applications of machine learning, where large datasets emerge naturally, and repeatable experiments may be conducted. For example, in the scope of image recognition, these types of algorithms have millions of images to train their models on, and as a result can leverage a large number of features in doing so.

Additionally, in these more standard applications of machine learning the data is generally stationary, whereas in finance the presence of non-stationary data and market regimes may create challenges for the underlying machine learning algorithms. Another aspect that makes it difficult to apply machine learning techniques is that, in finance, signal-to-noise ratios tend to be far lower than in other applications of ML. This is a function of the fact that markets are dynamic and highly efficient and any edge that is found is likely to be priced out of the market as more participants become aware of it, weakening the signal.

Finally, another challenge commonly associated with applying machine learning methods in finance is its interpretability. Many of the more complex ML algorithms have a black-box component to them, that is, they are difficult for practitioners to understand, even quants. This can be seen as an undesirable feature as many market participants, especially traders and portfolio managers like to understand the way their underlying models work, and the types of relationships it relies on. In some ML algorithms, however, it can be difficult to ascertain which features are most critical, and how they interact. When using standard regression methods, for example, the sign of the coefficient clearly tells us what an increase in the variable implies about our prediction. In some ML methods, this type of relationship can be more opaque, making even the directionality of a change in the prediction far less clear. Later in the chapter we discuss some tools for mitigating potential interpretability issues, and introduce the set of feature importance tools to aid feature selection.

Nevertheless, these challenges by no means imply that machine learning methods should not be applied within the context of financial modeling. Instead they suggest that the level of complexity in our model should match the available data. If we are estimating a model based on a single daily price series, then we should not attempt to throw it into a machine learning algorithm with hundreds of underlying features. Instead, we should distill the problem further based on some knowledge of the problem, such as our economic intuition. In this case machine learning may still be beneficial in identifying the complex, non-linear relationships between variables, but cannot be expected to sift through hundreds of features. In other cases, such as those related to alternative datasets rather than price/returns data, we may be able to apply a model that more closely mimics applications of machine learning in other fields, with a larger feature set.

23.3 SUPERVISED VS. UNSUPERVISED LEARNING METHODS

23.3.1 Supervised vs. Unsupervised Learning

A primary differentiating feature for machine learning algorithms is whether they fall into the category of supervised or unsupervised learning methods. Unsupervised learning algorithms, as the name implies, look to identify relationships in underlying data without the quant providing additional structure to the problem. Supervised learning, in contrast works with labels, and features created by the user. Said differently, in unsupervised learning algorithms there is no set output data. We are just trying to understand the underlying relationships better. There are several machine learning tools, such as clustering that can help in this regard. Supervised learning, however, requires us to specify a set of labels or output data that our features will then help us predict. Unsupervised learning techniques are often used for exploratory data analysis, and can help to narrow a large space of possible solutions into the most salient. Supervised learning, can help us generate forecasts, and better understand the relationships between the variables that we specify.

To take an example, consider the problem of using machine learning algorithms to identify market regimes. Unsupervised learning methods would work by processing a large market data set, with no knowledge of the regime, and break the market data into the most commonly occurring regimes. But importantly and fundamentally, we would not have to specify the regimes that we use to train the model. Instead the most likely regimes would be inferred strictly from the input dataset passed to the algorithm, and would be estimated in an unsupervised manner. In a supervised learning context, however, this problem would be formulated slightly differently. Most fundamentally, in this formulation we would need to label each piece of input data to a corresponding regime, and would then use a set of features to identify the properties of these regimes.

In some cases, it will be quite clear whether supervised or unsupervised learning techniques should be applied in a given situation, and both have their own strengths and weaknesses depending on the setting. Having said that, the first deciding factor in choice of a model in an ML context is whether to use a supervised or unsupervised approach. In the remainder of this section, we highlight some of the most common methods used for both supervised and unsupervised learning.

23.3.2 Supervised Learning Methods

As mentioned earlier in the section, supervised learning methods work by training themselves on a set of outputs, or labels given to the algorithm, in addition to a set of input data, or features. Supervised learning methods enable the user to provide more structure to a given model and have many applications in quant finance. In the following list, we highlight some of the most common supervised learning techniques, and the distinguishing features between them:

- **Linear Regression**: Linear regression is a technique that should be quite familiar to the reader, as it has been prevalent throughout the book, starting with its introduction in chapter 5. In fact, among other things, we saw how linear regression could be used to build an expected return forecast, or alpha model, or to calculate exposure to different risk factors. Along these lines, regression can be thought of as a machine learning technique, where we are trying to better understand the relationship between a dependent variables and one or more explanatory variables. Of course, when we think of incorporating machine learning techniques into quant finance problems, linear regression techniques are not necessarily the first thing that comes to mind. Instead we think of the more advanced techniques that we describe below. Nonetheless, the underlying set of concepts is the same and viewing these techniques from the same lens can help us understand how to extend our problems to more complex ML algorithms. Regression should be thought of as supervised learning because the dependent variable defines a set of output data, or labels for the problem.

- **Support Vector Regression**: Support Vector Regression (SVR) can be thought of as an extension of linear regression. Like linear regression, the goal is to predict a continuous output variable as a function of one or more underlying features. Unlike linear regression however, we can incorporate non-linear dynamics between the features and the output variables into SVR. Also unlike linear regression, SVR works by finding a separating hyperplane rather than minimizing a least squared difference. Fundamentally, this makes the algorithm less prone to outliers, a concept we will discuss with respect to Support Vector Machines later in the chapter.

- **Logistic Regression**: A classification model where we forecast the probability of a binary, or multi-classifying event. For example, we might use logistic regression to estimate the probability of a 0 or 1 event, such as default. Alternatively, we might use logistic regression to estimate the probability that one of a discrete set of events takes place. Logistic regression can be thought of as the equivalent to linear regression when the dependent variable is binary or from a finite group, although there are some subtle differences in the assumptions. In the next section, we highlight these differences and detail the framework for Logistic regression models.

- **Support Vector Machines**: Support Vector Machines (SVM) is another classification technique that improves on logistic regression by enabling more complex, non-linear relationships. Like SVR, SVM also works by finding a separating hyperplane, instead of minimizing the least squares distance, meaning that it is less prone to outliers. SVM techniques are discussed in more detail in the next section.

- **Nearest Neighbor**: K-nearest neighbor algorithms are simple machine learning techniques that rely on similar observations in order to generate a prediction for new data. These predictions can either be of continuous or binary variables. In the context of classification, a new observation would be classified based on the closest observations in the training set, with the idea being that if its neighbors were all classified with a certain tag, then the new observation is likely to have that tag as well, as it is similar. The k in the name K-nearest neighbor reflects the fact that the k nearest neighbors are used to inform the classification for each new piece of data. Nearest neighbor algorithms are discussed in more detail in the next section.

- **Neural Networks**: Neural networks can be both a supervised as well as unsupervised learning technique, depending on the algorithm. In the case of supervised learning, neural networks can be used to help classify labeled sets of data. In the case of unsupervised learning, they are a tool for clustering data. With respect to supervised learning, neural networks work by using a set of features, referred to as the input layer. This input layer is then transformed in a hidden layer that is responsible for formulating the problem such that it can produce an output label, which is referred to as the output layer. Neural networks can contain one or more hidden layers, and these hidden layers may incorporate linear or non-linear relationships between the chosen features and the output variable of interest. Neural networks with many hidden layers, referred to as Deep Neural Networks, are introduced in the next chapter. The theory behind Neural networks is not discussed in detail in this text, however, interested readers should consult [4] for a more detailed treatment.

- **Decision Trees**: Decision trees enable us to build a model with conditional control statements, nodes, and corresponding outcomes that may have a reward or a cost associated with them. Algorithms can then be tuned to maximize an objective over these control sequences, for example maximizing utility or minimizing cost. As such, these types of structures are commonly used in machine learning techniques. Random forest, for example, a popular machine learning technique, uses decision trees in order to build a classification or regression model. These techniques are not covered in detail in this text, however, interested readers should consult [74] for more information.

Later in the chapter, we explore several of these classification techniques, including K-Nearest Neighbor, Logistic Regression and Support Vector Machines. More detailed treatment of the additional regression and classification techniques not covered in this text can also be found in [141] or [74].

23.3.3 Regression vs. Classification Techniques

In the previous section, we discussed two main types of supervised learning techniques: regression and classification. In the case of regression, the output we were forecasting or otherwise working with was a continuous variable. For example, returns are inherently continuous variables so forecasting them at first glance seems to be a problem best solved via regression techniques. By contrast, in classification problems the output we were working with is either binary or from a fixed set of classifiers. For example, instead of using regression to forecast a return we could use classification techniques to forecast the sign of a future return. Alternatively, we could classify returns as good, bad and neutral, and train a classification algorithm to forecast which tag new observations are most likely to have.

The use of machine learning techniques, in particular opens up a wide array of classification techniques, many of which are introduced in this chapter, that we can leverage in our models. In many situations, these techniques can be quite helpful, and can overcome some of the potential challenges associated with forecasting a continuous variable in the presence of significant estimation error. Nonetheless, there are no doubt places where application of regression and classification problems may each be more appropriate. When conducting a supervised learning project, the choice between these regression and classification will be a defining feature of the model, and is often one of the first modeling decisions that we need to make.

23.3.4 Unsupervised Learning Methods

Unlike supervised learning techniques, in an unsupervised learning problem we ask the underlying algorithm to extract structure itself, without providing any labels or output variables. This type of technique can be useful to understand the underlying data better, and identify relationships among the underlying variables. As such, these techniques are commonly used in exploratory data analysis. [7] Additionally, unsupervised learning techniques can be used to help narrow the space of potential models. In the remainder of this section we discuss a few of the most common unsupervised learning methods.

- **Principal Component Analysis**: In chapter 5 we introduced PCA as a technique for transforming, or decomposing a matrix into a set of orthogonal factors.

[7] And we introduced many of these techniques in chapter 10 when introducing EDA.

PCA also helped us understand which components explained the majority of the variance, and which were possibly statistically insignificant. We then saw examples of how PCA can be used in the realm of fixed income in chapter 19, and trading, in chapter 27. Just like regression can be viewed as a machine learning technique, so too can PCA. In particular, it is a means of explaining the underlying structure of a matrix, often a covariance or correlation matrix. This structure is extracted itself, without our labeling or guiding the algorithm. As such, PCA can be naturally thought of as an unsupervised learning technique.

- **Autoencoder**: Roughly speaking, autoencoder can be thought of as an extension of PCA where we are able to incorporate non-linear relationships. Like PCA, autoencoder can be used for dimensionality reduction, and doesn't require labels or creation of a set of output variables. Instead, autoencoder is a type of neural network that helps to classify, or code a set of unlabeled data. Said differently, it is another unsupervised learning technique that can help us to understand the structure of the data that we encounter in quant finance problems. We provide more details on autoencoder algorithms in the next chapter, and provide an example of its use as an alternative to PCA. More information on autoencoder algorithms can also be found in [25].
- **Clustering**: Cluster analysis, or clustering is another unsupervised machine learning technique wherein we group different parts of our dataset based on their similarity to other observations. These groups are then referred to as clusters. These groups are formed without any predetermined labels. Instead, clustering algorithms examine the structure of the underlying data, usually using some distance metric, in order to extract commonality. There are different types of clustering algorithms based on different definitions of similarity of the underlying data. Clustering has many applications in quant finance. For example, it can be used to identify stocks or other assets whose properties are the most similar. It can also be used to find market periods that are most similar.

In the next section, we provide a broad overview of two of the more common clustering algorithms, K-Means and Hierarchical clustering. Additional coverage of clustering and other unsupervised learning techniques can be found in [141] and [74].

23.4 CLUSTERING

23.4.1 What is Clustering?

Clustering analysis, as previously mentioned, is an unsupervised learning technique that seeks to extract structure from a dataset. It is particularly helpful when there is some differentiation in the underlying data. For example, maybe the data comes from different regimes, and, within each regime, their statistical properties may be similar, but between regimes, may be highly differentiated. These regimes then would correspond to different clusters in the data. To see this, consider the following, hypothetical example of returns for a theoretical asset:

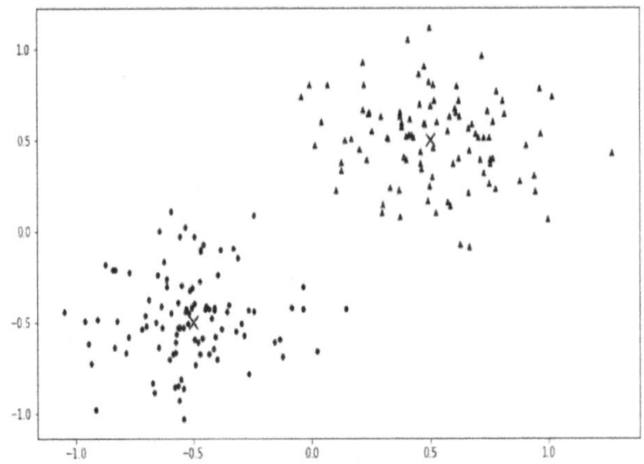

As we can clearly see in the chart, the returns data appears to come from multiple different underlying centers. Within each center, we can see a reasonably tight pattern of what the return expectation might be. But between these clusters, the data is fairly differentiated. Thus, if we were to simply use an average over the entire dataset, we might miss this multi-modal nature of the data, which might lead to suboptimal decisions.

We might also postulate that the same phenomenon exists across the cross section of assets. For example, stocks within the same industry, or sector, may display similar statistical characteristics but those across sectors may be more differentiated. Clustering algorithms can uncover these types of dependencies and similarities, including those that are more subtle and harder to identify subjectively. In the remainder of the section we discuss k-means and hierarchical clustering techniques, both of which are commonly used to break datasets into clusters, and in doing so find underlying relationships in the data.

23.4.2 K-Means Clustering

K-means is one of the simplest and also most commonly used clustering algorithms. It is used to break a dataset into k clusters, where each datapoint is uniquely identified as being from one of the k clusters. The datapoints are assigned to clusters such that the distance between the datapoints and the means, or centroids of the clusters, is minimized. This can be written mathematically as the following equation:

$$\min_{S} \sum_{i=1}^{k} \sum_{j=1}^{N} \mathbb{1}_{x_j \in S_i} \left(x_j - \mu_i\right)^2 \tag{23.1}$$

where k is the number of clusters, N is the number of observations in the dataset $X = \{x_1, x_2, \ldots, x_N\}$, μ_i is the mean in cluster i, and $\mathbb{1}_{x_j \in S_i}$ is an indicator function equal to 1 if the datapoint is within the cluster and 0 otherwise. Further, S is a vector of cluster assignments, S_i, corresponding to each observation in X. It can be shown that this can equivalently be written as a minimization against the sum of squared distances between all points within each cluster, as shown below:

$$\min_{S} \sum_{i=1}^{k} \left(\frac{1}{N_i} \sum_{x_j, x_k \in S_i} (x_j - x_k)^2 \right) \tag{23.2}$$

where as before k is the number of clusters, N is the number of observations in the dataset $X = \{x_1, x_2, \ldots, x_N\}$, and N_i is now the number of observations in cluster i. Further, S is a vector of cluster assignments, S_i, corresponding to each observation in X.

These minimization problems can be solved using the iterative techniques developed in chapter 14. However, in practice this is unnecessary as the sklearn package in Python has a built-in **sklearn.cluster.KMeans** function. This makes k-means an easy clustering technique to implement in practice. The function returns an output vector corresponding to S, which identifies the cluster associated with each observation, i in the dataset. It should be noted that these clusters are chosen purely statistically such that the intra-cluster variance is minimized. As such, in some cases the clusters may have a natural interpretation, however, in other cases interpreting them may be more challenging.

23.4.3 Hierarchical Clustering

Another clustering technique is hierarchical clustering where multi-level clusters are created based on their similarity, or the distance between them. At the most granular level, each item may be placed in its own cluster, with only items that are very close to each other clustered together. At the other extreme, all items may be clustered together. A hierarchical clustering algorithm defines how items will be clustered from their most granular to roughest levels. Readers may have come across dendrograms in practice, which are common visualizations of hierarchical clustering algorithms. In a dendrogram we depict the closeness of each item to each other, and show how the linkages are performed by the algorithm. In the following chart, we show an example of a dendrogram applied to the US equity market sector ETFs:

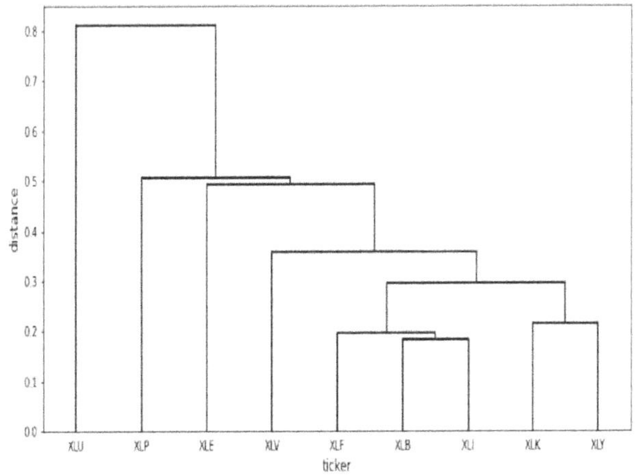

Here we can see that the dendrogram is consistent with the return correlation matrix. Specifically, the highly correlated assets, such as XLB and XLI, and XLK and XLY, appear next to each other on the dendrogram.

Hierarchical clustering algorithms are defined by their underlying distance metrics as well as their underlying linkage criteria. Distance metrics, as explored in the next section, define closeness of the underlying data. Linkage criteria, conversely, refers to how the items are placed into clusters based on those distance metrics. For example, we might use the minimum, maximum or average distance from a cluster in order to decide which cluster to link a particular item to. The actual clusters in a hierarchical clustering algorithm are then formed based on the specified number of desired clusters passed to the algorithm, which defines where on the y-axis of the dendrogram the clusters should be cutoff. In practice, choice of different linkage criteria or distance metric can cause notable changes to the resulting clusters. As such, these choices must be made judiciously. Sensitivity analysis, in particular, to different time periods and universes, can help to gauge stability of the clusters.

23.4.4 Distance Metrics

Clustering methods inherently work by finding elements in a set that are closest to each other according to some metric. In order to do this, the algorithm needs to be able to define closeness, which requires computing the distance between each element. Some techniques, such as k-means implicitly work with a certain distance metric, as we saw the algorithm was minimizing the variance within each cluster. Hierarchical clustering techniques, in contrast, can use different types of distance metrics, including a sum of squared distance approach, or a correlation-based metric. In practice, the choice of a distance metric can lead to significantly different results, and as such, it is important to choose a metric that is stable in the context of the problem we are trying to solve.

The following list details a few of the most common distance metrics used in hierarchical clustering algorithms:

- **Euclidean Distance**: Uses a sum of squared distance approach. Mathematically speaking, this is referred to as minimizing the L2 norm, and can be written as:

$$d_{\text{euclidean}} = \sqrt{\sum_{i=1}^{N} (x_i - y_i)^2} \tag{23.3}$$

- **Manhattan Distance**: Use the sum of the absolute deviations, which can equivalently be referred to as the L1 norm and written as:

$$d_{\text{manhattan}} = \sum_{i=1}^{N} |x_i - y_i| \tag{23.4}$$

- **Pearson Correlation Distance**: Pearson correlation is based on the correlation coefficients between the underlying elements, and can be expressed as:

$$d_{\text{pearson}} = 1 - \frac{\sum_{i=1}^{N}(x_i - \mu_x)(y_i - \mu_y)}{\sqrt{\sum_{i=1}^{N}(x_i - \mu_x)^2 \sum_{i=1}^{N}(y_i - \mu_y)^2}} \tag{23.5}$$

These are by no means the only distance metrics that exist in practice but are some of the most commonly used. Lopez de Prado, in his work on hierarchical risk parity [73], suggested using a correlation-based distance metric.

23.4.5 Optimal Number of Clusters

When using k-means, or hierarchical clustering, the number of clusters is a required input into the algorithm. That is, it is an internal model parameter. This means that we as quants must choose the number of optimal clusters. One method for helping make this choice is to leverage an Information criteria, such as AIC [53]. Readers may recall that this same approach was used in chapter 5 when determining the number of lags to include in ARMA models. Using this approach, we can generate a visualization like the following where we plot the AIC vs. the number of clusters:

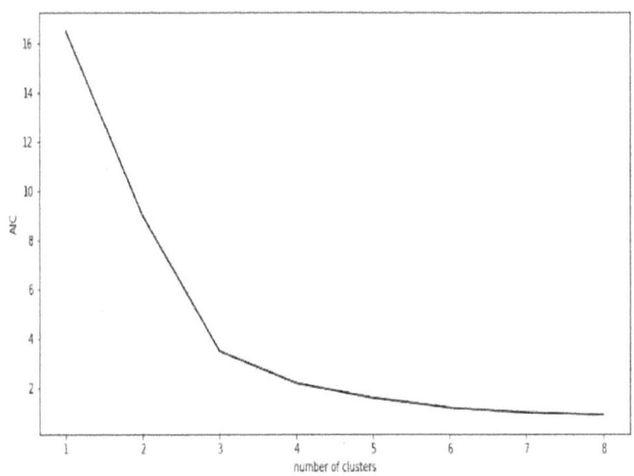

The optimal number of clusters, would then be chosen to be around the "elbow" of the visualization, which indicates a trade-off between model accuracy and generalizability. Other subjective components might also factor into our decision, such as if there are certain expected clusters that we expect to see manifest in the data. Cross validation techniques, alternatively, could be used to choose the optimal number of clusters.

23.4.6 Clustering in Finance

Applications of clustering analysis naturally occur in many contexts within quant finance. For one thing, they can be extremely beneficial in understanding the underlying data better. That is, they can be helpful in performing exploratory data analysis, as detailed in chapter 10. Similarly, they can be useful for anomaly or outlier detection, as we detailed in chapter 9. Clustering techniques also can be used to help us understand the underlying correlation structure across a set of assets, which is a fundamental problem in quant finance when it comes to building trading strategies and forming portfolios. For example, in chapter 27, we discussed pairs trading as a potential trading strategy that worked on highly correlated pairs. Along these lines, clustering can be a useful tool for identifying a universe to search for statistically significant pairs. Similarly, clustering can also uncover subtle dependencies between assets that could then be leveraged to find lead-lag relationships, or cross asset correlation, as detailed in chapter 27. More generally, we can use clustering techniques in this regard as a tool for narrowing the large, sometimes intractable space of possible signals into a smaller set.

Clustering techniques have also recently been employed in the context of portfolio optimization techniques, most notably by Lopez de Prado [76]. The premise behind this work is to allocate weights, or risk contributions, to different clusters. Intuitively, as each cluster is likely to be less correlated, it is sensible to think that we would want each distinct cluster to have similar weights, or in the context of risk parity, similar risk contributions. This approach enables us to incorporate correlation information, via the appropriate distance metric, without suffering from many of the challenges that we introduced in chapter 24 with respect to standard mean-variance techniques. Additionally, clustering techniques can be used as a regime detection tool, by splitting the historical data into clusters. These clusters then tell us the most common return patterns over time. For example, when doing this we often see risk-on and risk-off clusters emerge, which are defined by strong and poor returns for equities, and other related assets. In the next section, we show how to apply clustering techniques on a set of assets as a tool for understanding their underlying market regimes.

23.4.7 Clustering in Practice: Asset Class & Risk-on Risk-off Clusters

In this example, we use historical returns for a list of stocks and bonds, to cluster data for a set of ETFs spanning multiple asset classes [8]. In doing this, we leverage the clustering coding sample in the repository. We first cluster the assets into two distinct clusters. Doing so, we can see that all the stocks are clustered into one group, and all bond and credit instruments clustered into the other:

Cluster 0 (Bonds)	**Cluster 1 (Stocks)**
TLT	SPY
AGG	ACWI
EMB	IWM
HYG	EFA
GOVT	EEM

[8]Using the period 2021-01-01 through 2024-02-29.

Next, in the following table, we show the mean and standard deviation for the two clusters:

Cluster	Mean	Standard Deviation
0 (bond)	-0.02%	0.69%
1 (stock)	0.15%	1.18%

This is consistent with the fact that stocks tend to give higher returns over bonds, which comes at a cost of higher risk.

On the other hand, we can cluster the returns by date instead of asset. In this case we would input the returns for all assets, including both the stocks and bonds in the universe, and find the days with most commonality. When doing this, we again allow the k-means algorithm to cluster the data without any guidance, that is, we are using unsupervised learning. Once the clusters have been assigned, however, we can analyze them ex-post and define them in accordance with their properties. For example, we often find that so-called risk-on days cluster together, and, similarly, days that the market sells off tend to cluster together. Of course, mathematically speaking this doesn't have to be the case, as the extraction of clusters is purely statistical, meaning that the presence of an interpretable risk-on or risk-off cluster need not happen by necessity. Nonetheless, it is common to observe this pattern in markets, and it is what we find in our exercise as well. In the following visualization, we show the labels of the historical dates, identifying which are risk-on and which are risk-off:

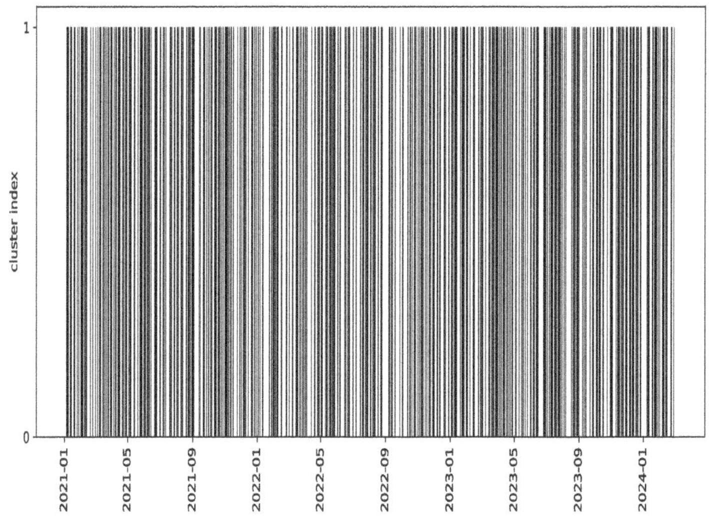

Similarly, in the following table we highlight the mean and standard deviation of all assets in the risk-on and risk-off clusters:

Cluster	Mean	Standard Deviation
0 ("risk-on")	0.47%	0.82%
1 ("risk-off")	−0.53%	0.86%

The reader should notice that the means in the two clusters are quite different, with one being very positive and the other being very negative. The reader may also notice that the standard deviation in the risk-off cluster is noticeably higher than in the risk-on cluster. This is a manifestation that volatility tends to be heightened in risk-off periods, and is consistent with many of the examples presented throughout the text.

Looking more closely, we might expect to find that different assets have different return profiles in risk-on and risk-off periods. For example, one might expect that the negative correlation between stocks and bonds many have observed would lead to strong positive returns for stocks in risk-on periods, and positive returns for bond instruments in risk-off periods, providing diversification to multi-asset portfolios. Interestingly, however, during this period of analysis, we observe that this potential diversification is less prevalent, and in fact, all asset returns have the same sign in risk-on and risk-off periods, albeit with different magnitudes. In fact, this was a well-known phenomenon over the period of analysis, that is, in the aftermath of Covid the correlation between stocks and bonds became positive, providing challenges for many investors expecting diversification from fixed income instruments. In the following table, we highlight the average return of each ETF in each type of market environment:

ETF	Risk-On Mean Return	Risk-Off Mean Return
ACWI	0.74%	−0.77%
AGG	0.08%	−0.12%
EEM	0.69%	−0.83%
EFA	0.71%	−0.75%
EMB	0.31%	−0.38%
GOVT	0.03%	−0.07%
HYG	0.31%	−0.34%
IWM	1.04%	−1.13%
SPY	0.75%	−0.74%
TLT	0.05%	−0.18%

Here we can indeed see this picture emerge, that is, consistent returns for all assets in risk-on and risk-off regimes. The reader is encouraged to leverage the coding repository to analyze the behavior of asset class returns in different regimes over different historical time periods.

23.5 CLASSIFICATION TECHNIQUES

23.5.1 What is Classification?

As mentioned previously in the chapter, classification problems are ones that involve estimation of a binary or ordinal set, rather than a continuous variable. They arise naturally when a set of input variables, or features, can help us identify which item in the ordinal set an item belongs to. In other cases, we can create a binary or ordinal set to use within a classification framework. To see this, consider the following hypothetical example of market returns on the y-axis by a theoretical single explanatory feature on the x-axis:

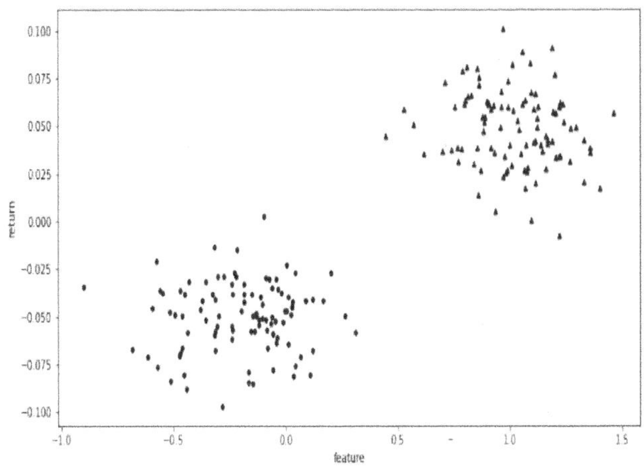

We can clearly see in the above chart that the behavior of the market is fundamentally different when the value of this hypothetical feature is above or below 50%. That is, they appear to come from different distributions or market regimes. This type of relationship of course would be quite beneficial to be able to identify in practice, and would logically lead to a strategy that is long the market when the quantity is above 50% and short the market otherwise. The quality of this signal expressed via a regression approach may, conversely, not be as strong because of the different behavior in each cluster. Instead, mathematically, this relationship could be formulated as a classification problem. We can label poor stock returns and strong market returns as a binary variable and then use this single explanatory feature to classify future data. Of course, although classification in this over-simplified example is trivial and quite clear, in practice we, unfortunately, encounter relationships much more complex. If relationships such as those in the figure above did exist, then predicting returns would be much easier. Instead the relationship in practice is much messier.

23.5.2 K-Nearest Neighbor

K-nearest neighbor is one of the simplest classification techniques, and, roughly speaking, it works by classifying new data using the data in the training set that it most closely resembles. To define similarity we use a set of specified features, with the idea being that observations with similar feature sets should result in similar outputs, or classifications. As such, k-nearest neighbor is a supervised learning technique where we need to specify a set of input features that aid the classification. Of course, to define the datapoints that most closely resemble a particular point also requires the use of a distance metric, such as Euclidean distance, Manhattan distance, or another distance metric more suited to binary values. For example, in the case of a Euclidean distance metric, we would have a distance equation of the form:

$$D(x, n) = \frac{1}{d} \sum_{i=1}^{d} (x_i - n_i)^2 \tag{23.6}$$

Once this distance metric has been defined, we can create the distance between our new data and all existing points, using the predetermined set of features. We can then sort by the distance metric, choosing only the k smallest values. The classification assignment is then done according to the classification of these k similar, neighbors. Python has built-in functionality for k-nearest neighbor algorithms, most notably in the sklearn package where **sklearn.neighbors.KNeighborsClassifier** can be used to perform classification via a nearest neighbor algorithm.

The number of neighboring datapoints, k that is used to estimate the output or classification is a model input, or, phrased differently, is a model hyperparameter. In the case of $k = 1$, the new data is classified solely based on the single, nearest observation in the training set. In the case of $k > 1$, then the new data is classified using multiple observations in the training set. To do this assignment, the modal, or most common classification among the neighbors is used as a classifier. As k is an internal model parameter, or hyperparameter, it is something that can be optimized using the cross-validation techniques introduced earlier in the chapter.

In addition to classification, a k-nearest neighbor approach can also be used to estimate a continuous rather than binary or ordinal value. In this case we would use k-nearest neighbor to perform a regression, and the estimated value of the dependent variable would be equal to the average of the values of the dependent variable in its k-nearest neighbors, as indicated below:

$$\hat{y}_{\text{knn}} = \frac{1}{k} \sum_{i=1}^{m} y_i \mathbb{1}_{y_i \in N} \tag{23.7}$$

where k is the number of neighbors and is a model parameter, m is the number of datapoints, and $N = \{n_1, \ldots, n_k\}$ are the set of defined nearest neighbors according to some distance metric such as those described in 23.4.4. Further, k-nearest neighbor algorithms can also be used to detect data anomalies or outliers. In this context, as we discussed in chapter 9, we would look for points with large distances from their neighbors, and examine the data quality of these points. In the coding repository, we provide an example that walks the reader through leveraging the K-Nearest Neighbor algorithm on country ETFs.

23.5.3 Probit Regression

Probit regression, contrary to the name, is another supervised learning classification technique that can be used to solve problems when the dependent variable is binary. In probit regression, we model the probability that the binary variable is true, or equal to 1. In probit regression, this probability is modeled using the CDF of a standard normal distribution, and is a function of a set of input, explanatory variables. Along these lines let's consider the case where each y_i can be either 1 or 0. When using probit regression, this is modeled mathematically via the following equation:

$$P(y_i = 1|X_i) = \Phi(X_i^\top \beta) \tag{23.8}$$
$$P(y_i = 0|X_i) = 1 - \Phi(X_i^\top \beta) \tag{23.9}$$

where Φ is the cumulative normal distribution, X_i is a vector of input, explanatory variables, or features, and β is a vector of coefficients that are in fact model parameters.

We could, of course, use other cumulative distribution functions. The salient point is that the CDF is monotone and maps $\mathbb{R}$ into $[0, 1]$, which is true for any CDF. This means that, for any CDF, it has a natural interpretation as the probability of a binary event occurring. In the next section, we will introduce logistic regression, which is another closely related technique that employs the same approach.

To calibrate a probit regression model, we need to find the coefficients, β that best fit our data. In practice this can be done by maximizing the likelihood of the data given our model parameters, β. For a given observation, using the PDF of the Bernoulli distribution we can write this likelihood as:

$$L(\beta)(X_i, y_i) = \left(\Phi(X_i^\top \beta)\right)^{y_i} \left[1 - \Phi(X_i^\top \beta)\right]^{1-y_i} \tag{23.10}$$

Assuming that all the observations are independent, we can similarly calculate the likelihood of the entire dataset given a set of model parameters, which can be written as:

$$L(\beta) = \prod_{i=1}^{N} \left(\Phi(X_i^\top \beta)\right)^{y_i} \left[1 - \Phi(X_i^\top \beta)\right]^{1-y_i} \tag{23.11}$$

It should be emphasized that, if the observations are not independent, then this likelihood function no longer holds. Thus, this is an important assumption of probit regression. That aside, we can maximize this log-likelihood function over the set of possible values for β using the optimization techniques introduced in chapter 14. Readers may also recall that this maximum likelihood approach was presented in chapter 25 in the context of estimated the parameters in a GARCH model.

Mathematically, it is more convenient to work with the log of the likelihood function as it enables us to switch the outer product with a more tractable sum. The log-likelihood for probit regression can then be written as:

$$\log L(\beta) = \sum_{i=1}^{N} y_i \log \left(\Phi(X_i^\top \beta)\right) + (1 - y_i) \log \left[1 - \Phi(X_i^\top \beta)\right] \tag{23.12}$$

Once we have estimated the set of model parameters, we can then estimate the probability of any new piece of data taking the values $y_i = 1$ or $y_i = 0$, respectively. In the next section we explore a close relative of probit regression, logistic regression, wherein we assume that a different function maps the explanatory variables to the probability of the event.

23.5.4 Logistic Regression

Like probit regression, logistic regression is another classification technique which can be used to estimate the probability of a binary event. Also like probit regression, logistic regression is a supervised learning technique that requires us to specify a set of explanatory variables, or features, which are then used to estimate the probability of the discrete or binary event. In fact, logistic regression is closely related to probit regression with the subtle difference in the function that we assume computes the probability of the underlying binary event. Logistic regression can also be extended to consider a larger set of discrete events. This is referred to as multinomial logistic regression.

Whereas probit regression uses the CDF for a standard normal to map from explanatory variables to the probability of a binary event, logistic regression relies on the logit function for this transformation. This logit function can be written as:

$$\text{logit}\,(z) = \frac{1}{1 + \exp\,(-z)} \tag{23.13}$$

It is further assumed that z is a linear function of the features, X_i, and the model parameters, β. That is:

$$z = X^\top \beta \tag{23.14}$$

In logistic regression, the probability that the binary event occurs, that is, the probability that $y_i = 1$ is simply the logistic function. This means that the probability distribution of y_i can be characterized by the following equation:

$$\begin{align} P(y_i = 1|X_i) &= \text{logit}\,(z) \tag{23.15} \\ &= \frac{1}{1 + \exp\,(-X^\top \beta)} \tag{23.16} \\ P(y_i = 0|X_i) &= 1 - P(y_i = 1|X_i) \tag{23.17} \end{align}$$

Notice that this has a similar form to probit regression, with a modified functional form on the right hand side of (23.15). Importantly, however, the logit function inherits the desirable properties that we had in the previous case, most notably that it is defined between 0 and 1 and is a monotonically increasing function of $X^\top \beta$.

It can be shown that, the probability in 23.15 can be re-written as [137]:

$$\begin{align} \log\left(\frac{Pr(y_i = 1|X_i)}{1 - Pr(y_i = 1|X_i)}\right) &= X_i^\top \beta \tag{23.18} \\ \frac{Pr(y_i = 1|X_i)}{1 - Pr(y_i = 1|X_i)} &= \exp\left(X_i^\top \beta\right) \tag{23.19} \end{align}$$

Some readers, especially those who play cards or gamble, may recognize the quantity $\frac{p}{1-p}$ in the equations above as the odds of the event $y_i = 1$. This means, that, a defining feature of logistic regression is that is implies a linear relationship between the features, X, and the log-odds.

Just as we did with probit regression, the coefficient estimates in the context of logistic regression can be estimated using maximum likelihood techniques. As the output variable y_i is still assumed to be binary, we can again leverage the PDF of a Bernoulli distribution in order to write the likelihood, and log likelihood, in terms of the logit function. As before, working with the log-likelihood function will be more convenient. The details of this log-likelihood approach are omitted here and instead left as an exercise for the reader.

In practice, a key feature of logistic regression is its interpretability. Like linear regression, logistic regression has easy to understand coefficients and corresponding t-statistics and p-values that are familiar to statisticians and econometricians. This is a significant advantage of logistic regression as it makes the underlying model easier to understand. In particular, it is easy to gauge the impact of each underlying explanatory variable, discern what the most significant features are and identify whether the variables increase or decrease the probability of the binary event occurring.

Conversely, the assumed relationship between the explanatory variables and the probability, or the log-odds, may be oversimplified and unrealistic in practical applications. In these cases, it may be preferable to leverage a technique such as SVM that is able to incorporate more complex non-linear relationships between the variables. Further, like standard regression techniques, logistic regression is prone to issues with multi-collinearity. As such, unlike some other machine learning techniques it is challenging to use logistic regression in the context of a large set of features that may have high levels of correlation. Additionally, recall that in our maximum likelihood setup we assumed that the observations were independent. In practice, this is often not the case, as we detailed in chapter 25. This could occur because of cross-sectional correlation, or time-series serial correlation. Time series serial correlation in particular, is introduced when using overlapping observations in our model. When using logistic regression, the presence of this type of correlation between the observations violates the underlying model assumptions, and may lead to biased t-statistics and standard errors if not accounted for properly.

While logistic regression is conceptually similar to the standard linear regression techniques that we introduced in chapter 5 as have relied on throughout the book, there are a few subtle differences in the assumptions in the two approaches. For one, while both assume a linear underlying relationship in the case of logistic regression this linear relationship is with respect to the log odds rather than the explanatory and dependent variables directly. Secondly, while both assume that the observations are independent, linear regression further requires that the observations are identically distributed, in particular meaning that they are homoscedastic. This assumption of homoskedasticity is not required when using logistic regression. Additionally, the assumption of normally distributed residuals in a linear regression framework is not necessary in the context of logistic regression. Notably, however, both logistic and linear regression assume a lack of correlation between the input, explanatory variables. This makes both techniques susceptible to issues associated with multi-collinearity.

23.5.5 Support Vector Machines

Support Vector Machines (SVM) is another classification technique that, like probit and logistic regression, can be used to predict which group new observations belong to. For example, SVM can naturally handle classification of binary variables, or those from a discrete, ordinal set. This is done using a set of input features, as it was for the previous techniques. Thus, SVM is another supervised learning technique. It can be thought of as an improvement on the previous techniques, however, in the sense that it allows us to incorporate complex, non-linear relationships into our classifications. As we will see, it is also built on a different type of optimization approach, that is based on a separating hyperplane instead of least squares or log-likelihood, making the technique less prone to issues with multi-collinearity and less sensitive to outliers.

In the context of SVM, we again assume we are working with a binary classification. That is, we assume that the output variable, y_i must take one of two values. In this case, however, we now assume that y_i can either be 1 or -1. Ideally, we want to classify the data such that we are able to create a line, or more formally a separating hyperplane such that all values above the line are tagged with $y_i = 1$ and all values below the line are tagged with $y_i = -1$. Note that the choice of ± 1 is made for convenience, as we will see shortly. To see this consider the following illustration of points that have a clear classifying criteria:

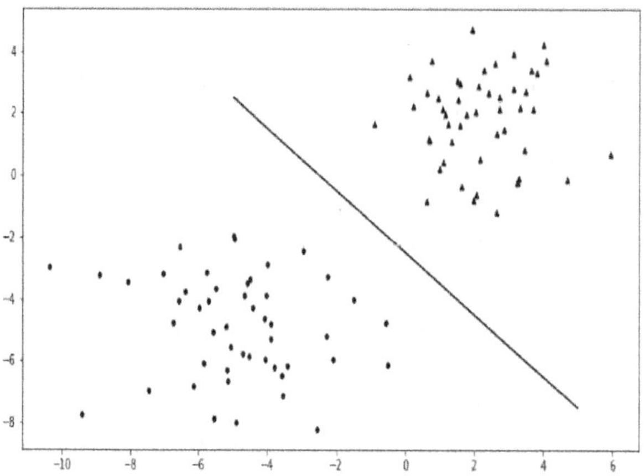

Looking at this visualization, we can see the data clearly separates into two categories based on the single, assumed explanatory feature. SVM works by creating a set of lines, such as the one shown in the chart, that separate the data. We can see above that there are many possible lines that do this, however, in SVM we want to choose the lines that are the furthest apart, as that will ensure that the classifiers are as far apart as possible.

Mathematically, this optimal line, or separating hyperplane, can be found by solving the following optimization problem:

$$\min_{\beta} \left\{ \|\beta\|^2 \mid \left(X_i^\top \beta + b \right) y_i \geq 1 \quad \forall i \right\} \tag{23.20}$$

where $(X^\top\beta + b)\, y_i$ is a convenient way of writing:

$$\begin{cases} X_i^\top\beta + b \geq 1 & y_i = 1 \\ X_i^\top\beta + b \leq -1 & y_i = -1 \end{cases} \tag{23.21}$$

In this context, new datapoints, X_i, can be classified according to the following crieteria:

$$X_i \mapsto \begin{cases} X_i^\top\beta + b \geq 1 & \text{Category 1} \\ X_i^\top\beta + b \leq -1 & \text{Category -1} \\ \text{otherwise} & \text{undecided} \end{cases} \tag{23.22}$$

Said differently, items that are above the line, or hyperplane, will be classified as $y_i = 1$. Similarly, items that are below will be classified as $y_i = -1$. Items that appear in between, in contrast, will be those that we do not know how to classify. This criteria meets our primary objective in SVM. That is, we classify data by finding the hyperplanes that separate the data and are furthest apart. We can similarly find the set of vectors X_i that define the hyperplane in the middle of the gap via the following condition:

$$X_i^\top\beta + b = 0 \tag{23.23}$$

Classification using this rule in (23.23) instead of (23.22) would result in a classification of every new piece of data, with no undecided observations. Conversely, it would involve applying classifications to data that we have less confidence in.

Remember that the parameter set, β is defined by the model. That is, we will choose values for β such that the optimization in (23.21) is minimized on a training dataset. Once this process is complete, the parameters, β will be fixed and can then be used to classify new observations. This is analogous to finding the β's in a regression model that best fit the underlying data. This process in the context of machine learning is referred to as training the model. In other parts of the text, we referred to this analogously as calibration, or estimation of the model parameters.

The vectors at the edges of the gap will be the ones that meet the following criteria:

$$X_i^\top\beta + b = \pm 1 \tag{23.24}$$

The data that lie on these edges, or hyperplanes are called support vectors and only they matter in the classification. Importantly, if new data appears that is outside the hyperplanes, it will not change any of the parameters of the SVM. This is a critical differentiating feature between SVM and linear, logistic or probit regression. An important caveat of this is that is makes SVM less prone to outliers, as, provided the outliers aren't on the hyperplane, they do not impact the calculation or model parameters. It is also an important characteristic for efficiency reasons, as it effectively leads to dimensionality reduction in that only a subset of points are used in the calculation.

Of course, in realistic problems, it will, generally speaking, not be possible to cleanly separate the two sets. As the data becomes messier, mis-classified items, or violations of the separating hyperplane, are inevitable. In this context, it is often convenient to allow some of these violations, rather than significantly altering the slope of the hyperplane. To do this, we can introduce slack variables, which are defined as:

$$\zeta_i = \max(0, 1 - y_i \left(X_i^\top \beta + b\right)) \tag{23.25}$$

This leads to an augmented optimization framework, which can be written as:

$$\min_{\beta,\zeta} \left\{ \frac{1}{N} \sum_{i=1}^{N} \zeta_i + c\|\beta\|^2 \;\middle|\; y_i \left(X_i^\top \beta + b\right) \geq 1 - \zeta_i \quad \forall i \right\} \tag{23.26}$$

This formulation allows a fraction of the training points to leak across the boundrary into the gap. Importantly, it also introduces a new internal model parameter, c. This model parameter determines the relative penalty of a violation of the separating hyperplane versus an increase in complexity of the model, which may lead to overfitting. In order to choose this cost parameter in the context of SVM, cross-validation is generally used.

Up until now, we have presented a linear classifier for SVM. While this is not without use, and can still enable us to benefit from certain advantages to SVM, in many cases it will not be an optimal classifier. For example, consider a two-dimensional problem where the classification cleanly separates between points that are inside a unit circle and those that are outside. Even though these points are cleanly classified, no linear classifier will be able to handle this type of classification. Instead, we need to apply a non-linear transformation to our problem. In practice, this can be done using what is known as the kernel trick [242]. Fundamentally, the kernel trick enables us to apply a different, non-linear function in our classification. Further, it enables us to do so without compromising efficiency. A brief list of the most common kernels is presented in 23.5.5.

To get at the Kernel Trick, we start by trying to solve the optimization problem (23.20). To see how the Kernel trick works, we will consider the formulation of SVM that cleanly separates the data, in (23.20) without introducing slack variables, however, it should be emphasized this is for convenience and the kernel trick can be applied to (23.26) as well. To proceed, as we learned in chapter 14, let's begin by formulating the Lagrangian. Recall that this involves transforming our optimization problem from one with constraints, to an unconstrained optimization of a transformed function.

The Lagrangian for (23.20) can then be written as:

$$L(\beta, b, \lambda) = \frac{1}{2}\beta^\top \beta - \sum_i \lambda_i \left[\left(X_i^\top \beta + b\right) y_i - 1\right] \tag{23.27}$$

with a further constraint that $\lambda_i \geq 0$ for all i. Note that the constant $\frac{1}{2}$ is added for convenience as it will simplify terms after the appropriate derivatives are taken and has no impact on the solution as it is a constant.

In order to solve this optimization problem, we can solve for the first-order conditions, as we detailed in chapter 14, which involves taking the gradient with respect to the variables β and b. Doing this yields:

$$0 = \nabla_\beta L = \beta - \sum_i \lambda_i y_i X_i \tag{23.28}$$

$$\beta = \sum_i \lambda_i y_i X_i \tag{23.29}$$

$$0 = \partial_b L = -\sum_i \lambda_i y_i \tag{23.30}$$

From (23.29), we can see that the normal to the separating hyperplane, β, is a linear combination of the features X_i.

Substituting (23.29) and (23.30) into the Largangian above in (23.27) and multiplying, rearranging and collecting terms leaves us with the following optimization problem, which is referred to as the dual problem:

$$L(\lambda) = -\frac{1}{2}\sum_{i,j} \lambda_i y_i \lambda_j y_j X_i^\top X_j + \sum_i \lambda_i \tag{23.31}$$

Importantly, in this dual problem formulation, the features, X_i, only appear inside the dot product $X_i^\top X_j$, and do not appear by themselves. Thus, we can redefine this dot product using another function, or transformation, which we refer to as the kernel, as long as the new dot product meets certain criteria[9]. In practice, this means that (23.31) can be generalized to the following function which includes an arbitrary kernel function:

$$L(\lambda) = -\frac{1}{2}\sum_{i,j} \lambda_i y_i \lambda_j y_j K\left(X_i, X_j\right) + \sum_i \lambda_i \tag{23.32}$$

where $K\left(X_i, X_j\right)$ is a kernel.

This Kernel Trick allows us to turn linear classifiers into non-linear ones and, no less importantly in big data, allows to speed up calculations quite a lot by helping to keep dimensionality of the problem low. As such, this is an important feature of SVM, and means that, a more general form of (23.22) classification is:

$$X_i \mapsto \begin{cases} K(\beta, X_i) + b \geq 1 & \text{Category 1} \\ K(\beta, X_i) + b \leq -1 & \text{Category -1} \\ \text{otherwise} & \text{undecided} \end{cases} \tag{23.33}$$

where $K(\beta, \mathbf{X}_i)$ is a pre-specified kernel, which is an input to the SVM model.

In practice, choosing this kernel can have a significant impact on the quality of our classification model. Proper choice of this kernel is problem specific and requires an understanding of the relationship between our features and our corresponding classification. This becomes increasingly difficult as the number of dimensions, or number of features, increases, as visualizing the multi-dimensional relationship and therefore knowing how reasonable the chosen kernel is, can be challenging. With the appropriate choice of a kernel, we may find a nice, cleanly separated hyperplane

[9]Notably that the new dot product is a positive definite function of the two vectors.

with few violations and high accuracy. On the other hand, this same problem with another kernel may have high amounts of error and large numbers of mis-classified observations. For example, in this text we considered linear classifiers. In real-world applications, however, this linear classifier is likely to suffer from many of the same challenges the other linear techniques detailed in the text suffer from. Instead, we may want to use a different, non-linear kernel that more accurately represents the potential relationship between the variables.

A few of the most common alternate, non-linear, kernels that are used in finance problems are the radial basis function, a sigmoid function or a polynomial, each of which is defined below:

- **Radial Basis Function**:

$$K(X,Y) = \exp\left(-c\|X-Y\|^2\right) \tag{23.34}$$

- **Sigmoid**:

$$K(X,Y) = \exp\left(-c\|X-Y\|\right) \tag{23.35}$$

- **Polynomial**:

$$K(X,Y) = \left(cX^\top Y + r\right)^d \tag{23.36}$$

where in each case c is a cost parameter and X and Y are vectors representing the terms in the dot product in (23.31). This list is by no means exhaustive, but provides the reader with exposure to the different types of kernels used in SVM algorithms. Python has built-in functionality for SVM, for example in the sklearn package where there is a **sklearn.svm.SVC** function, and within this package different kernel functions are allowed to be specified. In the coding repository, we provide an example using SVM to solve a classification problem using country ETFs.

23.5.6 Confusion Matrices

One way to judge the accuracy of a classification model is to examine the fraction of data that is classifies properly, as well as the number of misclassified items. These results are often aggregated into a matrix that summarizes the different types of classification error, such in the example below:

$$C = \begin{bmatrix} \text{Model}_0|\text{Actual}_0 & \text{Model}_0|\text{Actual}_1 \\ \text{Model}_1|\text{Actual}_0 & \text{Model}_1|\text{Actual}_1 \end{bmatrix} \tag{23.37}$$

where $\text{Model}_i|\text{Actual}_j$ is the number of observations that the model has classified with tag i and has actual tag j.

Here we can see that the two matrix dimensions are the Actual and Model classification, which the matrix then summarizes with the correct classifications on the diagonal, and the off-diagonal elements representing the mis-classified items. Elements in the lower left off diagonal represent false positives, or those which the model classified as positive despite not being labeled as positive in the data. Conversely, elements in the upper right off diagonal represent data which was labeled as positive but the model did not classify as positive. [10]

In many ML techniques we can specify different penalties for different types of errors, for example, if we want to discourage false classifications in the lower left off-diagonal relative to false classifications in the upper right off-diagonal. In practice adjusting this penalty can help balance the amount of type I (lower left) and type II (upper right) errors in the model.

Clearly, our goal when designing a classification algorithm should be high accuracy, both in the training set as well as the test set. As such, we should create confusion matrices for both datasets, and ideally would like to see a large fraction of the data fall into the diagonals, with relatively small off-diagonals. Confusion matrices are an important tool for helping to understand the predictive power and accuracy of a given classification model. As such, it is recommended that they are part of the standard output that is analyzed when building a classification model[11].

23.5.7 Classification Problems in Finance

Classifications problems arise in finance in many contexts. For one, we may use them in conjunction with non-traditional data sources to try to summarize them effectively. For example, we might use a classification technique to classify news articles about an asset or industry as either positive or negative. In this case, the output labels would be binary, assuming articles are either positive or negative. The underlying feature set, then would be a function of the underlying text. Similarly, we may use a classification model as a regime detection model, wherein we try to identify for example, high and low volatility market environments. It should be emphasized that, for supervised learning problems, these classification problems will require us to create labels for the data. This would mean marking articles in a large training set as positive or negative in the first example. In the second example, it would mean identifying which regime each datapoint comes from.

Another common use of classification within financial modeling is to forecast the sign of a return. This leads to an embedded classification of positive and negative returns, which we could predict given a set of relevant features of our choice, including previous returns in the spirit of momentum and reversal signals. In this context we could just as easily choose a differently delineated classification of returns, such as:

- **Very Poor**: $r_{i,t} \leq t_{\text{very poor}}$
- **Poor**: $t_{\text{very poor}} < r_{i,t} \leq t_{\text{poor}}$
- **Neutral**: $t_{\text{poor}} < r_{i,t} \leq t_{\text{neutral}}$
- **Strong**: $t_{\text{neutral}} < r_{i,t} \leq t_{\text{strong}}$
- **Very Strong**: $t_{\text{strong}} < r_{i,t} \leq t_{\text{very strong}}$

[10]Sometimes referred to as false negatives.

[11]In many cases, Python creates this output by default, meaning no additional work is needed.

where $t = \{t_{\text{very poor}}, t_{\text{poor}}, t_{\text{neutral}}, t_{\text{strong}}, t_{\text{very strong}}\}$ are a set of monotonically increasing return thresholds that define the various return classifications. Of course, this is just an example potential return classification, and the precise details of how to optimally classify returns in this context are left as an exercise to the reader. It should be noted, however, that some research has shown that classification techniques may be more effective than trying to estimate a continuous return variable. Estimation error and a low signal to noise ratio of the continuous return estimates are potential reasons this approach might be more effective. In the next example, we construct a simple example using classification in the context of forecasting the sign of an equity return, using SVM.

23.5.8 Classification in Practice: Using Classification Techniques in an Alpha Signal

In this example, we leverage the coding sample in the repository to show a naive approach for predicting SPY returns in a classification framework. To do this, we take the previous daily return as the only feature in order to predict the sign of the next day's return, via SVM. We then check the model performance on both the training set and an out-of-sample test period using built-in functions in sklearn. In the following chart, we show the probability of a positive return over time:

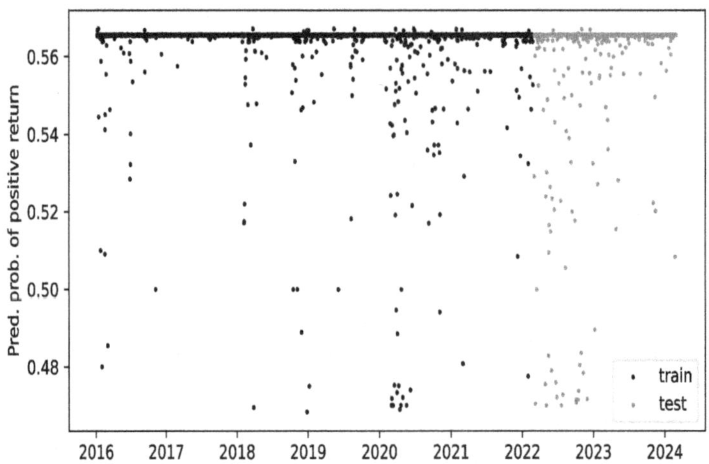

Unsurprisingly, in our naive model it is difficult to discern positive from negative returns and the result is positive return probability slightly greater than 50% in most time periods. The reader is encouraged to try to extend this simplified model with a more realistic feature set to try to improve the robustness of the predictions.

23.5.9 Other uses of Classification Techniques: Credit Risk Modeling

As we referenced in chapter 19 when discussing mortgage backed securities, one of the most natural places to fit machine learning algorithms in quant finance is when trying to determine the credit risk, or credit worthiness of a particular borrower. This is inherently a big data problem, as there are huge amounts of data on individual consumers that we can use to try to inform our credit risk model. Credit risk problems also often are naturally suited to be formed as classification problems, where we are trying to calculate the probability of an individual, or other entity, defaulting on a loan, for example. This means that the host of classification-based ML techniques that we introduced in this chapter, such as logistic regression and SVM, can be helpful in solving credit risk problems.

If we think about credit risk problems from the perspective of the entity extending the credit, they are tasked with figuring out whether it is worthwhile for them to give a loan to a given customer, and how much the loan should cost. This decision of course, is a direct function of the probability that a borrower will repay their loan, which is a function of their unique credit history. This probability then, is potentially a complex function of many underlying features, as indicated below:

$$P\,(\text{default}) = f(x_1, x2, ..., x_n) \tag{23.38}$$

where $x_1, \ldots, x_n$, are the data for the features. The features might be a function of a borrower's credit score, their income, and their other debts. As noted above, this probability can be modeled using machine learning techniques. In fact, in most cases, there is a plethora of available data on the borrower. Machine Learning is a natural choice to try to extract meaning behind the data and the ensuing default behavior, as these relationship are likely to be non-linear. The combination of the large number of borrowers/loans available, as well as the potential rich dataset for each borrower, make this type of model ideal for ML algorithms. In comparison with other approaches, ML techniques can make this problem more tractable, helping to identify key features and their potential non-linear relationship to each other and to defaults.

23.6 FEATURE IMPORTANCE & INTERPRETABILITY

23.6.1 Feature Importance & Interpretability

As mentioned earlier in the chapter, one of the main challenges associated with machine learning techniques, especially in a financial context, is the difficulty in interpreting the model. The need to truly understand the underlying workings of a model of course varies in different settings. In a high frequency trading algorithm, or a technique to understand alternative data, we may not require full knowledge of the model dynamics. In other situations, practitioners may be uncomfortable with black-box type algorithms where they can't understand the intuition or underlying relationships between the variables. In other words, in finance a certain value is placed on building models that are interpretable. This can be challenging to do in the context of many hundred features, or explanatory variables.

Regression models, comparatively, for all of their documented flaws, are relatively interpretable[12]. T-stats, for example, provide a natural way of gauging the statistical significance, and therefore the importance of a feature. As we incorporate more complex models and non-linear relationships, interpretability can become more challenging. In a non-linear model, there may be times when an increase in a certain input

[12]Which is perhaps an explanation for their prevalence in financial quant models.

feature increases the likelihood of an event, and other cases where an increase in the same input feature reduces the likelihood of the same event. Inherently, this creates challenges in interpreting our results and makes it harder to align these relationships with our economic intuition. It also means that a simple static coefficient and t-stat is no longer sufficient to gain intuition on the relationship, meaning we must rely on other tools for judging significance.

In this context, fortunately, there are certain things that we can do to help understand our models better. Sensitivity analysis, for example, can be helpful in determining both the significance of different features and also the direction of the relationship, at least locally. With respect to feature importance, we can see how much the fit is degraded when each feature is excluded. The more that excluding the feature degrades the fit of the model, clearly, the more significant the feature. Additionally, we can perform sensitivity analysis on our predictions to determine the relationship between the variables. It should be emphasized that this uncovers the local relationship, which may be dynamic and change as conditions evolve. As an example, we can shift an input feature, and then re-evaluate our prediction. This will implicitly tell us the impact on our prediction of a change to that specific feature, holding all other variables constant.

Python also has a set of built-in feature importance tools that greatly aid the process of machine learning models more interpretable. The reader is encouraged to consult the coding repository for more information on feature importance techniques, where we apply these tools in a simplified context to both a toy dataset as well as a classification model for predicting returns.

Feature importance analysis is important for many reasons. As mentioned, the first reason is that it can help us to understand our ML models better and to provide better intuition. Additionally, feature importance tools and techniques can be useful for narrowing down our feature universe. That is, they are a critical part of the feature selection process. In many ML algorithms we begin by casting an extremely large net of potential features. In a way this is the point of leveraging ML tools to begin with: they help us to identify relationships we may have otherwise missed. Often, however, it is desired to pare down this feature set prior to a final production model. This process is referred to as feature selection, and can prevent overfitting and lead to better out-of-sample performance of our models.

CHAPTER 24

Artificial Intelligence: Incorporating Deep Learning, Large Language Models and Working with Unstructured Data

24.1 OVERVIEW

IN THE LAST chapter we provided an overview of the broad landscape that is AI. As we mentioned, AI applications can range from simple regression models to Large Language Models [6] and Generative AI [94]. Within that broad spectrum, AI applications can also consist of what are referred to as expert or rules based systems [121]. In these cases, the AI models and logic behind them may be created by the model developers instead of trained by a machine algorithm. On the other end of the spectrum, generative AI and large language models are algorithms that can actually generate content themselves. This content may be text, code or even an investment strategy. We provide an introduction to the concepts behind LLMs and generative AI in section 24.4. Readers interested in a more in depth treatment may also consult [253].

AI models can also span many different types of underlying datasets ranging from structured, traditional datasets to unstructured data such as text. In fact, much of the innovation that we have seen with technological advances in AI & ML has been associated with its ability to unlock previously untractable data sources. This could include signals related to earnings transcripts, news articles or analysts research notes, or much more.

Up to this point, we have focused exclusively on machine learning applications that deal with structured data. For example, in the last chapter we considered examples where returns data was used to aid clustering and classification examples.

DOI: 10.1201/9781003530626-29

In this chapter, we introduce the concepts behind working with text based alternative datasets. One of the key tools for working with this type of unstructured data are natural language processing (NLP) algorithms which can help us analyze text data. Once the text data is processed, we can then, for example, attempt to associate it with positive or negative sentiment and package it into an investment signal. In section 24.3, we provide an introduction to NLP and working with alternative data.

Another defining feature of AI is the type of learning that is performed by the underlying model. In the cases that we explored in the last chapter, this learning tended to be either supervised, where structure was provided to the model in the form of labels, or the unsupervised learning was based on a simple representation or single transformation. In many machine learning applications, these types of classical machine learning algorithms are sufficient. In fact, as we explored in the last chapter, in many financial applications there is a preference towards simple, explainable supervised learning models. In some cases, however, there is a desire to incorporate more complex types of learning that may be based on multiple input layers. This is referred to as deep learning, and is detailed in section 24.2.

In the next section, we provide some context for how alternative data, natural language processing, deep learning, generative AI and large language models can be used in finance and investing. We then turn to introduce the underlying algorithms and techniques, beginning with deep learning.

24.1.1 Artificial Intelligence and the Investment Process

Just as AI is a broad field, so too are the potential applications in the world of finance and investing. This is evidenced through the plethora of recent research trying to use AI to parse and interpret alternative data [103], use deep learning to help find previously undetected relationships between variables for the purpose of alpha research [71], and much more. This trend of financial applications of AI has accelerated even further in recent years with the introduction of Large Language Models such as Chat GPT.

In the context of alpha research, we often leverage these techniques to help us find non-linear relationships. For example, instead of using linear regression to identify a relationship say between past and future returns for a given asset, we can train a deep learning model with the hope of capturing more nuanced relationships. This is an extension of some of the examples we saw in the last chapter, where we used classification techniques such as SVM when alpha modeling. Many researchers such as [82] have explored this type of approach to identifying alpha signals.

In the context of portfolio allocation or optimization, we might apply deep learning instead of standard portfolio optimization techniques like mean-variance optimization and risk parity [1]. Theoretically, this would allow us to build an allocation process that incorporates features beyond the mean and variance, and would allow us to incorporate non-linearity in the underlying building blocks into our model. To do this, we might train a deep learning algorithm to maximize some objective function

[1]Each of which were detailed in chapter 24.

such as a risk adjusted return, or sharpe ratio, on a set of input data by varying the weights of our underlying signals.

Lastly, as mentioned above one of the main triumphs of AI in the field of finance has been its ability to unlock alternative datasets that were previously inaccessible. These datasets are primarily text-based, including fundamental analyst reports, sell-side commentary, earnings transcripts and news feeds. NLP algorithms, a key subset of AI provide us the tools for understanding and quantifying these datasets. We can then transform these processed datasets into numeric signals that can then inform an alpha research or broader investment process. For example, we could summarize the entirety of available sell-side commentary or fundamental analyst notes to provide a bullish or bearish signal on a given asset or market.

24.2 DEEP LEARNING

24.2.1 Deep Neural Networks

In chapter 28 we briefly introduced Neural Networks as a technique for either supervised or unsupervised learning, where the model contained one or model hidden layers which were used to translate input data in order to produce an output label. We also saw that Neural Networks could be used in classification or regression models where the desired output is a continuous variable. In this section, we look to build off of this and discuss Deep Neural Networks (DNNs), where multiple hidden layers are used in the transformation process.

In the following chart, we show how a neural network works:

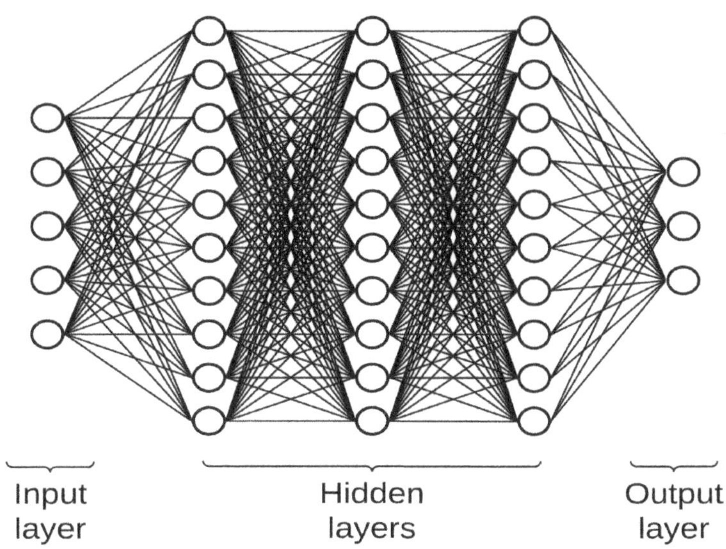

As the chart shows, a neural network is fed a series of input data, which is referred to as an input layer. This is then fed into one or more hidden layers. In the case of a Deep Neural Network multiple hidden layers will be used in this transformation process. Each layer will consist of one or more neurons which are actually responsible for performing the mathematical transformation. These neurons can be trained, on a large amount of input data, in order to learn the optimal transformation. This learning process is analogous to finding the optimal betas in a linear or logistic regression model. Once the data has been processed by these hidden layers, they are given to the output layer and given output labels. In the case of a classification problem, the labels may be 0 or 1, and the process for assigning the labels would be analogous to using a probability threshold in a logistic regression or SVM model, most commonly of 0.5, to decide what the most likely output label is.

The type of neural network shown in the above chart is known as a feed-forward neural network because the signals are processed by each hidden layer only once. In section 24.2.5 we consider another example called a Recurrent Neural Network (RNN) where the signal may be processed multiple times by the same hidden layer.

To take a simple example, each neuron within an neural network might use a function like the following in a hidden layer:

$$\mathbb{E}_{\phi}(x) = f(Wx + b) \tag{24.1}$$

where the function f is used to incorporate non-linearity in the relationships and is often referred to as an activation function. Similarly, W and b are weight vectors and a vector of constants to incorporate the linear relationship.

To understand how a neural network works, we can compare it to a simple technique like linear or logistic regression. In the case of linear regression, there would be no need to incorporate non-linearity or use an activation function, and each neuron would simply incorporate the intercept and multiply the coefficient times the training data. In the case of logistic regression, the activation function would be the logit function, $\frac{1}{1+\exp(-z)}$, however, in this case the logit function would be applied after the coefficients are multiplied by the input data and aggregated, and hence would take place in the output layer. A neural network, however, importantly, provides us with the framework to provide this type of activation function for each neuron.

For a neural network to be useful, it needs to be trained to a large input dataset. This training process chooses the parameters in the layers of the neural network such that the network has the highest accuracy on the training set. In other words, the neural networks parameters are optimized to minimize the training error. The general form of this optimization is done via a loss function that is minimized over the set of available parameters. A common implementation is to maximize the log likelihood [2] of the parameters to the data, however a cross entropy optimization or other approach could be similarly used. In each case, the optimization problem is generally solved using a gradient descent based optimization [3]. For a classification problem that leverages a log likelihood approach, the objective function for training a neural network can be written as:

$$L(\theta) = -\frac{1}{m}\sum_{i=1}^{m}\sum_{k=1}^{K}\left(y_k^i \log(h_\theta(x^i))_k + (1 - y_k^i \log(1 - h_\theta(x^i))_k))\right) \tag{24.2}$$

where y represents the training data, θ represents the parameters in the neural network, $h_\theta()$ represents the output from the neural network given a set of parameters

[2] Or minimize the negative of the log likelihood.

[3] More information on gradient based optimization techniques can be found in chapter 15.

and data and we are summing over both the elements in the training data and the nodes in the neural network.

To solve this optimization problem, efficient computation of the gradient becomes important, especially as layers and neurons are added and the dimensionality increases. A process called backpropagation, which leverages the chain rule and iterates backward through the neural network to reuse derivative computations efficiently, is used to perform this optimization. Interested readers can consult [40] for more information on backpropagation.

As is depicted in the chart above, in a deep neural network, every neuron in a given hidden layer is connected to every neuron in the next hidden layer. Each connection between the neurons also has its own weight which is optimized as part of the calibration process. This creates a rich parameter set for a deep neural network, however, a challenge in using a deep neural network is avoiding the potential for overfitting.

24.2.2 Autoencoder

In chapter 10 we introduced autoencoders as a dimensionality reduction technique and alternative to PCA. In fact, an autoencoder can be thought of as a non-linear version of PCA. As such, autoencoder is an unsupervised learning algorithm that is trained to identify the main features in a dataset through the use of a neural network. In this chapter, we build off of this and show how an autoencoder algorithm works, and how to apply it in different contexts.

An autoencoder is based on an encoding function, that translates the input data, and a decoding function, that translates the encoded data into output data. The process behind an autoencoder is highlighted on the following chart:

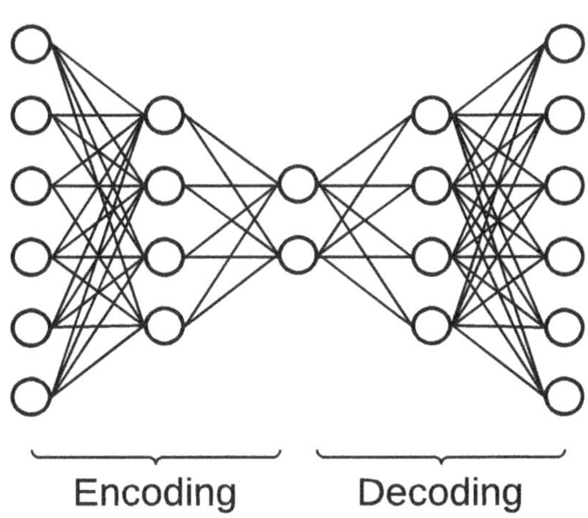

The goal of an autoencoder is to learn an efficient representation of the data, or said differently, to create an efficient encoding function. Autoencoders by construction, when calibrated seek to maintain a perfect representation of the data through the encoding and decoding process. This of course could be done trivially by duplicating the signal. However, autoencoders typically look to compress the data in the encoding layer, identifying only the important features. This makes autoencoder a natural tool for dimensionality reduction. When used for this, we can explicitly limit the number of neurons used in the encoding layer, which is analogous to only consider the first principal components when leveraging PCA for dimensionality reduction.

To take a simple example, an autoencoder might use a function like the following to perform the encoding:

$$\mathbb{E}_{\phi}(x) = f(Wx + b) \tag{24.3}$$

where the function f is used to incorporate non-linearity in the relationships and is often referred to as an activation function. Similarly, W and b are weight vectors and a vector of constants to incorporate the linear relationship. Note that this activation function mimics what we saw in the previous section for a neural network, and in fact, an autoencoder is a particular application of neural networks that is built on the same underlying neuron structure.

The autoencoder would then be trained on a large set of data to identify the optimal parameters. This optimization is typically based on minimizing a loss or cost function. For example, a loss function for an autoencoder might be something like:

$$L(\theta, \phi) := \mathbb{E}_{x \sim D}[d(x, D_{\theta}(E_{\phi}(x)))] \tag{24.4}$$

where we would minimize $L(\theta, \phi)$ over the parameters θ and ϕ, respectively, and μ_{ref} is the empirical or reference distribution. In chapter 14 we introduced the optimization techniques required to solve these types of optimization problems. In particular, we highlighted both gradient based and gradient free iterative optimization methods. Autoencoder training is usually done via a gradient descent approach, which is beyond the scope of this text. Interested readers should consult [40].

24.2.3 Autoencoder in Practice: Reducing the Dimensionality of a Covariance Matrix

In 8.5.3 we presented an example of using dimensionality reduction in the context of predicting ACWI returns via a set of country ETFs. In that example we saw that PCA would be leveraged to use the top PCs as explanatory variables, improving the stability of the problem and lowering the dimensionality. While that approach presented has these advantages, it still inherently relied on a linear relationship between the variables. Of course, PCA is not the only dimensionality reduction tool we can use, and in this example we extend the previous example to leverage an autoencoder. This will have the potential benefit of incorporating non-linear features in the country ETF returns data. Readers are encouraged to consult the coding repository to see how autoencoder was applied in this extended example.

To do this, in line with how we described autoencoder in the previous section, we begin by encoding the country ETF data while limiting the dimensionality of the encoded representation. In this example, we choose to allow three dimensions in the encoding layer, however, this is a parameter that the reader is encouraged to experiment with. The autoencoder can then decode the data to retrieve a compressed representation of the data, which is what we will then input into our predictive linear regression model. In the following chart, we show the fit of using autoencoder as a non-linear dimensionality reduction technique, and alternative to PCA:

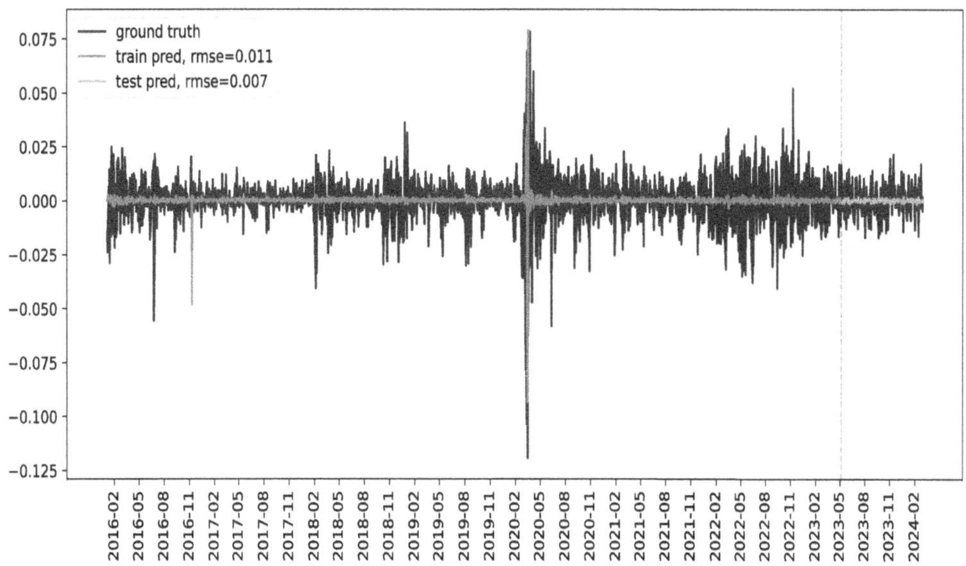

Here we can see that this alternative approach achieves similar results to those obtained in chapter 10. This is not surprising given the simplistic market example we rely on, however, importantly, in cases where non-linear relationships exist in the underlying data we are looking to reduce the dimensionality of, then a technique like autoencoder will be preferred compared to PCA. The reader is encouraged to think through how this can be extended to a more realistic prediction model for global equity indices, leveraging the other concepts discussed throughout the text.

An astute reader may question whether the data that is fed into our autoencoder algorithm should first be normalized. This is analogous to whether we want to perform PCA of a covariance matrix or a correlation matrix, and there are relative strengths and weaknesses to each approach. On the one hand, normalizing the data will help ensure that higher volatility inputs don't have a disproportionate impact on the results. Conversely, information may be lost in the normalization process.

24.2.4 Convolutional Neural Networks

In section 24.2.1 we introduced deep neural networks and saw that, often each neuron in a given layer was connected to each neuron in the subsequent layer. These connections also have separate weights associated with them to enable the model to uncover the interconnectedness between features in the model. While this rich structure has many advantages, and makes DNNs powerful, it also comes with the potential for the problem to be high-dimensional and to be overfit.

Convolutional Neural Networks (CNNs) are an alternative type of deep learning network that enable us to reduce the dimensionality of the problem through parameter sharing and the use of a different structure in the underlying hidden layers, convolutional layers.

In the case of a DNN as discussed in 24.2.1 the neurons are connected via separate weights that are assigned to them. This makes the operation of moving from one hidden layer to the next a matrix multiplication, and, as mentioned, results in high dimensionality. In contrast, a CNN uses convolution instead of matrix multiplication as the mathematical operation underpinning the hidden layers. These convolutional neurons, then, are able to share the same weights across a convolutional layer, resulting in parameter sharing and greater efficiency.

Mathematically, the convolution operation can be written as:

$$f(t) = (x * w)(t) = \sum_{a=-\infty}^{\infty} x(a)w(t-a) \tag{24.5}$$

As the equation shows, convolution is an operation that combines two functions, or signals and outputs a new transformed function or signal. In doing so, convolution expresses how a function overlaps with another as we shift the function over the other. In the case of a CNN, this operation (convolution) is applied to the input data, which consists of one or more input channels, producing feature maps that are passed as input signals to the hidden layers. The resulting feature maps from each hidden layer are referred to as output channels, which serve as input channels for the subsequent hidden layer. The function, w is the kernel that we use to transform the input function and in the context of CNN is generally referred to as a filter. Part of the calibration process of a CNN, then is to optimize this filter such that the CNN best fits the data. Importantly, this process of using a convolution to transmit the data from one hidden layer to the next allows for parameter sharing, enabling us to build models that are more parsimonious than a comparable DNN. Interested readers should consult [40] for more details on Convolutional Neural Networks.

24.2.5 Recurrent Neural Networks

The previously considered types of neural networks, DNNs and CNNs, are both feed-forward networks. While useful in many contexts, these types of networks have a few key underlying assumptions. First, DNNs and CNNs are defined by fixed lengths of input and output. Second, and perhaps most fundamentally, each input data point is processed independently of other data points. That is, each piece of data is input into the model independently with no temporal or other dependence.

Recurrent Neural Networks (RNNs) are a specialized type of deep learning neural network that works on sequences rather than independent observations. In finance, these types of networks are particularly useful as the most common datasets are time series data, which contains an inherent time dependence. In the following chart we show how an RNN works:

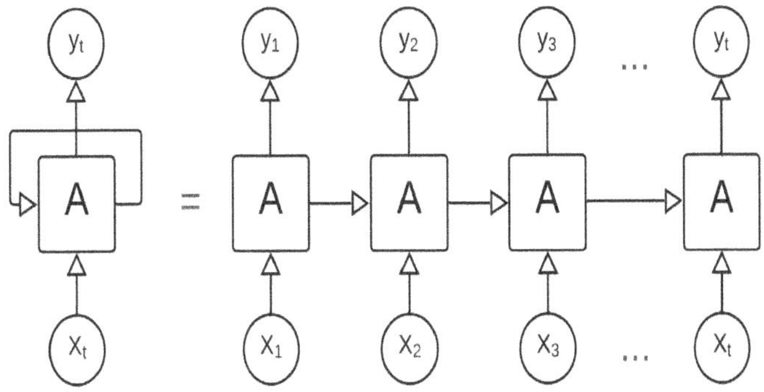

Importantly, as we can see in the chart that unlike a feed forward network where the output from a hidden layer is passed forward to the next hidden or output layer, RNNs have a recurrent feature that enables the same network to be visited multiple times with the information in the network being updated as the data is processed sequentially.

One particular type of RNN that is very commonly used in finance is Long short-term memory. The details of this algorithm are beyond the scope of this book, however, interested readers are encouraged to consult [102].

24.2.6 Generative Adversarial Networks

Generative Adversarial Networks (GANs) are a type of neural network used to generate data resembling a desired distribution. GANs are generative models, and an example of generative AI because they are able to generate content. In the case of GANs, the content that is created is the data itself, whereas in other cases the content might be code, text or another format. GANs are just one example of a generative model that is used to generate data. Others include Variational Autoencoder, which leverages the Autoencoder technique introduced earlier in the chapter, and Markov Chain Generative Stochastic Networks. These additional techniques are not covered in this text, however, interest readers are encouraged to consult [105].

GANs are built on top of two neural networks, which compete with each other in a zero-sum game to find the optimal distribution to generate from. These networks are referred to as a **generator network**, [140] which is trained to produce realistic looking samples, and a **discriminator network**, [140] which tries to discern whether a given sample came from the training dataset or the generator network. The generator network tries to create samples that are sufficiently high quality that the discriminator is fooled, whereas the discriminator is trained to be as accurate as possible in labeling the samples. This dynamic between the discriminator and the generator networks creates the zero-sum game nature of GANs.

GANs are referred to as implicit generative models, and begin by sampling from a known, tractable distribution. For example, we might sample from a multi-variate Normal distribution. The generator network is then trained to compute a transformation or mapping function between the sampling distribution and the desired output distribution. Once trained, the generator can take input samples from a known distribution and use this to generate samples that are indistinguishable from the distribution of the training data.

In the following flow-chart, we show the logic behind a GAN and the generator/discriminator process:

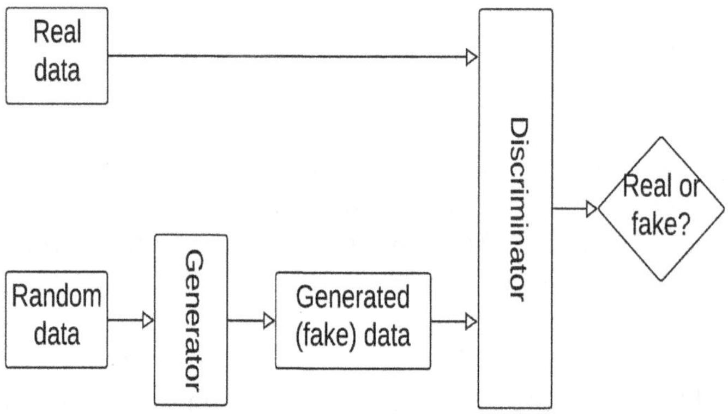

Mathematically, behind a GAN is an underlying optimization problem based on gradient descent [40]. In this optimization, we combine the objective of the generator and the discriminator through a cost function. The cost function of the discriminator involves minimizing the number of mis-classified samples. To see this, consider D, the discriminator's predicted probability of a sample being training data and not a generated sample. To do this, we minimize the cross-entropy loss for classifying training data vs. generated samples:

$$L(G, D) = \mathbb{E}_{x \sim D}[\log D(x)] + \mathbb{E}_z[\log(1 - D(G(z)))] \tag{24.6}$$

While the discriminator wants to minimize the number of mis-classified samples, and correspondingly the cost function, the generator wants to create samples that are not differentiable from the training data. As a result, the generator will maximize the same loss function L_D. The subsequent optimization function looks to maximize the cost function for the discriminator, and minimizes the cost function for the generator, as shown in the following optimization equation:

$$\max_G \min_D L(G, D) \tag{24.7}$$

This optimization is performed via gradient descent, meaning that information regarding the functions' derivatives will be used to update the parameter set. This mechanism is similar to the backpropagation algorithm described in section 24.2.1. Later in the chapter, in section 24.5.4, we show how GANs might be used in risk management and in simulation more broadly.

24.3 NATURAL LANGUAGE PROCESSING

24.3.1 Alternative Data Sources

As we have emphasized throughout the book, Natural Language Processing (NLP) is a key driver of innovation in quant finance due to its ability to unlock previously inaccessible datasets. These datasets are often text-based, but can come in many forms. For example, NLP might allow us to read company reports, news feeds or monitor social media for increased mentions of a given product or company. Presumably, this could then give investors insight into potential future sales and revenue data prior to it being reflected in other publicly available sources.

In the following list, we provide an overview of some of the most common alternative data sources leveraged within the finance industry, along with some context for why they might be useful data sources:

- **News**: These news sources could be text or video-based, and capture company specific or broad market sentiment. Parsing these news sources efficiently, then, might allow us to find investments that are over or underpriced, and trade on their ultimate convergence.

- **Economic Announcements**: Much like news, the market receives announcements related to economic activity periodically. Sometimes, the releases of these data are scheduled, such as a monthly payroll report, a quarterly GDP print or the result of a Fed interest rate decision. In some cases, like a Fed decision, this type of announcement might come with an accompanying statement. In either case, these announcements are another potential source of alternative data that investors might use to try to identify trends in the market or broader economy.

- **Analyst Ratings and Commentary**: Fundamental analysts dedicate significant amounts of time to due diligence processes for companies. This includes visiting the companies, speaking with leadership and trying to identify intangible information that might impact their profit. In a discretionary investment setting, this research is consumed by a portfolio manager who then uses it to inform his or her trades. Alternatively, however, this text-based content created by the underlying analyst could be consumed as alternative data by an NLP algorithm, and then processed as a quantitative signal.

- **Sell-Side Commentary**: Investment banks and dealers often possess large research teams that are trained to closely monitor the markets and provide commentary on recent developments, expectations and potentially attractive structures. As such, just as with other analyst content, it is natural to try to parse this as a text-based signal with the help of an underlying NLP algorithm.

- **Social Media Activity**: Investors may monitor Google Trends, or the number of times a product or company is mentioned on X [4], Facebook or another comparable site to look for clues related to sentiment and future growth potential.
- **Earnings Transcripts**: Investors may parse earnings call transcripts to look for indications beyond the reported earnings of likely future shifts in a firm's profitability.
- **Company Filings**: Companies are required to file certain documents periodically, such as 10-K filings. These filings could then be parsed to look for indications of changes in the firm's financial position that are not necessarily reflected in the headline numbers.
- **Geolocation Data**: Monitor foot traffic or images of parking lots to identify real-time sales trends.
- **Website & Mobile Device Usage**: Measure website clicks or usage rates of different companies mobile applications to identify potential increases or decreases in interest.
- **Product Reviews**: Analysis of product reviews, and any related changing tone might be used to gauge sentiment about a given product, theoretically helping to understand it's future sales trajectory.
- **Credit Card and Other Point-of-Sale Transactions Data**: Access to transactional data such as credit card transactions could be aggregated to give a sense of emerging trends in a firms sales or revenue, and give investors an opportunity to identify areas where this hasn't fully been reflected in the firm's share price.
- **Event Dates**: Trading the behavior of stocks around corporate events ranging from mergers to conferences to earnings releases has been a strategy pursued by many funds for years. At the heart of this, potentially, could be for an algorithm to search for changes in event dates and interpret any related sentiment.

We can see that the list of potential alternative data sources is quite large, and has grown significantly in recent years. As technological advances continue to allow for it, it is reasonable to expect this growth to continue, reinforcing the importance of NLP algorithms. In the next section, we provide a brief introduction into the main approaches used to process text via NLP.

24.3.2 Natural Language Processing Algorithms

In order to access any alternative data source, we need to be able to process and analyze the text programmatically. To do this, we need to find an appropriate numeric representation of the text, enabling us to compare and categorize the text with other

[4] Formerly known as Twitter.

pieces of text. This process is performed by Natural Language Processing (NLP) algorithms, which are introduced in the remainder of this section. The goal is to provide the reader with some context and intuition into how these algorithms work rather than to provide a full mathematical explanation. There is a significant amount of literature on the details of NLP algorithms, however, and interested readers are encouraged to consult [20].

In the following list, we highlight some of the core algorithms and techniques used to process and analyze text data via NLP:

- **Lemmatization and Stemming**: A pre-processing technique for NLP where we reduce words to their root or base form, limiting the inherent variability in the text an NLP algorithm is processing, resulting in a more efficient algorithm with a narrower dataset.
- **Word Embedding**: A representation of text, or words, as real-valued vectors in high dimensional space. The use of word embeddings [246] allows us to convert text to a space where comparison and categorization is easier, and thus the concept is core to NLP algorithms [5]. These word embeddings are trained on a large corpus of text using a set of underlying conversion algorithms such as Word2Vec [245].
- **Tokenization**: Another technique for converting pieces of text to numerically comparable representations is tokenization, where text is first broken into pieces, often individual words, and then encoded. These individual pieces are commonly referred to as tokens, and the process for breaking larger text data into a set of words, or smaller pieces is known as tokenization. Once the data has been broken into tokens, an encoding process then enables the algorithm to process the text. Models using tokenization then operate in the space of these tokens, before encoding and decoding these tokens back to text to create the desired output. Tokenization is used by Large Language Models, which we introduce in 24.4.
- **Bag of Words**: [182] An algorithm that represents text as an unordered collection of words, capturing the frequency but disregarding the order. The most common words can then be interpreted as the most important, and can be analyzed using Word Embedding or another technique.
- **Named Entity Recognition**: An algorithm that identifies entities, such as people or companies and classifies them among a predefined group.
- **Text Classification**: Classifying text into a set of predefined categories using techniques like Bag of Words and Word Embedding. Text classification is done using Machine Learning techniques like SVM, Decision Trees or Random Forests, where we might first standardize and pre-process the data using lemmatization and stemming. Once the text classification algorithm has been

[5] As are similar concepts.

trained, it can be used to judge the sentiment of the text, or classify the main subject as in topic modeling.

- **Topic Modeling**: An algorithm that extracts the main topics from a text or document, enabling the algorithm to break a set of documents or data into themes or groups. Topic Modeling is often done using Latent Dirichlet Allocation. Interested readers can consult [44] for more information on topic modeling and Latent Dirichlet Allocation.
- **Sentiment Analysis**: An NLP technique that captures the emotion in a given set of text. For example, we might use it to describe whether a news article about a company has a positive, negative or neutral tone. This is one of the most commonly used NLP algorithms and is often based on a Naive Bayes model for interpreting the statements and assigning an emotion. More details on sentiment analysis can be found in [196].

In this section we aimed to provide the reader with a broad overview of the principles and algorithms that underlie NLP algorithms and processing text data. In practice, much of the work on these algorithms has already been done, and there are in fact Python packages that can help analyze alternative datasets and implement NLP algorithms. In the next section, we explore some of these algorithms in Python and show how they can be used to parse news in the context of an attempted trading signal.

24.3.3 Natural Language Processing in Practice: Parsing News for Trading Signal

In this section we provide an example of how text data can be used in the context of a trading signal by analyzing whether the sentiment of text is positive or negative. Such sentiment analysis, and understanding whether the context of an article or other piece of text is positive or negative, has become an area of great interest in the algo trading space, and is a great way to incorporate alternative data into an investment strategy.

Once sentiment is extracted, an algorithm that buys on positive news and sells short on negative news can then be the basis of a trading strategy. The code for this sentiment analysis based NLP model is provided in the coding repository. To explore the sentiment of news data, we rely on the vaderSentiment built in package and import news data from reuters. We then extract the sentiment of each piece of news, resulting in a tag between -1 (max bearish) and $+1$ (max bullish). In this example, we focus on news stories with the word "trade" in them and calibrate our model to a large corpus of news.

To check the consistency of our model, we explore the sentiment tags of a few news stories. First, we select the story that was given the highest positive sentiment score, which was titled "FEBRUARY U.S. JOBS GAINS SHOW STRONGER ECONOMY." Secondly, we select the story with the most negative sentiment, which is titled "JAPANESE TARIFFS SEEN AS WORLDWIDE WARNING." In both cases, the sentiment of the news stories appears fairly unambiguous, and the sentiment algorithm appears to have characterized each news story correctly.

Finally, we examine a news story with neutral sentiment and a score closest to zero. This would be a news story that the algorithm couldn't classify as positive or negative. The title of this item in the news feed was "CANADA PLANS TO MONITOR STEEL IMPORTS, EXPORTS, TRADE MINISTER SAYS." In this case, the NLP algorithm was unable to determine much from the title and text, so it assigned a neutral sentiment score.

It should be emphasized that in this example while we have built an algorithm that extracts the sentiment of a given set of text, and performed diagnostics on the accuracy of the algorithm, we have not tested these sentiment scores against subsequent changes in markets. To do this, the reader can use the techniques introduced in chapter 27, such as examining the correlation between the sentiment score signal and future market movements. Readers are also encouraged to tweak the data source and filter criteria.

24.4 LARGE LANGUAGE MODELS

24.4.1 What are LLMs?

When Chat GPT [224] was launched for public use it marked a seismic shift in the perceived capabilities of AI and Machine Learning. The tool's ability, based on a Large Language Model, to generate text-based content that mimics human language and intelligence [6] was beyond what many had previously thought that AI was capable of. At heart, these tools work by leveraging NLP algorithms in order to form a sequence of text, where at each point the algorithm attempts to identify the next most likely piece of text.

To do this, an input set of text, known as a prompt, is fed to the LLM. We provide more background on these prompts in section 24.4.3. The text is then processed by the LLM using a process referred to as tokenization, where text is broken down into pieces and translated to tokens that can be compared numerically. Working with these tokens allows LLMs to rapidly search the set of input data for similar occurrences, and to model the probability of an ensuing token being next. This process of modeling the probability of the next token is based on an artificial neural networks, known as a transformer network, with a large number of neurons. A brief overview of how transformer networks are used by LLMs is provided in 24.4.2.

These tokens are then decoded back to text that a model user can interpret. LLMs continue this process sequentially to generate output in the form of paragraph text, and in doing so use what is referred to as an attention mechanism [21]. This allows the algorithm to assign weights to different parts of the text, paying more or less attention to the parts that are deemed more or less important in the training process. This allows for LLMs to account for context in a given piece of text. In the remainder of this section, we provide a brief introduction to LLMs and the technologies that underlie them. Readers looking for a more in depth treatment should consult [253].

[6] Not to mention code.

24.4.2 Transformer Networks

Transformer networks are a type of artificial deep neural network that are commonly used in Large Language Models. Transformer networks leverage tokenization and word embeddings to translate text into a numeric vector representation to be processed by the model. Importantly, this is done by comparing tokens using a large training dataset but also using an attention mechanism [87] to contextualize the content and generate likely sequences of text.

Transformer networks consist of the following underlying components:

- **Tokenization**: [158] Tokenization was briefly introduced in the last section and refers to the process of translating the input data into a set of tokens. This is done by breaking down the input text into small, parsable units and transforming these small pieces of text into tokens.
- **Embedding Layer**: A single layer that converts tokens and their location into a vector representation that can be processed efficiently by the layers in the transformer network.
- **Transformer Layers**: The heart of a transformer network and LLM. A transformer layer is responsible for extracting information from the vector representation of the text, doing so using a combination of a feed-forward neural network and an attention mechanism that helps to contextualize the input text.
- **Decoding Layer**: An optional layer that converts the transformed vector representation of the output data to a probability or likelihood of tokens. The likeliest tokens can then be converted back to text to form the standard text output for an LLM.

While the details of these stages of the model are beyond the scope of this text, they leverage the same concept developed in 24.2. Interested readers should may also consult [205].

24.4.3 Prompt Engineering

The heart of working with an LLM is asking it questions, or providing it prompts that it then responds to. It turns out that, due to the inner workings of LLMs, how a prompt is structured can determine how successful a user is at getting an effective answer. In some cases, an iterative process may be most effective in getting an LLM to generate the desired response.

As a result of this, being able to specify effective prompts with knowledge of how the LLM operates is likely to be an increasingly desired skillset in coming years. This is known as prompt engineering, which is the art of structuring prompts such that we get the best quality output. For example, one might use in-context learning [229] where we provide examples within the prompt to help the LLM refine its response. Additionally, chain of thought prompting [148] might be used, where we ask the model to provide incremental steps or a sequential flow to the answer in hopes of improving the output. More information on prompt engineering, and techniques that have been found effective can be found in [109].

24.4.4 Hallucinations

While LLMs have a tremendous ability to produce content that is plausible and is highly effective at completing many types of tasks [7], in some cases an LLM will produce something that is demonstrably false or not in line with common sense / simple logic. This phenomenon is known as a hallucination [269], which refers to this tendency for an LLM like ChatGPT to produce content that is blatantly wrong.

One reason that such a hallucination could occur is because the LLM is trained on poor data. In fact, no matter the sophistication level of the algorithm it will only be as good as its underlying data, and LLMs are no exception. Along those lines if data with incorrect conclusions is fed to the model, then it should be no surprise when the corresponding output has errors.

Additionally, as noted earlier in this section, LLMs work by converting input text into tokens, via an encoding and decoding process. As a result, in some instances errors in the encoding and decoding algorithm can lead to a hallucination. Further, because the comparison is then done on tokens, a hallucination may occur when two token representations are viewed as very similar by the model, but the underlying text or numeric data has a very different meaning. As an example an LLM might view two numeric based text strings as similar because of common digits, even if they are different lengths and thus have very different orders of magnitude.

As a user of an LLM, one of the main dangers is an undetected hallucination, which could lead to the completely wrong conclusion and model errors. As a result, to avoid a hallucination, it is crucial to verify the content that is produced. For example, if the output of an LLM references a publicly available paper, then the user of an LLM should perform some level of due diligence on the paper to confirm that it exists, has some degree of credibility, and is represented accurately by the LLM. Proper prompt engineering can also reduce the risk of such hallucinations, further reinforcing the potential importance of such skills going forward. More details on how hallucinations occur and what can be done to prevent them can be found in [269].

24.5 APPLICATIONS OF DEEP LEARNING & UNSTRUCTURED DATA

24.5.1 Delta Hedging Schemes & Optimal Execution via Reinforcement Learning

Earlier in the text, in chapter 16, we learned the importance of delta hedging schemes in the realm of options trading and discussed a few simple approaches. In particular, we saw that hedging schemes based on a set calendar or move based re-balance, could provide an effective hedging scheme. We also saw the inherent trade-off between hedging too frequently, and incurring large transactions costs, and waiting too long in between re-balancing our hedge, in which case undesirable exposure, and volatility, may arise in our portfolio. Of course, the problem of devising a hedging scheme,

[7] For example, summarizing large texts.

whether it be for delta or another Greek is a complex problem that may require an optimal solution beyond the scope of these two simple schemes. We must account for transactions costs, as well as the market impact of our traded hedges. Additionally, we must consider the dynamics of the underlying asset, including its volatility and its tendency to drift or revert to a set of equilibrium conditions. This makes the problem potentially path dependent, and dependent on exogenous market participants.

Within this context, the field of machine learning, in particular reinforcement learning, provides a framework for trying to solve the problem of finding an options hedging scheme, robustly. Similarly, it can be used as a tool to estimate an optimal execution policy. Reinforcement learning enables us to develop a dynamic environment where we can evaluate the cost or gain of different hedging decisions and develop an optimal strategy that minimizes a specified cost function over a set of these decisions. Many researchers such as Hull et al [149] have conducted research on this topic, and, in many ways, it is a natural application of machine learning within the field of quant finance.

24.5.2 Volatility Surface Calibration via Deep Learning

Calibration of a volatility surface was discussed in great detail in chapter 14. The approach that we discussed was based on specifying a stochastic model, and then running an optimization designed to find the parameters in that model that best fit the set of market data. Doing this effectively also required knowledge of the underlying stochastic model. For example, we learned that, in the Heston model, we were able to calibrate all expiries simultaneously. The Variance Gamma model, by contrast, required a separate calibration for each expiry. The SABR model, further required us to fix one of the four underlying model parameters. One of the challenges of this approach is that it is inherently computationally, and therefore time intensive.

Alternatively, we can try to view the volatility surface calibration process as a machine learning, or deep learning problem. That is, we can use a large training set and predict the next option price based on options with similar features, such as strike and expiry. That is, we would label a large dataset with option prices, and then use this dataset to predict prices for new observations based on similar characteristics. Halperin [123], for example, and others [145], have conducted research on this technique and found that it can be a way to speed up calibration algorithms significantly. As such, it is another natural use of deep learning.

24.5.3 Credit Risk Estimation via Deep Learning

In the last chapter, we discussed how classification techniques like SVM and logistic regression could be used to estimate default probabilities and credit risk. In that case, the risk embedded in a loan was modeled using a set of features, for example a credit score, debt and income levels for the customer, etc. Classification techniques then provided a natural way to frame the problem as the binary nature of a classification problem maps well to modeling whether or not the customer will repay their loan. It turns out that these problems can alternatively be solved using deep learning, where we could try to incorporate more complex relationships between the features and repayment decision of the consumer.

Additionally, while many of the features that we'd naturally want to use are well structured, numeric datasets, the techniques we learned in this chapter would also enable us to incorporate alternative data based features. For example, if the underlying loan were made to a company, then we could use an NLP algorithm to parse news articles on the company, identify keywords that would help flag news that would significantly impact their probability of repayment, and then translate this to a numeric feature to process by a classification technique or deep learning model [268]. In the coding repository, we provide an application of how deep learning, and machine learning models in general can be used in the context of estimating credit risk using a publicly available loan level dataset.

24.5.4 Leveraging GANs in Risk Modeling

In chapter 26 we learned about all aspects of risk management, from risk measurement to risk oversight and accounting for undefined risks. For an investment firm, a key part of the risk measurement process was identifying scenarios and metrics like VaR and CVaR that would help us understand the potential risk in a given portfolio. To do this, we discussed that scenarios could be constructed via a simulation approach or could be based on user defined adverse scenarios. Similarly, when calculating VaR or CVaR we relied on two types of simulation: historical simulation, or bootstrapping and simulating from a joint [8] distribution. GANs, as we learned in this chapter, provide another approach to simulation that we could leverage in the context of estimated scenarios or calculating risk metrics like VaR and CVaR. In this case, it would allow us to move beyond the standard distributional assumptions embedded in a standard Gaussian Copula type model but also enable us to embed more sophistication than a typical bootstrapping algorithm. Recently, many researchers have attempted to leverage GANs for estimate VaR, such as [72].

24.5.5 Generating Expected Returns via Generative AI & LLMs

As Large Language Models like Chat GPT were introduced, researchers naturally wondered whether they could form the basis for an investment strategy. While great at summarizing large amounts of data and producing content that mimics human intelligence, an open ended question remains about whether the intelligence that tools like this can create rises to the high bar needed to be successful in investing.

While this is still very much an open area of research, many researchers have already begun exploring how prompts could be fed to an LLM that would return data useful to us in building an investment strategy. As one example, we could try to train a LLM exclusively on sell-side [9] research and have it generate a sense of the sentiment of the aggregate text [89]. The LLM of course could be trained on a variety of input data, from news to commentary from industry experts to company reports.

[8] Often Gaussian.

[9] Or other.

Alternatively, we could structure our prompts to the LLM such that an expected return is provided by the model. This approach was explored by [267].

At heart, one of the challenges in pursuing this type of approach is no doubt the complexity of investing. Further, the non-stationarity of the underlying datasets as well as the lack of a sufficient number of observations to successfully train an LLM on will also prove problematic if not handled properly. Additionally, because by nature Large Language Models work essentially by sequentially generating the most likely next piece of text, enabling them to produce actionable investment decisions can also be a challenge.

24.5.6 Challenges & Limitations

While the use of AI in finance brings with it great promise, it also comes with significant questions, challenges and potential limitations. Some of these challenges, such as interpretability issues and questions about responsible use apply across all applications of AI regardless of the field. Others are unique to finance, such as the lack of robust data sets to train sophisticated AI models on. In this section we aim to highlight some of the potential obstacles to using AI in financial applications as well as some context on how we can build models responsibility against that backdrop.

As noted throughout the text, one of the defining features of finance is the lack of repeatable experiments. As a consequence, datasets tend to be limited and far smaller than what we would use to train a deep learning or generative AI model in other fields. This then calls into question whether in finance we even have the datasets required to justify use of such a model, and to train it accurately.

Partly as a result of this limited data, the field of finance has very few widely agreed upon laws or rules of nature. This means that even basic rules about how markets operate, such as whether or not they are efficient [10] can be controversial and lead to legitimately differing opinions. This creates challenges throughout the field identifying statistically significant relationships. As discussed, in many cases this means that we need to rely on a combination of statistical tests and our financial intution in order to judge whether a financial relationship is persistent.

In this context, interpretability, which is sometimes challenging to achieve as model complexity increases, becomes of the utmost importance. Linear regression, for example, comes with many well documented flaws and limitations, however, it also provides us with intuitively easy to understand tools for judging the relationships between variables. Deep learning or generative AI models struggle to offer this transparency, though the feature importance tools discussed in the last chapter can certainly aim to bridge some of the gap.

This problem of limited data then becomes amplified as we look to leverage AI algorithms which inherently work best when they can be trained on massive datasets that we in general are not given in finance. In fact, in chapter 24 we detailed the legitimate challenges just to estimate an expected return. In this context, we will certainly face challenges allowing an unconstrained AI algorithm to identify relationships without our oversight, as is often the case with deep learning or generative AI

[10] As detailed in chapter 5.

models. The challenges associated with lack of data are then compounded by the inherent non-stationarity of those limited datasets. This could include regime changes where the underlying relationships change, or are no longer significant. For an AI algorithm to succeed, it must then learn and incorporate this non-stationarity and these regime changes. [11]

Because of the need for large amounts of data to efficiently train a deep learning or other sophisticated AI model, a natural first place to look for trading applications is within the higher frequency space, where the datasets are naturally larger and we tend to be less concerned with interpretability. These types of strategies, however, need to act fast in order to succeed, creating a bit of a paradox where some of the potentially most intuitive use cases for these techniques are in applications where the computational load of the techniques creates barriers to adoption and may make them infeasible.

As discussed in the section on Large Language Models, another serious challenge to generative AI models is the potential for it to produce so called hallucinations. These hallucinations occur when Large Language Models appear to be tricked or produce results that are implausible or inconsistent with basic logic. While there are ways to attenuate the risk of such hallucinations, they create the need for an oversight process to ensure that the models are operated as desired and any potential inconsistencies don't result in suboptimal decision-making.

24.5.7 Responsible Use

As a result of all these challenges, it is extremely important for users of AI, especially those in finance, to leverage models responsibly with a keen awareness of the strengths and weaknesses inherent in the underlying techniques. At heart, whether we are leveraging a generative AI or linear regression model, we must remember that all models are approximations of reality. As George Box famously said "All Models are Wrong, Some are Useful" and this still needs to be remembered in the excitement of LLMs and deep learning. These models still rely on simplified representations of reality, and quants must be upfront about these embedded assumptions and limitations. Further, we must be diligent in mapping the model that we chose to the complexity justified by the underlying data and problem scope. As we are generally working with smaller datasets in finance, this needs to be reflected in the models we select, and how we train them.

It is also important to be explicit about what exactly we are doing in a given model. As we discussed in this and the last chapter, the scope of AI is actually quite broad, ranging from linear regression to classification techniques to deep learning to generative AI. Therefore, users claiming to use AI need to be upfront on what part of that broad spectrum they are operating in.

Finally, as model complexity increases, oversight of the model becomes more challenging. This includes model and data validation, which we discussed in great detail in chapter 11, but also includes ensuring proper model use. When we leverage the tools and techniques for working with alternative data discussed in this chapter,

[11] When some of these future regimes may not even be present in the data.

for example, it becomes much harder to validate and detect outliers or bad data. Nonetheless, it will still be crucial to success of the model and it is important for there to be accountability for each part of the process, from the data to the model design to the model directive.

If done properly, integration of AI has the potential to deliver great promise, increasing productivity and helping promote innovation. As such, working with AI will be of the utmost importance for quants to continue to stay at the forefront of the industry. That said, while AI has shown an ability to synthesize information effectively, such as a set of academic papers, analyst notes or news, it is less clear whether it can rise to the level of creating standalone investment processes. Against that backdrop, quants will need to increasingly be able to work with the machines and new technologies that emerge.

Bibliography

[1] Treasury Inflation Protected Securities (TIPS):https://treasurydirect.gov/marketable-securities/tips/.

[2] Why is it so hard to estimate expected returns? 2012.

[3] Adam Hayes. Sequential Pay CMO: What it is, How it Works: https://www.investopedia.com/terms/s/sequentialpaycmo.asp#:~:text=A%20sequential%20pay%20collateralized%20mortgage,works%20best%20for%20their%20strategy.

[4] Charu Aggarwal. *Neural Networks and Deep Learning*. Springer Publishing, New York, NY, USA, 2018.

[5] Irene Aldridge. *High Frequency Trading: A Practical Guide to Algorithmic Strategies and Trading Systems*. John Wiley & Sons, Hoboken, NJ, USA, 2013.

[6] Tim Salimans Alec Radford, Karthik Narasimhan and Ilya Sutskever. Improving Language Understanding by Generative Pre-Training. *arXiv*, 2024.

[7] Fausto Saleri Alfio Quarteroni and Paola Gervasio. *Scientific Computing with MATLAB and Octave*. Springer Publishing, New York, NY, USA, 2010.

[8] Steven Allen. *Financial Risk Management: A Practitioner's Guide to Managing Market and Credit Risk*. John Wiley & Sons, Hoboken, NJ, USA, 2003.

[9] Robert Almgren and Neil Chriss. Optimal execution of portfolio transactions. *Journal of Risk*, 2000.

[10] Leif Andersen. Efficient Simulation of the Heston Stochastic Volatility Model. 1985.

[11] Leif Andersen and Vladimir Piterbarg. *Interest Rate Modelling*. Atlantic Financial Press, New York, NY, USA, 2010.

[12] Chris Kenyon Andrew David Green and Chris Dennis. KVA: Capital Valuation Adjustment. *SSRN*, 2014.

[13] Andrew Lesniewski. Course Lecture Notes: http://lesniewski.us/papers/lectures/Interest_Rate_and_FX_Models_NYU_2013/.

[14] Andrew Matysin. Modelling Volatility and Volatility Derivatives.

[15] Anna Katherine Barnett-Hart. The Story of the CDO Market Meltdown: An Empirical Analysis: `https://www.hks.harvard.edu/sites/default/files/centers/mrcbg/files/Barnett-Hart_2009.pdf`.

[16] Swat Chandra Antti Illmanen and Nicholas McQuinn. Demystifying Illiquid Assets: Expected Returns for Private Equity. *The Journal of Alternative Investments*, 2020.

[17] Apache Spark Docs. Website: `https://spark.apache.org/`.

[18] Apache Spark Python API. Website: `https://spark.apache.org/docs/latest/api/python/index.html`.

[19] George Arfken. Taylor's Expansion. *Mathematical Methods for Physicists*, 1985.

[20] Rajesh Arumugam and Rajalingappaa Shanmugmani. *Hands-On Natural Language Processing with Python.* Packt, Birmingham, UK, 2018.

[21] Niki Parmar Jakob Uszkoreit Llion Jones Aidan Gomez Lukasz Kaiser Ashish Vaswani, Noam Shazeer and Illia Polosukhin. Attention is All you Need. *Advances in Neural Information Processing Systems*, 2017.

[22] Myron Scholes Ashwin Alankar, Michael DePalma. An Introduction to Tail Risk Parity: Balancing Risk to Achieve Downside Protection. *Alliance-Bernstein*, 2012.

[23] Cliff Asness. Stocks vs. bonds: Explaining the Equity Risk Premium. *Financial Analysts Journal*, 2000.

[24] Cliff Asness. The Illiquidity Discount?, 2019.

[25] Rowel Atienza. *Advanced Deep Learning with Keras.* Packt Publishing, Birmingham, UK, 2020.

[26] Marco Avellaneda and Stanley Zhang. Path-Dependence of Leveraged ETF Returns. *SSRN*, 2012.

[27] Marco Avelleneda and Jeong-Hyun Lee. Statistical Arbitrate in the U.S. Equities Market. *NYU Research Paper*, 2008.

[28] Sakis Kasampalis & Kamon Ayeva. *Mastering Python Design Patterns: A guide to creating smart, efficient, and reusable software.* Packt Publishing, Birmingham, UK, 2018.

[29] L. Bachelier. Theory of speculation. 1964.

[30] Suleman Baig and Moorad Choudhry. *The Mechanics of Securitization: A Practical Guide to Structuring and Closing Asset-Backed Security Transactions.* John Wiley & Sons, Hoboken, NJ, USA, 2013.

[31] Nicholas Barberis and Richard Thaler. A Survey of Behavioral Finance. *Handbook of the Economics of Finance*, 2003.

[32] Basel. Margin requirements for non-centrally cleared derivatives: `https://www.bis.org/bcbs/publ/d317.htm`.

[33] David Bates. Jumps and Stochastic Volatility: Exchange Rate Processes Implicitly in Deutsche Mark Options. *The Review of Financial Studies*, 1996.

[34] Michael Bauer and Thomas Mertens. Information in the Yield Curve about Future Recessions, 2018.

[35] Martin Baxter and Andrew Rennie. *Financial Calculus: An Introduction to Derivative Pricing.* Cambridge University Press, New York, NY, USA, 2001.

[36] Stan Beckers. The Constant Elasticity of Variance Model and Its Implications for Option Pricing. *The Journal of Finance*, 2009.

[37] Lorenzo Bergomi. *Stochastic Volatility Modeling.* Chapman & Hall / CRC Press, Boca Raton, FL, USA, 2015.

[38] Dimitri Bertsekas. *Constrained Optimization and Lagrange Multiplier Methods.* Academic Press, New York, NY, USA, 1982.

[39] Patrick Billingsley. *Probability and Measure, 3rd Edition.* John Wiley & Sons, Hoboken, NJ, USA, 1995.

[40] Christopher M. Bishop and Hugh Bishop. *Deep Learning: Foundations and Concepts.* Springer Publishing, New York, NY, USA, 2024.

[41] Tomas Bjork. *Arbitrage Theory in Continuous Time.* Oxford University Press, Oxford, UK, 2020.

[42] Fischer Black and Robert Litterman. Global Portfolio Optimization. *Financial Analysts Journal*, 1992.

[43] Fischer Black and Myron Scholes. The Pricing of Options and Corporate Liabilities. *Journal of Political Economy*, 1973.

[44] David Blei. *Practical Natural Language Processing.* O'Reilly, Boston, MA, USA, 2012.

[45] G Blundell and C Ward. Property Portfolio Allocation: a multi-factor model. *Land Development Studies*, 1987.

[46] Bob Prince. Risk Parity is about Balance: `https://www.bridgewater.com/research-and-insights/risk-parity-is-about-balance`.

[47] Ronald Bracewell. *The Fourier Transform and Its Applications.* McGraw-Hill, Boston, MA, USA, 2000.

[48] Douglas Breeden and Robert Litzenberger. Prices of state-contingent claims implicit in option prices. *Journal of Business*, 1978.

[49] Damiano Brigo and Massimo Masetti. *Counterparty Credit Risk Modeling: Risk Management, Pricing and Regulation.* Risk Books, New York, NY, USA, 2006.

[50] Damiano Brigo and Fabio Mercurio. *Interest Rate Models — Theory and Practice with Smile, Inflation and Credit.* Springer Publishing, New York, NY, USA, 2006.

[51] Nicholas Burgess. The Hull-White 1 Factor Convexity Adjustment and the Special Case when the Hull-White and Ho-Lee Models are Equivalent. *Journal of Computational Finance*, 2016.

[52] Nicholas Burgess. Interest Rate Swaptions: A Review and Derivation of Swaption Pricing Formulae. *Journal of Economics and Financial Analysis*, 2018.

[53] Kenneth Burnham and David Anderson. Multimodel inference: understanding AIC and BIC in Model Selection. *Sociological Methods and Research*, 2004.

[54] Merrill Liechty Campbell Harvey, John Liechty and Peter Muller. Portfolio Selection with Higher Moments. *Quantitative Finance Journal*, 2010.

[55] Peter Carr and Roger Lee. Robust Replication of Volatility Derivatives. *University of Chicago Whitepaper*, 2009.

[56] Peter Carr and Dilip Madan. Option valuation using the fast Fourier transform. *Journal of Computational Finance*, 1999.

[57] Peter Carr and Dilip Madan. Saddlepoint methods for Options Pricing. *The Journal of Computational Finance*, 2009.

[58] Antonio Castagna. *FX Options and Smile Risk.* John Wiley & Sons, Hoboken, NJ, USA, 2010.

[59] CDS Snac Docs. Website: `http://www.cdsmodel.com/cdsmodel/`.

[60] Ernest Chan. *Quantitative Trading: How to Build Your Own Algorithmic Trading Business.* John Wiley & Sons, Hoboken, NJ, USA, 2009.

[61] Ernest Chan. *Algorithmic Trading: Winning Strategies and their Rationale.* John Wiley & Sons, Hoboken, NJ, USA, 2013.

[62] Ludwig Chincarini and Daehwan Kim. *Quantitative Equity Portfolio Management: An Active Approach to Portfolio Construction and Management.* McGraw-Hill, Boston, MA, USA, 2006.

[63] Uta Pigorsch Christian Pigorsch and Ivaylo Popov. Volatility estimation based on high-frequency data. *Handbook of Computational Finance*, 2012.

[64] Kai Lai Chung. *A Course in Probability Theory, Third Edition.* Academic Press, New York, NY, USA, 2001.

[65] Andrea Frazzini Clifford Asness and Lasse Pedersen. Leverage Aversion and Risk Parity. 2011.

[66] Tobias Moskowitz Clifford Asness and Lasse Pedersen. Value and Momentum Everywhere. *The Journal of Finance*, 2013.

[67] James Cooley and John Tuckey. An algorithm for the machine calculation of complex Fourier series. *Mathematics of Computation*, 1967.

[68] James Cordier and Michael Gross. *The Complete Guide to Option Selling.* McGraw-Hill, Boston, MA, USA, 2005.

[69] Corinna Cortes and Vladimir Vapoik. Support-Vector Networks. *Advances in Neural Information Processing Systems*, 1995.

[70] John Crank and Phyllis Nicolson. A Practical Method for Numerical Evaluation of Solutions of Partial Differential Equations. *Advances in Computational Mathematics*, 1996.

[71] Matthias X. Hanauer David Blitzi and Tobias Hoogteijling. The term structure of machine learning alpha. *SSRN*, 2023.

[72] Shelton Peiris David E. Allen and Leonard Mushunje. GANs and synthetic financial data: calculating VaR. *SSRN*, 2023.

[73] Marcos Lopez de Prado. Building Diversified Portfolios that Outperform Out-of-Sample. *Journal of Portfolio Management*, 2016.

[74] Marcos Lopez de Prado. *Advances in Financial Machine Learning.* John Wiley & Sons, Hoboken, NJ, USA, 2018.

[75] Marcos Lopez de Prado. The 7 Reasons Most Econometric Investments Fail (Presentation Slides). *SSRN*, 2019.

[76] Marcos Lopez de Prado. *Machine Learning for Asset Managers.* Cambridge University Press, New York, NY, USA, 2020.

[77] Freddy Delbaen and Hiroshi Shirakawa. A Note on Option Pricing for the Constant Elasticity of Variance Model. *Asia-Pacific Financial Markets*, 2009.

[78] Emanuel Derman and Iraj Kani. The Volatility Smile and Its Implied Tree. *Risk Magazine*, 1994.

[79] Luc Devroye. *Non-Uniform Random Variate Generation.* Springer Publishing, New York, NY, USA, 1986.

[80] David Dickey and Wayne Fuller. Distribution of the Estimators for Autoregressive Time Series with a Unit Root. *Journal of the American Statistical Association*, 1979.

[81] Dmitry Pugachevsky. Wrong-way Risk: Regulatory Aspects and Computational Challenges: `https://www.garp.org/hubfs/Whitepapers/a1Z1W0000054x9fUAA.pdf`.

[82] Fabian Echterling Dominik Wolff. Stock picking with machine learning. *Journal of Forecasting*, 2023.

[83] Kevin Dowd. *Measuring Market Risk.* John Wiley & Sons, Hoboken, NJ, USA, 2005.

[84] DQ Nykamp. Introduction to Taylor's theorem for multivariable functions: `http://mathinsight.org/taylors_theorem_multivariable_introduction`.

[85] Bruno Dupire. Pricing with a Smile. *Risk Magazine*, 1994.

[86] Richard Durrett. *Stochastic Calculus: A Practical Introduction.* Chapman & Hall / CRC Press, Boca Raton, FL, USA, 1996.

[87] Kyunghyun Cho Dzmitry Bahdanau and Yoshua Bengio. Neural Machine Translation by Jointly Learning to Align and Translate. *arXiv*, 2014.

[88] Ronald Hua Edward Qian and Eric Sorensen. *Quantitative Equity Portfolio Management: Modern Techniques and Applications.* Chapman & Hall / CRC Press, Boca Raton, FL, USA, 2007.

[89] Zhiyuan Tu Edward Xuejun Li and Dexin Zhou. The promise and peril of generative ai: Evidence from gpt-4 as sell-side analysts. *SSRN*, 2024.

[90] Jiequn Han Ee Weinan and Arnulf Jentzen. Deep learning-based numerical methods for high dimensional parabolic partial differential equations and backward stochastic differential equations. *Communications in Mathematics and Statistics*, 2017.

[91] Engineering and Statistics Handbook. What is EDA?: `https://www.itl.nist.gov/div898/handbook/eda/section1/eda11.htm`.

[92] Robert Engle and Clive Granger. Co-integration and error correction: Representation, estimation and testing. *Econometrica*, 1987.

[93] Bjorn Eraker. The Volatility Premium. 2007.

[94] Danielle Li Erik Brynjolfsson and Lindsey R. Raymond. Generative AI at Work. *National Bureau of Economic Research*, 2023.

[95] Lawrence Evans. *Partial Differential Equations.* American Mathematical Society, Providence, RI, USA, 2010.

[96] Franke Fabozzi. *The Handbook of Mortgage Backed Securities: Seventh Edition.* Oxford University Press, Oxford, UK, 2016.

[97] Eugene Fama. Forward and spot exchange rates. *Journal of Monetary Economics*, 1984.

[98] Eugene Fama and Kenneth French. The Cross-Section of Expected Stock Returns. *The Journal of Finance*, 1992.

[99] Eugene Fama and Kenneth French. Size, value, and momentum in international stock returns. *The Journal of Financial Economics*, 2012.

[100] Eugene Fama and James MacBeth. Risk, Return, and Equilibrium: Empirical Tests. *Journal of Political Economy*, 1973.

[101] Fang Fang and Cornelis Osterlee. *SIAM Journal on Scientific Computing*, 2009.

[102] Nicol Schraudolph Felix Gers and Jurgen Schmidhuber. Learning precise timing with lstm recurrent networks. *Journal of Machine Learning Research*, 2002.

[103] Byoung-Hyoun Hwang Feng Chi and Yaping Zheng. The use and usefulness of big data in finance: Evidence from financial analysts. *SSRN*, 2023.

[104] Tony Finch. Incremental calculation of weighted mean and variance. *University of Cambridge*, 2019.

[105] David Foster. *Generative Deep Learning: Teaching Machines to Paint, Write, Compose, and Play.* O'Reilly, Boston, MA, USA, 2019.

[106] Dessislava Pachamanova Frank Fabozzi, Petter Kolm and Sergio Focardi. *Robust Portfolio Optimization and Management.* John Wiley & Sons, Hoboken, NJ, USA, 2007.

[107] Mark Garman and Michael Klass. On the Estimation of Security Price Volatilities from Historical Data. *Journal of Business*, 1980.

[108] Jim Gatheral. *The Volatility Surface: A Practitioner's Guide.* John Wiley & Sons, Hoboken, NJ, USA, 2011.

[109] Dina Genkina. Ai prompt engineering is dead: Long live ai prompt engineering. *IEEE Spectrum*, 2024.

[110] Robert Christian George Casella and Martin Wells. Generalized Accept-Reject sampling schemes. 2004.

[111] Mark Kritzman George Chow, Eric Jacquier and Kenneth Lowry. Optimal Portfolios in Good Times and Bad. *Financial Analysts Journal*, 1999.

[112] Igor Girsanov. On transforming a certain class of stochastic processes by absolutely continuous substitution of measures. *Theory of Probability and Its Applications*, 1960.

[113] Git Documentation. Website: `https://git-scm.com/doc`.

[114] Paul Glasserman. *Monte Carlo Methods in Financial Engineering.* Springer Publishing, New York, NY, USA, 2003.

[115] Gene Golub and John Welsch. Calculation of Gauss Quadrature Rules. *Mathematics of Computation*, 1969.

[116] Myron Gordon. Dividends, Earnings and Stock Prices. *Review of Economics and Statistics*, 1959.

[117] Allan Gottlieb and George Almasi. *Highly parallel computing.* Benjamin/Cummings, Redwood City, CA, USA, 1989.

[118] Andrew Green. *XVA: Credit, Funding and Capital Valuation Adjustments.* John Wiley & Sons, Hoboken, NJ, USA, 2015.

[119] Jon Gregory. *The xVA Challenge: Counterparty Risk, Funding, Collateral, Capital and Initial Margin.* John Wiley & Sons, Hoboken, NJ, USA, 2020.

[120] Richard Grinold and Robert Kahn. *Active Portfolio Management.* McGraw-Hill, Boston, MA, USA, 2000.

[121] Crina Grosan and Ajith Abraham. *Intelligent Systems: A Modern Approach.* Springer Publishing, New York, NY, USA, 2011.

[122] Charles Cao Gurdip Bakshi and Zhiwu Chen. Empirical Performance of Alternative Option Pricing Models. *The Journal of Finance*, 1997.

[123] Igor Halperin. QLBS: Q-Learner in the Black-Scholes(-Merton) Worlds. *SSRN Electronic Journal*, 2017.

[124] Larry Harris. *Trading and Exchanges: Market Microstructure for Practitioners.* John Wiley & Sons, Hoboken, NJ, USA, 2003.

[125] Linda Kaufman Alex Smola Harris Drucker, Chris Burges and Vladimir Vapoik. Support Vector Regression Machines. *Advances in Neural Information Processing Systems*, 1996.

[126] Espen Haug. *The Complete Guide to Option Pricing Formulas.* McGraw-Hill, Boston, MA, USA, 1997.

[127] Todd Hawthorne. Volatility as an asset class. *Pensions & Investments*, 2015.

[128] Fumio Hayashi. *Econometrics.* Princeton University Press, Princeton, NJ, USA, 2000.

[129] Lakhbir Hayre. *Salomon Smith Barney Guide to Mortgage and Asset Backed Securities.* John Wiley & Sons, Hoboken, NJ, USA, 2001.

[130] Steven Heston. A Closed-Form Solution for Options with Stochastic Volatility with Applications to Bond and Currency Options. *The Review of Financial Studies*, 1993.

[131] Geoffrey Hinton and Sam Roweis. Stochastic neighbor embedding. *Neural Information Processing Systems*, 2002.

[132] Geoffrey Hinton and Laurens van der Maaten. Visualizing data using t-sne. *Neural Information Processing Systems*, 2008.

[133] Ali Hirsa. *Computational Methods in Finance.* Chapman & Hall / CRC Press, Boca Raton, FL, USA, 2013.

[134] Salih N Neftci & Ali Hirsa. *An Introduction to the Mathematics of Financial Derivatives.* Academic Press, San Diego, CA, USA, 2014.

[135] Tin Kam Ho. Random decision forests. 1995.

[136] Stephen Hocht and Rudi Zagst. Pricing Distressed CDOs with Stochastic Recovery. *Review of Derivatives Research*, 2010.

[137] David Hosmer and Stanley Lemeshow. *Applied Logistic Regression.* John Wiley & Sons, Hoboken, NJ, USA, 2000.

[138] Hwei Hsu. *Fourier Analysis.* Simon and Schuster, New York, NY, USA, 1967.

[139] John C. Hull. *Options, Futures, And Other Derivatives, Seventh Edition.* Prentice Hall, Upper Saddle River, NJ, USA, 2009.

[140] Mehdi Mirza Bing Xu David Warde-Farley Sherjil Ozair Aaron Courville Ian Goodfellow, Jean Pouget-Abadie and Yoshua Bengio. Generative adversarial nets. *Proceedings of the International Conference on Neural Information Processing Systems*, 2014.

[141] Matthew Dixon Igor Halperin and Paul Bilokon. *Machine Learning In Finance: From Theory to Practice.* Springer Publishing, New York, NY, USA, 2020.

[142] Antti Ilmanen. *Expected Returns: An Investor's Guide to Harvesting Market Rewards.* John Wiley & Sons, Hoboken, NJ, USA, 2011.

[143] Rodney Sullivan Ing-Chea Ang, Roni Israelov and Harsha Tummala. Understanding the Volatility Risk Premium. *AQR Whitepaper*, 2018.

[144] Steven E. Shreve Ioannis Karatzas. *Brownian Motion and Stochastic Calculus.* Springer Publishing, New York, NY, USA, 1987.

[145] Andrey Itkin. Deep learning calibration of option pricing models: some pitfalls and solutions. *Risk.net*, 2019.

[146] Peter Jaeckel. *Monte Carlo Methods in Finance.* John Wiley & Sons, Hoboken, NJ, USA, 2002.

[147] Lloyd Han Jamil Baz and Mac-Antoine Loo. Three Models of the Liqudity Premium. *SSRN*, 2024.

[148] Dale Schuurmans Maarten Bosma Brian Ichter Jason Wei, Xuezhi Wang and Fei Xia. Chain-of-thought prompting elicits reasoning in large language models. *Advances in Neural Information Processing Systems*, 2022.

[149] John Hull Jay Cao, Jacky Chen and Zissis Poulos. Deep Hedging of Derivatives Using Reinforcement Learning. *University of Toronto Research*, 2020.

[150] Narasimhan Jegadeesh and Sheridan Titman. Returns to Buying Winners and Selling Losers: Implications for Stock Market Efficiency. *The Journal of Finance*, 1993.

[151] Soren Johansen. Estimation and Hypothesis Testing of Cointegration Vectors in Gaussian Vector Autoregressive Models. *Econometrica*, 1991.

[152] Andrew Lo John Campbell and Craig MacKinlay. *The Econometrics of Financial Markets.* Princeton University Press, Princeton, NJ, USA, 1997.

[153] Jonathan Ingersol John Cox and Stephen Ross. A Theory of the Term Structure of Interest Rates. *Econometrica*, 1985.

[154] Svetlozar Rachev John Hsu, Biliana Bagasheva and Frank Fabozzi. *Bayesian Methods in Finance.* John Wiley & Sons, Hoboken, NJ, USA, 2008.

[155] Mark Kac. On Distributions of Certain Wiener Functionals. *Transactions of the American Mathematical Society*, 1949.

[156] Daniel Kahneman and Amos Tversky. Advances in prospect theory: Cumulative representation of uncertainty. *The Journal of Risk and Uncertainty*, 1992.

[157] Zura Kakushadze. 101 Formulaic Alphas. *Wilmott Magazine*, 2015.

[158] Ayush Kaushal and Kyle Mahowald. What do tokens know about their characters and how do they know it? *arXiv*, 2022.

[159] Donald Keim. Size-Related Anomalies and Stock Return Seasonality: Further Empirical Evidence. *The Journal of Financial Economics*, 1983.

[160] Ralph Kimball. *The data warehouse ETL toolkit : practical techniques for extracting, cleaning, conforming, and delivering data.* John Wiley & Sons, Hoboken, NJ, USA, 2004.

[161] Peter Kloeden and Eckhard Platen. *Numerical Solution of Stochastic Differential Equations.* Springer Publishing, New York, NY, USA, 1992.

[162] Peter Lax. *Linear Algebra and its Applications.* John Wiley & Sons, Hoboken, NJ, USA, 2007.

[163] Oliver Ledoit and Michael Wolf. Improved Estimation of the Covariance Matrix of Stock Returns With an Application to Portfolio Selection. *Journal of Empirical Finance*, 2003.

[164] Oliver Ledoit and Michael Wolf. Honey, I Shrunk the Sample Covariance Matrix. *The Journal of Portfolio Management*, 2004.

[165] Cheng-Few Lee and John Lee. *Handbook of Quantitative Finance and Risk Management.* Springer Publishing, New York, NY, USA, 2010.

[166] Roger Lewenstein. *When Genius Failed: The Rise and Fall of Long Term Capital Management.* Random House, New York, NY, USA, 2001.

[167] Andrew Lo and Craig MacKinlay. When are Contrarian Profits Due to Stock Market Overreaction? *The Review of Financial Studies*, 1990.

[168] Francis Longstaff and Eduardo Schwartz. Valuing American Options by Simulation: A Simple Least Squares Approach. *The Review of Financial Studies*, 2001.

[169] Burton Malkiel. *A Random Walk Down Wall Street.* W.W. Norton, New York, NY, USA, 1996.

[170] Aaron Maltais and Bjorn Nykvist. Understanding the role of green bonds in advancing sustainability. *Journal of Sustainable Finance & Investment*, 2020.

[171] Danilo Mandic and Jonathon Chambers. *Recurrent Neural Networks for Prediction: Learning Algorithms, Architectures and Stability.* John Wiley & Sons, Hoboken, NJ, USA, 2001.

[172] Craig Friedman Nicolas Grandchamp Lukasz Kruk Marco Avellaneda, Robert Buff and Joshua Newman. Weighted Monte Carlo: A New Technique for Calibrating Asset-Pricing Models. *NYU Research Paper*, 1998.

[173] Markit CDX. Website: `https://www.markit.com/product/cdx`.

[174] Markit ITRAXX. Website: `https://www.markit.com/product/itraxx`.

[175] Harry Markowitz. Portfolio Selection. *The Journal of Finance*, 1952.

[176] George Marsaglia and Wai Wan Tsang. A simple method for generating gamma variables. *ACM Transactions on Mathematical Software*, 2000.

[177] V.A. Marcenko and Leonid Pastur. *Math USSR Sb.*

[178] MatplotLib Documentation. Website: `https://matplotlib.org/gallery.html`.

[179] Jan Mayle. *Standard Securities Calculation Methods: Fixed Income Securities Formulas for Price, Yield and Accrued Interest.* Securities Industry Association, Silver Spring, MD, USA, 1993.

[180] Sandip Mazumder. *Numerical Methods for Partial Differential Equations Finite Difference and Finite Volume Methods.* Academic Press, Orlando, FL, USA, 2015.

[181] Geoffrey J. McLachlan. *Discriminant Analysis and Statistical Pattern Recognition.* John Wiley & Sons, Hoboken, NJ, USA, 1989.

[182] Michael McTear. *The Conversational Interface.* Springer Publishing, New York, NY, USA, 2016.

[183] Robert Merton. On the Pricing of Corporate Debt: The Risk Structure of Interest Rates. *The Journal of Finance*, 1974.

[184] Attillio Meucci. *Risk and Asset Allocation.* Springer Publishing, New York, NY, USA, 2009.

[185] Richard Michaud and Robert Michaud. Estimation Error and Portfolio Optimization: A Resampling Solution. *Journal of Investment Management*, 2007.

[186] Richard Michaud and Robert Michaud. *Efficient Asset Management: A Practical Guide to Stock Portfolio Optimization and Asset Allocation.* Oxford University Press, Oxford, UK, 2008.

[187] Rik Lopuhaá Michel Dekking, Cor Kraaikamp and Ludolf Meester. *A Modern Introduction to Probability and Statistics.* Springer Publishing, New York, NY, USA, 2005.

[188] Monotonic Transformation of Utility Function. UKEssays: `https://www.ukessays.com/essays/economics/monotonic-transformation-utility-3581.php?vref=1`.

[189] Sheldon Natenberg. *Option Volatility and Pricing: Advanced Trading Strategies and Techniques, 2nd Edition.* McGraw-Hill, Boston, MA, USA, 2015.

[190] Whitney Newey and Kenneth West. A Simple, Positive Semi-definite, Heteroskedasticity and Autocorrelation Consistent Covariance Matrix. *Econometrica*, 1987.

[191] NumPy Documentation. Website: `https://docs.scipy.org/doc/numpy/reference/`.

[192] Dominic O'Kane. *Modelling singe-name and multi-name Credit Derivatives.* John Wiley & Sons, Hoboken, NJ, USA, 2008.

[193] Bernt Oksendal. *Stochastic Calculus for Finance I: Discrete-Time Models.* Springer Publishing, New York, NY, USA, 2013.

[194] Kim Oosterlinck. Sovereign debt defaults: insights from history. *Oxford Review of Economic Policy*, 2013.

[195] Pandas Documentation. Website: `http://pandas.pydata.org/pandas-docs/stable/`.

[196] Bo Pang and Lillian Lee. Seeing stars: Exploiting class relationships for sentiment categorization with respect to rating scales. *Proceedings of the Association for Computational Linguistics*, 2005.

[197] Michael Parkinson. The Extreme Value Method for Estimating the Variance of the Rate of Return. *Journal of Business*, 1980.

[198] Andrew Lesniewski Patrick Hagan, Deep Kumar and Diana Woodward. Managing Smile Risk. *Wilmott Magazine*, 2002.

[199] Andrew Lesniewski Patrick Hagan, Deep Kumar and Diana Woodward. An oil futures contract expiring Tuesday went negative in bizarre move showing a demand collapse. *CNBC*, 2020.

[200] Paul Wilmott. Wilmott Modeller's Manifesto: `https://wilmott.com/financial-modelers-manifesto/`.

[201] Jonas Palmquist Pavlo Krokhmal and Stanislav Uryasev. Portfolio Optimization with Conditional Value-At-Risk Objective and Constraints. *Journal of Risk*, 2002.

[202] Dilip Madan Peter Carr and Eric Chang. The Variance Gamma Process and Option Pricing. *European Finance Review*, 1998.

[203] Dilip Madan Peter Carr, Helyette Geman and Marc Yor. Stochastic Volatility for Levy Processes. *Mathematical Finance*, 2003.

[204] Peter Wentworth, Jeffrey Elkner, Allen B. Downey, and Chris Meyers. How to Think Like a Computer Scientist Learning with Python 3: `http://openbookproject.net/thinkcs/python/english3e/`.

[205] Mary Phuong and Marcus Hutter. Formal Algorithms for Transformers. *arXiv*, 2022.

[206] Vladimir Piterbarg and Marco Renado. Eurodollar Futures Convexity Adjustments in Stochastic Volatility Models. *Journal of Computational Finance*, 2004.

[207] Stanley Pliska. *Introduction to Mathematical Finance: Discrete Time Models.* Blackwell Publishers, Oxford, UK, 1997.

[208] Michael Pykhtin and Steven Zhu. A Guide to Modeling Counterparty Credit Risk. *GARP Risk Review*, 2007.

[209] Python Design Patterns Guide. Website: `https://python-patterns.guide/`.

[210] Python Expressions Docs. Website: `https://docs.python.org/3/reference/expressions.html`.

[211] Python Programming Database Wikibook. Website: `https://en.wikibooks.org/wiki/Python_Programming/Databases`.

[212] Edward Qian. Risk parity portfolios: Efficient portfolios through true diversification. *PanAgora Asset Management*, 2005.

[213] Quantifi. Funding Valuation Adjustments:`https://www.quantifisolutions.com/funding-valuation-adjustment-fva-part-1-a-primer/`.

[214] Lasse Pedersen Ralph Koijen, Tobias Moskowitz and Evert Vrugt. Carry. *Chicago Booth Research Paper*, 2016.

[215] Ashwin Rao and Tikhon Jelvis. *Foundations of Reinforcement Learning with Applications in Finance.* Stanford University, California, CA, USA, 2021.

[216] Florian Rehm and Markus Rudolph. KMV Credit Risk Modeling. 2000.

[217] Kenneth McKay Riccardo Rebonato and Richard White. *The SABR/Libor Market Model: Pricing, Calibration and Hedging for Complex Interest-Rate Derivatives.* John Wiley & Sons, Hoboken, NJ, USA, 2009.

[218] Kurt Friedrichs Richard Courant and Hans Lewy. On the partial difference equations of mathematical physics. 1967.

[219] Tyrrell Rockafellar and Stanislav Uryasev. Optimization of conditional value-at-risk. *Journal of Risk*, 2000.

[220] Tyrrell Rockafellar and Stanislav Uryasev. Conditional value-at-risk for general loss distributions. *Journal of Banking & Finance*, 2002.

[221] L. Rogers and Stephen Satchell. Estimating variance from high, low and closing prices. *Annals of Applied Probability*, 1991.

[222] Thierry Roncalli. *Introduction to Risk Parity and Risk Budgeting.* Chapman & Hall / CRC Press, Boca Raton, FL, USA, 2014.

[223] Bryan Kelly Ronen Israel and Tobias Moskowitz. Can Machines 'Learn' Finance? *The Journal of Investment Management*, 2020.

[224] Konstantinos Roumeliotis and Nikolaos Tselikas. ChatGPT and Open-AI Models: A Preliminary Review. *Journal of Business*, 2023.

[225] André Alves Portela Santos. The out-of-sample performance of robust portfolio optimization. *Brazilian Review of Finance*, 2010.

[226] Stephen Satchell and Alan Scowcroft. A demystification of the Black–Litterman model: Managing quantitative and traditional portfolio construction. *Journal of Asset Management*, 2000.

[227] Philipp Schönbucher. *Credit Derivatives Pricing Models: Models, Pricing and Implementation.* John Wiley & Sons, Hoboken, NJ, USA, 2003.

[228] William Sharpe. Capital Asset Prices: A Theory of Market Equlibrium Under Conditions of Risk. *The Journal of Finance*, 1964.

[229] Percy Liang Shivam Garg, Dimitris Tsipras and Gregory Valiant. What can transformers learn in-context? a case study of simple function classes. *arXiv*, 2022.

[230] Andrei Shleifer and Robert Vishny. The Limits of Arbitrage. *The Journal of Finance*, 1997.

[231] Steven E. Shreve. *Stochastic Calculus for Finance I: Discrete-Time Models.* Springer Publishing, New York, NY, USA, 2004.

[232] Steven E. Shreve. *Stochastic Calculus for Finance II: Continuous-Time Models.* Springer Publishing, New York, NY, USA, 2004.

[233] SOBOL Python Library. Website: `https://people.sc.fsu.edu/~jburkardt/py_src/sobol/sobol.html`.

[234] Pavel Solin. *Partial Differential Equations and the Finite Element Method.* John Wiley & Sons, Hoboken, NJ, USA, 2006.

[235] Dan Stefanica. *A Primer for the Mathematics of Financial Engineering.* FE Press, New York, NY, USA, 2011.

[236] Harvey Stein and Kin Pong Lee. Counterparty Valuation Adjustments. *SSRN*, 2012.

[237] Gilbert Strang. *Calculus.* Wellesley-Cambridge, 2010.

[238] Alan Stuart and Keith Ord. *Kendall's Advanced Theory of Statistics: Volume I Distribution Theory.* John Wiley & Sons, Hoboken, NJ, USA, 1994.

[239] Biliana Bagasheva Svetlozar Rachev, John Hsu and Frank Fabozzi. *Bayesian Methods in Finance.* John Wiley & Sons, Hoboken, NJ, USA, 2008.

[240] Rama Cont & Peter Tankov. *Financial Modelling with Jump Processes.* Chapman & Hall / CRC Press, Boca Raton, FL, USA, 2004.

[241] J.W. Thomas. *Numerical Partial Differential Equations: Finite Difference Methods.* Springer Publishing, New York, NY, USA, 1995.

[242] Bernhard Scholkopf Thomas Hofmann and Alexander Smola. Kernel Methods in Machine Learning. *The Annals of Statistics*, 2008.

[243] Thorsten Schmidt. Coping with Copulas: `https://web.archive.org/web/20100705040514/http://www.tu-chemnitz.de/mathematik/fima/publikationen/TSchmidt_Copulas.pdf`.

[244] Richard Crump Tobias Adrian and Emanuel Moench. Pricing the term structure with Linear Regression. *Journal of Financial Economics*, 2013.

[245] Greg Corrado Tomas Mikolov, Kai Chen and Jeffrey Dean. Efficient estimation of word representations in vector space. *arXiv*, 2013.

[246] Kai Chen Greg Corrado Tomas Mikolov, Ilya Sutskever and Jeffrey Dean. Distributed representations of words and phrases and their compositionality. *arXiv*, 2013.

[247] F. Barry Nelson Tracy V. Maitland and Daniel Partlow. *Convertible Securities: A Complete Guide to Investment and Corporate Financing Strategies*. McGraw-Hill, Boston, MA, USA, 2022.

[248] Lloyd Trefethen and David Bau. *Numerical Linear Algebra*. SIAM, Philadelphia, PA, USA, 1997.

[249] Troy Segal. Understanding Pass-Through Securities and Their Risks: `https://www.investopedia.com/terms/p/passthroughsecurity.asp`.

[250] Ruey Tsay. *Analysis of Financial Time Series*. John Wiley & Sons, Hoboken, NJ, USA, 2010.

[251] Bruce Tuckman and Angel Serrat. *Fixed Income Securities Tools for Today's Markets*. John Wiley & Sons, Hoboken, NJ, USA, 2012.

[252] John Tukey. *Exploratory Data Analysis*. Pearson, Hoboken, NJ, USA, 1977.

[253] Garrett Somers Uday Kamath, Kevin Keenan and Sarah Sorenson. *Large Language Models: A Deep Dive: Bridging Theory and Practice*. Springer Publishing, New York, NY, USA, 2024.

[254] Oldrich Vasicek. An Equilibrium Characterization of the Term Structure. *Journal of Financial Economics*, 1977.

[255] Ragav Venkatesan and Baoxin Li. *The Volatility Surface: A Practitioner's Guide*. Chapman & Hall / CRC Press, Boca Raton, FL, USA, 2017.

[256] Ganapathy Vidyamurthy. *Pairs Trading: Quantitative Methods and Analysis*. John Wiley & Sons, Hoboken, NJ, USA, 2004.

[257] Detlof von Winterfeldt and Ward Edwards. Decision trees. *Decision Analysis and Behavioral Research*, 1986.

[258] W3 Schools Python Operator Documentation. Website: `https://www.w3schools.com/python/python_operators.asp`.

[259] Jay Walters. The BlackLitterman Model: A Detailed Exploration. 2008.

[260] Alan Weir. Fubini's Theorem. 1973.

[261] Wikibooks Design Patterns. Website: `https://en.wikibooks.org/wiki/C%2B%2B_Programming/Code/Design_Patterns`.

[262] William T. Vetterling William H. Press, Saul A. Teukolsky and Brian P. Flannery. *Numerical Recipes in C++*. Cambridge University Press, New York, NY, USA, 2002.

[263] Jeffrey Woodbridge. *Econometric Analysis of Cross Section and Panel Data*. MIT Press, Cambridge, MA, USA, 2001.

[264] Uwe Wystup. *FX Options and Structured Products*. John Wiley & Sons, Hoboken, NJ, USA, 2007.

[265] Dennis Yang and Qiang Zhang. Drift-independent volatility estimation based on high, low, open, and closing Prices. *Journal of Business*, 2000.

[266] Yoshua Bengio Yann Lecun and Geoffrey Hinton. Deep Learning. *Nature*, 2015.

[267] Bryan T. Kelly Yifei Chen and Dacheng Xiu. Expected Returns and Large Language Models. *SSRN*, 2023.

[268] Luka Vidovic & Lei Yue. Machine learning and credit risk modelling. *S&P Global Market Intelligence*, 2020.

[269] Rita Frieske Tiezheng Yu Dan Su Yan Xu Etsuko Ishii Ye Jin Bang Andrea Madotto Ziwei Ji, Nayeon Lee and Pascale Fung. Survey of Hallucination in Natural Language Generation. *ACM Computing Surveys*, 2023.

Index